W9-AXH-145

THE
MIDDLE
EAST
A GEOGRAPHICAL STUDY

DS
49.7
.B36
1988

THE MIDDLE EAST

A GEOGRAPHICAL STUDY

SECOND EDITION

Peter Beaumont
Professor of Geography, St. David's University College, University of Wales

Gerald H. Blake
Senior Lecturer in Geography, University of Durham

J. Malcolm Wagstaff
Senior Lecturer in Geography, University of Southampton

HALSTED PRESS
a division of JOHN WILEY & SONS, Inc.
605 Third Avenue, New York, N.Y. 10158
New York • Chichester • Brisbane • Toronto • Singapore

SEP 2 6 1988

476873

Published in the Western Hemisphere and Japan by Halsted Press:
a division of John Wiley & Sons, Inc. New York

©Peter Beaumont, Gerald Blake & J. Malcolm Wagstaff 1988

All rights reserved. No part of this publication may be reproduced, stored in a retrieval system or transmitted in any form, or by any means, electronic, mechanical, photocopying, recording or otherwise, without the prior permission of the publishers.

LIBRARY OF CONGRESS
Library of Congress Cataloging-in-Publication Data

Beaumont, Peter
The Middle East: a geographical study/Peter Beaumont, Gerald H. Blake, J. Malcolm Wagstaff.—2nd ed.
p. cm.
Bibliography: p.
Includes index.
ISBN 0-470-21040-0 (Halsted)
1. Middle East—Description and travel. I. Blake, Gerald Henry. II. Wagstaff, J. Malcolm (John Malcolm), 1940- . III. Title.
DS49.7.B36 1988
956′.04—dc19 87-37148
CIP

ISBN 0-470-21040-0

Printed and bound in Great Britain

Preface

Eleven years ago, when our book first appeared, we little thought that a second edition would be called for. We are pleased that *The Middle East: A Geographical Study* has proved sufficiently useful to warrant revision and updating. We have adhered to the original plan but completely rewritten those sections and chapters where our information and views had been rendered obsolete by recent developments, notably the rises in oil prices (1973, 1979), the outbreak of civil war in Lebanon (1975) and the fall of the Shah (1979). In the process of rewriting we have been greatly impressed by the rate and extent of change in the region since 1976.

The second edition is the culmination of more than twenty years of research and teaching on the region, in the course of which we have received much practical assistance and generous hospitality from numerous friends and associates in various parts of the Middle East and North Africa. We hope that their kindness will be rewarded at least in some measure by any contribution which our book may make to a better understanding of the problems and aspirations of the people of the region. The need for such understanding is greater than ever before. We are equally indebted to many colleagues in this country, notably to the late Professor W.B. Fisher, formerly head of the Department of Geography, and Professor H. Bowen-Jones, formerly Director of the Centre for Middle Eastern and Islamic Studies, in the University of Durham. We are grateful to the Centre for Middle Eastern and Islamic Studies and other institutions which have enabled us to conduct fieldwork in the Middle East on many occasions. The Centre's Documentation Section has also assisted us with valuable primary source material, often difficult to obtain, in the region itself.

We wish to thank all those who have commented on parts of the text at various stages in its production, particularly Mr. C. G. Smith of Keble College, Oxford; Professor J. I. Clarke, Professor J. C. Dewdney and Dr. R I. Lawless of Durham University; Dr. M. El-Mehdawi of Garyounis University; Dr. S. A. Khater of Alexandria University; Dr. M. F. Davie of Université Saint-Joseph. Beirut; and Dr. B. S. Hoyle of Southampton University. Dr. A. J. Trilsbach and Dr. I. J. Seccombe of Durham University generously gave of their expertise in the revision of Chapters 19 and 5 respectively and we appreciate their enthusiastic collaboration. In addition, we owe a great deal to several generations of students, including many from the Middle East, who asked most of the questions that this book seeks to answer, and many more which we have been unable to tackle adequately.

In the production stage, Mr. A. S. Burn of the Cartographic Unit at Southampton University was responsible for drawing all the maps for both editions with skill and forbearance. We also wish to pay warm tribute to our publisher, Michael Coombs of John Wiley and Sons, who commissioned the book in the first place in the belief that it would be of long-term value. His faith has been repaid with the appearance of this second edition. We are grateful to David Fulton and Associates for having the confidence to take us over. Wendy Hudlass saw us through the press with much kindly advice and encouragement.

P. B.	Lampeter
G. H. B.	Durham
J. M. W.	Southampton
January, 1988	

Acknowledgements

Copyright material is reproduced by kind permission of the following whose co-operation we gratefully acknowledge: *Nature* (Macmillan Journals Ltd), (Figures 1.2 and 1.4); Oxford University Press, (Figure 1.3); Ministry of Agriculture, Republic of Turkey (Figure 1.7); United Nations Food and Agriculture Organization, (Figure 1.8); British Society of Soil Science, (Figure 1.9); Institute of Land Reclamation, Alexandria University (Figure 1.10); Iranian Meteorological Organisation, (Figure 2.1); The Geographical Association, (Figure 2.7); C. W. Thornthwaite Associates, Centerton, New Jersey, (Figures 2.12 and 2.15); United Nations (UNESCO–FAO), (Figure 2.13); National Water Well Association, (Figure 2.18); American Association for the Advancement of Science, (Figure 3.4); Koninklijk Nederlands Aardrijskundig Genootschap, (Figure 3.5); Wolf Hütteroth, (Figures 3.6, 4.3 and 4.4(a)); Department of Surveys, Tel Aviv, Israel, (Figure 3.8); World Meteorological Organization, (Figures, 4.1 and 13.2); American Geographical Society, (Figures 4.5 and 13.3); Douglas L. Johnson, (Figure 4.7); Editor, *Middle East Journal*, (Figure 4.8); Lars Eldblom, (Figure 4.9); Israel Exploration Journal, (Figure 4.10); David F. Darwent, (Figure 6.5); The Controller of Her Majesty's Stationery Office (Figure 6.7); University of London Press Limited, (Figure 6.8); M. M. Azeez, (Figure 6.8); Keter Publishing House, Jerusalem, Limited, (Figure 10.2); Associated Book Publishers, (Figure 10.5); University of Chicago Press, (Figure 12.6); Soil Conservation Society of America (Figure 15.3); Plan Organisation of Iran, (Figures 18.2 and 18.3).

We also thank the following for permission to quote from previously published tables: Iranian Meteorological Department, (Table 2.1); The Controller, Her Majesty's Stationery Office, (Tables 2.2, 2.3, 2.3, 2.4, 2.5 and 2.6); World Meteorological Organization, Geneva, (Table 2.7); The Geographical Association, (Table 2.10); American Philosophical Society, (Table 5.1); The British Petroleum Co Ltd, (Table 9.4); United Nations Educational Scientific and Cultural Organization, (Tables 12.2 and 12.3); and Echo of Iran, (Table 18.1).

All sources are fully cited in the chapter references and in the bibliography.

A Note on Names, Units and Measures

One of the problems encountered in studying a foreign region is the spelling of names. This is compounded where languages are written in the unfamiliar Arabic script and different systems of transliteration have been employed to turn names into European languages. Although T. E. Lawrence gloried in the variety of forms which could be produced, most readers are likely to be confused unless some standardization is achieved. Three principles have directed the choice for this book. As far as possible, place names have been spelt in the form used by *The Times Atlas of the World* since this is generally accessible, but where an English version is already in common use that spelling has been preferred. For example, Taurus Mountains is preferred to Toros Dağlari and Euphrates to al-Fūrat or Firat. Where reference is made to places not marked on the maps of *The Times Atlas*, the versions employed in the English language literature have generally been accepted, though for places in Turkey the modern Romanized spelling is used (J. C. Dewdney, *Turkey*, Chatto and Windus, London, 1971, pp 8–10). For other names, the practice adopted by the *Encyclopaedia of Islam* has been followed, even though this has sometimes created inconsistencies in the text. To avoid the inconvenient term 'Persian/Arabian Gulf', 'the Gulf' has been adopted throughout.

The metric system has been adopted for most quantities, in line with current British practice. For financial matters, however, a variety of currencies are used. Standardization over a long period of time is virtually impossible in a situation of inflation and rapidly changing exchange rates.

Contents

6. Towns and Cities 199

7. Problems of Economic Development 233

8. Industry, Trade and Finance 251

15. Jordan – The Struggle for Economic Survival 408

16. Israel and the Occupied Areas: Jewish colonization 433

17. The Industrialization of Turkey 460

18. Iran – Agriculture and Its Modernization 481

Introduction

0.1 Terms and definitions

The term 'Middle East' appears to have originated in the British India Office during the 1850s, in the early days of expansionist rivalry between Russia and Britain. It became current in the English-speaking world around 1900 when the American naval historian, A. T. Mahan, employed it in a discussion of British naval strategy in relation to Russian activity in Iran and a German project for a Berlin to Baghdād railway.[1] He was referring to a region centred on the Persian Gulf, for which the current terms 'Near East' and 'Far East' seemed inadequate. The term was also taken up by *The Times* correspondent in Tehrān, V. Chirol, for a series of articles on the lands forming the western and northern approaches to India,[2] the defence of which had been a sensitive issue for more than a century and became more and more crucial as the strategic centre of the British Empire, no less than British trade, became centred upon the subcontinent (Figure 0.1). 'Middle East' was given respectability when it was used in the House of Lords on 22 March, 1911 by Lord Curzon in opening a discussion of 'the state of affairs in Persia, the Persian Gulf, and Turkey in Asia, in relation . . . to the construction of railways . . . '[3] (Figure 0.1).

Clearly, the term 'Middle East' was one of strategic reference, developed in a Eurocentred world, just as the older terms 'The East', 'Far East' and 'Near East' had been. It was developed further during the First World War when the operational theatre of the Mesopotamia Expeditionary Force came to be distinguished as 'Middle East' from the 'Near East' of Palestine and Syria in which the Egyptian Expeditionary Force operated.[4] Although Curzon had already given the term a wider application than the lands centred about the Gulf, this only became permanent by a series of accidents in military organization. In 1932 the existing Royal Air Force Middle Eastern Command, in Iraq, was amalgamated with Near Eastern Command, in Egypt, but the new command retained the title 'Middle East'. When the Italian threat to the Suez Canal at the beginning of the Second World War led to the establishment of a military headquarters in Cairo, the army followed the RAF in calling this 'GHQ Middle East'. Between 1940 and 1943, the Cairo headquarters controlled British and Allied operations over a very wide region (Figure 0.1). The constant use of 'Middle East' to describe this region in communiqués and amongst military personnel made the term familiar to a large public. Continued political ferment in the region and its basic strategic importance have maintained the term in use, though not without some pleas for the retention of the old term 'Near East'.[5] Indeed, the term 'Middle East' has

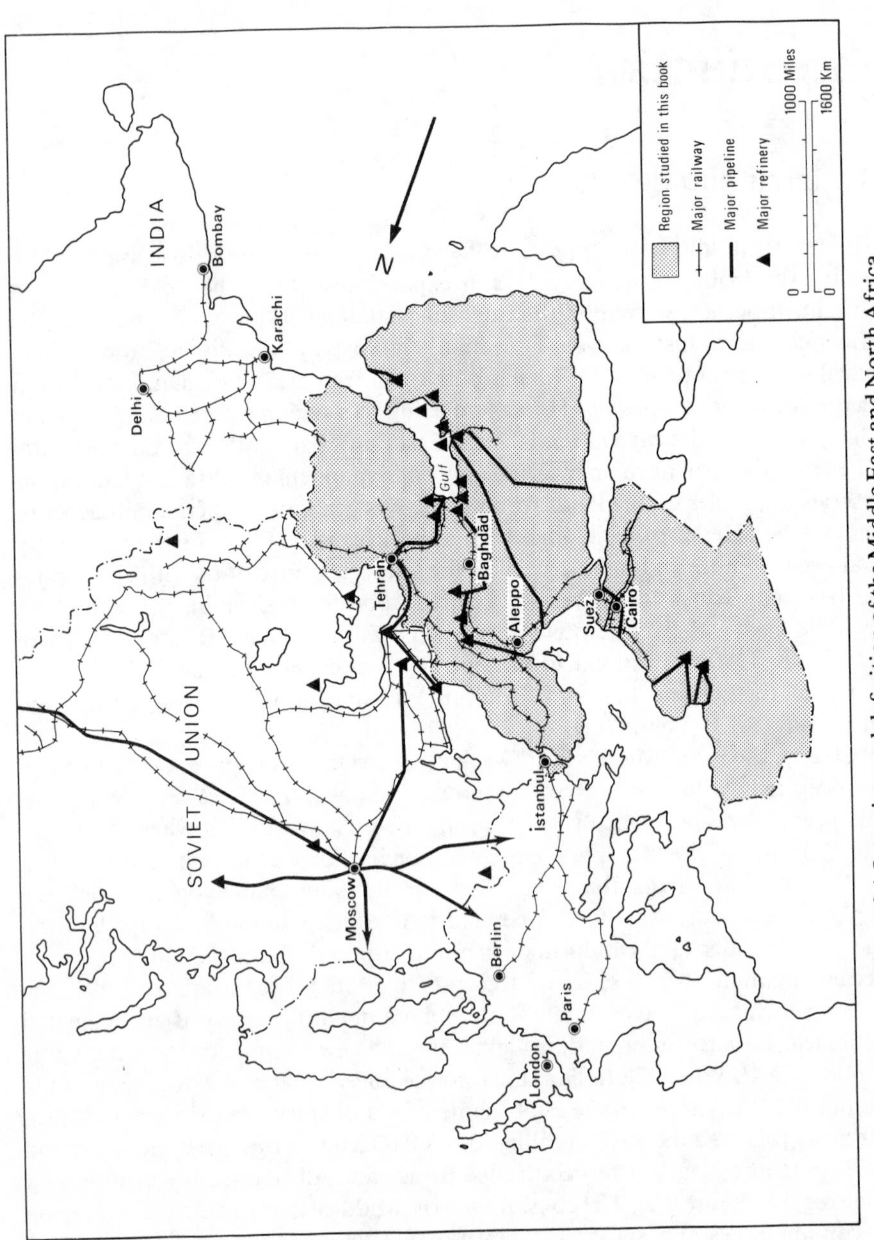

Figure 0.1 Location and definition of the Middle East and North Africa

become so useful that it is employed by the Russians and even the inhabitants of the region itself, though sometimes with reference to slightly different areas.

In an effort to gain clarity, two other terms are now widely used in the literature to refer to subregions of the 'Middle East' and will be so employed here. 'Southwest Asia' is a term originally devised by American commentators to cover that part of the region lying east of the Isthmus of Suez and north of the eastern Mediterranean Sea, thus excluding North Africa. 'North Africa' itself was used during the Second World War to designate the subregion of the 'Middle East' where fighting between Allied and Axis troops was actually taking place, particularly the Western Desert of Egypt and Libya. Later the term was extended to the whole of Africa between the Mediterranean Sea and the steppe lands of Sahel, including the Sahara Desert and the north-western corner of the continent which the Arab geographers had called *Jezira al-Maghreb* ('Island of the West') and Europeans had known as *Barbary*. Although the Maghreb states of Morocco, Algeria and Tunisia are not given detailed consideration in this book, some reference to them is made as part of the wider context necessary to certain of the thematic chapters. On these occasions, the term 'North Africa' may be understood in its widest sense, but, generally, usage will be restricted to Libya and Egypt.

The regional and subregional definitions employed in this book are thus essentially pragmatic. The macro-region itself is defined in terms of modern states, since today these may be regarded as constituting distinctive socio-economic systems, despite a number of shared characteristics. Iran and Turkey form the northern tier, with frontiers shared with the Soviet Union. To the south lie the Arab states of Lebanon, Jordan, Syria and Iraq, as well as the Jewish state of Israel. Further south is the heartland of Arabia, consisting of Saudi Arabia, a fringe of small states along the Gulf and the larger units of Oman and the two Yemens. Libya and Egypt form the southwestern corner of the region (Figure 0.1).

Apart from political frontiers, there are no clear boundaries around the region defined in this way. The sea, which penetrates deeply into the region, is as much a medium for movement as a barrier, and it is easily crossed at the Red Sea, between Africa and Arabia, and the Turkish Straits, between Europe and Asia. The upland which forms much of eastern Iran may retard movement, but well-defined corridors lead through it into central Asia and northern India. On the south, deserts interpose not so much an impassable barrier between the Middle East, on the one hand, and the Maghreb and the Sahel, on the other, as a difficult and exhausting zone of transition.

0.2 Unity and diversity[6]

Certain shared characteristics give a degree of unity to the region. The most fundamental is climate. The region is characterized by extremely arid summers and a winter-spring maximum of precipitation. Continental effects are so marked in the centre of the main land masses – Asia Minor, Iran, Arabia and

North Africa – and the precipitation so low that population is confined to oases or thinly scattered as nomadic groups. Human activity throughout the region is closely adapted to climatic conditions. The seasonal rhythm in farming and herding is dependent upon the incidence of precipitation, and all human life requires successful harvests of wheat and barley. Nomads and settled cultivators have been mutually dependent here for millenia, and though similar patterns are found in northern India, central Asia and sub-Saharan Africa, they do not appear to have been so well developed or so closely interdependent. In the Middle East, the towns, situated at nodes on natural route ways, have been the organizational centres of the region, closely influencing land use patterns and mediating high cultural developments.

The culture of the region today is fundamentally Muslim and deeply penetrated by the Arabic language. To Muslims Arabic is the language of God's revelation in the Qur'ān and of the suppliant's prayers. It spread as Islam itself spread throughout the region, partly by conquest but largely by the slower process of conversion.[7] The close identification of Islam with everyday life and administration has ensured the transference of Arabic words and phrases into the other languages of the region. Islam itself is a major integrating force. It stresses the equality of men before the mercy and power of God. Although the Middle East forms only a part of the *Dar al-Islam* ('House of Islam'), it constitutes the core about the major pilgrimage centre of Mecca, which all the faithful are commanded to visit at least once in their lives.[8] The Great Pilgrimage has bestowed considerable unity on the region, for until recently the lands closest to Mecca sent most people to the Ka'ba and the diffusion of Islamic influence backwards was correspondingly great. However, not all the peoples of the Middle East are Muslim, even after the population movements of the last sixty years. Nonetheless, the culture of Jews and Christians has been powerfully shaped by centuries of life as 'People of the Book' within the Muslim theocracy.

Diversity is as characteristic of the Middle East when examined at a detailed level, as unity appears to be at the regional scale. In many ways the lack of unity looms very large. In detail, a variety of terrain and climate exists in the region, and human response is not uniform. The presence of several races has been recognized. Three major languages – Arabic, Persian and Turkish – are spoken in different parts of the region, but their dominance was often slow in developing and a number of more ancient languages are still spoken by particular groups, whilst recent Jewish immigration has revived Hebrew as the language of Israel (Figure 0.2). The rise of nationalism has weakened many of the old sympathies and relationships, so that much of the cultural unity of the region appears to be disintegrating, whilst political regimes spread through a wide spectrum from patriarchal monarchy to people's democracy. States vary greatly in size, from the virtual city states of the United Arab Emirates to Saudi Arabia, which is about twelve times the size of England. Population densities range from minimal figures for this desert kingdom and much of Libya to over 800 per sq km in the Nile valley. Finally, the region does not yet constitute a unified trading bloc, since intra-regional trade is poorly developed.[9]

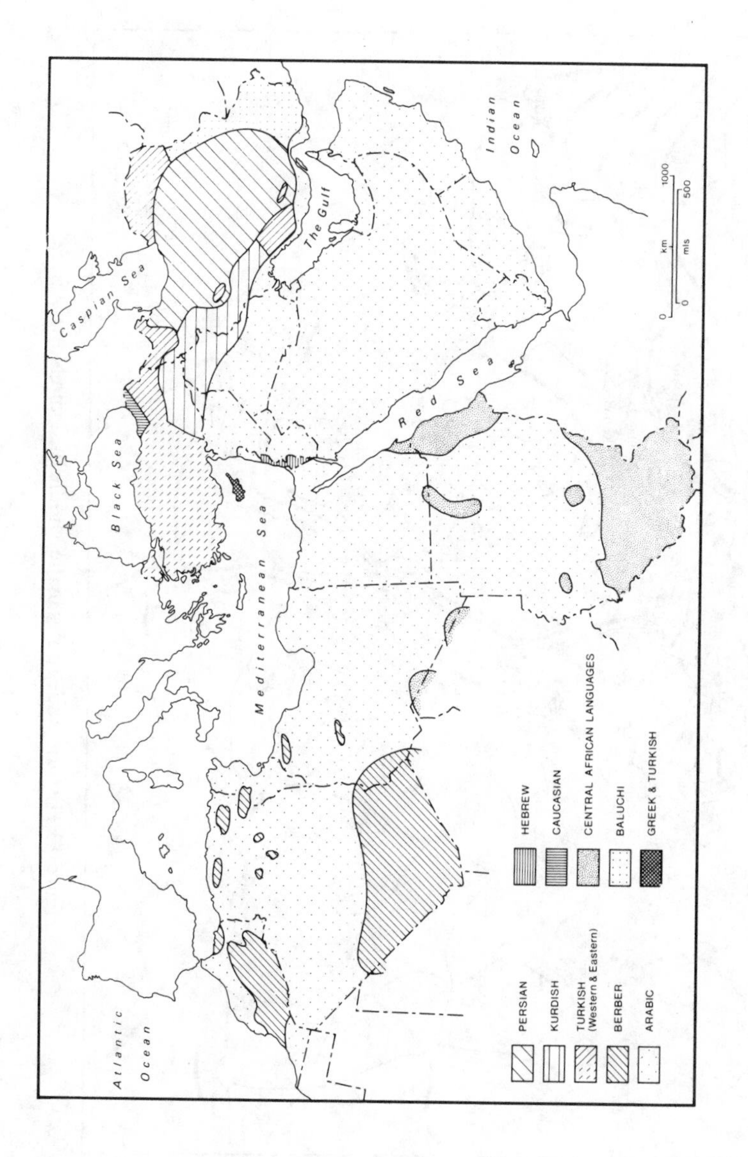

Figure 0.2 Major languages of the Middle East and North Africa

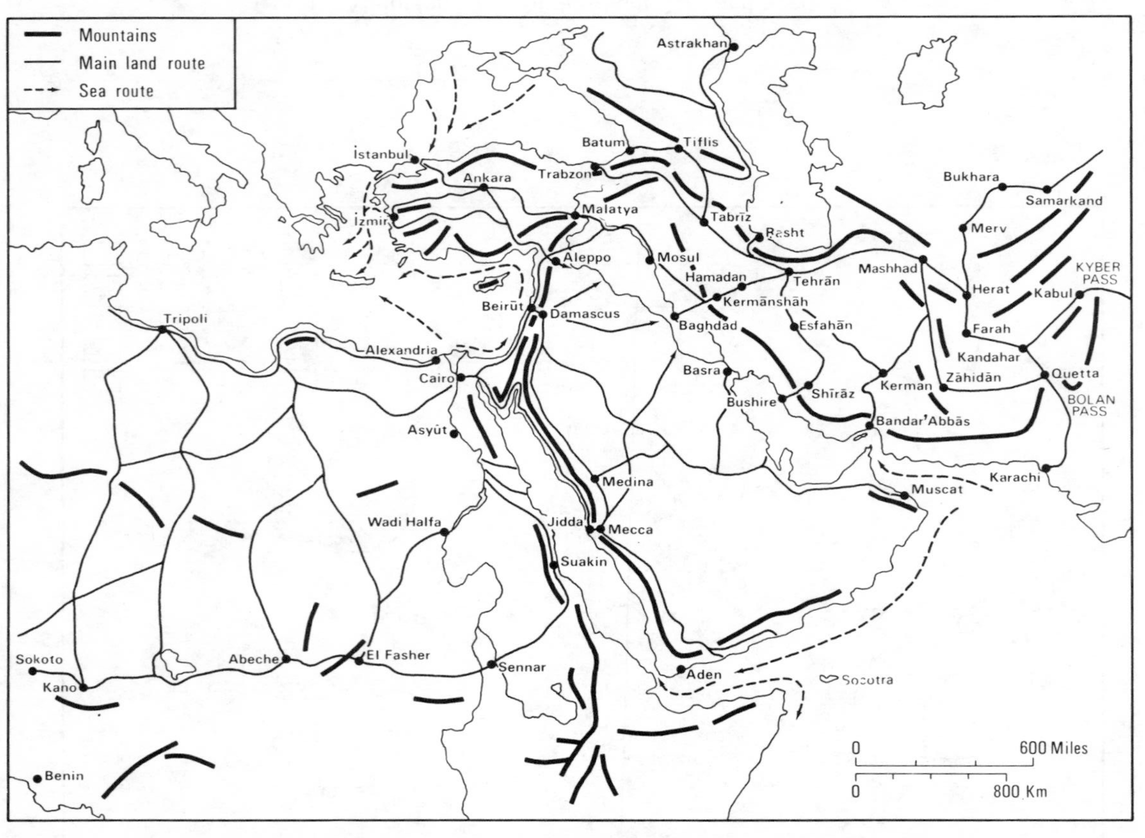

Figure 0.3 Historic trade routes in the Middle East and North Africa

0.3 Geopolitical significance

The region centres around the junction of Africa, Asia and Europe, where the land mass is deeply penetrated by the sea. Maritime trade has long been important and control of ports and constrictions in the seaways has alternately supported cities and disrupted commerce. The four great isthmuses which lie between the arms of the ocean and the inland Caspian Sea have been crossed by major routes for millenia (Figure 0.3), and their control has been of considerable geopolitical concern. This may be illustrated briefly with reference to the nineteenth century. In sailing ship days, when the sea passage from Britain to India via the Cape of Good Hope took anything between five and eight months, the quicker land routes were of vital importance. One route ran from Alexandria through Cairo to Suez and thence by ship down the Red Sea, but it was superseded in 1869 by the Suez Canal, so vital to British imperial strategy, particularly in two World Wars. An alternative route ran across Syria from the eastern Mediterranean to the Euphrates and followed the river as far as the latitude of Baghdād. There one branch continued to Basra and the Gulf, whilst the other struck across Persia and reached northern India via the Bolan or the Kyber passes. A comparatively little used route also ran through Persia, but came down through Tabriz from Batum or Trabzon on the Black Sea. Ease of movement between the Gulf, on the one hand, and the eastern Mediterranean or the Caspian Sea, on the other, made this area one of great concern to both Britain and Russia in the nineteenth century, as well as to the Allies and Germany during the First and Second World Wars. Other routes, of course, converged in the region. Mention might be made of the romantic-sounding but arduous Silk Road, which entered the region from Samarkand and ultimately China through the valleys of the Atrek and Kashaf rivers, and of the slave route which ran from the vicinity of Lake Chad through the Saharan oases to the Mediterranean coast at Tripoli in what is now Libya.

Land and sea routes tied the region into the rest of the world. Along them came people, ideas and plants, as well as the trade which nurtured cities and enriched empires. The region was one great transit zone, a major crossroads in the world. The Middle East flourished economically and politically as long as the ancient routes were used, but decayed when they were either closed, often by political change, or bypassed, as when the Dutch and English began to ship large cargoes from India and the Far East via the Cape. Although the region is now both the source and destination of important commercial and passenger movements, the ancient transit routes have retained some of their former significance. Caravan trails have been replaced by all-weather roads and railways, but both carry a surprising amount of international traffic. Their geopolitical value was emphasized during the Second World War, when sea routes were hazardous and Allied forces had to be supplied by land. Some of them flourished again during the closure of the Suez Canal between 1967 and 1979. Pipelines across the great isthmuses are also of major importance in the movement of oil from the Gulf region and northern Iraq (Figure 0.4); their significance was underlined during the years 1967–79 when the Suez Canal was closed and, since 1979, as a result of the

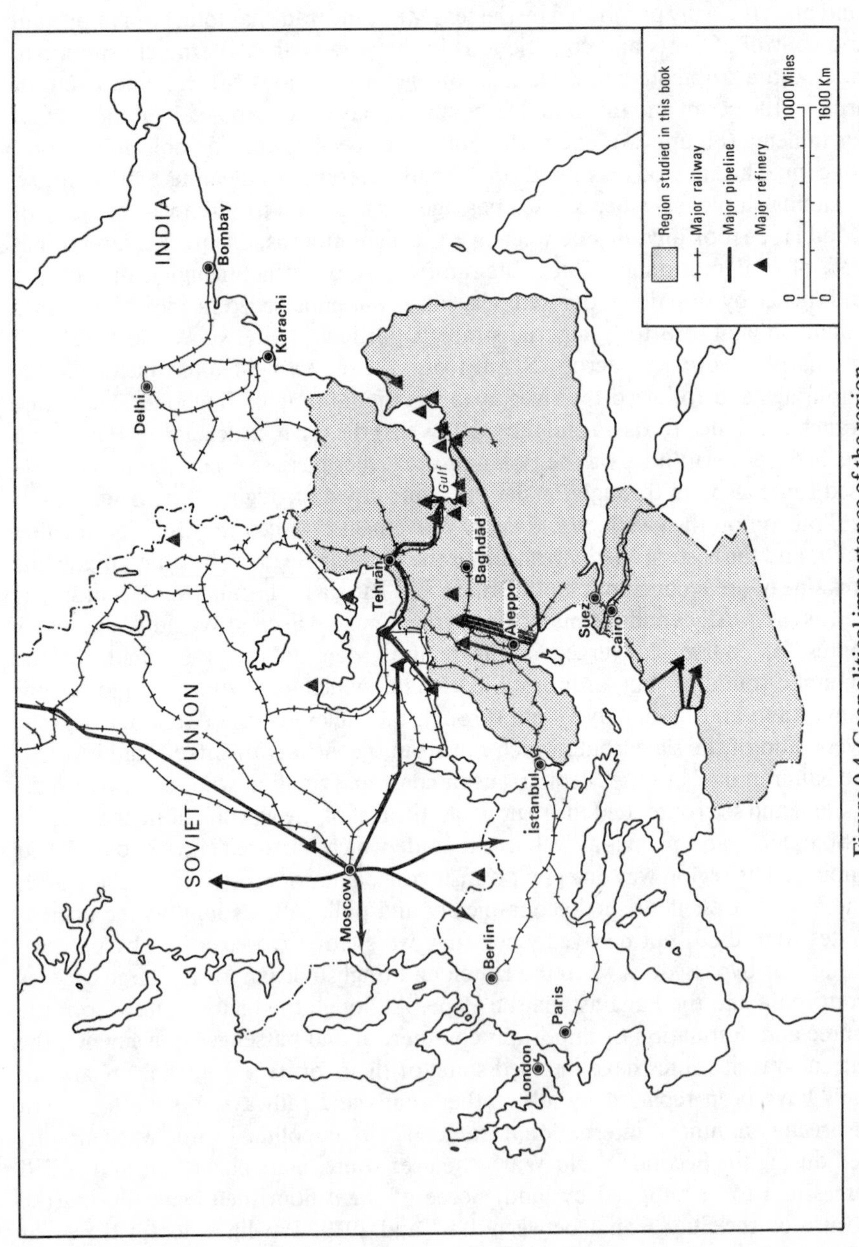

Figure 0.4 Geopolitical importance of the region

Iran–Iraq war which closed Iraq's ports. Meanwhile, tanker traffic in the Gulf, the Red Sea, the northern Indian Ocean and the eastern Mediterranean makes the peripheral seas among the most important highways in the world.[10] International air traffic between Europe and India, Southeast Asia, the Far East and Australia crosses the region, partly to take advantage of approximations to great circle routes, partly to benefit from clear air and fairly stable weather conditions, and partly because of the restrictions on flying over the Soviet Union. Cairo and Beirūt, for example, emerged as major airports. Military staging posts maintained at various times in Libya, Cyprus, Egypt, Oman, southern Turkey and the Gulf testify to the same basic facts.

Movement between different countries in the region increased markedly during the 1960s and 1970s, but two features are particularly important. The first is the growth in the numbers of Muslims going on pilgrimage to Mecca; the total probably exceeded two million in 1987, if Saudi citizens are included. The increase was facilitated by relatively cheap and easy transport, especially by air to Jiddah, and the generous provision of facilities by the Saudi government. Labour migration is the second important aspect of intra-regional movement. It also increased until the late 1970s or early 1980s. The main supplying countries were Egypt and Yemen AR, as well as Jordan and Syria (with Palestinians forming a large component of the migrants from these two). Saudi Arabia and the Arab states on the Gulf were the major recipients. Demand for labour from labour-scarce but capital-rich economies with ambitious development programmes was the main reason for the movement, but limited employment opportunities and low wages at home were also important (see Chapter 5). The slowing down of the oil-rich economies in the early 1980s caused a good deal of return migration which provoked a certain amount of stress.

The volume of tourists from abroad increased during the 1960s, but numbers have moved up and down according to the incidence, real or perceived, of violence in the region and its extension outwards in terrorism, including the hijacking of aircraft. The cumulative effect of this first hand knowledge could be a powerful force in the creation of better international understanding of the region. Real and lasting peace and cooperation will come to the Middle East only when technological developments and economic pressures together create powerful internal interaction. One optimistic forecast saw a common market, a network of international motorways, universities linked by computer systems and a high degree of scientific and cultural cooperation by the end of this century.[11] In the meantime, friction and strife continue in the region.

Military control of the region may be less important in the age of the intercontinental ballistic missile than it was in the 1940s or even 1950s, but, as 'Russia's back garden'[12] and the world's major source of oil, the Middle East retains a world significance (Figure 0.4). Eastward expansion in the eighteenth and nineteenth centuries gave Russia an interest in the northern parts of the region, and she was particularly concerned about control of the Turkish Straits and Persia. In large measure, this became the reverse side of the same coin as British concern with the approaches to India, once protection had been extended to the

Old and new urban architecture in Sharjah

native princes, and the Gulf had been brought under control. The causes of Russian concern have changed, but Turkey's involvement with NATO, as well as Iran's former alliance with the United States in CENTO, and the presence of units of the American naval squadrons in the eastern Mediterranean and the Indian Ocean, are seen by the Soviet authorities as a threat to national security. Like the United States, the Soviet Union has developed a forward defensive strategy which has sought to increase her influence in the region. To some extent Turkey and Iran have now been neutralized by economic links and internal upheaval (fall of the Shah 1979; *coup d'état* in Turkey, 1980) while Russian friendship with some Arab States has prospered on the bases of anti-western feeling, manipulation of the Arab–Israeli conflict and the supply of arms and some technical assistance. The

An international hotel in Beirūt before the outbreak of the Civil War in 1975 (National Council of Tourism in Lebanon. Photograph by Yetenegian)

growth of Soviet sea power and the presence of naval units in the Mediterranean Sea and the Indian Ocean have increased Russian influence, though it has diminished in Egypt since 1976.

Soviet interest in the Middle East may have increased during the 1970s by the need for oil, both for Russia's own needs and to supply to the other members of COMECON. Although the Soviet Union has vast resources of her own (Figure 0.4), demands are rising rapidly as industry and transport are modernized, whilst there is a continuing need to sell crude oil abroad to earn the foreign exchange now vital for continued economic development at home. Most Russian imports were formerly shipped from Iran via the Red Sea and the Eilat–Ashdod pipeline, but later the quantity imported from Libya increased. Russia also assisted in the development of Iraqi and Syrian oil reserves.

The United States, Western Europe and above all Japan are dependent upon the Middle East for supplies of crude oil. Following the Arab–Israeli war of October, 1973 technical and commercial agreements were reached which seem to guarantee the interdependence of industrial Europe and the oil-rich states for at least a decade.

Apart from importing Middle Eastern oil, Western Europe has become linked with the region in other ways. Bilateral agreements have been negotiated between the EEC and various Middle Eastern states, chiefly those around the eastern basin of the Mediterranean Sea. Trading links have strengthened. Turkey is a member of NATO and of the Council of Europe, as well as a long-standing Associate of the European Community which it applied to join in full early in 1987. It is important to remember that Europe is physically united with the Middle East by the Mediterranean Sea and that this facilitates seaborne trade, as well as providing the backdrop to the minuets performed by cruise ships. The Mediterranean itself is one of the seas seriously at risk from pollution, and the bordering states are beginning to cooperate in the management of its waters and resources, whilst competing in the provision of facilities for yet more beach holidays.[13] Another connection was forged in the 1960s and early 1970s by the large numbers of migrant workers employed in Western Europe, particularly on jobs requiring relatively low-level skills. Turks formed a large proportion of the influx. Although some Turks were found in practically every West European country, many settled in West Germany. Their remittances were of vital importance to the Turkish economy during the 1970s. As the recession of the late 1970s and 1980s deepened, return migration became important, but about three million Turkish families are more or less permanently resident in the German Federal Republic.

0.4 Middle East: image and reality

The understanding of mundane issues like these is complicated by the region's historic role in the clash of western and eastern civilizations. The West often sees the Middle East in terms of romantic, slightly unfocused stereotypes which have been fostered by childhood familiarity with some of the tales from the *Arabian Nights* and novels like John Buchan's *Greenmantle*. Knowledge of the region's languages, and thus of much of its history, literature and philosophy, has been confined to the few. Understanding of the region's peoples has been clouded by religious prejudice, stretching back at least as far as the Crusades, and befogged by the West's recently attained military and economic superiority.[14] While the Middle East's great social and economic leap forward is only just gathering momentum, many of the region's advances over the last decade have gone largely unnoticed in the western world. If this volume succeeds in dispelling some of the myths and misunderstandings and in introducing some of the changes taking place in the region today, its appearance will be justified.

Although the Middle East has had an enormous impact on the history and culture of Western Europe and features frequently in the news media, the region has received extraordinarily little attention in English-speaking schools and universities. Caricature has often passed for knowledge. However, the recent Islamic revival, expanding business interests in the region and the importance of Middle Eastern oil and finance have produced a demand for more information and given many people firsthand experience of the region and its peoples. More has been written and published about the Middle East in the last decade than ever before, at both research and popular level, even if this has not penetrated far into the world of formal education. A number of regional geographies and atlases are now available, but there is room for much more geographical research. Work at the local level has often been limited in scope and patchy in coverage; hundreds of topics remain untouched. In the field of human geography, for example, several of the priority areas for research indentified by the Economic and Social Science Research Council in 1971 have particular significance in the Middle East: perception studies; population and migration; regional taxonomy; and the processes of regional economic and social development.[15]

During the next two or three decades, new horizons will open up, as the offshore waters of the Middle East are further explored and exploited to a degree inconceivable a few years ago. The offshore oilfields of the Gulf and Red Sea already yield substantial quantities of oil, and output is likely to increase. Offshore exploration is also taking place in the Black Sea and the Mediterranean. The rich metalliferous muds of the Red Sea are already being exploited by the Sudan and Saudi Arabia Red Sea Commission.[16] The variety of marine life is considerable and most Mediterranean and Middle Eastern countries are making strenuous efforts to develop their fisheries. At the same time, the tourist potential of the coasts and seas is being developed for the benefit of an increasingly mobile and prosperous local population, as well as anticipated increases in overseas visitors. Parts of the eastern Mediterranean have already developed beach tourism, but other parts, notably in the Red Sea and the Gulf, could evolve entirely new types of holiday which might combine the desert safari with coastal cruises to places of historical interest. Developments in Yemen AR are pioneering in this regard. Interest in archaeology throughout the Middle East and North Africa gains momentum all the time, on account of the region's glorious past, the appeal of its superb ruins and the need to establish national identity. Significantly, the nascent discipline of underwater archaeology has developed in the waters of the Middle East where so much remains to be discovered. Several Arab countries in Southwest Asia have positively discouraged tourism, but it is to be hoped that one day visitors will be welcomed. Tourist revenues will be relatively insignificant, except pehaps in Israel, Jordan, Syria and possibly at some future date, Lebanon, so that many of the host countries have a unique opportunity to confine the number of visitors to the absolute capacity of the amenities they come to enjoy.

The Middle East is now experiencing some of the environmental problems associated with advanced industrial economies. Rapid growth of urban centres since the end of the Second World War has produced a few cities, often national capitals, which dwarf all others in the state. Accompanying this process has been a rise in urban standards of living, with the introduction of modern methods of industrial production and transportation. The number of cars and lorries in towns and cities has increased enormously, while sprawling industrial development is now characteristic of the outskirts of the large urban centres.

The serious effects of such rapid growth on the environment have not always been appreciated. Large areas of cultivated or potentially cultivable flat land have been swallowed up by the expansion of many cities. The growing concentrations of population have severely strained locally available resources, particularly of fresh water. This has necessitated large-scale capital investment in new projects in an attempt to overcome the severe water shortages often experienced during the summer. Linked to fresh water supply is the problem of sewage disposal. Until about 1960, most of the cities of the region did not possess any form of organized sewage removal or treatment system. Twenty-five years later the situation has been little improved in many countries, for while new and efficient sewerage systems are being built for the largest cities, most others are not likely to obtain sewage treatment facilities within the next ten years. It is increasingly likely that water supplies will be polluted.

In many places, especially where industrial development has become important, new pollutants are beginning to cause ecological damage. Petrochemicals and other chemical waste products present the greatest dangers to what are often very fragile ecosystems. Oil slicks covered huge areas of the Gulf at various times during the Iran–Iraq war when oil installations were damaged.[17] Pollution by solid waste from urban centres is another growing problem throughout the region. Waste disposal methods are generally primitive. The usual practice is either to throw domestic and industrial refuse into the nearest watercourse, whether dry or flowing with water, or to tip it indiscriminately just outside the city boundaries. Once again, the risk of disease is very real.[18]

Regrettably, atmospheric pollution is already characteristic of the Middle Eastern urban environment. Exhaust fumes from automobiles are probably the greatest single cause of this nuisance, but domestic and industrial consumption of hydrocarbons is also an important contributory factor. In winter, Ankara is frequently blanketed by a thick brown smog which collects in its enclosed basin as a result of temperature inversions and the use of lignite in central heating systems.

Photochemical smog is now apparent in the larger cities during the hot summer months, and it is likely to grow worse becasue of the lax regulations governing automobile exhaust emissions. Closely associated with automobile and air pollution is the growing amount of urban noise and its deleterious effect on the inhabitants. From this point of view, cities such as Cairo, Tehrān, Tel Aviv and İstanbul are now as unpleasant to visit and work in as London or New York. The tranquillity and charm which recently characterized Jerusalem and Eşfahān, for example, have been lost, possibly for ever. Old city centres have been devastated in the urge to modernize, but conservation policies, somewhat belatedly, are now beginning to take effect.

0.5 Perspective

The authors of this book cannot claim to be free from prejudice or to have attained perfect knowledge, but in discussing the geography of the Middle East they hope to display reasonable sympathy with the people of the region. Their object is to deal with the region as it is today. A historical perspective is used only where it has direct bearing on the present situation. The contention is that the region is as vital in world affairs towards the end of the twentieth century as it ever was. More than ever it is a central stage of the 'Great Debate' between West and East, and the source of considerable political friction, as well as a danger to world peace. But the Middle East is a region eminently worth studying in its own right. It was one of the major hearths of civilization and the birthplace of Judaism, Christianity and Islam. Man/land relationships are particularly close, and, whilst the physical environment has affected human activity profoundly, man has in turn left a deep imprint on the land in his struggle for survival. The settlement history of the region has demonstrated both man's destructive capacity and his ability to derive a reasonable living, when he adjusts to his environment with skill and ingenuity. The testimony of the Middle Eastern landscapes should be more widely known in an age of ecological concern. Although the region's social and economic life was once markedly more advanced than that of Europe, the Middle East is now, in many respects, one of the less-developed parts of the world. The long decay experienced by the region and its recent efforts to revive itself are a fascinating study. Some states in the region possess immense wealth which, wisely invested, can create a bright long-term future for their own people and the region at large.

The approach to regional geography used here is broadly conventional. The first part of the book deals with themes on a regional scale. It begins with discussions of the fixed and dynamic elements in the physical environment and man's interaction with them over time. This provides the setting for an analysis of socio-economic activity, including land use. General problems of population growth and economic development are discussed as a prelude to an outline of modern industrial development, contemporary trading and financial patterns, and an evaluation of the role of petroleum in the region. The final section, introduced by an essay on the political geography of the region, consists of ten studies which explore the geography of the modern states comprising the region as functional subregions. The approach in these chapters is thematic. Themes of human geography outlined in the first half of the work are studied in some detail and are chosen to illuminate the personality and evolution of the countries concerned. Physical geography is largely assumed in these chapters, but its role has frequently been crucial in this often ecologically fragile region.

References

1. (a) C. R. Koppes, 'Captain Mahan, General Gordon and the origin of the term "Middle East" ', *Middle East Studies*, **12**, 95–98 (1976).
   (b) R. H. Davison, 'Where is the Middle East?', *Foreign Affairs*, **38**, 665–675 (1960).
2. V. Chirol, 'The quest for the Middle East', *The Times*, beginning 14 October 1902, reprinted as *The Middle East Question, or Some Political Problems of Indian Defence*, John Murray, London, 1903.
3. *Parliamentary Debates, House of Lords*, 5th Series, **7**, col. 575.
4. (a) C. G. Smith, 'The emergence of the Middle East', *Journal of Contemporary History*, **3**, 3–17 (1968).
   (b) G. Macmunn and C. Falls, *History of the Great War. Military Operations: Egypt and Palestine*, vol. 1, HMSO, London 1928, 94–97.
5. (a) P. Lorraine, 'Perspectives of the Near East', *Geogr. J.*, **102**, 6–13 (1943).
   (b) L. Martin, 'The miscalled Middle East', *Geogr. Rev.*, **34**, 355–356 (1944).
6. (a) W. B. Fisher, 'Unity and diversity in the Middle East', *Geogr. Rev.*, **37**, 414–435 (1947).
   (b) R. Patai, 'The Middle East as a culture area', *Middle East Journal*, **6**, 1–21 (1952), reprinted in *Readings in Arab Middle East Society and Cultures*. (Eds A. M. Lutfiyya and C. W. Churchill) Mouton, Paris and The Hague, 1970, 187–204.
7. R. W. Bulliet, *Conversion to Islam in the Medieval Period: An Essay in Quantitative History*, Harvard University Press, Cambridge, Mass. and London, 1979.
8. *Qur'ān*, II, 192f.
9. G. H. Blake, J. C. Dewdney and J. K. Mitchell, *The Cambridge Atlas of the Middle East and North Africa*, Cambridge University Press, 1987.
10. A. D. Drysdale and G. H. Blake, *The Middle East and North Africa: A Political Geography*, Oxford University Press, New York and London, 1985.
11. The Association for Peace, *The Middle East in the Year 2000*, Tel Aviv, 1969, 1–40.
12. W. Laqueur, *The Struggle for the Middle East. The Soviet Union and the Middle East 1958–69*, Penguin, Harmondsworth, 1972, 221.
13. G. Luciani (Ed), *The Mediterranean Region*, Croom Helm, London and St Martin's Press, New York, 1984.
14. (a) N. Daniel, *Islam, Europe and Empire*, Edinburgh University Press, 1966.
    (b) B. Lewis, *The Middle East and the West*, Weidenfeld and Nicolson, London, 1964.
    (c) E. W. Said, *Orientalism*, Routledge and Kegan Paul, London, 1978.
15. M. Chisholm, *Research in Human Geography*, Social Science Research Council, Heinemann, London, 1971, 71–72.
16. A. M. Farid, *The Red Sea: Prospects for Stability*, Croom Helm, London and St Martin's Press, New York, 1984.
17. (a) 'Gulf states pool resources to combat slick', *Middle East Economic Digest*, 20 May, 1983, 14–17.
    (b) The Mediterranean Sea – special issue, *Ambio*, **6** (1977), No 6.

18. (a) P. Beaumont, 'The Middle East - environmental, management problems', *Built Environment Quarterly*, **2**, 104–112 (1976).
(b) P. Beaumont, 'Urban water problems', in G. H. Blake and R. Lawless (Eds), *The Changing Middle Eastern City*, Croom Helm, London, 1980, 230–50.

CHAPTER 1

Relief, Geology, Geomorphology and Soils

1.1 Relief

In a region as large and diverse as the Middle East, it is difficult to describe topographical conditions in a simple way. However, relief factors have played an important and often directly controlling role on the human occupancy of the region, and it is, therefore, essential that at least a brief sketch is given of the major features. For convenience the region can be divided into a northern mountainous belt, comprising the states of Turkey and Iran, and a southern zone made up largely of plains and dissected plateaus (Figure 1.1).

In Turkey, two major, though not continuous, mountain belts are usually recognized. The Pontus Mountains are an interrupted chain of highlands paralleling the Black Sea coast. They rise in altitude in an easterly direction to heights of more than 3000 m south of Rize. Inland from the southern coast of Turkey is the much more formidable range of the Taurus Mountains. Being less dissected by river systems than their northern counterparts, these uplands have always presented a considerable barrier to human movement, so focusing routes through passes such as the Cilisian Gates, to the northwest of Adana. Between the two ranges the central or Anatolian Plateau lies sandwiched. This is almost everywhere above 500 m in height and relatively isolated from the coastal regions.

In eastern Turkey the Pontus and Taurus ranges coalesce in a complex upland massif near Mount Ararat (5165 m) where crest elevations surpass 3000 m. From here eastwards into Iran the mountain chains divide once more. In the north along the southern shore of the Caspian Sea are the Elburz Mountains, which in Mount Damavand (5610 m) contain the highest peak of the region. Although these uplands are relatively narrow in a north to south direction, they present the greatest barrier to human movement anywhere in the Middle East. Southwards from Mount Ararat, overlooking the Tigris–Euphrates lowlands and the Gulf, stretch the broad Zagros Mountains. They attain a maximum height of 4548 m in Zard Kuh. These parallel ranges with their wide upland valleys have never presented the same degree of difficulty of movement as the Elburz.

In eastern Iran, on the borders of Afghanistan, a very complex pattern of mountain ranges, usually described as the Eastern Iranian Highlands, are found. These are lower than both the Elburz and Zagros Mountains, attaining only 2500 m in altitude. Surrounded by these highlands is the Central Plateau of Iran. This, like the Anatolian Plateau, is almost everywhere above 500 m in height, and

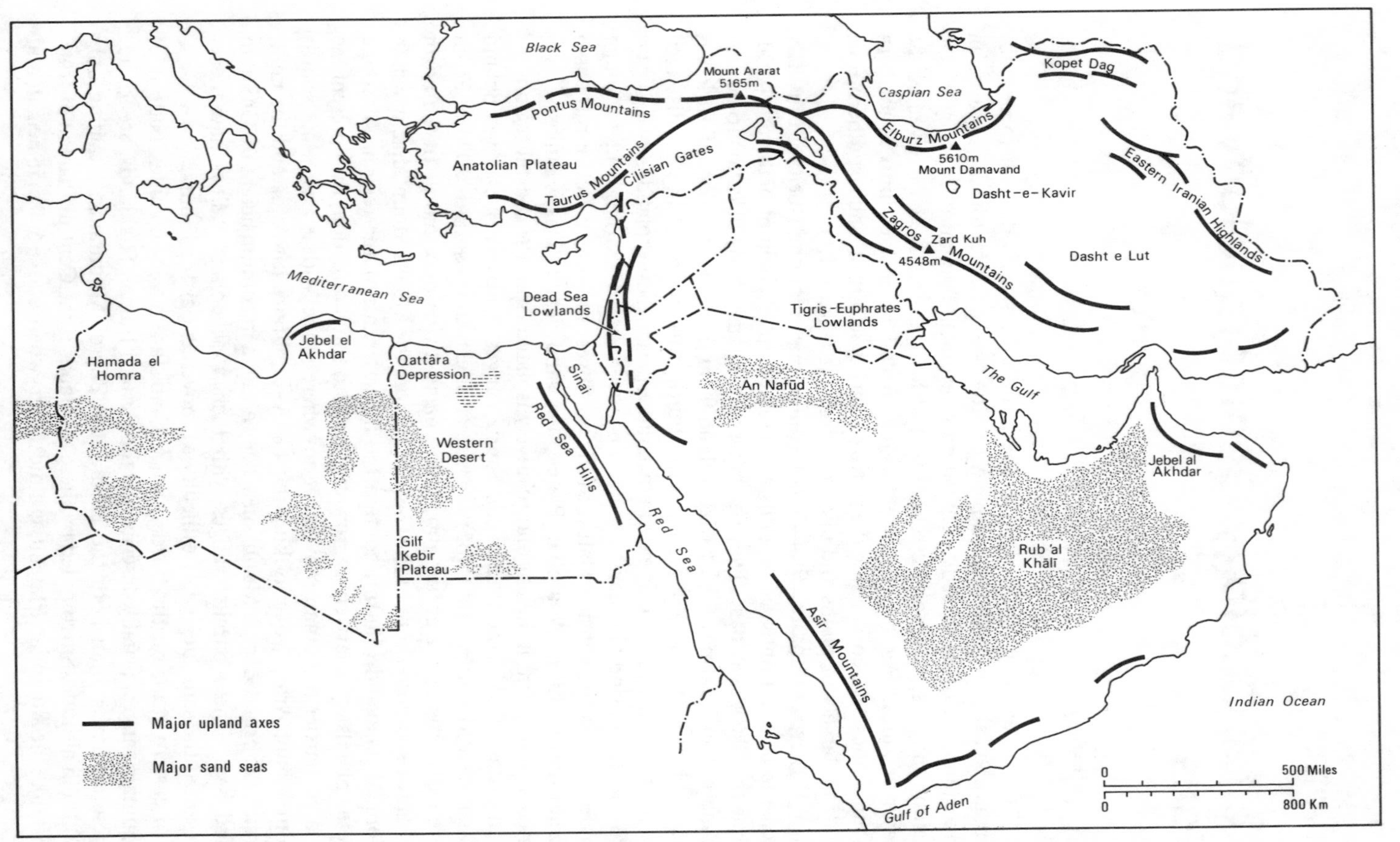

Figure 1.1 Major relief features of the Middle East and North Africa

is subdivided into two major basins both of inland drainage. The Dasht-e-Kavir forms a huge salt desert in the north, while the term Dasht-e-Lut is used for the southern basin.

In the southern region of plains and dissected plateaus a useful east/west division can be made along the line of the Red Sea and the Suez Canal. In eastern Egypt the Red Sea Hills are the major upland area, while the Nile delta forms the major lowland. To the west, a relatively simple topographical picture can be drawn, with narrow lowlands along the coast rising inland to upland plateaus along the southern margins of Libya and Egypt. Even on these interior plateaus heights of more than 1000 m are rarely surpassed. Large sand seas form important landscape features in this zone. An important exception to this general description is the existence of a small upland zone, the Jebel el Akhdar, in north-east Libya. Although only some 1000 m in height this region has played an important role in the human settlement of the region.

To the east of the Red Sea the highest land in this zone is found at the south-western corner of the Arabian peninsula. Here in Yemen altitudes of more than 3700 m are attained. Highland also occurs along the whole of the western part of Arabia, with the general level of the land declining to the north and east. In central Arabia, characteristic features of the relief are a series of westward facing escarpments, in arc-like form around the main highland mass of the west coast. Although none of these landforms are particularly high they have concentrated the routes across the peninsula towards the regions of most easy access. In the Levant, upland areas are found in proximity to the coast, with a gradual decline in altitude towards the interior. Heights here too can be considerable, reaching almost 3000 m at Mount Hermōn. The pattern of relief in this region is complicated by the existence of the north-south fault zone of the Dead Sea Lowlands, which has dissected the upland belt to form a trough-like region descending to 300 m below sea level.

Stretching from northern Iraq to the coast of the Indian Ocean in Oman is the largest lowland belt of the region. In Iraq it is crossed by the large rivers Tigris and Euphrates and in the southern parts relief is minimal. Some of the oldest human settlement in the world is found here. The lowland belt continues as an attenuated zone along the western shore of the Gulf to broaden into an extensive plain in southeastern Arabia. In this latter zone the largest sand sea in the world, the Rub'al Khālī, is situated. Here some of the dunes are more than 200 m in height. At the easternmost tip of Arabia, a belt of uplands, also called Jebel al Akhdar ('Green Mountain') reach a maximum height of more than 3000 m. This region has always been an extremely isolated part of the region cut off by both sand and water from adjacent areas.

1.2 Geology

In recent decades our knowledge of the evolution of the continental masses has increased tremendously owing to detailed geological and geophysical investigations in many parts of the world. As a result of this work new theories of sea floor spreading and plate tectonics have been put forward, and widely accepted by most earth scientists.[1] The basic idea of these theories is that the continental masses are embedded in huge plates which move over the denser material beneath the earth's crust. These plates, and the continents on top of them, travel across the surface of the earth probably as the result of convectional currents acting deep within the earth. This movement can lead to the plates coming into contact with one another, so producing crush zones, or mountain ranges. In such contact zones, parts of one or other of the colliding plates are dragged down towards the centre of the earth, along what are termed Benioff or subduction zones. In contrast, where two plates are moving away from one another, upwelling of magma occurs usually beneath the ocean floors, to produce sea floor spreading.

Geologically speaking, the Middle East and North Africa is a particularly complex region, as a number of different continental plates have come into contact here. North Africa and Arabia represent the remnants of an ancient continental landmass in the southern hemisphere known as Gondwanaland. During the Mesozoic period this landmass, composed mainly of Palaeozoic and older rocks, began to split up and drift northwards. Eventually these moving continental plates made contact with a similar landmass in the northern hemisphere, known as Laurasia. When this occurred, probably during the Tertiary period, the younger sediments which had formed between and overlapped onto the ancient and stable continental platforms in the Tethyan Sea, buckled and contorted as the result of compressive stresses to produce the mountain ranges which stretch through the region from the Alps to the Himalayas. Although this simple picture of events gives a reasonable idea of what happened, a glance at a detailed geological map illustrates just how complex the real situation is. Indeed, it is still true to say that the exact inter-relationships of the different continental plates in the region are not known with any degree of certainty.

In the Mediterranean and Middle Eastern region three major plates can be identified. These are the African, Eurasian and Arabian plates, and the boundaries between them are the Azores–Gibraltar ridge and its extension across North Africa, the Red Sea, and the Alpide zone of Iran (Figure 1.2).[2] Seismic activity in Yugoslavia, Greece and Turkey is much more pronounced than in the western Mediterranean. The reason for this appears to be the existence of two small, but rapidly moving plates, named the Aegean and Turkish plates respectively. Observations reveal that the Aegean plate is moving towards the southwest relative to both the European and African plates. As a result of its motion, it is overthrusting the Mediterranean Sea floor south of the Cretan arc and thus causing it to sink beneath the Aegean Sea.

The Turkish plate is moving almost due westwards with respect to the Eurasian and African plates. Its northern boundary is the North Anatolian fault, but the southern limit is not fully defined. The existence of these two small plates and the role they play in the regional tectonics is possibly explained by differences in behaviour between continental and oceanic lithosphere. Material forming the

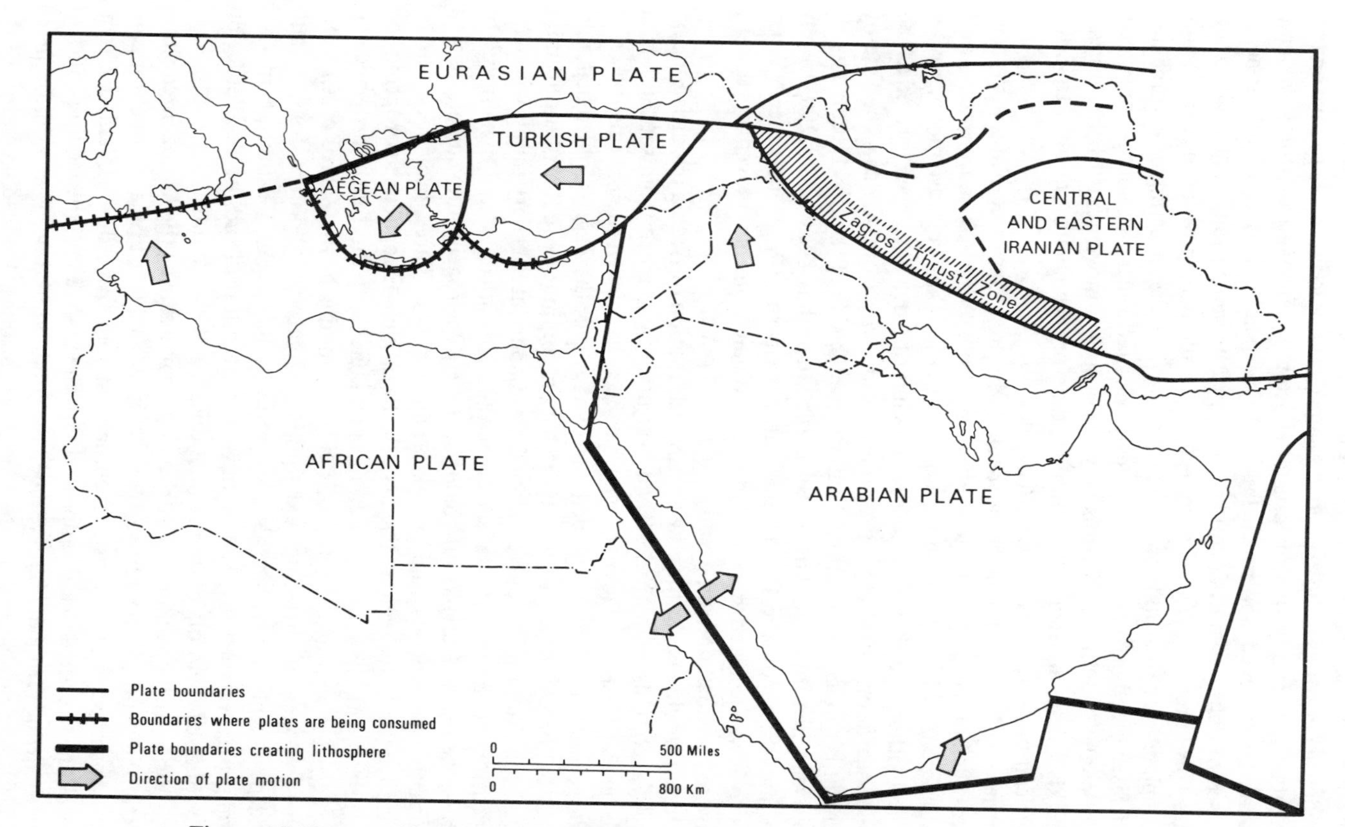

Figure 1.2 Structural units in the Middle East (modified by permission of *Nature* 1970, 1970)

continents is light in density and, therefore, cannot be drawn down into the denser mantle of the earth. When two plates of such material come into contact tremendous energy has to be expended against gravitational forces to thicken the continental crust and so permit crustal shortening. In contrast, the same end result can be achieved and much less energy expended if oceanic lithosphere, composed of denser material, can be consumed by being dragged down into the mantle. This latter is indeed what appears to be happening with the Aegean and Turkish plates, for their motion is such that further overthrusting of continental material is avoided in Turkey, and, instead, Mediterranean Sea floor is consumed in front of the Cretan arc.[3] In this way the observed movements of the Aegean and Turkish plates are such as to minimize the work which must be done to move the African plate towards the Eurasian one.

Further south in the region it has been discovered that three plates also meet towards the southern end of the Red Sea. These are the African, Arabian and Somalian plates.[4] In this area it would seem that all the plates are moving away from each other, and that the Red Sea and Gulf of Aden have been formed as the result of sea floor spreading as Arabia moved away from Africa. The rate of movement is thought to be about 1 cm/yr on each side of the rift. Detailed work in the Dead Sea lowlands has revealed that there has been about 100 km left lateral movement on the Dead Sea Fault system since Miocene times.[5] What seems to have occurred here is that the Arabian plate has moved northward relative to the small and apparently fixed Sinai plate.[6]

In Iran it would appear that the northwards motion of Arabia towards Eurasia has been accomplished by widespread overthrusting in a belt from southern Iran to the central Caspian. The net result has been to thicken the continental crust over large areas. Iran can be divided into two regions: the Zagros folded belt, and the rest of the country. In the Zagros region continuous sedimentation under tranquil conditions has occurred from Cambrian to late Tertiary times, when the sediments were folded into a series of parallel anticlines and synclines.[7] In contrast, the rest of Iran has suffered more severe epeirogenic movements, as well as considerable igneous and metamorphic activity. Three provinces can be identified in this latter region. The first, the Reza'īyeh–Eşfandegheh orogenic belt runs parallel with the Zagros mountains and unites with the Taurus orogenic belt of Turkey. It is separated from the Zagros Mountains by the Zagros crush zone, which is an area of thrusting and faulting. Central and eastern Iran, a fault bounded, roughly triangular shaped region with its apex in the south, forms the second province. The Elburz Mountains of northern Iran and the parallel region to the south of them make up the final division of the country.

The northward movement of Africa and Arabia during the Mesozoic caused a reduction in width of the Tethyan Sea. This was achieved by a subduction zone which consumed oceanic crust. Eventually at some time during the late Cretaceous all the oceanic crust disappeared into the mantle and the leading edges of the African and Arabian plates reached the subduction zone. When this occurred ophiolites were emplaced along the Zagros crush zone at the leading margin of the Arabian plate.[8] The Zagros sedimentary basin, the present Zagros Mountains, continued as the shelf of the old Afro-Arabian continent, with continuing sedimentation, mostly of a carbonate nature.

Although relatively little drift occurred in Eurasia during the Mesozoic period, it was sufficient to cause a partial break-up of the northern continents. This led to the formation of micro-continents, such as central and eastern Iran, which were separated by narrow oceanic areas of the Red Sea type. Continued northerly drift of the African and Arabian plates during the early Tertiary era resulted in the closure of these basins and the emplacement of melange complexes along their former axes. Eventually, therefore, the continental plates of Eurasia, Africa and Arabia became connected by a series of micro-continents, with only a few shelf seas still existing in the region. At this time, to the east, the northward movement of the Indian plate was underthrusting Eurasia, and giving rise to the Himalayan ranges.

Yet another period of compression occurred in the late Tertiary period, associated with the formation of the Red Sea and Gulf of Aden, and also the southeastwards movement of Eurasia. These movements, which are continuing at the present day, led to the underthrusting of Iran by the Arabian plate and resulted in the complex folding of the Zagros Mountains, together with folding and faulting in other parts of the country.

Bearing the above outline in mind, the distribution of rocks of different ages is relatively simple to explain, at least in general terms. The oldest rocks, of Pre-Cambrian and Palaeozoic age, are found on the stable masses of both North Africa and Arabia, with occasional smaller outcrops occurring in other places throughout the region. These basement complex rocks are only exposed, however, on a large scale in eastern Egypt and western Saudi Arabia (Figure 1.3).

Northwards, progressively younger rocks, mostly of sedimentary origin, which have overlapped onto the basement complex, as it was downwarped during its northwards passage, are found. In general the older sedimentary rocks of both North Africa and Arabia tend to be of continental origin and are represented by such rock types as the famous Nubian Sandstone. Further north still, as one moves into the Levant region and also southern Iran, marine sediments, in particular limestones and marls of Mesozoic age, make up a much larger proportion of the outcrop. These sediments, it is believed, were deposited in the Tethyan Sea, between the respective remnants of Gondwanaland and Laurasia. Calcareous sediments such as these have tremendous importance in the economic life of the region, for it is in these rocks that the oil reserves are concentrated. For example, in Iran, the Asmari Limestone is the most important reservoir rock for petroleum accumulation.

Rocks of Tertiary age, mostly marine sands, clays, marls and limestones, although some continental deposits also do occur, tend to be confined to the lowest lying areas of the region. In particular, sediments of this period are found extensively along the North African coast in Libya and Egypt, as well as along a wide zone paralleling the Gulf. More restricted outcrops of Tertiary rocks also occur in the inland basins of Iran and Turkey.

Thick Quaternary sediments, almost all of which are unconsolidated, are

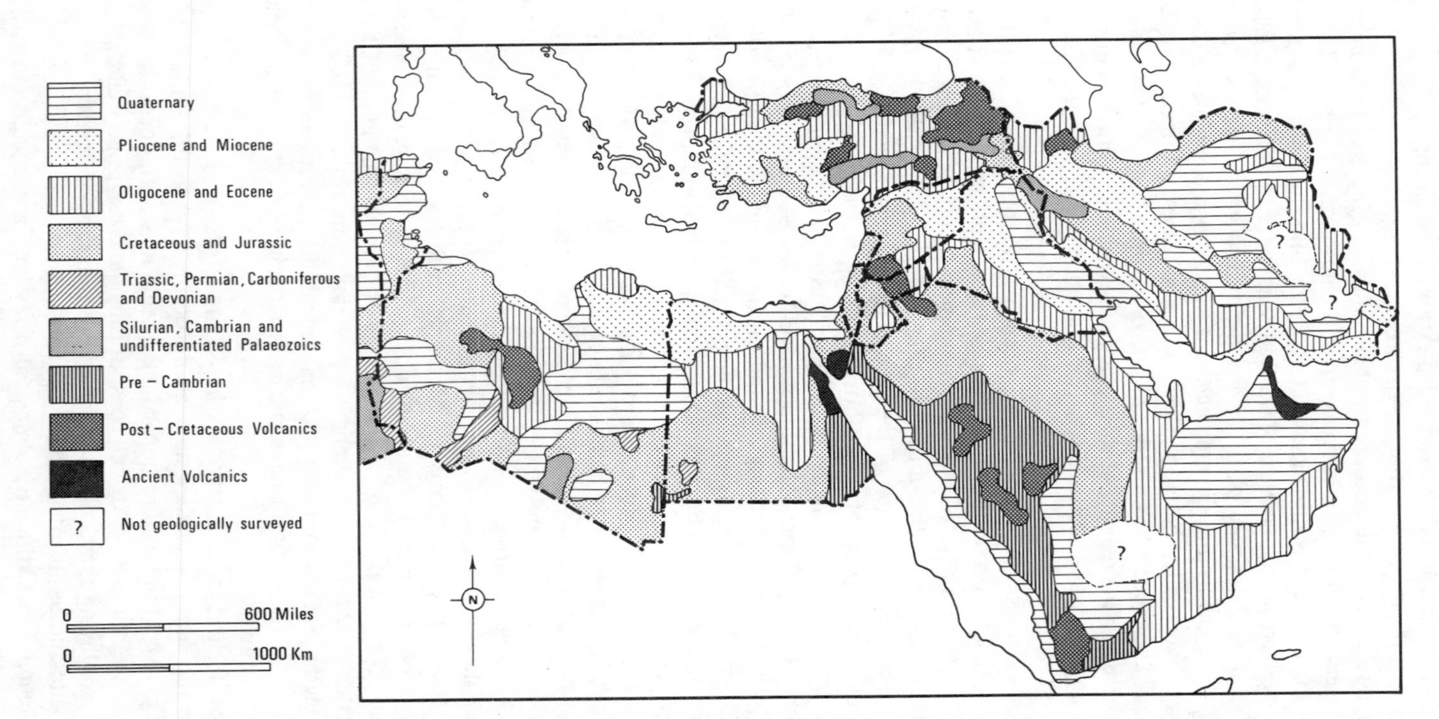

Figure 1.3 Geology of the Middle East (reproduced by permission of Oxford University Press)

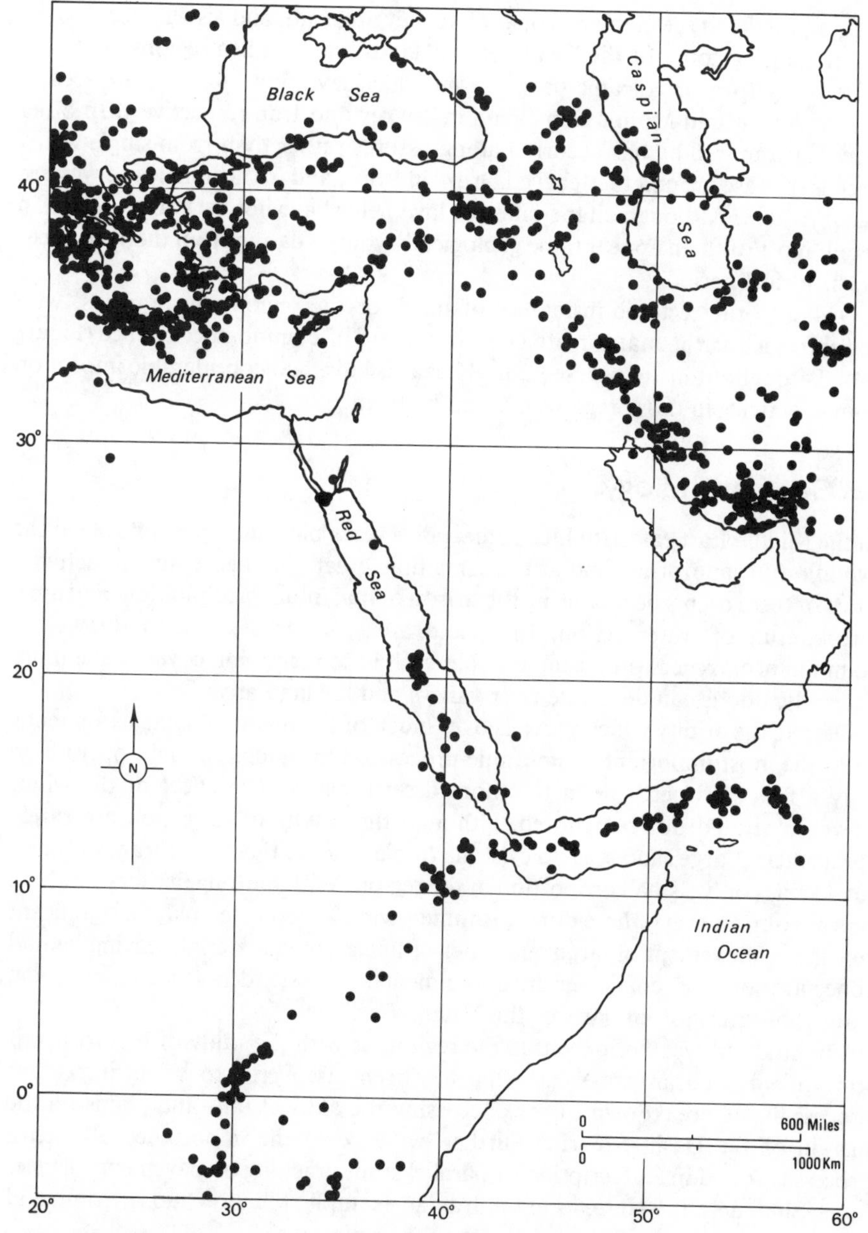

Figure 1.4 Epicentres of earthquakes (reproduced by permission of *Nature*)

confined to the upland basins of the highland zone, and the major valley systems of such rivers as the Tigris–Euphrates and the Nile.

A characteristic feature of the region is the widespread occurrence of eruptive rocks, chiefly basalts. They are found associated with zones of structural weakness, especially in the highland zones of Turkey and Iran, and also adjacent to the major faulting zones of the Dead Sea Lowlands and Red Sea regions.

Perfectly formed volcanic peaks, from which lava flows radiate, are seen in Mount Ararat and Mount Damavand in Turkey and Iran respectively. In other places lava upwelling has occurred along fissures rather than from single vents. Such is the case around the Jebel ed Druze in Syria, and in the uplands of western Saudi Arabia. Although all the different lava fields have not been investigated in detail, most of them appear to be geologically young, dating from the Tertiary to the historical period.

Owing to the tectonic instability of much of the region, earthquakes, with epicentres along the major plate boundaries, are of common occurrence (Figure 1.4). Throughout history these natural hazards have had considerable impact on human activity in those regions affected by them.

1.3 Geomorphology

In the Middle East catastrophic geomorphic events play an important role in the evolution of landforms, and at the same time greatly influence human activity. Most of these events occur during the season of maximum precipitation, and are a direct result of water action. In upland areas, landslides and mudflows are common occurrences on unstable slopes with little vegetation cover, while more generally, floods can devastate river valleys and lowland areas.

Despite the aridity which prevails over much of the region, fluvial action tends to be the most important geomorphic process, even though it might operate on only a few days per year in the more desertic parts. The effect of the wind, however, should not be ignored, although the results of its action are rarely spectacular. Large sand seas do exist in Arabia, Libya, Egypt and Iran, but their total area is only a small proportion of the region. With the long dry season which prevails almost everywhere during summer, wind action often plays a significant role in the movement of large quantities of fine grained material, leaving behind concentrations of coarse grained sediments. Loessic deposits are common throughout the eastern parts of the Middle East.

The study of weathering within the region, as with the study of landforms, is still limited. In Iran salt weathering has been discovered to be an important process in the breakdown of rocks crossing the alluvial fans and plains on the margin of the Dasht-e-Kavir.[9] Further evidence of the importance of such a process is found in a description of parallel stone cracking on pavement surfaces in the Sinai desert.[10] Here, as in the Iranian example, the growth of gypsum and salt crystals in fine crevices seems to be the prime cause.

The complex physical nature of the Middle East makes it extremely difficult to generalize about landform types. It is, however, useful to employ the concept of a simple basin model to provide a framework with which to look at both the physical and the human environment. In its simplest form the basin is divided into four zones (Figure 1.5). These are upland, alluvial fan, alluvial plain, and salt lake or salt desert.

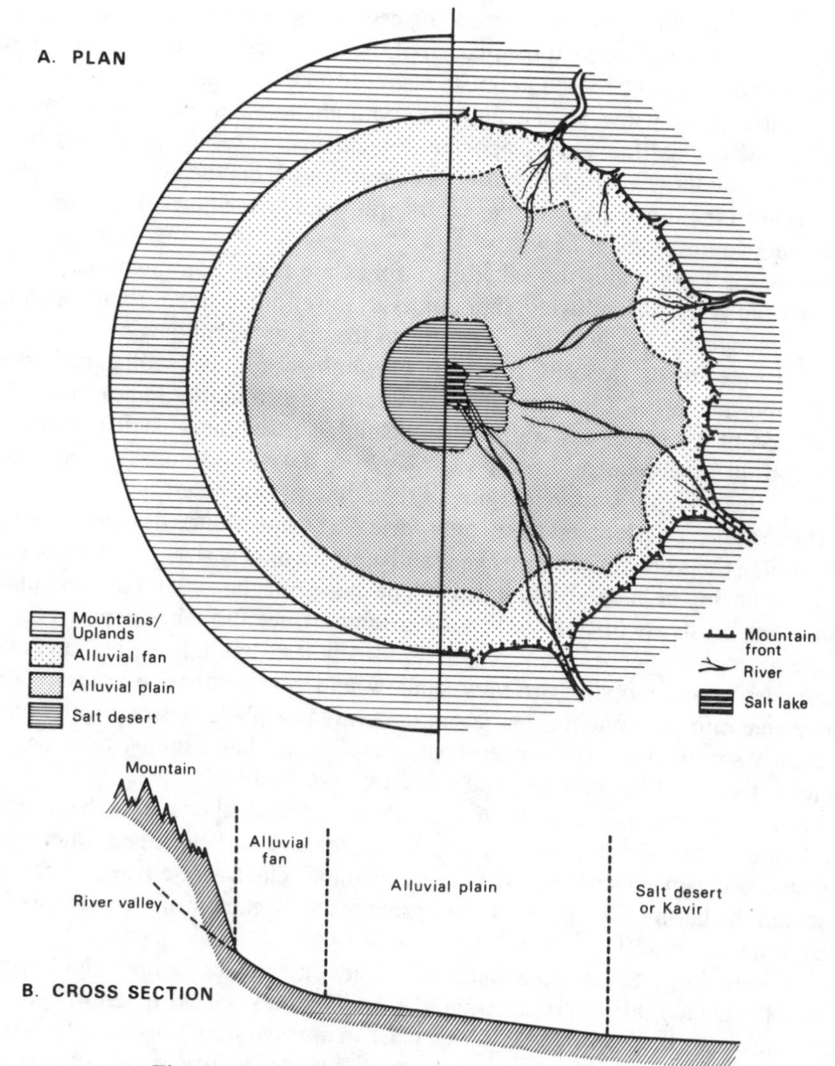

Figure 1.5 Basin model with four environmental zones

The upland zone is characterized by the outcrop of usually resistant rock types and very steep slopes. The amplitude of relief may be hundreds or even thousands of metres. Weathering profiles are thin or absent, and coarse scree-like material mantles the lower slope segments. The river valleys tend to be flat floored, owing to the deposition of material eroded from the adjacent slopes. Freeze–thaw action is an important weathering phenomenon, and consequently angular debris is common. In general, this is a zone of maximum erosive activity as the result of severe weathering conditions, mass movement on the steep slopes, and considerable fluvial activity along the major valley systems.

The three other morphological zones are essentially depositional in nature. The alluvial fan zone is situated at the margin of the upland zone and is characterized by slopes with angles between about 15 degrees and three degrees. Amplitude of relief is usually very low. Coarse material, both angular and rounded, makes up most of the alluvial fan zone. Deposition of sediment is brought about by both fluvial and mudflow activity. In most cases, mudflows from the upland areas only penetrate to the upper parts of the fan region, while river flows often continue onto the two lower zones as well.

The alluvial plain and the salt desert zones are really continuations of the alluvial fan zone. The alluvial plain possesses lower angle slopes than the fans, and, in many parts of the Middle East, forms the major morphological zone. It is chiefly formed by the depositional action of running water, but aeolian activity is also important here. The finer surface material is commonly removed by the wind, leaving a stone or desert pavement behind. The sediment carried or moved by the wind is subsequently redeposited as loess or as sand dunes in the lower parts of this zone, or on the margins of the adjacent salt desert.

The salt lake, or salt desert zone, only occurs in basins with no drainage outlets. Material in this zone is normally fine grained, of silt or clay size, with the coarser material having been deposited during its passage over the alluvial fan and plain zones. Salt crusts are often found on the ground surface and saline deposits occur in the soil profile. Slopes are very gentle, usually less than one degree, and as a consequence water bodies are very shallow and cover extensive areas. Water movement into this zone from adjacent uplands takes place mostly in late winter and early spring as the snow pack begins to melt. By late summer all standing water has normally evaporated and salt crusts are formed.

In any one region, not all of the zones outlined above may be found. Undoubtedly the model can be best applied in Iran and Turkey, but elsewhere also the ideas and concepts generated by it are of value. The Dasht-e-Kavir of central Iran and the basin of the River Jordan provide excellent environmental units for illustrating the model.

Our knowledge of environmental and climatic changes in the Middle East during the Quaternary era is still fragmentary. This makes it extremely difficult to correlate events there with those taking place in northwestern Europe at the same period. In this latter region, an accumulation of evidence from many sources has shown that the last glacial period, known as the Wisconsin–Würm–Weichsel, occurred between about 70,000 years and 10,000 BP (Before Present), and was followed by the Holocene Period.

During the latter part of the Wisconsin–Würm–Weichsel period, between 25,000 and 15,000 years ago approximately, a major ice advance occurred, which, at its maximum, covered a large proportion of the northern hemisphere. The nature of the changes which occurred beyond the ice front, in places such as the Middle East, still remain largely unknown, owing to the paucity of research work which has been carried out there.

Early workers postulated climatic changes on a large scale, which they claimed had a fundamental effect on the geomorphic processes operating in the region.

Pluvial periods, that is times with higher precipitation than at present, were considered to have occurred in the Middle East contemporaneously with glacial activity in northwest Europe. More recent work, however, has shown that the pattern is much more complex than this, and that perhaps different conditions prevailed in different areas at the same period.[11] The lack of evidence from large parts of the region makes it difficult, though, to be precise in drawing detailed conclusions.

Information on the Nile Valley has been greatly enhanced over the last few years as the result of work on the Aswân High Dam as well as by continued archaeological investigations.[12] Studies of alluvial deposits in Egypt have been used for two main purposes. Small wadi deposits provide an indication of local climatological conditions, while flood-plain sediments of the Nile record environmental events, particularly summer monosoonal rainfall, in Ethiopia. Evidence of pluvial conditions in southern Egypt, possibly about 60,000 years ago, is provided by the Wadi Floor Conglomerate, which overlies bedrock along many wadi floors. At the time of the accumulation of this deposit there is no evidence for higher Nile flood levels, which would be indicative of pluvial conditions in Ethiopia.

The Korosko Formation, consisting of Nile silts, was deposited between 50,000 and 25,000 years ago. During the first part of this cycle, pluvial conditions still continued in southern Egypt, while at the same time increased precipitation in Ethiopia produced greater floods along the Nile. Later, pluvial activity in Egypt declined, although even higher flood levels were recorded down the Nile river system. This lack of complete agreement between environmental conditions in south Egypt and the source of the Blue Nile in Ethiopia seems characteristic of the latter part of the Pleistocene Period.

During the time of maximum Wisconsin–Würm–Weichsel glaciation in northwest Europe, the climate in southern Egypt appears to have been arid, as it is today, while at the same time comparatively wet conditions prevailed in Ethiopia. In the 10,000 years following 15,000 years BP, the discharge of the Nile apparently inceased, as too did wadi incision in southern Egypt. This latter fact is taken to indicate the return of pluvial conditions in that region.

Nile flood levels began to decrease in the third millenium BC, and local wadi activity also declined in intensity. Since this period the climate of southern Egypt has remained intensely arid. The lowest Nile floods appear to have been recorded between 2,350 BC and 800 BC, with slightly less arid conditions prevailing since that time.

Although similar sequences of alluvial deposits exist throughout the Tigris–Euphrates basin, their chronological significance has not yet been deciphered in any detail. So far the most detailed work on the Euphrates comes from Syria, where late Pleistocene and Holocene terraces have been described.[13] Terrace suites on the Tigris above Baghdād are also known to exist.[14]

Pleistocene stratigraphy has been particularly well studied in the basin of the River Jordan.[15] Here, the main sequence of sediments, known as the Lisan Marls, are believed to be late Pleistocene in age and to have been deposited in a huge lake over 300 km in length, which possessed a water level about 200 m above that of the present Dead Sea. Following the maximum development of the lake, more arid conditions occurred, and the lake decreased in size. With a lowering of water surface elevation a sequence of shorelines were cut in the Lisan Marls at many different heights down to the present level of the Dead Sea.

Glacial activity at the present day is confined to a few small corrie glaciers and permanent snow patches in the Pontus and Taurus ranges and the Armenian Plateau of Turkey, as well as on Mounts Savalan, Suleiman and Damavand in the Elburz Mountains of Iran. In Turkey the contemporary snowline reaches a minimum height of about 3200 m in the northern part of the Pontus Range, rising inland to more than 4000 m, while in Iran it appears to be slightly higher at between 4000 and 4300 m.[16] During the Pleistocene it seems that glaciers grew larger and that new ones came into being in other upland regions. The actual amount by which the snowline was depressed during this period is still the subject of considerable controversy. Values of 800 to 1200 m have been reported for Turkey,[17] while in parts of the Zagros it is claimed to have been at least 1200 m.[18] Morphological evidence for such advances remains in the form of numerous retreat moraines along the higher valleys of Turkey[19] and of the Elburz range of Iran.[20]

In the Zagros Mountains, Pleistocene shorelines have been identified around Lake Rezā'īyeh up to 20 m above the present lake level.[21] From hydrological studies it was concluded that a 5°C lowering of mean annual temperature would be sufficient to explain a lake volume 10 times that found today, without the necessity of any increase in precipitation.

One of the most important works of regional geomorphological significance in an upland region concerns the origin of transverse drainage lines in the Zagros Mountains.[22] In this work it was postulated that the transverse gorge systems are the result of normal fluvial development in certain structural–lithological environments. The critical type of environment seems to be a succession of limestones and flysch deposits which have been subjected to orogenic activity to produce *en echelon* anticline groups. With this view older ideas suggesting single mechanisms operating on a regional scale, such as superimposition of a drainage pattern, or the development of antecedent drainage, are considered incorrect.

In western Saudi Arabia a well-developed pediplain, the Najd, developed on Pre-Cambrian crystalline rocks with isolated inselbergs rising above it, has been mapped.[23] As parts of the pediplain are covered with early Tertiary lavas, the surface must predate this period. Parts of it may well be a Pre-Cambrian surface which has been buried and subsequently exhumed. An important period of erosion and planation seems also to have occurred in middle Tertiary times and this has caused modifications to the forms of the earlier surfaces.

Erosion surfaces have also been identified further north along the rift of the Dead Sea Lowlands.[24] Three high level surfaces at 1,650 m, 1,200 m and 900 m are found, with the lowest, the Sinai or Arabia surface, extending far into central Arabia. This lowest surface is thought to have been formed prior to the formation of the main rift system, and to be of most probably Upper Oligocene age. A series of much smaller surfaces and terraces are described below 600 m, mostly identifiable along the margins of the rift system.

Salt deserts have been studied in detail in Iran by a number of workers. The classic work in this field is on the Dasht-e-Kavir of the Central Plateau.[25] In this it is claimed that the Kavir is not a relict landform, but one which is actually forming under present day environmental conditions. To prove this assertion a number of examples of where kavir fill of silt and clay sized material is transgressing over the adjacent eroded surfaces of *dasht* (gravel desert) are quoted. This rise in the level of the kavir fill, it is suggested, has been caused by a climatic change which has resulted in more humid conditions. The data of such a change is unknown, but it is believed to have been preceded by earlier drier periods, when erosional processes dominated. Early views on the origin of the Dasht-e-Kavir considered it to have been an enormous lake or closed-in part of a Tertiary sea, which was later reduced to a mudfilled basin as hydrological conditions changed.[26] More recent research, however, revealed that most of the marginal surfaces of the Kavir were erosional in origin with series of anticlines and synclines truncated by the processes of erosion. Further work in the same region confirms the idea that no widespread lakes existed in the Dasht-e-Kavir during the Pleistocene period.[27] It is also concluded that the Pleistocene climate of northern Iran was similar to that of the present day, with the exception that the precipitation evapotranspiration ratio was higher owing to lower summer temperatures.

A considerable body of research has been carried out in the Middle East on alluvial deposits of Tertiary and Quaternary age fringing the major upland regions. These deposits are often of very great thicknesses, upwards of 300 m, and would seem to indicate continued continental sedimentation over very long periods of time. In a series of papers dealing with alluvial deposits along the southern slopes of the Elburz Mountains three major alluvial formations of post-Upper Miocene age have been identified.[28] The oldest, between 100 and 120 m in thickness, is named the Hezardarrah Formation. It has often been subjected to folding and is correlated with the Mio-Pliocene or Pliocene Bakhtiari Formation of the main oil producing region of Iran. Overlying this is the Kahrizak Formation. This is much thinner, rarely more than 60 m in thickness, but it too has been subjected to folding and faulting. Its age is estimated to be mid-Quaternary. Finally, the youngest formation, the Tehrān Alluvium, is generally less than 35 m in thickness and has been relatively unaffected by orogenic activity.

The dating of recent alluvial material, and the land forms to which they give rise, still poses considerable problems. Two phases of alluvial deposition are widely recognized in the Middle East.[29] In Iran, the earlier phase began no more than 50,000 years ago and had probably ended by the fourth millenium BC. This was followed by a period of erosion, after which a second phase of deposition occurred during the Middle Ages. At the present time this deposit is being eroded. Detailed work on a number of alluvial fans near Tehrān suggests that sedimentary deposition over the last 750 years has been of only minor importance, and that optimum conditions for fan formation probably occurred during the glacial phases of the Quaternary.[30] Almost all the fans, particularly the larger ones, show evidence of fan-head trenching. This has meant that the upper portions of the fans cannot be alluviated under present conditions. Some deposition is currently occurring on the lower parts of the fans but the dominant processes appear to be largely erosional in nature.

In the Konya basin of central Turkey, research suggests that present environmental conditions are not responsible for the major geomorphological features.[31] Such features, including abandoned shorelines, wave-cut cliffs, sandpits and deltas, all testify to lacustrinal conditions during the recent geological past. Studies of the fauna of the lacustrinal sediments suggest that the water was fresh, even though it had no outlet. Archaeological evidence shows that the basin was largely dry by 8500 years ago. At present depositional processes appear to be confined to alluvial fan formation along the margins of the basin. The total thickness of sediments within the basin is unknown, but boreholes have revealed that in some places it is at least 400 m.

Sea level changes also occurred during the Pleistocene period leaving behind well marked raised beaches in the Mediterranean Sea, Red Sea, and the Gulf. As yet few attempts have been made to correlate the differing levels which have been described in studies of local significance. An exception is found in the Mediterranean, where published work in the region has been reviewed in the early 1970s by the Mediterranean and Black Sea Shorelines Subcommission of INQUA (International Quaternary Association), with the objective of trying to separate out reliable observations from those of a more doubtful character.[32]

In the Black Sea region eight distinct Quaternary shorelines have been re-recognized, ranging in height from 105 m to the present sea level.[33] A much more complex situation exists in the Mediterranean with no general agreement about correlation between the different areas, owing to the absence of diagnostic faunal assemblages. Shorelines have been described ranging from 200 m to 2 m above the present sea level, with many of these being recognized in the eastern Mediterranean region. In Lebanon, high-level shorelines have been identified at 180–190 m, 110–120 m and at 60 m, together with a series of lower ones.[34]

1.4 Soil

Soil is one of the most vital natural resources of any region, for without it agricultural activity is not possible. It is also relatively easily destroyed by the careless actions of man. Certain basic factors can be thought of as controlling soil formation.[35] These are climate, parent material, vegetation, slope, hydrology, micro-organisms, man, and time. In a given region any one of these factors may have a dominating influence on the soil type produced, but, in general, parent material and climatic conditions are usually the most significant.

In a region like the Middle East, where arid conditions and in particular summer drought prevail almost everywhere, climate produces a characteristic stamp on the soils. One of the crucial climatically controlled factors in soil formation is the predominant direction of moisture in the soil profile. When it is downwards, the soluble products of weathering are carried through the soil and into streams and rivers. Such soils are known as pedalfers. In contrast, in arid climates these same weathering products accumulate in the upper layers of soil, as water evaporates from the surface, to form salt, gypsum and calcium carbonate layers or nodules. The soils produced under these conditions are termed pedocals. In the Middle East, pedocal soils predominate in the more arid regions of the south, whilst pedalfers are commonly developed in the northern highland regions of Turkey and Iran, as well as in the Black and Caspian Sea lowlands where high annual precipitation totals occur.

Soil classification is still the subject of much controversy, even though increasing standardization is being introduced. Until recently, the Russian genetic classification into great soil groups, with subsequent modifications by North American workers, has been most used in the western world. As new information became available it was obvious that the modified Russian genetic classification, often known as the Great Soil Group classification, with its three major orders of zonal, intrazonal and azonal soils, was inadequate for detailed survey and research work. It was therefore decided in the United States that a new scheme should be established, and so there appeared in 1960 a publication entitled *Soil Classification*: a *Comprehensive System* (*7th Approximation*).[36] This system, which is now the chief one employed in many parts of the world, is based on the morphological characteristics of the soil profile, particularly diagnostic horizons, and the sequence of occurrence of different horizons. The aim of the classification is to ensure that soils from any part of the world can be placed in their proper relation to each other. To facilitate this, an entirely new nomenclature has been introduced and ten major global orders recognized.

The final classification which has to be mentioned in the Middle East context is that established by FAO/UNESCO as the basis for their 1 : 5,000,000 soil map of the world.[37] This scheme includes ideas and information from both the major classifications already outlined, together with data on the suitability of a particular soil as a production resource. In this respect it has similarities with other methods employing land capability approaches. In all 103 soil units are recognized, which are grouped into 23 higher categories.[38]

All the above mentioned classification systems have been utilized at some time or other in the Middle East. By far the majority of the earlier soil studies made use of the Great Soil Group classification, but later work has tended to utilize either the 7th Approximation system or the FAO/UNESCO approach. For exploratory surveys in unknown areas land capability studies have often been the only feasible method.

In the Middle East, a number of soil types stand out as being of exceptional importance with regard to either their total area of occurrence or their agricultural productivity. In terms of the Great Soil Group classification these are Red Desert soils (including lithosols and sand); Sierozems (including lithosols and sand); Reddish Prairie, Reddish Chestnut and Reddish Brown soils; Terra Rossa, and Rendzinas; Chestnut and Brown soils; and alluvial soils (Figure 1.6).

The most widely occurring soils in the region are without doubt, the Red Desert

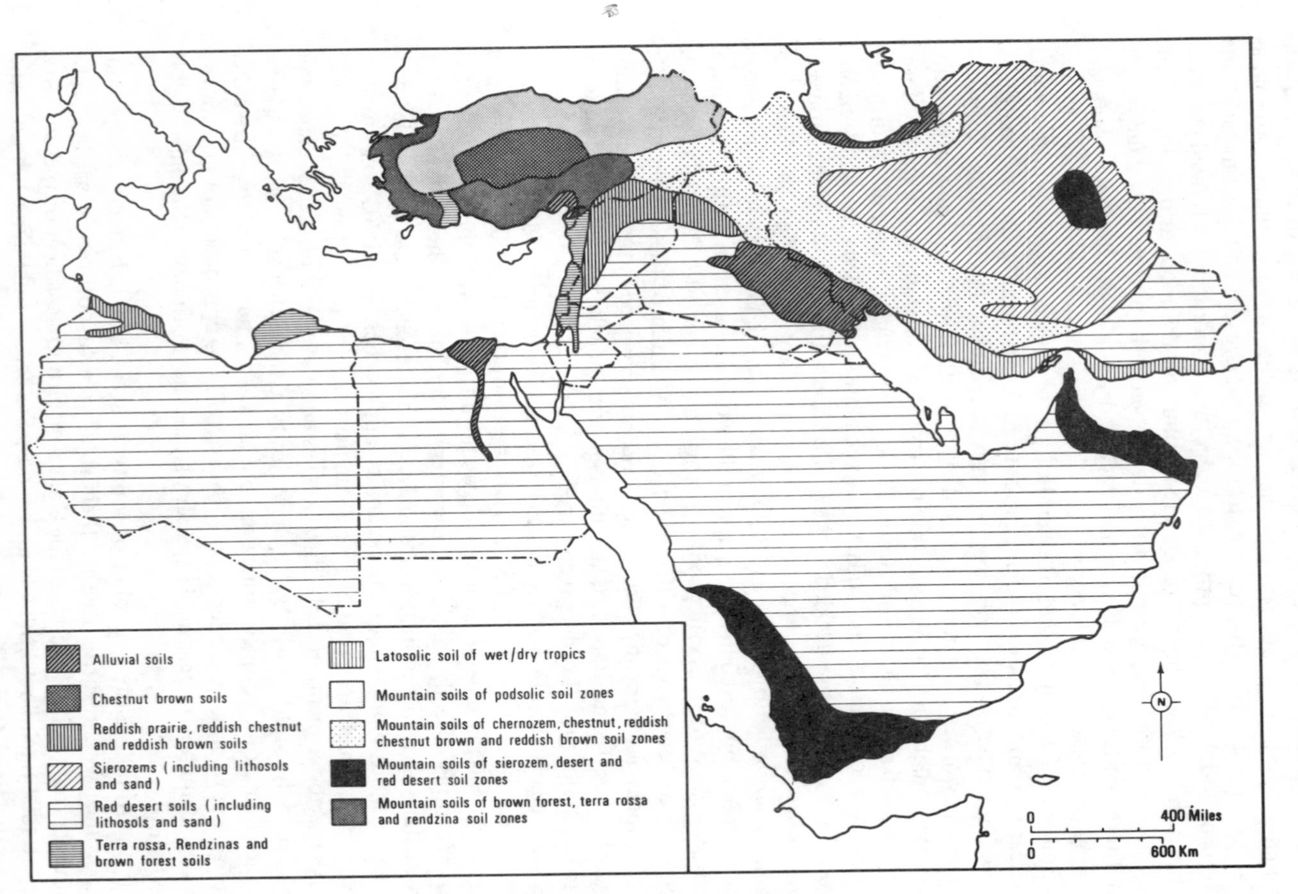

Figure 1.6 Soil groups of the Middle East

soils, found throughout North Africa, much of Arabia and southern Iran. Owing to the dryness of the regions in which these soils are found, animal and plant growth within them is minimal and, as a consequence, the humus content is very low. Soil horizons are only poorly developed and the texture is coarse with many unweathered fragments of rock present.

Closely related to the Red Desert soil are the Sierozems, or Grey Desert soils. These soils are of more limited occurrence than their red counterparts, and are confined mainly to the northern and central parts of Iran where temperature conditions are cooler. They lack a marked profile, but are usually characterized by the accumulation of calcium carbonate just below the surface. In some cases, this has built up into a resistant layer known as caliche. Neither the Red Desert nor the Grey Desert soils are usually of great agricultural significance.

Reddish Prairie, Reddish Chestnut and Reddish Brown soils are found in the region of the 'Fertile Crescent' of northern Iraq and Syria. The Reddish Prairie soils represent a transitional type between the pedocals and pedalfers. They appear to have been developed under grassland conditions and are usually extremely productive agriculturally, particularly for cereals. In slightly drier areas Reddish Chestnut and Reddish Brown soils occur. These seem to form under a range of vegetation conditions, but all show the development of a calcium carbonate horizon at depth.

Of exceptional agricultural importance are Terra Rossa soils. These develop on hard, pure crystalline limestones under a Mediterranean climatic regime, and occur extensively in Turkey, Syria, Lebanon, Jordan, Israel and Libya. Although found on limestone parent material, they possess little or no free lime and their pH values are between seven and eight. The deep red colour of these soils is thought to be due to the presence of sesquioxides of iron in considerable quantities. The clay content is commonly above 50 per cent and, because of this, their moisture holding capacity is good. The structure of the soil is particularly unstable when water is added, breaking down to a paste-like consistency, which greatly reduces the infiltration capacity of the underlying soil layers. These soils are, therefore, very susceptible to erosion when the vegetation cover is removed. Following long continued erosion, bare limestone hills remain in many parts of the region, with only traces of soil being found in the deeper joints between the limestone blocks.

The actual soil-forming processes associated with Terra Rossa development are still not fully understood.[39] Chemical weathering of the limestone to produce an insoluble residue is the fundamental process, and this occurs mainly during the wet winter season, along cracks and fissures in the rock. Soil water charged with carbon dioxide in the form of carbonic acid is believed to be the major chemical agent. A characteristic feature of Terra Rossa soils is the very sharp weathering front between the base of the soil profile and the underlying limestone.

Rendzina soils possess dark grey or black surface layers overlying lighter coloured material which is highly calcareous. Such soils are found only on soft limestones and marls, usually in association with Terra Rossa soils which form on more resistant calcareous formations.

In the transitional zone between the true prairie or grassland soils and the arid or desert soils, Chestnut and Brown soils are found. They are best developed in central Anatolia. The Chestnut soils are similar in type to Chernozem soils with a well-developed A horizon, but one which lacks the high humus content of a true black earth. This horizon grades down into a lighter brown B horizon, followed across a sharp line of demarcation by a light coloured C horizon. In more arid regions, the Chestnut soils are replaced by Brown soils, which are similar in profile characteristics, but contain less humus and consequently are of a lighter colour. Both these soils tend to be found in regions which are marginal for dry-land farming.

Alluvial soils are amongst the most fertile soils found in the Middle East. They occur along all the major river systems, and are especially well developed along the Nile Valley and in the Tigris–Euphrates lowlands. The texture of this type of soil varies considerably, depending upon local sediment supply conditions. In the larger river basins silty-clays, silty-loams, and silty-clay-loams tend to predominate. Where poor drainage occurs, salinity can be a problem. As these soils are often added to annually, their profiles are not well developed. Occasionally, a horizon of calcium carbonate deposition can be distinguished.

1.4.1 Soil surveys

For illustrative purposes summaries of soil surveys carried out in four countries of the region are included. The countries selected, Turkey, Iran, Israel and Egypt, were chosen with a view to providing details of the differing soil types which commonly occur throughout the Middle East and North Africa.

1.4.1.1 Turkey

The only comprehensive survey of soils in Turkey is that made by Oakes in 1957, in which 18 major soil types were recognized.[40] Over about one-fifth of the area of the country, particularly the mountains of the east, it did not prove possible to classify the soils adequately. In these regions the soils are usually thin and discontinuous, wth igneous and metamorphic rocks forming the parent material (Figure 1.7).

Pedalfers cover approximately one third of the country and are represented by Red and Grey–Brown Podsolic soils, and Brown Forest soils. These soil types tend to occur in the wetter upland regions, with the Brown Forest soils developed on calcareous strata. The soil profiles are often thin and truncated, owing to the erosion of the steep mountain slopes. In total these three soil types cover almost 40 per cent of Turkey.

An intermediate grouping of soil types, which are difficult to classify as either pedalfers or pedocals, are the non-calcic Brown soils, Rendzinas and Grumusols. The non-calcic Brown soils are similar to the Red Podsolic soils, but less strongly leached. Rendzinas and Grumusols are both azonal types found in similar climatic zones to those of the non-calcic Brown soils. The former are highly

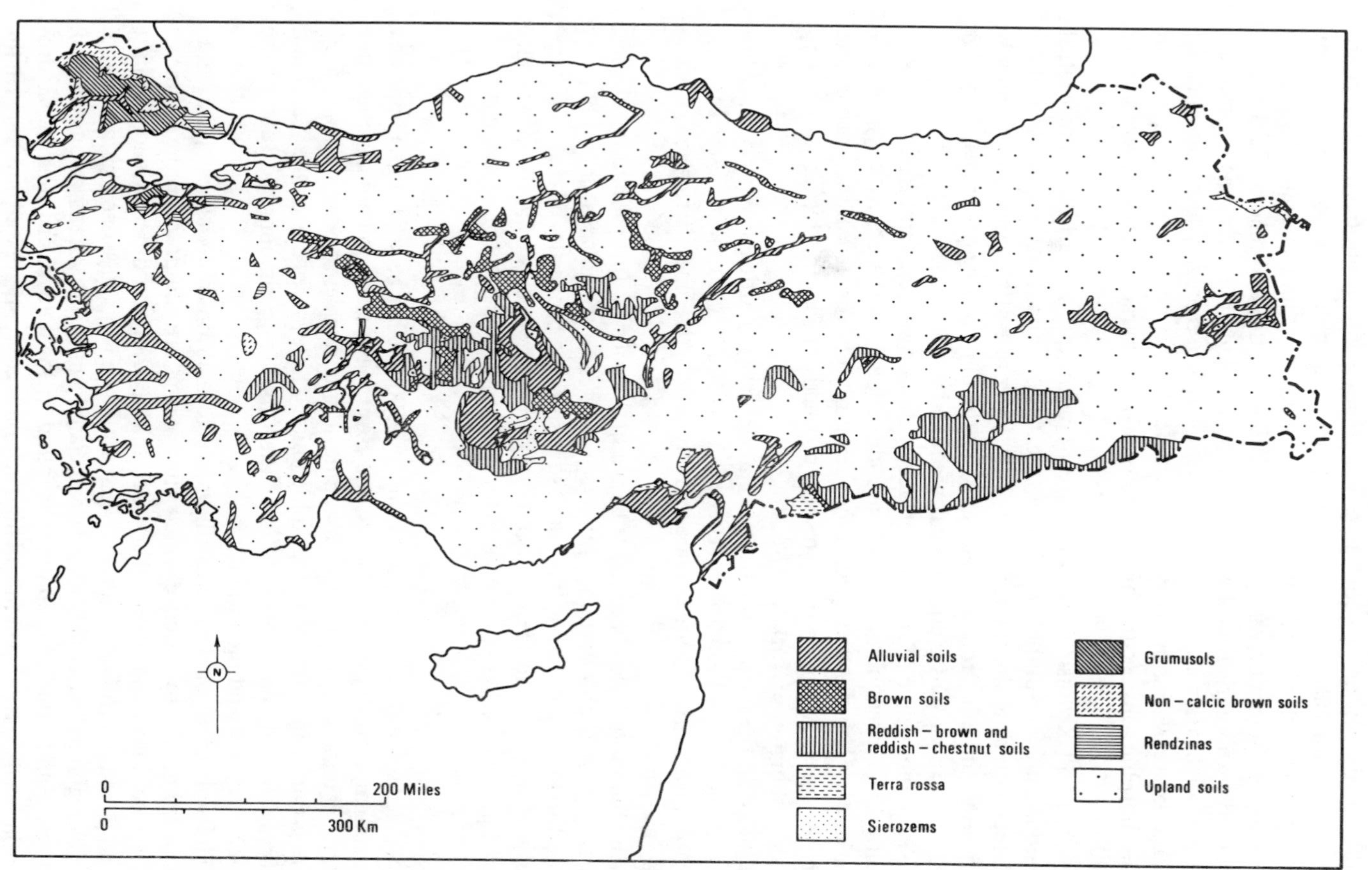

Figure 1.7 Soil map of Turkey (modified by permission of the Ministry of Agriculture, Turkey)

calcareous and are usually developed on soft or marly limestones, while the latter tend to be deeper, and to contain a higher clay content, although they too are associated with calcareous parent material. The remaining soil types can be broadly classed as pedocals, with the exception of the alluvial soils. These latter occur along the major valley systems of the country, and tend to be the most productive soils agriculturally. They exhibit considerable variations in lithology and are all characterized by a relatively immature profile development, and marked lateral variations over short distances. Where drainage is poor, alluvial hydromorphic soils occur. Some of these soils are characterized by severe salinity build-up.

Brown and Reddish Brown soils are found in the semi-arid areas of central Anatolia and the southeast of Turkey near the Syrian border, and are usually developed on a calcareous substrate. They cover one-fifth of the country, mostly rolling terrain, and are important soils for the dry farming of cereals. In wetter areas the Brown soils grade into calcareous Chestnut soils.

Within the regime of the Mediterranean climate proper, Terra Rossa and Red Prairie soils predominate. They are developed as the result of limestone weathering, with the Red Prairie soils tending to be located in the warmer and damper areas. Although they are very distinctive soils, they cover only a very small proportion of Turkey.

Finally, in the most arid parts of central Turkey, particularly in the southern part of the Konya basin, Sierozem soils are found. These semi-desert soils are highly calcareous, but are often extremely productive under irrigation, if carefully managed.

An attempt has also been made to group the soils of Turkey in terms of their suitability for arable farming. Fifteen groups are distinguished on the basis of altitude, slope, and drainage as well as the intrinsic properties of the soils themselves. Eighty per cent of the total area of the country is classed in groups 11 to 15, which are considered as unsuitable for arable farming.

1.4.1.2 Iran

A detailed soil map of Iran at a scale of 1 : 2,500,000 has been prepared jointly by the Iranian Ministry of Agriculture and FAO.[41] Nineteen soil associations were identified and used as a basis of mapping. For convenience these mapping units were grouped into four physiographic units. These were soils of the plains and valleys, soils of the plateau, soils of the Caspian Piedmont, and soils of the dissected slopes and mountains.

The soils of the plains and valleys are formed on material which has been deposited mainly by water or wind. Alluvial soils are composed of the sediments of terraces and flood plains. They lack marked horizon differentiation and are usually medium to fine textured. Where drainage is poor, saline soils are found. Coarser grained alluvial soils are associated with alluvial fans along foothill regions. Sand dunes are commonly found in lowland basins in Iran. Hydromorphic soils, dark in colour and high in humus content, are well developed in the paddy lands of the Caspian Lowlands.

Solonchak and Solonetz are saline and alkali soils found in the drier areas of the country. They are either poorly drained or have developed under impeded drainage conditions, contain large quantities of soluble salts and are low in organic matter. Solonetz soils are produced by the partial leaching of Solonchak soils when irrigation is practised without proper drainage. They have a light coloured surface layer over a heavy dark coloured subsoil. Both soil types are commonly found in the basins of central Iran, where they can also be associated with salt marsh soils.

The soils of the plateau have been formed under modified continental climatic conditions with marked aridity at heights of around 1,000 m. Grey Desert and Red Desert soils are common in this region. They are characterized by a thin surface crust of compacted or cemented material which produces the typical desert pavement. These soils are always alkaline in reaction, possess a high calcareous content, and are very low in humus. Sierozem soils are light grey in colour and extremely calcareous. The surface layer is usually powdery and

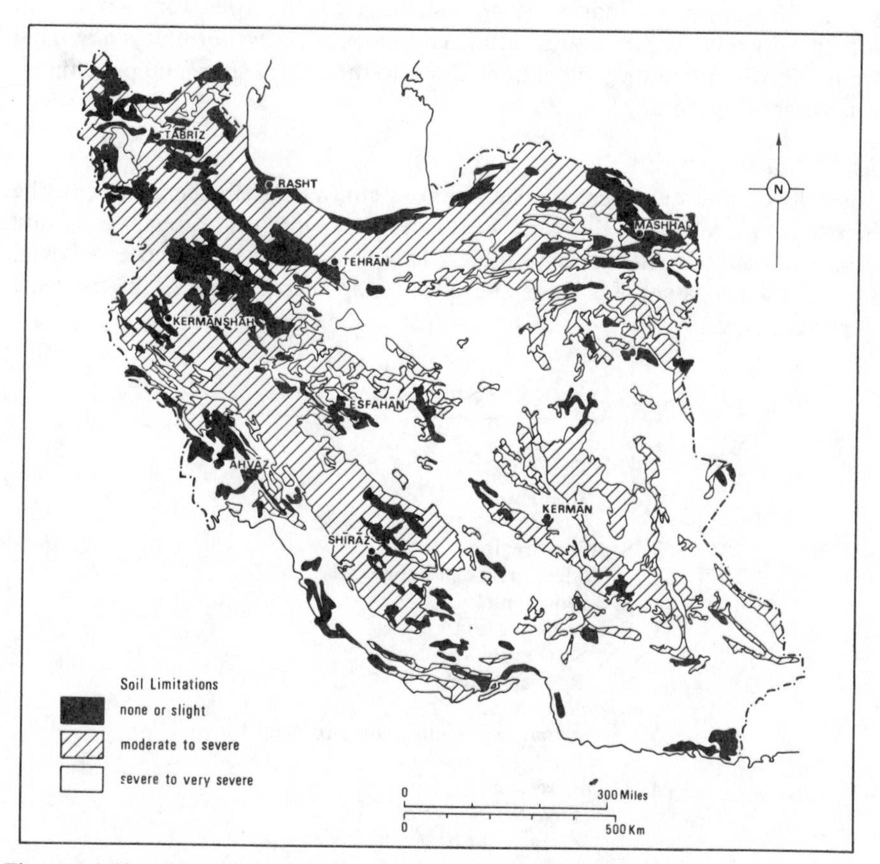

Figure 1.8 Simplified soil potential map of Iran (modified from Dewan and Famouri, *The Soil of Iran*, FAO, Rome, 1964)

organic matter content is minimal. Such soils are common in desert steppe areas. Closely related to the Sierozems are the Brown Steppe soils, which are probably the most widespread soils in Iran. These are brown in colour and usually overlie calcareous horizons. They have been developed beneath grass vegetation under semi-arid climatic conditions.

The soils of the Caspian Piedmont have climatic conditions different from any other part of Iran, being characterized by precipitation throughout the year and an almost subtropical temperature regime. Vegetation is abundant, and chemical weathering of the outcropping Mesozoic sediments intense. Many different soil types, including Brown Forest, Red–Yellow Podsolic, Grey–Brown Podsolic, and Red–Brown Mediterranean soils, have been described. The total area of these soils, however, is only 0.2 per cent of the country.

The soils of the dissected slopes and mountains are stony, shallow and lack profile development. They are described mainly as lithosols. In total they occupy almost half the area of the country.

To make the soil map more useful for planning purposes it was decided to group the classes to indicate the limitations of the soils for agricultural productivity. Five broad groups with ten divisions in all, were employed ranging from soils with no or only slight limitations, to those with almost no potentiality whatsoever (Figure 1.8).

1.4.1.3 Israel

The soils of Palestine were the subject of a study by Reifenberg prior to the Second World War.[42] In this work he divided the country into four regions and described the soils within each (Table 1.1). For agricultural purposes, alluvial soils, Terra Rossa soils, and Mediterranean Steppe soils were by far the most important types.

TABLE 1.1
The Soils of Palestine

1. Arid region
   - Desert soils
   - Lisan Marl soils
   - Loess
2. Semi-arid region
   - Mediterranean steppe soils
   - Dune sands
3. Sub-humid region
   - Kurkar soils
   - Red sandy soils
   - Nazzaz soils
   - Alluvial soils and Aclimatic Black Earths
   - Peat
4. Humid region
   - Terra Rossa
   - Red earths on volcanic rocks

Mountain marl soils

Source: After Reifenberg A, 1947

This work has now been superseded by more detailed Israeli surveys.[43] For descriptive purposes the country has been divided into three terrain types: a coastal region, a mountain and hill region, and a valley, plains and plateau region. Each of these is then further subdivided with respect to moisture availability, and lithology (Figure 1.9).

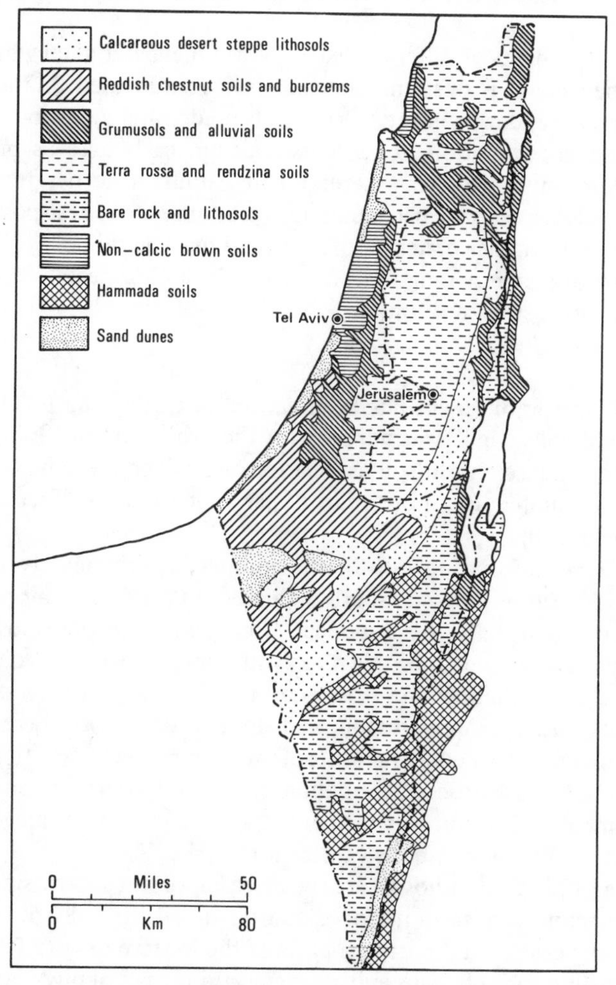

Figure 1.9 Soil map of Israel (modified from Dan and Koyumdjisky 1963, by permission of the British Society of Soil Science)

The soils of the coastal region are formed on sandy sediments, and belong to three great soil groups. Non-calcic Brown soils are found in the north under Mediterranean climatic conditions. Further south the soils become calcareous. In the semi-arid zone Reddish Chestnut soils occur, while in the arid zone proper coarse textured Burozems (arid brown soils) are common.

In the mountain and hill region, the parent material on which the soils have developed consists mainly of hard limestone, dolomite, and marl. Under the Mediterranean regime, Terra Rossa soils have formed on the more resistant calcareous strata with brown Rendzinas on the marly limestones. With increasing aridity, brown Rendzinas first become the predominant soil on all parent materials, to be replaced under even drier conditions by calcareous desert-steppe lithosols.

Quaternary sediments of differing textures form the material on which most of the soils of the valleys, plains, and plateaus have been formed. In the wettest or Mediterranean zone, dark coloured Grumusols predominate, giving way to more silty soils with accumulation horizons of calcium carbonate as precipitation decreases. Burozems also occur here. Finally, under true desert conditions, usually gravel plains, soils termed hammadas predominate. These possess a stone cover, beneath which finer material, often with a marked saline layer, is seen. In areas of poor drainage a range of soil types, including peats, Planosols and Hydrohalomorphic soils, are found.

1.4.1.4 Egypt

A recent soil survey of Egypt has been carried out using the FAO/UNESCO classification of soils which was adopted for the project on the Soil Map of the World.[44] Altogether, eighteen soil associations were identified, but of these, six associations accounted for more than 85 per cent of the surface area of the country (Figure 1.10).

Lithosols cover a large portion of the country, approximately 17 per cent, especially in the Eastern Desert, south Sinai and on the Gilf Kebir plateau. The parent material is usually the Basement Complex, consisting mostly of Pre-Cambrian igneous and metamorphic rocks with some more recent volcanic rocks. The soil profiles are shallow and stony, and possess only a weakly developed A horizon. Rock outcrops are common, and slopes nearly everywhere steep.

A new soil unit, termed Ermolithosol, was introduced for this survey to describe the characteristic desert formations usually referred to in the literature as ‘desert pavements’. The unit was further subdivided into three categories, lithic, gravelly, and argillic, on the basis of parent material.

Limestone Lithic Ermolithosols are the most important subdivision within the Ermolithosol unit, and account for a quarter of all the soils of the country, especially in the central and northern parts of the western desert. These soils are developed on limestone plateaus and are merely thin crusts of physical weathering which have been smoothed and polished by aeolian activity. Some fine material occurs beneath the weathered stones, making the surface very level.

Sandstone Lithic Ermolithosols are developed on the Nubian Sandstone and cover some 20 per cent of the country, mainly in the southern part of the western desert. These soils produce a very bare and smooth form of desert pavement which has been intensely affected by wind action. Again no profile development is seen.

Another new unit, Ergosols, was introduced for the mapping of sand dune

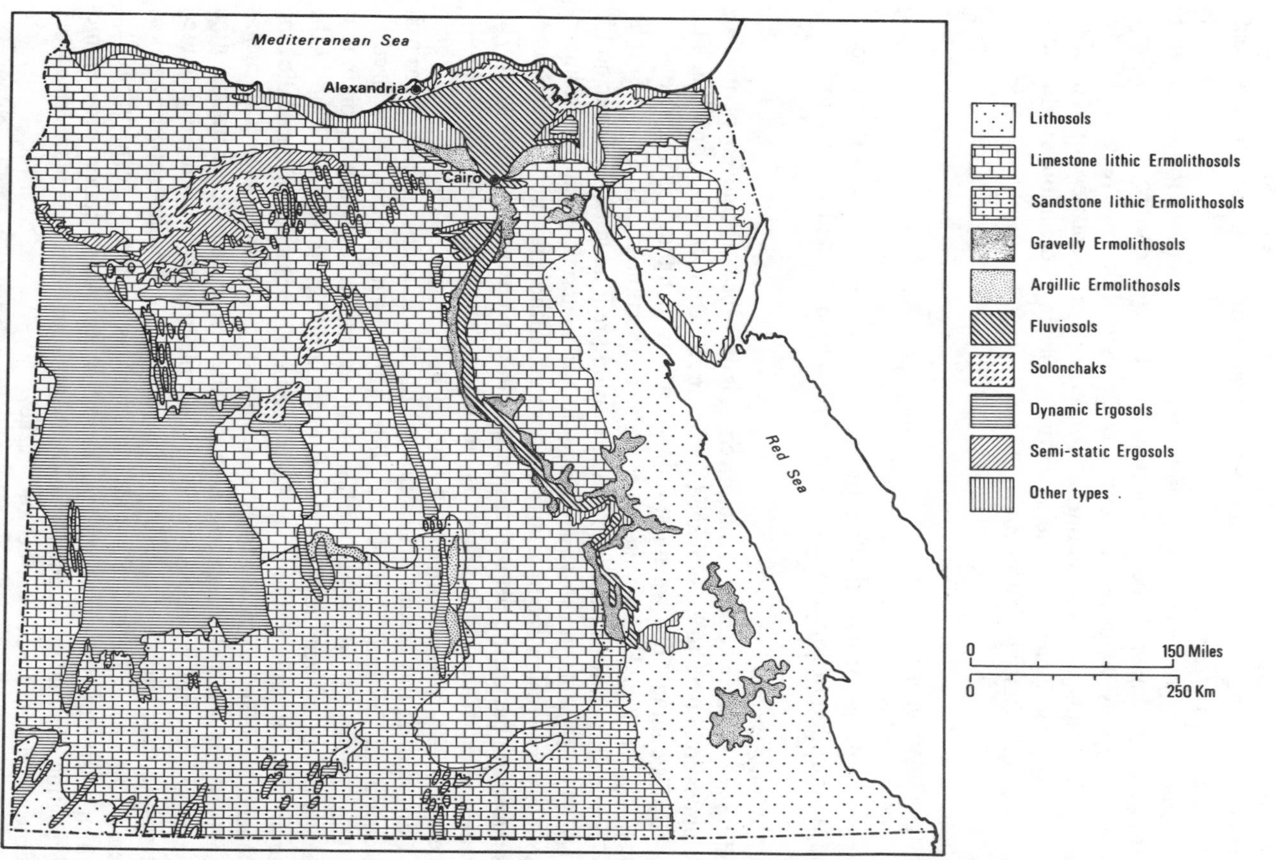

Figure 1.10 Soil map of Egypt (reproduced by permission of the Institute of Land Reclamation, Egypt)

complexes. Shifting sand dunes are classified as Dynamic Ergosols, while semi-stabilized dunes are known as Semistatic Ergosols. Dynamic Ergosols cover 15 per cent of the country in a series of scattered zones throughout the main plateau region. The best known occurrence of these soils is undoubtably the great sand sea of the Western Desert.

All other soil types cover only very small parts of the country. However, it must be realized that almost all the cultivated area is restricted to one of these groups, namely Fluviosols, which occur extensively throughout the Nile Valley and Delta region. These soils, which cover only 2.5 per cent of Egypt, and developed on Nile silt, and, as a consequence, are heavy textured. Gleying is common, and in badly drained areas salinity is a problem.

1.4.2 Soil management

In most countries of the Middle East two major problems exist with regard to man's utilization of soil resources. These are soil erosion and soil salinity.

1.4.2.1 Soil erosion

Soil erosion occurs when the vegetation cover is disturbed or removed, permitting the upper soil layers to be subjected to the direct action of water or wind. The problem is compounded when the soil structure is such that it is unstable when water is present, and also when slopes are steep (Section 15.3.1).

Throughout most of the region water erosion predominates. Gully erosion is especially spectacular on soft and unconsolidated materials, while almost imperceptible sheet erosion occurs nearly everywhere. Although water action reaches its maximum during the wet winter months, the converse is true of the wind. The greatest effect of this is felt in the dry summer, when surface soils become dust-like and so are easily transported in strong winds.

Control measures against soil erosion are usually difficult to implement as well as expensive. Two major approaches, often carried out together, are utilized.

The first, and most simple method is to attempt to establish a continuous vegetation cover over the soil which is being subjected to erosion. One of the major causes of soil erosion in the Middle East is the practice of bare fallowing the land between the cereal harvest and the first rains of the winter. This means that large quantities of soil are lost as a result of sheet erosion from what are often the most fertile and productive regions. A simple, though as yet not widely used remedy, is to strip plant grasses in belts parallel to the contours between areas of cereal cultivation. Deforestation also accentuates erosion by removing foliage which dissipates the effects of rain drop impact. Afforestation programmes throughout the region are helping to combat this problem with pioneering work being carried out by both Israel and Turkey.

The second method of erosion control is the construction of physical barriers which reduce the speed of surface runoff. Throughout the Middle East the most widely used soil conservation technique is terracing. Terraces are constructed by hand labour and have been a feature of rural agriculture in many parts of the region for millenia. They are particularly well developed in the eastern Mediterranean littoral.

Since the Second World War modern conservation methods have been increasingly introduced. These include gradoni terracing, contour walling, and gully-plug or check-dam construction (Section 15.3.1).

1.4.2.2 Soil salinity

The major effects of soil salinity are most clearly seen in lowland and riverine areas where irrigated agriculture is practised and drainage systems are inadequate (Chapter 12). The major cause of soil salinity is the presence of ground-water, often though not necessarily saline, close to the surface. Under these conditions water is drawn upwards to the surface by capillary action, and from there it evaporates into the atmosphere. When it does so, it leaves behind any dissolved chemicals to form a surface crust, or saline horizon, in the top few centimetres of the profile. Salt can also be added directly to the soil through poor quality irrigation water, even though satisfactory drainage systems are in existence, if insufficient water to permit full leaching is not utilized.

The major salts found in soils are chlorides and sulphates of sodium, calcium and magnesium. In arid soils, sodium ions tend to present the greatest problem as they cause the breakdown of soil structure and a great reduction in permeability. These ions are also toxic to most plant species when present in large quantities. The reclamation of saline soils is a straightforward process if sufficient water is available for leaching away the soluble salts, and provided that an adequate drainage systems exists (Section 12.2)

References

1. R.S. Dietz and J.C. Holden, 'The Breakup of Pangaea', *Scient. Am.*, **233**, 30–41 (1970).
2. D.P. McKenzie, 'Plate tectonics of the Mediterranean region', *Nature, Lond.*, **226**, 239–43 (1970).
3. D.P. McKenzie, 'Plate tectonics of the Mediterranean region', *Nature, Lond.*, **226**, 242 (1970).
4. D.P. McKenzie, D. Davies and P. Molnar, 'Plate tectonics of the Red Sea and East Africa', *Nature, Lond.*, **226**, 243–8 (1970).
5. A.M. Quennell, 'The structural and geomorphic evolution of the Dead Sea Rift', *Q. Jl. geol. Soc. Lond.*, **114**, 1–24 (1958).
6. D.P. McKenzie, D. Davies and P. Molnar, 'Plate tectonics of the Red Sea and East Africa', *Nature, Lond.*, **226**, 247 (1970).
7. M. Takin, 'Iranian geology and continental drift in the Middle East', *Nature, Lond.*, **235**, 147–50 (1972)
8. M. Takin, 'Iranian geology and continental drift in the Middle East', *Nature, Lond.*, **235**, 149 (1972).
9. P. Beaumont, 'Salt weathering on the margin of the Great Kavir, Iran', *Bull. geol. Soc. Am.*, **79**, 1683–4 (1968).

10. D. H. Yaalon, 'Parallel stone cracking, a weathering process on desert surfaces', *Geological Institute Technical and Economic Bulletins, Series C, Pedology, Bucharest*, **18**, 107–11 (1970).
11. K. W. Butzer, *Quaternary stratigraphy and climate in the Near East*, Bonner Geog. Abh., **24**, 1958, 157 pages.
12. K. W. Butzer, and C. L. Hansen, *Desert and River in Nubia*, University of Wisconsin Press, Madison, 1968, 562 pages.
13. W. J. Van Liere, 'Observatons on the Quaternary of Syria', *Berichten, Rijksdienst Oudheidkundig Bodemonderzoek*, **10–11**, 1–69 (1961).
14. P. Buringh, *Soils and soil conditions in Iraq*, Ministry of Agriculture, Republic of Iraq, Baghdād, 1960, 322 pages.
15. K. W. Butzer, *Quaternary stratigraphy and climate in the Near East*, Bonner Geog. Abh., **24**, 1958, 157 pages.
16. (a) B. Messerli, 'Die eiszeitliche und die gegenwärtige Vergletscherung im Mittelmeeraum', *Geographica helv.*, **3**, 105–228 (1967).
(b) B. Freznel, 'Die Vegetations und Landschaftszonen Nord-Eurasien während der letzten Eiszeit und während der postglazialen Wärmezeit', *Abh. Akad. Wiss. Liter. (Mainz) Math.—Naturw. Kl.*, **13**, 164 pages (1959) and **6**, 167 pages (1960).
17. B. Messerli, 'Die eiszeitliche und die gegenwärtige Vergletscherung im Mittelmeeraum', *Geographica helv.*, **3**, 105–228 (1967).
18. H. E. Wright, 'Pleistocene glaciation in Kurdistan', *Eiszeitalter Gegenw.*, **12**, 131–64 (1962).
19. J. H. Birman, 'Glacial reconnaissance in Turkey', *Bull. geol. Soc. Am.*, **79**, 109–126 (1968).
20. H. Bobek, 'Die Rolle der Eiszeit in Nordwestiran', Z. *Gletscherk, Glacial geol.*, **25**, 130–83 (1937).
21. H. Bobek, 'Nature and implications of Quaternary climatic changes in Iran.' in *Changes in Climate*, UNESCO, Paris, 1963, Arid Zone Research, **20**, 403–13.
22. T. M. Oberlander, *The Zagros Streams: A New Interpretation of Transverse Drainage in an Orogenic Zone*, Syracuse Geographical Series No. 1, Syracuse, 1965, 168 pages.
23. G. F. Brown, 'Geomorphology of Western and Central Saudi Arabia', *Report of XXI International Geological Congress, Copenhagen*, **21**, 150–59 (1960).
24. A. M. Quennell, 'The structural and geomorphic evolution of the Dead Sea Rift'. *Q. Jl. geol. Soc. Lond.*, **114**, 1–24 (1958).
25. H. Bobek, *Features and Formation of the Great Kavir and Masileh*, Arid Zone Research Centre, University of Tehrān, Publication No. 2, Tehran 1959. 63 pages.
26. W. T. Blandford, 'On the nature and probable origin of the superficial deposits in the valleys and deserts of Central Persia', *Q. Jl. geol. Soc. Lond.*, **29**, 495–503 (1873).
27. D. B. Krinsley, 'Geomorphology of three kavirs in Northern Iran', in *Playa Surface Morphology: Miscellaneous Investigations* (Ed. J. T. Neal), USAF, Office of Aerospace Research, Environmental Research Papers, No. 283, 1968, 105–30.
28. (a) E. H. Rieben, *Geological Observations on Alluvial Deposits in Northern Iran*, Geological Survey, Iran, Report 9, 1966, 41 pages.
(b) E. H. Rieben, *Les Terrains Alluviaux de la région de Tehrān*, Publication No. 4, Arid Zone Research Centre, Tehrān, 1960, 41 pages.
(c) E. H. Rieben, 'The geology of the Tehrān Plain', *Am. J. Sci.*, **253**, 627–639 (1955).
29. (a) C. Vita-Finzi, 'Late Quaternary alluvial chronology of Iran', *Geol. Rdsch.*, **58**, 951–73 (1968).
(b) C. Vita-Finzi, *Mediterranean Valleys*, Cambridge University Press, 1969, 140 pages.
30. P. Beaumont, 'Alluvial fans along the foothills of the Elburz Mountains, Iran', *Palaeogeography, Palaeoclimatology, Palaeoecology*, **12**, 251–73 (1972).

31. N. A. De Ridder, 'Sediments of the Konya Basin, Central Anatolia, Turkey', *Palaeography, Palaeoclimatology, Palaeoecology*, **1**, 225–54 (1965).
32. R.W. Hey, 'Quaternary shorelines of the Mediterranean and Black Seas', *Quarternaria*, **XV**, (VIII Congrés INQUA–Les Niveaux Marins Quaternaires), II–Pleistocene), 273–84 (1971).
33. P. V. Federov, 'The marine terraces of the Black Sea coast of the Caucasus and the problem of the most recent vertical movements'. *Dokl. Acad. Nauk*. USSR, **144**, 431–4 (1969), (in Russian).
34. (a) P. Sanlaville, 'Sur les niveaux marins quaternaires de la région de Tabarja (Liban)', *Comptes Rendues Somm. Soc. Geol. Fr.*, 157–8 (1967).
(b) P. Sanlaville, 'Sur le Tyrrhènien libanais', *Quaternaria*, 15 (VIII Congrés INQUA–Les Niveaux Marins Quaternaires, II–Pleistocene), 239–48 (1971).
35. H. Jenny, *Factors of Soil Formation*, McGraw-Hill Book Co, New York, 1941, 281 pages.
36. Soil Survey Staff, Soil Conservation Service, United States Department of Agriculture, *Soil Classification: A Comprehensive System (7th Approximation)*, US Government Printing Office, Washington DC, 1960.
37. FAO, *Definitions of Soil Units for the Soil Map of the World*, FAO/UNESCO Project, World Soil Resources Office, Land and Water Development Division, FAO, Rome, 1968, 72 pages.
38. FAO, *Key to soil units for the Soil Map of the World*, Soil Map of the World, FAO/UNESCO Project, Soil Resources, Development and Conservation Service, Land and Water Development Division, FAO, Rome, 1970, 16 pages.
39. K. Atkinson, 'The dynamics of Terra Rossa soils', *Bulletin of the Faculty of Arts, University of Libya*, **3**, 15–35 (1969).
40. H. Oakes, *The Soils of Turkey*, Republic of Turkey, Ministry of Agriculture, Soil Conservation and Farm Irrigation Division, Ankara, Division Publication No 1, 1957, 180 pages.
41. M.L. Dewan, and J. Famouri, *The Soils of Iran*, Food and Agricultural Organization of the United Nations, Rome, 1964, 319 pages.
42. A. Reifenberg, *The Soils of Palestine* (translated by C.L. Whittles). Thomas Murby, London, 1947, 179 pages.
43. J. Dan and H. Koyumdjisky, 'The soils of Israel and their distribution', *J. Soil Sci.*, **14**, 12–20 (1963).
44. M. M. Elgabaly, I.M. Gewaifel, N.N. Hassan, and B.G. Rosanov, *Soil map and land resources of U.A.R.*, Institute of Land Reclamation, Alexandria University, Research Bulletin, Alexandria, No. 22, 1969, 14 pages.

CHAPTER 2

Climate and Water Resources

2.1 Climate

2.1.1 Introduction

No single climatic regime prevails throughout the Middle East. Indeed, as in so many other respects the region is a transitional zone, in this case between equatorial and mid-latitude climates. A characteristic feature of all subtropical latitudes, of which the Middle East forms a substantial part, is the prevalence of aridity, with a marked precipitation minimum, over both the oceans and continents, centred on 30 degrees of latitude. The fact that aridity is found over the oceans as well as the land, suggests the operation of some mechanism which inhibits the precipitation process, as there is obviously no shortage of water for replenishing atmospheric moisture.[1] In oceanic regions, annual evaporation can attain values of more than 200 cm/annum, making these areas one of the most important sources for atmospheric water recharge.[2] Given these facts there can be little doubt that the aridity of subtropical latitudes can only be accounted for by dynamical factors related to the general circulation of the atmosphere. Although it is possible to outline a number of phenomena which help to explain the arid zone, it is difficult, if not impossible, to point to a single factor which can be regarded as the ultimate control of the whole system.

Much of the subtropical region is characterized by divergent air flow in the atmosphere at low levels. In turn, this implies the presence of converging and subsiding air aloft, which being subjected to dynamical warming, will produce a lowering of relative humidities and the creation of stable atmospheric conditions. Under these circumstances, convectional activity, capable of producing precipitation, is reduced to a minimum.

Closely related with both the surface divergent flow of air and widespread air subsidence, is the occurrence of a normally well developed high pressure zone near the 30th parallel, which separates the tropical easterlies from the circumpolar westerlies. In both the regimes of the tropical easterlies and mid-latitude westerlies, a large proportion of the precipitation is the result of moving wave disturbances, which play such a crucial role in the distribution of heat, water vapour and momentum. The intensity of these disturbances, and hence the amounts of precipitation, decline as the subtropical high pressure belt is approached from either a northerly or a southerly direction. Although clearly detectable at sea level over the oceanic areas, the high pressure belt is usually only well developed over land above heights of two to three kilometres owing to surface heating effects.

A belt of strong westerlies, known as the subtropical jet stream, is found in the upper atmosphere in the Middle East, above the light and variable winds associated with the subtropical high pressure system which occurs at the earth's surface. The position of the core, or line of maximum velocity, of the jet stream varies over 10 to 15 degrees of latitude from summer to winter, especially over the eastern part of the region (Table 2.1). In July it is centred over the Caspian Sea, while by January it has moved southwards to lie over the northern part of the Gulf.[3] In the west over the Mediterranean Sea the jet stream core reveals a much more restricted latitudinal movement (Figure 2.1).

TABLE 2.1
Average speed and latitude of subtropical jet stream along 45 E

Month	J	F	M	A	M	J	J	A	S	O	N	D
Knots	120	120	100	90	90	80	70	60	50	60	70	90
Latitude	30	29	29	29	T	39	40	40	39	T	30	30

T = transitional period

Source: Weickman, J. 1961 'Some characteristics of the subtropical jet stream in the Middle East and adjacent regions'. Meteorological Publications Series A-No. 1 Ministry of Roads, Iranian Meteorological Department, Tehrān, p.4

This seasonal shift of the subtropical jet stream in the Middle East parallels the movement of the major climatic belts between their summer and winter positions. Observation of the jet stream has shown that the movement between the two positions is normally abrupt. The winter position, over the northern part of the Gulf, is maintained for approximately six to seven months from mid-October to April. This is followed by a rapid shift to the summer position over the Caspian Sea, which is held for three to four months from June to the end of September.

The reasons for the rapid shift of the jet-stream core position are not known with certainty. It has been discovered, however, that it appears to be related to atmospheric conditions generated by the existence of the high mountain belt of the Himalayas. Observations have revealed that the change in position of the jet-stream, together with the related sudden climatic change from winter to summer and vice versa, is associated with a jet-stream movement from one side of the Himalayan range to the other. The Himalayas apparently retard the movement of the jet-stream core in some direct or indirect manner, until a certain threshold value is exceeded. There then follows a sudden movement of the jet-stream core to the opposite side of the Himalayan chain, and the subsequent establishment of a new equilibrium position.

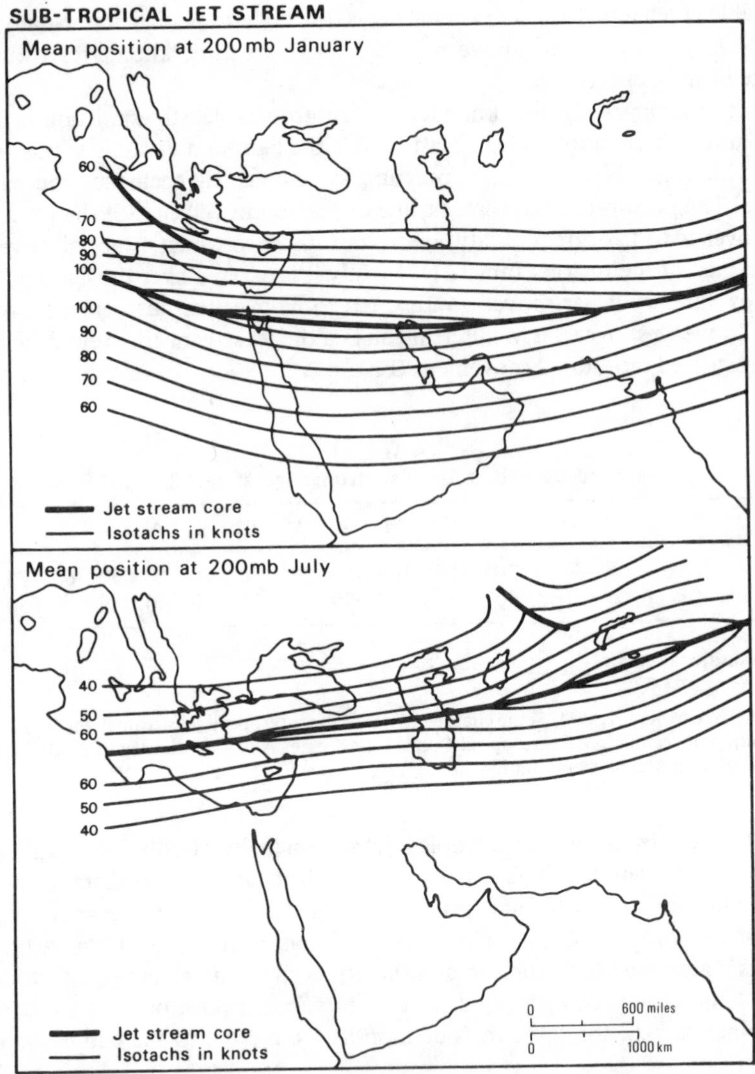

Figure 2.1 Position of the jet-stream core in winter and summer (reproduced by permission of the Iranian Meteorological Organization)

2.1.2 Weather systems

The weather patterns of the Middle East can best be explained by reference to the succession of cyclones which pass over the region. Cyclones are wave disturbances generated along the polar front, separating polar and tropical air masses. They develop over the North Atlantic and especially the Mediterranean Sea, and travel westwards over the Middle East.[4]

During summer the paths of the cyclones tend to pass north of the Pontus and Elburz Mountains, and, therefore, their effect on the climate of the Middle East, with the exception of northern Turkey and the Caspian littoral of Iran, is minimal. In winter a quite different pattern occurs, with the paths of the cyclones displaced southwards over the Mediterranean Sea and the Middle East (Figure 2.2). Even at this time, however, very few depressions penetrate south of 30 degrees north and their frequency of occurrence increases markedly as one travels northwards. As winter progresses, very cold air masses are developed over the high Anatolian and Iranian plateaus and these tend to exercise a steering effect on the cyclones, causing them to pass either northwards or southwards of the main mountain masses. For much of the Middle East the most important cyclone track is across northern Syria, Iraq, and southern Iran, as it is disturbances moving along this route which account for a very large proportion of the total precipitation of the region.

Three types of cyclonic disturbance have been recognized in the region.[5] The first are shallow waves moving rapidly in the upper troposphere, which cross the area along a west-east corridor between southern Turkey and Jordan. Rain associated with these disturbances reaches a maximum during the winter months, and falls mainly in Lebanon, Syria, Jordan and along the Zagros Mountains.

Cold troughs in the upper atmosphere, creating almost stationary cyclones in certain regions, form the second type of the group. The best known area for this

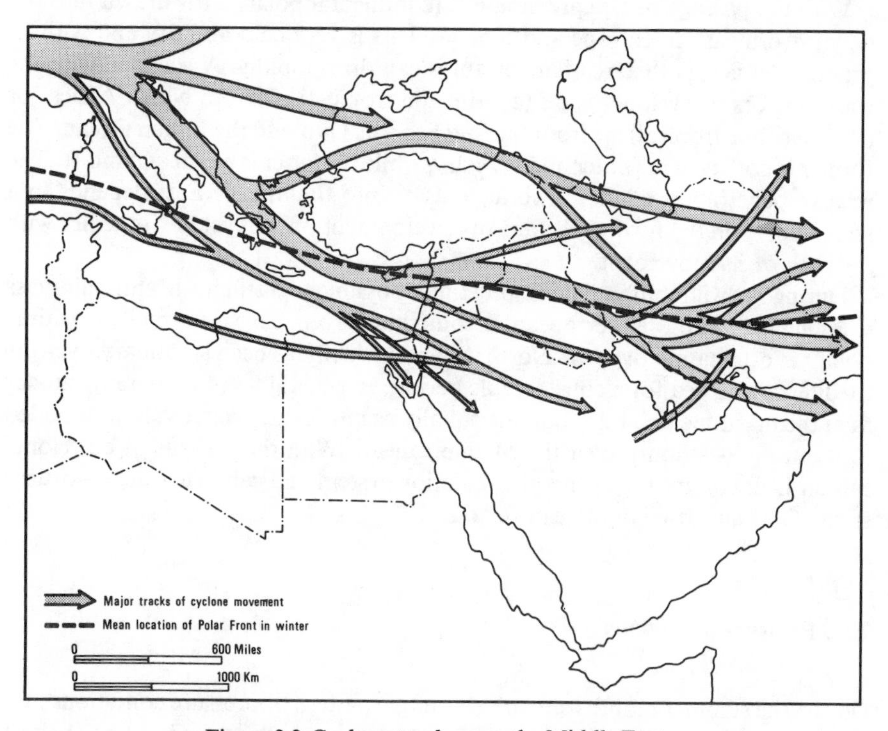

Figure 2.2 Cyclone tracks over the Middle East

kind of disturbance is centred on the island of Cyprus, where cyclones moving through the Mediterranean slow down, and often become part of a semi-permanent low pressure system. Such slow moving cyclones bring considerable precipitation to the lands fringing the Levant coastline. Moving further east these disturbances tend to become semi-stationary once again over the steppe region of northern Iraq, or over the Turkish mountains to the north, where they draw in moist air from the Gulf region and cause precipitation along the line of the Zagros Mountains and northwestern parts of the Iranian Plateau. Maximum precipitation associated with these systems occurs during spring.

The final type of cyclone, sometimes known as the 'Khamsin type', originates over the northern Sahara and moves eastwards over the southern part of the Middle East. Occasionally they come into contact with a cold upper air trough and under such conditions can give rise to appreciable rainfall amounts.

The Middle East comes under the influence of five air masses at different times of the year. During the summer months there is relatively little air movement throughout the region, and a cT (continental tropical) air mass, which is very hot and dry, prevails. A very different situation occurs in winter when air masses from four major source areas can enter the region as the result of cyclonic activity. The most important air mass is mP (maritime polar) air, originating over the North Atlantic. This contains large amounts of moisture and provides most of the precipitation experienced throughout the Middle East.

With the passage of a depression cP, (continental polar) air is drawn into the region from eastern Europe and Siberia. This is very cold and dry and with its progression across the area temperatures can drop rapidly. Associated with the warm sectors of cyclones is cT (continental tropical) air from North Africa, or mT (maritime tropical) air from the Red Sea, the Gulf and the Indian Ocean. The former is hot, even in winter and very dry, while the latter is warm and moist. The relative penetration of these two air masses into the Middle East depends to a large degree on the form and stage of development of the cyclone, together with the path of its movement.

During the winter months it is sometimes possible to distinguish a fifth air mass to which the term 'Mediterranean' is sometimes given. This is normally maritime polar air originating over the North Atlantic which has been stationary over the Mediterranean Sea for a considerable period. In general it is considerably milder than the maritime polar air, but it is equally as moist since water vapour is added to it during its sojourn over the Mediterranean. With the passage of a cyclone, this air is drawn into the general circulation system and can penetrate eastwards to the Gulf and Iranian plateau region.

2.1.3 Pressure and winds

During the summer months a relatively simple pattern of pressure conditions predominates throughout the region. The main feature at this time is a large belt of low pressure formed over the Gulf and the adjacent lowlands of Iraq. Smaller and less permanent low pressure centres also occur over Anatolia, central Iran and the southern Red Sea. Although these low pressure zones can often be identified on synoptic charts, the pressure gradients tend to be slight throughout the whole region and winds gentle. To the southeast of the region, very low pressure develops over the northern India as the monsoonal circulation is established in mid-summer. This system, however, only affects the extreme southern fringe of the Middle East.

Despite the gentle pressure gradients, remarkably persistent local winds can develop at this time of year, blowing towards the low pressure centres. In the Tigris–Euphrates lowlands the *shamal* is a dry north or northwesterly wind which blows throughout much of the summer. A similar wind in the Sīstan basin of southeast Iran is known as the *sad-ou-bistbad* (120 day wind). This, too, blows from the north and northwest, is extremely hot, and, owing to its high velocities, heavily dust laden.

Along the coastal areas of the Mediterranean, and in particular in western Turkey, land–sea breezes are well developed during the summer months. These blow onshore during the day, especially in the afternoon, bringing cooler but more humid conditions to narrow coastal areas. At night, these winds decline markedly in intensity and are sometimes replaced by offshore winds. Where they are funnelled along valley systems they can often attain high velocities. In summer, locally very strong pressure gradients can also be developed as the result of intense solar heating, producing a range of small dust storms characteristic of the interior deserts of Iran and Saudi Arabia.

In winter, it is much more difficult to speak of average pressure conditions, for this is the period when the region is crossed by a succession of cyclones. In general, one can identify a tendency for low pressure conditions to be dominant over the eastern Meditteranean, centred on Cyprus, and also over the Black Sea. Similarly, high pressure systems develop commonly over Anatolia and central Iran, as a result of the very cold winter temperatures.

With the passage of depressions locally steep pressure gradients can result, and, if associated with favourable topographic conditions such as a sharp altitudinal change over a short distance or else a funnelling effect, these can produce the various types of local winds for which the Mediterranean and Middle East region is so well known.

Similar local winds are also caused by particular air masses being drawn into a region with the passage of a cyclone and its associated frontal systems. For example, hot dry winds, with differing local names, such as *khamsin* in Egypt; *ghibli* in Libya; *shlour* in Syria and Lebanon: *shargi* in Iraq; and *simoon* in Iran, develop when tropical continental air from over North Africa and Arabia, is brought into a region in the warm sector of a travelling cyclone. Such winds are often strong, and, therefore, frequently give rise to dust and sand storms in desert regions, especially during autumn and spring. As they are associated with moving weather systems, these winds last only for three or four days, at most. A sudden rise in temperature is associated with their arrival, while an equally rapid fall in temperature and increase in humidity witnesses the passage of the following cold front and the subsequent invasion of cooler polar maritime air.

Pronounced adiabatic warming of air masses is also a common phenomenon in the Middle East in areas where high mountains are found in close proximity to lowlands. Warm winds produced by this mechanism are common in spring along both the Mediterranean and Black Sea shores of Turkey, in the Tigris–Euphrates and Khuzestan lowlands bordering the Zagros Mountains, and along the Caspian Sea coast of Iran.

During the winter months, the intense cold which develops over Anatolia and central Iran, means that outbursts of air from these regions still reach adjacent lowland regions as very cold winds, despite the adiabatic warming effect. This is particularly true in western Turkey where very low temperatures can be experienced in the İzmir region as the result of cold air movement down the valleys of the Gediz and Büyük Menderes rivers.

2.1.4 Temperatures

Throughout the Middle East summer temperatures are high almost everywhere. The highest mean daily temperatures at this period, of more than 30°C, are

TABLE 2.2
July temperatures at stations with mean daily temperatures greater than 30°C

	Average daily		Mean	Absolute	
	max	min	daily range °C	max	min
Coastal stations					
Quseir (Egypt)	34	26	8	40	23
Jiddah (Saudi Arabia)	38	26	12	42	21
Aden (South Yemen)	36	28	8	40	23
Muscat (Oman)	36	30	6	45	25
Bahrain (Bahrain)	37	29	8	44	24
Kuwait (Kuwait)	40	30	10	48	26
Būshehr (Iran)	35	29	6	44	23
Bandar 'Abbās (Iran)	38	29	9	45	26
Interior stations					
Ghadames (Libya)	43	22	21	54	15
Kufra (Libya)	38	24	14	43	17
Aswân (Egypt)	42	26	16	51	21
Riyadh (Saudi Arabia)	42	25	17	45	19
Baghdād (Iraq)	43	24	19	50	17
Mosul (Iraq)	43	22	21	51	15
Tehrān (Iran)	37	22	15	43	15
Sīstan (Iran)	40	26	14	48	17

Source: Meteorological Office, London. 1966. 'Tables of Temperature, Relative Humidity and Precipitation for the World', Part V–Asia, HMSO. Meteorological Office, London. 1967. 'Tables of Temperature, Relative Humidity and Precipitation for the World', Part IV–Africa, the Atlantic Ocean south of 34°N and the Indian Ocean, HMSO.

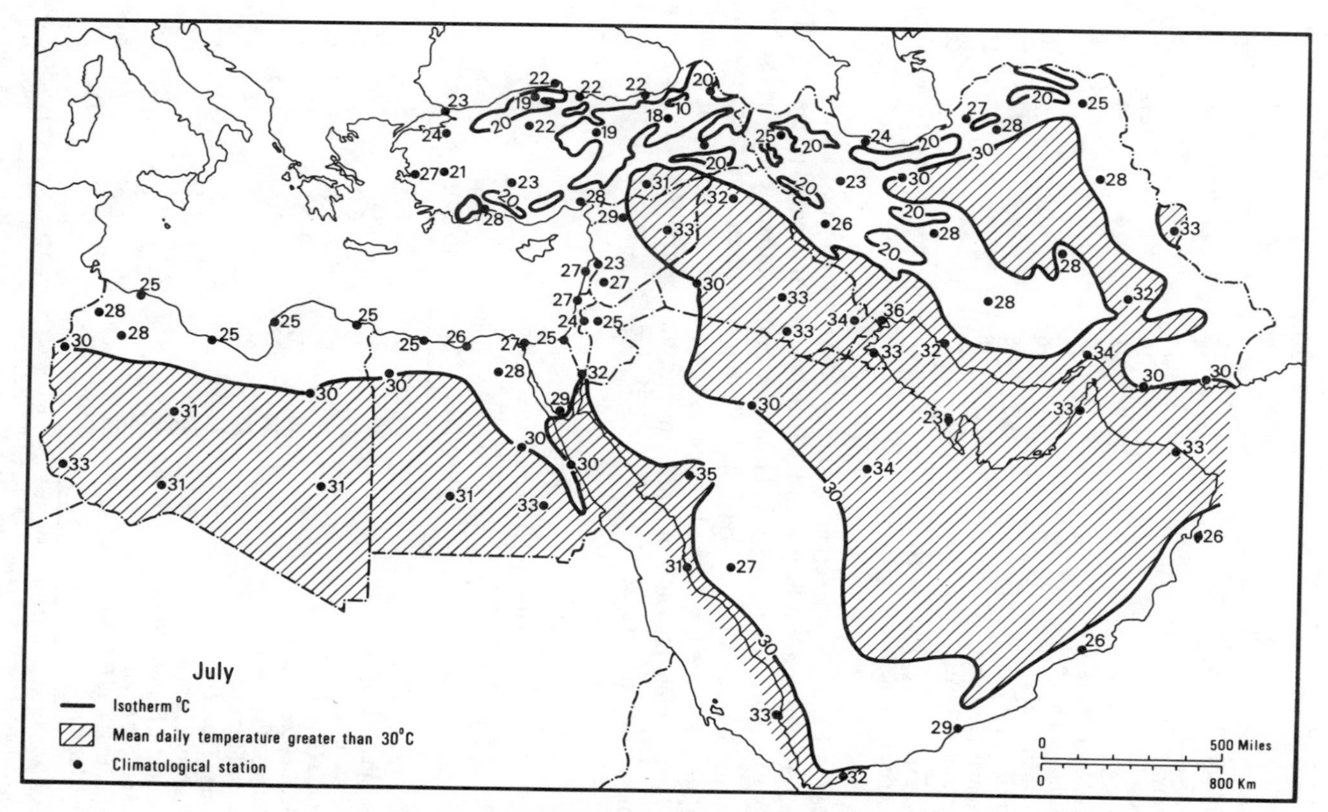

Figure 2.3 Mean daily temperatures for July

TABLE 2.3
July temperatures of stations in northern highland regions °C

	Average max	daily min	Mean daily range °C	Absolute max	Absolute min
Ankara (Turkey)	30	15	15	38	7
Uşak (Turkey)	29	14	15	38	3
Konya (Turkey)	30	15	15	38	7
Sivas (Turkey)	27	10	17	37	3
Erzurum (Turkey)	25.5	11.5	14	34	3
Mashhad (Iran)	33	17	16	41	8
Kermānshāh (Iran)	37	13	24	42	3.5
Eşfahān (Iran)	38	19	19	42	9
Damascus (Syria)	35.5	18	17.5	42.5	13
Amman (Jordan)	31.5	18	13.5	40	13
Jerusalem (Israel)	30	17	13	38	10

Source: Meteorological Office, London, 1966. 'Tables of Temperature, Relative Humidity and Precipitation for the World', Part V–Asia, HMSO.

recorded in the southern desert areas of Libya and Egypt, the Red Sea coastlands, the Gulf coastlands and adjacent lowlands and parts of the central plateau of Iran (Figure 2.3). In the coastal regions diurnal temperature ranges tend to be smaller than in the interior stations (Table 2.2). During the summer months very few places, however, are free from the relentless heat during the middle of the day. Even on the high plateaux of northern Iran and central Anatolia day-time maxima can often surpass 35°C (Table 2.3). Indeed, it is only in the highest

TABLE 2.4
July temperatures in coastal locations °C

	Average max	daily min	Mean daily range °C	Absolute max	Absolute min
Tripoli (Libya)	29.5	21	8.5	45	15
Benghazi (Libya)	29	21.5	8.5	40	11
Alexandria (Egypt)	29	23	6.0	40	17
Haifa (Israel)	31	24	7.0	35.5	17
Beirūt (Lebanon)	30.5	23	7.5	40	13
Antalya (Turkey)	34	22.5	11.5	43.5	15
İzmir (Turkey)	33	20.5	12.5	42.5	11
Sinop (Turkey)	24	18	6.0	34	13
Trabzon (Turkey)	25.5	19	6.5	32.5	15
Bandar-e Pahlavī (Iran)	29.5	21	8.5	—	—
Bābol Sar (Iran)	29.5	22	7.5	—	—
Masīra Island (Oman)	31.0	23	7.0	38	20.5
Salālah (Oman)	27.5	24	2.5	32	21

Source: Meteorological Office London. 1966. 'Tables of Temperature, Relative Humidity and Precipitation for the World'. Part V–Asia. HMSO. Meteorological Office, London. 1967. 'Tables of Temperature, Relative Humidity and Precipitation for the World'. Part IV–Africa, the Atlantic Ocean south of 35°N and the Indian Ocean. HMSO.

upland areas of Turkey, Iran and the coastal Mediterranean region where daytime maxima do not normally rise above 30°C at this time of year. Along the narrow strip bordering the Black, Caspian, and Mediterranean Seas, together with a similar zone along the Indian Ocean in southern Arabia, summer temperatures are moderated by the proximity of large water bodies (Table 2.4). A few miles inland this influence quickly declines and temperatures soar to values approaching those found in inland regions.

Much stronger contrasts in temperatures are noted between different regions in winter. In general, a regional gradient, with temperatures declining in a northerly direction, tends to dominate the picture, and this pattern is intensified by the location of the major mountain masses in the northern parts of the region running through Turkey and Iran. During the coldest month it is possible to divide the Middle East into two major zones, north and south respectively of the 10°C mean monthly January isotherm (Figure 2.4).

To the south of this line mean daily temperatures are above 10°C, and, with the exceptions of the southern coastal areas of Arabia and southeast Egypt, less than 20°C. Strong temperature gradients, therefore, are absent from this region. The mean daily values, however, do tend to conceal variations in the magnitude of diurnal ranges which exist between coastal and interior location (Table 2.5). Nearly everywhere on the coast this range is much smaller than inland.

TABLE 2.5
January temperatures in the southern part of the region °C

	Average daily		Mean daily	Absolute	
	max	min	range°C	max	min
Coastal					
Tripoli (Libya)	16	8.5	7.5	28	1
Benghazi (Libya)	17	10	7.0	24.5	3.5
Alexandria (Egypt)	18	10.5	7.5	28	3
Beirūt (Lebanon)	16.5	10.5	6.0	25	−0.5
Kuwait (Kuwait)	16	9.5	6.5	27.5	0.5
Būshehr (Iran)	17.5	10.5	7.0	26.5	0
Bandar 'Abbās (Iran)	23	13	10.0	28	3
Sharjah (United Arab Emirates)	23	12	11.0	29.5	2.5
Interior					
Sebra (Libya)	17.5	5	12.5	28.5	−2
Kufra (Libya)	20.5	6	14.5	32	−3.5
Dakhla (Egypt)	21	5	16.0	35.5	−0.5
Aswân (Egypt)	23	10	13.0	38	3
Riyadh (Saudi Arabia)	21	8	13.0	30	−7
Hā'il (Saudi Arabia)	16.5	4	12.5	26.5	−2.5
Bam (Iran)	16.5	5	11.5	—	—

Source: Meteorological Office, London. 1966. 'Tables of Temperature, Relative Humidity and Precipitation for the World'. Part V–Asia. HMSO. Meteorological Office, London. 1967. 'Tables of Temperature, Relative Humidity and Precipitation for the World'. Part IV–Africa, the Atlantic Ocean south of 35°N and the Indian Ocean, HMSO.

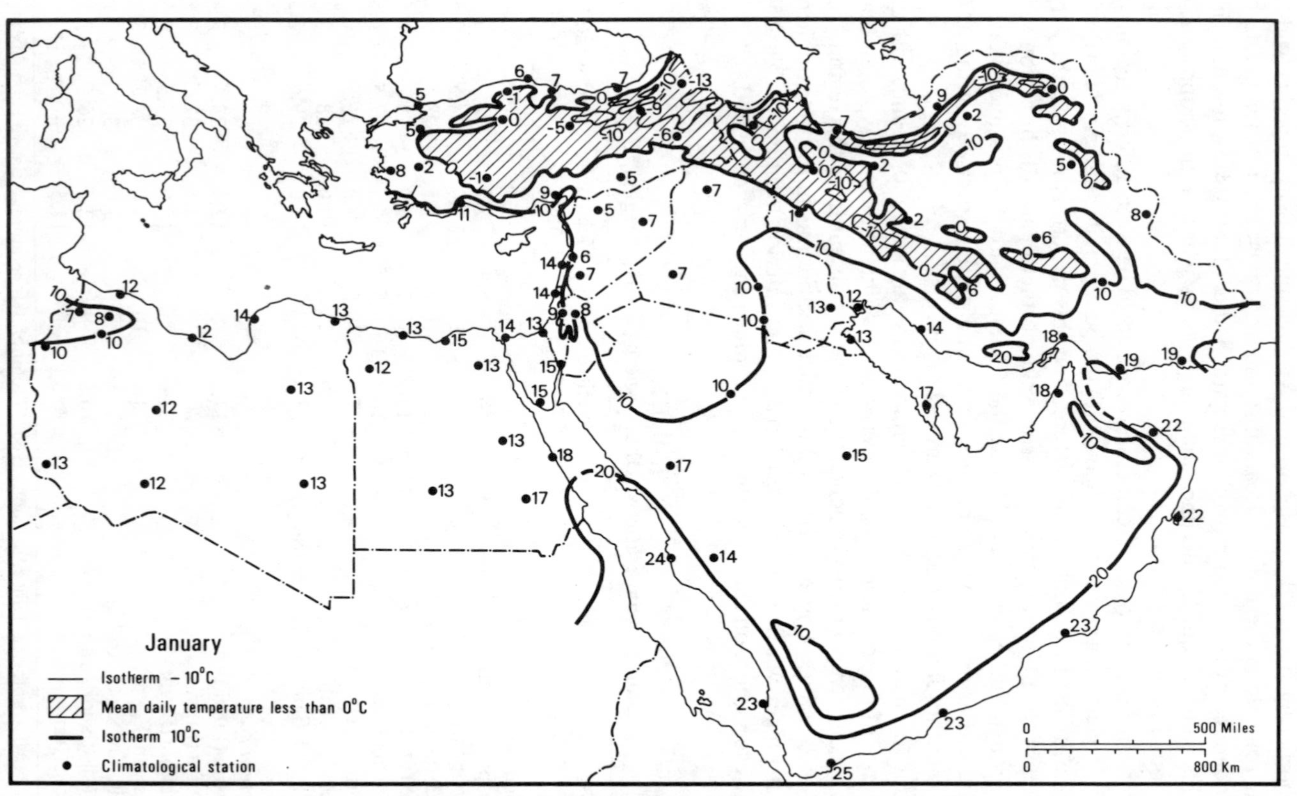

Figure 2.4 Mean daily temperatures for January

North of the 10°C mean January isotherm, daily temperatures decline rapidly across the Taurus and Zagros Mountains to give mean daily minimum values of less than −10°C over parts of the plateau regions of eastern Turkey, and in smaller pockets along the Elburz and Zagros chains. Few people live in such areas and, as a consequence, climatological stations are rare. However, a number of the high plateau and mountain stations in both Turkey and Iran clearly indicate the severity of conditions during winter (Table 2.6). For example, in the interior of eastern Turkey the mean maximum daily temperatures of January are below freezing point.

As a consequence of these regional differences in temperature between the summer and winter months, the mean annual range of temperatures reveals marked variations between the northern and southern limits of the area. The greatest mean annual temperature range, of more than 30°C, is found in the mountainous areas of eastern Turkey (Figure 2.5). Much of eastern Turkey and northern Iran experiences a range of more than 25°C, particularly over the enclosed plateau regions. To the north of this zone, a sharp downward gradient is experienced toward both the Caspian and Black Sea coastlands, while a much more gentle reduction in the mean annual range is noted to the south. The lowest mean annual temperature range of less than 5°C is found along the southeastern coastline of Oman. Most of the Red Sea coastline, together with the coastal area of South Yemen, have an annual range of less than 10°C.

The temperature regimes of individual stations can best be illustrated in terms of diagrams showing mean maximum and mean minimum monthly temperatures (Figure 2.6). These clearly reveal both the annual as well as the monthly temperatures ranges. The ameliorating influence of coastal locations, and the extreme severity of winters in the interiors of Anatolia and Iran stand out markedly.

Features which are difficult to illustrate, owing to the lack of data, are the sharp meteorological contrasts which can occur over relatively short distances at a given

TABLE 2.6
January temperatures in the northern part of the region °C

	Average daily		Mean	Absolute	
	max	min	daily range° C	max	min
Eşfahān (Iran)	10	−2	12	20	−16
Tehrān (Iran)	9	−1	10	19	−16
Mashhad (Iran)	8	−2	10	24	−24
Kermānshāh (Iran)	8	−4	12	19	−21
Tabrīz (Iran)	4	−5	9	17	−25
Ankara (Turkey)	4	−5.5	9.5	15	
Konya (Turkey)	4	−5	9	16	−27
Sivas (Turkey)	−0.5	−9.5	9	10.5	−31
Erzurum (Turkey)	−5.5	−13	7.5	5.5	−30
Kars (Turkey)	−4	−18	14	5	−36

Source: (a) Meteorological Office, London. 1966. 'Tables of Temperature, Relative Humidity and Precipitation for the World'. Part V–Asia. HMSO.
(b) Ganji, M. H. 'Climate'. Chapter 5, in *The Land of Iran*, (Ed W. B. Fisher), Vol. 1, Cambridge History of Iran, Cambridge University Press.

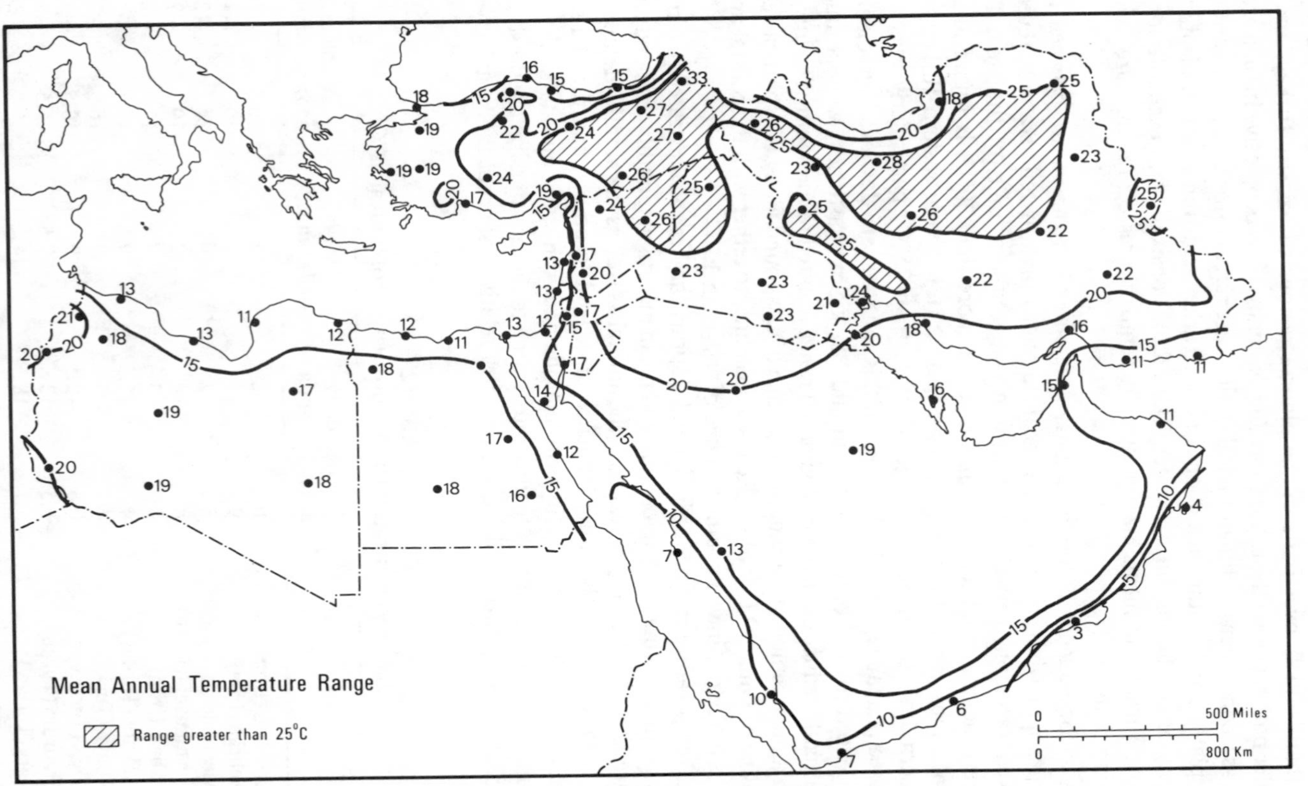

Figure 2.5 Mean annual temperature range

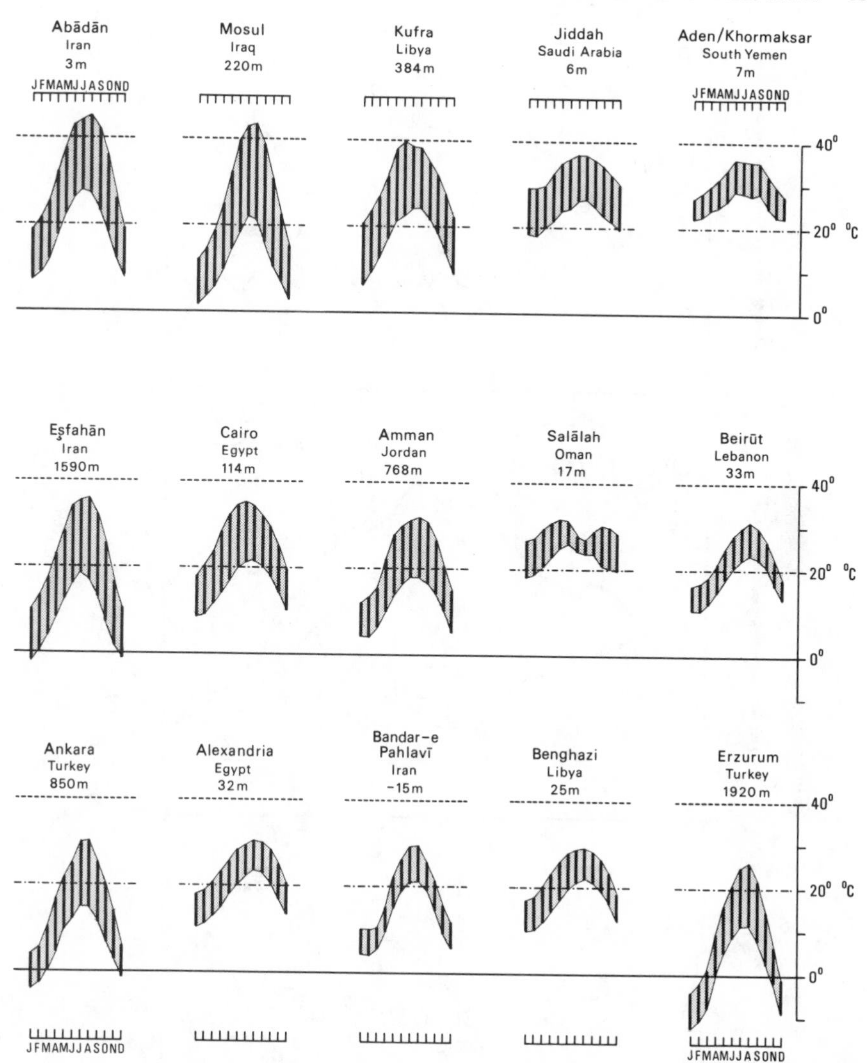

Figure 2.6 Diagrams of mean maximum and mean minimum monthly temperatures for selected stations

time. An illustration of such a contrast is given by two temperature traverses on consecutive days made by one of the authors across the Dead Sea Lowlands from Amman to Jerusalem and back again.[6] Along this route, a distance of 80 km, the road descends from about 1000 m above sea level near Amman to almost 400 m below sea level, the lowest point on earth, and then ascends to 700 m above sea level in the vicinity of Jerusalem. The temperature and humidity changes along this route are registered in Figure 2.7. Variations in absolute temperature of 11.5° and 12.5°C, and in relative humidity of 38 per cent respectively were recorded on the two days. Such studies clearly reveal the considerable local climatological

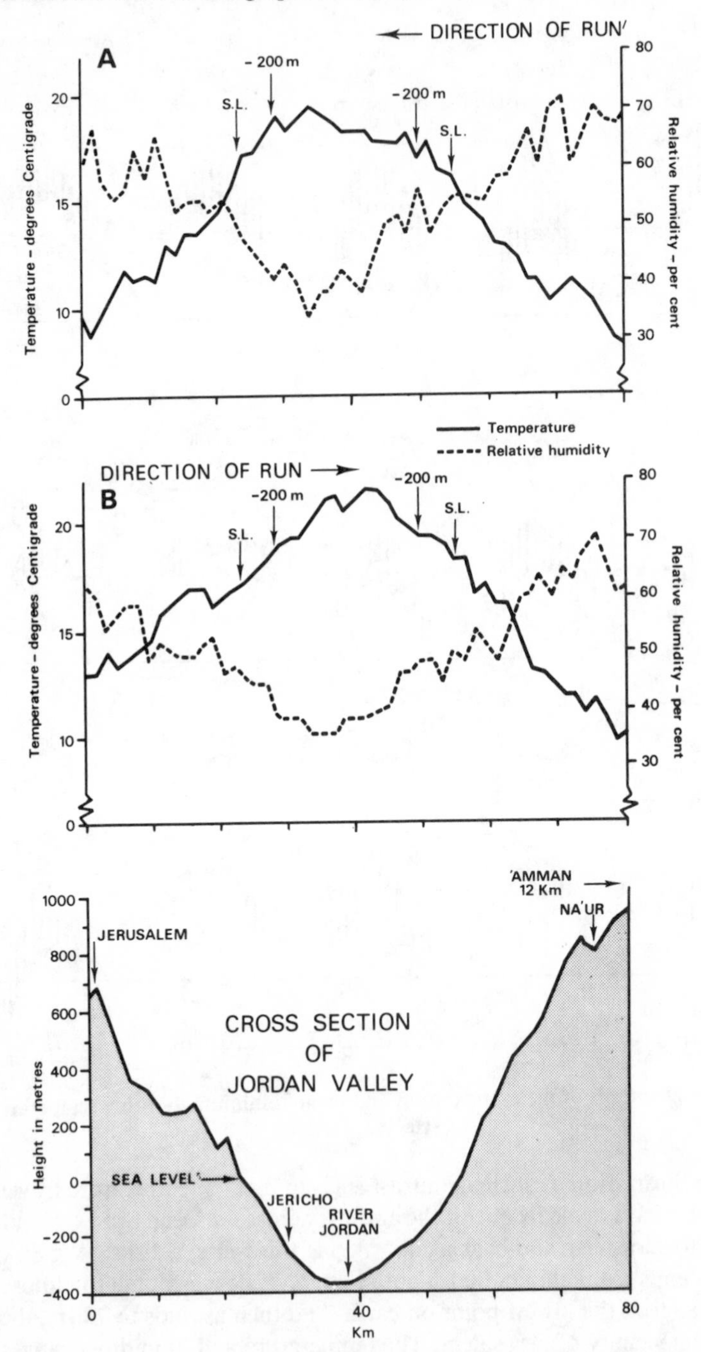

Figure 2.7 Temperature and relative humidity variations in a transect across the Dead Sea lowlands (reproduced from *Geography* 53, 1968)

differences which exist within the Middle East in areas which possess considerable amplitude of relief. Perhaps the greatest contrast of all in the Middle East is to be found in northern Iran along a traverse from the well-watered Caspian Sea lowlands, over the summits of the Elburz Mountains which attain crest heights of more than 3000 m, and down to the arid salt desert of central Iran.

2.1.5 Precipitation

Precipitation occurs when moist air is lifted above its condensation level and water droplets are formed. The major lifting mechanisms are of an orographic, cyclonic or convectional nature. Orographic lifting is usually brought about by the presence of highland regions over which the moving air masses have to rise. As the position of mountain ranges is fixed, this lifting mechanism is in operation all year round, though whether it actually produces precipitation or not depends upon dynamic factors in the atmosphere.

Lifting associated with cyclones is the result of the interaction and intermixing of air masses with different thermal characteristics along frontal boundaries. The net result is that warm air masses, especially those associated with the warm sector of a cyclonic disturbance, are uplifted to the condensation level. Cyclonic lifting obviously occurs over all types of terrain, both lowland and highland, but its effects, in terms of precipitation amounts, are often greatly enhanced when further lifting is produced by the movement of a cyclone over a highland barrier. The individual paths or tracks of cyclones tend to be highly variable. However, if one studies statistically the average paths of a large number of cyclones, certain preferred routeways appear to exist, and along such routeways annual precipitation amounts reach their maximum values.

The final type of lifting is that associated with convectional, or thermal activity. This occurs chiefly during spring and summer in lowland and plateau regions subjected to strong solar heating. It often gives rise to thunderstorm formation and extremely intense precipitation.

Precipitation amounts are related to two major factors, first, proximity to the major moisture sources of the Mediterranean, Black and Caspian Seas, and, secondly, altitude. In general, precipitation totals decline in an easterly and southerly direction away from the Mediterranean Sea. Over much of the region most of the precipitation is caused by frontal lifting of moist air masses, a process which is intensified by orographic lifting, when highland zones are crossed. In the extreme north of the region, along the Black Sea and Caspian Sea coasts, precipitation occurs all the year round since this region always falls under the influence of mid-latitude cyclonic disturbances. Not surprisingly, therefore, this part of the Middle East receives some of the highest precipitation totals within the region (Figure 2.8).

Few areas in the Middle East receive more than 600 mm/annum of precipitation. With only one major exception, they are confined to narrow coastal belts backed by mountain ranges. In Turkey both the northern, or Black Sea coast and

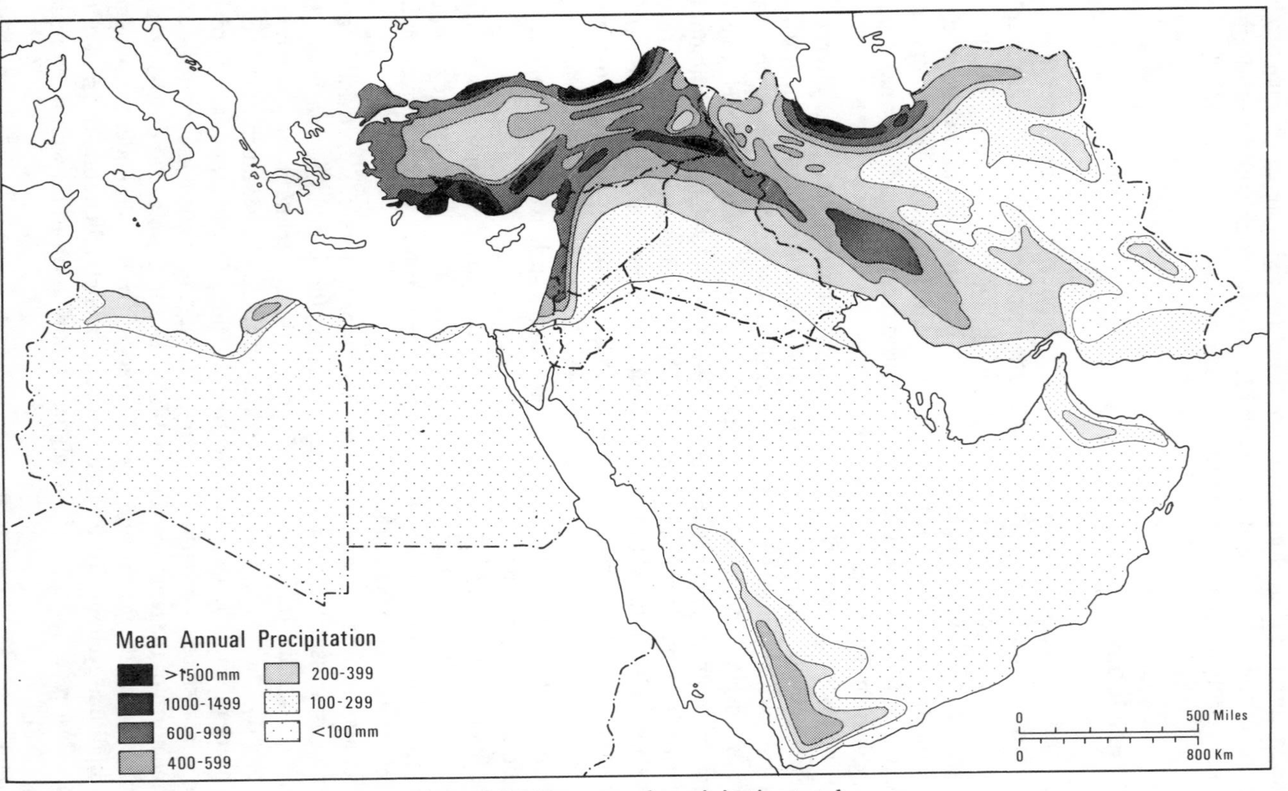

Figure 2.8 Mean annual precipitation totals

the southern, or Mediterranean coast possess regions where precipitation totals are in excess of 1000 mm/annum. Even in these areas, though, the belt of high rainfall is not continuous. The eastern shoreline of the Mediterranean, in Syria, Lebanon and Israel, together with the mountains behind it, is another favoured region in terms of precipitation amounts. Maximum values are recorded in the northern highlands, and precipitation totals tend to decline southwards.

Along the Caspian Sea lowlands of Iran high precipitation values, commonly over 1000 mm/annum in the western area, are recorded. A very marked gradient is found in an easterly direction along the coast with annual values dropping at the eastern margin of the lowlands to less than one half of those recorded in the west. A similar and perhaps somewhat surprising decrease in precipitation amounts is seen as one ascends southwards from the Caspian lowlands towards the crest line of the Elburz Mountains.

The final zone of high precipitation is a long and discontinuous belt commencing in eastern Turkey and running southeastwards parallel with the crest of the Zagros Mountains. Only rarely do precipitation totals rise above 800 mm/annum, but the sheer size of the area gives it tremendous importance in the water balance of the region. Although the highland regions continue eastwards to the Pakistan border, precipitation totals begin to fall off rapidly south of 30 degrees north.

Of all the countries of the Middle East, only two, Turkey and Lebanon, do not possess areas of extreme aridity, where precipitation falls below 200 mm/annum. In contrast, well over three-quarters of the total areas of Libya, Egypt and Arabia receive annual precipitation totals of less than 200 mm. All other states experience a range of variations between these two extremes.

The seasonal distribution of precipitation in the Middle East, reflects, as one might expect, the passage of winter cyclones (Figure 2.9). Winter precipitation maxima predominate almost everywhere with the exception of parts of the southern coastlands of Arabia where summer monsoon conditions often penetrate. Along the Black and Caspian Sea littorals summer rain occurs, but here also an autumn/winter maximum is normally observed.

Throughout the upland areas of the Middle East, particularly in Turkey, northern Iran and the higher parts of the Levant, a considerable proportion of the precipitation falls as snow. This means that large quantities of water are stored in the highlands as snow pack during the winter period to be released, often as flood discharges, when temperatures begin to rise in spring and early summer.

Even over relatively short distances, relief and rain shadow influences can produce considerable variations in precipitation. For example, in the eastern Mediterranean region of Israel and Jordan daily precipitation totals for selected stations forming a west to east transect from the coast to the interior desert, clearly reveal both these effects (Figure 2.10). Analysis of the precipitation records show rain day sequences of two to five days duration associated with the passage of depressions separated by longer periods of dry weather. At the drier stations, the number of rain days per annum is fewer, and the daily amounts of precipitation less than at the wetter stations.

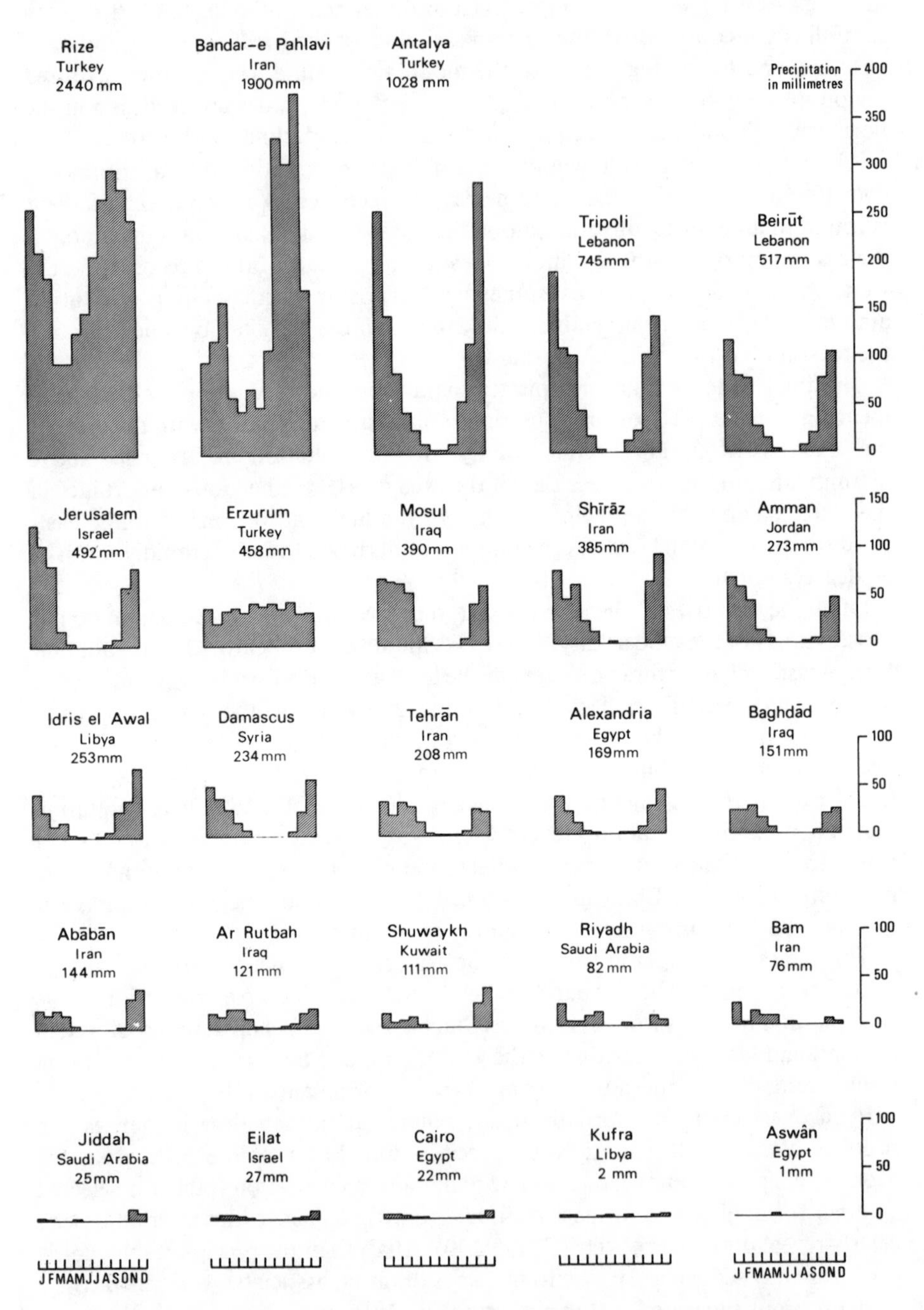

Figure 2.9 Diagrams of mean monthly precipitation for selected stations

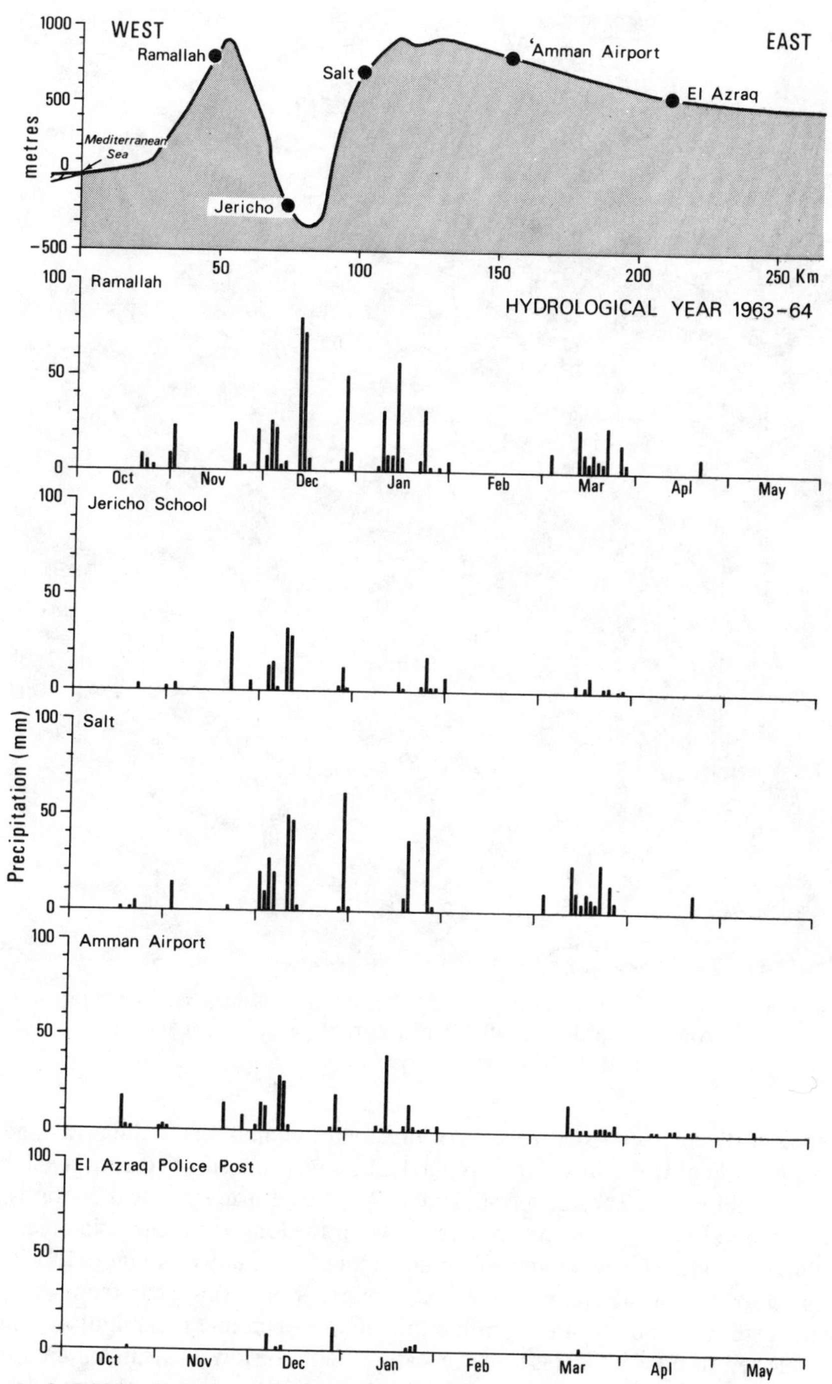

Figure 2.10 Variations in precipitation along a transect from the Mediterranean Sea to the Jordanian desert

Snowfalls can be heavy in some areas; winter conditions in a refugee camp near Amman, Jordan (UNRWA photograph by John Bonar)

One of the characteristic features of precipitation amounts in the arid zone is the high annual variability. Unfortunately, however, the short records available for most climatological stations make it difficult to utilize statistical methods to assess this variability with any accuracy. Two of the longest precipitation records within the Middle East, for Jerusalem and Tehrān, are shown in Figure 2.11, and these clearly illustrate the very marked changes which can occur from year to year. A study of the maximum, minimum and mean annual precipitation values for selected stations indicates that the stations with the lowest mean annual rainfall show the greatest difference between the maximum and minimum annual recorded fall (Table 2.7).

TABLE 2.7
Annual precipitation in millimetres

Region	Station	Mean	Maximum	Minimum	Ratio of maximum to minimum
Egypt	Alexandria	169	313.6	33.2	9.45
	Port Said	63	129.6	13.0	9.97
	Cairo	22	63.4	1.5	42.27
	Asyût	5	25.0	0.0	—
	Dakhla	Trace	11.0	0.0	—
Turkey	Rize	2440	4045.3	1757.5	2.30
	Antalya	1028	1644.9	560.8	2.93
	İzmir	695	1116.5	441.2	2.53
	Erzurum	458	829.6	253.7	3.27
	Ankara	362	500.8	247.5	2.02
	Konya	316	500.5	143.7	3.48
Israel	Jerusalem	529	957.7	273.1	3.51
	Eilat	27	96.9	5.5	17.62
Jordan	Amman	273	476.5	128.3	3.71
Iraq	Mosul	390	585.2	208.2	2.81
	Baghād	151	336.0	72.3	4.65
	Ar Rutbah	121	269.9	46.9	5.75
Arabia	Bahrain	76	169.4	10.1	16.77
	Aden	39	93.0	7.6	12.24

Source: World Meteorological Organization 1971, 'Climatological Normals (Clino) for Climate and Climate Ship Stations for the Period 1931–1960'. WMO/OMM-No. 117 TP52.

More sophisticated types of analysis, involving calculations of relative inter-annual variability of precipitation, have been made as part of a study of agroclimatology in the Near East.[7] This work showed that the relative interannual variability of annual precipitation in per cent within the region studied increased as the mean annual total decreased. Using these data, together with mean annual precipitation values, it was concluded that a mean annual rainfall of 240 mm, with a relative interannual variability of 37 per cent, was the normal minimum requirement for dryland farming in the region. In mountainous areas, where the relative interannual variability was particularly high, a greater mean annual rainfall total was necessary to ensure regular dryland farming than in the lowland regions.

Information on the intensity of rainfall in the Middle East is also very meagre since recording rain gauges have only been introduced on a large scale within the last few years. The most generally available data which can be used to gain some indication of precipitation intensity are daily precipitation figures which are normally collected from all precipitation stations. To date, relatively little of this

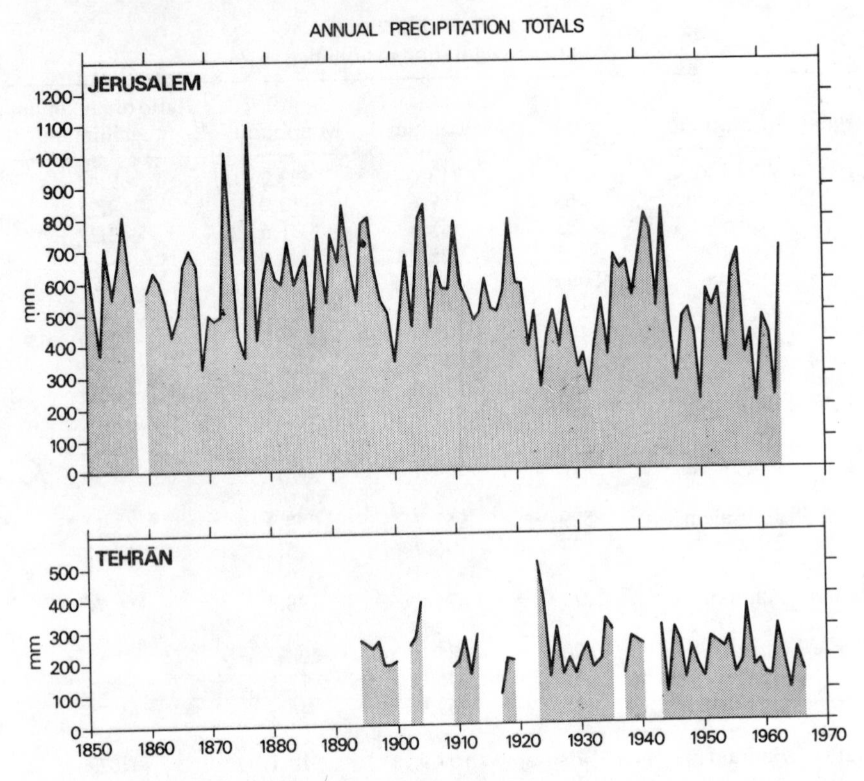

Figure 2.11 Annual precipitation variability at Jerusalem and Tehrān

information has been subjected to analysis. At Tehrān, however, daily precipitation data going back to 1898 have been analysed to discover the return period of falls of a given magnitude.[8] For the winter season this work revealed that a daily fall of 35 mm of precipitation would be expected to occur once in every ten years on average, whilst a daily fall of 75 mm would be expected only once in 1000 years (Table 2.8).

TABLE 2.8
Maximum daily precipitation – Tehrān

	Daily precipitation (mm) likely to be equalled or exceeded once in:						
	10 yrs	20 yrs	50 yrs	100 yrs	200 yrs	500 yrs	1000 yrs
Winter season	35	41	49	55	61	69	75

Source: Gordon, A. H. and Lockwood, J. G. 1970.

These results clearly indicate that torrential downpours over long periods of time, that is of 24 hours or longer, are of relatively rare occurrence in the Tehrān situation.

2.1.6 Evapotranspiration

In any climatic regime, temperature and precipitation data often do not provide a clear picture of the prevailing moisture conditions as experienced by plants. To obtain such information a number of empirical formulae have been devised which attempt to measure the combined effects of evaporation and transpiration, and thus permit a detailed water balance for a region to be drawn up. The best known formulae are those of Penman and Thornthwaite, both of which measure potential evapotranspiration (PET).[9] This is defined as the maximum amount of evaporation and transpiration which would occur from a vegetated surface (grass) if an abundant and continuous supply of moisure is available in the upper soil layers. Because a continuously available supply of water is assumed, potential evapotranspiration is really a measure of the available energy in the atmosphere. In contrast, actual evapotranspiration is the amount of moisture which is lost from a vegetated or ground surface under normal conditions. This is always less than, or equal to, the figure of potential evapotranspiration, and is governed by the amount of soil moisture storage, as well as by air temperature conditions. Of the two methods, the Penman formula produces results which agree most closely with field measurements of potential evapotranspiration. However, the amounts of meteorological data required for its calculation mean that the formula can only be applied at relatively few climatological stations in the Middle East, where detailed records are available. As a result, the simpler Thornthwaite method, relying only on mean monthly temperature and precipitation data, together with information on latitude, has been most widely used throughout the region.

By using calculated potential evapotranspiration and precipitation data it is possible to draw up water balance diagrams for climatological stations which reveal periods of water surplus and water deficit in the annual cycle. Water surplus occurs when precipitation and available soil moisture reserves exceed potential evapotranspiration for a given month, while water deficit results when precipitation and available soil moisture reserves are less than potential evapotranspiration.

Detailed work on potential evapotranspiration amounts in the Middle East revealed marked variations throughout the region (Figure 2.12).[10] The highest values of mean annual potential evapotranspiration, of more than 1140 mm, are confined to the lower lying areas south of 30 degrees north. Medium values, between 570 and 1140 mm/annum, occur throughout much of Turkey, the eastern Mediterranean coastal region, Syria, northern Jordan, northern Iraq and in the highlands of Iran, while figures below 570 mm are found only in the highest upland regions of Turkey and in northern Iran.

2.1.7 Climatic classification

In the geographical study of the environment of a region, it is always useful to have some simple framework within the confines of which a number of related

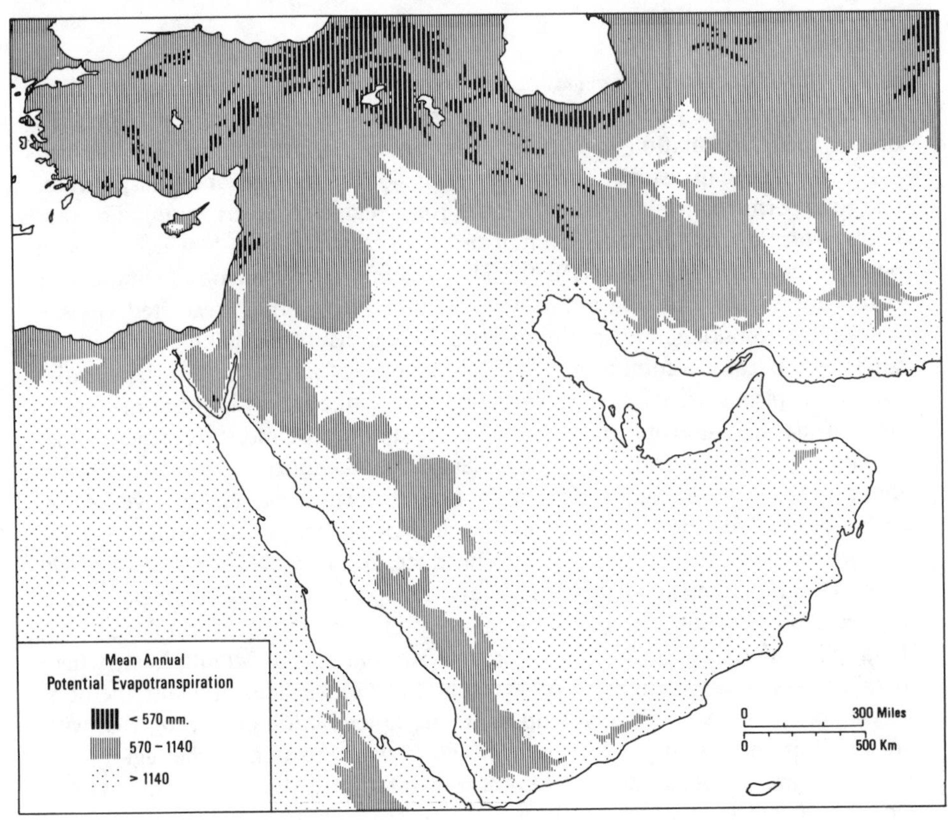

Figure 2.12 Mean annual potential evapotranspiration in the Middle East (after Thronthwaite, Mather and Carter, 1958)

factors can be studied. Climatic classifications provide such a framework, in so far as they integrate information concerning, in particular, temperature and precipitation. In turn, these parameters are important controls of soil development, vegetation patterns and man's use of the environment.

One of the better known climatic classifications is that proposed by Köppen, and later elaborated by Geiger.[11] Köppen realized that the major vegetation belts were determined by important climatic controls, and, therefore, placed his climatic boundaries at the margins of the major vegetation regions. These lines were then empirically defined in terms of mean monthly temperatures and precipitation amounts and their seasonal distribution (Table 2.9).

In the Middle East the only climatic types which occur on a large scale are the B and C types. Much of the region, however, does possess mountainous regions which do not fit easily into a climatic classification owing to the very large changes which can occur over relatively short distances.

Throughout North Africa, Arabia and much of Iran, BW (arid) climates predominate. BS (semi-arid) climates occur in the Levant and in an arc paralleling

TABLE 2.9
Köppen climatic classification

A –	humid climates – no winter
B –	dry climates (subtypes BS – semi-arid; BW – arid)
C –	climates with mild winter
D –	climates with severe winter
E –	polar climates – no true summer.

the Taurus and Zagros Mountains, as well as in the northwestern margin of the Iranian plateau, while C climates are confined largely to Turkey, parts of northern Syria and the Caspian littoral of Iran.

A more comprehensive division of the Middle East into climatic zones has been carried out by FAO/UNESCO, as part of their arid zone research programme.[12] This classification, based on temperature, precipitation totals, number of rain days, and amounts of atmospheric humidity, mist and dew, was designed to synthesize the effects of the climatic factors of particular importance for living organisms.

The initial division of climates is into three broad groupings on the basis of the mean temperature of the coldest month (t). With hot, warm temperate, and temperate climates the mean temperature of the coldest month is always more than 0°C, whereas for cold and cold temperate climates the mean temperature of the coldest month is less than 0°C. Finally in glacial climates mean monthly temperatures for all the months of the year are less than 0°C.

A second division, again into three groups, is then made on the basis of the distribution, nature and intensity of the drier periods, making use of a 'xerothermic index'. This term denotes the number of 'biologically' dry days during the dry season and seeks to integrate information on precipitation amounts, number of rain days and atmospheric humidity conditions. Using the xerothermic index, 31 separate climatic types have been distinguished within the major divisions of hot and warm temperate climates. A simplified map of the differing climatic types in the Middle East using this classification is seen in Figure 2.13.

The actual human response to a climatic regime is still very difficult to quantify with any accuracy, yet it is obviously of vital importance in a developing country when new industrial and urban development is being planned. Although some general works do exist, as yet little research appears to have been carried out on this topic in the Middle East.[13] Recently, interesting works of a biomedical nature have appeared dealing with Libya and Kuwait, which have attempted to access all factors affecting human life, including local climate conditions.[14] One of the simplest methods for outlining comfort conditions for humans in differing climatic regimes is the use of climographs.[15] With these diagrams, monthly plots are made of relative humidity against wet bulb temperatures, and they do permit at least a rough classification of the major cities within the region in terms of their general suitability for human activity (Figure 2.14). From this figure it can be seen that very few of the large cities possess what would be classified as an 'ideal' climate, except for just a few months per year. Indeed, some coastal stations, such

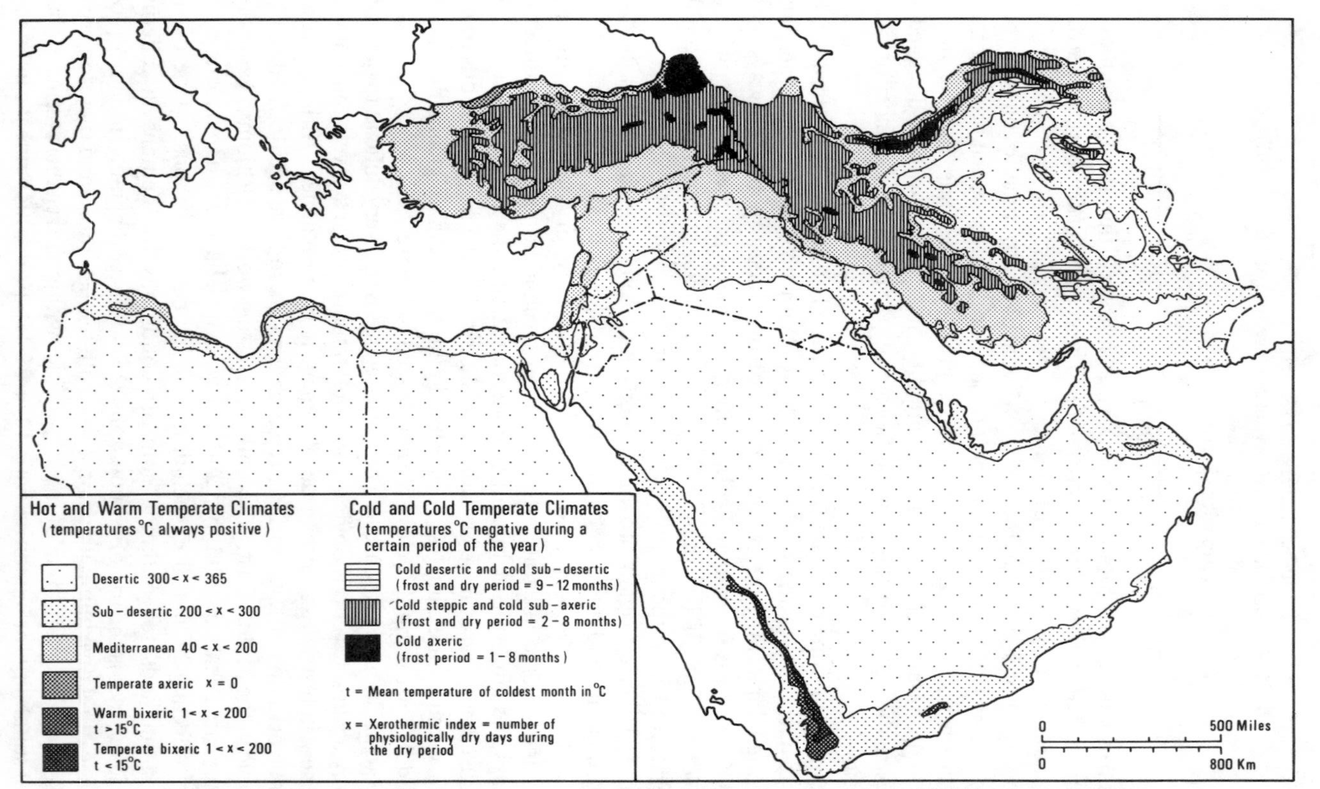

Figure 2.13 Simplified bioclimatic map of the Middle East (modified from UNESCO–FAO, 1963)

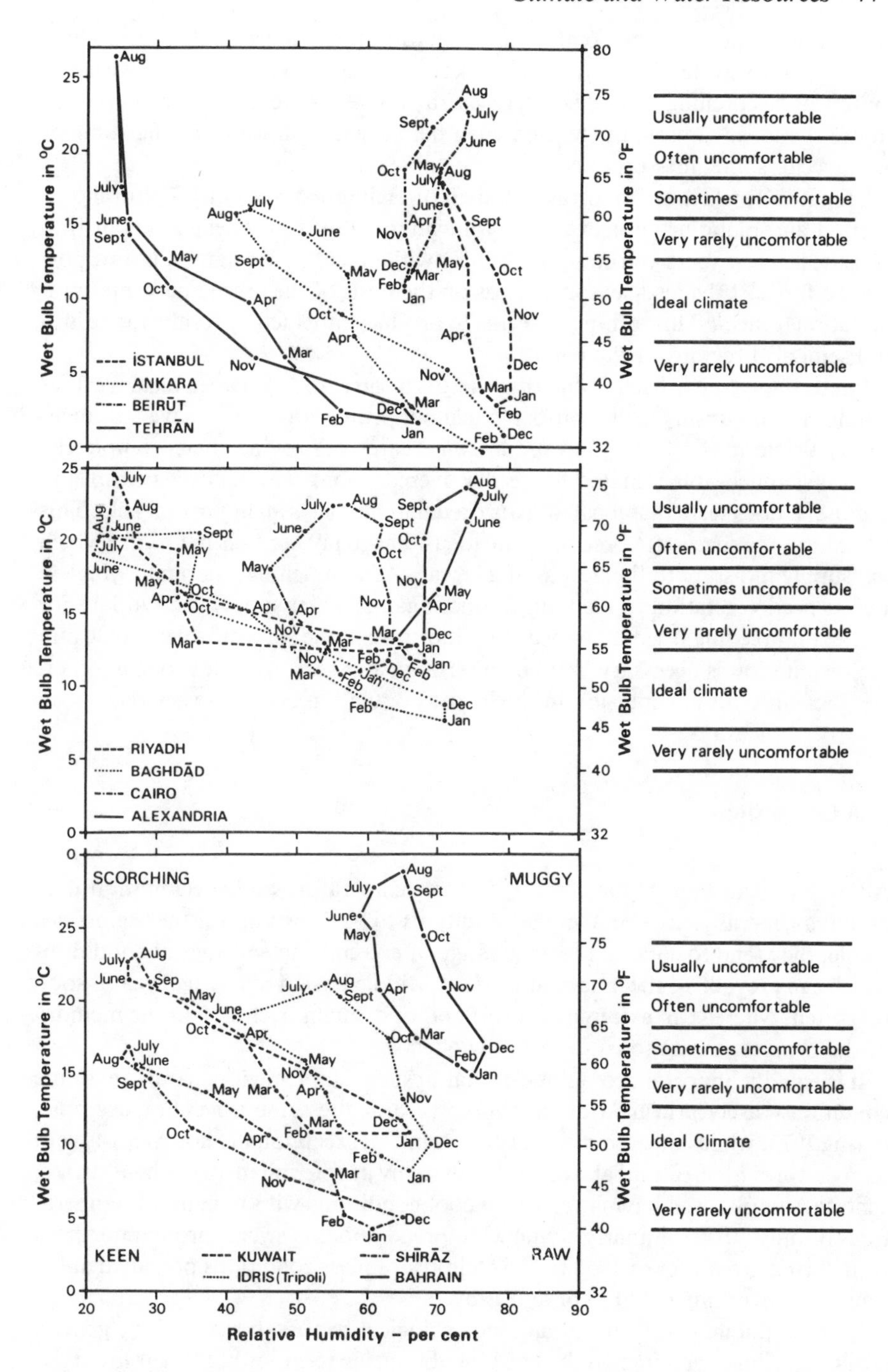

Figure 2.14 Climographs for selected cities in the Middle East

as Bahrain, experience 'uncomfortable' conditions for most of the year, whilst interior stations, such as Tehrān and Ankara, reveal marked extremes, with raw winters and scorching summers. Only Shīrāz possesses a climate which approximates the ideal climatic range, and even here the temperatures of the summer months can sometimes produce uncomfortable conditions.

Climate also plays an important, though often indirect role in agricultural production and planning. For example, following analysis of wheat yields in Iran over the last few years and comparing these with precipitation data, it is now possible to forecast the next season's yields on the basis of the preceding winter precipitation totals.[16] This technique seems to produce satisfactory results in the arid and semi-arid regions of the country.

Interestingly, it has been found that there is a correlation between local weather conditions and upsurges of locusts.[17] Such happenings obviously have a tremendously deleterious effect on crops; in some cases destroying them completely. Although much still remains to be discovered about why locusts swarm and migrate, it does seem that local moisture availability plays an important role. This is because, for successful breeding, the locust has to lay her eggs in ground which will supply moisture to the developing larvae. After hatching, the small wingless locusts need vegetation to feed on, so once again a water supply is essential. For successful breeding it would appear that moisture in the soil equivalent to 20 mm of precipitation is necessary.[18] When a series of wet seasons follow one another, the possibility of an explosion in the locust population becomes very real.

2.1.8 Conclusions

Although mean monthly and mean annual temperature and precipitation data permit a general picture of average conditions to be drawn up for the region as a whole, they tend to conceal the wide range of extreme meteorological conditions which can prevail. To the inhabitants of the Middle East, an 'average' year is soon forgotten, whereas an extreme year of flood or drought lives long in the memory of the peoples, owing to its devastating effects.

It is equally important to point out that in such a large region, similar extreme conditions rarely prevail over the whole area at the same time. For example, during 1973, while Khuzestan was experiencing an exceptionally hot August, with temperatures of 50°C and above, and with many people collapsing of heat stroke, central Anatolia was having relatively cool conditions with maximum temperatures of only 30°C. Similarly, while widespread flooding swept through northern Iran during the winter of 1968 to 1969, almost rainless conditions prevailed at the same time over much of Egypt and Libya.

Perhaps the most difficult of all meteorological factors to assess is the general effect of a climatic regime on the local inhabitants in terms of both their life styles and an adaption of their living conditions to meet the requirements of a harsh environment. It should be admitted, however, that it is often impossible to separate what is a cultural trait of a people developed in another region from a direct response of a group to the local physical environment.

The flat roofed, courtyard house so characteristic of the southern part of the region, with its thick whitewashed brick or mud walls, small windows and high-ceilinged rooms, provides cool indoor conditions as a relief from the scorching desert conditions outside (Chapter 6). In southern Iran, lowland Iraq and the Gulf region, underground rooms ventilated from wind towers are often constructed in houses, to protect the inhabitants from the fierce oppressive heat of summer. High precipitation conditions have resulted in the construction of house types with pitched roofs in western and northern Turkey and along the Caspian coast of Iran. In Iran, such roofs are often thatched, whereas in Turkey the red-tiled roof so characteristic of the Mediterranean region predominates.

Modern buildings, now so prevalent in Middle Eatern cities, with their extensive use of concrete, steel, and glass, often do not appear to be designed with any particular climatic regime in mind. Occasionally one finds tower blocks with balconies which shade the windows beneath from the midday sun, but mostly the internal climates of these large buildings are controlled, at considerable expense, by sophisticated air conditioning units. Indeed, the buzz and whine of air conditioners during summer is a new urban noise characteristic of many Middle Eastern cities. In the large cities which are situated in dry inland locations, simple evaporative coolers have become tremendously popular over the last 20 years. These work on the principle that as water evaporates, energy is given up from the air, producing a cooler temperature. Relative humidity is increased, however, and so these coolers can only be utilized successfully where summer humidities of less than 30 per cent prevail. In Tehrān, there was an explosion in the sale of these coolers in the early 1970s and this began to give rise to concern owing to the very large quantities of water being consumed.

A meteorological phenomenon of increasing importance in the Middle East is air pollution. To some extent this occurs in all the larger cities of the region, although its cause and season of maximum incidence varies. For example, serious air pollution in Ankara is largely a winter phenomenon. It occurs because the city is situated in a large basin which experiences very cold still conditions during the winter period. To combat the cold, the urban population, which now numbers more than 5 million, utilize numerous small and generally inefficient lignite and oil burning stoves. As the quality of the lignite which is used is often very poor, the amount of pollution is considerable and visibility is often reduced to a few tens of metres. This problem has become so severe that the government, with the aid of the Middle East Technical University, Ankara, is now attempting to devise a strategy using better quality coal and more efficient combustion methods which will at least alleviate the pollution problem. The fact that the population of Ankara is expected to rise to about five million by year 2000 clearly illustrates the potential seriousness of the issue in future.

Tehrān, like Ankara, also has a serious air pollution problem, though here the situation is most pronounced during the hot summer period. The basic cause is the accumulation of exhaust gases from the many vehicles which crowd the streets of this city of more than three million inhabitants. These gases, and in particular the reactive hydrocarbons, undergo photochemical reactions under the strong sunlight of the summer period, which result in the conversion of nitric oxide into nitrogen dioxide. This gas may have adverse effects on plants, animals and humans if high concentrations occur. Other products reduce visibility and some can cause eye irritation. These phenomena are usually termed 'smog', and their effects are best known from the city of Los Angeles, USA. The first symptom of this form of pollution is visibility reduction, followed by damage to plants and finally eye irritation. In central Tehrān, on hot summer days, the eye irritation stage has now been reached. The problem here is compounded also by the fact that, unlike California, no emission control regulations as yet exist for automobile exhaust systems. Consequently, badly tuned engines, burning incorrect octane fuels and spewing out noxious gases are a common sight in the streets of Tehrān.

Although Tehrān, on account of its huge size, has been singled out to illustrate the photochemical air pollution problem created by automobiles, it is by no means the only large city in the area to suffer in this way. Indeed Beirūt, Tel Aviv, Baghdād, Cairo, and İstanbul all suffer this same type of pollution to varying degrees. With the expected growth of all these cities, together with the associated rise in the standard of living of the inhabitants, there can be little doubt that the problem will become much worse in the future, unless regulations controlling exhaust emissions are introduced quickly.

Finally, mention must be made of one of the most controversial issues relating to the environment of the Middle East, which is the question of whether the region has been subjected to marked changes of climate during the Quaternary era (Chapter 1). Although climate change during the Pleistocene period is now widely accepted as having occurred in the Middle East, the nature of climatic change during the Holocene is still a matter of considerable dispute and the subject of a considerable research effort. A number of works have appeared treating the issue of climatic change on a world scale.[19] These have emphasized the fact that any major breakthrough in our knowledge of the climatic changes which may have occurred during the last 10,000 years, is more likely to come from our increasing understanding of the general circulation of the atmosphere, than from detailed local and regional works. Certainly, the present range of literature dealing with Holocene climatic changes in the Middle East reveals a dearth of information which can be interpreted in a number of different ways.[20] Unfortunately, it would seem that man's impact on the Middle East has been so great, that it is probably now impossible, from the evidence remaining, to separate environmental changes caused by variations in climate from those induced by human activity.

2.2 Water resources

2.2.1 Introduction

A reliable source of water for both domestic and agricultural use is one of the prerequisites of human survival. As a result, perennial water sources have played a crucial role in the siting of settlement and the growth of economic activity in the Middle East. In the past, water resource development has taken place largely at the local level, with the aim of supplying immediate agricultural and domestic needs from either surface or groundwater sources. With increasing populations, and in particular the rapid growth of large cities, local water sources are often totally inadequate to supply the new demands. As a result, individual countries have had to resort to the implementation of a number of large water resource projects, often at least partially financed from overseas sources.[21]

2.2.2 Available water resources

The two major dimensions of water are quantity and quality. The former can usually be adequately described with the aid of precipitation and runoff data, but the later, owing to the lack of detailed observations, is often a neglected aspect of water resource studies. Water quality is normally assessed in terms of its physical, chemical and bacteriological characteristics. The quality requirements for water depend largely upon the use to which it is to be put. Highest standards are associated with drinking water, while many industrial processes can utilise water of low quality. For irrigation purposes, a detailed knowledge of water quality is also essential, if serious soil problems are to be avoided. A classification of irrigation waters is normally made in terms of the salinity hazard, and the sodium (alkali) hazard.[22]

The total water supply within a region is best studied by means of a map of annual precipitation. Such a map, however, is not necessarily a good indicator of water availability, in terms of the water supply which can be utilized by man for beneficial purposes. In the hot and arid areas of the world, a large proportion of the precipitation is lost through evapotranspiration, from water and vegetation surfaces, with little or no economic gain. To obtain more accurate assessments of water availability, maps of water surplus have to be constructed for parts of the Middle East[23] (Figure 2.15). These maps, constructed from monthly data, show the annual water surplus which represents the sum of the monthly differences between precipitation and potential evapotranspiration. Water surpluses are often transported from their places of origin by river flow. As a result the mean annual discharge from a drainage basin is equivalent to the mean annual water surplus. Although most of the water surplus normally occurs within a restricted period of the year, its export is often delayed as the result of infiltration and subsequent discharge as groundwater, or by snow storage. Water surpluses, whether abstracted from surface water sources, or from groundwater, permit irrigated agriculture throughout much of the Middle East.

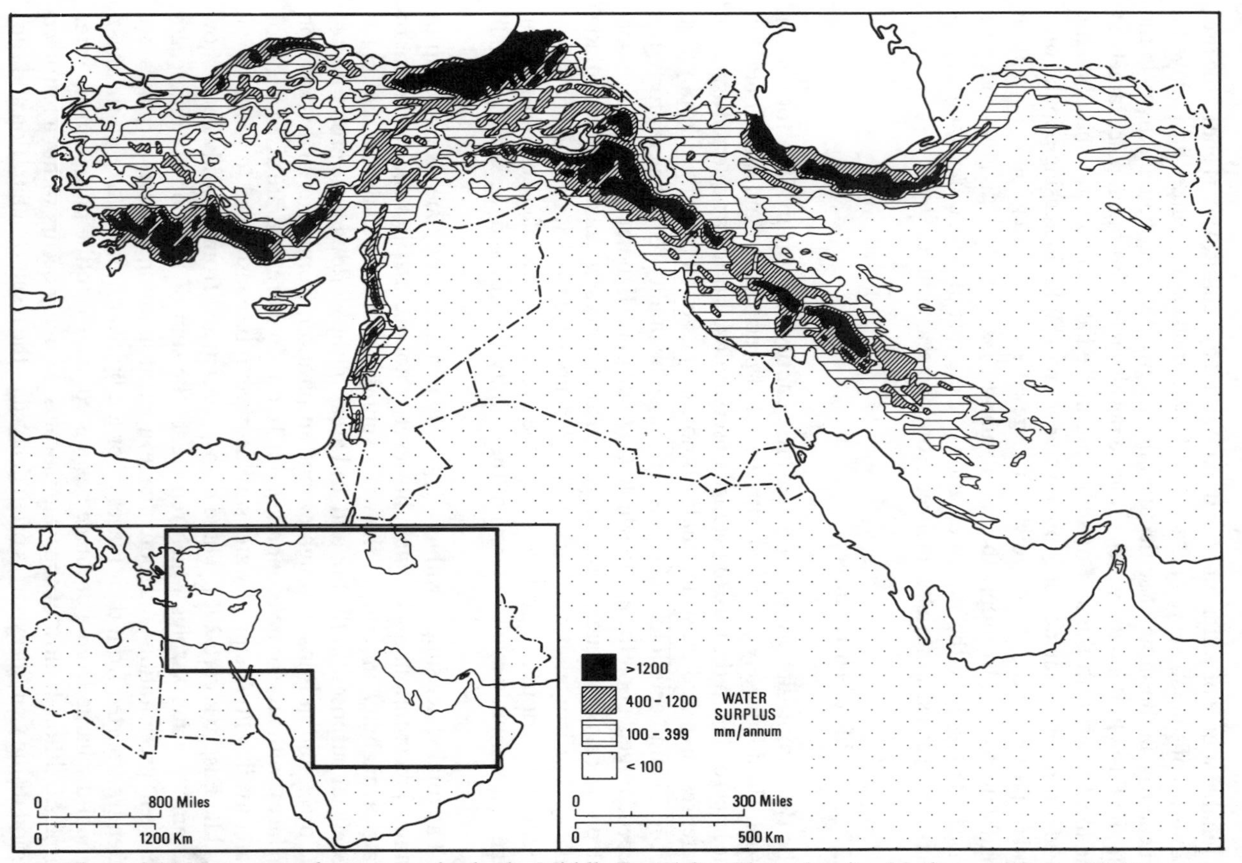

Figure 2.15 Mean annual water surplus in the Middle East (after Thornthwaite, Mather and Carter, 1958)

The map of water surpluses, constructed by use of the Thornthwaite method, clearly illustrates that surpluses are confined to the northern part of the region, and are closely associated with upland areas. The greatest water surplus of 2,400 mm/annum is found at the eastern edge of the Black Sea in northeastern Turkey. Other large surplus regions occur in mountains fringing the Black Sea and the Mediterranean coasts of Turkey, in the coastal uplands of Syria and Lebanon, in the Elburz mountains along the margin of the Caspian Sea, in the mountain region of eastern Turkey, and in scattered regions throughout the Zagros Mountains. Egypt, much of Syria, Jordan, Iraq and Arabia, together with central and western Iran, are characterized by the absence of a water surplus.

Despite the great size of the Middle East, there are only three rivers, the Nile, Euphrates and Tigris, which can be classed as large by world standards. Of these, the Nile receives most of its discharge from precipitation falling well outside the Middle East on the upland plateau of East Africa and the highlands of Ethiopia. In contrast, the watersheds of both the Euphrates and Tigris are situated within the Middle East, dominantly in the countries of Turkey, Syria and Iraq. Elsewhere, perennial river systems are confined to the more northern upland regions of Turkey and Iran, together with the coastal highlands of the Levant.

Owing to the general aridity within the region, a very large proportion of the total area consists of endoreic or inland drainage. Included in this zone is most of North Africa south of the coastal fringe, almost all of Arabia, with the exception once again of a narrow coastal strip, a large part of central Iran and, finally, smaller zones in Turkey, Iraq and the Levant.

Even though maximum precipitation occurs during the winter period, almost all the larger rivers of the northern part of the region are characterized by regime hydrographs with maximum discharge in spring and early summer. This is the result of the superimposition of a large snow-melt component onto the direct runoff discharge pattern.[24] The larger the proportion of upland within a catchment, the more pronounced the spring and early summer discharge peak tends to be. Further south in the region and also at lower altitudes the regime hydrograph peak more loosely reflects the pattern of precipitation distribution throughout the year. From June through to September, owing to the absence of precipitation over most of the region, the rivers are fed almost entirely from groundwater reserves and, as a result, the rivers reveal well developed recession curves.

The total amounts of river water which are available in the countries of the Middle East are still not known with any degree of certainty, although attempts have been made to tabulate existing data[25] (Table 2.10). The favoured position of Egypt, Turkey and Iraq is clearly seen. No comparable data are yet available for Saudi Arabia or Libya, but in both these countries river discharges are small and unreliable.

Average data on the amounts of runoff per unit area are still scarce within the Middle East, since many river gauging stations have only been established since the 1960s. In Iran, a country with a wide range of climatic conditions, work has revealed runoff figures of up to 550,000 cu m/sq km/annum, or the equivalent of a water depth of 550 mm/annum, for some of the river systems.[26] Amongst the

TABLE 2.10
River water available in Middle Eastern countries

Country	Estimate of total mean annual flow of major rivers (in billion cu m)	
Egypt	84	18.5 billion cubic metres of this can be used by Sudan
Turkey	80	About 40 billion cu m of this is in the Euphrates and the Tigris
Iraq	76	Of which only 20 to 30 billion cu m originates in Iraq
Syria	28	Of which 24 billion cu m is in the Euphrates
Iran	42	22 billion cu m in Karun and Dez systems in Khuzestan
Israel	1	Includes Jordan flow
Jordan	0.5	Excludes main Jordan
Lebanon	1	Excludes the upper Orontes (0.5 billion cu m)

Source: Smith, C.G., 'Water resources and irrigation development in the Middle East', *Geography*, **55**, p 424, 1970.

larger rivers of Iran, for which reliable data are available, the Kārūn catchment has by far the largest runoff potential, with a value of 255,000 cu m/sq km/annum. In Turkey even higher annual runoff rates of more than 450,000 cu m/sq km have been recorded in small rivers flowing into the Black Sea and Mediterranean Sea.[27]

Many of the rivers of the Middle East, particularly those flowing into basins of inland drainage, exhibit a decrease in discharge in a downstream direction. The reasons for this are complex. Some water is lost through evaporation and percolation, but large volumes too are often diverted for irrigation purposes. A good example of this seen on the Zaiandeh River which waters the Eşfahān oasis in Iran. Here, four gauging stations along the river reveal a considerable reduction in flow (Figure 2.16).

In the Middle East it is possible to distinguish two major types of aquifers.[28] Along river valleys and beneath alluvial fans and plains there are shallow alluvial aquifers. These are generally unconfined, small in area, and have water tables which respond rapidly to local precipitation conditions. The second type are deep rock aquifers, usually of sandstone and limestone. These are often confined systems, sometimes of considerable areal extent, and contain water which can in part be many thousands of years old.

The shallow alluvial aquifers are recharged along broad gravel floored valleys in the highlands, or on the upper part of alluvial fans, where the rivers leave the highland zone and divide into a number of distributary channels. In such areas water percolation can be extremely rapid, with all the waters of a river disappearing while crossing an alluvial fan zone. Most of this recharge occurs during spring and early summer at the time of maximum river discharge.

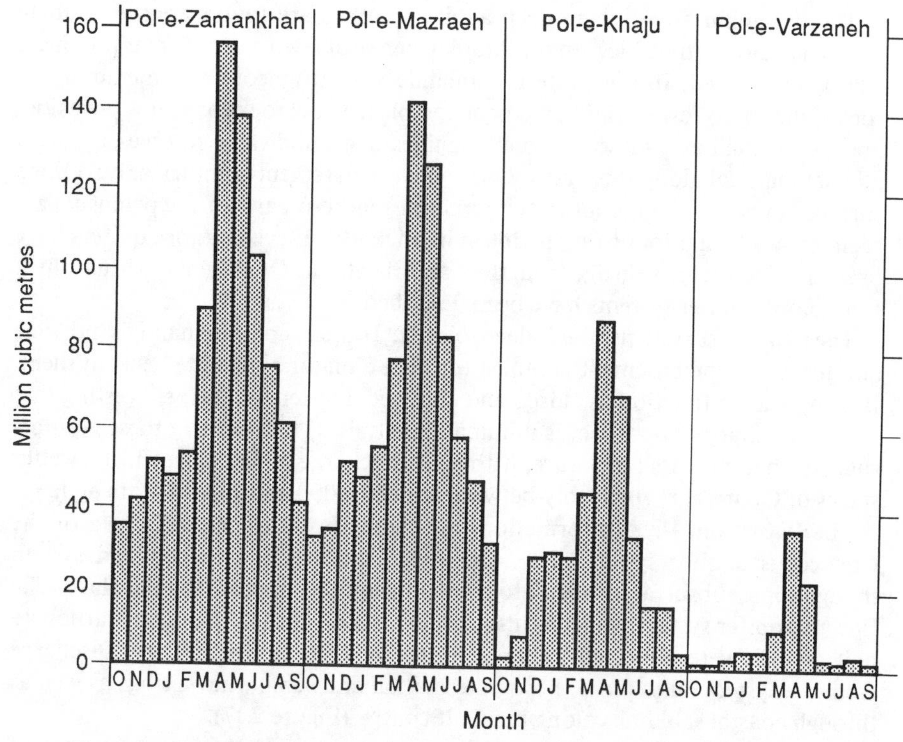

Figure 2.16 Downstream variations in discharge along the Zaindeh River, Iran

The deep rock aquifers often cover many thousands of square kilometres in area, with natural recharge occurring in upland and foothill zones where the rocks have surface outcrops. The mode of recharge is uncertrain, but seems to be achieved by the concentration of storm runoff into river channels followed by percolation through the bed into the underlying aquifer. In this way the drainage of a large area can be collected and recharged into a rock unit which may have only a limited surface outcrop.

With these large rock aquifers there is still considerable uncertainty as to the degree to which recharge is taking place at the present day, partly because little is known about how much runoff is generated during the rare, but often intense, local storm events. Some researchers have suggested that recharge can take place under very arid conditions. For example, in the Ad Dahnā sand dunes of Saudi Arabia, it has been claimed that even with an annual rainfall of only 80 mm about one quarter of this total may percolate downwards to recharge the aquifer systems.[29] However, in other semi-arid areas, such as Texas it has been clearly shown that average recharge rates are often about 5 mm/annum.[30] This clearly suggests that most of the water in these large rock aquifers is fossil in origin and that once it is taken out it will not be replenished. Exploitation of these deep aquifer systems can, therefore, only continue for a few decades in most cases.

The two main areas where deep aquifer systems are found are the Arabian peninsula and north Africa. In the Arabian peninsula two groundwater provinces can be recognized. In the western highlands ancient igneous and metamorphic rocks outcrop to form a rigid basement complex. These rocks have low porosities and permeabilities and so the only significant groundwater reserves occur in alluvial material along the wadis. Overlying the basement complex and forming surface outcrops throughout the eastern and southern parts of the peninsula are sedimentary formations of up to 500 m in thickness. Eleven major aquifers have been identified ranging in age from the Cambrian to the Quaternary, within which four major aquifer systems have been described.[31]

The two oldest systems, the Palaeozoic and Triassic, contain mainly sandstone aquifers, with the basement complex forming a major aquiclude beneath them. Both appear to function as closed and confined aquifer systems, suggesting that natural discharge from them is minimal. Analysis of the water in these aquifers suggests that it is fossil in nature, with major recharge taking place during wetter stages of Quaternary, probably between about 35,000 and 16,000 years ago.[32]

The Wasia and Biyadh formations are the main water-bearing strata of the Cretaceous aquifer system. This also functions largely as a confined system though some subsurface transfer to overlying aquifers appears to take place. The Eocene aquifer system, which is also confined, is composed mainly of carbonate units, which were subjected to karstic conditions during the Tertiary. Discharge from the system takes place through terrestrial and submarine springs as well as through coastal sabkhahs along the Gulf coast[33] (Figure 2.17).

The Nubian aquifer system underlines much of north Africa. It consists of sands, sandstones, clays and shales and reaches a maximum thickness of 3500 m. Although a number of different aquifer units are present within the Nubian series it appears to function as a single multi-layered artesian system covering about 2.5 million sq km.[34,35] Natural discharge from the aquifer occurs in a series of depressions across the northern part of the region. The largest is the Qattâra depression, but also includes the Siwa, Farafra, Bahariya and Dakhla oases. Recharge to the aquifer is thought to be runoff generated in the highlands of northeast Chad, the western uplands of the Sudan and the Ennedi and Tibesti plateau. Recharge may also occur directly from the Nile in the area south of Khartoum, as well as through the bed of Lake Nasser.[36] The dominant direction of water movement in the Nubian sandstone aquifer is towards the northeast.

Development of the Nubian sandstone aquifer has taken place in the New Valley project in Egypt. This consisted of the agricultural development of the major oasis areas using deep wells to provide irrigation water from the Nubian aquifer. At first much of the water flowed under artesian conditions, but with growing use pressure quickly fell and ever greater amounts of pumping had to be employed. In some cases, too, saline water began to be obtained from the wells, so limiting the crops which could be grown. The projects also suffered from human problems in that many of the managerial staff did not like living in such isolated areas. Overall the project cannot be considered a success.

Similar schemes were also tried in Libya where agricultural developments were

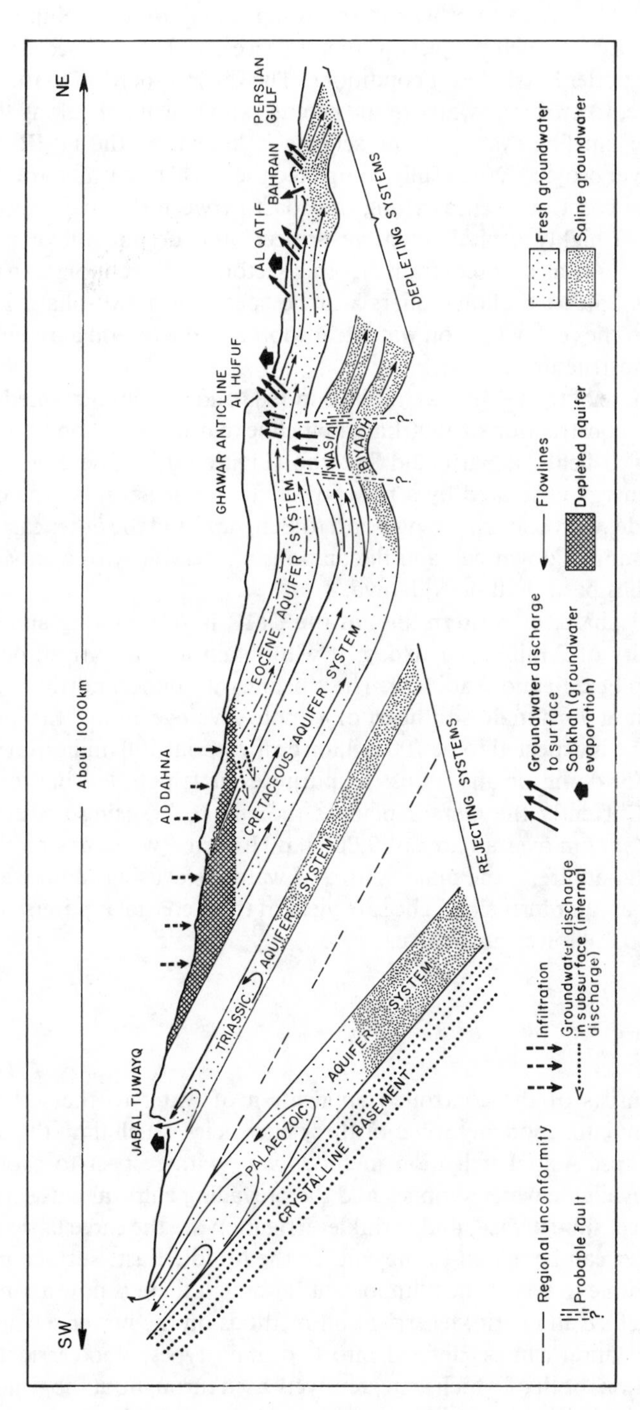

Figure 2.17 Major aquifer systems of Saudi Arabia

carried out at Al Kufra and other centres, using water from the Nubian aquifer system. As in Egypt, such projects were not successful because people were not willing to live under harsh desert conditions. The Libyan solution to the problem was, therefore, to move the water to the people, who live mainly along the coast, through huge pipeline systems. The scheme is known as the GMR or Great Manmade River project. When fully completed there will be a water grid from the interior to the coast, as well as along the coast between the major population centres. It is a highly capital intensive project and, despite Libya's huge oil revenues, has put considerable strain on other sectors of the economy. In total it is expected that 25,000 million dollars will be spent on a two-phase project.[37] Three-fifths of the cost will go on water transport and the rest on agricultural and irrigation infrastructures.

The first phase of the GMR was initiated in 1983 and will be completed by 1990. It involves the construction of 1900 km of pipeline to bring 2 million cu m a day of water from well fields at Sarir and Tazerbo (Figure 20.2). These wells will be pumped by energy generated by a new power station at Sarir. At the coast the pipelines divide, with one arm going north to Benghāzī and the other east to Sirte. In order to dampen down demand fluctuations a reservoir with a capacity of 4 million cu m has been built at Ajdābiya.

The second phase, or western line, of the GMR is 700 km long and will also have a capacity of 2 million cu m/day. It will obtain its water from desert well fields at North Fezzan and Wadi Aril. The pipeline will connect northwards with a large reservoir at Tarhūnah, southeast of Tripoli. Average annual throughput of the western line is expected to be 700 million cu m. About 100 million cu m of this will be used for domestic and industrial purposes in Tripoli, leaving the rest for agriculture. Currently the Gefara plain is thought to be using about 650,000 million cu m/annum over an area of 9,000 ha. However, water abstraction from shallow coastal aquifers, the main source of water, is causing rapid water table decline and sea water intrusion. The safe yield of these coastal aquifers is thought to be only 200 million cu m each year.

2.2.3 Irrigation

Irrigation consists of the controlled application of water to the soil with the objective of making good any soil moisture deficiencies which limit the optimum growth of crops. Actual irrigation methods vary with respect to topography, soils, crops, available water supplies and the prevailing cultural pattern within a region. Surface, sub-surface, and sprinkler irrigation are the three basic methods by which water can be added to the soil. In the Middle East, surface irrigation methods continue to be the most important, although there is now an increasing use of sub-surface and sprinkler irrigation methods on the newer schemes.

Surface irrigation can be divided into two main types, flood irrigation and furrow irrigation, both of which are extensively used throughout the region. With flood, or basin irrigation as it is sometimes known, fields are levelled, and surrounded by earth dykes. Water is then rapidly applied to the basin and held there until it soaks into the ground. This type of irrigation is widely used for cereal cultivation. In furrow irrigation, water is led in trenches, either along or down the slope between individual row crops. It requires a considerable input of labour to maintain the furrow systems, and, consequently, tends to be utilized only for high value vegetable and fruit crops.

One of the earliest known sites in the Middle East where irrigation appears to have played an important role in the establishment of human settlement, is to be found in the town of Jericho adjacent to the Dead Sea. Here an arid climate prevails, with precipitation totals of less than 150 mm/annum. The water source at this site is a perennial spring, *'Ain es Sultan*, situated at the foot of the western escarpment of the Dead Sea lowlands. The earliest occupation of the site was apparently by Mesolithic hunters and food gatherers in the period around 9000 BC.[38] From this date, almost continuous occupation seems to have occurred leading to the establishment of a walled settlement by about 8000 BC, during the Pre-Pottery Neolithic A period. This settlement is believed to have covered an area of about four hectares and to have probably contained a population of between 2000 and 3000 inhabitants. By the time the walled city came into being, it would seem that an organized community with a strong leadership must have existed in Jericho. Equally, it seems likely that to feed this relatively large population in such an arid environment, some form of irrigated cultivation of the Jordan valley, using water from the adjacent spring, must have been practised.

The distribution of water in irrigation systems has traditionally been achieved by the use of hand-dug canals, usually in alluvial materials. These canals, some of which, for example the Nahrawān canal in lowland Iraq, were many metres in width, formed the lifelines on which the local agricultural economies relied. In many cases, these canal systems were extremely complex and designed to irrigate areas of many hundreds of square kilometres (Chapter 12). Even today intricate irrigation systems, almost exclusively hand constructed, are found. One of their main problems is that water losses through seepage can result in the loss of up to 50 per cent of the water passing through the canal intake. Although such wastage sounds excessive, it should be remembered that in these regions groundwater is often also widely utilized, and that the canal seepage losses are sometimes a very useful method of groundwater recharge.

At the present time it is the usual practice when new irrigation systems are being built, or traditional ones modernized, to construct concrete-lined, trapezoidal, primary canals, as the main arteries of the system. Much more variety exists at the secondary and tertiary distribution levels, including such variates as concrete canalets, underground PVC pipes or, in the majority of cases, still the hand-dug canal.

In the more arid parts of the region, water supply in the past was largely obtained from groundwater sources, either from springs in foothill regions or from hand-dug wells along river valleys or on gravel plains. For centuries, Mecca was dependent upon water from the well of Zam Zam.[39] Later, new wells were sunk and then, as the population of the town expanded and the numbers of

Two types of irrigation: (top) traditional, using mud walls for diversion purposes (Libyan Embassy, London); (bottom) sprinkler system (Libyan Embassy, London)

Modern river regulator under construction on the Zaindeh River near Esfahān for irrigation purposes (Peter Beaumont)

people visiting Mecca on pilgrimage increased, water was led into the town along aqueducts from spings adjacent to the town. The most famous of these, *Aine Zubeda*, was about 16 km in length, and brought water from the springs of Zubeda, in the Arafat Hills northeast of the city. It was built in the early part of the ninth century, under the instigation of Queen Zubeda, the wife of Hārūn-al-Rashīd. At Medina, too, water was largely obtained from a shallow aquifer, with about seventy wells in existence during the lifetime of the Prophet Muhammad.[40] A number of aqueducts brought water from springs and wells on the margins of the town.

Further east in Arabia, groundwater has also played an important role in rural life. Between Al Kharj, some 100 km to the southeast of Riyadh, and Al Aflaj, are a series of spectacular solution cavities in calcareous strata. Some of the pits at Al Kharj are 100 m in diameter and more than 130 m in depth, with pools of fresh water forming their floors.[41] At Al Aflaj, the largest pit is almost 0.75 km in diameter. Here it has been claimed that the remains of three irrigation ditches indicate that the water table has been lowered nine metres in the last millenia, perhaps owing to climatic changes. Water from these pits is utilized for the irrigation of date gardens and cereals.

Al Hufūf, which is the largest oasis in Saudi Arabia, is situated about 100 km west of the Gulf coast. Here groundwater reaches the surface by way of nine prolific springs at a rate of more than 5.6 cu m per second.[42] Over the years huge date gardens have been planted close by to make use of this water. The sheer volume of available water, however, has led to a serious drainage problem, and in a number of places, salinity build-up within the soil has taken place.

Comparisons of the proportion of irrigated land within the countries of the Middle East are extremely difficult to make owing to the differing interpretations

TABLE 2.11
Cultivated and irrigated areas in the countries of the Middle East

Country	Cultivated area (Temporary and permanent crops) (ha)	Irrigated area (ha)	Irrigated area as a per cent of cultivated area
Turkey	26,390,000	2,120,000	8.03
Iran	13,700,000	4,000,000	29.19
Iraq	5,450,000	1,750,000	32.10
Syria	5,607,000	580,000	10.34
Lebanon	298,000	86,000	29.90
Israel	437,000	220,000	50.34
Jordan	416,000	38,000	9.13
Saudi Arabia	1,135,000	405,000	35.68
Kuwait	2,000	2,000	100.00
Egypt	2,471,000	2,471,000	100.00
Libya	2,097,000	230,000	11.00

Source: FAO Production Yearbook 1984, FAO (UN). Rome, Tables 1 and 2.

as to what constitutes cultivated land, and indeed, what is implied by the term irrigation. Nevertheless, accepting that these problems do exist, it is possible to note major differences in the areas under irrigation within the countries of the region (Table 2.11). In terms of the absolute size of the irrigated areas, Iran, Iraq, Egypt and Turkey stand out as the most important countries, whilst in Egypt and Kuwait cultivation is not possible without irrigation. The relatively low proportion of irrigated lands in the very arid countries of Libya and Saudi Arabia is largely caused by the use of a very broad definition as to what constitutes the cultivated area.

2.2.4 The Qanāt

Throughout much of Iran, groundwater is abstracted by means of an unusual engineering construction known as a *qanāt*. This consists of a gently sloping tunnel which conducts water from an infiltration section beneath the water table to the ground surface by gravity flow. In the construction of a *qanāt* a shaft or well is initially sunk to prove the presence of groundwater at depth. An outlet point for the water is selected, and then a tunnel is dug back into the hillside to link up with the original shaft, or mother well (Figure 2.18). To aid in construction a series of vertical shafts are sunk along the line of the tunnel. These permit the extraction of spoil and provide a measure of air circulation to the workers below. Eventually, the tunnel will intersect the water table, but construction continues beyond this point, as it is only in this section that water penetrates into the *qanāt*.

The construction of *qanāts* is carried out by a team of workers known as *muqannis*. All work is carried out by hand and so progress, even in unconsolidated sediments is slow. As a consequence, it normally takes many years to construct a single *qanāt*.[43] Almost all *qanāts* are constructed in alluvial material where the water table is relatively close to the surface. This means that tunnel collapse is often a serious problem, which necessitates the lining of the tunnel with baked clay rings, known as *kavulls*. The average tunnel dimensions are 1.2 m in height by 0.8 m in width. Frequent caving of the roof and sides of the gallery in unlined *qanāts* means that actual dimensions vary considerably.

The great advantage of the *qanāt*, is that once constructed, it will continue to suppply water for long periods with little energy input, apart from annual cleaning operations. Its largest drawback is that the water discharge is uncontrollable. This means that water runs to waste during the winter season when irrigation is not required. Even during summer months, water flow during the night is unused, unless a storage reservoir is constructed downslope from the point where the tunnel reaches the ground surface.[44]

Large *qanāt* systems, often with more than 100 individual *qanāts*, are found throughout Iran[45] (Figure 2.19). These are commonly located on huge alluvial fans in foothill regions, where precipitation totals average between 100 and 300 mm/annum. Groundwater recharge is believed to occur in the better watered adjacent high mountain regions.

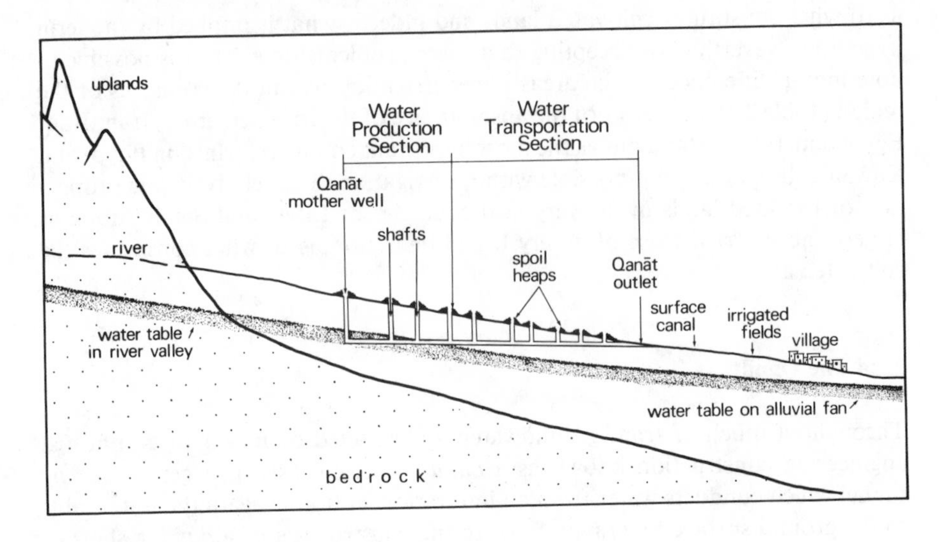

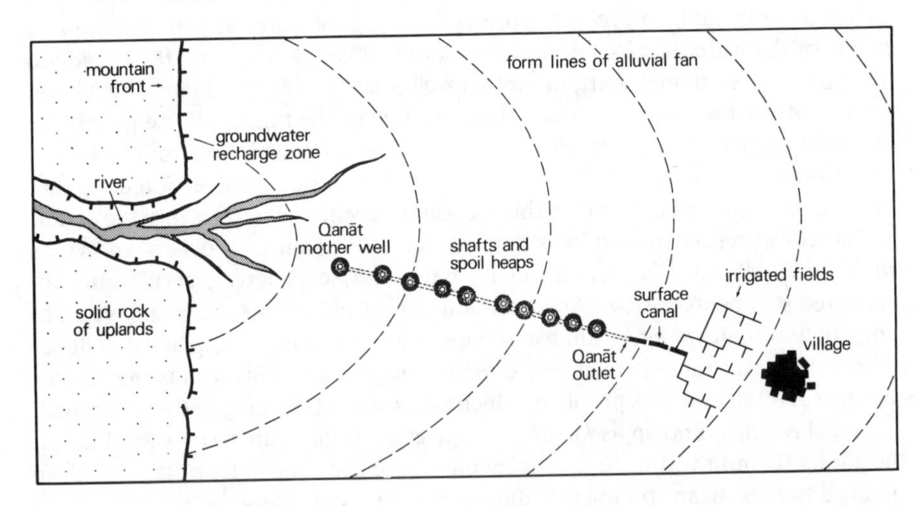

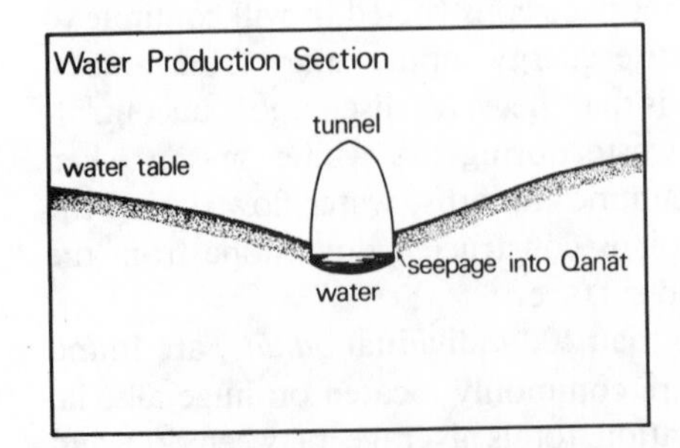

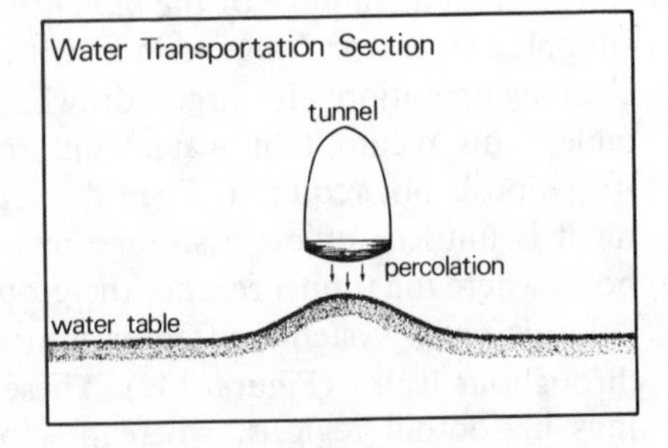

Figure 2.18 A *qanāt*: cross-section and plan (reproduced by permission of the National Well Water Association)

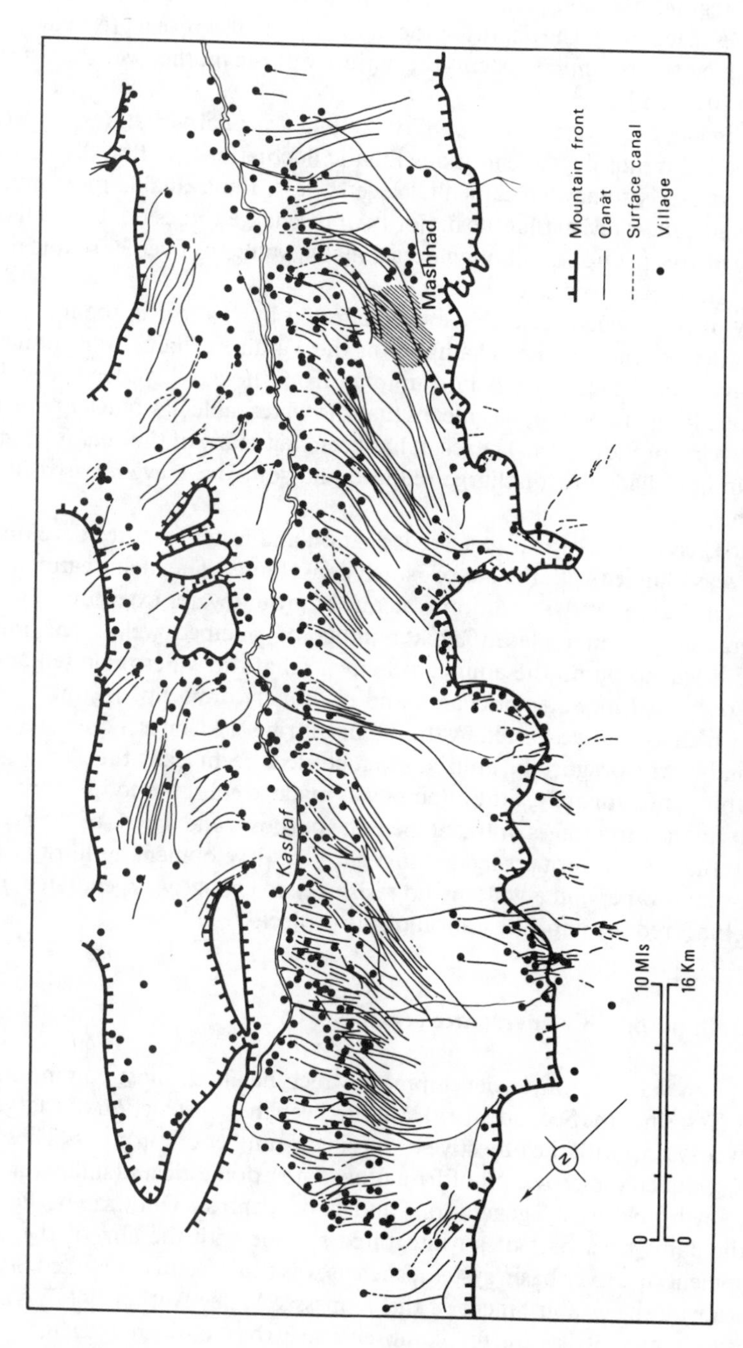

Figure 2.19 A *Qanāt* system around Mashhad, Iran

Although *qanāts* with lengths of more than 50 km have been described from the Kermān region, it would seem that the vast majority are between one and five kilometres in length.[46] The depths of the mother wells also reveal great variations, though in most large *qanāt* systems the majority of the mother well depths range between 10 and 50 metres.[47]

The discharge of water from *qanāts* exhibits seasonal variations, as well as longer period trends dependent upon climatic fluctuations, so that discussion of average values is somewhat difficult. Nevertheless, the available measurements for Iranian *qanāts* reveal that the majority of discharge values fall between 0 and 80 cubic metres per hour, although occasionally, values of over 300 cubic metres per hour have been recorded.[48]

Many of the *qanāt* systems have been effected by changes in the methods of groundwater extraction. For example, on the Varāmīn Plain, Iran, numerous wells have been drilled since 1955. Pumping from these wells, together with the effects of a long dry period, has meant that the water table has fallen appreciably in many parts of the region. The result has been that many of the *qanāts* irrigating the plain have had their discharge reduced and some have even ceased flowing altogether.[49]

In the Qasvīn area of Iran, a large irrigation and water resource development project was planned with the aid of Israeli consultants. The main feature of this project has been that the traditional means of groundwater extraction by *qanāt* have gradually been replaced by a number of pumped wells. The method employed was to pump the aquifer heavily so that the water table fell and the *qanāts* decreased in water discharge and eventually dried up. In this way, the waters which used to be wasted by the *qanāts* during the winter months remained in groundwater storage to be utilized when needed. Technically the plan was very sound, but, unfortunately, little attempt was made to enlist the cooperation of the peasants in the early stages or to tell them the purposes and details of the plan. As a result, the peasant farmers tended to regard the development authority as akin to a new landowner and a widespread resistance to change was generated, which greatly hindered the implementation of the project.[50]

2.2.5 Multi-purpose water resource schemes

The type of water resource development which has been most common in the Middle East since the Second World War has been the construction of a large dam on a river system with the objective of serving a number of purposes. These have usually included the provision of irrigation water, domestic and industrial water supply, hydro-electricity generation, and flood control. Only rarely, however, have the dams been part of an integrated scheme with the aim of the unified development of a river basin system, such as was achieved in the United States of America with the establishment of the Tennessee Valley Authority (TVA). Even within a country, it is rare for comprehensive river basin management to be implemented. All too often in the past, a dam has been constructed with the sole

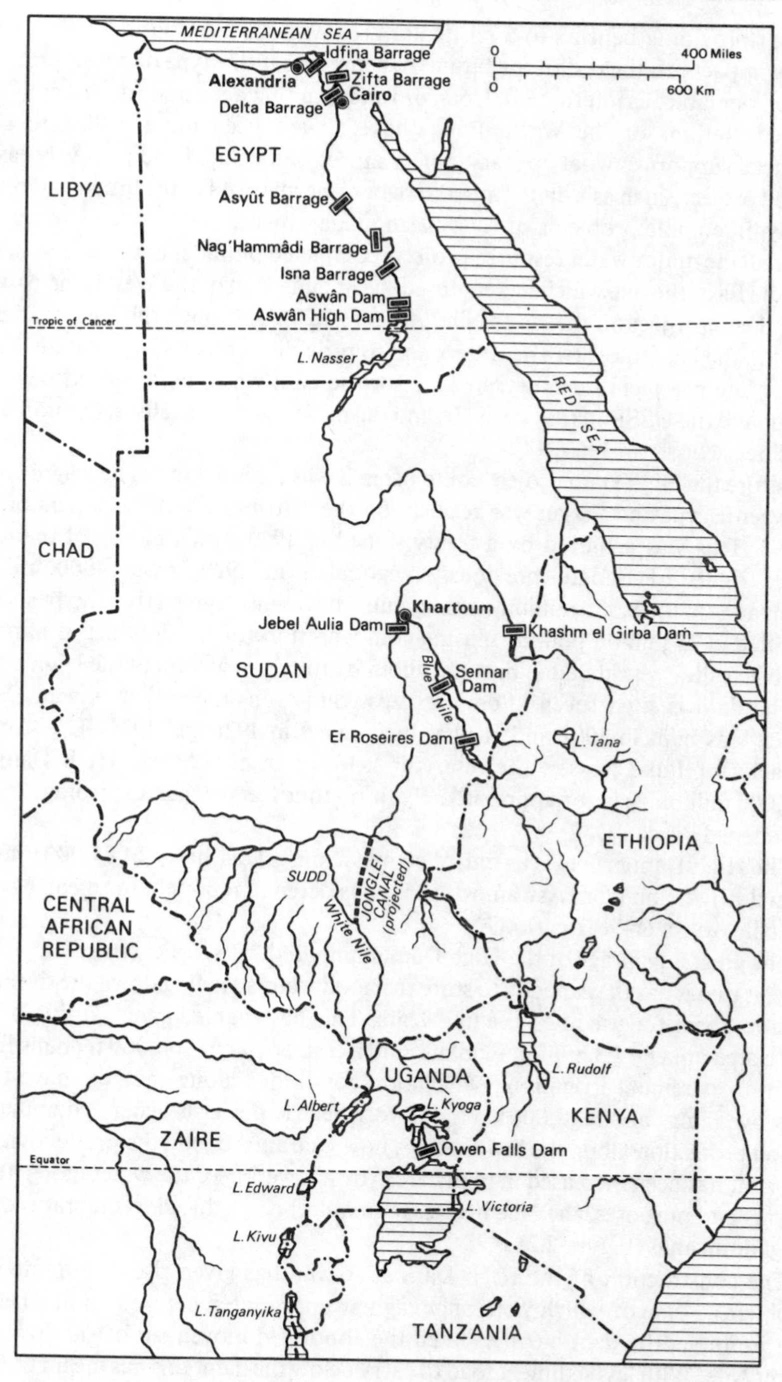

Figure 2.20 Hydraulic works along the Nile

aim of providing benefits to a particular area without any consideration of the likely impacts on the rest of the basin. To some extent this type of development is understandable, as international loan or fund granting agencies, for example, the United Nations or the World Bank, have always been more willing to give financial support to what appears to be a single, well defined and relatively easily costed project, such as a dam, rather than become involved in the broad and often costly, intangible problems of river basin management.

Of all the major water resource projects completed or under construction in the Middle East, the one which has captured the imagination of the West is the Aswân High Dam project on the River Nile[51] (Figure 2.20). Although the idea of a large dam on the Nile to control its waters and to provide 'century storage' dates back to the late nineteenth century, finance for the dam was only arranged between Egypt and the USSR in the late 1950s and the project was officially inaugurated by President Nasser in 1960.[52]

Before the High Dam project could begin it was essential that a new agreement between Egypt and Sudan was reached on the distribution of the waters of the Nile.[53] This was achieved by a treaty signed in 1959 which increased the total water rights of Sudan, previously negotiated in 1929, from 4000 million cu m/annum to 18,500 million cu m/annum. It was also agreed that Egypt should pay Sudan 15 million pounds sterling compensation for the flooding of parts of the Nile valley within Sudan brought about by the construction of the High Dam.

The dam is sited seven kilometres upstream from the earlier Aswân Dam, which was built in 1902 and further heightened in 1921 and 1933. The storage capacity of Lake Nasser, the reservoir held up by the Aswân High Dam, is 164,000 million cu m, or approximately thirty times larger than the storage of the heightened Aswân Dam.[54]

The High Dam provides a controlled mean annual discharge of 84,000 million cu m downstream from Aswân, which is considered to represent the mean flow of the Nile over a 60-year period.[55]

The great advantage of the High Dam, compared with earlier structures on the Nile, is that it has the capacity to store the flood waters moving down the river and to store them for use in succeeding years. This has enabled an expansion of the cultivated area by 1.3 million feddans, and a conversion of 700,000 feddans from basin to perennial irrigation. Navigation conditions along the Nile have been improved, and about 10,000 kWh of hydroelectric power is generated annually. Flood protection along the lower Nile is now guaranteed, and the productivity of the land has been increased in many areas by a lowering of the water table. To all intents and purposes, the Nile below Aswân has been reduced to the status of an irrigation canal (Figure 2.21).

The construction of the High Dam at Aswân has given rise to a number of problems, some of which were only vaguely appreciated before building began. For example, the dam has prevented the continued movement of silt down the River Nile. With less sediment load to carry below the dam, this has meant that the water has become more erosive, and has begun to undermine the foundations of some of the older bridges and hydraulic structures.[56] The absence of silt in the lower

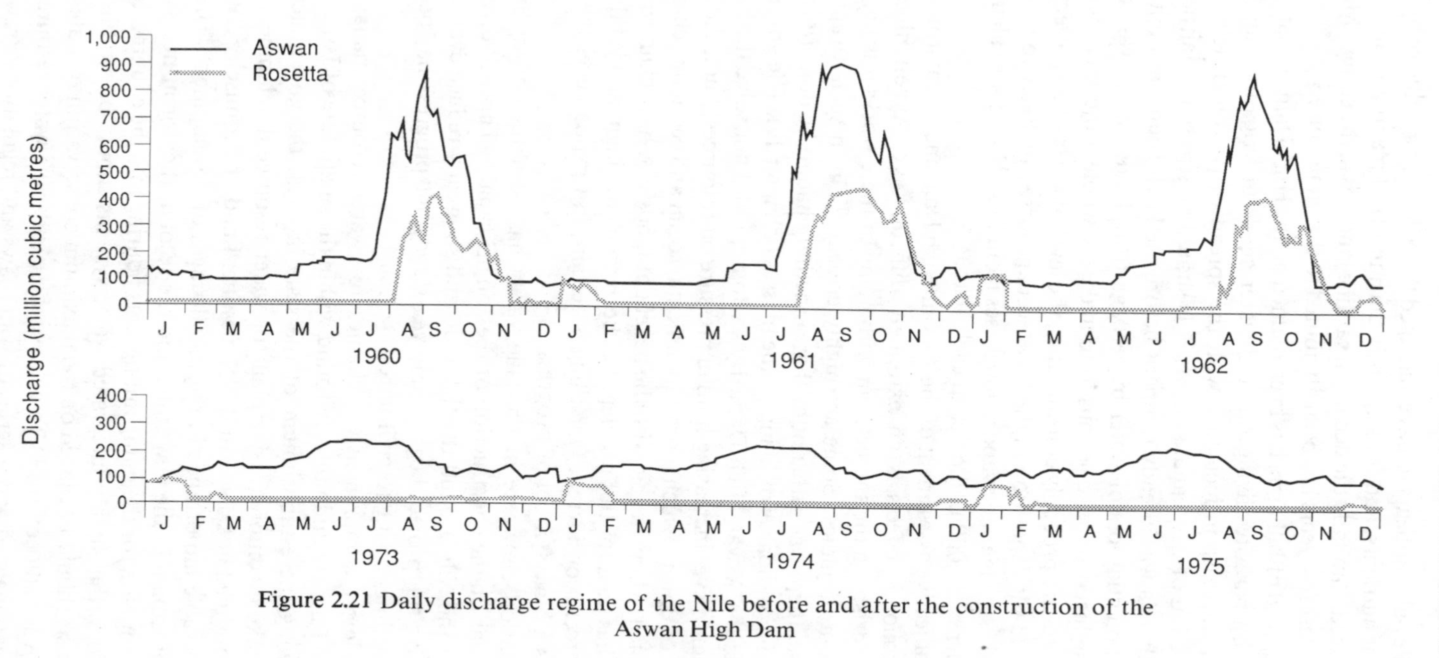

Figure 2.21 Daily discharge regime of the Nile before and after the construction of the Aswan High Dam

Nile valley also appears to have damaged the Nile fishing industry by reducing the supply of nutrients, and consequently the number of plants and animals on which the fish feed. Similarly the decline in sardine landings in the eastern Mediterranean over the last few years has been attributed to the same cause.[57]

The spread of bilharzia has been blamed on the High Dam as well. This disease, carried by parasites living on freshwater snails in irrigation canals, will, it is claimed, increase in incidence with the spread of perennial irrigation and the construction of new canal systems.[58] It should be stressed in fairness, however, that bilharzia was already rampant before the High Dam was built.

The potential loss of water by seepage from Lake Nasser was considered by some engineers to be so serious as to make the whole project a failure. Although such losses are probably considerable, the level of the lake rose as expected. With continued silt deposition in the lake, seepage losses are likely to be progressively reduced. Seepage and evaporation losses from Lake Nasser are calculated to be of the order of 15,000 million cu m each year.[59]

An objective assessment of the Aswân High Dam and its environmental, social and economic effects is an extremely difficult task. A voluminous literature already exists on the subject, but with many of these works it is hard to separate fact from prejudice. Some are highly critical.[60] Given such a rapid population increase, there was little doubt that something had to be done to try to alleviate the falling standards of living of the masses (Chapter 19). The government chose to build the Aswân High Dam. By so doing, it did not solve the basic Egyptian problem of ever increasing human pressure on resources, but it did at least buy a period of time in which more fundamental social and economic changes could be attempted. Many people with the benefit of hindsight, have claimed that the price Egypt has paid for this privilege has been much too high, and yet the alternatives are scarcely considered. One thing is certain, in Egyptian eyes the Aswân High Dam is a huge symbol of progress.

Attention elsewhere in the Nile valley has been focused in Sudan, and in particular on the construction of the Jonglei canal.[61] This will divert part of the flow of the Nile around the the Sudd and so greatly reduce evaporation losses. Work on the project began some years ago, but progress has been delayed at various times owing to civil unrest.

The most comprehensive schemes for water resource management in the Middle East are undoubtedly found within the small state of Israel (Figure 2.22). Following the establishment of the state in 1948, the government decided to undertake a comprehensive plan for water resource development based on the ideas outlined in Lowdermilk's 'Palestine: Land of Promise'.[62] Two factors had considerable importance in the initial stages of development. The first was the lack of capital in the new state, and the second, the urgent necessity to provide water supplies for the many immigrants pouring into the country. It was decided, therefore, that highly integrated water systems should be established, and that emphasis should be placed on the maximum conservation of water, as well as on strict water allocations.[63] To achieve these ends, the water resources of the state were nationalized and a series of water laws and regulations drawn up.

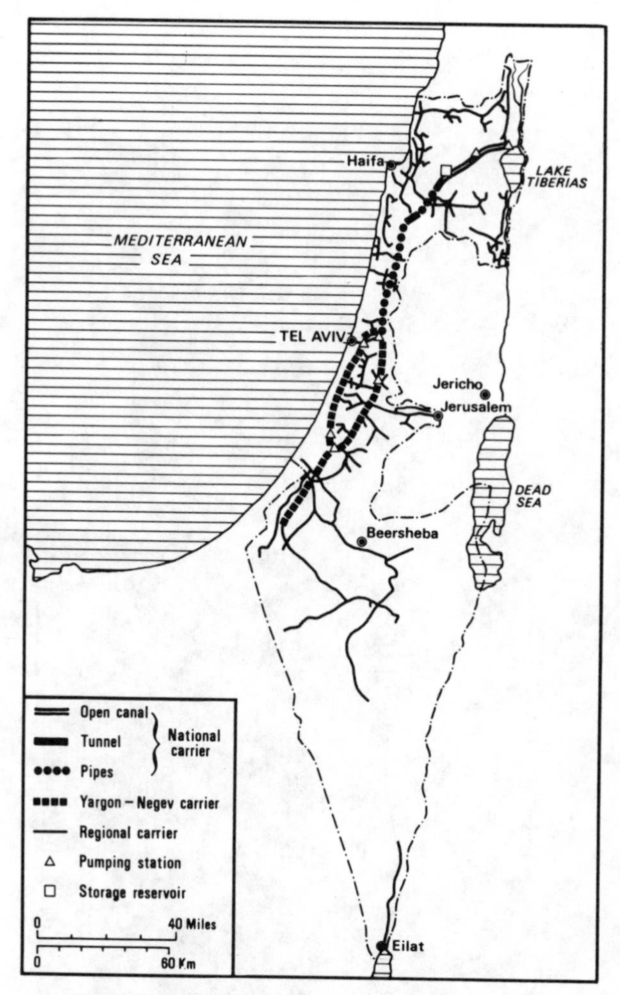

Figure 2.22 Water distribution networks in Israel

Initially, attention was concentrated on low cost projects, such as the drilling of wells, which produced quick results. These pumped wells permitted the irrigation of new lands in the coastal plain and northern Negev.

With medium term development projects, schemes were chosen which provided the minimum investment per unit of water supplied, which were not technically complex, and which were capable of having the investment divided into a number of stages. At the same time, the idea evolved that every project within the country, no matter what its size, should be capable of being integrated into a nationwide hierarchical water supply system.

A number of long term projects, which possessed a regional, rather than local significance, were also implemented. One of the largest of the early schemes was the Yarqon–Negev Project, which diverts water from the River Yarqon at the

Desalination plant on the Gulf of Aqaba at Eilat, Israel (Israel Embassy, London)

Rosh Ha'ayin springs near Tel Aviv southwards towards the Negev desert. With the output of numerous wells integrated into the project, the total water delivered is 270 million cu m/annum.[64]

Another large development was the Western Galilee–Kishon Project. In this, water from western Galilee was transported to the fertile, but dry Jezreel Plain. In this, 85 million cu m/annum are carried by this system. This is made up largely of surface water supplies during the winter months, when these are relatively abundant, and groundwater during the drier summer period.

An unusual project is located in the Beit She'an Valley. Here abundant water is available from saline springs, but by itself, it is too salty to be used for irrigation purposes. However, it has been found possible to utilize this water by diluting it with purer water obtained from Lake Tiberias.

The largest water resource development project in Israel is the National Water Carrier, which is a huge aqueduct and pipeline network carrying the waters of the River Jordan southwards along the coastal plain region. This scheme stems from earlier ideas and concepts for the integrated development of all the waters of the River Jordan for the mutual benefit of the states of Lebanon, Syria, Jordan and Israel. In the earlier 1950s, discussions took place between Israel and the adjoining Arab states in an attempt to reach an understanding as to how the waters of the River Jordan might be most fairly allocated amongst the four states. This plan, which was drawn up for the United Nations, is usually referred to as the 'Main Plan'.[65] After prolonged negotiations, modifications to the original plan were made and this new version became known as the 'Johnston Plan', named after the American mediator, Eric Johnston. This gave Israel 36 per cent of the utilizable water of the Jordan, estimated at 1380 million cu m/annum, compared with 52 per cent for Jordan, 7 per cent for Syria and 3 per cent for Lebanon. It is widely assumed that the technical experts of the various countries involved agreed upon the details of this plan, although soon afterwards the governments rejected it for political reasons.

With the failure of these negotiations, both Israel and Jordan decided to proceed with water projects situated entirely within their own boundaries. As a result Israel began work on the National Water Carrier in 1958. The main storage reservoir, and also the starting point of the scheme, is Lake Tiberias. From here water is pumped through pipes from 210 m below sea level, to a height from where it flows by gravity to a reservoir at Tsalmon. After a further lift, the water flows via a canal to a large storage reservoir at Beit Netofa, which forms a key part of the system. South of Beit Netofa, the water is carried in a 270 cm pipeline to the starting point of the Yarqon–Negev distribution system at Rosh Ha'ayin. In the initial stages 180 million cu m/annum of water were carried. This capacity was increased to 360 million cu m/annum in 1968, and it is now believed that the maximum capacity approaches 500 million cu m/annum.[66] This has, however, not yet been attained owing to water salinity problems. At the present time, the national water grid interconnects all the major water demand and supply regions of the country, with the exception of a number of desert regions in the south. In total, it supplies approximately 1400 million cu m/annum, or about 90 per cent of all Israel's water resources. This allows more than 116,000 ha of land to be irrigated compared with a figure of 28,000 ha when Israel was established in 1948.[67] More than half of the water is obtained from the River Jordan and its tributaries, with a further 14 per cent from the River Yarqon. The largest groundwater contribution is produced from coastal aquifers and amounts to around 30 per cent of the total. Approximately 5 per cent is obtained by the reclamation of waste waters from the Tel Aviv metropolitan area.

Israel provides an excellent case study of water resource management, as it has already had to face problems which other countries of the region are just beginning to experience. In the early years of its development, the major efforts of water resource management schemes in Israel were directed towards the agricultural sector. This was partly for strategic reasons to ensure that the countryside was occupied, but also to provide food and employment for the rapidly growing population. As a result it has given the agricultural lobby considerable power in the state decision making process.

Up to about 1965 and the completion of the National Water Carrier, there was enough water awaiting development to satisfy all needs.[68] All that was required was new schemes to tap the resources and make efficient use of them. From the late 1960s onwards it became extremely difficult to make any extra water supplies available, and so emphasis had to be shifted to making more efficient use of available supplies. In agriculture there have been spectacular achievements and today in Israel almost all irrigation is carried out by sprinkler, drip or sub-surface systems. It has meant that a given irrigated area can now be watered with much less water than previously. At the same time it does mean, however, that little future water savings can be made by agriculture by increasing efficiency, as irrigation in Israel is as economical of water use as any in the world.

In the late 1970s and early 1980s Israel has faced growing demand for water from urban and industrial sectors of its economy. Experiments have been made to reuse urban waste waters through the Dan Wastewater Recovery project, but success has been less than had been hoped for owing to difficulties to remove contaminants within the waste waters.[69] Similarly research into different water desalination systems has concluded that currently such water is too expensive, except for specific projects.

The result has been that Israel has been faced with the fact that the only way to obtain water for the growing cities is to divert water from one use to another. As agriculture still accounts for more than three-quarters of Israel's total water use, it is inevitable that this sector of the economy will suffer, despite protests from the powerful agricultural lobby. What seems likely to happen increasingly in Israel, as has hapened in states such as Arizona in the USA, is that irrigated land adjacent to large urban centres will be taken out of cultivation and the water diverted to urban and industrial uses. Economically this makes good sense too, for a cubic metre of water devoted to the production of Uzi submachine guns will generate more profit than the same water used to grow Jaffa oranges.

In future other countries in the Middle East will have to face the issue of diverting water from the agricultural to industrial sectors of the economy. Many will be able to delay the decision somewhat by the introduction of more efficient irrigation techniques. At the moment basin and furrow irrigation is almost universally used for water application in all the countries of the region, and so major water savings will be able to be made if sprinkler and related systems can be introduced. However, by the early decades of the twenty-first century almost all countries of the region will be facing severe water shortages in urban centres as populations continue to grow. It seems inevitable, therefore, that water will have to be diverted away from irrigation to urban/industrial uses. Indeed, it is worth noting that, to some extent, it has already occurred. For example, in Iran the construction of dams on the Jaji and Karaj rivers to supply Tehrān with water, has meant that less water is available for irrigation on the alluvial plains downstream. This was rarely admitted in planning documents. Instead it was usually claimed that the new water storage facilities would permit a greater proportion of the river's flow to be used effectively. While this is correct, the net effect is the diversion of water from agricultural to urban uses.

2.2.6 Industrial and domestic demands for water

The growth of large cities in the Middle East has imposed a tremendous burden on the water supply facilities of the urban centres for domestic and industrial use. In the past, drinking water was obtained from local sources such as wells and streams. The incease in population numbers of the last 40 years or so has meant that such supplies have now become totally inadequate, and the water catchment region have had to be continuously enlarged in an attempt to cope with water demands. Even with these tremendous efforts, it is true to say that almost every large city in the Middle East has water supply problems, and that these are likely to increase before the end of the century.

Over the last few years, Tehrān, the capital of Iran, has experienced growing problems with both domestic water supply and sewage disposal. Since 1933, the population of greater Tehrān grew from 210,000 to an approximately 3.4 million in 1971.[70] By 1985 it was estimated that Tehrān had a population of at least 6 million and perhaps one as high as 9 million. It has been this phenomenal growth which has strained the water resources of the area to their limit. With no perennial streams close by, Tehrān has been dependent on 34 local *qanāts* for its water supply until recent years.[71]

With growing water demand during the 1920s, a canal was constructed to carry water from the River Karaj, in the west, to Tehrān. This project was completed in 1930, and had a capacity of 1.3 cu m/sec.[72] By 1950, water was once again in short supply, and so a well drilling programme in the city was initiated. At the same period, the installation of a piped water system was commenced to replace the prevailing open ditch distribution system, but the rapid growth of the city meant that large areas still received their supply by traditional methods.

In the early 1960s, the severity of the Tehrān water shortage was alleviated by the construction of the Karaj dam and of two pipelines for transporting the water to the city. This scheme provided Tehrān with an extra 144 million cu m of water, to make a total available supply of 184 million cu m/annum.[73] With the continued expansion of the capital still further water was needed. This time it was provided by the Latian dam on the River Jaji to the east of Tehrān, which was opened in 1967, and supplied an extra 80 million cu m each year.[74] In the earlier 1970s it was obvious that still further water would be required. It was decided that the best project would be to divert the waters of the River Lar, which flows into the Caspian Sea, southwards in a huge tunnel beneath the Elburz Mountains and into the River Jaji[75] (Figure 2.23). The water then flows into the reservoir behind the Latian dam from which it can be distributed to the Tehrān network through existing pipelines.

Related to the problem of water supply is that of sewage disposal. Until very recently, Tehrān has possessed no integrated sewerage system, and human wastes were discharged untreated into the ground or into the nearest water course. As a result, the shallow aquifers have become contaminated, but, fortunately, the deeper and at present more important ones, have remained unpolluted. With the construction of a modern sewage disposal system, it is hoped that the problem of disease has been minimized and water pollution controlled.

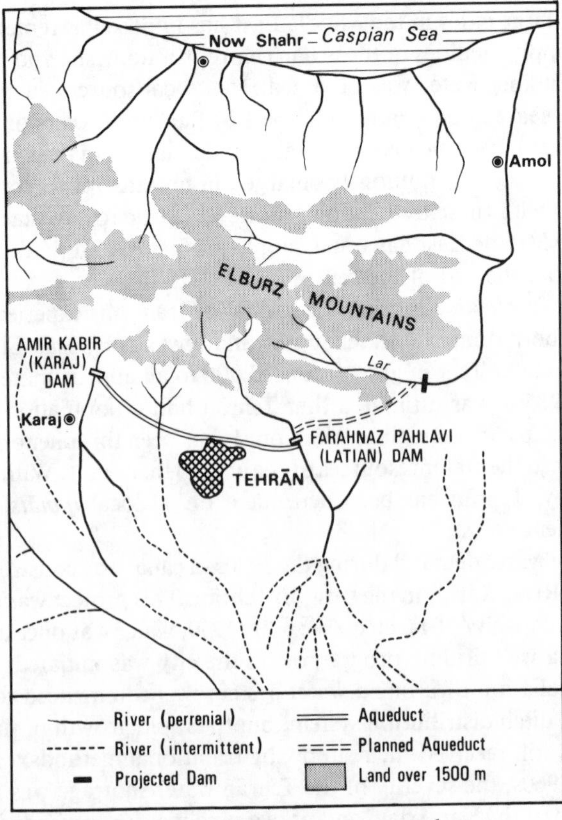

Figure 2.23 Tehrān: water supply

An unexpected aspect of transporting all the extra water to Tehrān has been that on disposal, much of the water has entered the groundwater system. From the early 1970s onwards this has led to an increase in the height of the water table in the lower part of the alluvial fan on which Tehrān is situated. Many old *qanāts* which had dried up as the result of the well drilling programme began to flow once again, often with very high discharges. It is hoped that this new water resource will be able to be tapped by wells and used for irrigation on the western edge of the Varāmīn Plain.

İstanbul is another large city which is suffering from severe water supply and waste disposal problems at the present time. Parts of the city are not provided with a public water supply and industrial expansion has been hampered by the lack of water. Groundwater has been contaminated locally due to an inadequate sewerage system and beaches and waterways around the city have been grossly polluted by domestic and industrial effluents. Here, as elsewhere, the major problem has been population growth, coupled with a rise in the per capita demand for water. By the early 1970s the population of the Greater İstanbul region reached three million and official estimates predict a figure of nine million by the year 2020. In the 1970s water supplied to the region was approximately 120 million cu m/annum, though actual demand was estimated to be nearer 170 million cu m/annum. This clearly illustrates the deficiencies of both the present distribution system and also the resources available.

Detailed work carried out by, and for, the Turkish government has shown that an adequate water supply for the İstanbul city region can be supplied from within a radius of 100 km to meet the anticipated demands of the early twenty-first century. However, it does entail the construction of a number of large reservoir projects to control surface runoff, and, naturally, involves a very large capital investment programme.

Waste and effluent disposal within the area is also causing concern, as the waters of the Sea of Marmara, the Bosphorus and İsmit Bay, all of which possess great value from aesthetic, recreational and other utilitarian standpoints, become increasingly polluted. Throughout much of the region sewage works are rare, and the waste waters are usually discharged untreated into the nearest water course. Until recently the only sewerage system in İstanbul was one dating back to the Ottoman and Byzantine administrations which only served about one quarter of the city's area. Now there is an integrated sewerage system with separate storm and foul water sewers. The problem of efficient disposal is made particularly difficult by the fact that İstanbul is a linear coastal city, and, therefore, sewage has to be collected over a very wide area. Outfalls for the new systems are into the Sea of Marmara and the Bosphorus.

In some of the more arid parts of the Middle East, in particular the Gulf states, the problem of urban water supply has been solved by the desalination of sea water. Kuwait was the first state to develop such a programme, linking electricity generation stations to desalination plants. In this way the low pressure steam from the generators can be used to provide energy for the desalination process. As a result costs are minimized. Kuwait began desalinated water production in

1957 when 3.1 million cu m of water were produced. By the mid 1970s this figure had risen to 55 million cu m.[76] It now has a sophisticated distribution system whereby saline waters are piped through one pipeline system for toilet flushing and related uses, while drinking water is transported through a separate system. The costs of desalinated water are high and Kuwait is now looking to provide extra supplies through an agreement with Iraq to establish a pipeline to transport waters from the Shaṭṭ al' Arab to Kuwait City. Although general agreement has been reached between the two countries, nothing can be implemented until the Iraq–Iran war is over.

In Qatar, too, an intensive programme of desalinated water production has been started, which should be supplying about 150 million cu m of water per year by the year 2000.[77] This is believed to be about three-quarters of the total water need, with the rest being supplied from groundwater sources. About half of the country's demand will be generated from the urban/industrial centres.

Saudi Arabia entered the desalinated water field much later than Kuwait, with the first plant not beng commissioned until 1970.[78] It has, however, gone in for an ambitious programme of desalination plant construction on both the Red Sea and Gulf coasts.[79] To date more than 20 plants have been built, including a one million cu m/day facility at Al Jubayl, which is currently the largest in the world.[80] Along both of the coasts the plants are connected by a water grid. The water is used largely for urban consumption, but some industrial use is also planned. As in Kuwait there is increasing government concern about the production costs of desalinated water and every effort is being made to ensure that use is as efficient as possible.

All of the cities of the region face, to a greater or lesser degree, the sort of problem outlined in the case studies for Tehrān, İstanbul and the Gulf states. Indeed, it seems that the supply of water to cities, and the disposal of waste and contaminated water is likely to cause much greater problems in the Middle East that those generated by the provision of irrigation water to supply agricultural needs. Whether they can be efficiently and economically overcome in the immediate future is still uncertain. Major attempts are being made and now even Cairo is to have an integrated sewerage system.

References

1. F.K. Hare, 'The causation of the arid zone', in *A History of Land use in Arid Lands*, Arid Zone Research, **17**, UNESCO, Paris, 1961, 35–50.
2. M.I. Budyko, N.A. Yefimova, L.I. Aubenok and L.A. Strokina, 'The heat balance of the surface of the earth'. *Soviet Geogr.*, **3**, 3–16 (1962).
3. L. Weickman, *Some characteristics of the sub-tropical jet-stream in the Middle East and adjacent regions*, Meteorological Publications, Series A, No. 1, Ministry of Roads, Meteorological Department, Tehrān, Iran, 1961, 29 pages.
4. Meteorological Office, *Weather in the Mediterranean*, **1** (*Second edition*), *General Meteorology*, London, Her Majesty's Stationery Office, 1962, 32.
5. G. Perrin de Brichambaut, and C.C. Wallén, *A Study of Agroclimatology in semi-arid and arid zones of the Near East*, World Meteorological Organisation, Technical Note No. 56, Geneva 1963, 8.

6. P. Beaumont, 'The Road to Jericho–A Climatological Traverse across the Dead Sea Lowlands', *Geography*, **53**, 170–4 (1968).
7. G. Perrin de Brichambaut and C.C. Wallén, *A study of Agroclimatology in semi-arid and arid zones of the Near East*, World Meteorological Organisation, Technical Note No. 56, Geneva, 1963, 10 et seq.
8. A.H. Gordon and J.G. Lockwood, 'Maximum one day falls of precipitation in Tehrān', *Weather, Lond.*, **25**, 2–8 (1970).
9. (a) H.L. Penman, 'Natural evaporation from open water, bare soil and grass', *Proc. R. Soc. Series A.*, **193**, 120–45 (1948).
   (b) C.W. Thornthwaite and J.R. Mather, *Instructions and tables for computing the potential evapotranspiration and the water balance.* Publs. Clim. Drexel Inst. Technol., **X**, 1957, 311 pages.
10. C.W. Thornthwaite, J.R. Mather and D.B. Carter, *Three water balance maps of southwest Asia*, Publs. Clim. Drexel Inst. Technol., **XI**, 1958, 57 pages.
11. J.E. Van Riper, *Man's Physical World*, McGraw-Hill Book Company, New York, 1971, 627–30.
12. UNESCO–FAO, *Bioclimatic Map of the Mediterranean Zone*, Arid Zone Research, **XXI**, UNESCO, Paris, 1963, 58 pages.
13. UNESCO–FAO, *Environmental Physiology and Psychology in Arid Regions–Review of Research*, Arid Zone Research, No.**21**, UNESCO, Paris, 1963, 345 pages.
14. (a) H. Kanter, *Libya*, **1**, Geomedical Monograph Series, Geomedical Research Unit of the Heidelberg Academy of Sciences, Springer–Verlag, Berlin, 1967, 188 pages.
    (b) G.E. Ffrench and A.G. Hill, *Kuwait*, **4**, Geomedical Monograph Series, Geomedical Research Unit of the Heidelberg Academy of Sciences, Springer–Verlag, Berlin, 1971, 124 pages.
15. G. Taylor, *Australia*, Methuen and Co Ltd, London, 1955, 72–4.
16. J. Lomas, 'Forecasting wheat yields from rainfall data in Iran', *World Meteorological Bulletin*, **XXI**, 9–14 (1972).
17. D.E. Pedgley and P.M. Symmons, 'Weather and the locust upsurge'. *Weather, Lond.*, **XXIII**, 484–92 (1968).
18. D.E. Pedgley and P.M. Symmons, 'Weather and the locust upsurge', *Weather, Lond.*, **XXIII**, 485 (1968).
19. (a) H.H. Lamb, *Climate: Present, Past and Future, Vol. 1. Fundamentals and Climate Now*, Methuen & Co Ltd, London, 1972, 613 pages.
    (b) Royal Meteorological Society, *World Climate from 8000 to 0 B.C.*, Proceedings of the International Symposium held at Imperial College, London, 18 and 19 April 1966, Royal Meteorological Society, London, 1966, 229 pages.
    (c) W.C. Sherbrook and P. Paylore, *World Desertification: Cause and Effect*, Arid Lands Resource Information Paper No. 3, University of Arizona, Office of Arid Lands Studies, Tucson, Arizona, 1973, 168 pages.
20. UNESCO-FAO, *Changes of Climate*, Arid Zone Research, **XX**, UNESCO, Paris, 1963, 488 pages.
21. C. Gischler, *Water Resources in the Arab Middle East and North Africa*, Middle East and North African Studies Press, 132 pages, 1979.
22. L.V. Wilcox, *classification and Use of Irrigation Waters*, US Dept. Agric. Circular No. 969, Washington, 1955, p 7.
23. C.W. Thornthwaite, J.R. Mather and D.B. Carter, *Three Water Balance Maps of Southwest Asia*, Publs. Clim. Drexel Inst. Technol., Vol 11, 1958, 57 pages.
24. P. Beaumont, *River Regimes in Iran*, Occasional Publications (New Series) No 1, Department of Geography, University of Durham, 1973, 29 pages.
25. C.G. Smith, 'Water resources and irrigation development in the Middle East', *Geography*, **55**, p 424, (1970).
26. P. Beaumont, *River Regimes in Iran*, Occasional Publications (New Series) No 1, Department of Geography, University of Durham, 1973, pp 10–13.

27. *Hydrological Yearbook*, General Directorate of State Hydraulic Works, Turkey, 1968.
28. P. Beaumont, 'Water resources and their management in the Middle East', in *Change and Development in the Middle East*, (Eds, J.I. Clarke and H. Bowen-Jones), Methuen, London, 1981, pp 40–72.
29. T. Dincer, A. Al-Mugrin, and U. Zimmerman, 'Study of the infiltration and recharge through the sand dunes in arid zones with specific reference to the stable isotopes and thermonuclear tritium', *Journal of Hydrology*, **23**, pp 79–109, 1974.
30. P. Beaumont, 'Irrigated agriculture and ground-water mining on the High Plains of Texas, USA', *Environmental Conservation*, **12, (2)**, pp 119–30, 1985.
31. D.J. Burdon, *Groundwater resources of Saudi Arabia, Groundwater Resources in Arab Countries*, ALESCO Science Monographs, No 2, 1973.
32. L. Thatcher, M. Rubin, and G.F. Brown, 'Dating desert groundwater', *Science, N. Y.*, **134,** pp 105–106, 1961.
33. J.G. Pike, 'Evaporation of groundwater from coastal playas (Sabkhah) in the Arabian Gulf', *Journal of Hydrology*, **11**, pp 79–88, 1970.
34. H.Y. Hammad, *Ground Water Potentialities in the African Sahara and the Nile Valley*, Beirut Arab University, 1970.
35. I.H. Himmida, 'The Nubian Artesian Basin, its regional hydrogeological aspects and palaeohydrological reconstruction', *Journal of Hydrology, New Zealand*, **9**, pp 89–116, 1970.
36. T.A. Wafer and A.H. Labib, 'Seepage from Lake Nasser', in *Man-Made Lakes; their Problems and Environmental Effects*. (Eds. W.C. Ackerman, G.F. White, E.B. Worthington and J.L. Young), American Geophysical Union, Geophysical Monograph, **17**, pp 287–291, 1973.
37. M. Ritchie, 'Libya: taking the plunge with the GMR', *Middle East Economic Digest*, **29, (29)**, pp 14–16, 1985.
38. K. Kenyon, 'The origins of the Neolithic', *Advmt. Sci., Lond.*, **26,** p 155, (1969).
39. N.A. Jiabajee, 'Saudi Arabia – water supply of important towns', *Pakist. J. Sci.*, **9,** p 192, (1957).
40. N.A. Jiabajee, 'Saudi Arabia – water supply of important towns', *Pakist. J. Sci.*, **9,** p 197, (1957).
41. K.S. Twitchell, 'Water resources of Saudi Arabia', *Geogrl. Rev.*, **34,** p 380, (1944).
42. K.S. Twitchell, 'Water resources of Saudi Arabia', *Geogrl. Rev.*, **34,** p 384, (1944).
43. P.W. English, 'The origin and spread of qanāts in the Old World', *Proc. Am. phil. Soc.*, **112,** p 174, (1968).
44. D. J. Flower, 'Water use in north-east Iran', in 'The Land of Iran' *The Cambridge History of Iran*, (Ed W.B. Fisher), Vol 1, Cambridge University Press, 1968, Chapter 19, p 603.
45. P. Beaumont, 'Qanāt systems in Iran', *Bull. int. Ass. scient. Hydrol.*, **16,** pp 39–59, (1971).
46. (a) P. W. English, 'The origin and spread of qanāts in the Old World', *Proc. Am. phil. Soc.*, **112,** p 170, (1968).
    (b) P. Beaumont, 'Qanāt systems in Iran', *Bull. int. Ass. scient. Hydrol.*, **16,** p 43, (1971).
47. P. Beaumont, 'Qanāt systems in Iran', *Bull. int. Ass. scient. Hydrol.*, **16,** p 45, (1971).
48. P. Beaumont, 'Qanāt systems in Iran', *Bull. int. Ass. scient. Hydrol.*, **16,** p 47, (1971).
49. P. Beaumont, 'Qanāts on the Varamin Plain, Iran', *Trans. Inst. Br. Geogr.*, **45,** p 177, (1968).
50. A.K.S. Lambton, *The Persian Land Reform*, Clarendon Press, Oxford, 1969, p 281.
51. D. Whittington and K.H. Haynes, 'Nile water for whom? Emerging conflicts on water allocation for agricultural expansion in Egypt and Sudan', in *Agricultural Development in the Middle East*, (Eds P. Beaumont and K.S. McLachlan), Wiley, Chichester, pp 125–149, 1985.

52. United Arab Republic, Ministry of the High Dam, Aswân High Dam Authority, *Aswân High Dam – Commissioning of the First Units – Transmission of Power to Cairo*, Ministry of the High Dam, Aswân, Egypt, 1963, p 6.
53. J. Waterbury, *Hydropolitics of the Nile valley*, Syracuse University Press, New York, 1979.
54. Arab Republic of Egypt, Ministry of Culture and Information, *The High Dam*, State Information Office, Cairo, Egypt, 1972, p11.
55. United Arab Republic, Information Department, *The High Dam: Bulwark of our future*, Information Department, Cairo, Egypt, p 6.
56. C. Hollingworth, 'Egypt's Aswan balance-sheet', *The Times*, London, 15 January, (1971).
57. Arab Report and Record, 'Dam devastates Mediterranean fishing', *Arab Report and Record*, Issue 23, 1–15 December, p 677, (1970).
58. J. McCaull, 'Conference on the ecological aspects of international development', *Nature and Resources*, UNESCO, **5**, (1969).
59. T. Little, 'Why cry havoc at Aswan?', *Middle East International*. June, No 3, p 5, (1971).
60. M. Lavergne, 'The seven deadly sins of Egypt's Aswan High Dam', in *The social and environmental effects of large dams*, (Ed E. Goldsmith and N. Hildyard), Wadebridge Ecological Centre, 1986, pp 181–183.
61. K.E. Haynes and D. Whittington, 'International management of the Nile: Stage three?', *Geographical Review*, **71, (1),** pp 17–32, 1981.
62. W.C. Lowdermilk, *Palestine – Land of Promise*, Victor Gollancz, London, 1944, 167 pages.
63. (a) A. Wiener, *The Role of Water in Development*, McGraw–Hill, New York, 1972, p 403.
    (b) A. Wiener, 'Coping with water deficiency in arid and semi-arid countries through high efficiency water management', *Ambio*, **6**, pp 77–82, 1977.
64. Y. Prushansky, *Water development*, Israel Digest, Israel Today, No 11, Jerusalem, 1967, p 24.
65. C.T. Main, *The Unified Development of the Water Resources of the Jordan Valley Basin*, Boston, Massachusetts, 1953.
66. (a) C.G. Smith, 'The disputed waters of the Jordan', *Trans. Inst. Brit. Geogr.*, **40,** p 122, (1966).
    (b) E. Orni and F. Efrat, *Geography of Israel*, Israel Universities Press, Jerusalem, 1971, p 451.
67. A. Wiener, *The Role of Water in Development*, McGraw–Hill, New York, 1972, pp 404–405.
68. I. Galnoor, 'Water policy making in Israel', in *Water Quality Management under Conditions of Scarcity – Israel as a Case Study*. (Ed H. I. Shuval) Academic Press, New York, 1980, pp 287–314.
69. H.I. Shuval, *Water Renovation and Reuse*, Academic Press, New York, 1977.
70. Echo of Iran, *Iran Almanac 1971*, Echo of Iran, 1971, p 101.
71. X. De Planhol, 'Geography of Settlement', in *The Land of Iran*, (Ed W.B. Fisher), Vol 1, The Cambridge History of Iran, Cambridge University Press, Cambridge, 1968, p 452.
72. Plan Organisation, *Dam Construction in Iran*, Bureau of Information and Reports, Tehrān, 1969, p 69.
73. Plan Organisation, *Dam Construction in Iran*, Bureau of Information and Reports, Tehrān, 1969, p 69.
74. M. Vahidi, *Water and Irrigation in Iran*, Plan Organisation and Bureau of Information and Reports, Tehrān, 1969, p 69.
75. R. Marwick and J.P. Germond, 'The River Lar multi-purpose project in Iran', *Water Power and Construction*, **27**, pp 133–141, (1975).

76. P. Beaumont, 'Water in Kuwait', *Geography*, **62,** pp 187–197, (1977).
77. (a) M.J. Hall and N.A. Hill, 'A master water resources and agricultural plan for the State of Qatar. Part 1: Physical setting and resources'. *International Journal of Water Resources Development*, **1,(1)**, pp 15–30, 1983.
    (b) T.J. Annesley, M.J. Hall and N.A. Hill, 'A master water resources and agricultural plan for the State of Qatar. Part 2: the systems model and its application', *International Journal of Water Resources Development*, **1,** no. 1, pp 31–49, 1983.
78. P. Beaumont, 'Water and development in Saudi Arabia', *Geographical Journal*, **143,** pp 42–60, 1977.
79. Saudi Arabia Central Planning Organization, *Development Plan 1395–1400*, Riyadh, Saudi Arabia, 1975.
80. *Journal of the Arab-British Chamber of Commerce*, London, 'Water technology and irrigation', p 34, April/May, (1985).

CHAPTER 3

Landscape Evolution

3.1 Introduction

Over millenia wind and weather have shaped a variety of landscapes from the rocks of the Middle East, though locally tectonic movements and volcanic eruptions have been important. Man's basic demands for food, fuel and raw materials played their part, particularly through their effects upon vegetation.[1] Palaeolithic man no doubt had some effect upon his environment, especially through the use of fire.[2] However, it was with the domestication of plants and animals, what Childe called the 'Neolithic Revolution', that man began to clear vegetation on an increasing scale and to initiate far-reaching, even irrevocable changes in his environment. Crucial developments after initial domestication were the emergence of peasant farming and nomadic pastoralism, together with the diffusion of their characteristic techniques throughout the region. Farming and pastoralism have gradually worn away the surface of the natural landscape, leaving only its bare bones behind in many districts. On the margins of the region's deserts, man's work has been assisted by irregular fluctuations of precipitation, for in such marginal zones any interference with the delicate natural balance can produce havoc, though, as Lamb has pointed out, climatic and human effects are often inextricably confused in this sensitive zone.[3]

The result of physical and human interaction by the closing decades of the twentieth century has been the production of a variety of landscapes. Some of these are mapped in Figure 3.1, but particular emphasis has been given to the 'humanized' landscapes. The most obvious 'natural' landscapes include the mountains above 3000 to 4000 m in Iran and Turkey, where corrie glaciers and permanent snow fields are found, and the great sand seas of the Nafūd and the Rub'al Khālī in Arabia. At least as impressive are the blackened lava tracts and barren, stone-strewn surfaces of Arabia and Egypt, as well as the shimmering salt flats of central Iran. Towns, especially those above 500,000 in population, contain the clearest manmade landscapes in the region, but the gold and green landscapes of dry and irrigated farming are equally human creations, as are the burnt and overgrazed pastures of many districts. But man, plants and animals are sustained by water, and its availability is a fundamental constraint in this arid region, while relief even today exerts a profound effect upon spatial organization and distributional systems. Olive, fig and other tree plantations sometimes constitute veritable manmade forests, just as much as deliberately afforested areas, but both cover only a small proportion of the region which, over the centuries, has been either completely denuded of vegetation or severely ravaged by human

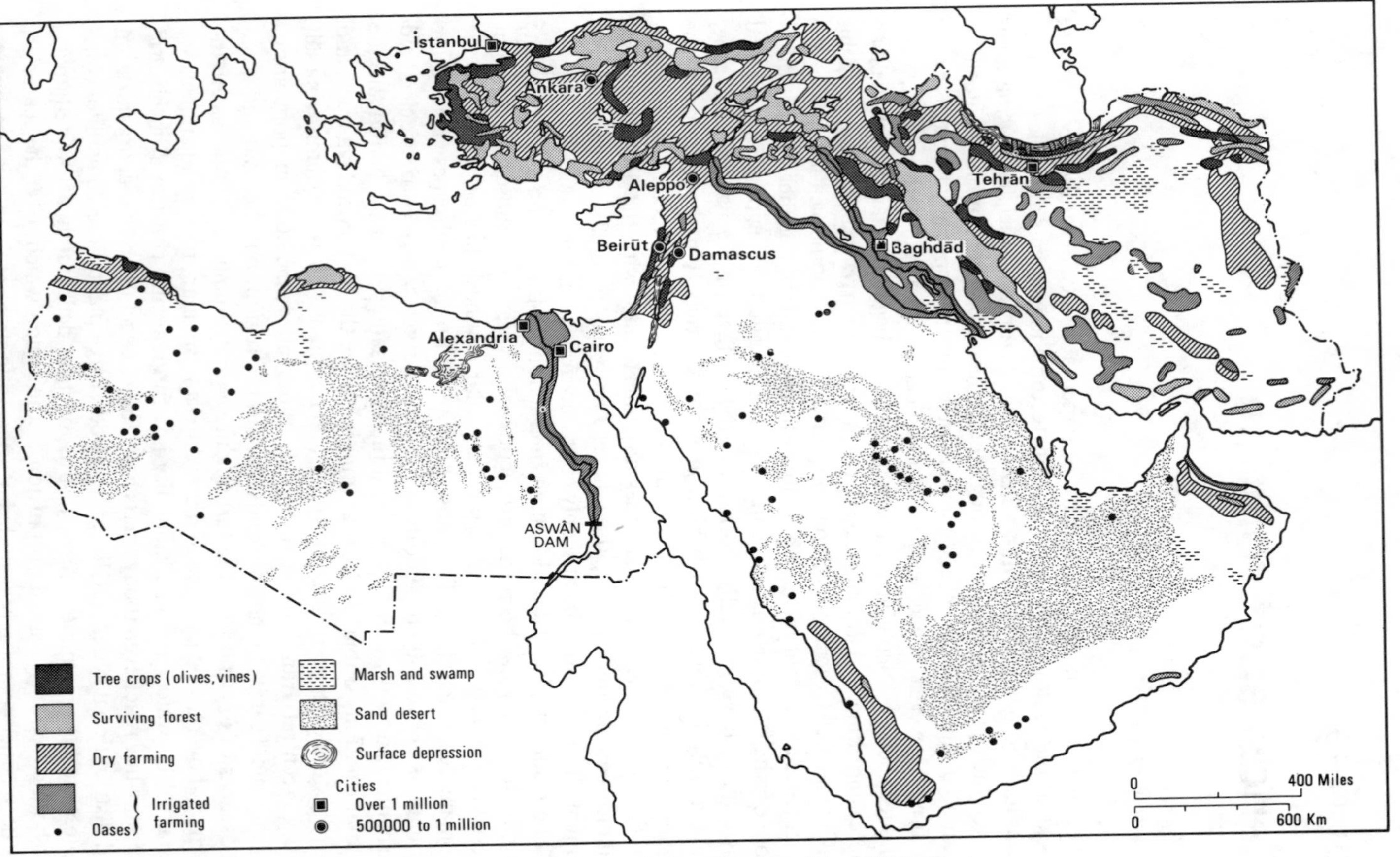

Figure 3.1 Contemporary landscapes of the Middle East

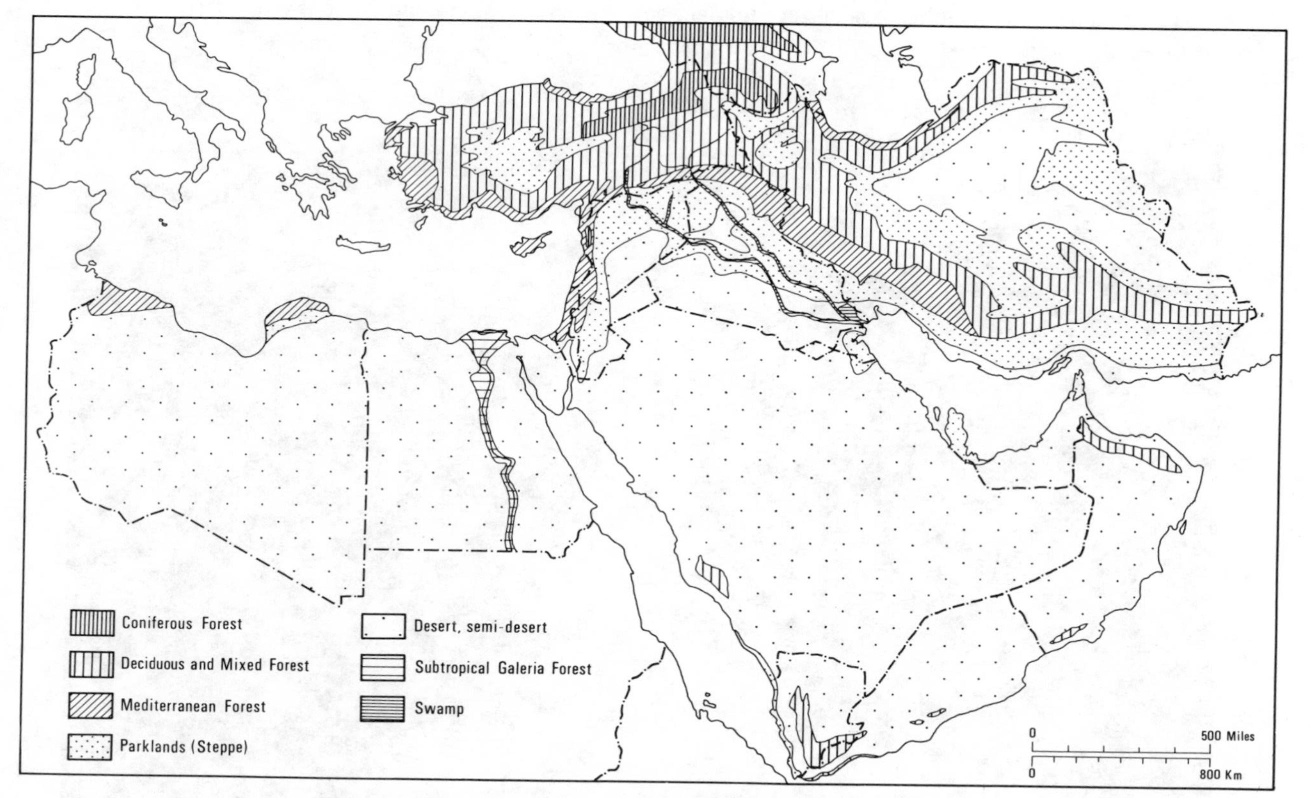

Figure 3.2 Generalised natural vegetation c.8000 BC

Dense forest along the southern margin of the Plain of Mazanderan, Iran (Peter Beaumont)

activity. Very little 'natural' vegetation survives. A reconstruction of the 'natural' vegetation associated with present climates will provide a base from which to gauge the effects of human activity upon the landscape. The effects of man on the 'natural' environment provide the theme for this chapter.

3.2 'Natural' vegetation

Vegetational assemblages are generally related to physical conditions, particularly the prevailing climate. Variations in the types of deposit (aeolian or alluvial) in certain areas of the Middle East have been interpreted as indicating drier or wetter conditions than those prevailing at the present. The period around 3000 BC, for example, may have been moister (or cooler) than the present while a climate similar to today's may have developed by 1000 BC.[4] Certainly, within the period of direct evidence from instruments, precipitation across the region was between 2 and 11 mm lower during the 1920s and 1930s than during the period 1891 to 1910 and temperatures were up to 0.75 °C higher.[5] However, Butzer has argued that the main precipitation and temperature characteristics of the present were established by 8000 BC, though more recent work suggests it might have been later.[6] 'Natural' vegetation over the region must be adjusted to these conditions (Figure 3.2).

Butzer stated that about 8000 BC the high mountains of Kurdistan were covered predominantly with coniferous forest,[7] but the accuracy of this view has been contested. Wright has suggested that the Alpine vegetation characteristic there today was already in existence about 10,000 years ago.[8] At lower elevations, where conditions were less severe but precipitation comparatively heavy, as over the Taurus and western Zagros, a mixed forest of deciduous and coniferous species was found in which black pine (*Pinus nigra*), juniper (*Juniperus communis*) and the holm or evergreen oak (*Quercus ilex*) were predominant. The mixture varied from area to area so that, for example, the deciduous Persian oak (*Quercus persica*) was common in the western Zagros, while in southwestern Turkey the main components of the forest were juniper and pine.[9] In Lebanon and parts of the maritime Taurus, stands of cedar were found (*Cedrus libanus*). At still lower altitudes lay a zone of Mediterranean forest consisting of evergreen, drought-resistant species, such as the Aleppo pine (*Pinus halpensis*), valonia, holm and kermes oaks (*Quercus macrolepis/aegilops, Q. ilex, Q. coccifera*), lentisc (*Pistacia lentiscus*), carob (*Ceratonia siliqua*), wild olive and wild vine. The whole assemblage was probably fairly open and may have contained a variety of grasses, including the early domesticates – two-rowed barley (*Hordeum distichum*), emmer and einkorn wheat (*Triticum dicoccum, T. monococcum*).

Precipitation over central Asia Minor, northern Arabia, much of Iran, and along the Libyan and Egyptian coasts was insufficient to support forest; steppe grasslands punctuated by occasional trees were probably characteristic. With a further decline in precipitation over Arabia and the Sahara even grassland was impossible. It was replaced by grass tufts, shrubs and low bush, which became wider and wider apart until vegetation virtually disappeared as annual precipitation faded away to nearly nothing and the expected frequency became extremely low. Even so, an irregular shower would produce a miraculous flush of vegetation.

Cutting across the desert and semi-desert were linear oases along the valleys of the Nile, the Tigris and the Euphrates. Before their occupation by agricultural communities, these valleys were probably marked by subtropical galeria forest containing species such as tamarisk (*Tamarix gallica, T. mannifera*), aspen (*Populus tremula*) and oleander (*Nerium oleander*). Swamps probably existed in low-lying areas where floodwater accumulated, especially in the southern parts of the Tigris–Euphrates lowland and in the Nile delta. Indeed, swamps so impressed the early population of Egypt that such an environment became the epitome of the underworld through which the souls of the dead were required to pass on their way to reincarnation on the eastern horizon. Islands and lines of riverine woodland must also have existed in desert and semi-desert areas wherever a spring, stream or permanent pool occurred. The abundant food supplies of these watery environments were so attractive to early man that agriculture was slow to penetrate, once the initial stages of domestication had been accomplished elsewhere.

3.3 Domestication

Domestication of plants and animals represented a considerable change in the ways by which men secured their subsistence. It is perhaps best summarized as the introduction of deliberate management and control into food production, though, as Flannery has pointed out, the subsistence base was not necessarily more stable or more reliable than a hunting-fishing-gathering economy under Middle Eastern conditions.[10] The chief advantages of farming were that it greatly increased the carrying capacity of a given unit of land[11] and made more intensive use of time and space.[12]

Several theories have been advanced to account for domestication.[13] Most of them are unsatisfactory and proof is difficult, especially since the available archaeological evidence is so equivocal. Explanation based solely on climatic change has been popular in the past, but now seems unacceptable. Man's food supply is unlikely to have been so insecure for famine to be the spur to domestication, while the study of other mammals suggests that, far from starvation acting as the principal regulator of human populations, social behaviour may be a mechanism for maintaining human populations in harmony with the carrying capacity of their inhabited space. Whatever the true explanation, at the time of writing there is a concensus of opinion that the final steps towards recognizable domestication in the west Eurasian tradition were taken at the end of a gradual, almost imperceptible process of developing human control over selected plants and animals within a context provided by an original symbiosis of plants, animals and man. These final steps appear to have been taken within the uplands of, in

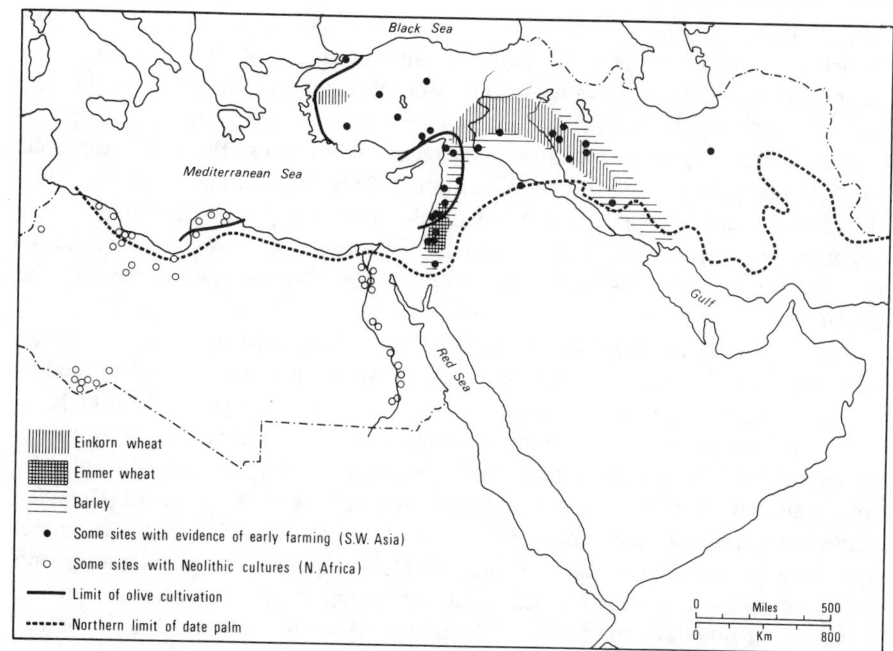

Figure 3.3 Natural distribution of early domesticates: early farming sites

Darlington's words, 'a single connected region'.[14] Certainly, a fully developed agricultural complex, involving the cultivation of wheat and barley, as well as the keeping of sheep and goats, was widespread by the end of the sixth millenium BC in an arc from Khuzestan through the intermontane valleys of the Zagros and the Taurus mountains into central and western Asia Minor, on the one hand, and southwards into Syria and Palestine, on the other (Figure 3.3). Virtually nothing is known at present of the cultivation techniques employed by the first farmers, though the recovery of flint sickle blades from early contexts in Palestine indicates a specialized range of implements. Even if its use may be considerably older, the appearance of the plough in the archaeological record about 3000 BC marks the final transition to dry farming of the type which has existed in the Middle East for several millenia down to the present and which seems so well adjusted to the region's soil and water conditions. Wheat and barley, however, were not the only plants which were domesticated. Fruit trees were probably brought into regular cultivation at later dates, but by the fourth millenium the growing of vines and olives had created the triad so characteristic of the Mediterranean fringes of the region down to the present,[15] and so basic to the development of a type of civilization markedly different from those found in the interior parts of the Middle East.

Farming spread from the uplands of the Middle East into the Tigris–Euphrates lowland, the Nile valley and parts of North Africa. Progress was slow. Part of the reason was the existence of resilient collecting economies based on the comparatively rich resources of the rivers and their banks, or, in the case of North Africa, of the comparatively damp mountain areas of the interior. Also important, however, was the need to master techniques of irrigation so that the lack of adequate rainfall could be compensated. Irrigated farming had spread into the Tigris–Euphrates lowland by the late fifth millenium BC and throughout the Nile valley during the fourth millenium BC. Cultivation may have spread to the coastal areas of North Africa independently of Egypt and dry farming was apparent in several places before 5000 BC. Another 2,000 years passed, however, before cultivation was established in the mountain valleys and oases of the interior.[16]

Pastoralism appears to have developed later than cultivation, though goats were probably domesticated before 7000 BC. Although cultivators continued to keep animals, specialist forms of pastoralism soon developed (before 6000 BC in North Africa), perhaps as a means of using different, but adjacent environments to meet the needs of settled cultivators for cheese, hides and draught animals. Important developments were the domestication of the Bactrian camel (*Camelus bactrianus*) and the dromedary (*Camelus dromedarius*). The Bactrian camel appears to have been domesticated in central Asia between the third and second millenium BC, but was introduced to the northern parts of the Middle East much later. The dromedary may have been first broken in southern Arabia before 2500 BC, but its effective use as a beast of burden and the basis of 'desert power' came with the adaptation of horse saddles in northern Arabia some time before the ninth century BC when the first evidence appears.[17] Use of the dromedary allowed fully nomadic pastoralism to develop in Arabia and adjacent areas and to appear in North Africa. Various forms of pastoralism and cultivation have co-existed in the Middle East and North Africa for several millenia, alternately expanding and contracting their domains in response to political and economic conditions, as well as to fluctuations in precipitation. Both economies have slowly transformed the 'natural' landscapes of the region.

3.4 Effects of domestication

Domestication itself initiated widespread, radical and perhaps irreversible changes in the uplands. Woodland was thinned and even cleared in various ways to allow cereals to grow. The extent of the cultivated area required would depend upon the size of the community which had to be fed, but the support of even quite small numbers could result in extensive clearance. For example, to support the 125 to 175 people at the famous site of Jarmo, which flourished around 6000 BC, it has been estimated that a crop area of between 45.0 and 91.2 ha would have been necessary.[18] Forest and bush fallowing techniques, which may have been used by early cultivators, would have required a larger area. Regeneration of woodland on these cleared plots may have been made difficult partly by the browsing and grubbing of domesticated animals and partly by changed soil and water conditions, so that in time quite wide areas of woodland were devastated. The net result of cultivation, then, was to provide open space in the 'natural' forest. Open ground must have become gradually more extensive around the various farming communities. Population increase, albeit slow, meant that more land had to be cleared and eventually may have brought in a short-fallow system of working, a development which would have expanded the area of permanently open ground. The unrelenting and steadily mounting demands of the domestic hearth must have also consumed large areas of irreplaceable woodland. Pressure and clearance undoubtedly increased when pottery began to be fired and metals worked.[19] Demand for constructional timber became immense, especially from the evolving kingdoms and empires of the region, and resulted in the deforestation of the more accessible parts of the uplands, such as the upper Euphrates catchment in southeastern Turkey from which timber was sent as rafts to southern Iraq for several thousand years,[20] and the virtual elimination from Lebanon of stands of cedar, a wood valued for its colour and perfume.[21]

Removal of the woodland and its slow regeneration had several important effects upon the landscape. Clearance of undergrowth may have diminished the number, and perhaps the variety of plants (berries, roots), previously used as food or else forced them into peripheral locations with respect to settlements. Wild game, initially important to early farming villages, almost certainly must have been forced further and further away to become characteristic of remote districts.[22] Forest, too, survived in something approaching its original state only in the more distant and physically difficult parts of the region, such as the Taurus mountains and the High Atlas. Where woodland did not regenerate quickly enough, any forest and bush-fallowing practices must have come to an end and been replaced by cultivation techniques, involving the plough and fallowing, which themselves kept land open and largely free of 'natural' vegetation. Although dry farming techniques may help to conserve soil moisture and to some extent preserve natural fertility, the existence of large areas of bare plough soil in winter must have led to the destruction of soils by wind erosion, sheet wash and gullying. The long-term effects were almost certainly changes in soil character, a lowering of the water-table and the gradual reduction of the cultivable area. River flow probably became more erratic, too, exposing the lower reaches of valleys to greater floods and the deposition of silt.[23]

Domestication and then peasant farming brought about changes on the human as well as on the physical side of the man–land equation. In the first place, the land area from which people lived was reduced. This may be demonstrated with reference to Iran.[24] In 1956, about 65 per cent of the surface area of Iran was classified as uninhabitable and marginal for exploitation. The remaining 35 per cent was available for some form of exploitation, but only 10 per cent was considered as arable land, actual or potential. A similar situation prevails in other parts of the Middle East (Table 4.6). Domestication, thus, meant that man became increasingly dependent upon a smaller area for subsistence and one which was being diminished by erosion initiated and perpetuated by man's own activities. A second result of domestication may have been an increase in population by reducing the frequency of miscarriages and increasing the rate of conception.[25]

Either way, the impact of man on the land would have increased. A third change resulting from domestication and the development of peasant farming was also cumulative. This was the gradual emergence of an open, regulated landscape in the uplands. It contained ploughed fields, terraces and treeless pastures, all of which were linked by trackways and crossed locally by irrigation channels.

Although development began in the uplands, great transformations occurred in the major lowland areas of the region as a result of the spread of cultivation. Not only was 'natural' galeria forest removed, but marshes and swamps were at least partially reclaimed. They were replaced by a regulated landscape composed of basins used for irrigation and networks of irrigation canals. Irrigation, however, was confined to relatively narrow bands along the river channels or to favoured localities on the edges of the surviving marshes. Beyond the immediate vicinity of the rivers lay untamed land. In the Tigris–Euphrates lowland this was semi-arid steppe used as grazing, but scrub vegetation may have survived for quite a long time between the river and the steep valley sides marking the edge of the desert in Egypt. Patterns in the hydraulic landscape changed through time, especially in the Tigris–Euphrates lowland where natural water channels were unstable, meandering and braiding.[26] Developing salination also had effects in southern Mesopotamia leading to the spread of barley growing (relatively salt tolerant) and sheep rearing, as well as to the abandonment of the most severely poisoned land (Chapter 12). The Nile valley did not suffer to the same degree because of the natural flushing of the annual flood, but some salination seems to have developed in parts of the delta by Byzantine times. In Egypt, more than in the Tigris–Euphrates lowland, cultivation responded to the annual fluctuations in the height of the flood, since this conditioned the amount of land which could be cropped in the pre-dam era.[27]

Semi-permanent or permanent settlements predate domestication, but their stability and number grew after domestication because farming increased the carrying capacity of land by as much as 200 times.[28] The density and spacing of permanent settlements in Khuzestan, a region in southwestern Iran which has been well explored by archaeologists, had achieved their modern characteristics by about 3500 BC, though the details of the patterns have changed considerably over the intervening time. The same region has also produced evidence for the emergence of towns (Figure 3.4). Around 3500 BC many settlement sites in Khuzestan varied from one to two hectares in extent, but some had areas of four to five hectares, while a few covered as many as 20 ha. The size hierarchy strongly suggests variations in population and the possibility that the largest settlements performed some urban functions, perhaps arising not only from local marketing needs, but also from some long-distance trade.[29] Although most towns in the Middle East may have resulted from organic growth, some appear to have been created, even during the fourth millenium BC, by the deliberate concentration of population as an administrative and military device,[30] while their internal layout shows clear evidence of structural planning. Once they had appeared, towns became increasingly characteristic of the Middle East. Many continued to result from organic growth at some convenient central place and their impor-

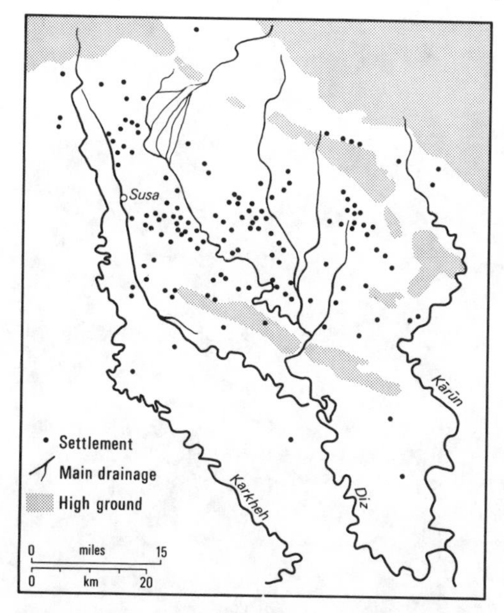

Figure 3.4 Settlement pattern in Khuzestan before about 3500 BC (Adams, R. M., 'Agriculture and urban life in early southwestern Iran', *Science* 136, 109–122, Figure 3 (13 April 1962). Copyright 1962 by the American Association for the Advancement of Science)

tance, even their existence, depended upon the general economic health of their immediate hinterland and, in some cases, upon the flow of trade along long-distance caravan and sea routes. Other towns, however, were artificial creations, generally built for some administrative or political purpose, but they were able to perpetuate themselves because of location at a natural focus, like Baghdād or Cairo,[31] or because their administrative functions survived long enough for them to become the centres of a sustaining web of socio-economic relationships. Whatever their origins, from early times the towns of the region constituted a distinctive built environment, in which large numbers of relatively densely packed dwellings were dominated by large public buildings. They were also organizational and marketing centres, so that their demands for food, fuel and raw materials had significant effects upon the landscape of surrounding regions. In particular, the fluctuating demands of the town economy controlled the rhythm of exploitation in the countryside and thus the pace of landscape change.[32]

3.5 Spatio-temporal variations[33]

So far in this chapter a case has been made for gradual but relentless change in Middle Eastern landscapes as the result of domestication and the subsequent

Aerial view of the city of Rezaiyeh, Iran in 1956, showing the abrupt transition from the built-up area to the irrigated fields, with arid land beyond (National Cartographic Centre, Iran)

spread of farming and pastoralism. However, over time the direction and pace of landscape change have varied regionally. A frequent theme in the history of the whole region has been the struggle between the desert and the sown, the advance and retreat of the pastoral domain alternating with the contraction and expansion of the cultivated area. Related to this rhythm has been one of change in agricultural specialization in those areas which remained under cultivation. Such spatio-economic changes may be regarded as responses to forces, internal and external to the region, which were as much political as economic. The next few pages attempt to outline some of the major changes and their causes.

Roman rule over North Africa and much of Southwest Asia was sufficiently strong and efficient over several centuries for cultivation to expand, though in some districts at considerable cost to the physical environment. Within the bounds of modern Libya, agricultural expansion during the first and second centuries AD took place within the zone at present suitable for dry farming, particularly the area near Tripoli and the plateau of Cyrenaica, though the coastal plain near the modern town of Sirte was also intensively developed. Careful attention was given to the control, storage and distribution of winter stream flow, leaving behind massive concrete dams at the heads of various wadis as evidence for this activity. More extensive, though, are systems of terraces running down wadi beds. These were obviously designed to trap silt, soak up runoff, irrigate land and prevent erosion. During the third century AD, following a series of military campaigns against the nomads of the interior, marginal land beyond the effective limits of dry farming was brought into cultivation, largely through widespread use of water-conserving techniques. For example, the great wadi systems of Sofejin and Zemzem south of modern Misurata were intensively farmed by soldier-farmers living in closely spaced but small, tower-like farmsteads which were part of a deep zone of frontier defence incorporating a few very strong fortresses. Both phases of development were associated with an expansion of olive growing, partly to meet the domestic needs of a growing population and partly to supply the public baths with oil. Oleiculture was already established around the old Greek colonies in Cyrenaica before the Romans took over the territory, but under Roman rule it spread further, particularly across the hilly hinterland of Leptis Magna, where the remains of several oil presses attached to substantial farmsteads may still be seen. In neighbouring parts of modern Tunisia, oil-growing was carefully organized within a grid division of land (*centuriation*) based upon a square with a side of 710 m.[34]

Oleiculture also expanded along the western coast of Asia Minor, where the remains of ancient presses are common. Behind the coasts, though, forests in this region were ruthlessly cut for constructional and fuel purposes and much cleared hill land was brought under the plough or intensively grazed for the first time. The medium-term effects of this 'extractive economy', developed in Hellenistic times but perpetuated under Roman rule, were disastrous. Not only was some land completely ruined by erosion, but increased sedimentation led to the silting up of coastal bays (Figure 3.5), and assisted the gradual extinction of once flourishing ports such as Ephesus and Miletus.[35] Aggradation and increased runoff modified the coastal valleys, producing derelict and marshy land.

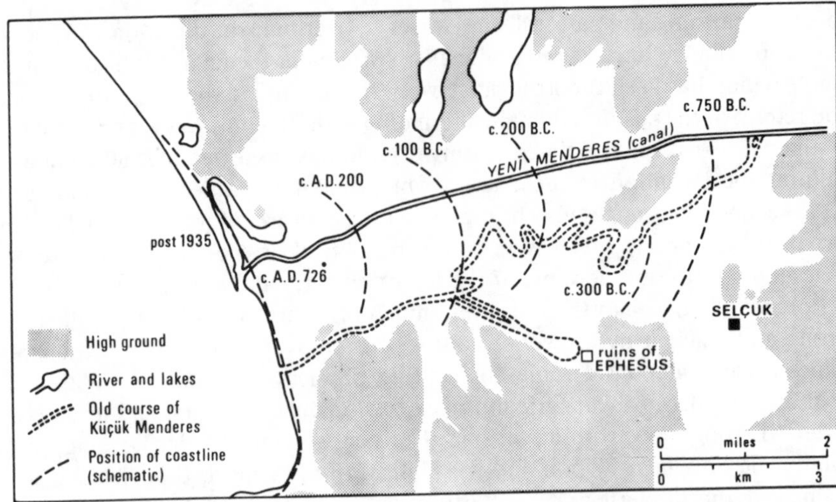

Figure 3.5 Advance of the coastline at the mouth of the Küsçük Menderes (reproduced by permission of Koninklijk Nederlands Aardrijkskundig Genootschap)

The profound economic crisis which affected the Roman and Byzantine Empires between the third and sixth centuries resulted in some release of the exploitive pressure, and may have initiated a progressive contraction of the cultivated area, especially in districts where precipitation was low and uncertain. Contraction has been explained as the result of denudation and climatic change, but, though there may be an element of truth in these views, an alternative explanation is favoured here. Contemporaries firmly believed that the contraction of arable land resulted from oppressive taxation. The land tax was the basic source of revenue to the state, and it had to be progressively increased in order to pay for an expensive series of wars against Barbarian invaders in the west and Persia in the east, as well as to maintain a bureaucracy expanded as a result of Diocietian's reforms in imperial administration. The effects were to reduce the profitability of estate farming in marginal areas and to concentrate cultivation where it remained profitable, that is, in the main dry-farming areas.[36]

Although partial recovery had taken place by the end of the sixth century, and further local expansions of cultivation were to occur in subsequent centuries, the period from roughly the end of the sixth century down to the end of the eighteenth century saw, on the regional scale, a marked contraction of arable land and pulsating expansions of the pastoral domain, as nomads were allowed to spill out of Arabia and the Sahara, or occasionally swept down irresistibly from central Asia. Contrary to popular opinion, the rate of deforestation may have slackened with the spread of nomadism, except close to certain coasts from which shipbuilding and other timber were taken,[37] but regeneration was retarded by the grazing of animals and radically changed soil and water conditions. Irrigation works became too expensive to maintain both in North Africa and Syria, for example, and were allowed to decay, though in Mesopotamia some canals were breached in the frequent military campaigns. Although some towns were founded by new rulers, whether Arab, Mongol or Turk, generally towns and villages lost population. Many villages were deserted altogether, while the ruins of once magnificent public buildings dwarfed the black tents of nomads in the vicinity of once important towns like Baalbek and Jarash in Southwest Asia and Leptis Magna in North Africa.

Various explanations have been offered for such widespread and long-lasting decay. The situation obviously varied locally, but the keys very often appear to have been the degree of administrative control and long-term investment which a particular government and a particular society were capable of making. Thus, the earlier Caliphs restored prosperity to the former frontier provinces of Iraq and Syria, which had been devastated by a series of bitter wars between the Roman and Persian empires. 'New' crops, such as rice, sugar cane and citrus, spread across the region and with them mechanical methods of raising water.[38] From the ninth century onwards, however, falling tax receipts suggest declining agricultural production, especially in the Tigris–Euphrates lowland where cultivation was precariously dependent upon the maintenance of an elaborate system of flood control and irrigation works (Chapter 12). Contemporary evidence suggests that the system was simply neglected. Consequently, canals and distributaries silted up, while floods created extensive damage which was seldom made good. At the same time, the peasants were squeezed by an inefficient tax farming system to support a luxurious court and to fight interminable wars, which could no longer pay for themselves in booty. Invasions and ruthless exploitation by foreign rulers completed the process of decay.[39] A similar story can be told for Palestine and Transjordan, which appeared so prosperous when the Crusaders invaded, but which already had acquired a desolate character by 1516, when the territory was incorporated into the Ottoman Empire. Turkish rule did not improve conditions, and the desertion of villages was quite considerable over the next 300 years, especially east of the Jordan (Figure 3.6), where nomadism was allowed to expand.[40]

Western travellers to the Middle East during the eighteenth and early nineteenth centuries painted sad and sorry pictures of the lands which they crossed. Although their views may have been exaggerated, travellers were struck by the widespread devastation and desertion which they saw, as well as the dominant position of the ubiquitous nomad. In Iraq, settled life had ceased altogether along the Tigris north of Baghdād and along the Euphrates north of Al Hillah, while cultivation in Syria had shrunk back to the wetter parts of the region, largely lying west of the road from Aleppo to Damascus. Palestine seemed particularly deserted and devastated to many travellers, familiar with its milk-and-honey landscapes from their reading of the Bible. They came probably expecting too much, but from their accounts it is clear that settled life had contracted into the hills, where a largely subsistence economy based on cereal-growing was maintained. The plains were generally empty and left as grazing to the nomadic tribes which had penetrated from the east during the seventeenth and early eighteenth centuries. Neglect of springs and steams lead to swamp development and its expansion in several areas, notably the Hula Valley.[41]

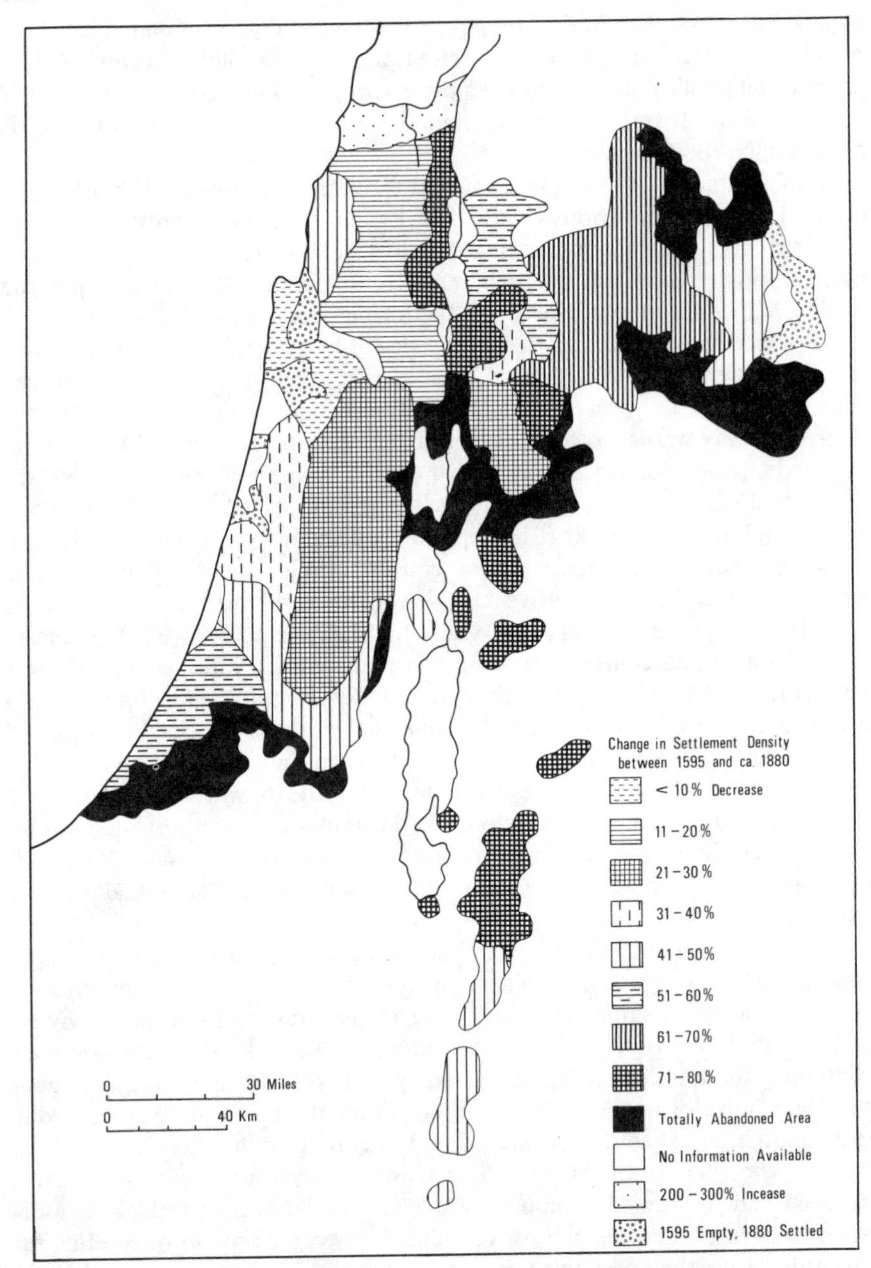

Figure 3.6 Desertion of settlements in Palestine and Transjordan, 1595 to 1880 (reproduced by permission of Wolf Hütteroth, 1969)

Change, however, was already underway. In favoured retreat areas, especially Mount Lebanon and northern Palestine, agricultural expansion began in the eighteenth century (Chapter 14), while Druze colonists were settling in the upland area further east which subsequently became known as Jebel ed Drūze (Chapter 13). Rising demands for food and raw materials from the industrializing countries of western Europe produced a steady, if spatially scattered expansion in the cultivated area of the Middle East during the nineteenth century. Coastal areas were affected first, but the frontier of recolonization moved inland as roads were constructed, railways built and river navigation improved. Thus, the building of the road from Beirūt to Damascus between 1858 and 1863 stimulated cereal growing in the valley of El Beq'a, while the Ankara branch of the Anatolian Railway carried five times as much grain in 1911 as it had in its first full year of operation (1893), an increase which almost certainly resulted from an expansion in the cultivation area in districts accessible to the line. Completion of a narrow gauge railway from Beirūt to Damascus in 1895 further stimulated grain production in the Hauran district of Syria, though the wave of agricultural expansion had begun some sixty years before during the stable period of Egyptian rule (Chapter 13).[42] The introduction of steam navigation to the Tigris in the 1830s cut freight rates and allowed the country to supply markets in India with wheat and barley much more easily than in the past, thereby making it worthwhile for small-scale irrigation projects to begin and extend the cultivated area. Larger irrigation schemes were attempted in the 1850s in Iraq, but without very much success. The expansion of irrigated farming was much more marked in Egypt during the course of the nineteenth century. It began in the 1820s with the construction of small barrages on river channels in the delta to hold back some of the annual flood and use the water for extending the temporal and spatial limits of cultivation. The first major step in developing irrigation took place in 1843 when work began on the construction of barrages at the bifurcation of the two main channels of the Nile. The delta barrages were completed in 1861 and subsequently improved. They allowed extensive reclamation in the delta and the extension of perennial irrigation. Barrages were subsequently built further and further up the valley, but traditional forms of basin irrigation tended to persist here well into the inter-war period. Towards the end of the nineteenth century, however, it was realized that effective use of the barrage system really depended upon storing as much of the flood as possible and gradually using it as the flow of the river dropped through the year. The most obvious site for a dam to control the irrigation system of the Nile valley was at the First Cataract. The first Aswân Dam was constructed at this point in 1902 and was subsequently raised in 1912 and 1933. The building of the dam had tremendous effects, but in particular it initiated the penultimate stage in the transformation of Egypt's irrigation system from one dependent upon the direct use of the flood in basins to a perennial system in which water was made available through canals and distributaries. Intensive cropping thus became feasible.[43]

Much of the new land made available in Egypt was put down to cotton, which gradually came to dominate both the economy and the landscape of Egypt during the nineteenth century. Commercial cropping also expanded elsewhere in the Middle East during the century. Cotton growing came to dominate the reclaimed lands of the Çukorova in southeastern Asia Minor. Silk production expanded in

Lebanon, particularly in the 1850s, and mulberry trees became very characteristic of Mt Lebanon itself. In western Asia Minor, railways opened up the valleys near Smyrna from 1863, changed the orientation of communications systems to the railheads, and reinforced the development of local specialisms in dried grapes, dried figs and tobacco.[44]

Parallel with these developments ran improvements in administration and political control. Their clearest effect lay in the control which was gradually extended over the nomads and which either assisted in sedentarization or forced nomadic pastoralism into topographically difficult and climatically hazardous districts. Security and stability allowed the growth and the spread of settlements. Old sites were occupied, notably in Syria and Transjordan,[45] and numbers of completely new settlements were established in many districts. A few villages grew organically into towns as the rural economy prospered, as in the case of Zahle which became a collecting centre for the grain of the Beq'a valley as well as a stopping placing on the Beirūt to Damascus road. Some towns, however, were deliberately planted, like Elâziğ, the lowland successor to the citadel-town of Harput (Kharpout) in eastern Turkey.[46] New ports, like Haifa, appeared and old ones, such as Beirūt and Smyrna, expanded considerably in population and extent. Many other towns grew beyond their ancient walls and spatious suburbs were sometimes laid out on European lines. Sedentarization of the nomads and the expansion of permanent, cultivating settlements and market towns were responsible in turn for a renewed assault upon the surviving forests of the region, especially where access was improved by railways and modern roads. A rapid retreat of the forest seems to have resulted.[47]

The landscape changes described so far may be represented as the indirect effects of growth in the European market during the nineteenth century. European influence was much more direct in parts of North Africa. A large French territory was carved out of the Maghreb in the decades following the capture of Algiers in 1830, and French colonists were introduced from 1838 onwards, with the greatest flow in between 1860 and 1900.[48] French experience influenced the Italian government after its conquest of Libya in 1911. Much of Tripolitania at that time was seen as empty or under-used land which was ripe for development using techniques already in use in Italy and reviving Roman irrigation systems. Colonization began in an experimental way in 1914, only to be ended by the First World War. A more active colonization and development policy was pursued after 1922, initially by establishing large estates worked with local sharecroppers but, after 1928, by attracting peasant farmers from Italy. By 1940, when colonization again stopped, some 15 per cent of the total productive land of Libya was being used by Italians. In both the Maghreb and Libya, Europeans were largely interested in cash crops. By experimenting with new methods of dry farming and then introducing machines, the French were able to increase the area devoted to cereals, though extensively eroded land was often the long-term result in the steppe region of the Maghreb.[49] Vines and olives were grown in Maghreb and Libya on a larger scale than at any time since the Roman period. Irrigation was extended in both regions by the construction of dams and the partial restoration of ancient flood control works, but the rate of progress was slow, mainly because of the technical problems and the vast expense involved. The net effect was the production of enclaves of European landscape characterized by rectangular field patterns and alien architecture. Except in the case of a few large estates, which followed the European example, indigenous farming was affected chiefly by the appropriation of land which, in some cases at least, upset traditional extensive land use practices and increased poverty. Pastoralism was completely disrupted by increasing mechanization, the ploughing up of traditional pastures and continual attempts at containment and sedentarization. Restrictions on the range of pastoral activities was one reason for the rapid reduction in the forest cover of the Maghreb under French rule, though the demand for constructional timber by the *colons* and for export to France was also important.

The Middle East has experienced profound political and economic changes since the end of the First World War. The inter-war period saw the establishment of mandates over Syria, Palestine, Transjordan and Iraq, and the consequent aggrandizement of European power and influence. After the end of the Second World War came gradual retreat. Independence resulted in the emigration of European colonists from the Maghreb and Libya, as well as the expropriation of European property in many countries, though not the total eradication of European imprint from local landscapes. Population increased considerably, especially after the Second World War, bringing greater pressure than ever before on natural resources. The World Wars and the depression were major causes of shifts in world commodity prices to which land use in the Middle East responded, often painfully. However, in the history of landscape evolution, the period since the end of the First World War may be visualized largely as one in which previous developments continued, but at an accelerating pace. The only major new departures have been the addition of distinctive industrial buildings to the rapidly expanding towns, especially to the port and capital cities, and the appearance of derricks, flares, pipes and storage tanks in all the places where petroleum is produced, refined and exported, but with an important concentration on the Gulf (Chapter 9). Elsewhere, continued development has produced less weird landscapes. These are discussed systemmatically in Chapters 4 to 8, and in more detail in the regional essays (Chapter 11 to 20). Only a brief general sketch need be given here.

In the age-old struggle between the desert and the sown, victory seems declared for the cultivator. Sedentarization of the nomads has proceeded rapidly, often as a result of changed economic and land-use conditions, rather than the application of military force, though this has not been wanting on occasion. Nomadic populations are now very small. In Iraq, for example, the proportion of nomads in the population fell from about 35 per cent in 1867 to 1.1 per cent in 1947.[50] Although numbers have fallen dramatically, and continue to decline throughout the region, there is a growing demand for the traditional nomadic products of wool, hair and cheese. Migratory herds can now be kept mainly in the high mountains of eastern Turkey and the Zagros and some parts of Arabia, with the result that concentration and overstocking leads to overgrazing, further pressure on woodland, destruction of the grazing and an increase in erosion (Chapter 4).

As pastoralism has retreated, so cultivation has expanded its area (Chapter 4). Much of the effort has been devoted to the recolonization of territory which was once cultivated, but has been used largely by nomads for centuries. In some districts, though, advance has been across land which is, in effect, new. In the plains and valleys especially, the land now being brought into cultivation is frequently composed of silts laid down in cycles of erosion, storage and release triggered by land clearance and the neglect of conservation measures such as hill terraces.[51] Reclamation and recolonization, however, have combined with population emigration to ensure that once attractive hill land is being abandoned; fields lie idle and terraces are allowed to collapse, threatening low-lying land with flooding and siltation. Some of the upland, of course, is now being managed as forest in an attempt to contain erosion, as well as to conserve and extend local timber resources. (Figure 3.7).

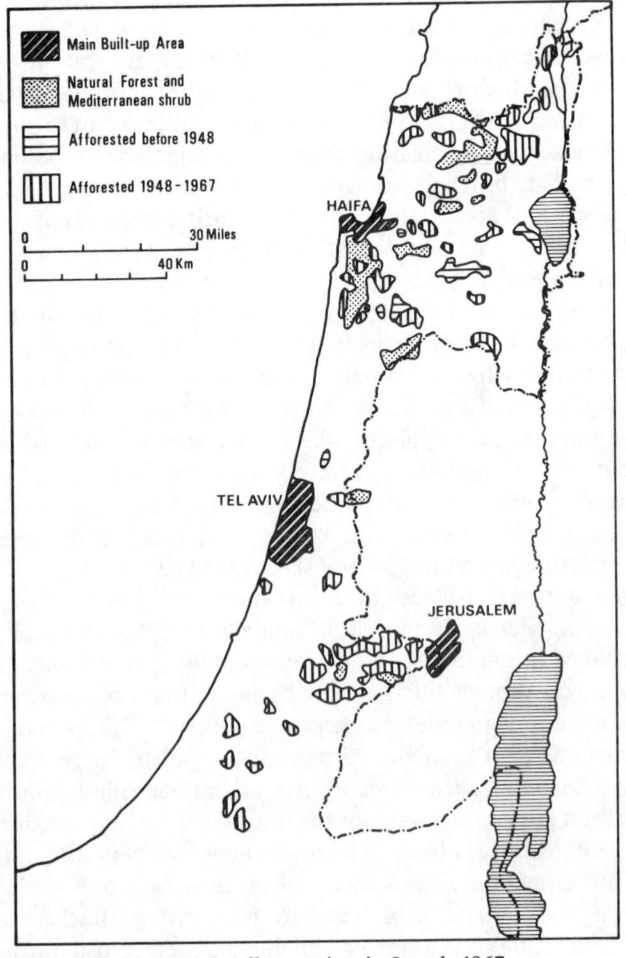

Figure 3.7 Afforestation in Israel, 1967

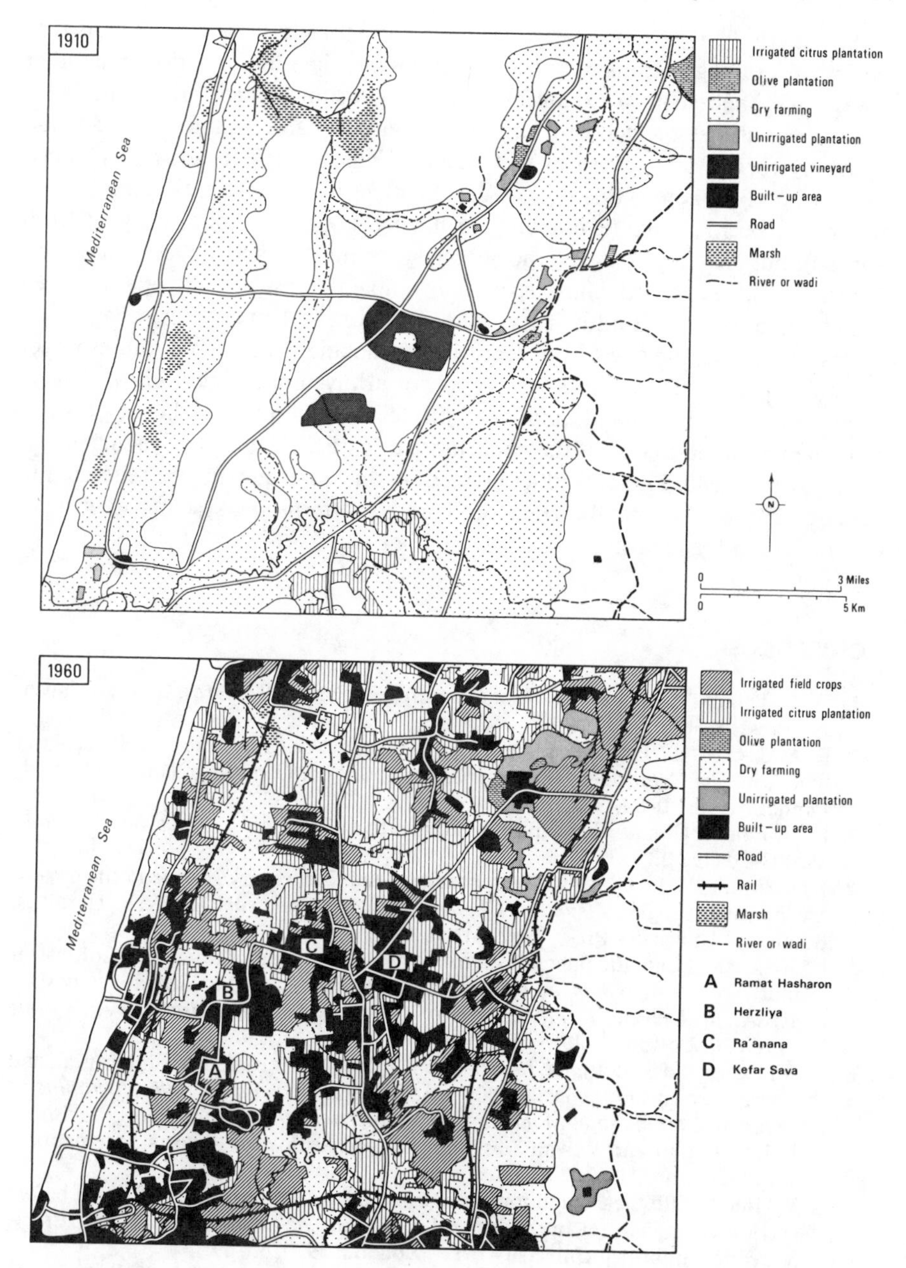

Figure 3.8 Transformation of landscapes of the southern plain of Sharon, 1910 to 1960 (modified by permission of Department of Surveys, Tel Aviv, Israel)

Much of the valley and plain land is suitable for irrigation. In consequence, large dams with their associated hydraulic networks and continuously cropped fields are very much a part of modern landscapes in the Middle East.[52] In some countries, though, as land is expensively gained for cultivation in one district, neglect of drainage means that a certain amount is lost by salination in another area, whilst the relentless expansion of towns consumes much prime farmland. Irrigation is also associated with large-scale regional planning and the production of highly formal landscapes of rectangular fields, straight roads and planned settlements. Nowhere is this type of development more apparent than in Israel, which has been fortunate not only in having sufficient capital (largely from abroad), but also in possessing the necessary technology and skilled manpower. Figure 3.8 shows a particularly striking example of how a landscape has been transformed over a period of fifty years. But even in Israel, where change has been so complete, the scars from centuries of human misuse of natural resources are still apparent. They are expensive, if not always impossible to heal. Everywhere in the region badlands and bare hillsides are reminders of the long history of human occupation and of the importance of care and conservation in a precarious environment. The difficulties created by that environment are reiterated in the next chapter (Chapter 4), which outlines the patterns and systems involved in using the countryside today.

References

1. W. L. Thomas (Ed), *Man's Role in Changing the Face of the Earth*, Chicago University Press, 1956.
2. D. A. E. Garrod, 'Primitive man in Egypt, western Asia and Europe in Palaeolithic times', in *The Cambridge Ancient History*, **1**, 3rd ed, Cambridge University Press, London, 1970, 70–89.
3. H. H. Lamb, 'Climatic background to the birth of civilisation', *Advt. Sci. London.*, **25**, 103–20 (1968).
4. J. L. Bintliff, 'Climatic change, archaeology and Quaternary science in the eastern Mediterranean region', *Climatic Change and Later Prehistory* (Ed A. F. Harding), Edinburgh University Press, 1982, 143–61.
5. N. Rosenam, 'Climatic fluctuations in the Middle East during the period of instrumental record'. In *Changes of Climate, Proceedings of the Rome Symposium organised by UNESCO and the World Meteorological Organisation, Arid Zone Research 20*, UNESCO, 1963, 67–73.
6. W. van Zeist and S. Bottema, 'Vegetational history of the eastern Mediterranean and the Near East during the last 20,000 years', in *Palaeoclimates, Paleoenvironments and Human Communities in the Eastern Mediterranean Region in Later Prehistory*, (Eds J. L. Bintliff and W. van Zeist), British Archaeological Reports, International Series 133, Oxford, 1982, 277–321.
7. K. W. Butzer, 'Physical conditions in eastern Europe, western Asia and Egypt before the period of agricultural and urban settlement'. In *The Cambridge Ancient History*, **1**, 3rd edn, Cambridge University Press, London, 1970, 35–69.
8. H. E. Wright, 'Natural environment of early food production north of Mesopotamia', *Science, N. Y.*, **161**, 334–9 (1968).
9. W. van Zeist, 'Reflections on prehistoric environments in the Near East', in *The Domestication and Exploitation of Plants and Animals* (Eds P. J. Ucko and G. W. Dimbleby), Duckworth, London, 1970, 35–46.

10. K. V. Flannery, 'Origins and ecological effects of early domestication in Iran and the Near East', in *The Domestication and Exploitation of Plants and Animals* (Ed P. J. Ucko and G. W. Dimbleby), Duckworth, London, 1970, 73–100.
11. W. Allan, 'The influence of ecology and agriculture on non-urban settlement', in *Man, Settlement and Urbanism* (Eds P. J. Ucko, R. Tringham and G. W. Dimbleby), Duckworth, London, 1972, 211–26.
12. T. Carlstein, *Time Resources, Society and Ecology*, George Allen and Unwin, London, 1982, **1**, 162–75.
13. (a) B. Bender, *Farming in Prehistory: From Hunter-gatherer to Food Producer*, John Baker, London, 1975.
(b) J. Hutchinson, G. Clark, E. M. Jope and R. Riley (Eds), *The Early History of Agriculture*, Oxford University Press, 1977.
(c) C. A. Reed (Ed), *Origins of Agriculture*, Mouton, The Hague and Paris, 1977.
(d) R. Stigler, R. Holloway, R. Solecki, D. Perkins and P. Daly, *The Old World: Early Man to the Development of Agriculture*, St Martins Press, New York, 1974.
(e) J. M. Wagstaff, *The Evolution of Middle Eastern Landscapes: An Outline to A. D. 1840*, Croom Helm, London, 1985, 39–43.
14. C. D. Darlington, 'The silent millenia in the origin of agriculture', in *The Domestication and Exploitation of Plants and Animals* (Eds P. J. Ucko and G. W. Dimbleby), Duckworth, London, 1970, 67–72.
15. (a) J. M. Renfrew, *Palaeoethnobotany: The Prehistoric Food Plants of the Near East and Europe*, Methuen, London, 1973, 125–36.
(b) D. Zohary and P. Spiegel-Roy, 'The beginnings of fruit growing in the Old World', *Science, N. Y.* **187**, 319–27 (1975).
16. (a) K. W. Butzer, *Environment and Archaeology*, 2nd ed., Methuen, London, 1972, 563, 585–95.
(b) J. D. Clark, 'The spread of food production in sub-Saharan Africa', *Journal of African History*, **3**, 211–28 (1962).
(c) J. D. Clark, 'The prehistoric origins of African culture', *Journal of African History*, **5**, 161–83 (1964).
(d) J. D. Clark, *Atlas of African Prehistory*, University of Chicago Press, Chicago and London, 1967.
17. R W. Bulliet, *The Camel and The Wheel*, Harvard University Press, Cambridge, Mass, 1975.
18. R. J. Braidwood and C. A. Reed, 'The achievement and early consequences of food-production: a consideration of the archaeological and natural historical evidence', *Cold Spring Harbor Symposia on Quantitative Biology*, **22**, 19–31 (1957).
19. F. R. Matson, 'Power and fuel resources in the ancient Near East', *Advmt. Sci., Lond.*, **23**, 146–153 (1966).
20. M. B. Rowton, 'The woodlands of ancient Asia', *Journal of Near Eastern Studies*, **26**, 261–277 (1967).
21. M. W. Mikesell, 'The deforestation of Mount Lebanon', *Geogrl. Rev.*, **58**, 1–28 (1969).
22. K. V. Flannery, 'Origins and ecological effect of early domestication in Iran and the Near East', in *The Domestication and Exploitation of Plants and Animals* (Eds P. J. Ucko and G. W. Dimbleby), Duckworth, London, 1970, 73–100.
23. (a) G. W. Dimbleby, 'Climate, soil and man', in *The Early History of Agriculture* (Eds J. Hutchinson, G. Clark, E. M. Jope and R. Riley), Oxford University Press, 1977, 197–208.
(b) V. B. Proudfoot, 'Man's occupance of the soil', in *Man and his Habitat. Essays Presented to Emyr Estyn Evans* (Eds R. H. Buchanan, E. Jones and D. McCourt), Routledge and Kegan Paul, London, 1971, 8–37.

24. K. V. Flannery, 'Origins and ecological effects of early domestication in Iran and the Near East', in *The Domestication and Exploitation of Plants and Animals* (Eds P. J. Ucko and G. W. Dimbleby), Duckworth, London, 1970, 73–100.
25. (a) D. R. Harris, 'Alternative pathways towards agriculture', in *Origins of Agriculture* (Ed C. A. Reed), Mouton, The Hague and Paris, 1977, 179–243.
(b) P. E. L. Smith, 'Land-use, settlement patterns and subsistence agriculture: a demographic perspective', in *Man, Settlement and Urbanism* (Eds P. J. Ucko, R. Tringham and G. W. Dimbleby), Duckworth, London, 1972, 409–25.
26. (a) R. M. Adams, 'Factors influencing the rise of civilisation in the alluvium illustrated by Mesopotamia', in *City Invincible* (Ed C. H. Kraeling and R. M. Adams), University of Chicago Press, 1960, 24–34.
(b) R. M. Adams, 'Patterns of Urbanism in early southern Mesopotamia', in *Man, Settlement and Urbanism* (Eds P. J. Ucko. R. Tringham and G. W. Dimbleby), Duckworth, London, 1972, 735–49.
27. (a) R. M. Adams and H. J. Nissen, *The Uruk Countryside. The Natural Setting of Urban Societies*, Chicago University Press, Chicago and London, 1976.
(b) K. W. Butzer, *Early Hydraulic Civilisation in Egypt: A Study in Cultural Ecology*, Chicago University Press, Chicago and London, 1976.
28. (a) K. V. Flannery, 'The origins of the village as a settlement type in Mesoamerica and the Near East: a comparative study', *Man, Settlement and Urbanism* (Ed P. J. Ucko, R. Tringham and G. W. Dimbleby), Duckworth, London, 1972, 23–53.
(b) W. Allan, 'The influence of ecology and agriculture on non-urban settlement', in *Man, Settlement and Urbanism* (Eds P. J. Ucko, R. Tringham and G. W. Dimbleby), Duckworth, London, 1972, 211–26.
29. (a) R. M. Adams, 'Agriculture and urban life in early southwestern Iran', *Science, N. Y.*, **136**, 109–122 (1962).
(b) H. E. Wright and G. A. Johnson, 'Population, exchange and early state formation in southwestern Iran', *American Anthropological Journal*, **77**, 267–89 (1977).
30. R. M. Adams, *Heartland of Cities*, Chicago University Press, Chicago and London 1981.
(a) J. M. Wagstaff, *The Evolution of Middle Eastern Landscapes: An Outline to A.D. 1800*, Croom Helm, London, 1985, 97–105.
31. (a) Abu-Lughod, *Cairo: 1001 Years of the City Victorious*, Princeton University Press, Princeton, 1972.
(b) C. H. Kraeling and R. M. Adams (Eds), *City Invincible. A Symposium on Urbanisation and Cultural Development in the Ancient Near East*, University of Chicago Press, 1960.
(c) J. H. G. Lebon, 'The site and modern development of Baghdād', *Bull. Soc. Géogr. Egypte*, **2**, 7–32 (1956).
(d) P. Marthelot, 'Bagdād; notes de géographie humaine', *Annls Géogr.*, **74**, 24–37 (1965).
32. (a) P. W. English, *City and Village in Iran. Settlement and Economy in the Kirman Basin*, University of Wisconsin Press, Madison, 1966, 87–110.
(b) G. E. von Grünebaum, 'The Muslim town and the Hellenistic town', *Scientia*, **90**, 364–70 (1955).
(c) P. Lampl, *Cities and Planning in the Ancient Near East*, Studio Vista, London, 1968.
33. (a) J. Despois, 'Development of land use in northern Africa', in *A History of Land Use in Arid Regions* (Ed L. D. Stamp), *Arid Zone Research*, **17**, UNESCO, 1961, 219–237.
(b) R. O. Whyte, 'Evolution of land use in south-western Asia', in *A History of Land Use in Arid Regions* (Ed L. D. Stamp), *Arid Zone Research*, **17**, UNESCO, 1961, 57–118.

34. (a) *Atlas des Centuriations romaines de Tunisie*, Institut Géographique National, Paris, 1954.
(b) J. Baradez, *Fossatum Africae*, Arts et Métiers Graphiques, Paris, 1949.
(c) A. Caillemer and R. Chevalier, 'Les centuriations romaines de l'Africa vetus', *Annales, Economies, Sociétés, Civilisations,* **9**, 433–460 (1954).
(d) A. Caillemer and R. Chevalier, 'Centuriations romaines de Tunisie', *Annales Economies, Sociétés, Civilisations,* **12**, 275–286 (1957).
(e) R. G. Goodwood, 'Farming in Roman Libya', *Geogr. Mag.*, **25**, 70–80 (1952).
(f) R. G. Goodwood, 'The mapping of Roman Libya', *Geogr. J.*, **118**, 142–152 (1952).
35. (a) W. C. Brice, *Southwest Asia*, University of London Press, 1966, 96–98.
(b) W. C. Brice and A. N. Balci, 'The history of forestry in Turkey', *Orman Fakültesi Dergisi, İstanbul Universitesi*, **5**, 19–42 (1955).
(c) R. T. Marchese, *The Lower Meander Flood Plain: A Regional Settlement Study*, British Archaeological Reports, International Series 292, Oxford, 1986.
(d) R. J. Russell, 'Alluvial morphology of Anatolian rivers', *Ann. Ass. Am. Geogr.*, **44**, 363–391 (1954).
36. A. H. M. Jones, *The Decline of the Ancient World*, Longmans, London, 1966, 304–310.
37. (a) M. B. Rowton, 'The woodlands of ancient Asia', *Journal of Near Eastern Studies*, **26**, 261–77 (1967).
(b) M. Lombard 'Les bois dans la méditerranée musulmane', *Annales, Economies, Sociétés, Civilisations*, **14**, 234–54 (1959).
38. A. M. Watson, *Agricultural Innovation in the Early Islamic World*, Cambridge University Press, 1983.
39. R. M. Adams, *Land Behind Baghdād*, University of Chicago Press, Chicago, 1965, 69–71.
40. (a) M. Benvenisti, *The Crusaders in the Holy Land*, Israel Universities Press, Jerusalem, 1970.
(b) U. Heyd, *Ottoman Documents on Palestine, 1552–1615*, Oxford University Press, London, 1960.
(c) W. Hütteroth, 'Schwankungen von Siedlungsdichte und Siedlungsgrenze im Palästina und Transjordanien seit dem 16. Jahrhundert', in *Deutscher Geographentag Kiel, 21–26 Juli, 1969*, 463–75.
(d) G. Le Strange, *Palestine Under the Muslims*, 650–1500, Palestine Exploration Fund, London, 1890.
41. (a) M. A. Hachicho, 'English travel books about the Arab Near East in the eighteenth century', *Die Welt der Islam*, **9**, 1–206 (1964).
(b) A. Hourani, 'The changing face of the Fertile Crescent in the eighteenth century', *Studia Islamica*, **8**, 89–122 (1957).
(c) H. Margalit, 'Some aspects of the cultural landscapes of Palestine during the first half of the nineteenth century', *Israel Explor. J.*, **13**, 208–23 (1964).
(d) R. Owen, *The Middle East in the World Economy, 1800–1914*, Methuen, London, 1981.
(e) M. C. F. Volney, *Voyage en Syrie et en Egypte pendant les années 1783, 1784 et 1785*, Paris, 1786.
42. (a) R. Owen, *The Middle East in the World Economy, 1800–1914*, Methuen, London, 1981.
(b) D. Quataert, 'Limited revolution: the impact of the Anatolian Railway on Turkish transportation and the provisioning of İstanbul, 1890–1908', *Business History Review*, **51**, 139–60 (1977).
(c) D. Quataert, 'The commercialisation of agriculture in Ottoman Turkey, 1800–1914', *International Journal of Turkish Studies*, **1**, 38–55 (1980).
43. (a) Z. Y. Hershlag, *Introduction to the Economic History of the Middle East*, Brill, Leiden, 1964.

(b) C. Issawi (Ed), *The Economic History of the Middle East, 1800–1914*, University of Chicago Press, 1966.
(c) D. S. Lander, *Bankers and Pashas. International Finance and Economic Imperialism in Egypt*, Heinemann, London, 1958.
(d) E. R. J. Owen, *Cotton and the Egyptian Economy, 1920–1914. A Study in Trade and Development*, Oxford University Press, London, 1969.

44. (a) O. Kurmuş, 'The Role of British Capital in the Development of Western Anatolia, 1850–1913' (unpublished PhD thesis, University of London, 1974)
(b) R. Owen, *The Middle East in the World Economy*, 1800–1914, Methuen, London, 1981.
45. (a) N. E. Lewis, 'Malaria, irrigation and soil erosion in central Syria', *Geogrl. Rev.*, **39**, 278–290 (1949).
(b) N. E. Lewis, 'The frontier of settlement in Syria, 1800–1950', *International Affairs*, **31**, 48–60 (1955).
46. V. Cuinet, *La Turquie d'Asie. Géographie administrative, statistique, descriptive et raisonnée de chaque province d'Asie Mineure*, **2**, Paris, 1892, 355–56.
47. M. B. Rowton, 'The woodlands of ancient Asia', *Journal of Near Eastern Studies*, **26**, 261–77 (1967).
48. (a) S. Amin, *L'économie du Maghreb*, **1**, *Colonisation et la Décolonisation*, Editions de Minuit, Paris, 1966.
(b) N. Barbour (Ed), *A Survey of North-West Africa*, 2nd ed, Oxford University Press, London, 1962.
(c) J. Despois, 'Development of land use in northern Africa', in *A History of Land Use in Arid Regions* (Ed L. D. Stamp), *Arid Zone Research* **17**, UNESCO, 1961, 219–37.
(d) M. M. Knight, 'Economic space for Europeans in French North Africa', *Economic Development and Cultural Change*, **1**, 360–75 (1952–1953).
49. (a) G.L. Fowler, 'Italian colonisation of Tripolitania', *Ann. Ass. Am. Geogr.*, **62**, 627–40 (1972).
(b) R. G. Hartley, 'Libya: economic development and demographic responses', in *Populations of the Middle East and North Africa* (Eds J. I. Clarke and W. B. Fisher), University of London Press, London, 1972, 316–18.
50. M. S. Hasan, 'Growth and structure of Iraq's population, 1867–1947', *Bull. Oxf. Univ. Inst. Statist.*, **20**, 339–52 (1958).
51. J. M. Wagstaff, *The Evolution of Middle Eastern Landscapes: An Outline to A. D. 1840*, Croom Helm, London, 1985, 258–60.
52. (a) P. Beaumont, 'Water resources and their management in the Middle East', in *Change and Development in the Middle East* (Eds J. I. Clarke and H. Bowen-Jones), Methuen, London, 1981, 40–72.
(b) C. G. Smith, 'Water resources and irrigation development in the Middle East', *Geography*, **55**, 407–25 (1970).

CHAPTER 4

Rural Land Use: Patterns and Systems

4.1 The importance of cultivation and pastoralism

Although agriculture and pastoralism are not the only types of land use in the Middle East, their role is of fundamental importance in the life of the region. Table 4.1 gives some of the indicators for the role of agriculture, which is more measureable than pastoralism and indeed essential to its support. Over the whole region, agriculture contributed directly 9 per cent of GDP about 1980, compared with 2 per cent in the case of the United Kingdom and 3 per cent in that of the United States. The variation across the region was considerable. A small contribution, for example, was made by agriculture in Israel, where manufacturing industry is particularly developed, and an even smaller one in Libya and Saudi Arabia, where the oil industry almost completely dominates the economy. A significantly greater than average contribution to GDP was made by agriculture in Egypt, where irrigated farming has reached a high level of development, and in the largely rainfed agricultural countries of Syria and Turkey.

Agriculture produced many of the basic raw materials for the region's manufacturing industry (Chapter 8). However, its direct contribution by value to non-petroleum exports from the region about 1982 was only 17 per cent, though the range was considerable. It was less than 1 per cent by value in the case of Bahrain, Kuwait, Libya, Qatar and Saudi Arabia, but well above the regional figure for Iran, Syria and Turkey. Egypt and the UAE were close to the regional figure, while Israel, Jordan, Lebanon and Yemen AR had significant agricultural exports (and virtually no petroleum or petroleum products). About 45 per cent of the region's economically active population reputedly worked in agriculture, but the figures are suspect because of the differing labour contribution of women and children. In Egypt, Saudi Arabia, Turkey and Yemen AR the regional proportion was exceeded. Elsewhere, though, the proportion of the labour force in agriculture was lower than the regional figure, considerably lower in the case of Iraq, Israel, Jordan and Lebanon. Taken together, the various indices reveal that productivity in agriculture is low. Whilst this is closely related to adverse physical conditions for farming over much of the region, it can also be attributed to the relative neglect of agricultural development by most of the region's governments over a long period and, in some cases, in violation of their published development plans.

TABLE 4.1
Importance of Agriculture

Country	Percentage contribution to GDP (year)	Percentage contribution to exports about 1982	Percentage economically active population in agriculture, 1980 estimates
Egypt	22 (1979)	19.2	45.7
Iran	17 (1980)	1.0 (1977)	36.4
Iraq	7 (1980)	1.0*	30.4
Israel	6 (1980)	10.6	6.2
Jordan	6 (1980)	17.4	10.2
Kuwait	na	0.3	1.9
Lebanon	7 (1977)	16.2 (1977)	14.3
Libya	2 (1980)	0.0	18.2
Qatar	1 (1980)	0.1 (1978)	na
Saudi Arabia	1 (1979)	0.1	48.4
Syria	21 (1980)	16.9 (1979)	32.3
Turkey	22 (1980)	36.5	58.4
UAE	1 (1980)	0.9*	na
Yemen AR	na	8.6	na
Yemen, PDR	na	na	41.1

* estimate
Sources: United Nations FAO *Production Yearbook, 1985*, Rome 1986, Table 2; *International Trade Statistics Yearbook 1983*, New York 1985; *Statistical Yearbook, 1982*, New York 1985.

4.2 Influence of physical conditions

Water availability is the most important constraint on rural land use in the region and the effects on cultivation are very marked. Except in southern Arabia, precipitation regimes are characterized by winter or spring maxima and a long summer drought (Chapter 2). This pattern largely determines the time available for crop growth, though low temperatures in winter and high evapotranspiration in summer restrict rapid growth to the period from February to May in the so-called Fertile Crescent and from April to July in the high central parts of Turkey and Iran. Wheat and barley are the most important crops in terms of the area devoted to them as well as their large part in the diet, and, as indigenous plants, they are well adapted to local conditions. So are the fodder crops clover, lucerne (alfalfa) and vetch, upon which draft animals – still important in the region – are dependent. The date at which the long drought is broken is crucial, for the arrival of the rain conditions the amount of land which can be sown in any one year. If the rains are early, for example, then more time is available for ploughing and a greater area can be sown. Unfortunately, the rains arrive at a time when the vegetation has been dried out, and they continue over the period of ploughing. Erosion is severe and valuable soil is lost every year, while the fields may become ravaged by gullying. Run-off is rapid, resulting not only in the loss of vital soil moisture but also in the frequent flooding of the better crop land in the valleys and plains. Spring rain is vital since it falls towards the end of the growth cycle, when timing and abundance greatly affect yields. If the rain is late, or light, yields will be low and famine may threaten; it is never very far away in some parts of the region. Yields vary quite markedly from year to year (Chapter 18) and, because of the basic importance of field crops to national economies, partly explain the curious interannual fluctuations in GDP characteristic of the region.

Successful long-term cereal growing without irrigation (*dry farming* or *rainfed farming*) depends upon the relationships between the minimum amount of precipitation required for growth and the variability of precipitation from year to year. The critical conditions are met only in parts of the region where mean annual precipitation exceeds 240 mm and relative interannual variabilities are less than 37 per cent (Figure 4.1).[1] Most of the region cannot support dry farming and permanent cultivation is dependent upon the availability of water for irrigation, and, as we have seen (Chapter 2), this is restricted.

Water availability affects livestock rearing, as well as cultivation. Although herding takes place within the same areas as rainfed farming, it is especially associated with the margins of the dry farming zones. Herbage here varies according to the seasonal and annual availability of precipitation. In consequence, herding has historically been fully or partially nomadic, while a run of unusually dry years in the past often assisted an extension of the nomadic domain deep into the core areas of dry farming and encouraged settling. The broad temporal and spatial movements of nomadic groups were adjusted to the observed and expected regularities in the precipitation, which determined the distribution of grazing and drinking water. Indirectly, precipitation also affects the herds themselves. The type of animals kept in particular areas is related to the amount of precipitation and drinking water to be expected there. Herd size is closely affected by the availability of these items, and thus by interannual variation in precipitation. Numbers increase in wet years and decrease in dry ones, creating a kind of dynamic equilibrium which is easily upset. For example, serious overgrazing is now widespread in the region as a result of concentrating larger numbers of animals in diminishing areas, a situation created, on the one hand, by the ploughing up of former grazing land, and, on the other hand, by increasing demand for animal products. The effects of drought in such a situation can be devastating.

If water availability exerts a dominant influence on the broad patterns of rural land use, cultivation and herding are also affected in detail by topography and soils. Extremely rocky areas may not be used at all. Steep, bare slopes can be grazed only by the agile goat, while slopes in excess of a few degrees cannot be irrigated. Slopes greater than 35 degrees cannot even be ploughed. Cultivation is possible in such circumstances only with the help of terraces, but in the past terracing was a feature of districts like Mount Lebanon, the Yemens and the Judaean hills where cultivable land was scarce but, for historical reasons, the population became particularly dense. The soils of the region have already been described (Chapter 1), but it is important to emphasize certain of their characteristics for agriculture. They are generally deficient in humus and nitrogen so that,

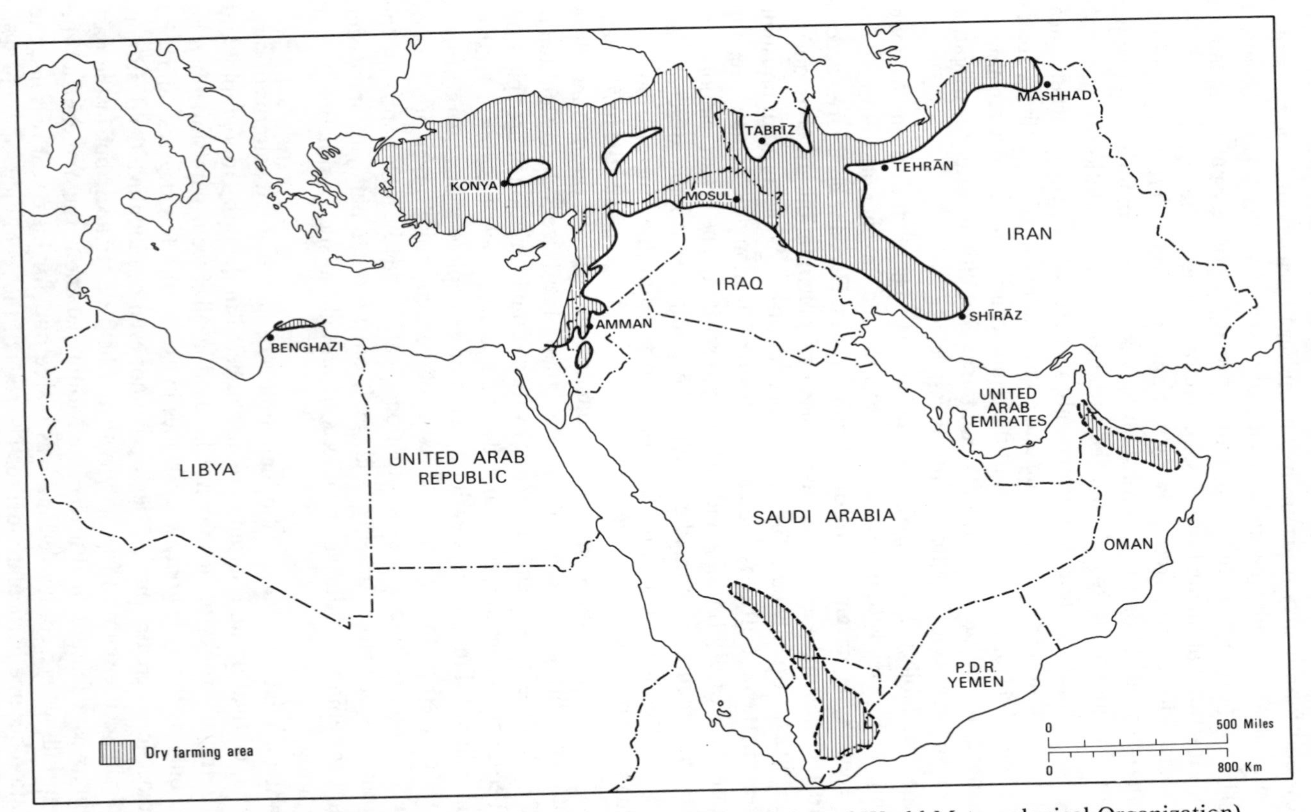

Figure 4.1 Theoretical limits of dry farming (reproduced by permission of World Meteorological Organization)

without the application of fertilizer, yields are low over a long period of time, though the ploughing up of old pasture has given high yields in areas like the Jezira of northeastern Syria, in the short term. In many districts, natural fertility has been reduced to a stable minimum, determined by the mineral composition of the soil, as the result of centuries, or even millenia, of almost continuous cultivation. Some soils are rich in bases and so become susceptible to salination, especially where irrigation water is carelessly applied and drainage is inefficient. Plant growth is inhibited in these conditions, and may even become impossible, as in parts of southern Iraq (Chapter 12). Partly as a result of the slow development of soils under the present climatic regimes of the region, and partly as a result of centuries of ill-treatment and mismanagement, soils away from the plains and valleys are shallow and stoney, and moisture is not retained for any length of time, thus imposing another limitation on the growing season. Water percolation is often retarded and runoff increased by the tendency to puddle at the surface, which results from both precipitation characteristics and the high clay content of most soils. Puddling is also unfavourable to the early growth of cereals, while ploughing is rendered difficult by the tendencies of clayey soil to bake hard during the summer and become heavy and sticky after rain. Altogether, the farmer's task is not an easy one.

4.3 Land use organization

4.3.1 Permanent settlements

Cultivation is generally organized from, and carried on around, permanent settlements, though some completely nomadic groups in Libya grow crops. The stereotype settlement is the shapeless agglomeration of flat-roofed, single-storey, rectangular houses. Streets are narrow and winding, whilst culs-de-sac are frequent. The inhabitants are grouped spatially by kinship ties and sometimes by religious affiliation, but the entire population is often divided into at least two different factions between which the normal hostility still occasionally flares into violence of such magnitude that the stability of the community seems something of a puzzle.[2]

In fact, the basic monotony of settlement type is broken in a variety of ways. Variation, for example, is quite marked in house type (Figure 4.2). The simple single-storey house is by no means universal, and two storeys are characteristic of some regions, while tower-houses are particularly characteristic of North Yemen and Ḥaḍhramawt. The colour and texture of local stones and muds, the display or not of whitewash and external decoration, and the use of timber, all provide subtle variations from place to place. But the most obvious variations are in roof type. Common is the heavy flat roof, built of layers of insulating materials resting on heavy beams and topped with a layer of salted mud, but other forms are also found. Pitched roofs are characteristic of districts where precipitation is heavy, particularly the Pontic Mountains and the Caspian lowlands. Roofs consisting of

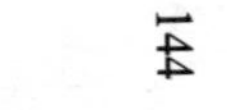

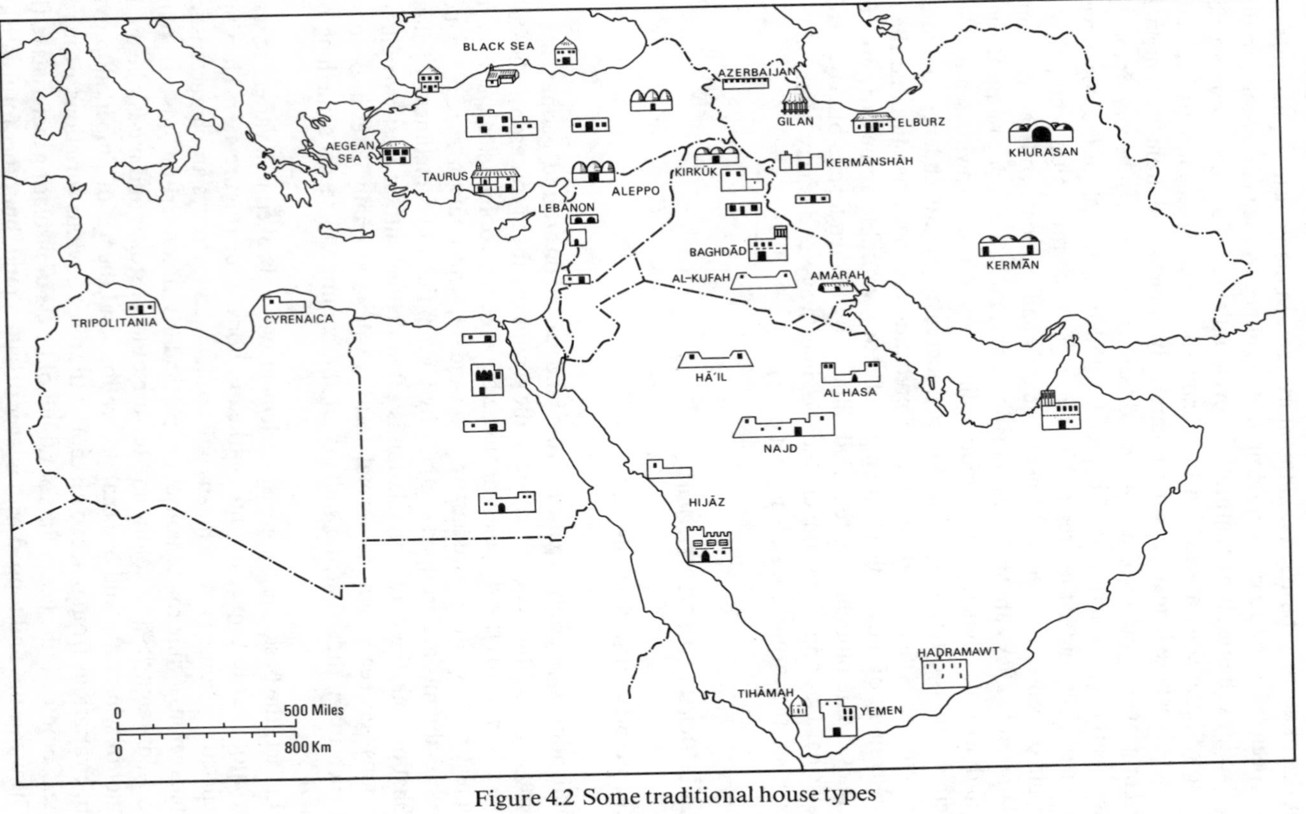

Figure 4.2 Some traditional house types

several corbel-built cones are typical of the largely treeless steppe of Syria near Aleppo, but they are also found in adjacent parts of southern Turkey and occasionally in Upper Egypt. Conical roofs are also characteristic of the round huts of the coastal plain in the 'Asir and Yemen AR districts of southwestern Arabia, whilst barrel-vaulted roofs are a marked feature of traditional reed architecture in the marshes of southern Iraq. Almost everywhere concrete and brick are now replacing traditional local materials and village houses are becoming more standardized in appearance, as well as more comfortable to inhabit.

Settlement form displays a basic dichotomy between the mountains and the lowland. Compact agglomerations of 100 to 2,000 houses (perhaps 400 to 8,000 people) are characteristic of lowland areas (Figure 4.3). Most are unfortified, but defensive walls were found in districts, like central Arabia (Najd), Iran and the Tigris–Euphrates lowland, which were particularly exposed to small-scale

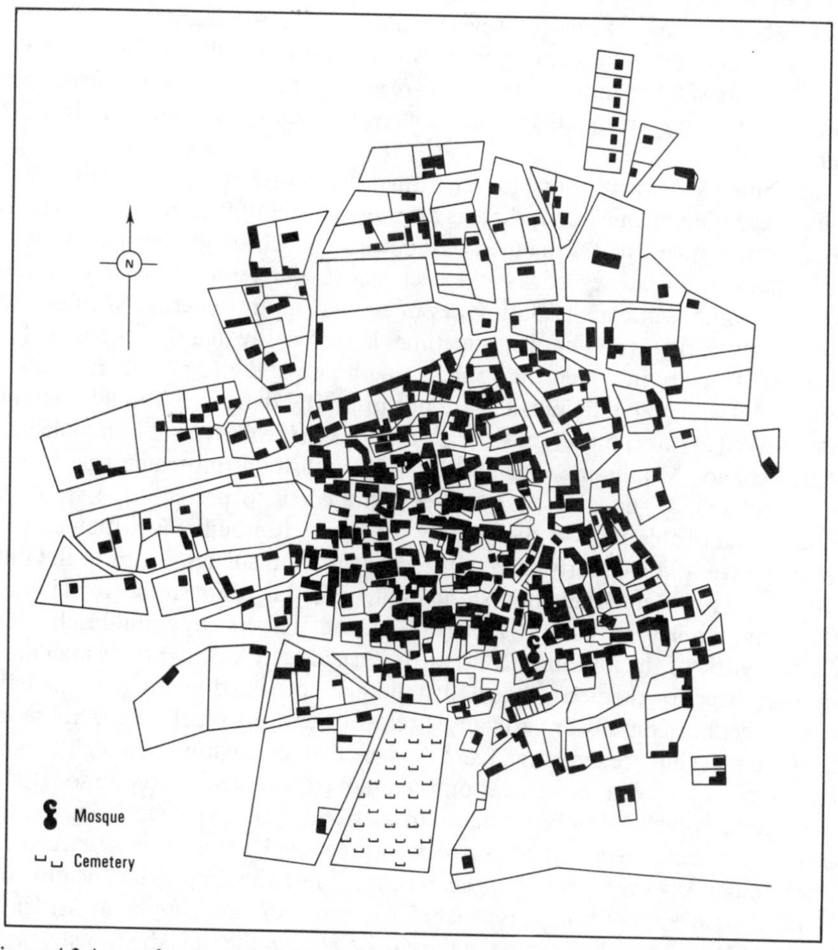

Figure 4.3 An agglomerated plains settlement: Karagedik in central Turkey (reproduced by permission of Wolf Hütteroth, 1971)

mauruding. In the open plains, settlements appear to be scattered almost at random, but they frequently occupy a mound (Arabic *tell*; Turkish *hüyük*) formed from the erosion and rebuilding of generations of mudbrick houses, though it must be pointed out that continuous occupation of the same site over several millenia is likely to be exceptional rather than the rule. In central Turkey, settlements are generally situated below the general level of the undulating plateau, so that they are almost invisible from a distance, but they are located above the floors of valleys and depressions where floods normally appear in the winter. The same situations are preferred in the Beq'a valley of Lebanon. In Iran, however, settlements are generally found around the outlet of a *qanāt* or along the side of a water course. Linear settlements also stretch along the levees of the Tigris and Euphrates in Iraq and beside the Nile in Egypt, though desert-edge settlements are also found in Egypt, along with polynuclear forms which have developed from the growing together of once distinct settlements as a result of population increase.[3] Practically everywhere the mudbrick houses are built in the corner, or at one side, of a courtyard in which many domestic activities are carried out in warm weather and around which are gathered stores and byres. Settlements are built, in effect, from units of such courtyard houses, integrated by kinship relationships.

Mountain settlements have looser forms, though two basic types may be distinguished. On the one hand, there is the loose cluster of perhaps 10 or 20 houses and, on the other, the fragmented settlement in which smaller groups of houses are separated by ravines, rocks and steep slopes, but yet lie closer to each other than any other settlement. Both types of settlement are generally sited on steep slopes, spurs, ridges and knolls, sometimes in such a way that the settlement has a stepped structure in which the roof of one house is the terrace in front of the house above. These difficult sites are not primarily defensive, though they may have served in this capacity on several occasions in the past, as at Maaloula in the Anti-Lebanon Mountains which successfully withstood full-scale sieges in 1850, 1860 and 1925.[4] The sites were chosen more often to preserve the scarce but valuable cultivable land in the vicinity. In the high mountains settlements frequently face in a southerly direction to benefit from solar warmth in the bitter winters. Houses are frequently stone-built, with flat roofs supported either by transverse arches or vaulting, instead of the timbers and mudbrick pillars characteristic of the lowland. Courtyards are generally lacking; there may not be enough space for them. Storage and stalling are provided by dividing the house itself either horizontally or vertically. In the first case people share the house with their animals, but are separated from them by their occupation of a slightly raised platform. Where vertical separation is employed, people are accommodated on the second floor, while stores and animals are kept below, often in deep cellars and even artificial caves, where the rock is soft enough to permit easy excavation.

Although Arab villages in Israel can be fitted into the simple classification outlined, Jewish agriculture is associated with very different types of settlement (Chapter 16). The earliest type, the *moshavot*, started the trend by transferring to Palestine a basically central-European settlement form, in which houses with

Contrasting village types: Pitched roofs made of thatch, Caspian Sea Lowlands, Iran (top); flat-roofed stone built houses, northern highlands Jordan (bottom) (Photographs: Peter Beaumont)

pitched roofs were arranged on either side of a village street (Figure 16.1). *Kibbutzim* were a later introduction. They are communal settlements of 300 to 1700 people, often originally located in regions of political insecurity or exceptional environmental difficulty. The central focus is a group of communal buildings, characterized by the separation of farm functions from the living quarters. *Moshavim* (villages of cooperative smallholders) are somewhat smaller, with populations of 200 to 450 people, but are rigorously planned. The earliest *moshavim* of the 1920s were laid out around an elliptical street with services located at the centre (Figure 16.1). This wasted valuable agricultural land and produced awkwardly shaped fields. To meet these problems, *moshavim* built after 1948 were generally laid out along a single street, or on 'T' or 'L' shaped patterns. There was also an increasing tendency to make each *moshav* a member of a completely planned pattern of settlement designed to facilitate the provision of services in a hierarchy of the type described by Christaller and Lösch. Thus, groups of up to six *moshavim* are generally served by a single rural service centre accommodating only the families of non-agriculturalists to maintain a wide range of social and economic services.

4.3.2 Land tenure

The permanent settlement is the organizational centre of the farm, but its composition, layout and pattern of working are controlled by the way in which the land itself is held. Land tenure is an important organizational structure underlying present patterns of land use.

Ultimate ownership of most of the arable land in Muslim countries is vested in the state, and state-owned land (*miri* in the Arab countries and Turkey; *khāliseh* in Iran) has been carefully distinguished from privately owned property (*mulk* or *milk*). The classic distinction was made originally for tax and administrative purposes, but it is of little practical significance today, since the legal tenants of the state pay no rent and their titles are inherited, mortgaged and sold, provided only that the land is not allowed to go out of cultivation for more than five years, when it can be claimed by anyone who will undertake to work it. Of more practical importance today are the existence of four other characteristic features of land tenure: mortmain land (Arabic *wakf*; Turkish *evakf*), large estates, share-cropping and the fragmentation of holdings.

Wakf is property, particularly land, which has been dedicated to God and its income allocated to some religious or charitable purpose, such as the maintenance of a particular mosque or bathhouse, or to the support of the poor of a particular locality.[5] However, it became customary for the income to be used for the maintenance of the dedicator's family over several generations before it actually devolved upon the ultimate beneficiary. This was one way in which family property could be maintained intact, since it could not be alienated by the state nor foreclosed upon in the event of debt. The management of *wakf* was put into the hands of an overseer, who was often the head of the grantor's family in successive generations, or a respected individual, or, where other managers lapsed, the ruler. Management was not very efficient, so that much *wakf* land was unproductive and the income often badly applied. The inalienable nature of *wakf* meant that consolidation and improvement schemes were often thwarted.

Much has been made of these facts by commentators, especially since it has been estimated that in the closing years of the Ottoman Empire about three-quarters of all the arable land was in *wakf*.[6] The situation has changed considerably since then. Independent governments have attempted to regulate *wakf* by establishing special ministries or agencies, and have reduced its extent bu confiscation (illegal under the Sharīah), and reallocation. *Wakf* has not been totally abolished, but even by 1950, on the eve of the great land reform movements in the region, it had been reduced to about 7.0 per cent of all registered land in Egypt, 1.2 per cent in Iraq and 0.3 per cent in Jordan.[7]

On the eve of the land reforms of the 1950s and 1960s, the ownership of land was far from equal (Table 4.2). Very large estates, sometimes consisting of the whole arable area of several villages, were found even as late as the 1950s, particularly in the dry farming areas of Syria, the irrigated districts of southern Iraq and the province of Kermān in Iran. By contrast, smallholdings were especially characteristic of two rather different subregions – mountainous terrain, like Lebanon, and the most densely populated provinces of Egypt. The situation has changed since the 1950s with the spread of effective land reform measures, but not drastically.

Development of large estates varied according to the different histories of land holding in particular subregions, but a number of common origins may be noted.

TABLE 4.2
Figures illustrative of Traditional Land Ownership

	Size of holding (ha)	No. of holdings ('000)	Percentage of total holdings	Percentage of total area
EGYPT 1952	<2	2642	94.3	35.4
	2–4	79	2.8	8.8
	4–8	47	1.7	10.7
	8–21	22	0.8	10.9
	21–42	6	0.2	7.3
	>42	5	0.2	17.0
LEBANON 1961	<2	92	35.0	4.0
	2–4	11	12.0	15.0
	4–20	16	13.0	38.0
	>20	2	1.0	37.0
TURKEY 1963	<5	2132	68.8	24.8
	2–20	853	27.5	41.7
	20–100	110	3.6	23.4
	>100	4	0.1	10.1

Source: Central Organisation for General Mobilisation and Statistics, *Annual Statistical Abstract 1952–1966*, Cairo, 1967, pp. 50–54; Ministry of Agriculture, *Census of Agriculture, 1961*, Beirūt, 1965. B. Kayser, 'Tendances de l'économie Turque', *Infoprmation Géographique*, **36**, 26 (1972).

Purchase was often the way in which estates were built up, particularly in recent times. Another very frequent origin, but one of greater age, was the granting of state land by the ruler (sultan, khedive or shah) to an individual who was then supported by the rents, dues and labour services of the people working the land. This arrangement is often loosely referred to as *feudal*, but it differed in a number of important respects from the classic feudalism developed in early medieval Europe. Although fiefs were granted as a way of maintaining troops (as late as the early twentieth century in Iran), the land grant in the Middle East during Islamic times was essentially a device for administering territory and raising taxes. It was not characterized by the bonds of personal loyalty and contract normal in Europe, and tokens of overlordship were few or non-existent. Moreover, the grant could not strictly be inherited, though this often happened, and could be revoked or resigned at any moment. Purchase also became characteristic. Other ways in which a large estate could be created was by the reclamation of swamp and marsh, as in the Nile delta during the second half of the nineteenth century, or by the seizure and subsequent colonization of any previously unoccupied land lying at a distance from an inhabited settlement (*mevat*),[8] a device which was particularly common during the expansion of the cultivation frontier in Syria (Chapter 13). In these cases, cultivation provided title to the land. Some large estates developed from the registration of land under the provisions of the Ottoman Land Code (1858) by farmers in the name of a few important persons. To some extent this was the direct result of Article Eight of the Code, which prevented a community from registering its collectively owned land (*musha*') as such, but recognized only registration by individuals.[9] Under a tribal system, it was natural for the farmers to register their collective possession in the name of their chief, who was simply *primus inter pares*. Many chiefs, however, came to abuse their position of trust and, with legal right and government support, claimed the land as their own and forced the tribesmen to work it for them as tenants. At the same time, the Code allowed unscrupulous notables to ensure, by bribery and coercion, that land was registered in their name.[10] Rather similar to this was estate building by the seizure of properties to pay off debts contracted with urban merchants or wealthy farmers (the *ağas*). Most farmers could not be anything else but debtors, given their need for ready cash for such things as the brideprice, dowry, and to pay for the rituals which mark the life crises, to meet the high rates of interest charged on loans, and to counter the vagaries of the weather. The situation was exacerbated by the weak bargaining position of sharecroppers. Credit and loans from government agencies have helped to relieve the situation, but it still exists. The small farmer is often refused loans and must turn elsewhere, even if it means eventual foreclosure on his property. In some parts of the region it is even possible to speak of a new wave of estate building.

Whatever their ultimate origins, the large estates of the region have generally been run purely as tenancy units. Outside the Nile delta, advantage was rarely taken of the possibilities for direct exploitation or of the economies of scale offered by such large units. Land was let out as small, scattered plots in return for a share of the harvest. The arrangement is not confined to large estates, but is used commonly in the region whenever a piece of property is let, while the principle of shares is widespread and found amongst seamen as well as farmers. The proportion of the harvest taken by the lessor depends upon the relative provision of the other factors of production apart from land (labour, tools, seed, fertilizer, draught animals, and in irrigated areas, water), the type of crop grown, and the probability of a successful harvest. Thus, in Syria, the landlord's share was higher in the core areas of dry farming than it was towards the margins of the desert, while a tree crop normally gives the tenant more rights than cereal monoculture. In Egypt, provision of labour only on the part of the tenant entitled the landlord to four-fifths or three-quarters of the harvest.

Sharecropping has the advantage that the landlord's share is related to the actual state of the harvest, and is not a fixed payment. On the larger estates, it also means that the tenant does not have to find the capital to supply seed, tools or even draught animals. The landlord frequently acts as an intermediary between his tenants and the government, protecting and helping them. As Weulersse observed,[11] the relationship of the estate owner to his tenant was normally that of the *patron* and *client*, rather than the *proprietor* and *métayer*. Unfortunately, like all leasing systems, sharecropping has been abused and exploited, particularly by the larger estate owners. The lessor's share is frequently so large that the lessee is left with barely enough to support himself and his family. Accordingly, many tenants lack capital to improve the land or increase yields, and are forced to borrow to meet even bare subsistence. The system is thus self-perpetuating. Contracts are not normally written down, with the result that they cannot be enforced at law, but may easily be exploited by the landlord. They are renewed every year, generally on terms unfavourable to the tenant, who must have land to survive and who is already in debt to the *ağa* for previous loans. Although one family might have farmed the same land for generations, annual contracts mean that there is little incentive to improve the land, even if the capital could be raised. In the past, neglect was reinforced in some districts by a tradition of periodically redistributing the holdings, sometimes by lot. The larger landlords inevitably gained control over the pattern of cropping in village territories, and are able to direct the organization of farm work, though power is usually exercised through an agent, since the more important *ağas* live in the towns and cities of the region.

Although landlords were content to draw their rents and use their tenants to support any political ambitions they might hold, they were in a position to embark upon modern commercial farming and carry out innovation when the right combination of incentives developed. The opportunity was seized in the Nile delta of Egypt following the early attempts to regulate the seasonal flow of the river, but little happened elsewhere until the tractor made its appearance; in many districts this was after the Second World War. In the Çukorova of southeastern Turkey, for example, the effect of introducing tractors was to curtail, or completely end, the sharecropping system, since manual labour is no longer needed in large quantities for large-scale cultivation. In consequence, farmers could no longer patch together a subsistence holding from their personal properties and the land they were able to lease. They are forced either to work as wage labourers, often on a seasonal basis, when they are in competition with cheaper migrant labour from mountain villages, or to emigrate to the industrial towns of the region or to Ankara and İstanbul.[12] The situation elsewhere has not been so well documented, but throughout the region the impression gained is that machines seem to be ending the sharecropping system. In some districts, it is being replaced by money rents and written leases, but the general effect is still to assist in forcing people off the land. Other factors encouraging people to leave the land include prices for agricultural products which are deliberately kept low by governments more orientated towards industry and the towns than to the land; the oppression of sharecropping (where it still survives); and the uneconomic size of farm holdings which has resulted from a combination of lack of surplus land to reclaim (at least cheaply), rapid population increase and the custom of equal division of property between at least the male heirs. Principles of equity combine with accidented terrain and local variations in soil and water to scatter holdings in irregular plots over village territories, thus creating a mosaic of tiny fields (Figure 4.4). Holdings may be collectively worked by members of the same family, but after several generations the degrees of kinship widen to such an extent that collectivization will no longer operate. Individual properties and shares of their joint produce become so small as to be insufficient for the support of individual families. Land may be left fallow in these circumstances, sold to some more fortunate neighbours or enter the pool of sharecropped land. Efficient use of time in modern terms is often difficult where journeys between plots become long, while consolidation and mechanization may be impossible to introduce effectively. Frustration, as well as poor living standards, is often a real impulse to migration.

Israel and the 'Occupied Territories' provide contrasts with the rest of the region. In the territories occupied by Israel since 1973, lack of written title has often provided the opportunity for evicting Arab proprietors and tenants and replacing them with Israelis. Within Israel itself, 78 per cent of all the land was state owned in 1957 and a further18.0 per cent was in the hands of the Jewish National Fund. Land was rented to individual farmers or groups of farmers for nominal sums and on hereditable leases.[13] Around *kibbutzim* cultivated land is worked in large, block fields which are ideal for the efficient development of machines. Strips or long wedges in individual ownerships are more characteristic of *moshavim*, but holdings cannot be subdivided or enlarged. The situation is different around the earliest Jewish *moshavot* on the citrus-growing coastal plain. Much land here is still in private ownership and, because of the way it was acquired and divided, it is not as regularly arranged as around the other types of Jewish settlement. The situation is more akin to that found in neighbouring Muslim countries.

4.3.3 Land reform

In the Muslim countries, the difficulties engendered by farm fragmentation, the evils of sharecropping, the uneven distribution of land and the social, economic

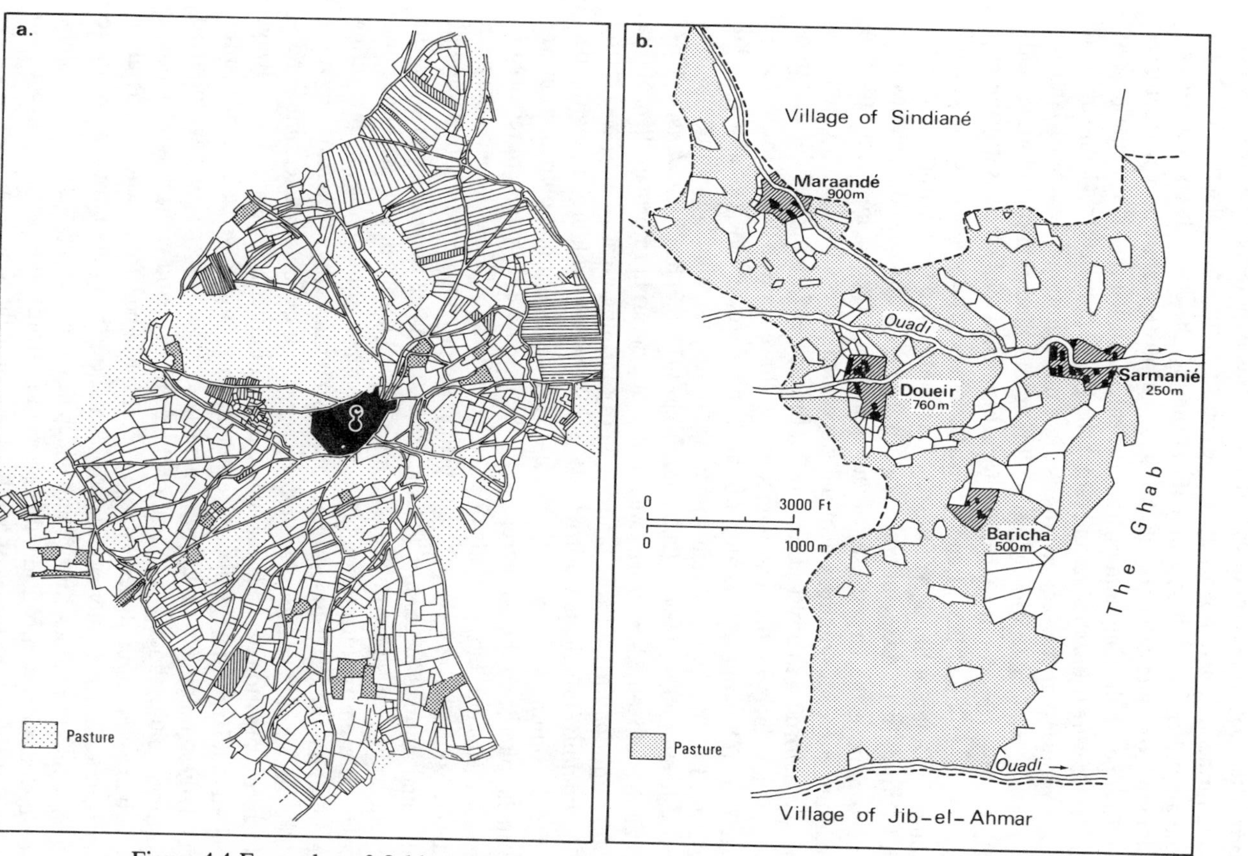

Figure 4.4 Examples of field patterns:
(a) Dry farming, plains-type: Runkus, central Turkey (after Hütteroth, 1971)
(b) Dry farming, mountain-type: Doueir el Akrad, western Syria (after Weulersse, 1946)

and political dominance of a comparatively small group of large estate owners all produced pressures for land reform. Before the main wave of effective land reform after 1950, most countries had already made sporadic allocations of state land to farmers, though not always to the benefit of the smallholders or of the increasing numbers of landless people. For example, the distribution of some state land by the Republic of Turkey in the 1920s and 1930s mainly benefited refugees from Europe. In the Kingdom of Iraq, the chief beneficiaries of a similar process in the 1930s were tribal sheikhs and established large proprietors. Confiscation of foreign property in the region provided a small additional source of land for distribution, while alienation of *wakf* allowed the allocation of still more land.

Despite these developments, land reform began in a really serious way in Egypt after the revolution of 1952. Land reform has not been attempted everywhere, for instance in Oman or Saudi Arabia, and was confined to the territory of the East Ghor Project in Jordan. But where reform has been introduced, the aims have been to remove the largest estates and redistribute their component lands. A maximum size of a holding was usually specified. This varied from country to country, according to physical conditions, the type of farming and cultural traditions. One village was the maximum holding allowed in Iran during the first stage of reform following the passing of the Land Reform Law in January, 1962 (Chapter 18). In Egypt, the maximum size of an individual's holding was progressively reduced from 84 ha (200 *feddans*) in 1952 to 21 ha (50 *feddans*) in 1969, but all the cultivated land is irrigated. Following agrarian reform in 1958, 500 ha (2000 Iraqi *donums*) was the limit in the dry farming areas of Iraq and 250 ha (1000 *donums*) in irrigated districts. In every country, holdings above the permitted maximum were seized by the state for distribution. Compensation was normally offered to the landlords, while the new state tenants were required to pay for their land, but generally at prices considerably below those on the open market. Cooperatives were normally planned to provide farmers with the credit, seeds and animals necessary for successful farming, as well as to offer the direction and assistance traditionally given by the landlord. Attempts were also made to prevent fragmentation of the new holdings, for example, by preventing the division of holdings of less than about two hectares (five *feddans*) in Egypt.[14]

The success of the various land reform schemes is difficult to determine, especially since evaluation depends very much upon the criteria chosen. It is not always realized that large estates still survive and that the reforms were aimed only at the very largest holdings. In Egypt, where land reform seems to have been so thorough-going, about 17 per cent of the cultivated land was actually involved, though in Syria about 40 per cent of the arable will be effected when confiscation and reallocation is finally completed.[15] Landless labourers have tended not to benefit from the reallocation of land, which has often gone to the smaller tenants. Non-cooperation and considerable opposition from landowners, together with uncertainty about the future of landholdings, brought considerable disruption to Iraqi's agriculture so that output of wheat, barley and rice, for example, fell by 25 per cent over the first three years following confiscation (Chapter 12). This was an important factor in the sudden flood of emigration from the marsh provinces to

Baghdād. The attempt to establish cooperatives is proving successful in Egypt, and has resulted in something like collective farming on the Soviet model,[16] but in Iraq and Syria, where Egyptian precedent was closely followed, the failures were considerable, partly because of the lack of trained staff but partly also because of different administrative traditions from those of the Nile valley. Credit is neither adequate nor easy to obtain, so that the old problem of debt remains, and the moneylender can still build up a moderate-sized estate by foreclosure. Attempts to curtail land fragmentation were confined to reallocated land, while share-cropping continues as the normal means of leasing land, though in Egypt money rents are now well established. Agricultural output and market responsiveness may have increased as a result of land reform, particularly in Egypt,[17] but the growth of productivity has been modest and there even seems to have been a decline in Iran and Iraq.[18] On the more positive side, though, the various land reforms have improved the agrarian situation to some extent. The rate of emigration from the countryside may have been moderated below what it might otherwise have been. The power of the landlords has been curtailed, although not everywhere broken. Tenure is now more secure than before, and rents may be lower in real terms.

4.3.4 Water rights[19]

Land is not the only aspect of cultivation which has required organization. The ownership and distribution of water for irrigation is crucial, especially beyond the limits of successful dry farming and if any summer cropping is to be practicable. Indeed, so scarce and so vital is water that access has customarily given title to land in some parts of the region, and in villages near Eşfahān the size of land holdings is reckoned in shares of water, as measured on a time basis. In Israel water provision is highly subsidized by the state and elaborate means have been developed to secure and distribute it, but the overall shortage of this vital commodity has established a system of allocation which has been said to be similar to that which would exist if an economic land rent prevailed.[20] Ownership, access to and use of water in the Muslim countries of the region are governed by customary and codified law, often based on the Qur'ānic teaching in favour of sharing and equity.[21] Practice varies considerably from country to country and district to district, depending upon hydraulic conditions. With the exception of Egypt, where a single system prevails and there is a long tradition of government control, the irrigation systems are fragmentary and independent of each other. Generally speaking, the scarcer the water is, the more complicated are the regulations. However, two broad methods of distribution have been observed. Water is frequently allocated on a volume basis, and the recipient takes a set share of the flow from a continual supply, as in the oases of the Fezzan in Libya. A time allocation is used more widely in the region. It provides for the entire supply from a particular source to be available on a rotational basis, though the actual method of calculating time, and hence the shares, varies quite considerably. Attempts have been made to standardize practice within national boundaries, as well as to protect both users and owners, but reforms similar to those carried through for land holding are probably impossible to implement. Modifications, rather than radical changes, are characteristic also of the way in which farming itself is carried on.

4.3.5 'Peasant economy'

Farming in the Middle East may be represented as a form of *peasant economy*.[22] Most of the region's farmers own or lease only small holdings. They and their ancestors have been exploited and oppressed, contributing to the high cultural developments of the region only through their rents and their labour. The land, however held, is worked by the family as an economic unit under the direction of its oldest male. Historically, paid labour was employed only in exceptional circumstances (in picking cotton or tea, for example), but it has become more common in recent times. The basis of subsistence is the family holding, and the product available to the farmer after the deduction of rent and other payments is largely consumed directly. The whole enterprise is more concerned with securing the subsistence of the family group than with making a profit in a capitalist sense, though the need for certain basic commodities (salt, iron), and the expenses of the rituals marking life stages, such as circumcision and marriage, require periodic sales.

In the past, a balance was maintained traditionally between subsistence needs and a distaste for manual labour, a distaste probably felt more keenly where a periodic reallocation of holdings was normal, and hence little opportunity to develop a sense of attachment to the land. The family would make an effort to produce more for enlarged consumption or investment only if the drudgery of labour was not pushed beyond a critical point where the increased output was outweighed by the irksomeness of the extra work. This balance occurred at different points according to the size of the family and the ratio of its working and non-working members (very young children, old adults). Most members of the family naturally worked on the land, making an increased labour contribution as they grew from infancy to maturity. The labour inputs would then stabilize for a while before declining as the mature adults became older and less capable of manual labour. The amount of labour available controlled the amount of land which could be cultivated. Additional land might have been secured by bringing under the plough more of the family holding, if they had a surplus, or by reclamation, where waste land still survived, or by acquiring a larger holding, where communal systems of land tenure continued and periodic redistribution was normal, or by entering into sharecropping contracts. Thus, the optimum amount of land which a family could work would vary through its biological cycle, as would the pivot between need satisfaction and drudgery of labour.

Several of the conditions under which *peasant economy* flourished have changed over the last twenty years. Communal tenure has all but vanished. Popu-

Pastoralism remains important in spite of the decline in nomadism; sheep grazing among date palms near Tripoli, Libya
(An Esso photograph)

lation growth and mechanization have meant that little surplus land is available for family-scale reclamation. Sharecropping is in decline as the needs for labour decline with the advance of mechanization. Labour shortages are actually acute in some districts. People have left the land in large numbers and the proportion of the active population engaged in farming has fallen slightly. Yet aspects of *peasant economy* linger, perhaps more in Iran and Turkey than elsewhere in the region. *Peasant economy* has different objectives from those of capitalist farming. It is hardly surprising, therefore, that it reacts differently to certain market forces, or that in some places and at certain times the 'peasant' has seemed reluctant to adopt new crops and new techniques, whilst in other districts and in other times he is eager to modernize. What has often been interpreted in the past as resistance based on ignorance and conservatism may be regarded in this light as perfectly rational responses to innovations which threaten a satisfactory life-style by invoking goals which may be little appreciated or which can be purchased at too high a social price. Change has come with the acceptance of different sets of social objectives, mediated through cooperatives, closer contact with the towns and greater exposure to the mass media. Considerable changes have taken place, though they have been less far reaching than those experienced in the pastoral sector of the economy, the organization of which is outlined below.

4.3.6 Pastoralism[23]

Although permanently settled communities of cultivators keep draught and herd animals, animal husbandry and cultivation are not normally integrated into a mixed farming system similar to that found in northwestern Europe. Until recently, mixed farming with fodder grown to supply stall-fed cattle was developed only in Israel. Everywhere else herd animals were not normally fed on fodder crops (which were saved for winter feed and for draught animals) and were kept away from the cultivated areas, as far as possible. Spatial separation of the two activities still continues, as does a certain amount of traditional group specialization.

Animal husbandry in the Middle East is organized from permanent bases where food and water can be obtained in the difficult season, whether that is winter, as in the mountains with their frosts and snow, or summer, with its extreme heat and drought, as on the edges of the Saharan and Arabian deserts. Until very recently, some herders were completely mobile after they had left their base areas. In following the available grazing they produced spatial patterns which will be described below. The basic grazing unit was usually the *tent* (of perhaps five people), each moving with a group of five or six others, but in no particularly stable or permanent association, only for mutual company and help. The individual movements of a *tent* in any year were conditioned by the state of the grazing and the availabilty of drinking water; they were closely affected by the spatial and temporal pattern of precipitation. Over a period of years, though, they tended to follow traditional routes which themselves led through tribal

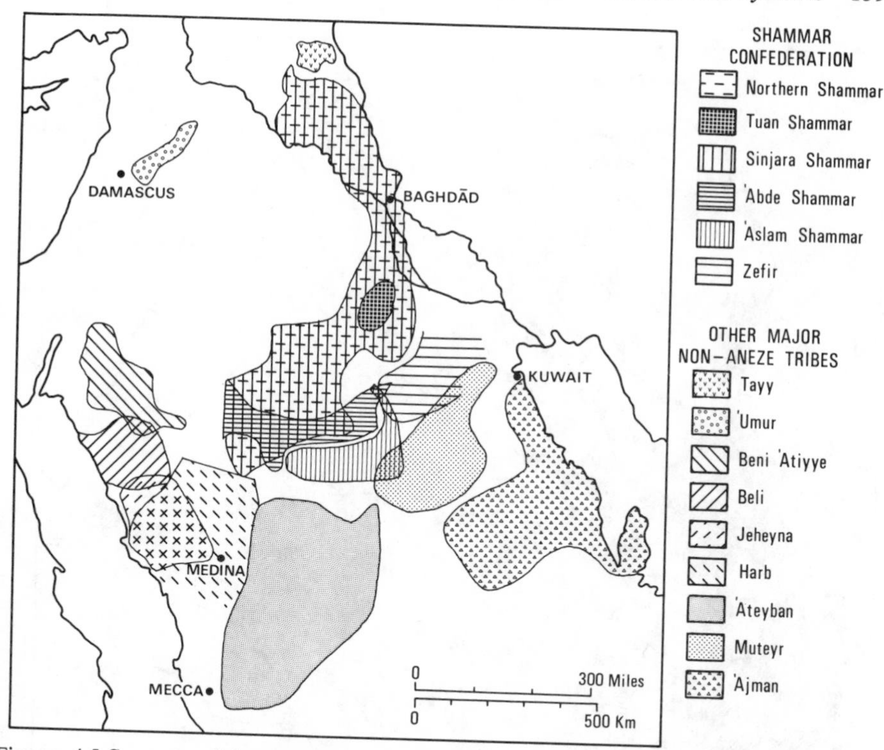

Figure 4.5 Some traditional tribal territories in northern Arabia (reproduced by permission of the *Geographical Review*)

areas. These are territories with well recognized if flexible limits, where a tribe has traditional rights at certain times of the year (Figure 4.5). Only in dire emergencies were the boundaries crossed, and then the pattern and extent of the trespass depended upon traditional agreements and alliances.

Animals were kept for sale, or for the sale of their products, so that the staple foods (dates and cereals) and other necessities could be bought. Supplies were also obtained by ownership of, or control over, cultivated land, while cash was often secured as 'protection money' from settled communities and caravans. Large herds meant prosperity and prestige, so that formalized stealing and raiding were part of the traditional way of life.

Other groups of herders have been more restricted in their movements. They have moved from bases in their own permanent, cultivating villages to seasonal grazing land at a distance, but on which often stood a second settlement. The *shieling*, to use a Scottish term, was generally less substantial than the permanent settlement, and was used by only a section of the group for part of the year, but it provided a focus around which the herds grazed on traditional pastures according to their known advantages for different kinds of animals[24] and a place where liquid milk was made into cheese and yoghurt. In addition, a few crops might be grown. As population pressure grew, the shielings were frequently converted into

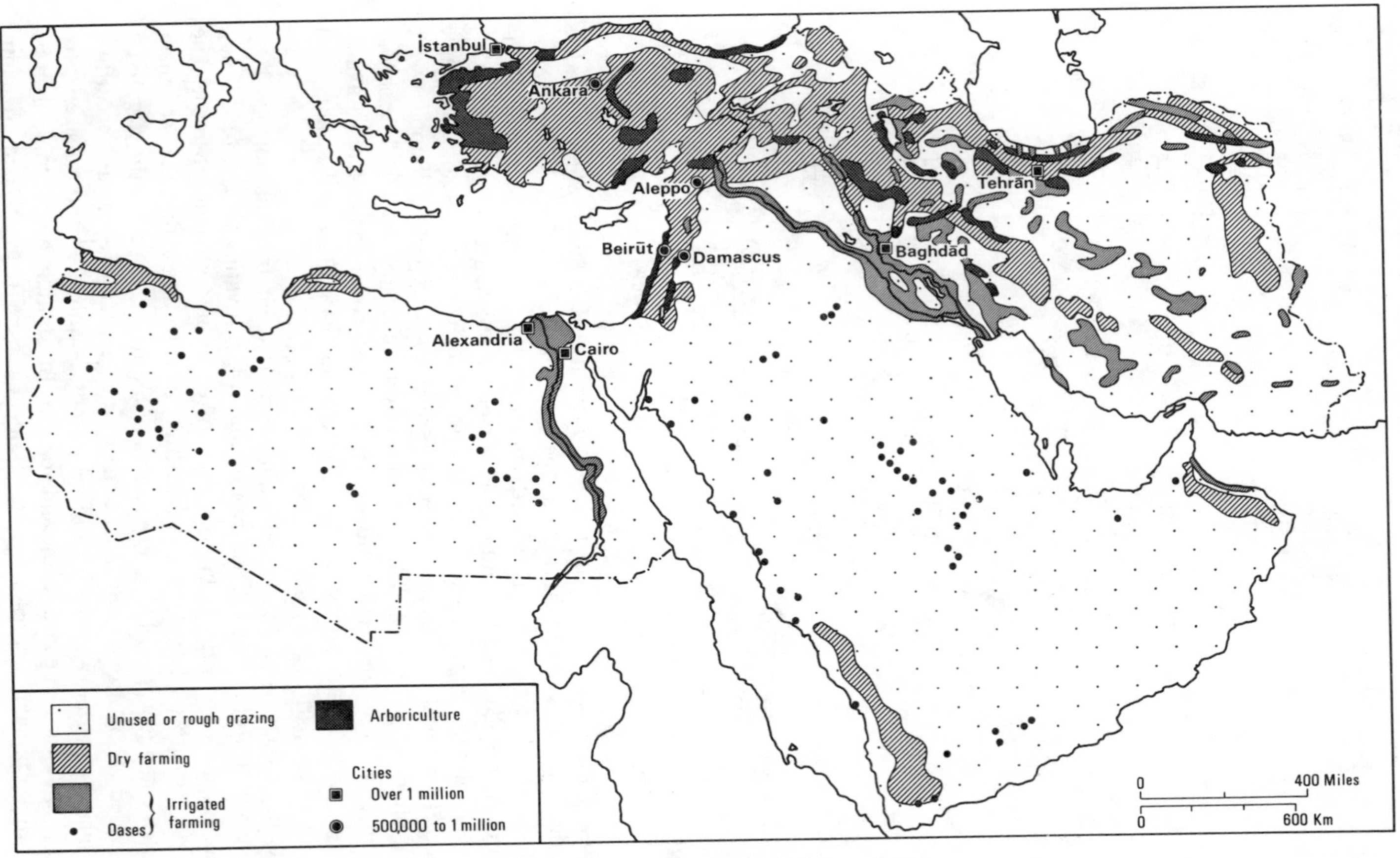

Figure 4.6 Generalised land use in about 1980

permanent cultivating villages, as the many *yayla* suffixes testify in Turkey today. Modern pressures, in a situation approaching one of land shortage, have produced severe overgrazing in many parts of the region, which has been exacerbated by the decline of more extensive forms of nomadism (Chapter 11).

4.4 Patterns of land use (Figure 4.6)

A large part of the Middle East is little used by man. The mountains are too high in places for comfortable use, and in winter they are too cold and snow-covered for permanent occupation, though Uludağ (near Bursa) is proving a joy to the skier. Small glaciers still exist in parts. Tracts of sand and stone desert provide a physical contrast, but human use is rare and limited to exceptional seasons when a flush of herbage has been produced by an occasional shower. Both the mountains and the true deserts have been the traditional haunts of outcasts, brigands and *jinns*, all characters in the region's folk stories. Nomadic pastoralists were the main organized groups to enter these apparently negative subregions, since they have traditionally operated across the margins of the permanently settled districts, but some of them were traversed by historic caravan routes.

4.4.1 Herding

Herding has been an essentially mobile way of life induced by seasonal variations in the availability of grazing and drinking water, and by the type of animal herded. Where the grazing and water resources have been good, sheep and goats have been kept, either in mixed herds or sometimes in separation. Sheep are rather fussy about what they eat and cannot graze far from permanent water, since they need to drink at least every ten days, even when the vegetation is fresh and green, and every two days, when it is old and more desiccated. Goats have reputations for being almost omniverous, and consequently very destructive of vegetation,[25] but they need less water than sheep. Their other attractions include an ability to climb were other herd animals cannot go, which means that they can graze in otherwise unusable country. Goats also have a higher rate of reproduction than sheep and a lactation period which is 50 to 100 per cent greater than that of sheep. Further deterioration in both water and grazing brings the dromedary (*Camelus dromedarius*) into its own. Not only can it eat parched grass and desiccated shrub, but it will exist on such a diet for several days without being watered. However, the quantity of water consumed is closely related to the quality of the vegetation. In the extremely arid conditions of summer, dromedaries may require water every day, while just after rain, when the grazing is succulent but surface water may still be scarce, they may need water only once in a month. This ability to last without water is the great advantage possessed by the dromedary over the donkey as a transport animal for, while they will eat much the same type of vegetation, the donkey requires watering every second day. Dromedary rearing had been characteristic of nomadic pastoralists over much of Arabia, Syria, Palestine and Iraq for centuries, probably spreading from these subregions into North Africa from the sixth century BC. It was replaced by rearing of the Bactrian camel (*Camelus bactrianus*) in Iran and Turkey. The two animals are similar in many ways, but the Bactrian camel is more adapted to the cold, snow and rocks of the high plateaus and mountains where it is required to work. Horses were traditionally bred by some nomads, and Arab stallions were famous, but they were always difficult to rear and often kept as a prestige symbol.

A minimum number of herd animals is needed to support a *tent* at subsistence level by producing a sufficient number of offspring and other products to pay for food. Estimates of the critical number have varied, but range from 25 to 60 for sheep and goats and from 10 to 25 for dromedaries. There is a tendency to build up the herds in years of good grazing in an attempt to provide a safeguard against the years of drought, when the number of stock will be reduced to a level from which it might be difficult to build up again.[26] This is especially true in the case of dromedaries, whose reproduction rate is low, and largely explains the importance of formalized raiding once common amongst north Arabian nomads.

The patterns of movement produced by relatively long-range animal herding have been classified by Johnson into two main types, horizontal and vertical. Horizontal movements are created by areal variations in the availability of water and grazing, and were characteristic of what are now Saudi Arabia, with extensions into neighbouring Jordan, Syria, Iraq and Kuwait. Mixed herds were common, but traditionally dromedaries were the most important animals and allowed the very long migrations associated with some nomadic groups. Movements had a marked seasonality to them, involving summer clustering about permanent wells, frequently on the fringes of cultivated areas, and winter migration from pasture to pasture. Johnson recognized two subtypes of horizontal nomadism (Figure 4.7a).[27] *Pulsatory* nomadism involves movement out from a dry season base on an oasis or group of wells into the surrounding desert and then a return back along almost the same line when the grazing fails. This is the pattern followed by sheep and camel nomads in Cyrenaica, in Libya. *Elliptical* nomadism, the second type, begins as a similar movement outward from a dry season base, but the return is made along a markedly different route and generally along a valley where water lasts longest, even though its permanency cannot be relied upon. The movements of the Ruwala of north Arabia provide a classic example. They formerly spent the dry season near Damascus, but from October onwards worked southwards towards al-Jawf, and then returned between late April and June down the Wādī es Sirhān, through modern Jordan, but their movements have been greatly curtailed in modern times.

Vertical nomadism makes use of altitudinal variations in the seasonal availability of pasture and water. It is associated, of course, with mountain areas and in some districts is similar to the classic transhumance of the Alps. Again two basic subtypes have been recognized (Figure 4.7b). *Constricted Oscillatory* nomadism is very widespread where mountains and lowlands are juxtaposed. Sheep and goats are the main herd animals, with camels and donkeys for trans-

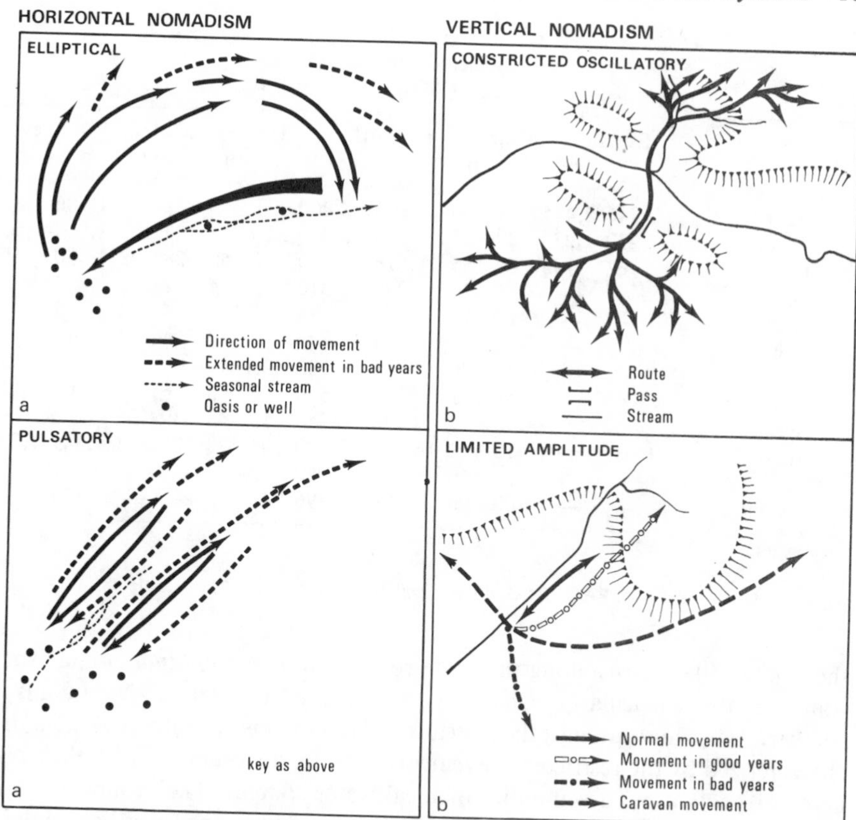

Figure 4.7 Schematic patterns of nomadic movement
(a) horizontal nomadism
(i) elliptical pattern (ii) pulsatory pattern
(b) vertical nomadism
(i) constricted oscillatory pattern (ii) limited amplitude pattern
(reproduced by permission of Douglas L. Johnson, 1969)

port. The herders usually have a firm agricultural base, either in their own fields and villages or through a close relationship with cultivators whose territory they cross during the migrations. Movement takes place from winter pastures in the lowlands as desiccation sets in, and is directed at upland pastures. Rights of passage and grazing are held along the migratory route, which may, in fact, be anchored by a permanent settlement at either one or other end. The distances covered vary considerably. The Yürüks of Turkey and the Kurds of the Zagros Mountains generally cover relatively short distances, but in historical times the Bakhtiari of Iran have travelled over 500 km each way over mountains and rivers in their annual migration from the plains of Khuzestan to upland grazings in the vicinity of Eşfahān. Movements are constricted because they follow the valleys and therefore have a linear form, but at the terminal area the members of the group spread out and go their separate ways. Movements are oscillatory, because

TABLE 4.3
Selected animals
('000)

Country	Buffaloes a	Buffaloes b	Camels a	Camels b	Cattle a	Cattle b	Horses a	Horses b	Mules a	Mules b	Donkeys a	Donkeys b
Egypt	2,346	2,415	84	160	1,906	2,800	11	9	1	1	1,706	1,850
Iran	220	230	29	27	7,800	8,350	316	316	123	123	1,800	1,800
Iraq	162	145	60	55	1,616	1,500	52	50	28	28	450	450
Israel	—	—	11	10	298	310	4	4	2	2	3	5
Jordan	—	—	13	15	38	35	3	3	4	4	19	19
Lebanon	—	—	—	—	64	45	2	2	4	4	10	10
Libya	—	—	134	170	164	200	25	41	—	—	60	60
Saudi Arabia	—	—	158	171	391	540	3	3	6	6	106	111
Syria	2	2	6	7	778	740	52	45	39	35	240	194
Turkey	1,031	544	12	3	16,499	17,300	804	623	306	213	1,349	1,226
Yemen AR	—	—	58	60	883	950	3	3	—	—	531	520
Yemen, PDR	—	—	100	90	89	96	—	—	—	—	164	170

a = average 1979–81
b = 1985
Source: United Nations, FAO, *Production Yearbook, 1985*, Rome 1986, Table 88.

the whole tribe returns along the same route. *Limited Amplitude* nomadism is found in the particularly harsh environments of the Tibesti Mountains in southern Libya and amongst the plateaus and mountains of southern Arabia. It is characterized by the restricted movements of individual families with their own herds of goats around a number of local water sources. The groups are fully mobile and do not appear to possess bases in the same sense as other nomads, though catch crops, sown after an exceptional shower, may tie them to a particular locality for longer than usual in any one year.

Everywhere in the region nomadism is in retreat as a form of land use, and the number of people involved appears to have declined appreciably (Table 4.3), though precise figures cannot be secured (Chapter 5). In Saudi Arabia, the 'true' desert nomad is already hard to find, and within a decade or two there may be few nomads anywhere in the region. The decline in numbers of animals in most countries, however, is confined to camels and horses (73 and 35 per cent respectively since around 1950) (Table 4.4). They are no longer required as draught animals or to provide military might. The caravan has been replaced largely by the truck, the dromedary and horse by the car, bus and tank. Pumps have replaced various types of animal-powered lifting devices in most districts, while ploughing and farm haulage is increasingly being done by tractors. The ploughing up of traditional grazing land, and the replacement of seasonal and annual fallow by rotational crops has diminished the grazing in the vital base areas, as well as in the seasonal pastures. Nomads find it increasingly hard to rent the grazing they need, and many have given up herding, especially after a disaster to their herds. Even in good seasons, the nomadic life is hard, while status has deteriorated as the relative wealth and military power of the tribes have declined. Some former

TABLE 4.4
Sheep and Goats
('000 head)

Country	Sheep		Goats	
	a	b	a	b
Egypt	1,590	2,500	1,451	2,650
Iran	33,833	34,500	13,500	13,600
Iraq	9,408	8,500	2,253	2,350
Israel	240	230	132	128
Jordan	978	990	510	500
Lebanon	137	135	413	450
Libya	5,046	5,500	1,400	900
Oman	114	200	240	700
Saudi Arabia	2,904	3,800	2,270	2,454
Syria	9,311	13,665	1,028	1,060
Turkey	46,199	40,391	18,755	13,100
Yemen AR	1,688	1,850	2,063	2,230
Yemen, PDR	892	940	1,237	1,380

a = average 1979–81
b = 1985 estimates
Source: United Nations, FAO, *Production Yearbook, 1985*, Rome 1986, Table 89.

nomads attach themselves as paid herders to established cultivating communities, where they are at the bottom of the social hierarchy. Others find various forms of paid employment. Still others take advantage of various government schemes to settle on the land and be transformed into cultivators. Governments act from mixed motives. There is undoubtedly a philanthropic, even a religious motive involved, but political factors have been important also, for the strong kinship ties vital to nomadic life, tribal loyalties and skill with weapons have meant that the nomads have frequently been seen as a threat to security and ordered administration.

Animal husbandry, however, is in a somewhat paradoxical situation. Against the decline in nomadism must be set a rising demand for meat, hair, wool and dairy products from within the region and which the nomads have traditionally supplied. The number of buffaloes, cattle, sheep, goats and donkeys in the region increased between about 1950 and 1980 (Table 4.3; Table 4.4). The increase in mules is probably more apparent than real since it is entirely accounted for by Yemen AR for which no data were available about 1950. Although the overall number of donkeys rose (40 per cent) the increase actually took place in Saudi Arabia and the two Yemeni states, and everywhere else there was a decline. The difference of experience is perhaps explicable in terms of the stage of development reached, for those states which experienced an overall decline did show increases in the period from around 1950–70. Donkeys provide a cheap means of transport and more were possibly needed in an early phase of land use intensification and in conditions where many roads were still inadequate for vehicles. At a later stage of development donkeys are replaced by light trucks and cars. The numbers of buffaloes and cattle rose by 47 and 67 per cent respectively between about 1950 and 1980. Whilst both types of animal are still used for ploughing and haulage, the main reasons for their increase is the demand for fresh milk and, to a lesser extent, for hides and different varieties of meat. The stall-feeding of cattle on specially grown fodder is increasing throughout the region, especially near large towns.

The traditional herd animals in much of the region, sheep and goats, have experienced different trends (Table 4.4). The number of sheep grew by 15.6 per cent over the thirty years, 1950–80, whilst the number of goats fell by 9.2 per cent. This reflects both a long-standing preference for lamb and mutton in the region and growing affluence. Iran and Turkey have much the largest national herds, and export of livestock has long been important to them. Sheep and goats are still ranged in traditional ways, though the movements are being curtailed as outlined above. In consequence, animals are being concentrated in areas formerly considered marginal for cultivation. Grazing has seriously deteriorated as a result. This has been helped to some extent by government philanthropic measures, deliberately or casually applied, to secure water for the surviving nomads. New wells, as along Tapline in Saudi Arabia, have focused and concentrated grazing, leading to overgrazing and the destruction of the pastures (Chapter 11). Without serious consideration of measures to conserve and improve traditional grazing lands, a great deal of land will go out of use. In that event, animal husbandry will have to be integrated more fully into the region's cultivation systems which are so fundamental to the socio-economic life of the region. Alternatively, large commercial enterprises, based on irrigated fodder, as at Kufra in Libya and 11 centres in Saudi Arabia, are practicable in capital-rich states.

4.4.2 Forestry

Traditional forest communities shared the characteristics of nomadism and sedentary cultivation. Forest generally survived to any extent only in the mountains, in a zone between the plains and lower slopes, on the one hand, which were the domain of the settled cultivator, and the upland pastures, on the other, where the pastoralists spent the summer (Figure 4.5). Livestock were herded in the forest or beyond and some crops were grown, sometimes under systems reminiscent of bush fallowing. The major activities in the forests, though, were cutting timber to provide lumber and firewood, producing charcoal for cooking and metal working, and extracting resins for adhesives, gums and paints. The specialist forest communities, like the Tahtaci of the southwestern parts of the Taurus Mountains, steadily destroyed the resources on which they depended and were thus forced to live in camps away from home and, from time to time, to shift their settlements completely. Forest destruction was aided by the activities of nomadic pastoralists who, on occasion, wilfully burnt forest to extent the open grazing and frequently had no alternative but to graze their animals amongst the trees as they passed through, thereby hindering regeneration.

TABLE 4.5
Surviving forest, 1980

Country	Area ('000 ha)	Percentage area
Egypt	2	0.0
Iran	18,000	11.0
Iraq	1,900	4.4
Israel	116	5.7
Jordan	38	0.4
Lebanon	85	8.3
Libya	600	0.3
Oman	1,000	4.7
Saudi Arabia	1,200	0.5
Syria	466	2.5
Turkey	20,199	26.2
Yemen AR	1,600	8.2
Yemen, PDR	1,600	4.8

(Forest is not reported from elsewhere in the region.)
Source: United Nations, FAO, *Production Yearbook, 1985*, Rome 1986, Table 1.

Only a small amount of forest survives today (Table 4.5; Figure 4.6), and much of that consists of little more than scrub. Forest removal increased steadily after the First World War as the demands for wood grew and as more land was cleared for cultivation but may have eased on the regional scale in recent years. The consequences of clearance have not only been the loss of a valuable resource, but extensive erosion in the mountains and an increase in flooding in the valleys and plains. Governments have acted to preserve forests, prevent coastal flooding, and curtail the activities of their traditional inhabitants, whose numbers are almost certainly in decline. In some countries, notably Israel, afforestation programmes are in progress (Figure 3.7). Conservation efforts elsewhere may be entering a successful phase as farming pressure on marginal land decreases in remote, more mountainous areas when people move away and as alternative sources of energy (electricity, gas) become more widely available.

4.4.3 Cultivation

Farming is still basic to the socio-economic life of the Middle East, despite recent attempts at industrialization (Chapter 8). Arable land, however, covers a very small part of the surface area of the region. FAO estimates for 1984 put it at only 6.9 per cent of the entire land surface area of the region. The range from country to country, however, was considerable (Table 4.6).

Tiny areas were cultivated in most of the states of peninsular Arabia, with the exception of the Yemen AR, while about 31 and 35 per cent of the total area was cultivated in Syria and Turkey respectively. Relatively high proportions of the surface area were estimated to be cultivated in Lebanon (29.1 per cent) and Israel

TABLE 4.6
Agricultural land use, 1984
(estimates)

Country	Total land area '000 ha	Arable Area '000 ha	Arable Percentage total area	Permanent crops Permanent area '000 ha	Permanent crops Percentage arable
Bahrain	62	2	3.2	1	50.0
Egypt	99,545	2,474	2.5	164	6.6
Iran	163,600	14,830	9.1	730	4.9
Iraq	43,397	5,450	12.5	200	3.7
Israel	2,033	437	21.5	92	21.0
Jordan	9,718	378	3.9	37	9.8
Kuwait	1,782	2	0.1	n/a	n/a
Lebanon	1,023	298	29.1	88	29.5
Libya	175,954	2,115	1.2	335	15.8
Oman	21,246	43	0.2	28	65.1
Qatar	1,100	3	0.3	n/a	n/a
Saudi Arabia	214,969	1,156	0.5	76	6.6
Syria	18,406	5,654	30.7	550	9.7
Turkey	78,058	27,411	35.1	2,911	10.6
UAE	8,360	15	0.2	8	53.3
Yemen AR	19,500	1,351	6.9	90	6.7
Yemen PDR	33,297	167	0.5	20	12.0
Total	892,050	61,786	6.9	5,330	8.6

Source: United Nations, FAO, *Production Yearbook, 1985*, Rome 1986, Table 1.

(21.5 per cent), but only small proportions of the national area in the remaining countries of the region.

Whilst Israel, Lebanon, Syria and Turkey appear to have experienced some expansion in their cultivated area during the 1970s, the other countries of the region have seen a contraction, following in some areas a period of expansion which started in the 1950s. Contraction has been most dramatic in Iraq. Whilst some of this may be attributed to the long-drawn out war with Iran, the principal reasons are probably the same as elsewhere. They include the abandonment of marginal land as populations left the remoter and more physically difficult rural areas; low prices for farm products; the rapid expansion of urban areas, often at the expense of some of the most fertile land in the region; the devastation of formerly productive land through salination and poor drainage; and a slowing down in the rate of reclamation of uncultivated land which had earlier compensated for some of the losses.

Cropping in the mid-1980s continued to be almost totally dominated by annual crops (91 per cent of the cropped area, Table 4.6). Wheat and barley were still the most important in terms of the proportion of the cultivated area devoted to them. Together they occupied an estimated 63 per cent of the arable area in 1985 (Table 4.7). The proportion exceeded 84 per cent in Libya, almost reached 80 per cent in Iran and Iraq and passed 70 per cent in Jordan and Saudi Arabia, but occupied only 13 and 15 per cent of the cropped area in PDR Yemen and Yemen AR

TABLE 4.7
Field crops, 1985 (000 ha)

COUNTRY	Wheat	Barley	Maize	Millet & Sorghum	Rice	Sugar Cane	Sugar Beet	Potatoes	Onions	Tomatoes	Cabbages	Cauli-flowers	Beans	Broad Beans	Peas	Cucumber	Melons	Water Melons	Chick Peas	Lentils	Pulses	Ground Nuts	Cotton	Sesame	Sun Flower	Flax	Tobacco	Total Area*
EGYPT	498	56	900	175	422	115	15	72	25	135	17	4	122	140	10	20	21	60	10	8	193	12	1175	22	6	17	—	4250
IRAN	6100	1850	45	27	390	30	—	110	42	38	—	—	160	—	50	—	68	115	75	73	379	2	390	3	19	—	15	9981
IRAQ	1000	850	20	7	40	4	158	7	10	43	1	—	6	6	—	33	30	46	16	6	35	—	17	13	13	—	15	2376
ISRAEL	65	20	5	3	—	—	—	6	3	6	1	1	—	—	3	2	3	6	3	—	8	5	227	—	8	—	1	376
JORDAN	94	55	—	—	—	—	—	1	1	11	—	1	1	—	—	4	2	3	2	6	12	—	—	—	—	—	3	196
LEBANON	14	4	—	—	—	—	2	8	2	5	1	—	1	—	1	3	1	2	2	4	9	1	—	—	—	—	4	64
LIBYA	274	130	1	3	—	—	—	16	6	15	1	—	—	—	1	—	4	10	1	—	9	7	—	—	—	—	—	478
OMAN	—	—	—	—	—	—	—	—	1	—	—	—	—	—	—	—	—	—	—	—	—	—	—	—	—	—	—	1
P.D.R. YEMEN	10	2	6	50	—	—	—	2	—	1	—	—	—	—	—	—	—	1	—	—	—	—	19	5	—	—	—	96
QATAR	—	—	—	—	—	—	—	—	—	—	—	—	—	—	—	—	—	—	—	—	—	—	—	—	—	—	—	—
SAUDI ARABIA	550	6	2	171	—	—	—	1	5	18	—	—	—	—	—	1	2	21	—	—	4	—	—	2	—	—	—	783
SYRIA	1146	1386	45	12	—	—	15	23	7	40	3	3	13	6	2	27	36	95	79	66	215	9	485	42	7	—	14	3776
TURKEY	9000	3250	580	11	64	—	323	245	80	130	37	4	164	43	8	42	—	260	350	620	1339	23	1448	68	600	7	210	18906
U.A.E.	1	—	—	—	—	—	—	—	—	2	1	—	—	—	—	1	1	1	—	—	—	—	—	—	—	—	—	7
YEMEN A.R.	60	52	35	500	—	—	—	12	—	—	—	—	—	—	—	—	—	—	—	—	70	—	10	10	—	—	7	756

N.B. Data very incomplete, and mostly F.A.O. estimates

* = Sum of areas under reported crops

Source: United Nations, F.A.O. Production Yearbook, 1985, Rome, 1986.

TABLE 4.8
Yields of Wheat and Barley in 1985
(kg/ha)

Country	Wheat	Barley
Egypt	3,763	2,679
Iran	984	892
Iraq	650	824
Israel	1,692	750
Jordan	1,060	545
Lebanon	1,071	1,250
Libya	546	n/a
Oman	3,774	n/a
Qatar	3,333	3,966
Saudi Arabia	3,091	2,000
Syria	1,496	748
Turkey	1,892	2,000
UAE	1,238	2,000
Yemen AR	667	962
Yemen, PDR	1,500	1,567
World	2,217	2,266

respectively. Wheat and barley are generally sown after the first rains in October or November. Despite their areal importance, however, yields were generally below the world average of 2266 kg per ha for barley and 2217 kg per ha for wheat, though these were exceeded in a few countries, notably Egypt (Table 4.8). Wheat was the most important single crop in terms of area, except in Syria where its place was taken by the more drought resistant barley, and in the two Yemens where the hardy grains, millet and sorghum, covered the largest proportion of the cropped area. Millet and sorghum ranked second to wheat in Saudi Arabia and were even more important there before the government's campaign during the early-1980s to promote wheat growing. They are normally planted as the summer rains end. Maize also is a summer crop, but it does well on the deep soils of the Nile valley and delta.

Despite their predominance, cereals are rarely grown as the only crops in any district today (Table 4.7). Traditionally, vegetables have been grown on a small scale in gardens in and around the settlements, but their production as field crops has steadily expanded since the Second World War. Improved communications have been largely responsible, since perishable commodities can now quickly reach the distant urban markets, where the demand has been rising, by fast truck. Potatoes, onions, tomatoes, cucumbers, melons and water melons, as well as legumes (lentils, chick peas and beans), have expanded their areas considerably in many districts. Nonetheless, they are most important in Israel, Jordan and Lebanon where export markets have been developed in addition to the urban home market. Total areas, however, remain very small. Vetch is widely grown as a fodder crop, but its place is taken in Egypt and parts of Arabia by *berseem* (clover). Unfortunately, FAO does not include fodder crops in its statistics, though the area is proportionately large, expecially when it is remembered that a portion of the cereals grown are intended for animal food, generally for the winter. The extreme is probably reached in Egypt, where up to one fifth of the entire cropped area is devoted to clover.

Some cereals and most of the vegetables are sold off the farm, but a good deal of the income is now obtained by growing specific industrial crops. The most important in terms of area is cotton. It has long been grown in the region, but its rapid expansion came in two phases. The first came during the nineteenth century, and was stimulated by rising demand from industrialized Europe. The second phase came after the Second World War, and was consequent upon world-wide shortages and high prices created by the Korean War (1950–53). By 1985 cotton occupied an estimated 9 per cent of the crop area of the region, with the largest contributions to the total regional area under the crop being made by Turkey (38.4 per cent), Egypt (31.1 per cent), Syria (12.9 per cent) and Iran (10.3 per cent). In terms of its share of the national crop area, cotton was most important in Israel (60 per cent), Egypt (27.6 per cent), PDR Yemen (19.8 per cent) and Syria (12.8 per cent). Practically everywhere the crop is grown under irrigation.

Other cash crops are of minor importance on the regional scale, though some are significant users of land at the national level. All the region's sugar cane is grown in three countries – Egypt (77.1 per cent in 1985) and where production has recently expanded in the Valley, Iran (20.2 per cent) and Iraq (2.7 per cent) – whilst sugar beet is largely grown in the cooler dry farming districts of Turkey (63.0 per cent) and Iraq (30.5 per cent) in rotations involving cereals. Tobacco has been an important industrial crop in the region since the eighteenth century. The leading growing areas are the Aegean and Black Sea coastlands of Turkey (78.0 per cent of the regional area under tobacco was found in this country in 1985), the Latakia district of Syria and various districts in Iran. Only minor contributions are made by Jordan, Lebanon and Yemen AR. Most of the region's area under sunflowers was found in Turkey (91.8 per cent in 1985). The other oil-producing crops in the region (sesame and groundnuts) were more widely grown but with the largest areas of seame in Turkey (41.2 per cent) and Syria (25.4 per cent), and the greatest proportion of groundnuts in Turkey (39.0 per cent), Egypt (20.3 per cent) and Syria (15.2 per cent). Flax was cultivated only in Egypt (70.8 per cent) and Turkey (29.2 per cent). Small areas of opium poppies are grown (legally and illegally) in several parts of the region, notably in Turkey, Iran, Iraq and Lebanon.

By contrast with the area under field or annual crops, that devoted to permanent crops, principally fruit trees, is small (Table 4.6) and in some countries agriculture has shown surprisingly little response to increased demand. Olives, vines and figs are the most numerous trees since they have the advantage of being deep-rooted and so able to draw moisture from depth in the long, arid summers characteristic of the region. However, the extreme aridity of the interior and the winter cold of the mountains and high plateaus make them a characteristic combination only in areas adjacent to the Mediterranean, chiefly Israel, Jordan, Lebanon and coastal Turkey. Undercropping with cereals is possible here, but with distance inland and falling precipitation, olives must be planted further apart and soil moisture is not sufficient to permit undercropping, unless irrigation is applied, as in the oasis of Damascus. Vines have a wider distribution than the other two tree crops, but, until recently, the area devoted to them had been restricted by the Muslim ban on wine-drinking. Tropical fruits, such as bananas, are grown on a small scale in Egypt, Israel, Lebanon, along the south coast of Turkey and in Cyprus. Oranges, which are both a subtropical and Mediterranean fruit, are grown in most countries, though production is particularly high from the coastal plain of Israel. Grapefruit and peaches are expanding in similar areas to supply the export market in western and eastern Europe. Temperate fruits (apples, pear and cherries) are widely found, but on a fairly small scale still. They are particularly associated with upland areas, where there are local specialities, such as the apples of Lebanon and the pears of Ankara. Sales are largely restricted to local or national markets, though Lebanon built up a trade in apples with some of its more southerly neighbours. Date groves, as commercial propositions, are more restricted still. The main commercial growing area is along the Shaṭṭ al'Arab in southern Iraq, but dates are also an important crop in the oasis of Arabia, Egypt and Libya. Dates are not only an important local food but a useful export. The palm trees are valuable also as a source of building materials in otherwise largely treeless areas and provide essential shade for the successful growing of cereals, vegetables and other crops under desert conditions.

Everywhere the commercial orientation of cropping patterns is increasing. Demand continues to rise, both within the region and beyond, as populations grow and incomes rise. Accessibility has been improved for many farming communities by the introduction of the truck and the construction of motor roads. Marketing has become easier and quicker. The mechanization of farm work and the extension of irrigation have also been important in affecting the changes, since the necessary investments, which are still comparatively low, can only be recouped by growing industrial or high value commercial crops. The price of these changes in cropping has often been the ending of traditional share-cropping contracts and the destruction of 'peasant economy', with the consequent appearance of cash rents and unemployment in the rural areas.

Cropping patterns, however, have not only responded to commercial pressures. Considerable influence has been exerted by governments on the national scale and, at the local level, by distance from the operational centre, the village. Distance from the nucleated village exerts a profound influence on the local pattern of land use and von Thünen's location theory seems to apply.[28] Land uses which require the application of relatively large amounts of labour, such as vegetable growing for home consumption, are generally situated close about the settlement, though some distortion occurs where excentrically placed and small-scale irrigation is practised. Cereal growing is less labour intensive, and is carried out further away from the settlement (Figure 4.8). With traditional techniques the cereal and fallow belt extended for a maximum distance of between three and four kilometres from the settlement, and it seems that beyond this distance the economic return did not normally exceed the drudgery involved in

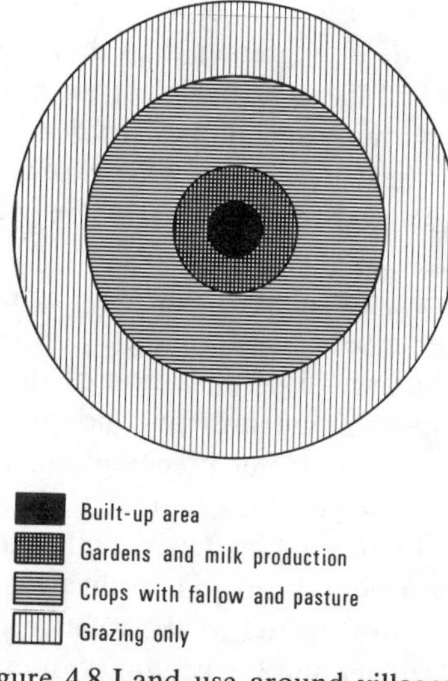

Figure 4.8 Land use around villages in central Turkey (reproduced by permission of *Middle East Journal*, 1971)

cultivating plots so far away from each other. Where tree crops are important, they appear to be grown in a zone more than three or four kilometres from the settlement, since they require less labour and the fewest visits in the year. Where trees are not of commercial importance, however, they might often be seen growing in and immediately around the village to give shade and an atmosphere of coolness. Grazing land lies beyond the commercial tree-crop zone, since the herding of animals requires comparatively little effort compared with other forms of land use, while the animals have to be kept away from the cropped areas as much as possible to prevent destruction of the vital harvest. Mechanization, of course, has extended the radius of cereal growing and, in some places, brought into existence a second cereal zone beyond the belt of tree crops, while increased demand has brought under the plough occasional tracts of common grazing found closer to the village. Such areas are divided into the long strips shown in Figure 4.4a.

Locational changes are associated also with the introduction of irrigated farming to dry farming districts, where it is usually accompanied by a change in crop type, often in response to government plans, as well as to the needs of capital investment programmes. In addition to undertaking reclamation and irrigation schemes, providing credit and supplying new strains, government has exerted other controls over land use. Restrictions are normally imposed on the area which can be planted with cotton and tobacco in an attempt to manage the market and stabilize prices, while poppy growing is controlled as much as possible to prevent illegal drug trafficking with the West. Production of other crops is not completely free either. Cereals, for example, are often supported by guaranteed prices which are higher than those on the world market and, in consequence, the areas under wheat and barley are considerably greater than they would be if market forces were allowed something like full play. Inefficient forms of land use are thus perpetuated, as are archaic farming techniques. In Saudi Arabia, by contrast, a remarkable expansion of wheat growing (>700 per cent) was achieved during the early 1980s using machines and sprinkler irrigation on newly reclaimed land in the Hail and Qassim districts.

Techniques of cultivation vary quite considerably across the region, but a basic distinction may be made between dry farming and forms of irrigated farming. Dry farming uses only the annual precipitation of the area, and the system is geared to conserving limited soil moisture in conditions of high summer evapotranspiration and relatively low winter precipitation. Fallowing of part of the cultivated area for at least one year was characteristic of the system, and is still retained widely. This was often explained by the mistaken belief that at least two seasons moisture could be accumulated and produce better crops. In fact, the use of fallow was probably more related to the relative abundance of land until recently, the shortage of fodder and the adequacy of extensive production, with traditional labour inputs, to support the population at an acceptable standard. Population growth and market demands have reduced the extent of fallowing by introducing a year of vegetables, fodder crops, sugar beet or groundnuts into a rotation with the basic cereals. Other features of the system can be more definitely explained as attempts to use available precipitation wisely.

Ploughing begins as the rains start, but after a sufficient interval has elapsed for the baked soil to soften. The traditional plough is a simple ard made of wood, frequently with a fire-hardened point, rather than an iron share. It is pulled by a pair of animals, frequently but not always, a yoke of oxen.[29] Such an implement merely scratches the surface of the ground, but when employed with simple harrows and mattocks, it is very effective in producing a fine tilth which aids water conservation. Seeds are then sown on to this seedbed, traditionally by broadcasting, and being harrowed or trodden in. The whole process might take two to three weeks under favourable conditions on an average-sized holding. Soil is still prepared in such a way in some districts, generally the remoter and more mountainous ones, where steel ploughs and tractors are not only relatively expensive but often ill-adapted to difficult terrain, awkward terraces, and thin, stoney soils. Steel ploughs, designed to turn a furrow, are almost universal, while in the plains and larger valleys multiple ploughs are commonplace. Haulage is frequently by tractors now, a fact which helps to explain the decline in the number of horses in the region since 1950. The number of tractors, in fact, grew by almost six times between 1963–64 and 1981, whilst the number of tractors per 1000 ha of cropped land has risen in every country. Israeli farming is the most mechanized in the region with more than 67 tractors per 1000 ha, but Kuwait ranked second and Turkey, a long way behind, third (Table 4.9). By comparison, agriculture in Iran, Iraq, Jordan, Oman and Saudi Arabia was considerably less mechanized with fewer than five tractors per 1000 ha. The other countries occupied a middling position.[30] Tractors and steel ploughs have allowed more land to be ploughed

TABLE 4.9
Tractors and Fertilizers, 1981

Country	Tractors (number)	Tractors per 1000 ha of cultivated land	Consumption of fertilizer per ha of cultivated land
Egypt	25,500	8.9	247.5
Iran	59,000	3.7	42.3
Iraq	22,100	4.1	14.1
Israel	28,335	67.8	199.6
Jordan	4,570	3.3	5.3
Kuwait	37	37.0	500.0
Lebanon	3,000	8.6	100.6
Libya	15,000	7.2	37.5
Oman	109	2.7	39.5
Qatar	na	na	280.0
Saudi Arabia	1,300	1.2	60.2
Syria	31,387	5.5	23.2
Turkey	457,425	16.1	45.4
UAE	na	na	281.2
Yemen AR	2,050	0.7	4.3
Yemen, PDR	1,265	6.1	8.8

Source: United Nations, FAO, *Production Yearbook, 1985*, Rome 1986.

with less labour. In addition, they are efficient at opening up the soil and allowing plants to get full benefit from the nutrients in the ground. But, unless contour ploughing is employed – and this is still exceptional – the soil is more dangerously exposed to erosion, especially gullying, than with traditional techniques.

In the past, artificial fertilizers were seldom applied and dunging was often by chance rather than deliberate application. Chemical fertilizers are used practically everywhere today in an attempt to maintain or improve yields. Rates of application are high in Egypt and the wealthy oil states of Kuwait, Qatar and the UAE, but exceptionally low in Jordan and the two Yemens (Table 4.9). This is partly because their value is not always appreciated, but it is largely because fertilizers are not readily available to the farmer. Prices are high, distribution is often inadequate to the requirement, and national output is frequently low.

Harvest time varies according to the type of crop, date of sowing and elevation, but the cereal harvest begins in May and ends in August across the region. Cereals were traditionally cut with sickles, though the use of scythes has been reported from parts of Iran. These methods are still employed in difficult terrain, on irregularly-shaped fields and by poor families, but combine-harvesters operated by contractors are now widespread. A similar use of mixed techniques is apparent at threshing. If hand cut, the cereals are taken to threshing floors, often located in windy places near the village. There they are prepared for winnowing by means of a threshing sledge or similar device, consisting of heavy pieces of wood shod with flints or pieces of sharpened steel, dragged round and round the circle of the floor.

Winnowing, to separate grain and chaff, then takes place by hurling spadefuls or forkfuls of the chopped material in to the wind – a skilled and tiring operation. In some districts, simple box-threshing machines, driven by tractor engines, are employed. It is at the threshing floors, of course, that sharecropping contracts are fulfilled with the division of the bagged grain.

Similar implements and techniques are employed in irrigated areas, but irrigated farming as a system of land use is dominated by water supply methods. The various sources of the vital water, thus allow a four-part classification of types of irrigated farming. These are oasis, terrace, *qanāt* and river types.

Various subtypes of oasis may be recognized, but all of them are situated in areas which receive little or no regular precipitation, and grow dates as their principal crop. Some oases depend primarily on surface water and have a compact or linear form (Figure 4.9), depending upon the amount of water available and the type of source actually tapped.[31] Water may be supplied by perennial streams, as in the case of the famous oasis of Damascus (Chapter 13) and the numerous humble oases on the desert fringes of the Omani and Yemeni mountains. Other oases depend upon irregular flash floods (*seils*) which are used extensively and successfully in the wadi systems of southern Arabia. Another type of oasis makes use of springs, the water from which is fed along systems of aqueducts fanning outwards from the source. One of the largest and most complex of such oases is al Hasa in eastern Arabia. The type image of an oasis, however, is dependent upon subsurface water reached by shallow wells. Traditionally, water was lifted in one of two ways. It might be raised in great leather bags by animals pulling on a rope stretched over a pulley as they descended a ramp. The alternative was to use the *noira* (Arabic *na'oura*), a waterwheel geared in such a way that it could be turned by an animal walking in a circle around the well head. The physical effort required in lifting comparatively small quantities of water produced an essentially fragmented pattern of fields or gardens, each centred upon its well (Figure 4.9). This pattern has been largely maintained, even though diesel pumps have replaced traditional lifting devices and produced greater flexibility through their ability to raise more water. The contiguous circles of irrigated land centred on sprinkler pivots are striking on satellite images of the newly-reclaimed areas in Saudi Arabia. The sheer efficiency of pumps, however, has caused local water tables to fall, and created serious problems for the continuation of irrigated farming in a few districts.

The irrigation of systems of terraces is probably only a fragment of what it may once have been. Terrace irrigation consists essentially of leading water on to the top of a flight of terraces and allowing it to flow downwards by gravity (Figure 4.10). The original source might be a spring, a seasonal stream diverted by dams or weirs, or even a system of cisterns storing winter runoff from a wide area. Each source, water conveyors, channels and levelled terraces constitute a separate hydrological system which must have been visualized as an entity before construction started.[32] Each system is tiny compared with the hydraulic system of the Nile valley in Egypt, yet each represents a considerable investment of labour, not only in the initial construction, but also in maintenance. Neglect easily ruins the

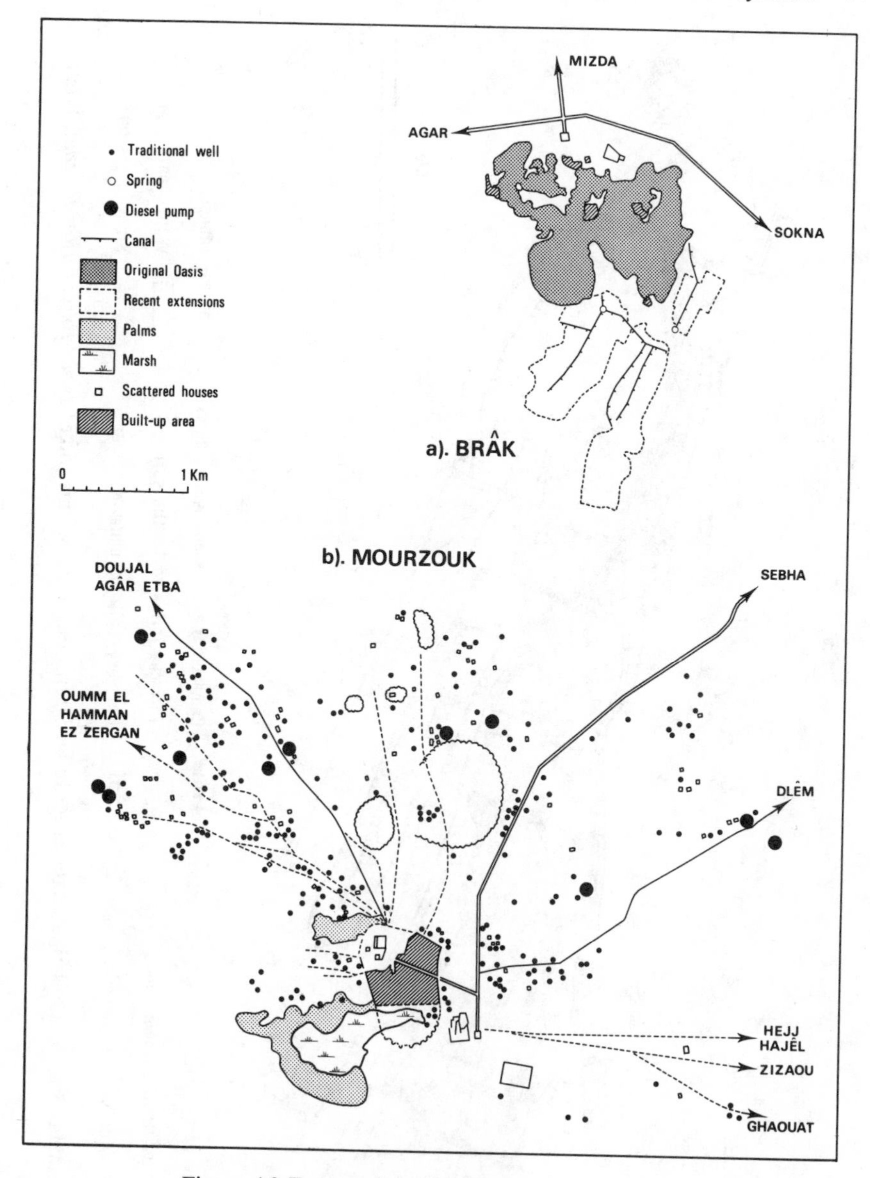

Figure 4.9 Forms of oasis in Libya
(a) Brâk and (b) Mourzouk
(reproduced by permission of Lars Eldblom, 1961)

entire system and may cause serious flooding, as recent experience in Lebanon and Yemen AR has shown. Terrace irrigation is characteristic either of once densely settled areas, like ancient Judaea and parts of Yemen, where as much of the sloping ground as possible must be cultivated, or of districts where labour might have been coerced, as in parts of Libya under Roman rule (Chapter 3).

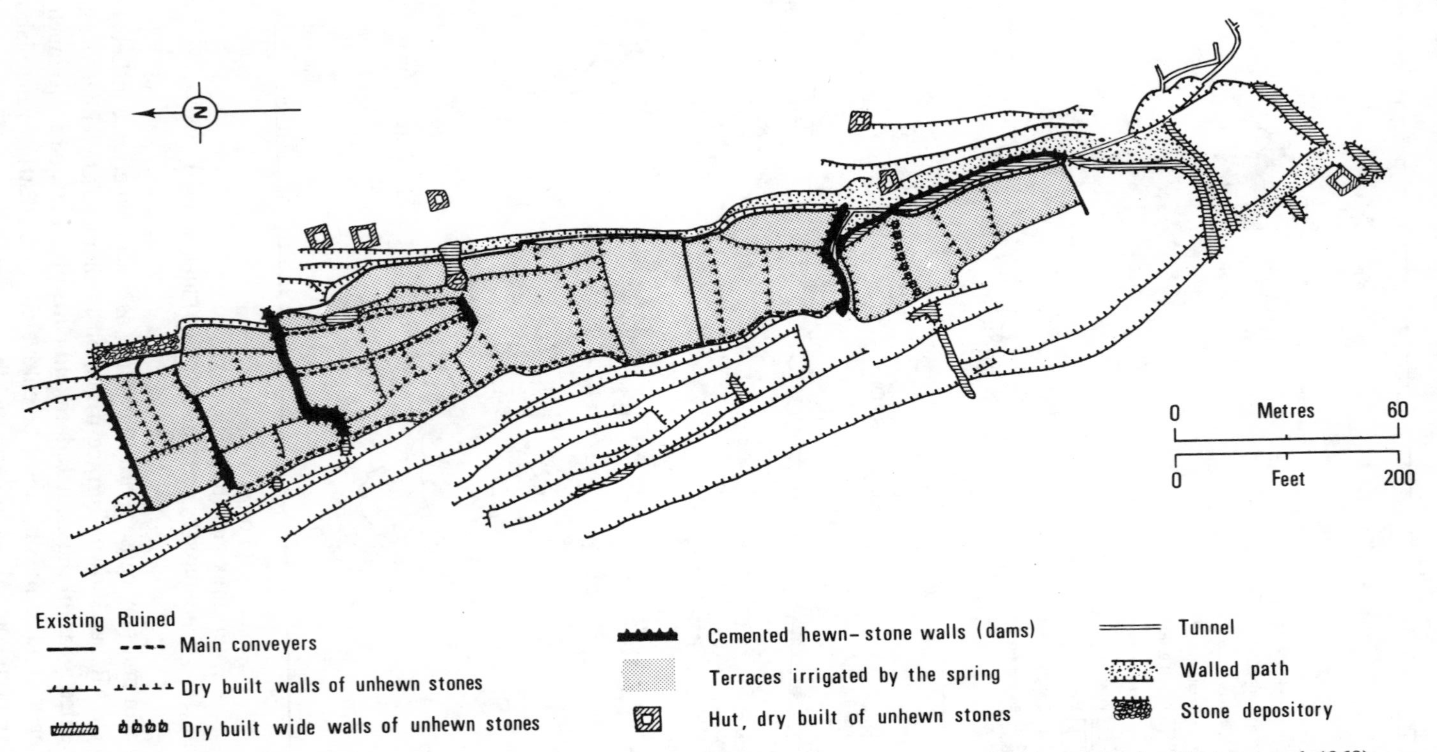

Figure 4.10 Ein Khandak system of irrigated terraces, Judaea (reproduced by permission of *Israel Exploration Journal*, 1969)

Labour costs today, however, mean that the systems are expensive to maintain, and must be used to produce high value vegetables and tree crops if they are to continue in use, while labour shortages, often resulting from emigration, put whole systems at risk.[33]

Qanāt is the Persian term for a subterranean gravity canal or aqueduct leading water from an aquifer in an alluvial fan situated up slope from the area to be irrigated (Chapter 2). Although probably developed in Iran,[34] the 'horizontal well', as it is sometimes called, is widespread in the region under a variety of local names. It is an expensive way of securing water, and it has been calculated that a median length of *qanāt* of about nine km gives a return in crops and water sold of only about 10 per cent.[35] Once provided, the water is allocated and used in similar ways to the water from other flow sources. Security of supply, however, is generally more uncertain. Excessive rainfall, leading to floods, can cause a roof-fall or choke the gallery with material washed in from the surface. Aquifers may dry up or the flow be reduced, either through natural events, like earth tremors, or through sinking deep bores in areas crossed by the *qanāt*. In any case, in many long settled areas, there is competition for underground water from other *qanāts* running in a variety of directions and at different depths. In recent years the viability of *qanāts* has been threatened by the sinking of deep wells for pumped irrigation, while emigration has often brought neglect.[36]

River irrigation is becoming more important in the region as a result of the construction of large dams across the major rivers, though its largest extent is still in the Nile valley and delta and in the Tigris–Euphrates catchment. Water is not obtained in the same way in all districts irrigated from rivers, and basic distinctions can be made between flow and lift-based types of irrigation, and between the perennial and seasonal availability of water. Until the recent advent of the motor pump, perennial irrigation was normally dependent upon lifting water out of the rivers and canals by various devices. The *shaduf* (a weighted beam with a bag on a rope at one end) is still used in Egypt. Waterwheels were once commonly used, but remained characteristic of the Orontes in the vicinity of Homs and Hama until recently (Chapter 13). The use of the *tambūr* (hand-operated Archimedean screw) was virtually confined to Egypt. Practically everywhere, these devices have been replaced and supplemented by pumps, which are not only more efficient in raising water, but, with the use of pipes, allow a much wider and more continuous strip of flood plain to be irrigated.

Flow irrigation, although found on other rivers in the Middle East, has been particularly characteristic of seasonal irrigation in the Nile valley and the Tigris–Euphrates lowland. Different systems were employed in the two subregions until the nineteenth century. Basin irrigation, using the annual flood, was characteristic of Egypt, while the use of canals and distributaries was preferred in the Tigris–Euphrates lowland, particularly where the rivers are closest together in the vicinity of Baghdād. Since the early nineteenth century, the irrigation systems in the two subregions have been transformed and made more alike. Perennial irrigation is now normal in Egypt, where greater use is made of canals and distributaries than before, while in Iraq the problems of uncontrolled and destructive flooding are being ameliorated by the use of reservoirs and diversionary works. These developments are described fully in Chapters 12 and 19, while the historical situation is outlined in Chapter 3. In Egypt, change in the irrigation system has brought more intensive forms of land use. Even in 1965, before the sluices in the Aswân High Dam were closed, an average of 1.6 crops was being taken from every *feddan*, while three crops in a year were common.[36] Intensification has been more limited in Iraq, and the traditional cereals are still dominant in extensive systems of land use.

Irrigation is being extended everywhere throughout the region,[37] as the regional chapters show. Popular devices include the sinking of deep bores and the damming of seasonal streams and perennial rivers. Further dam-building on the Euphrates by both Syria (Tabqa Dam) and Turkey (Keban, Karakaya and Atatürk) have reduced the amounts of water available for use in Iraq and made it more saline. Serious political problems appear to be looming as a result.

Water extraction from rivers is usually associated with flood control and frequently with the generation of electricity. Increased use is being made of more efficient methods of applying water. Sprinkler systems and drip-feeding are now widespread, while large polythene greenhouses are being introduced. Mechanization is usually a feature of these developments and is integrated with large rectangular fields. The projects are increasingly expensive, and frequently require financing from abroad, with important political and economic consequences, as Egypt discovered with the Aswân High Dam and subsequently Syria and Turkey with their Euphrates projects. The high levels of investment also mean that valuable industrial crops and relatively high-priced vegetables and fruits are the preferred crops, rather than low-priced but basic subsistence cereals. Local deficiencies are made good by imports.

Food imports increased appreciably everywhere, with the exception of Turkey, during the 1970s and continued into the early 1980s. To some extent this was the result of improved nutrition and the emphasis on industrial crops, as well as special circumstances such as the civil war in Lebanon and the war of attrition between Iran and Iraq, but the food deficit of the region as a whole also reflects the inefficiencies of Middle Eastern agriculture and the relative lack of attention given to this sector of the economy by the various national governments. Agricultural production has failed to keep pace with the demand created by rapid population increase. Controlled prices and low wages in industry and commerce have not provided incentives to local farmers. Levels of investment have been insufficient to meet the needs of a modernizing, market-responsive activity. Too much investment has gone into prestige projects like large dams and large scale irrigation schemes of doubtful long term benefit. Labour has become scarce. Worried by their dependency on outside sources, the wealthy Arab states, notably Saudi Araba, launched various programmes of food self-sufficiency during the 1980s, and the Islamic government in Iran stressed the spiritual and economic importance of agriculture. Saudi Arabia has been extremely successful and by 1985 was more than self-sufficient in wheat and dairy products, but the cost in financial terms has been high and the waste of soil and water resources considerable. With the exception of Turkey again, agricultural exports from the region do not compensate for the high levels of food imports, despite the importance attached to them by both farmers and national governments. Indeed, agricultural exports from the region generally fell during the 1970s, except for Turkey and, to a lesser extent, Syria. The fall was dramatic in the case of Egypt, from $982 million in 1974 to $610 million in 1979. In these circumstances, heavy reliance is placed on success in other lines of the economy. This gives added incentive to the need to maintain oil prices, for those states fortunate enough to export large quantities of petroleum, and to develop manufacturing industry, especially for those countries without large petroleum reserves. Considerable problems have to be faced. The whole field of economic development is explored in Chapter 7, while industrialization and petroleum exploitation are discussed in Chapters 8 and 9.

References

1. G. Perrin de Brichambaut and C. S. Wallén, *A Study of the Agroclimatology in Semi-Arid and Arid Zones of the Near East*, Technical Note No. 56, World Meteorological Organisation, Geneva, 1963.
2. C.A.O. Van Nieuwenhuijze, 'The Near Eastern village: a profile', *Middle East Journal*, **16**, 295–308 (1962).
3. (a) A. Ibrahim, 'Classification and characteristic patterns of rural settlements (Egypt)', *Mediterranea*, **23–24**, 332–45 (1968).
   (b) X. de Planhol, 'Geography of settlement', in *The Cambridge History of Iran, The Land of Iran* (Ed W.B. Fisher), Vol 1, Cambridge University Press, London, 1968, 409–467.
4. J. Weulersse, *Paysans de Syrie et du Proche Orient*, Gallimard, Paris, 1946, 281.
5. *Encyclopaedia of Islam*, **4**, E.J. Brill, Leyden, and Luzac and Co, London, 1934, 1096–103.
6. *Encyclopaedia of Islam*, **4**, E.J. Brill, Leyden, and Luzac and Co, London, 1934, 1096–103.
7. G. Baer, *Population and Society in the Arab East*, Routledge and Kegan Paul, London, 1964, 142 and note.
8. R.C. Tute, *The Ottoman Land Laws*, Jerusalem, 1927, 15–16.
9. R.C. Tute, *The Ottoman Land Laws*, Jerusalem, 1927, 17–19.
10. (a) J. Weulersse, *Paysans de Syrie et du Proche Orient*, Gallimard, Paris, 1946, 99–108.
    (b) Latron, *La Vie Rurale en Syrie et au Liban,* Beirūt, 1936, 213–15.
11. J. Weulersse, *Paysans de Syrie et du Proche Orient*, Gallimard, Paris, 1946, 116.
12. J. Hinderink and M.B. Kiray, *Social Stratification as an obstacle to Development. A Study of Four Turkish Villages*, Praeger Publishers, New York, Washington, London, 1970.
13. (a) M. Clawson, H.H. Landsberg and L.T. Alexander, *The Agricultural Potential of the Middle East*, American Elsevier Publishing Co, New York, London and Amsterdam, 1971, 60.
    (b) A. Drysdale and G.H. Blake, *The Middle East: A Political Geography*, Oxford University Press, 1985, 296–306.

14. (a) E. Eshag and M. A. Kamal, 'Agrarian reform in the United Arab Republic (Egypt)', *Bulletin of the Oxford Institute of Economics and Statistics*, **30**, 73–104 (1968).
(b) A. K. S. Lambton, *The Persian Land Reform, 1962–66*, Oxford University Press, 1969.
(c) D. Warriner, *Land Reform and Development in the Middle East*, 2nd edn, Oxford University Press, London, 1962.
15. Z. Keilany, 'Land reform in Syria', *Middle Eastern Studies*, **16**, 209–24 (1980).
16. F. Fattah, 'Farming cooperatives in Egypt', *World Marxist Review*, **15**, 96–9 (1972).
17. H. Askari, J. T. Cummings and B. Harik, 'Land reform in the Middle East', *International Journal of Middle East Studies*, **8**, 437–51 (1977).
18. K. McLachlan, 'The agricultural development of the Middle East: an overview', in *Agricultural Development in the Middle East* (Eds P. Beaumont and K. McLachlan), John Wiley, Chichester, 1985, 27–50.
19. D. A. Caponera, *Water Laws in Moslem Countries*, FAO Development Paper No 43, Rome, 1954.
20. M. Clawson, H. H. Landsberg and L. T. Alexander, *The Agricultural Potential of the Middle East*, American Elsevier Publishing Co., New York, London and Amsterdam, 1971, 60.
21. D. A. Caponera, *Water Laws in Moslem Countries*, FAO Development Paper No 43, Rome, 1954, 14–18.
22. T. Shanin (Ed), *Peasants and Peasant Societies*, Penguin Books, Harmondsworth, 1971.
23. (a) C. Gurdon, 'Livestock in the Middle East', in *Agricultural Development in the Middle East* (Eds P. Beaumont and K. McLachlan), John Wiley, Chichester, 1985, 85–106.
(b) D. L. Johnson, *The Nature of Nomadism: A Comparative Study of Pastoral Migrations in Southwestern Asia and Northern Africa*, Department of Geography, Research Paper No 118, Chicago, 1969.
(c) C. Nelson (Ed), *The Desert and the Sown: Nomads in the Wider Society*, University of California Press, Berkeley, 1973.
24. A. Tanoğlu, 'The geography of settlement, *Rev. geogr. Inst. Univ. Istanb.*, **1**, 3–27 (1954).
25 J. Kolars, 'Locational aspects of cultural ecology: the case of the goat in non-western agriculture', *Geogrl. Rev.*, **56**, 577–84 (1966).
26. T. R. Stauffer, 'The economics of nomadism in Iran', *Middle East Journal*, **22**, 284–302 (1965).
27. D. L. Johnson, *The Nature of Nomadism: A Comparative Study of Pastoral Migrations in Southwestern Asia and Northern Africa*, Department of Geography, Research Paper No 118, Chicago, 1969, 158–76.
28. W. A. Mitchell, 'Turkish villages in interior Anatolia and von Thünen's "Isolated State"; a comparative analysis', *Middle East Journal*, **25**, 355–69 (1971).
29. (a) L. Turkowski, 'Peasant agriculture in the Judaean Hills', *Palestine Exploration Quarterly*, **101**, 21–33, 101–12 (1969).
(b) H. E. Wulff, *The Traditional Crafts of Persia*, MIT Press, Cambridge (Mass) and London, 1966, 260–77.
30. P. Beaumont, 'The agricultural environment: an overview', in *Agricultural Development in the Middle East* (Eds P. Beaumont and K. McLachlan), John Wiley, Chichester, 1985, 3–26.
31. L. Eldblom, *Quelques Points de Vue Comparatifs sur les Problemes d'Irrigation dans les Trois Oases Libyennes de Brâk, Ghadamés et particuliément Mourzouk*, Lund Studies in Geography, No 22, Lund, 1961.
32. Z. Ron, 'Agricultural terraces in the Judaean mountains', *Israel Explor. J.*, **16**, 33–49, 111–22 (1969).

33. J.S. Birks, 'The reactions of rural populations to drought: a case study from south-east Arabia', *Erdkunde*, **31**, 299–308 (1977).
34. (a) P.W. English, 'The origin and spread of qanāts in the Old World', *Proceedings of the American Philosophical Society*, **112**, 170–81 (1968).
    (b) H. Goblot, 'Dans l'ancien Iran, les techniques de l'eau et la grande histoire', *Annales, Economies, Sociétés, Civilisations*, **18**, 400–520 (1963).
35. (a) P.H.T. Beckett, 'Qanāts around Kerman', *Royal Central Asian Journal*, **40**, 47–57 (1953).
    (b) E. Noel, 'Qanāts', *Royal Central Asian Journal*, **31**, 191–202 (1944).
    (c) J.C. Wilkinson, *Water and Settlement in South-East Arabia. A Study of the Aflaj of Oman*, Oxford University Press, 1977.
36. (a) J.A. Allan, 'Irrigated agriculture in the Middle East', in *Agricultural Development in the Middle East* (Eds P. Beaumont and K. McLachlan), John Wiley, Chichester, 1985, 51–62.
    (b) P. Beaumont, 'Qanāts in the Varamin Plain', *Trans. Inst. Br. Geogr.*, **45**, 169–80 (1968).
    (c) J.S. Birks, 'The reactions of rural populations to drought: a case study from south-east Arabia', *Erdkunde*, **31**, 299–308 (1977).
37. (a) M.E. Adams and J.M. Holt, 'The use of land and water in modern agriculture: an assessment', in *Agricultural Development in the Middle East* (Eds P. Beaumont and K. McLachlan), John Wiley, Chichester, 1985, 63–83.
    (b) J.A. Allan, 'Irrigated agriculture in the Middle East', in *Agricultural Development in the Middle East* (Eds P. Beaumont and K. McLachlan), John Wiley, Chichester, 1985, 51–62.

CHAPTER 5

Population

5.1 Historic perspective

Possibly the most tantalizing gaps in our knowledge of Southwest Asia and North Africa relate to population. Many attempts have been made to estimate population sizes for parts of the region at particular periods of history, but results are inconclusive. In the region as a whole, it is clear that the population was relatively small until the demographic upsurge of the last two or three hundred years. It is also clear that the first century population was larger than at any time until the eighteenth century. Table 5.1 (which does not correspond precisely with the region discussed in this book) gives some idea of population trends based on a variety of evidence.

The AD 1500 populations shown in Table 5.1 are consistent with calculations derived from the sixteenth century Ottoman censuses of Suleiman the Magnificent.[1] Estimates for Arabia and the Tigris–Euphrates region are unfortunately not shown. Arabia is estimated to have had a population of one million in AD 600 and the Tigris–Euphrates valley 9.1 million,[2] although there is more disagreement concerning the former populations of Iraq than any other part of the region. One estimate puts the population of Iraq as high as 20 million in the eighth to thirteenth centuries.[3] In general Arabia and the Tigris–Euphrates region undoubtedly experienced the same pattern of demographic decline and stagnation shown in Table 5.1. The most important causes of this decline were the disastrous plagues of the second, sixth, and fourteenth centuries which ravaged populations, though possibly on a smaller scale than in Europe. There were also local factors, usually arising from political instability. The long border struggle between the Arabs and Byzantines had a devastating effect on the populations of Syria and Iraq, and the Mongol invasion of Iraq in the thirteenth century marked the climax of a period of destruction and insecurity from which the rural areas have scarcely recovered. The most striking changes occurred in regions where high population densities could only be sustained by the painstaking upkeep of terraces or irrigation installations which were vulnerable to attack or deteriorated rapidly once neglected. Urban populations appear to have fluctuated even more markedly. Unlike today, towns were formerly subject to higher mortality than rural areas since they were more susceptible to epidemics of cholera, smallpox, and typhoid, as well as the plague.

From about the middle of the eighteenth century, for reasons that are not yet fully understood, populations in widely separated parts of the world began to show a simultaneous upward trend in their growth rates. What was the popu-

TABLE 5.1
Population estimates AD1–AD1500
(in millions)

	AD1	AD350	AD600	AD800	AD1000	AD1200	AD1340	AD1500
Asia Minor	8.8	11.6	7.0	8.0	8.0	7.0	8.0	6.0
Syria	4.4	4.0	4.0	4.0	2.0	2.7	3.0	2.0
Egypt	4.5	3.0	2.7	3.0	3.0	2.0	3.0	2.0
North Africa	4.2	2.0	1.8	1.0	1.0	1.5	2.0	3.5
	21.9	21.0	15.5	16.0	14.0	13.2	16.0	13.5

Source: J. C. Russell, 'Late ancient and medieval population', *Trans. Amer. Phil. Soc.*, **48**, Appendix B (1958).

lation of the Middle East on the eve of this unprecedented population increase? By backward extrapolation from the highest and lowest estimated 1920 populations of the region (including Afghanistan and Cyprus but not Egypt and Libya), J. C. Durand arrived at a 'low' estimate of 14 million for 1750 and a 'high' estimate of 44 million.[4] It now seems clear that the lowest figure for 1920 was much more accurate* and allowing for the addition of Libya and Egypt and the exclusion of Afghanistan and Cyprus, a 1750 population of 20 million is a fair estimate. By 1950 the same region contained approximately 44 million people,[5] having more than doubled in two hundred years. The significance of this increase becomes clear when it is realized that for the previous 10,000 years the world population had probably taken about one thousand years to double itself.[6] After the Second World War growth rates accelerated still further, largely in response to the widespread use of modern drugs and insecticides, and the population of the region can now double itself in 25 years or less.

5.2 Population size

The Middle East is widely regarded as having an inadequate and unreliable demographic data base. At the end of 1986 there were still three countries (Lebanon, Oman and South Yemen) which had never conducted a full national census. Only Egypt (since 1897), Turkey (since 1927) and Kuwait (since 1957), have a long series of census records.

Several factors account for the general paucity and poor quality of demographic information. There are the practical problems of census taking where a high proportion of the population are illiterate or live in remote rural areas – factors which also mean that civil registration is generally inadequate. There is also deep-seated suspicion since enumeration has traditionally been the prelude to

*The United Nations for example revised the 1950 estimated population of Southwest Asia *downwards* from 60 million to 44 million in the light of increasing census information. *UN Demographic Yearbook*, p 124 (1962), p 105 (1970).

taxation or conscription.[7] Political factors have also discouraged enumeration. In Lebanon, for example, a census would doubtless reveal radical changes in the delicate balance between Christians and Muslims.[8] In other cases results have remained unpublished because the enumerated population was either substantially smaller than previous estimates (for example the 1962 Saudi census), or revealed an unacceptably large foreign population (for example the 1970 Qatari census). Preliminary results from the October 1985 Turkish census had to be revised downwards by almost one million persons because village leaders had inflated their returns in order to support their claims for a greater share of public funds.[9] Nevertheless the broad characteristics of the region's population can be interpreted from available figures, while for individual countries, such as Kuwait and Bahrain, the statistics are generally reliable.

In 1985 the twenty countries of the Middle East region had a combined population of 264 million, roughly equivalent to that of North America. Table 5.2 ranks these countries by population, showing the enormous range in size from 50.7 million in Turkey to 0.29 million in Qatar. The range in size is strikingly illustrated by that fact that the annual increment of population in Egypt is greater than the total population of Oman.

TABLE 5.2
Populations of the Middle East in 1985

Country	Number	Per cent per annum	Growth Rate
Turkey	50,664,558	19.2	1.9
Iran	49,728,000	18.8	3.1
Egypt	46,637,980	17.7	2.7
Morocco	22,280,963	8.4	2.6
Algeria	20,203,666	7.7	3.1
Iraq	15,620,783	5.9	3.5
Saudi Arabia	11,215,477	4.2	3.8
Syria	10,452,516	4.0	3.7
Yemen AR	9,274,173	3.5	3.0
Tunisia	7,174,152	2.7	2.5
Israel	4,198,400	1.6	1.9
Libya	3,779,350	1.4	4.3
Lebanon	2,699,524	1.0	1.9
Jordan	2,689,572	1.0	5.2
Yemen, PDR	2,174,379	0.8	2.6
Kuwait	1,697,301	0.6	4.5
United Arab Emirates	1,600,000	0.6	7.6
Oman	1,218,804	0.5	4.7
Bahrain	427,271	0.2	4.2
Qatar	294,906	0.1	5.1
TOTAL	264,031,775	100.0	

Source: Based on data collated in *Demographic and related socio-economic data sheets 1984*, UN Economic and Social Commission for West Africa, Baghdād, (1985); *World Development Report 1985*, World Bank, Washington DC, (1986), and various national censuses.

TABLE 5.3
National and non-national populations in the labour-importing countries, 1985

Country	Total Population	Nationals	Non-nationals	Per cent Non-nationals
Saudi Arabia	11,215,477	8,609,923	2,605,554	23.2
Libya	3,779,350	3,363,622	415,728	11.0
Kuwait	1,697,301	681,288	1,016,013	59.9
United Arab Emirates	1,600,000	341,570	1,258,430	78.7
Oman	1,218,804	878,634	340,170	27.9
Bahrain	427,271	277,319	149,952	35.1
Qatar	294,906	121,115	173,791	58.9
TOTAL	20,233,109	14,273,471	5,959,638	29.5

Source: as Table 5.2

Three countries (Turkey, Iran and Egypt) account for almost 56 per cent of the region's total population. Moreover, the two largest populations in the region, Turkey and Iran, are both non-Arab. These two countries account for 38 per cent of the region's total population. The three largest Arab populations (Egypt, Morocco and Algeria) are all in North Africa rather than Southwest Asia. Overall North Africa accounts for 37.9 per cent of the region's population, compared to 10.5 per cent in the Arabian Peninsula.

The populations of the six Gulf Co-operation Council (GCC) countries (Bahrain, Kuwait, Oman, Qatar, Saudi Arabia, United Arab Emirates) have been considerably increased by the influx of immigrant workers and their families since the oil price increases of the early 1970s.[10] In the smaller Gulf states nationals are a minority in the total population (Table 5.3). The proportion of non-nationals is 59 per cent in Qatar, 60 per cent in Kuwait and almost 79 per cent in the United Arab Emirates. Even in Saudi Arabia almost one quarter of the population are foreign.

5.3 Birth rates

Crude birth rates are generally high throughout the region (Table 5.4). Half the countries of the region have a crude birth rate exceeding 40 per thousand. In the smaller Gulf States crude birth rates are biased downwards by the presence of large, and predominantly male, migrant populations. In Qatar, for example, the crude birth rate for nationals is much higher, at 45.8 per thousand, than that of non-nationals, at 29 per thousand. Among the *national* populations of the Gulf States crude birth rates range from 42.3 per thousand in Kuwait, to 46.5 in Bahrain. These countries have pro-natalist population policies which include the payment of substantial family allowances, the provision of free health and education services, as well as restrictions on contraceptive advertising and sales.[11] The promotion of rapid population growth among nationals is regarded as an important step in reducing dependence on non-national manpower. The two countries with the lowest crude birth rates in the region, Lebanon and Israel, both have substantial non-Muslim populations with significantly lower total fertility rates.

In most countries crude birth rates have fallen only marginally over the last twenty years. The persistence of high birth rates can be attributed to a number of factors, in particular to early and universal marriage. Moreover, high fertility fosters security by increasing the number of relatives and ensuring that parents will not be abandoned in their old age.[12]

There is evidence of significant fertility differentials, for example, between different religious communities in the Lebanon[13] or between rural and urban areas in Egypt and Syria.[14] There is some evidence that the increasing educational attainment, and formal labour force participation of women, is leading to a decline in total fertility.[15] In Jordan, for example, the total fertility rate of women is seen to decline from 8.3 for those with no formal education, to 7.6 with primary education and 3.4 with secondary education completion.[16]

5.4 Death rates

There is a very wide range in crude death rates in the region (Table 5.4). Mortality is lowest in the oil-rich Gulf countries, ranging from 3.7 per thousand in Kuwait and 3.8 in the UAE to 6.6 per thousand in Qatar. These rates are however distorted by the presence of migrant worker populations. With their age distribution skewed towards the 20–49 age cohorts, mortality rates tend to be lower among non-nationals than nationals. In the UAE, for example, the crude death rate for non-nationals is 2.7 per thousand, compared to 7.1 per thousand for nationals.

Higher death rates prevail in the poorer countries, particularly those with a large and dispersed rural population. These range from 22 per thousand in the Yemen Arab Republic (YAR) to 19 per thousand in South Yemen (PDRY) and 14 per thousand in Morocco. Mortality rates also tend to vary considerably within countries. Remarkable reductions have been affected in some urban areas where better housing and sanitation are combined with access to medical services.

Over the last twenty years mortality rates have fallen significantly in several countries. In Syria, for example, crude death rates have fallen from 16 per thousand in 1965 to 8.3 per thousand in 1985. In the Gulf States the reduction in crude death rates has been even more marked, falling by over 60 per cent in the case of Kuwait.[17]

The reduction in mortality is primarily a result of a sharp decline in infant deaths.[18] These are lowest in the Gulf States (for example, infant deaths in the first 12 months of life are 44 per thousand live births in Bahrain) where the highly urbanized population has ready access to medical services. In the poorer countries with a large rural population, infant mortality remains high, ranging from 82 per thousand in Iraq to 104 in Egypt, 137 in South Yemen and 164 per

TABLE 5.4
Demographic parameters 1985

Country	Crude Birth Rate (per 000)	Crude Death Rate (per 000)	Per cent under 15 years	Infant Mortality Rate (per 000)	Natural Increase
Yemen AR	49.6	22.1	47.4	164.0	2.8
PDR Yemen	49.3	18.7	48.2	137.0	3.1
Oman	47.5	15.3	44.1	116.5	3.2
Libya	45.0	11.0	47.0	91.0	3.1
Iraq	43.8	7.9	45.7	81.8	3.6
Syria	43.0	8.3	49.2	60.0	3.5
Saudi Arabia	41.8	12.6	44.7	110.0	2.7
Jordan	41.5	6.7	48.1	63.1	3.5
Morocco	40.0	14.0	46.0	98.0	2.6
Iran	40.0	10.0	45.3	100.0	3.1
Algeria	39.5	8.9	47.1	107.0	3.1
Egypt	37.4	10.9	39.0	104.5	2.7
Bahrain	36.8	5.9	33.0	44.1	3.1
United Arab Emirates	34.9	3.8	28.4	42.2	3.1
Tunisia	33.0	9.0	40.0	83.0	2.5
Kuwait	32.9	3.7	39.9	23.8	2.9
Turkey	31.0	9.0	39.7	82.0	1.9
Qatar	30.2	6.6	36.2	46.6	2.4
Lebanon	29.8	8.8	49.6	44.4	2.1
Israel	24.0	7.0	33.2	14.0	2.3

Source: As Table 5.2

thousand in North Yemen. Mortality statistics are notoriously unreliable in the Middle East and real levels, particularly in the poorer countries, may well be much higher than these official figures suggest.

The decline in death rates has been the chief cause of rapid population increase in the region as a whole. Given the room for further improvements in the mortality figures, population growth rates are likely to remain high and may increase.

5.5 Growth rates

The combination of falling death rates and high birth rates has resulted in rapid population growth in the Middle East. Between 1950 and 1960 the average annual increase was 2.5 per cent; it is now around 2.8 per cent per annum. Natural increase ranges from 1.9 per cent per annum in Turkey to 3.5 per cent in Jordan and Syria, and to 3.6 per cent in Iraq. High rates of immigration into the oil-producing GCC states has given them the highest rates of total population growth. These range from 3.8 per cent per annum in Saudi Arabia to 5.1 per cent in Qatar and 7.6 per cent in the UAE.

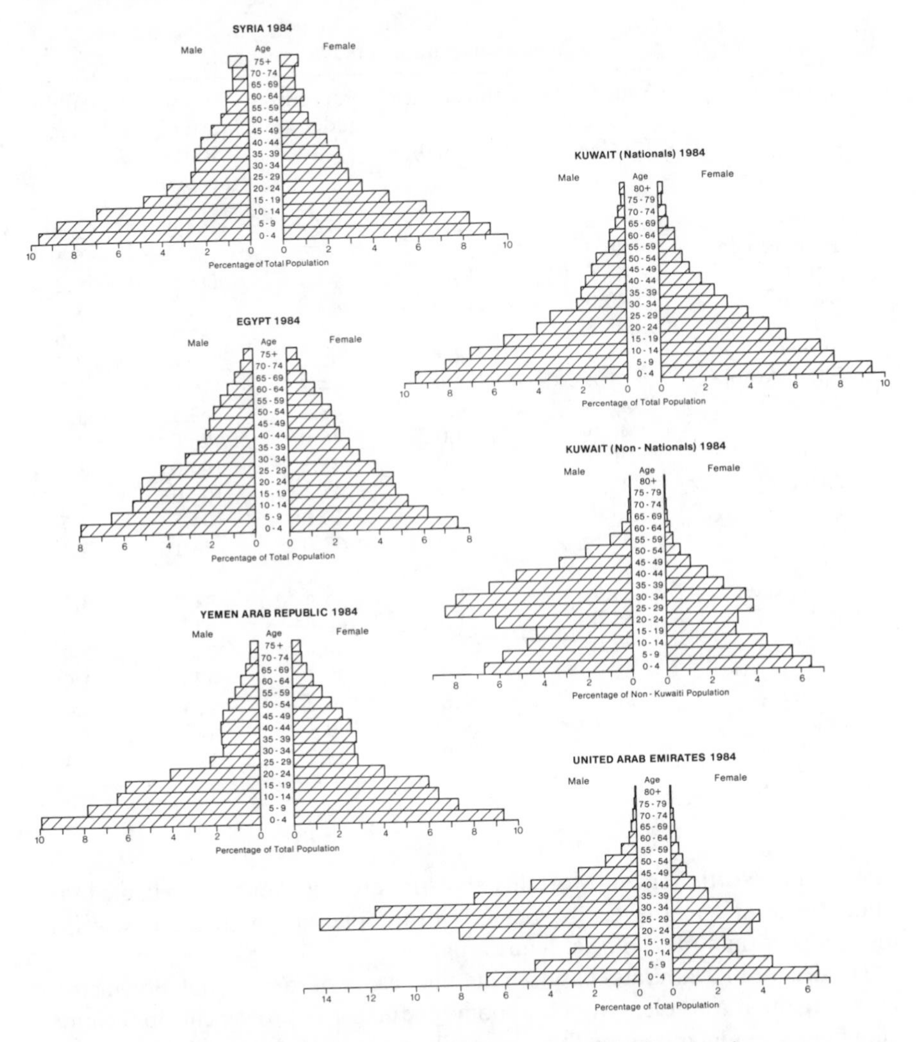

Figure 5.1 Age–sex pyramids for selected Arab countries

One characteristic of rapidly increasing populations is their youthful age structures. Age structure is important because it indicates the ratio of dependent groups to the active population, and it is also a major determinant of future population growth rates. Figure 5.1 illustrates the age-sex pyramids of selected countries. With the exception of the UAE and of Kuwait's non-national population, these all display very broad bases associated with extreme youthfulness. The proportion of the population under the age of 15 ranges from 39 per cent in Egypt to 47 per cent in North Yemen, and 49 per cent for Syria and Kuwaiti nationals. In contrast the proportion aged 65 and over is very small, ranging from 2 to 4 per cent. The age-sex pyramid for the UAE clearly shows the influence of its large migrant population in which male migrants aged 25–44 form the majority.

Kuwait has a longer history of immigration. Many migrant workers are accompanied by their dependants and a large proportion of the non-national community were born in Kuwait.[19] Overall some 34 per cent of non-nationals are aged under 15. Nevertheless, males aged 20–44 still form a prominent majority.

5.6 International migration

International labour migration has had a significant impact on the population geography of the Middle East. Today more than 5.5 million workers are employed outside their home countries. Until the early 1970s labour emigration was dominated by outflows to Western Europe from Turkey and the Maghreb states. With the oil price increases of 1973–74 new demands for expatriate manpower were stimulated within the region itself. By 1975 the six Gulf oil exporters, together with Libya, had a foreign labour force of about 1.6 million, representing some 17 per cent of their total workforce.[20] Recent estimates suggest that by 1985 the immigrant labour force in the Arab World exceeded 4.3 million, accounting for some 28 per cent of the total workforce.[21]

The greatest number of foreign workers are employed in Saudi Arabia and Libya. However, the relative dependence on non-nationals in the labour force is greatest in the smaller Gulf States such as Kuwait where immigrants account for 66 per cent of the total workforce, Qatar (86 per cent) and the UAE (90 per cent).[22]

In the mid-1970s the majority of these migrant workers came from surrounding Arab states, in particular Egypt, North Yemen and Jordan. Between 1975 and 1985, however, the Arab share of labour inflows has fallen from 65 to 48 per cent as the stock of Asian workers has risen to almost two million. The initial boom in Asian labour flows came from South Asia, particularly Pakistan and India. In the later 1970s these were joined by countries such as Korea, Thailand, the Philippines and the People's Republic of China.[23]

The growth of the expatriate workforce has been accompanied by the development of significant immigrant communities in the settlement of migrant workers' dependants.[24] Table 5.5 compares the size of the migrant worker population in 1975 and 1985. In the case of the UAE the majority of immigrants, and hence of the total population, are now non-Arab in origin.

The demographic evolution of these immigrant communities poses several crucial policy issues for the host countries. As well as adding to the costs of infrastructural and service provision, immigrant communities are increasingly seen as posing a threat to national culture and identity.[25]

Another state to have experienced high immigration rates is Israel, which received 1,407,000 Jewish immigrants between 1948 and 1971.[26] About half of these were from Muslim countries in North Africa and Southwest Asia. During the first four years of the state, the average annual rate of population increase was 23.7 per cent. At times in 1949 as many as 20,000 immigrants were arriving in Israel each month.[27]

TABLE 5.5
Nationals, non-national Arabs and other non-nationals in the labour-importing countries, 1975 and 1985

Year 1975 Country	Nationals number	Per cent	Non-national Arabs number	Per cent	Other non-Nationals number	Per cent	Total number	Per cent
Saudi Arabia	4,592,500	74.6	1,429,000	23.2	136,000	2.2	6,157,500	100.0
Libya	2,223,700	80.7	487,425	17.7	44,050	1.6	2,755,175	100.0
Kuwait	472,000	48.4	402,400	41.3	100,100	10.3	974,500	100.0
United Arab Emirates	200,000	30.5	120,200	18.3	335,800	51.2	656,000	100.0
Oman	650,000	83.1	21,000	2.7	111,250	14.2	782,250	100.0
Bahrain	214,000	79.3	12,200	4.5	43,800	16.2	270,000	100.0
Qatar	67,900	41.2	31,800	19.3	65,200	39.5	164,900	100.0
TOTAL	8,420,100	71.6	2,504,025	21.3	836,200	7.1	11,760,325	100.0
Year 1985 Country	Nationals number	Per cent	Non-national Arabs number	Per cent	Other non-Nationals number	Per cent	Total number	Per cent
Saudi Arabia	8,609,923	76.8	2,052.432	18.3	553,122	4.9	11,215,477	100.0
Libya	3,363,622	89.0	294,789	7.8	120,939	3.2	3,779,350	100.0
Kuwait	681,288	40.1	699,288	41.2	316,725	18.7	1,697,301	100.0
United Arab Emirates	341,570	21.3	244,800	15.3	1,013,630	63.4	1,600,000	100.0
Oman	878,634	72.1	24,376	2.0	315,794	25.9	1,218,804	100.0
Bahrain	277,319	64.9	22,645	5.3	127,307	29.8	427,271	100.0
Qatar	121,115	41.1	85,228	28.9	88,563	30.0	294,906	100.0
TOTAL	14,273,471	70.5	3,423,559	16.9	2,536,079	12.5	20,233,109	100.0

Source: J. S. Birks, 'The demographic challenge in the Arab Gulf', *Arab Affairs*, **1**, 1986, 84.

Two major displacements of population may be mentioned. In 1948 some 700–800,000 Palestinians, approximately half the Arab population of Palestine, became refugees as a result of fighting between Arabs and Jews. The majority fled from Jewish controlled areas into what became the Gaza Strip and the West Bank. In June 1967 another 350,000 Palestinians were again uprooted as Israeli forces took the West Bank and Gaza. The majority of refugees fled initially to the East Bank of Jordan, and subsequently moved on to Lebanon, Syria and the Gulf States following the 1970 civil war in Jordan.

A population movement of a rather different kind occurred between 1922 and 1924 when approximately 1.6 million Greek and Armenian-speaking peoples left Turkey, the former in exchange for some 400,000 Turkish speakers from Greece.[28] After the Second World War many Muslims also left Bulgaria to settle in Turkey. The most important exodus of Europeans from the Arab world in recent years was the large French settler community of the Maghreb, but Libya has also witnessed the departure of well over 100,000 Italians (about 15 per cent of the population in 1940) since the outbreak of the Second World War.

5.7 Distribution and density of population

This large region remains overall one of the least densely populated in the world, in spite of rapid rates of population increase which have created marked local increases in density during recent decades. The chief explanation is in the large tracts of arid and semi-arid land, and the more limited areas of high altitude unsuitable for permanent settlement. Southwest Asia has an average density of 17 persons per sq km, and North Africa, including the Maghreb, 10 persons per sq km. Such figures are of limited value since most states possess large tracts of almost uninhabited land, and over one third of the total population are urban dwellers. Lebanon's average density of 268 persons per sq km compares with the densely populated states of western Europe largely on account of the city of Beirūt. A more useful index is obtained by considering the density of agricultural populations on arable land, though there are problems arising from different definitions. The highest densities are associated with irrigated lands of river valleys and desert oases. By far the highest densities are encountered in parts of the Nile valley and delta where entire rural governorates record over 800 persons per sq km. Every country, however, displays considerable contrasts in population density; even Turkey which has less obvious signs of sparsely populated regions has rural densities ranging from seven to 127 per sq km. In general, physical controls, notably the availability of water, account for these contrasts, but historical factors may sometimes be responsible.

Figures 5.2 and 5.3 illustrate the distribution of population and the broad range of densities. Apart from the large areas of uninhabited semi-arid land, the abrupt changes in density are a notable feature, reflecting the rainshadow effect of some of the mountain ranges, and the margin between irrigated lands and desert particularly along the Nile valley and delta. It should be noted that the population

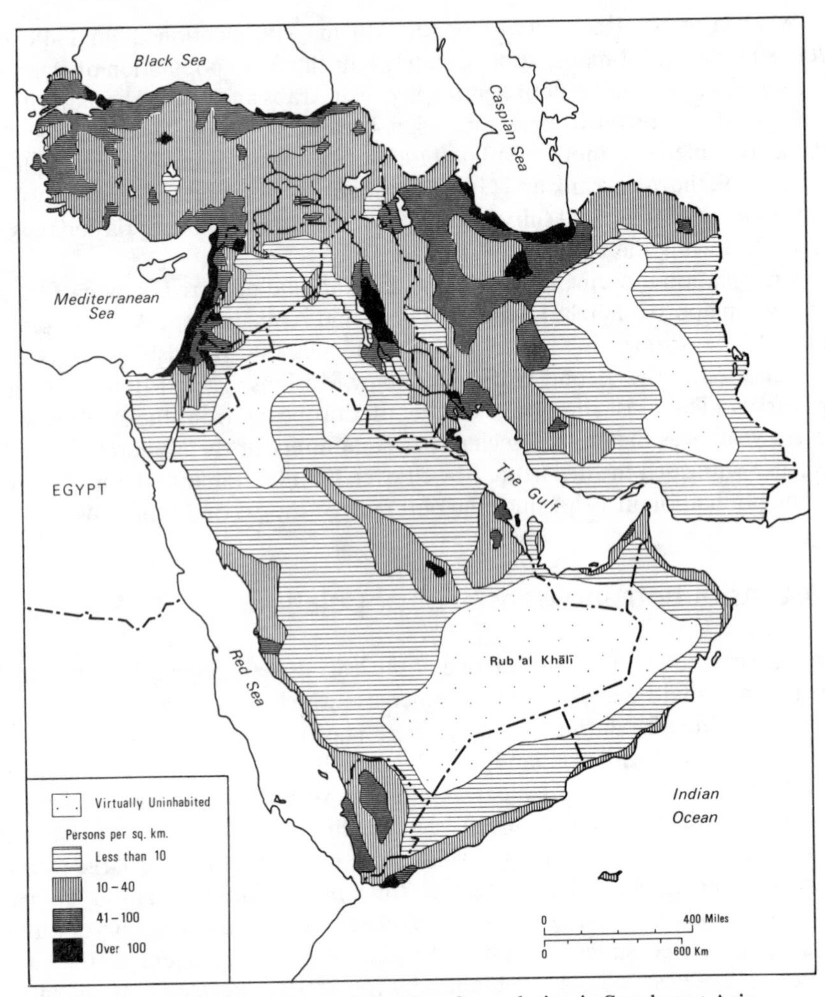

Figure 5.2 Distribution and density of population in Southwest Asia

distribution of the region has never been static. In certain regions the frontier between the desert and the sown has migrated in response to political and economic change. One famous example is the northern Negev, which was an area of nomadism in Old Testament times; under the Nabateans and their immediate successors a dozen flourishing cities grew up which had been largely abandoned by the seventh century AD. Today, modern technology and the needs of national security have again stimulated desert colonization in southern Israel. Of far more significance, however, are a number of large dams and barrages recently constructed in Turkey, Syria, Iraq and Iran which have brought large irrigated areas under the plough.

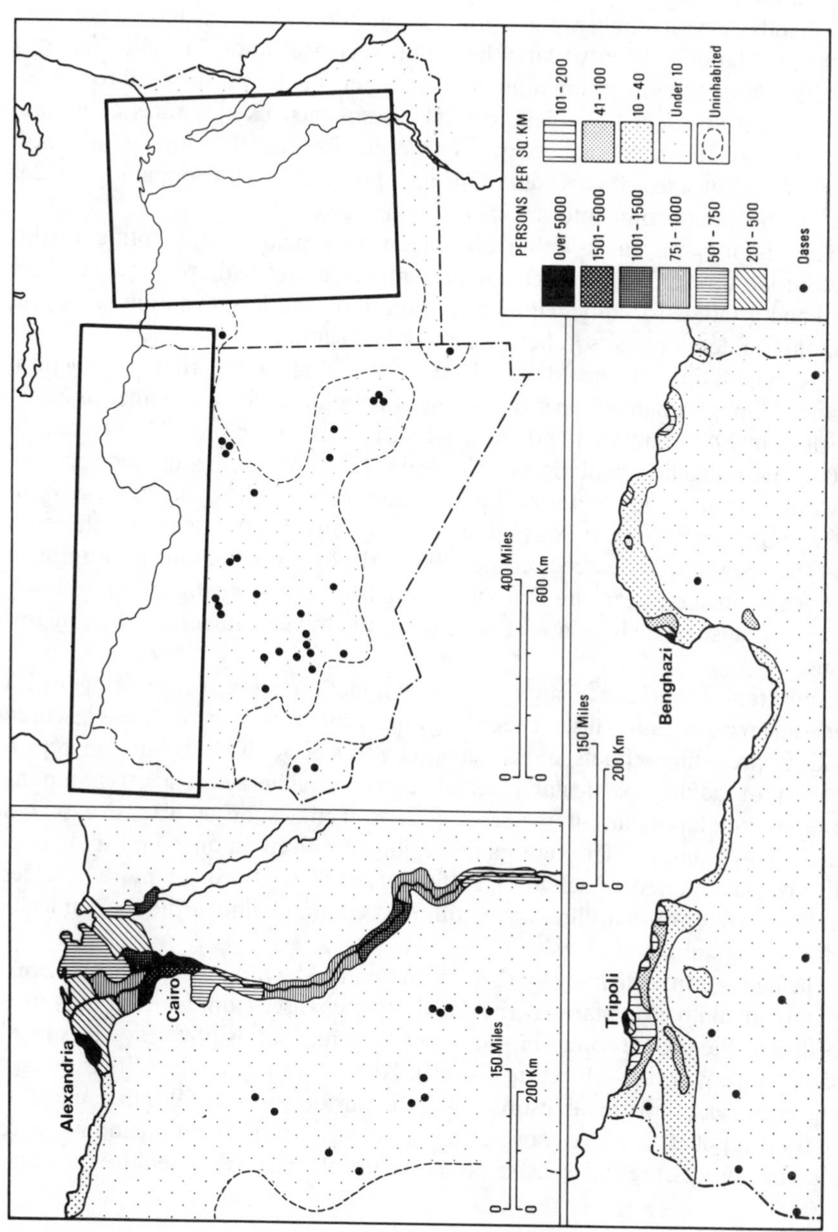

Figure 5.3 Distribution and density of population in Egypt and Libya

5.8 Conclusion: population growth and change

Population growth rates of around 3 per cent per annum which prevail in the region today have a number of important implications. These include: a doubling of the population in twenty years; the need to increase national production by at least the same rate just to maintain per capita income levels; the need to double employment and social overhead capital investments. Even if rates of natural increase were drastically reduced at once, population growth would not diminish over the next two decades because of the high proportion of the population under age 15 who would come into the child-rearing age cohorts.

The majority of states within the region have paid little attention to this Malthusian threat. While direct population control has had little impact, there are good grounds for anticipating a medium term fall in population growth in response to developments in health care and education.

The population of the Middle East is also undergoing radical structural changes. Approximately 40 per cent of the population could be classified as urban dwellers in 1985 compared with 35 per cent in 1970 and perhaps 10 per cent in 1900. The nomadic population of the region, in contrast, is experiencing considerable pressure to sedentarize, both indirectly and by direct action of governments.[29] Estimates of their exact number vary greatly, but there can be little doubt that they now scarcely exceed one per cent of the total population. Two-thirds of the people in the region are still village dwellers, but the proportion is slowly falling because of higher natural increase in the towns and rural–urban migrations.

Important qualitative changes in the population are also occurring which generally receive far too little recognition. In recent years, great progress has been made in providing schools and educational services of all kinds but a high proportion of adults, particularly females, remain illiterate, and even among children rapid population increases have sometimes offset extensive school building programmes. Throughout the region it is still common to find about one-third of the children of school age illiterate, and two-thirds or more of older adults illiterate, so that illiteracy is still most common among the economically active population.

The fight against illiteracy has been accompanied by progress in other spheres, notably in combating many of the debilitating diseases from which a high proportion of the people commonly suffered. Malnutrition is now less widespread than it once was, though recent work in Arab countries indicates that dietary deficiencies and nutritional disorders are still far too common. Besides augmenting food supplies, a whole range of measures need to be taken, including more local food processing and enrichment, pest control, and general health education.

References

1. O.L. Barkan, 'Essai sur les données statistiques des registres de recensement dans l'Empire Ottoman aux XVe et XVIe siècles', *J. Econ. Soc. Hist. Orient*, **1**, 9–36 (1958).
2. J.C. Russell, 'Late ancient and medieval population', *Trans. Amer. Phil. Soc.*, **48**, 89 (1958).
3. R.I. Lawless, 'Iraq: changing population patterns', in *Populations of the Middle East and North Africa* (Ed J.I. Clarke and W.B. Fisher), University of London Press, London 1972, 97.
4. J.D. Durand, 'The modern expansion of world population', *Proc. Amer. Phil. Soc.*, **III**, 151 (1967).
5. United Nations, *Demographic Yearbook*, New York, 1962, 124.
6. J.D. Durand, 'The modern expansion of world population,' *Proc. Amer. Phil. Soc.*, **III**, 137 (1967).
7. J.C. Hurewitz, 'The politics of rapid population growth in the Middle East', J. Int. Affairs, **19**, 27 (1963).
8. J.M. Wagstaff, 'A note on some nineteenth-century population statistics for Lebanon,' *Bull. Brt. Soc. Middle East Stud.*, **13**, 1986, 36–44.
9. *Middle East Economic Digest*, 30 (25 October 1986), 50.
10. J.S. Birks and C.A. Sinclair, *International Migration in the Arab Region*, International Labour Office, Geneva, 1980, 137.
11. J.S. Birks, 'The demographic challenge in the Arab Gulf,' *Arab Affairs*, **1**, 72–86 (1986).
12. J.C. Caldwell and P. Caldwell, 'Fertility transition with special reference to the ECWA region,' in *Population and Development* in the Middle East, ECWA, Baghdad, 1982, 92–118.
13. J. Chanie, 'Religious differentials in fertility: Lebanon,' *Population Studies*, **31**, 365–82 (1977).
14. M.L. Samman, *La population de la Syrie. Etude geo-démographique*, ORSTOM, Paris, 1978, 260.
15. R.P. Shaw, *Mobilizing Human Resources in the Arab World*, Kegan Paul, London, 1983, 105.
16. H. Rizk, (a) 'Fertility trends and differentials in Jordan,' in *Women's status and fertility in the Muslim World* (Ed J. Allman), Praeger, New York, 1978, 113; (b) *Jordan Demographic Survey 1981*, Department of Statistics, Amman, 1983, 64.
17. M. Ryan, *Health Services in the Middle East*, Economist Intelligence Unit, London, 1984.
18. K. Abu Jaber (Ed), *Levels and Trends of Fertility and Mortality in Selected Arab Countries of West Asia*, University of Jordan, Amman, 1980.
19. M. Galaleldin, 'Demographic consequences of alternative types of migration in the Middle East,' in *Proceedings of the Workshop on the Consequences of International Migration*, IUSSP, Canberra, 1984.
20. J.S. Birks and C.A. Sinclair, *Arab Manpower*, Croom Helm, London, 1980.
21. R. Owen, *Migrant Workers in the Gulf*, Minority Rights Group, London, 1985, 18.
22. I.J. Seccombe, 'Economic recession and international labour migration', *Arab Gulf Journal*, **6**, 43–52, 1986.
23. F. Arnold and N.M. Shah, 'Asian labour migration in the Middle East,' *International Migration Review*, **18**, 302, 1984.
24. J.S. Birks and C.A. Sinclair, 'Demographic settling amongst migrant workers,' IUSSP, Manilla, 1981.
25. J. Abu-Lughod, 'Social implications of labour migration in the Arab world,' in *Arab Resources: the Transformation of a Society* (Ed I. Ibrahim), Croom Helm, London, 1983, 237–66.

26. State of Israel, Statistical Abstract, **23**, Jerusalem, 1972, 127.
27. G. H. Blake, 'Israel: immigration and dispersal of population,' in *Populations of the Middle East and North Africa* (Eds J. I. Clarke and W. B. Fisher), University of London Press, London, 1972, 185.
28. J. C. Dewdney, 'Turkey: recent population trends,' in *Populations of the Middle East and North Africa* (Eds J. I. Clarke and W. B. Fisher), University of London Press, London, 1972, 42.
29. A. R. George, 'Processes of sedentarisation of nomads in Egypt, Israel and Syria,' *Geography*, **48**, 167–9 (1973).

CHAPTER 6

Towns and Cities

6.1 Introduction

The importance of towns and cities in the contemporary life of Southwest Asia and North Africa is demonstrated by the fact that almost half the population are urban dwellers, and the proportion is rapidly increasing. Chapter 8 discusses the modern role of towns as centres of economic and social change, but the influence of urbanism has been strong throughout the region for many centuries, even at times when the actual level of urbanization has been quite low. It is true that relations between towns and country have sometimes been weak, and the two have had little in common politically, socially and culturally. Yet successive political regimes have governed their territories from the towns, and almost every major movement to have spread through the region (Christianity, Islam, westernization, and nationalism) has done so through the medium of the town. Economically, the fortunes of town dweller and countryman have probably been rather more closely interdependent than is sometimes recognized. Long-distance trade between Europe and Asia and Europe and Africa gave many of the towns of the region considerable prosperity, but their day-to-day survival often depended on the success of local agriculture.

This chapter is largely devoted to an examination of the morphology and regional distribution of towns, rather than with the functions and processes of urbanization in individual countries. In the centuries following the death of Muhammad until the middle of the nineteenth century, Islam rebuilt and refashioned the towns of Southwest Asia and North Africa from their foundations, giving the region one of its most distinctive elements. For this reason some attention is given to the Maghreb in this chapter. With rapid urban growth over the last hundred years and the creation of a number of new towns, it has become misleading to talk of 'the Islamic town', though the impress of Islam survives to the present day. The region may still be distinguished from other developing regions of the world by its high level of urbanization, rapid rates of urban growth, and the relative significance of inland towns, in spite of the development of many ports between about 1850 and 1930. In common with other developing regions however, a high degree of urban primacy and a scarcity of medium-sized towns is often manifest, together with the first signs of the emergence of metropolitan core areas.

6.2 Pre-Islamic towns

6.2.1. Ancient towns

In ancient times the largest concentrations of population were along the valleys of the Tigris–Euphrates, the Nile, and Indus rivers. Until recently, the earliest cities were assumed to have arisen in these regions sometime during the fourth millenium BC, when improved irrigation techniques combined with stock rearing and fishing began to yield surpluses of food. The availability of river transport supplemented by wheeled vehicles facilitated the concentration of surplus at a few centres, while the need to canalize water and protect habitations from flooding also encouraged the aggregation of populations. When and where the first cities grew up is not certain, but Eridu is still popularly regarded as the oldest city on earth.[1] The Neolithic town of Çatal Hüyük in Anatolia, discovered in 1961, may have a stronger claim to this distinction,[2] and it may yet be proved that the emergence of urban life in Anatolia actually *preceded* the beginnings of agriculture, which could have originated in cities.[3] It was certainly in southern Mesopotamia that a network of walled towns appeared at an earlier date than elsewhere; some are shown on Figure 6.1. Although many of their inhabitants were farmers or fishermen, these ancient towns were genuinely urban, supporting many non-agriculturalists such as priests, administrators, traders and artisans. With populations of 7,000 to 25,000 they were at least ten times larger than any contemporary villages.[4] Ornate public buildings constructed on a monumental scale have been excavated, including temples, workshops, and granaries, but knowledge of domestic building is less complete. The houses of most ordinary citizens were probably mud constructions which have entirely disappeared, but some impressive domestic buildings have been unearthed. At Ur, for example, a variety of house types dating from 2000 BC have been discovered, including two-storey constructions based on the courtyard principle complete with highly efficient fresh water and sewage systems. Yet more elaborate houses have been found in the Indus cities of Mohenjo-Daro and Harappa. Some possessed two storeys and an interior staircase, while wells and sophisticated drainage systems appear to have been standard. Building materials included kiln-fired bricks for the thick exterior walls, cedarwood from the Himalayas, and a damp course of bitumen.[5]

Some ancient towns, including Ur, were partially planned. A remarkable example is the Egyptian town of Kahun, laid out in 2600 BC as a colony for workmen engaged in pyramid construction. Streets were arranged in gridiron pattern, with a standard house type consisting of three rooms and an open court-yard; a few double storey houses were included for foremen and officials.[6] Such comprehensive planning is not evident in the large cities, although it seems clear that at least in Sumeria all towns were laid out along basically similar lines. Three main elements can generally be identified; a walled city; suburbs, including more houses, but also fields, groves and cattle folds; and a commercial quarter. Trade between Egypt, Mesopotamia and the Indus has long been recognized as an

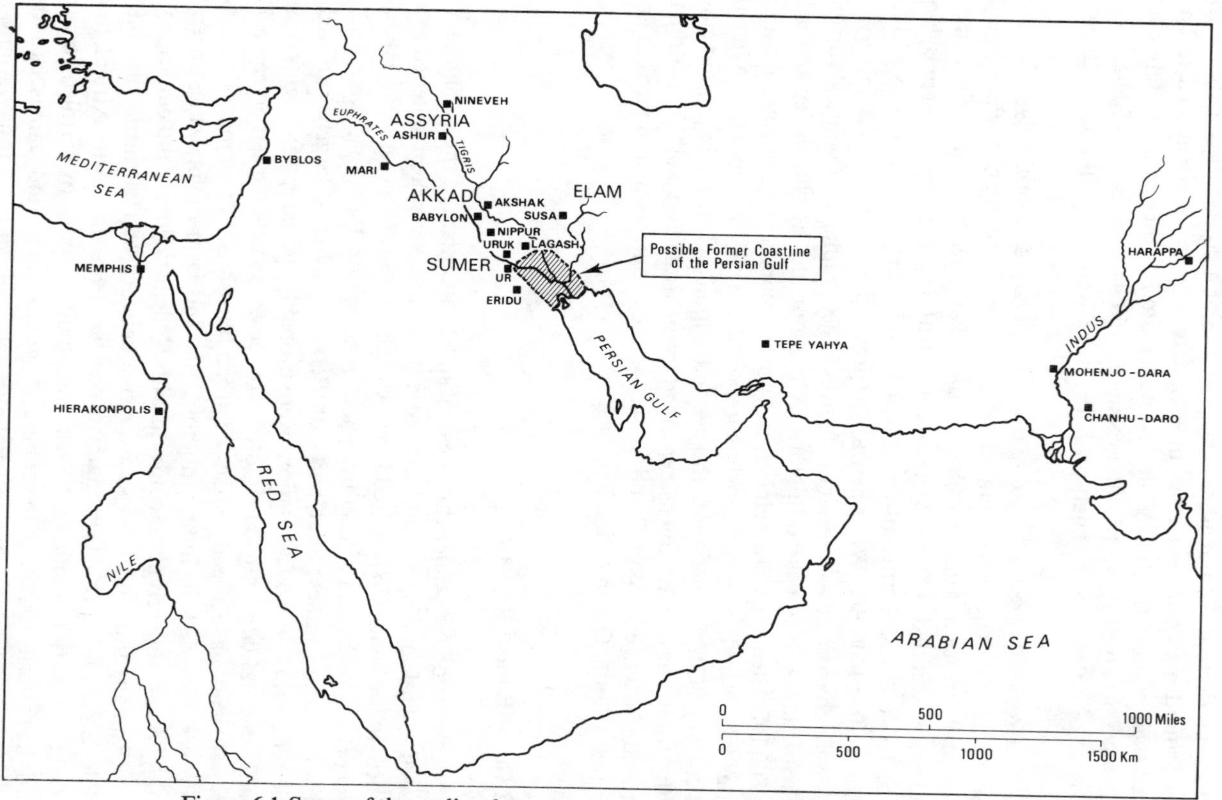

Figure 6.1 Some of the earliest known centres of urban civilisation, 4500 BC to 3000 BC

important factor in town building in the three great river valleys civilizations. Excavations at Tepe Yahya in southern Iran suggest the existence of another urban, literate civilization just as early, linked with Mesopotamia and the Indus by trade, but derived from neither. The existence of a green belt of gardens and houses around many ancient cities may explain why some are alleged to have covered such vast areas. During the first millenium BC for example, Babylon is said to have occupied some 1,000 ha, Nineveh 750 ha, and Erech 450 ha.[7] Such cities were of course exceptional; probably most were more like 10 to 20 ha in area.[8]

The network of pre-Greco-Roman towns changed continually. The twin requirements of defence and commerce stimulated the growth of towns throughout the fertile crescent, but particularly in the Levant. The eastern Mediterranean coast became studded with ports engaged in maritime trade, among them Byblos, Tyre and Sidon. The Hittites also created a few small urban centres in Asia Minor sometime after 1400 BC. While new towns grew up in Southwest Asia many old towns were abandoned as a result of changes in geographical or political fortune. The first true towns appeared in North Africa outside the Nile valley and delta after 800 BC when the Carthaginians took over a number of Phoenician trading stations which had been previously garrisoned but never colonized. After the sixth century BC Carthage itself, Tangier and Tripoli all became important cities, but the total number of towns remained comparatively small. The Romans destroyed Carthage in 146 BC and by AD 42 they controlled an unbroken zone from the Atlantic Ocean to the Red Sea.

6.2.2 Greco-Roman towns

The conquests of Alexander the Great from 334 BC marked the beginning of a phase of town-founding in Southwest Asia which has never been matched before or since. In less than two hundred years Alexander and his successors established a network of towns and roads throughout Asia Minor, Palestine, Syria, and the Tigris–Euphrates valleys and parts of Persia which has largely survived to modern times. *De novo* foundations were probably the exception, rights of self-government and city status being often conferred on existing settlements which may or may not have already possessed urban functions. Rather more than eighty towns were founded, including, notably, Alexandria in Egypt, while dozens of existing towns were rebuilt to conform with Hellenistic ideals and culture.[9]

The Romans thus inherited an extensive network of Hellenistic cities in the Middle East, and a few Carthaginian coastal towns in North Africa. In the provinces of Mesopotamia and Syria, they built a series of fortified towns to guard the eastern frontier of the empire; Aqaba, Amman and Damascus and the great cities of Jarash, Busra and Palmyra which lie in ruins today. Many new towns were built in Asia Minor for a variety of functions besides defence. Another region where extensive Roman urbanization occurred was in the North African provinces of Africa (modern Tunisia) and Numidia Tripolis. Ports, route centres, and fortified frontier towns were established, sometimes on the site of Carthaginian settlements. Both in North Africa and Southwest Asia a large number of Roman foundations were subsequently abandoned including rich and magnificent cities such as Leptis Magna and Jarash. Today, it is difficult to identify more than two dozen or so inhabited towns of genuine Roman origin in the Middle East and North Africa. Most of these are in Algeria and in Turkey, where two (Adana and Diyarbakir) have over 250,000 inhabitants. Persia never became part of the Roman empire, but several towns were founded there during the Sassanid dynasty (AD 226 to AD 651) most of which survive, including Ahvāz, Kermānshāh and Shīrāz.

Many Greco-Roman towns and cities declined in importance or where abandoned altogether in later centuries. W.C. Brice,[10] writing of Southwest Asia, attributes this to six major factors. First, in Asia Minor, Syria and Palestine, exploitation of soil, pasture and forest occurred in response to overseas demands for timber, grain, wool, horses, and other products of the land. By the third century AD soil erosion and overgrazing were so severe that some cities once dependent upon agriculture found themselves at the heart of an impoverished hinterland. Second, silting occurred along the coasts as a result of excessive losses of soil, leading to the abandonment of several ports such as Ephesus and Miletus. Third, hydrological changes affecting water supply through upstream diversion or silting spelt ruin as in the case of Harran in Turkey and Timgad in Tunisia. Similarly changes in the course of a stream could change the value of a site. Seleucia, for example, was deprived of its harbour with the eastward migration of the Tigris river. Fourth, many ancient towns depended on elaborate aqueducts, canals and tunnels for their water supply, and such installations were easily destroyed by raiders. The Vandals in North Africa in the fifth century, the Beni Hillal in the eleventh, the Seljuks in Asia Minor in the twelfth, and the Mongols in Iraq in the thirteenth century were all responsible for the death of towns. The sporadic raids of desert nomads also disrupted communications, destroyed effective administration, and ruined settlements. Fifth, political changes rendered Roman garrison towns obsolete both in North Africa and along the eastern frontier of the empire in Southwest Asia. Sixth, earthquakes probably dealt the final blows to some towns where economic foundations were already weak, particularly in Asia Minor where over 200 destructive earthquakes were chronicled between AD 33 and 1900.[11]

By far the most important factor however, was the changing pattern of trade and commerce following the break-up of the Roman empire. For centuries Southwest Asia in particular had derived great prosperity from transporting luxury goods between Europe and Asia, while many ports engaged in local trade as well. The importance of the overland routes ceased as the demand for goods declined in Europe, and much of Southwest Asia was plunged into years of savage warfare between Byzantium and the Persian Sassanids, depriving ports and inland towns alike of their livelihood. Thus Palmyra, Antioch, and the remarkable Nabatean cities of the Negev desert fell into ruins. The great Islamic conquests following the death of Muhammad in AD 632 further stimulated changes in the existing pattern of urbanization. While existing towns were rarely destroyed in the fighting because of the willingness of their inhabitants to come to terms with the Arab armies,[12] some declined rapidly as political and economic fortunes changed, notably in North Africa. New towns were founded, some of which took over the role of more ancient towns as regional centres; Kairouan finally displaced Carthage; Cairo rivalled Alexandria; and the rise of Baghdād brought ruin to Ctesiphon. The Arab empire was primarily land based, and all the new centres of power – Baghdād, Damascus, Cairo and Mecca – were at inland locations.

6.3 Islamic towns: seventh to eighteenth centuries

The first three centuries of Islamic domination were a period of marked urban revival and prosperity throughout Southwest Asia and North Africa. Although the early Islamic conquerors were largely desert nomads, Islam was from the beginning a religion of the towns. The religious precepts of Islam encourage the close association of believers and in many ways presuppose the existence of an urban society. The Arabs themselves added a number of new towns to the existing network, particularly in the Maghreb where nearly half the new foundations were located (Figure 6.2). Most of these towns are still prosperous, while a few which enjoyed unusual geographical advantages of site and location are among the leading cities of the region. The Islamic conquests led directly to the creation of a number of walled towns for military purposes; Kufra (AD 638), Basra (AD 637), Fustāt (old Cairo, AD 642) and Kairouan (AD 670), for example. Some new towns were centres of pilgrimage or religious learning such as Najaf (AD 791) and Karbela (AD 680) in Iran. Others were founded later for defence in wars between rival sultans or as the new political capital of a rising dynasty. Marrakech (AD 1062), for example, was originally founded as capital of the Almoravids, and Sāmarrā' (AD 836) was the court city of the Khalif of Baghdād. Several centres also arose in response to commercial needs, particularly in regions like the Maghreb and the Red Sea where trade had not previously flourished; several small towns even sprang up in the Sahara at this time.

In addition to these new towns a large number of existing urban centres found themselves favourably placed to participate in the resurgence of trade which followed the spread of Islam and continued until the eleventh century. Southwest Asia again became the corridor between the storehouses of the east and the markets of Europe, and the number of trans-Saharan caravan routes suggest that North Africa also indulged in considerable transit trade (Figure 6.2). There was also highly developed trade within the Arab world, as illustrated by Maurice Lombard's work on the movement of timber.[13] The Islamic civilization created unprecedented demands for wood. It was used extensively in great buildings, particularly mosques; the famous mosque at Kairouan even contained timber from India. Ships, fishing boats and river barges on the Nile and in Mesopotamia were constructed of wood. Arab craftsmen made exquisite goods of many non-

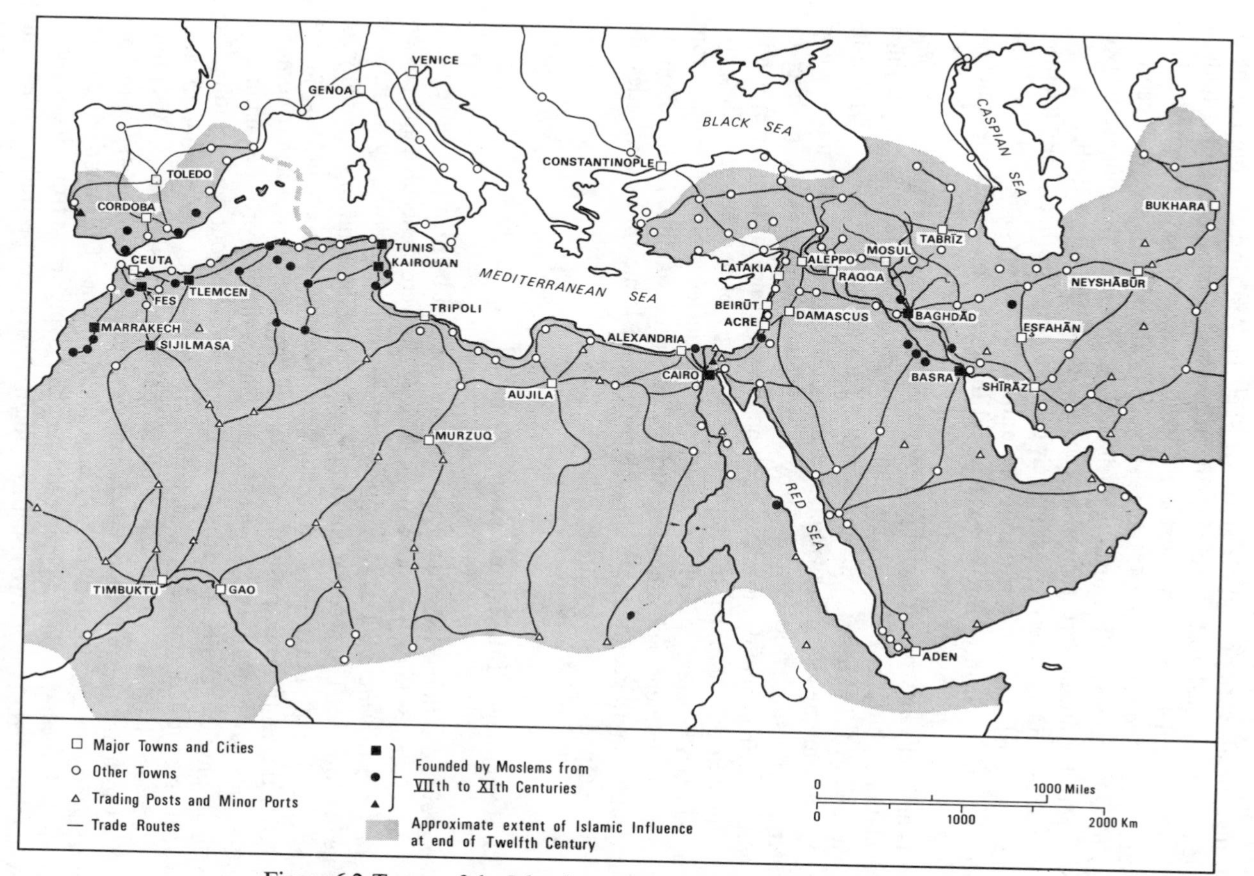

Figure 6.2 Towns of the Islamic world in the tenth to fourteenth centuries

Mediterranean woods like maple, teak and ash. Wood was used in the construction of waterwheels and mills and above all for fuel. Apart from domestic fuel requirements, wood was consumed in industries like sugar refining, charcoal burning, metal working and glassblowing. Yet the Arab world from northern Syria to Tunis was almost entirely lacking in forest resources, and timber had to be imported chiefly from the northern Mediterranean and the Maghreb. The failure of the Arabs to secure adequate supplies of timber after the tenth century may have been one of the causes of their decline, but until that time some two dozen Mediterranean ports thrived on the import and export of timber.

It has been widely observed that in their basic anatomy, form and architecture, towns of the Islamic world resemble one another in many respects.[14] The most striking affinities occur within the Arab world and Persia, but similarities can also be traced in the towns and cities of Asia Minor. The extent to which the shape of the Islamic city has been determined by its being Islamic is a matter of some speculation,[15] but it is arguable that Islam as much as any other single factor has given towns of Southwest Asia and North Africa a basic similarity of form in spite of the uniqueness of individual sites and differences of wealth and building material. The similarities displayed by these towns were also the product of common environmental influences, such as the frequency of dust-laden winds, the stifling heat of summer, and large diurnal ranges of temperature. They also evolved from processes of culture diffusion along ancient trade routes and from other shared influences of which insecurity is the prime example. In the following paragraphs some geographical characteristics of Islamic towns and cities are briefly outlined.

6.3.1 Morphology of the Islamic town

With few exceptions, Islamic towns were surrounded by high protective walls, often rectangular in their ground-plans. Most walls were crenellated and possessed a series of watchtowers. Even ports were sometimes enclosed by walls, as at Salé in Morocco. The need for elaborate defences of this kind was originally for the protection of the Muslims among a population of unbelievers, but they persisted over the centuries because of dynastic quarrels, the raids of desert nomads, and piracy along the North African coast. City walls also provided climatic as well as physical protection from conditions round about by excluding the scorching sand-laden winds characteristic of so much of the region. In some towns, waste land or pasture was included inside the walls, but their effect more often was to induce uniform building densities. Cemeteries – Jewish or Muslim – were nearly always situated outside the walls, near one of the main gates of the town. Each wall was generally pierced by a gate, giving access from all directions, while inside the town two or three main thoroughfares linking the gates converged on a single point, dividing the town into a number of quarters. Apart from these major axes, roads were generally narrow and winding, suitable only for pack animals, but affording welcome shade from the sun and some defence

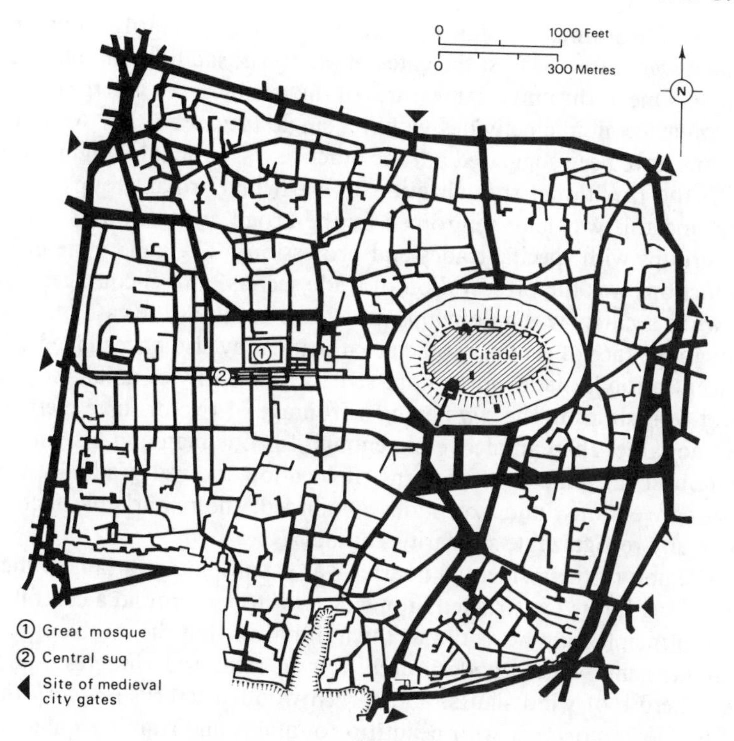

Figure 6.3 Old Aleppo in Syria, showing street pattern in 1941; Roman–Hellenistic influences are evident around the central *sūq*

against invaders. Blind alleys were numerous, and in small towns in North Africa, roads sometimes pass through 'tunnels' beneath houses. Nevertheless, in some cities of Southwest Asia, gridiron street patterns sometimes survived from Hellenistic or Roman times, as in the southwestern quarter of Aleppo (Figure 6.3).

The focal points of towns varied. In the large town it was sometimes a citadel or *Kasba* embraced by its own wall and gates, containing the palace and gardens of the sultan with his officials and administrative officers; Rabat, Tangier and Fèz, in Morocco are all examples. In other towns, a large mosque or market place or even the ruins of an ancient citadel, as occurs in Turkey, provided the focus of main roads into the towns.

A notable feature of Islamic towns was the segregation of residential and commercial areas. The centre of the commercial quarter was usually the great mosque or Friday mosque, which in turn influenced the disposition of many activities round about. Generally there was a religious school or *medersa* associated with it. Nearby were trades serving the needs of mosque and *medersa* - bookbinders, booksellers, furnishers of the sanctuary and so on. Next came cobblers, carpenters, tailors, carpet makers and beyond them, noisy and smelly pursuits such as metal working, tanning and dyeing. Crafts primarily of interest to countrymen were often nearest the gates of the town; saddlers and blacksmiths, for example. One of the universal features of the Islamic town was the concentration of similar economic activities within a single *bazaar* or *sūq*. A number of explanations have been suggested but the practice was probably originally associated with the traditional specialization of particular tribal groups in certain activities, and this was later reinforced by the broad association of ethnic and religious groups with specific trades and professions. The guild system, one of whose functions was the preservation of trade secrets, also encouraged concentration, while security was an obvious practical reason, as the case of jewellers. Solidarity in the face of arbitrary official extortion may also have played a part.[16] This concentration is made more impressive by the density of units, most shops, stores and workshops in the *sūq* having a frontage of two or three metres only; some are no larger than an alcove big enough for one man and his goods. It is worth emphasizing that these patterns of location evolved gradually through unwritten conventions, and exceptions abounded; the remarkable fact is that there were any recognizable similarities at all.

The basic unit of the residential quarter was usually some variant of the traditional dwelling house constructed in one or two storeys around a central courtyard. The principle may be as old as town life itself, but the Arabs introduced several refinements which modern architects in semi-arid climates could well emulate. The use of wind shafts, and carved wooden screens or *mashribiya* in place of windows, together with beautiful fountains and running water created cool and pleasant living conditions. In Iraq and Iran, extremes of heat and cold also led to the adoption of an underground room suitably ventilated for use in summer. The courtyard is a simple but effective way of securing thermal comfort in hot dry climates, and is widely used throughout North Africa and as far east as the plains of India. Open courtyards are less suitable for conditions in Turkey where they tend to be rare. Since only a small courtyard can afford effective protection from the sun, large houses were often constructed with two or more courtyards, each with a set of rooms with large doors and windows opening on to them. Houses are turned inwards towards the courtyard, with openings on to the heat, dust and noise of the street reduced to a minimum. Such an arrangement incidentally conformed with the Islamic emphasis on the family unit and the seclusion of women. Wherever possible it was also the custom for successive generations to build their houses contiguously, thus creating clusters of dwellings comprising extended kinship groups. From the air, therefore, the residential quarter of an Islamic town displays a cell-like structure (Figure 6.4). A fair proportion of the built-up area was private space open to the sky. Public space on the other hand was more limited; the narrow streets which wind their way between blocks of houses sometimes give way to small open spaces suitable for groups to congregate. The cooling effect of the streets was assisted by uneven building lines, and sometimes by overhanging windows and upper storeys, resulting in compartmentalized space, full of aesthetic interest.

The residential areas of the older Islamic towns today present a formidable

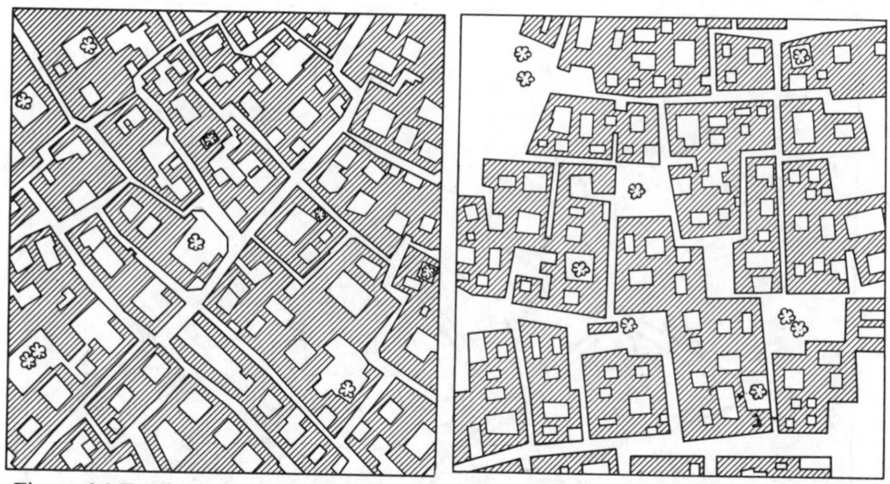

Figure 6.4 Traditional building patterns in the old city of Kuwait (left) and Misurata in Libya (right)

challenge to the urban planners seeking to provide water, electricity and sewerage, and above all access for vehicles.[17] In Iran and elsewhere the construction of broad straight avenues, sometimes crudely superimposed on a totally unrelated street pattern has been attempted in the major cities.[18] Mashhad is a particularly striking example of what has occurred in Iran since 1925 (Figure 6.5). Such broad streets may be necessary for modern traffic, but they are clearly less well adapted to the environment than the narrow streets of the old towns.

Residential areas were in turn divided fairly strictly into quarters (Arabic *mahalla*, or colloquially, *hara*) for peoples of differing ethnic origins and religions. Very few North African towns were without a Jewish community until their exodus following the birth of Israel, and the Jewish quarter was also a common element of most towns in Southwest Asia. Occasionally they were separated from other quarters by walls and most were able to close their streets at night with massive wooden doors. Originally, Jewish communities probably lived in very compact, congested quarters as in Shīrāz.[19] In the large cities however there is evidence of increasing dispersion of Jewish families around the fringes of the old *mahalla*, possibly since medieval times. Some *Mahalla* had slightly narrower streets than in neighbouring quarters. The location of the *mahalla* was sometimes near the palace of the ruler as in Fèz, Tlemcen, Constantine and Marrakech. Numerous other minority groups existed besides the Jews. In the coastal towns of North Africa in recent centuries, French, Italian, Greek, Spanish and Maltese communities might be found. The importance of Christian quarters in many towns in Lebanon is discussed in Chapter 13. Throughout Southwest Asia, minorities included not only groups from within the region, but many from outside, notably the Armenians, and Indians and Somalis and other Africans in the ports of the Red Sea and the Gulf. Even small communities of foreign merchants had their own quarters, so that many quarters could exist in one city.

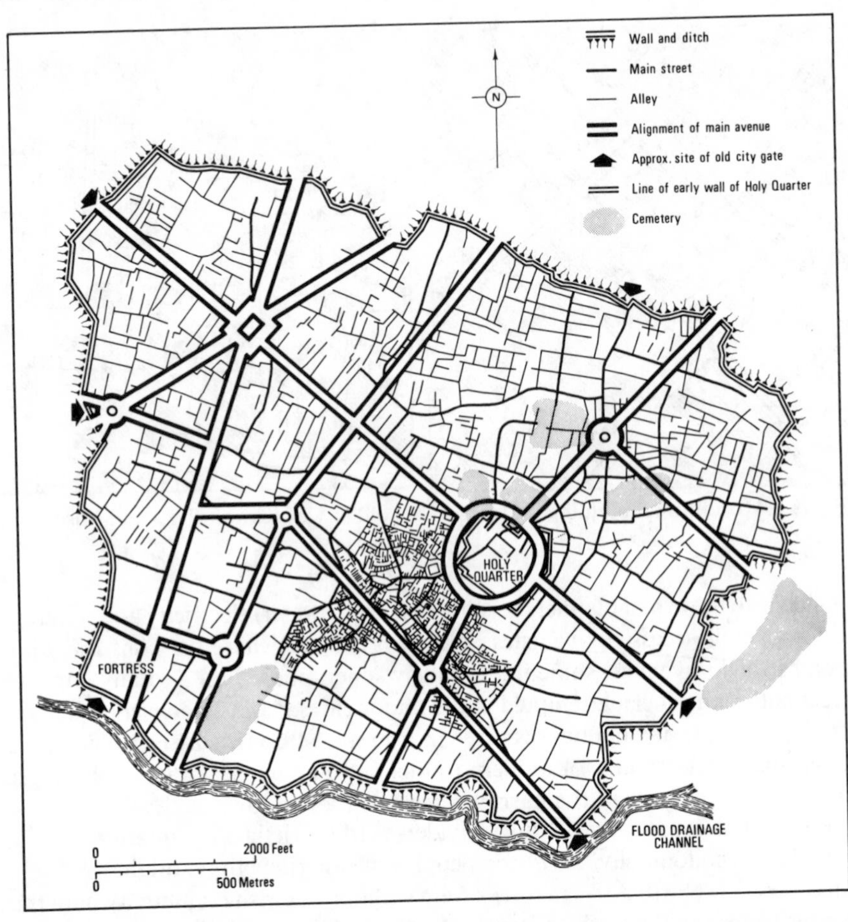

Figure 6.5 Mashhad in Iran: the superimposition of twentieth century avenues on the ancient street pattern (reproduced by permission of D. F. Darwent, 1965)

This emphasis on kinship and ethnic groupings reflected a vertical structure of society in which the clan was all-important as opposed to distinctions based on class or income.

The Islamic city had atmosphere, harmony and functional unity. Its public buildings were often fine examples of Islamic architecture, exhibiting superb craftsmanship in wood and tiles and stone, reflecting the golden ages of Persian and Arab civilization. Both Persian[20] and Arab town builders of the early Islamic period dealt skilfully with problems of water supply for houses, mosques, and gardens. In old Cairo every house had its own running water and drains in the ninth century.[21] To this day in several Iranian towns, *qanāts* can be found terminating in the *bazaar* or in a mosque, or sometimes in the house of the owner. Some underground rooms in houses are kept cool by the presence of running water from a *qanāt*. Although a high concentration of houses and narrow streets created a compact ground-plan in which the movement of people and pack animals was sometimes difficult, numerous small squares and gardens and courtyards with fountains and running water offered relief from the thronging streets. Here, cool air lingered long after sunrise. The renowned urban ethos of the Islamic town was derived as much from the subtle juxtaposition of buildings as from their quality as individual structures. The whole city had evolved organically over the centuries in an expressly Islamic environment and, in many ways, it was an embodiment of an Arab–Islamic culture. The centrality of the great mosque, the high status accorded to craftsmen and merchants, tolerance of minorities, and the importance of family and kinship group all arose out of the precepts of Islam and more or less directly affected the morphology of the town.

In spite of their structural similarities, Islamic towns possess a rich variety of townscapes, some of extraordinary beauty. Set in contrasting geographical environments, ranging from desert and coast to river valley and mountain, many types of traditional building material are available. Oven-fired or sun-dried bricks and mud were commonly used, creating a range of colour as varied as the parent materials. In southern Morocco light brown and red predominate; in the Nile valley every shade from grey to dark brown can be seen, while the old city of Baghdād was soft yellow in colour. More striking contrasts are evident in the stone-built towns. Diyarbakir in Turkey has massive black basalt walls; Mosul is extensively constructed of speckled grey alabastrine stone, and the old city of Jerusalem is renowned for its sparkling white limestone. Building styles also differ markedly from region to region. The sloping red tiled roofs of Turkey contrast strongly with the flat roof tops of most Arab and Persian towns. Some of the dazzling whitewashed settlements of the North African coast seem to have more in common with southern Europe than the Islamic world. On the other hand the centuries old apartment buildings of the Wadi Ḥaḍhramawt, constructed without steel or concrete are peculiar to a small part of the Arabian peninsula. They average some six storeys in height, but appear higher because each floor has a double set of windows.

The one visible element common to every Islamic town is the mosque. Surprisingly, the mosque did not originate from any specific directive in the Qur'ān. In the earliest years of the faith collective prayer was probably performed in an open space rather than within a specific edifice. Mosques were first built during the Islamic conquests for the purpose of excluding unbelievers, and minarets were originally added not to call to prayer but as symbols of the presence of the people who built them.[22] The puritanical Wahhābi sect of central Saudi Arabia still frown on tall minarets, and Riyadh while a strongly Islamic city, lacks this characteristic element. Later, when the call to prayer became common practice it seems likely that dwellings were constructed within *muezzin*-call of the mosque, and when this became impossible, further mosques were established. It is certainly evident that most parts of an Islamic town are served by a local mosque, the total number running into several hundred in the larger cities. Some of the important mosques of Southwest Asia, particularly in Turkey and Iran, possess magnificent blue or golden domes, another unmistakable feature of the Islamic skyline.

6.3.2 Municipal organization

An unusual aspect of the Islamic city was the form of government. Until the late nineteenth century, there were no councils charged with overall responsibility for town management. Occasionally the whole population might combine for defence, but civic feeling was generally weak among town dwellers compared with western Europe. The chief reason lay in the vertical structure of society and the importance of the residential quarter in which men of high and low estate were united on the basis of religion or blood. Groups cutting across quarters were rare and ineffective, and cooperation between quarters was consequently difficult. Most towns and cities had a governor or ruling sultan who appointed a local sheikh to raise taxes, maintain order, and represent the quarter.[23] The custom was reinforced under the Ottomans by the *millet* system which gave a certain degree of autonomy to religious minorities. Another instrument of administration was the trade corporations or guilds which flourished from the ninth century until their collapse ten centuries later. Their chief functions were the preservation of trade secrets, the maintenance of standards, and regulation of prices. By discouraging initiative and limiting expansion the guilds may ultimately have contributed to the economic stagnation of certain crafts in some cities. Whether or not this is so, the guilds provided links through their leaders, with a large proportion of the adult male population, since most were members of some guild; even beggars and thieves had their own.[24] While this style of urban government probably functioned reasonably effectively at the local level, the implementation of schemes for the town as a whole was rendered almost impossible. Thus with very few exceptions, chiefly in Persian cities, the Islamic period was one of unplanned organic growth in which individualism played an important part. The physical consequences are evident in the intricate morphology of the older towns and cities of the region today. The first serious attempts to draw up master plans for even the largest cities were made only in recent decades. The master plan drawn up for Aleppo in 1932, though not generally adopted, was probably the first in Southwest Asia[25], though the French had begun planning their towns in the Maghreb rather earlier in the century.

6.3.3 Urban decline: fifteenth to eighteenth centuries

The full flowering of Islamic urban civilization was reached in about the eleventh century. From then until the fifteenth century a period of stagnation set in, followed by demographic and economic decline until the end of the eighteenth century. It is not possible to give precise figures, but the reality of this decline is supported by evidence from many sources. While some towns disappeared altogether, others dwindled in size and importance. The population decline of a few leading cities is illustrated in Table 6.1. The earlier figures cannot be regarded as wholly reliable; nevertheless, the differences are sufficiently large to be worth noting. Similar processes were undoubtedly widespread, though even notional

TABLE 6.1
Population change of some leading cities
(ninth to nineteenth centuries)

	Year/Period	Estimated population in thousands	Year	Estimated population in thousands
Fèz	13th century	400	1900	95
Tunis	1517	200	1881	120
Alexandria	860	100	1800	7
Cairo	14th century	300	1800	250
Damascus	14th/15th century	100	1820	120
	1520/1530	57	1840	80
Baghdād	10th century	500	1800	50–100
			1831	20
Tehrān	17th century	600	1807	45

Source: See note 26 under references.

figures are non-existent for the smaller towns. Asia Minor, which for centuries constituted the favoured heartland of the Ottoman empire, appears to have been an exception, the largest cities at least showing population increase from the sixteenth century. Large discrepancies frequently occur in the estimated populations of individual towns, but these are not always entirely due to inaccurate guesswork. Wars, floods,[27] fire, earthquake and epidemics in former years periodically reduced urban populations, creating enormous fluctuations over short periods. In 1669 for example, 150,000 people are estimated to have died of plague in Aleppo, and the 1831 death toll of 40–50,000 in Baghdād partially accounts for the sharp decline shown in Table 6.1 between 1800 and 1831.[28] Cairo has also been fearfully devastated by plague on several occasions, as in 1492, when 12,000 corpses are said to have been carried out of the city in one day.[29]

The decline of urban populations from the fifteenth to eighteenth centuries was almost certainly accompanied by a general demographic decline. A combination of complex local factors usually accounted for such decline, but three common causes may be suggested. First, the period was marked by a succession of wars and political convulsions. In Southwest Asia the Crusades were followed by the repeated destructive incursions of the Mongols in the thirteenth and fourteenth centuries. During the sixteenth and seventeenth centuries the Ottomans conquered most of Southwest Asia except Persia and part of North Africa, while Persia and Turkey engaged in a succession of long and exhausting wars which left much of eastern Asia Minor, Iraq and western Persia in a state of ruin. Secondly, the Ottoman regime, and that of the oppressive Mamluks which preceded it in Syria and Egypt failed to revive industry, agriculture and trade. In the late seventeenth and eighteenth centuries the Ottomans imposed crippling taxation and arbitrary levies on merchants in the towns, thus inhibiting development. The discovery of the Cape route to the East by the Portuguese in 1498 was followed by disruption of trade in the Red Sea, the Persian Gulf, and the Indian Ocean which robbed Southwest Asia of its prosperity as the world's premier trade route. The declining standard of living and the increase of violence implicit in these events might themselves have lowered birth rates and raised death rates, but it also seems likely that during the period in question famines and epidemics became more frequent as irrigated agriculture contracted and the quality of urban life deteriortated, particularly housing and sanitation.

6.4 The nineteenth and twentieth centuries

6.4.1 Level of urbanization

Since nobody really knows the precise size of the total population, it is impossible to say whether the proportion of people living in towns rose or fell before the nineteenth century. If the level remained roughly the same as in early Islamic times, it could only have done so with the periodic influx of rural migrants to compensate for some towns abandoned altogether and to make up for high urban mortality sometimes in excess of births. C. Issawi stresses the importance of such rural–urban migration[30] but Z. Y. Hershlag concedes little or no movement from the villages during these centuries.[31] The truth may be that it was the desert nomads who settled from time to time in the city rather than the peasant cultivator. Certainly the level of urbanization in Southwest Asia and North Arica was high during the nineteenth century, comparable with much of western Europe. In 1800 at least 10 per cent of the populations of Turkey and Egypt lived in towns of over 10,000 inhabitants; in Syria and Iraq it was probably 15 to 20 per cent. Towards the end of the century the urban population of Iran was put at 25 per cent,[32] while figures of eight to nine per cent have been suggested for Morocco and Tunisia and 16 per cent for Algeria.[33] It should be noted that many urban dwellers were agriculturalists living inside the city walls, and in this sense comparisons with Europe are not altogether valid. Nevertheless it is clear that the urbanization of the last hundred years took place from a well-established demographic base.

Estimates of the degree of urbanization in the Middle East and North Africa today tend to differ largely because there is no agreed definition of 'urban'. Israel for example takes any settlement with over 2,000 inhabitants as urban, even when more than one third of the labour force is engaged in agriculture. In Saudi Arabia the urban threshold is put at 4,000, in Iran 5,000 and in Turkey 10,000. Other countries such as Egypt use administrative status as the yardstick. Furthermore there are often confusing differences between urban populations quoted for the political city (defined by municipal boundaries) and the geographical entity (defined by the built-up area). In spite of such difficulties a good impression of the present levels of urban population is given in Table 6.2. By 1984 53 per cent of the population of Southwest Asia and 42 per cent of North Africa were urban dwellers. More than half the states of the region record over 50 per cent of their populations as urban. Several record some of the highest levels in the world.[34]

TABLE 6.2
Urbanization

	Urban population				Percentage of urban population			
	As percentage of total population		Average annual growth rate (%)		In largest city		In cities of 500,000 persons	
	1965	1984	1965–73	1973–84	1960	1980	1960	1980
Algeria	32	47	2.5	5.4	27	12	27	12
Egypt	40	45	3.0	3.0	38	39	53	53
Libya	29	63	8.9	7.9	57	64	0	64
Morocco	32	43	4.0	4.2	16	26	16	50
Tunisia	40	54	4.1	3.8	40	30	40	30
Bahrain	—	81	—	—	—	—	—	—
Iran	37	54	5.4	5.0	26	28	26	47
Iraq	51	70	5.7	5.5	35	55	35	70
Israel	81	90	3.8	2.7	46	35	46	35
Jordan	47	72	4.7	4.7	31	37	0	37
Kuwait	75	93	9.3	7.7	75	30	0	0
Lebanon	49	77	6.2	2.8	64	79	64	79
Oman	4	27	10.8	17.6	—	—	—	—
Qatar	—	87	—	—	—	—	—	—
Saudi Arabia	39	72	8.4	7.3	15	18	0	33
Syria	40	49	4.8	4.3	35	33	35	55
Turkey	32	46	4.9	4.0	18	24	32	42
United Arab Emirates	56	79	16.7	10.4	—	—	—	—
Yemen AR	5	19	9.7	8.8	—	25	0	0
Yemen PDR	30	37	3.4	3.5	61	49	0	0

Source: World Development Report 1986. Oxford University Press, New York, 1986, pp 240–241. (Figures for Qatar and Bahrain from *Population Reference Bureau*, New York, 1984).

6.4.2 Rate of urban growth

The rate of urban growth has shown almost unchecked acceleration throughout the twentieth century. In 1900 the level was probably around 10 per cent. Fifty years later it had reached 25 per cent, rising to 45 per cent in 1984. The average annual growth rate of urban population in recent decades has invariably exceeded the average annual rate of natural increase of between 2.6 and 3.0 per cent (Table 6.2). Since natural increase tends to be higher in towns because of better living conditions, it often constitutes the largest component in urban population growth. In the oil-rich states however where migrant labour from overseas has been recruited on a large scale, the migration component is clearly dominant. Here, annual growth rates averaging over 10 per cent have been registered in Oman and the United Arab Emirates. These exceptional growth rates should not obscure the generally high rates shown in Table 6.2, most of which are at levels creating considerable strains on the urban housing and infrastructure. It has been calculated that if current rates of urbanization are maintained in the Arab world it will be necessary to build more houses in the next 25 years than in all previous history.[35]

The rapid increase of urban populations and the persistently high level of rural–urban migration has sometimes led to the conclusion that Southwest Asia and North Africa are 'overurbanized'. In the sense that migrants are often unemployed or at best undertake low productive employment the idea is valid. On the other hand, unemployment and underemployment are also common in rural areas and most migrant families enjoy higher incomes in towns than previously. While a proportion have been driven from the land by crop failure or acute population pressure, most migration occurs in response to both 'push' and 'pull' factors, the chief attraction being the steady expansion of employment opportunities since the turn of the century. Although the oil industry is capital-intensive and not labour-intensive, it has created substantial demands for labour. Oil revenues have generated major development projects. Several states, not necessarily oil producers, have begun to establish an increasingly wide range of manufacturing industries and these have also created jobs. The political fragmentation of Southwest Asia following the First World War and the era of independence following the Second World War also created enormous opportunities in government service and in service industries of all kinds associated with the concentration of high spending power in the new capital cities.

A further consequence of rapid urbanization has been described as 'underurbanism', the tendency for many rural migrants to retain their non-urban linkages, value systems and traditions long after arriving in town, living in effect as urban villages. The great majority of immigrants settle in one or two large cities, usually in the same general location, which encourages 'underurbanism'. In a number of cities rural migrants and their offspring constitute half or two thirds of the total population, so the phenomenon is widespread. While the retention of rural ways is a comfort to the migrant, it also delays the adoption of new ideas and attitudes and civic consciousness on the path to a genuine urban lifestyle.[36]

6.4.3 Concentration of urban populations

The concentration of urban population of individual countries in one or two large cities or city-regions is an unmistakable feature of most of North Africa and Southwest Asia. In several of these, there is probably overconcentration of urban population, with harmful social and economic consequences of congestion, pollution and land speculation. Overconcentration may have advantages for industrial and commercial activities but may deprive less favoured regions of industrial investment and professional personnel. The concentration of populations in the largest or primate city and in cities of over 500,000 inhabitants shown in Table 6.2 is very striking. Altogether about 12 per cent of the region's inhabitants live in such cities, though in individual countries the proportion is very much higher. A high degree of primacy is noted in Iran, Iraq, and Lebanon where the primate cities completely outclass their nearest rivals in size. Thus in 1976 Tehrān

(4,589,000) was followed by Eşfahān (842,000), Baghdād (2,969,000 in 1970) by Basra (371,000), and Beirūt (702,000 in 1980) by Tripoli (175,000).[37]

Because urban census areas and urban agglomerations rarely coincide, changes in the administrative boundaries or of census areas can give misleading results, such as those which show an apparent decline in urban concentrations between 1960 and 1980 (Table 6.3). In a few instances this may reflect successful attempts to encourage the growth of other cities, but most often it represents a statistical quirk. The figures also tend to under-represent the extent of population concentration which is often in a large and expanding metropolitan region comprising several urban centres. So far, the most heavily urbanized regions are emerging in Morocco, Tunisia, Egypt and Israel. In Morocco a zone no more than 150 km long and 50 km wide extending from Casablanca to Kenitra embraces half the population of the country. Some 50 years ago the population of this small area was insignificant compared with the inland lowland regions. Similarly in Tunisia, Tunis is the heart of a strongly urbanized region extending to a radius of some 80 km, including nearly half the medium-sized towns in Tunisia. The most impressive examples however are in Egypt and Israel. Greater Cairo embraces a region more than 50 km across, including 11 towns, two of them with over one million inhabitants. In Israel, nearly half of the population inhabit the Tel Aviv district, 20 km long and 8 km wide. Recent studies have drawn attention to the emergence of metropolitan core are around Amman in Jordan and Riyadh in Saudi Arabia.[38] These areas often enjoy geographical and historical advantages which give them political, economic and social supremacy over the rest of the country. It is arguable that these advantages should be exploited rather than resisted; even the most strenuous efforts by Middle Eastern governments to decentralize economic activity and foster the growth of small and medium-size towns has met with only modest success.

6.4.4 Number of towns

The number of towns in Southwest Asia and North Africa has greatly increased since the beginning of this century, both through the expansion of existing small settlements and the creation of new urban centres. It is a fair guess that in 1900 there were fewer than 100 towns with 20,000 or more inhabitants in North Africa and perhaps 150 in Southwest Asia. Table 6.3 shows 215 such towns in North Africa and 242 in Southwest Asia in about 1970. If anything, these figures are underestimates. The increase in the number of towns with over 100,000 inhabitants is remarkable. Table 6.3 shows 31 cities in this category in 1970 in North Africa and 56 in Southwest Asia; by 1980 the total number had reached 145. In 1900 there were four towns of this size in North Africa, and fewer than a dozen in Southwest Asia. The greatest number of 100,000-plus cities are in Turkey (26), Egypt (21), Iran (21) and Morocco (18). Table 6.3 also indicates 10 million-plus cities in 1980, whereas in 1900 there were none. Population estimates for these cities in 1987 are: Cairo 8.9 million; Tehrān 7.5 million; Alexandria 2.5 million; Baghdād 3.5 million; İstanbul 5.5 million; Casablanca 2.6 million. Algiers and Ankara had also achieved 'millionaire' status by 1970, while the Tel Aviv conurbation also contains over a million people in spite of the modest size often attributed to the political city of Tel Aviv. The exceptionally rapid growth of these cities has been an outstanding feature of the last eighty years. They now embrace over one tenth of the entire population of the region and one third of the urban population. All are growing more rapidly than the urban populations of their respective countries and the populations of several major cities will double in fifteen years. Greater Cairo already ranks about ninth in the world, and greater İstanbul about seventeenth.

The complexities of definition renders column I in Table 6.3 of limited value; many of the Egyptian settlements in this category are large villages with a high proportion of the population engaged in agriculture, while other figures represent strict census definitions or 'urban'. Nevertheless, the large number of small towns may be noted in Turkey, which has a relatively low level of urban population (Table 6.2).

TABLE 6.3
Number of towns in 1970
(1980 figures for columns IV, V and VI shown in brackets)

	I 10,000– 19,999	II 20,000– 49,999	III 50,000– 99,999	IV 100,000– 499,900	V 500,000– 999,999	VI 1 million and over	1970 Total
Algeria	24	52	12	4 (7)	1 (1)	— (1)	93
Libya	8	1	—	2 (2)	— (2)	—	11
Morocco(a)	22	12	5	6 (14)	1 (3)	1 (1)	47
Tunisia	22	14	3	— (1)	1 (1)	—	40
Egypt	200	65	20	13 (18)	— —	2 (3)	300
Bahrain	—	1	1	— (1)	— —	— —	2
Iran	48	41	15	11 (18)	2 (2)	1 (1)	118
Iraq	19	12	3	8 (7)	— —	1 (1)	43
Israel(b)	24	15	7	4 (8)	— —	— —	50
Jordan(b)	7	6	1	2 (2)	— (1)	— —	16
Kuwait	4	3	—	1 (2)	— (1)	— —	8
Lebanon	4	2	1	1 (1)	1 (1)	— —	9
Oman	1	—	—	— —	— —	— —	1
Qatar	1	—	1	— (1)	—	—	2
Saudi Arabia	12	5	3	3 (5)	— (2)	—	23
Syria	15	4	1	3 (3)	2 (1)	—	25
Turkey	134	77	21	17 (18)	1 (4)	2 (3)	252
United Arab Emirates	—	2	1	— (4)	—	—	3
Yemen AR	1	—	1	1	—	—	3
Yemen, PDR	—	2	—	1 (1)	— —	— —	3
TOTAL	546	314	96	77 (116)	9 (19)	7 (10)	1049

(a) With Spanish towns of Melilla and Ceuta
(b) Pre-1967 territories
Source: National censuses; *The New Geographical Digest*, George Philip, London, 1986, pp 14–22.

6.4.5 New towns

Several factors explain the surprising number of new towns in the region. For the second time in its long history, the area was gradually drawn into the sphere of European commercial and political influence from the mid-nineteenth century. With the opening of the Suez Canal in 1869 the historic function of acting as the crossroads of the world was resumed, and maritime traffic received a great impetus. The export of primary products, aided in part by the development of railways and pipelines, became increasingly important and new outlets were needed to meet changing patterns of trade. Until the outbreak of the Second World War, many of the fastest growing towns and cities were in coastal locations. Besides the revival of many ancient ports, several new ones were constructed, notably in North Africa. The colonial era in North Africa also stimulated a spate of town building. Many of the minor towns of the Maghreb were founded during the Fench occupation. A few were directly associated with mineral extraction, notably phosphates and coal, and other towns developed initially as small ports associated with mineral exports and a few were founded as resorts. A large number of towns, particularly in Algeria, were originally garrison towns with administrative functions, such as Sidi Bel Abbes. The largest category however, were the village and market centres established mainly in regions of European colonization; Souk Ahras, Boufarik and Saida, for example, in Algeria; Kasba Tadla, Khemisset and Souk El Arba in Morocco. Deliberately located at the foci of discrete geographical and agricultural regions, many of them are now effective centres for the surrounding countryside in areas far removed from the large towns. One reason why so many new towns had to be created was the scarcity of indigenous centres suitable for expansion. It seems likely that the operation of weekly tribal markets at any convenient geographical foci obviated the need for permanent market towns.[39]

Israel's contribution to the number of new towns may be noted; out of a total of 77 urban centres, half are new towns.[40] Tel Aviv, which now forms the heart of a conurbation of one and a half million people, began in 1909 as a Jewish garden suburb of Jaffa. Most other new towns have grown quite slowly. The majority of Israel's new towns were part of a national strategy to decentralize population from the old centres of Tel Aviv, Jerusalem and Haifa.

Some of the most important new town developments in Southwest Asia have been directly or indirectly, associated with the oil industry. The Iranian oil port of Abadan (1979 population 400,000) was founded in 1910 as an oil refining and exporting centre, laid out largely on European lines. Dhahran in Saudi Arabia is another early example, founded as the headquarters of Aramco and catering for large numbers of expatriate workers, many from the United States. The town (Figure 6.6) was like an urban transplant from North America with regular streets, open spaces, gardens, and recreational facilities including a golf course, swimming pool, tennis courts and bowling alley. Churches were not in evidence, however, being forbidden by law in Saudi Arabia.

Saudi Arabia has also been responsible for the two most spectacular new towns

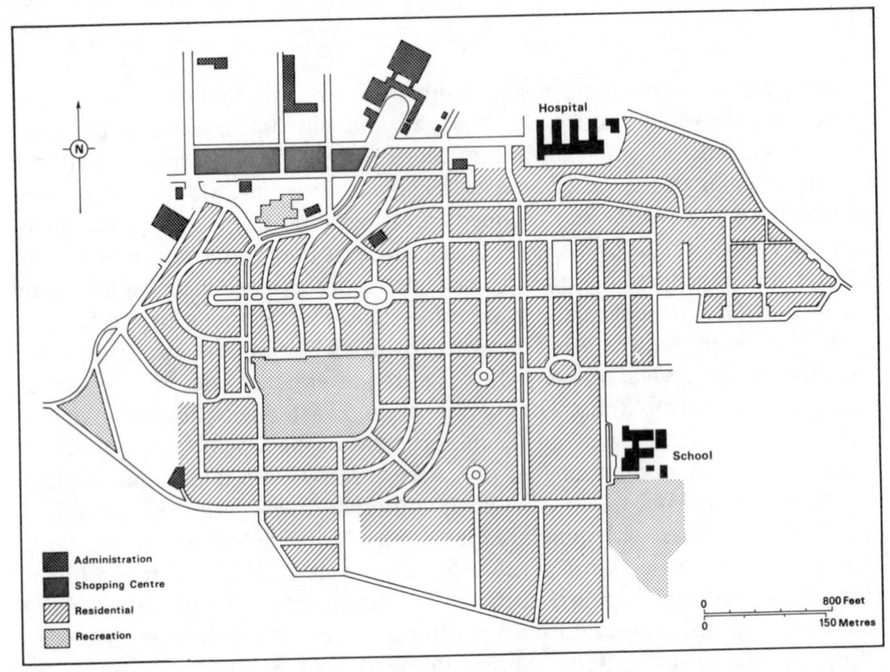

Figure 6.6 Dhahran, an oil town in Saudi Arabia in the early 1960s

in the region at Jubail on the Gulf coast and at Yanbu on the Red Sea. A Royal Commission for Jubail and Yanbu was set up in 1975 to begin work on two huge new industrial cities with population projections of 350,000 (Jubail) and 150,000 (Yanbu) by the end of the century. Both depend heavily upon oil refining and petrochemicals and have large associated ports. An equally ambitious project is the grandiose new town at Jebel Ali in the United Arab Emirates, planned to accommodate over half a million people by 2007 based on diverse industrial activities. There are a number of other new towns in Southwest Asia, though none on the scale of Jebel Ali. Many of these are relatively small satellite developments designed to relieve population congestion in the largest cities. Umm Said in Qatar was planned for this purpose with great sophistification, but has met with little success.[41] The best example are the new urban centres of the Nile delta region in Egypt, established in an attempt to slow down the growth of Greater Cairo.

6.4.6 Changes in urban form in the twentieth century

New towns contain only a fraction of the total increase of urban dwellers in recent decades, the majority having been absorbed into expanded pre-nineteenth century foundations. The effect of rapid expansion on the structure and morphology of traditional towns and cities has varied. In the largest cities redevelopment has often been so extensive that only small parts of the old town

A leafy suburb in Ankara looks much like its counterpart in a north European city (Malcolm Wagstaff)

The Nile at Cairo (The Times, Photograph by W. Harrison)

survive, as in Algiers, Cairo, Baghdād and Tehrān. In Iran, the construction of broad avenues through most large towns stimulated changes in the location of commercial activities and land values so that some degree of fusion between old and new occurred. Physical change has however, been quite moderate in some towns, expansion taking place adjacent to the original core. Even city walls have often remained substantially intact, although some have been removed to make way for much needed ring roads as in Aleppo (Figure 6.3). Within the old city or *medina* many buildings have changed their functions – hotels have replaced the old *caravanserais* and most synagogues have closed – but in general the outward appearance and atmosphere is not unlike that of a hundred years ago or more. The contrast between the old and new towns can be seen most obviously in street patterns, the one organically evolved, confused and dense, the other ruler-planned and orderly. Tetouan provides a graphic example of a colonial town of this kind (Figure 6.7). There are also less tangible implications of this duality. Population densities in the *medinas* have been found to be generally higher, and different patterns of retailing and movement of people and goods have been identified. Other research has revealed higher rates of natural increase in the *medinas*, lower rates of literacy, a lower proportion of active females, and the prevalence of social structures based primarily on kinship and clan groupings. Economic activity in the old town is generally unrelated to land values, while in the new town familiar processes are at work generating high values centrally, reducing towards the urban fringe.

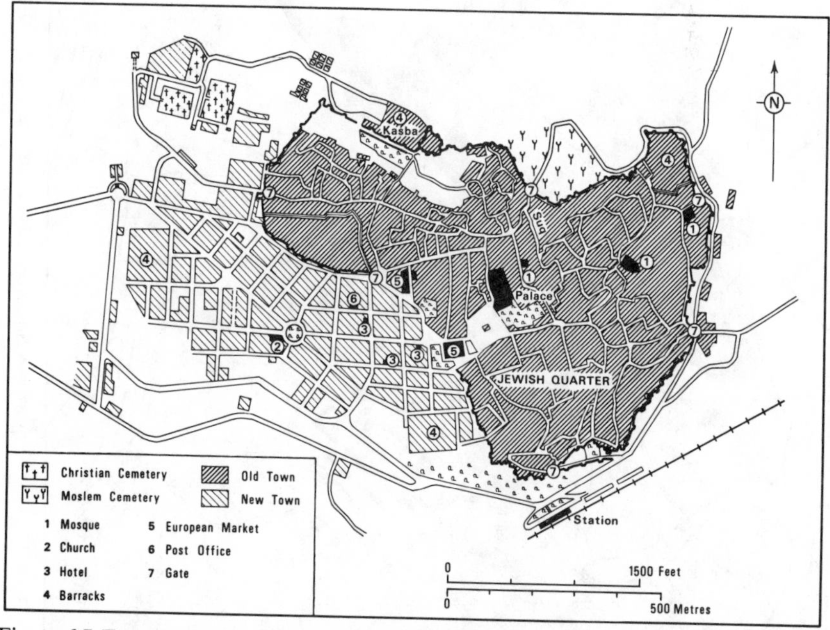

Figure 6.7 Tetouan in Spanish Morocco in about 1938, showing the old town and the new Spanish-built town (reproduced by permission of the Controller, Her Majesty's Stationery Office)

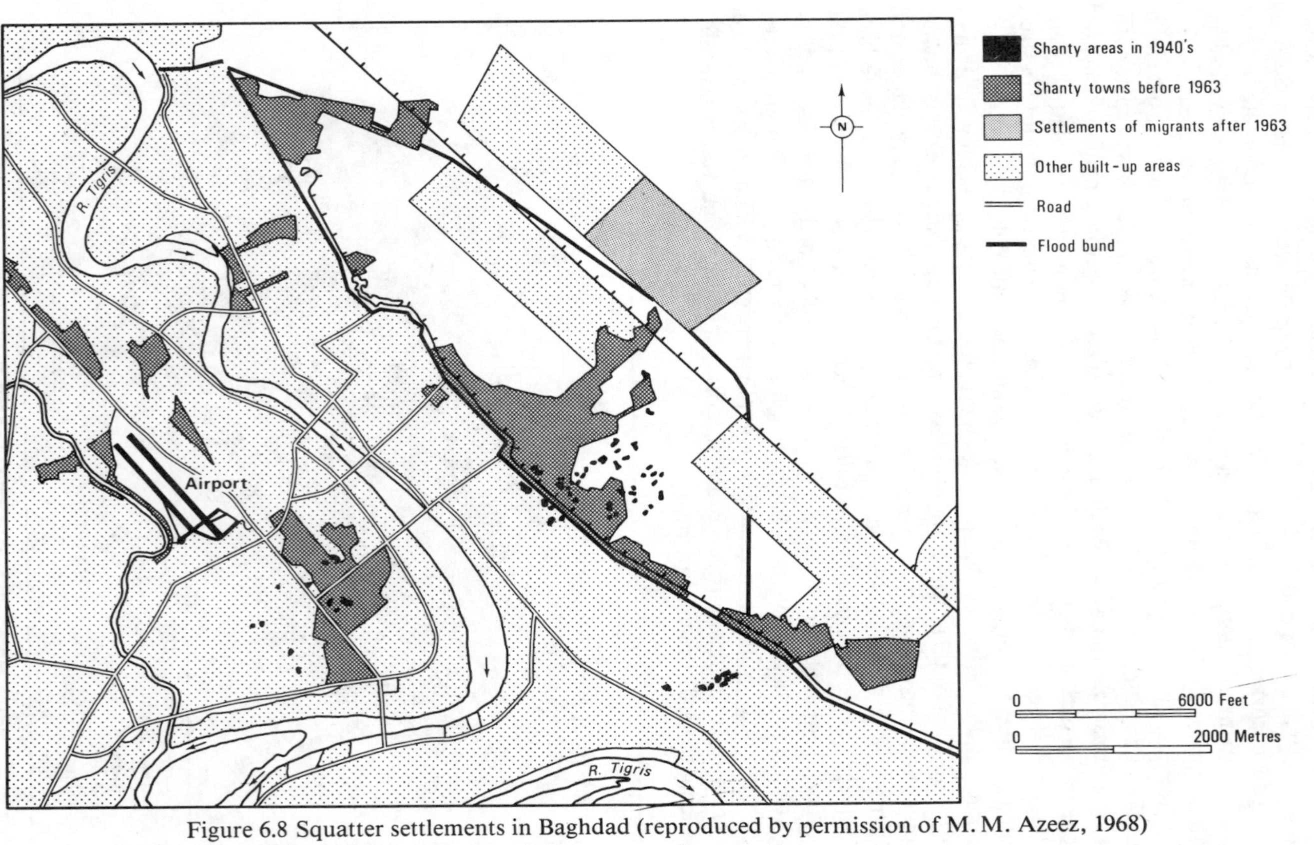

Figure 6.8 Squatter settlements in Baghdad (reproduced by permission of M. M. Azeez, 1968)

Even where the original town has largely survived, it now represents a mere fraction of the built-up area, usually accommodating a small share of the total urban population. New and distinctly alien elements have been added to the townscape – villa development, flats, shops, garages, industrial estates, railway sidings, government offices, schools and hospitals. Around the largest cities unsightly shanty towns or *bidonvilles* have sometimes been erected by poor rural immigrants. Casablanca and Baghdād (Figure 6.8) have had extensive *bidonvilles*, though partially successful efforts have been made to replace them with good housing. In most towns and cities today the characteristic Islamic skyline of domes and minarets is overshadowed by one characteristic of almost any western city with high rise apartments, chimneys and office blocks. Stone and mud have been largely replaced by bricks and concrete as building materials, and indigenous architectural forms have almost entirely disappeared. The centre of gravity has shifted from the old town to the new, and as urbanization proceeds an increasingly high proportion of the population are living in a western-style urban environment, no longer distinctly Islamic, Persian, Turkish or Arab.

6.5 The distribution of towns and cities

Figure 6.9 shows towns and cities with more than 25,000 inhabitants according to census data and other estimates largely made during the 1980s. Their distribution broadly corresponds with the network of medieval towns of all sizes (Figure 6.2). The largest number of towns are located in regions of over 400 mm of annual rainfall, where perennial streams and groundwater resources provide reasonably adequate water supplies and a wide choice of sites. The coast of North Africa, most of the Levant, Asia Minor and parts of northern Iraq and western Iran may be included in this category. Several towns even in these areas experience problems of water supply. A second important category with generally adequate water resources are the towns of the Nile valley and delta, and the Tigris–Euphrates lowlands where rainfall is generally less than 130 mm. Both regions possess large urban populations, but the network of towns in the Nile delta is unequalled elsewhere in North Africa or Southwest Asia. A third category of towns exist in regions with less than 400 mm of rainfall, and outside the Nile and Tigris–Euphrates valleys. In the largest of these towns the provision of an adequate water supply often presents acute problems, and in all of them the presence of a water supply in some form or other was decisive in the choice of site. Several of the most famous were caravan cities on the margins of the desert. Damascus, for example, has been described as 'the gift of the Barada', a river which rises in the mountains to the west and supplies the city with abundant water through its seven distributaries before being lost in the desert to the east.[42] Eşfahān is another classic example. Like most Iranian oases, it lies within sight of high mountains from which radiate scores of short streams, but the Zayandeh Rud brings water to the city and its surrounding oasis from a great distance. Not all towns in regions of rainfall deficiency are as favoured, particularly in the Arabian peninsula. Water

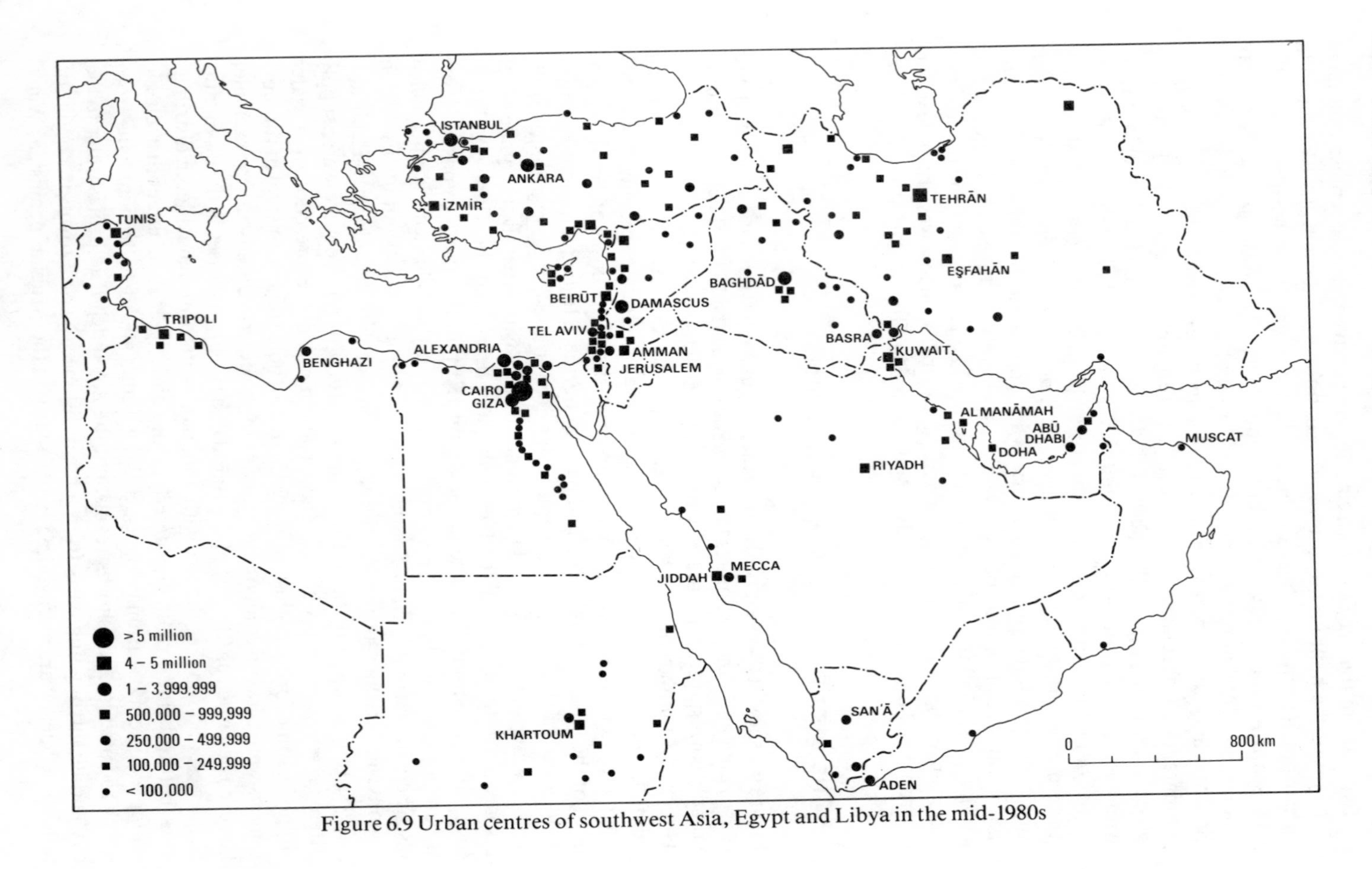

Figure 6.9 Urban centres of southwest Asia, Egypt and Libya in the mid-1980s

for Jiddah has to be brought 80 km by pipeline, and about half that distance to Riyadh. Kuwait, Abū Dhabi and Doha, the capital of Qatar, rely on desalination of sea water, and plants are also in operation at Jiddah, Eilat and elsewhere. Desalination plants will certainly multiply in future.

6.5.1 Capital cities

The location of capital cities deserves some comment. In the Maghreb and Libya all capital cities are on the coast, in keeping with the fashion in the rest of the continent. Rabat, Algiers and Tunis, while not geographically central in their respective countries are well placed with regard to the most productive and populous regions. Tunis is particularly well favoured in this respect. Libya, on the other hand, suffers from a polarization of population both physically and politically between the old provinces of Tripolitania and Cyrenaica. Thus Tripoli and Benghazi were formerly regarded as joint capitals, though an ambitious scheme to build a new capital at Al Bayda' in Cyrenaica was attempted and subsequently abandoned. Since 1969, the decision has gone in favour of Tripoli. Cairo is perhaps the most rationally situated capital in the world, linking the populous delta and valley regions, standing at the natural east–west crossing point of the Nile, and at the natural focus of river traffic and north–south communications by road and rail. Baghdād and Beirūt are the only capital cities in Southwest Asia which enjoy anything like the same advantages as Cairo, and significantly both are unrivalled in size in their own countries. Tehrān, which became capital of Persia at the end of the eighteenth century, dominates other cities in Iran despite its eccentric geographical location. Many towns have in turn fulfilled the function of capital city of Persia because of the difficulty of finding a good central site on account of the sparsely populated interior.

Ankara was a town of 30,000 inhabitants in 1923 when it was chosen as capital of Turkey in preference to İstanbul with its cosmopolitan associations. Its present population is around one million. Ankara enjoys a measure of centrality and is not unfavourably located in other respects, on an ancient east–west route, and at the focus of complementary geographical regions – forest, steppe, and cultivated land. Damascus and Aleppo had very similar geographical advantages as potential capitals for Syria after the First World War. Damascus may have been chosen partly because of its more favourable position with regard to Lebanon and the rest of the Arab world. Lebanese independence and the advent of Israel may have reduced these advantages. Certainly Aleppo remains a close rival of Damascus in size. The selection of an ideal capital for Saudi Arabia is problematic because of the marginal distribution of much of the country's population and resources. The choice of Riyadh as the Royal capital and Jiddah as administrative capital was an obvious compromise. The Yemen Arab Republic also had two capitals until 1969, but for different reasons; a Royalist capital at Ṣan'ā and a Republican capital at Ta'izz. Ṣan'ā is now the capital of the Yemen AR. Jerusalem was proclaimed capital of Israel in 1950, in spite of the wishes of the United

Nations that it should be internationalized. It is the seat of Israel's parliament, but most foreign embassies and certain government departments remain in Tel Aviv. It is worth noting that before June 1967, Jerusalem was a divided city, reached by a vulnerable corridor flanked by Jordanian territory.

6.5.2 Relations between town and country

If towns with fewer than 25,000 inhabitants could be shown on Figure 6.9, a positive relationship between rural population densities and the density of small towns would be indicated. There are perhaps 600 to 650 centres in the 10,000 to 25,000 range, the majority of which are small market towns. Many fulfil local administrative functions and possess a limited range of industrial activity, such as cotton ginning or olive pressing, and some craft industries such as leather-work or weaving. The symbiosis of town and country characteristic of these small centres may be illustrated by the functioning of Misurata in Libya, a town with less than 20,000 inhabitants in 1966. Misurata is located at the geographical centre of a small oasis, to the south of which lies a fairly extensive region of dry cereal cultivation, and enormous tracts of rough steppe pasture beyond. The total population of the surrounding region was less than 80,000 in the 1960s, yet three times a week, each market day, about 7,000 rural dwellers, predominantly men, entered the town to buy and sell their goods. Individuals made regular visits from 30 to 40 km away, and occasional visits from as far as 300 km to the south. The entire centre of the town was devoted to open air and covered markets, and more than 800 small retail stores and craftmen's shops were counted in the town in 1966. Some countrymen were found to own workshops in Misurata while a number of townsmen also farmed land within the oasis and beyond.[43]

It is therefore appropriate to conclude by considering relationships between the countryside and the larger towns and cities of North Africa and Southwest Asia. A widely accepted view is that from early times until the nineteenth century the townsman was isolated politically, socially and culturally from his counterpart in the countryside, whom he generally despised. Furthermore, since the town was often the seat of an alien administration, antipathy between the two groups was strong. These ideas are probably exaggerated. Town and country were poles apart in wealth and culture, but they were also frequently part of a single eco-system in which both were mutually interdependent. This was particularly true of the great cities like Marrakech, Damascus, Eṣfahān and others surrounded by large oases. The peasant looked to the town for protection and as a market for his produce, while the townsman depended largely upon local agriculture for food. Many farmers and landowners actually lived inside the city, and in small towns farmers may well have owned small stores and workshops as is customary in North Africa today. Some cities possessed extensive walled suburbs for gardening, such areas of green having both economic and amenity value. In fact, in many situations no absolute distinctions between urban and rural habitats could be drawn. The towns themselves were socially divided into quarters no larger than village communities, and in Egypt and Iraq canals and rivers sometimes physically separated parts of the town. Moreover, there was undoubtedly a steady trickle of rural migrants into even the meanest towns, if not a steady stream, and rural attitudes lingered in urban society. On the other hand some large villages were almost urban in character, fortified, with their own non-agriculture specialists, public baths and mosques.[44] It is probably true that those ancient towns and cities which survived the vicissitudes of interregional trade most successfully were those enjoying strong economic relations with a surrounding population of cultivators.

6.6 Urban planning

The necessity for urban planning in the Middle East is clear enough. Rapid urban growth stimulated by in-migration and high rates of population increase has created a pressing need for housing and services on a massive scale in the major cities. In addition many states have embarked upon industrial development, much of which finds the market forces of the major cities irresistible. At the same time modern transport has permitted rapid urban sprawl, with densely populated residential districts located far from the city core. This is exacerbated by the high dependence of the fringe on the centre for employment, much of which is in tertiary activities, and for basic services. Maps and air photographs reveal the astonishing spatial growth of many cities, particularly since the Second World War. Figures for Libya for example suggest that urban land in the rainy zone rose from 250 hectares in 1911 to 35,100 in 1970, a 140-fold increase.[45] Much of this expansion was at the expense of cultivable land. In Egypt, where good soils are even more precious, similar processes used up 622,000 hectares of agricultural land between 1952 and 1977, almost 70 per cent of land reclaimed in the same period.[46] The other consequences of such large scale urban expansion include severe congestion in certain large cities, and the general neglect of small and medium-size towns as well as peripheral regions.[47]

The planning response to these problems has been patchy, and in many cases largely ineffective. Most Middle Eastern countries have suffered from lack of indigenous planning expertise and the absence of adequate planning machinery.[48] Apart from the oil-rich states, financial problems also impeded the implementation of planning policies. The most common approach has been to tackle the immediate symptoms of congestion without evolving a national policy to encourage decentralization. While a number of states have national settlement policy guidelines including Iran, Iraq, Egypt and Libya, they do not translate easily into practical action. Iraq for example seeks to curb urban growth, maintain a minimum level of rural population, and develop mountain and desert areas.[49] Measures to relieve congestion have included building new highways, sometimes in place of ancient walls or through old cities, as in Figure 6.5. In Kuwait the wholesale demolition of the old city and reconstruction along modern lines was undertaken, while Tehrān embarked on a major scheme to move government offices to a site three kilometres north of the town.[50] The creation of satellite settlements beyond the fringe of the metropolitan core is another strategy, but the results are dubious. Around Cairo for example, four satellite towns may eventually add to the attractiveness of the Greater Cairo region rather than diminish regional congestion. Significantly alternative plans for new towns in the desert appear to have been abandoned. The truth is that decongestion and decentralization are both notoriously difficult to achieve in the face of powerful economic forces, often underpinned by the geographical advantages of a specific region.

During the 1960s and 1970s most of the larger Middle Eastern cities had physical master plans prepared which advocated decentralization among other measures. Unfortunately the planners had neither the authority nor the resources to pursue decentralization. Even the physical plans themselves dealing primarily with land-use zoning were often neglected because of weak planning structures.[51] Furthermore many of these plans were prepared with limited understanding of the social and economic context, or the wishes of the people themselves. Sadly too, many opportunities were lost to preserve the great historic and architectural inheritance of such cities, and far too much western experience commonly influenced the plans. On the other hand there have been some remarkable planning success stories, with the Saudi city of Jiddah a prime example of a blending of the old with the best of the modern, to standards which would have been impossible without almost unlimited funds.

References

1. M.E.L. Mallowan, 'The development of cities from Al'Ubaid to the end of Uruk 5', *Cambridge Ancient History*, **1**, Cambridge University Press, 1967, 6–7.
2. J. Mellaart, *Çatal Hüyük: a neolithic town in Anatolia*, Thames and Hudson, London, 1967.
3. J. Jacobs, *The Economy of Cities*, Jonathan Cape, London, 1969, 5–38.
4. V. Gordon Childe, 'The urban revolution', *Town Plann. Rev.* **21**, 3–17 (1950).
5. C. Daryll Forde, 'The ancient cities of Indus', *Geography*, **17**, 186 (1932).
6. F.R. Hiorns, *Town-building in History*, Harrap, London, 1956, 11–14.
7. A.L. Oppenheim, 'Mesopotamian cities', in *Middle Eastern Cities*, (Ed I.M. Lapidus), University of California, Berkeley and Los Angeles, 1969, 5.
8. R. McAdams, *Land behind Baghdād*, University of Chicago, 1965, Figures 2, 3 and 4.
9. A.A.M. Van der Heyden and H.H. Scullard (Eds) *Atlas of the Classical World*, Nelson, London, 1959, 84.
10. W.C. Brice, *Southwest Asia*, University of London Press, 1966, 96–101.
11. G.B. Cressey, *Crossroads*, J.P. Lippincott Co, International University Edition, New York, 1960, 81.
12. J.B. Glubb, *The Great Arab Conquests*, Hodder and Stoughton, London 1963, 158–159, 202, 230–1.
13. M. Lombard, 'Une carte du bois dans la Méditerranée musulmane (VIIe-IXe siècles)', *Annales-Economies-Sociétés-Civilisations*, 234–54 (April–June 1969).

14. (a) G. Shiber, *The Kuwait Urbanisation*, Kuwait Planning Board, Kuwait, 1964, 15–39.
    (b) G. Marçais, 'L'urbanisme musulman', in *Mélanges d'histoire et d'archaéologie de l'occident musulman,* **1,** Gouvernement Géneral d'Algérie, Algiers, 1957, 219–31.
15. A. H. Hourani, 'The Islamic city in the light of recent research'. In *The Islamic city*, (Ed A. H. Hourani and S. M. Stern), Bruno Cassirer, Oxford, 1970, 9–24.
16. W. B. Fisher, *The Middle East*, 5th edn, Methuen, London, 1963, 132.
17. J. I. Clarke and B. D. Clark, *Kermānshāh: an Iranian Provincial city*. University of Durham, Centre for Middle Eastern and Islamic Studies, 1969, 97–104, 127–134.
18. X. de Planhol, 'Geography of Settlement' In *Cambridge History of Iran; The Land of Iran*, **1** (Ed W. B. Fisher), Cambridge University Press, 1968, 438–40.
19. J. I. Clarke, *The Iranian City of Shīrāz*, University of Durham, Department of Geography, 1963, 48–51.
20. E. E. Beaudouin and A. U. Pope, 'City Plans' in *A Survey of Persian art* (Ed A. U. Pope), Oxford University Press, 1939, 1398.
21. G. Marçais, 'L'urbanisme musulman', *Mélanges d'histoire et d'archéologie de l'occident musulman*, **1,** Gouvernement Géneral d'Algérie, Algiers, 1957, 224–6.
22. O. Grabar, 'The Mosque', in *Middle Eastern Cities* (Ed I. M. Lapidus), University of California, Berkeley and Los Angeles, 1969, 39.
23. I. M. Lapidus, 'Muslim cities and Islamic societies', in *Middle Eastern Cities* (Ed I. M. Lapidus), University of California, Berkeley and Los Angeles, 1969, 49–50.
24. H. A. R. Gibb and H. Bowen, *Islamic Society and the West*, Royal Institute of International Affairs, Oxford University Press, London, 1950, 281–99.
25. N. Chechade, 'Aleppo', in *The New Metropolis in the Arab World* (Ed M. Berger), Allied Publishers, New York, 1963, 89.
26. Table 6.1 was compiled from:
    (a) G. Hamdan, 'The pattern of medieval urbanism in the Arab world', *Geography*, **47,** 121–33 (1962).
    (b) C. Issawi, 'Economic change and urbanisation', in *Middle Eastern Cities* (Ed. I. M. Lapidus), University of California, Berkeley and Los Angeles, 1969, 102–21.
    (c) I. M. Lapidus, *Muslim Cities in the Later Middle Ages*, Harvard University Press, Cambridge, Mass. 1967, 79–87.
    (d) D. Noin, 'L'urbanisation du Maroc', *Information Géographique* **32,** 69–81 (1968).
27. J. H. G. Lebon, 'The site and modern development of Baghdād', *Bull. Soc. Géogr. Egypte*, **24,** 7–33 (1956).
28. C. Issawi, 'Economic change and urbanisation', in *Middle Eastern Cities* (Ed I. M. Lapidus), University of California, Berkeley and Los Angeles, 1969, 106.
29. K. Baedeker, *Egypt and the Sudan*, 6th edn Baedeker, Leipzig, 1908, 40.
30. C. Issawi, 'Economic change and urbanisation', in *Middle Eastern Cities* (Ed I. M. Lapidus), University of California, Berkeley and Los Angeles, 1969, 105–6.
31. Z. Y. Hershlag, *Introduction to the Modern Economic History of the Middle East*, E. J. Brill, Leiden, 1964, 21.
32. 'Persia' *Encyclopaedia Britannica*, 10th edn, Edinburgh and London, 1902, 616.
33. S. Amin, *L'économie du Maghreb*, **1,** Editions de Minuit, Paris, 1966, 21–6.
34. *World Population Data Sheet 1984* Population Reference Bureau, New York, 1984.
35. S. Khalaf, 'Some salient features of urbanization in the Arab world', *Ekistics* **50,** (300) (1983), 219–22.
36. S. Khalaf, (1983), *Ibid*, 221.
37. *The New Geographical Digest*, George Philip, London, 1986, pp 14–22.
38. N. Kliot and A. Soffer, 'The emergence of a metropolitan core area in a new state – the case of Jordan'. *Asian and African Studies* **24** (2), (1986), 217–32.
39. W. Fogg, 'The sūq: a study in the human geography of Morocco', *Geography*, **17,** 257–67 (1932).

40. E. Spiegal, *New Towns in Israel*, Praeger, Stuttgart and Bern, 1967.
41. H. Roberts, *An urban Profile of the Middle East*, Croom Helm, St Martin's Press, New York, pp 109–19.
42. J. Garrett, 'The site of Damascus', *Geography*, 21, 283–96 (1936).
43. G. H. Blake, *Misurata: a Market Town in Tripolitania*, University of Durham, Department of Geography, 1968, 1–34.
44. I. M. Lapidus, 'Muslim cities and Islamic societies,' in *Middle Eastern Cities* (Ed I. M. Lapidus), University of California, Berkeley and Los Angeles, 1969, 60–4.
45. A. A. Mahmoud-Misrati, 'Land conversion to urban use: its character and impact in Libya', *Ekistics* **300** May/June (1983) 183–94.
46. J. Smit, 'Which future for Alexandria?', *Geoforum*, **8** (3), 1977, 135–40.
47. G. H. Blake, 'The small town', in *The Changing Middle Eastern City*, (Eds G. H. Blake and R. I. Lawless) Croom Helm, London, 1980, pp 209–29.
48. A. M. Findlay and R. Paddison, 'Planning the Arab City: the cases of Tunis and Rabat', *Progress in Planning* **26** (1) (1986), 3–82.
49. R. P. Misra and O. El Agraa, 'Urbanisation and national development: the quest for appropriate human settlement policies in the Arab world', *Ekistics* **300** May/June (1983), 210–18.
50. V. F. Costello, 'Tehran' in *Problems and Planning in Third World Cities*, (Ed M. Pacione) Croom Helm, London, 1981, pp 156–86.
51. B. D. Clark, 'Urban planning: perspectives and problems', in *The Changing Middle Eastern City*, (Eds G. H. Blake and R. I. Lawless) Croom Helm, London, 1980, pp 154–77.

CHAPTER 7

Problems of Economic Development

7.1 Introduction

The position of the Middle East in the development spectrum is far from clear. GDP per head (Table 7.1) simply reflects the extremely high incomes enjoyed by a handful of oil-exporting countries (Kuwait, Libya, Qatar, Saudi Arabia and the UAE) and the relative impoverishment of those either lacking in oil altogether (Jordan, Yemen AR) or still to enjoy the full benefits of its exploitation (PDR Yemen). High incomes from oil, moreover, do not reflect 'advanced' economic status or even a capacity for rapid economic development. Rather they indicate dependence on one resource and reveal the shortness of the period covered by development programmes in the oil-rich states. Growth in average per capita incomes over the decade 1970–80 show a more balanced and more modest picture of attainment, country by country. As a whole, the region fared better than Africa and the advanced western economies of Sweden, the UK and the USA, but not quite as well as the Caribbean and Latin America. Energy consumption was high in world terms in Bahrain, Kuwait, Qatar and the UAE, reflecting in large measure an energy-based solution to their water problems. Most countries in the region were worse off than the Caribbean and Latin American region, whilst Yemen AR fell below even the average energy consumption for Africa. Economic development has not reached the level of diversification, particularly through manufacturing industry, which requires large energy inputs. Rates of change in energy use were largely positive and often high in the Middle East over the decade 1970–80. This can be seen as the result of the progressive 'modernization' of its socio-economic patterns in certain directions, including industrialization, but also growing import-based consumerism (refrigerators, televisions, air conditioning and motor cars). Only Israel and Kuwait conformed to the 'western' pattern of falling levels of energy consumption. In western countries this was a response to the vast rise in the price of oil effected in 1973 and the consequent adoption of conservation measures and the search for alternative sources of energy. Israel's alignment with the western countries in this characteristic indicates the relative maturity of its social and economic structures. Other countries in the region have a long way to go before they achieve this degree of flexibility.

Rapid economic growth has been characteristic of Middle East states since the end of the Second World War. Despite the problems of comparison,[1] the region progressed more rapidly than some other less-developed parts of the world, and, as the figures for GDP suggest, most Middle Eastern countries are wealthy com-

TABLE 7.1

Crude indices of economic development for the Middle East and some comparisons

Country	Total 1979/80 US$	GDP per Capita Average Annual Growth 1970–80 (per cent)	Energy Consumption Per Capita 1980 kg of coal equivalent	Energy Consumption Percentage Change 1970–80
Bahrain	8,164	—	12,651	373.3
Egypt	435	8.2	488	77.4
Iran	2,382	5.9[2]	1,177	15.4
Iraq	-	5.2[3]	807	27.5
Israel	5,431	1.6	2,273	−11.2
Jordan	1,327	−2.3[4]	690	9.0
Kuwait	20,143	−3.6	5,019	−46.7
Lebanon	1,293	—	951	43.0
Libya	11,825	10.8[1]	2,456	357.3
Oman	7,056	—	725	262.5
Qatar	20,824e	—	26,667	42.3
Saudi Arabia	13,354	6.0	2,745	224.8
Syria	1,437	6.1	835	66.3
Turkey	1,266	3.5	718	49.9
UAE	40,587	6.6[5]	17,188	152.4
Yemen AR	533	7.7	95	400.0
Yemen, PDR	319	—	529	14.2
REGION	5,140	5.2e	4,471	143.0
CARIBBEAN and LATIN AMERICA	2,230	5.8	1,046	45.5
AFRICA	900	1.8	408	32.0
Japan	8,873	3.4	3,726	350.6
Sweden	14,882	1.6	5,376	− 16.4
United Kingdom	9,351	1.9	4,850	− 9.1
United States	11,416	3.0	10,386	− 5.7

e = estimate; 1 = 1974–78; 2 = 1970–77; 3 = 1971–75; 4 = 1971–76; 5 = 1975–80.

Sources: United Nations, *Yearbook of National Accounts Statistics, 1981*, New York, 1983;
United Nations, *World Energy Supplies, 1950–1974*, New York, 1976;
United Nations: *Energy Statistics Yearbook, 1983*, New York, 1985.

pared with African and Latin American countries[2] (Table 7.1). This achievement rests on foundations laid during the nineteenth century, when much of the Middle East was drawn firmly within the trading orbit of northwestern Europe as demands for food and raw materials grew. Considerable investment was made in the infrastructure during the otherwise largely stagnant period of the 1920s and 1930s,[3] and this paid dividends later. The Second World War marks something of a watershed, for it provided an important boost to local economies by its pressures for national self-sufficiency and import substitution. High prices created by the subsequent Korean War (1950–53) helped to sustain growth. Political independence created important stimuli, while additional but localized advantages were realized from oil exploitation and the aid offered by great powers. The sharp rises in oil prices of 1973 and 1979 increased the incomes available to the oil exporters, transforming the major producers overnight into capital surplus countries well able to embark upon ambitious programmes of social and economic development.

Favoured countries naturally wish to maintain their economic growth, while the more backward states of the region feel a need to increase their growth rates, to improve general living standards, as well as secure a greater measure of political independence and world influence. Improved living standards, however, depend upon a more equitable distribution of national wealth, while further economic growth, in many instances, can be achieved only by change in the structure of the economy, away from agriculture and towards more manufacturing industry. Structural changes are not easily secured, and the problems involved are the subject of this chapter. Some problems are peculiar to the Middle East, but many are shared with other developing regions of the world, albeit with a Middle Eastern flavour.

7.2 General problems of economic development

Economic development is promoted by a complex of interacting but poorly understood forces. Much depends upon the roles and relative importance of capital accumulation, past and present, and total factor productivity. Rates of saving and investment are crucial, especially where incomes are low (Figure 7.1). Equally important are the choices of techniques and methods employed to achieve the goals. While the temptation is always to borrow answers from more developed countries, problems should be formulated with Middle Eastern conditions specifically in mind. The quality of administrators becomes vital here, as in the rest of the development process, especially since Middle Eastern states are almost invariably forced to initiate, finance and control socio-economic change themselves. More basic elements in the development complex, though, are the role of agriculture, the influence of population growth, and the existence of regional inequalities.

7.2.1 Agriculture

Agriculture is of considerable importance in the economies of most Middle Eastern states, including the oil giants, Iran, Libya and Saudi Arabia, especially when petroleum earnings are removed from their national accounts (Table 4.1). Despite growing imports, local farming still produces much of the food consumed in the region, and generally employs the largest part of the labour force. In addition, agriculture provides much of the raw material for manufacturing industry and a great deal of locally-raised development capital. The export of agricultural products is an important source of foreign exchange.

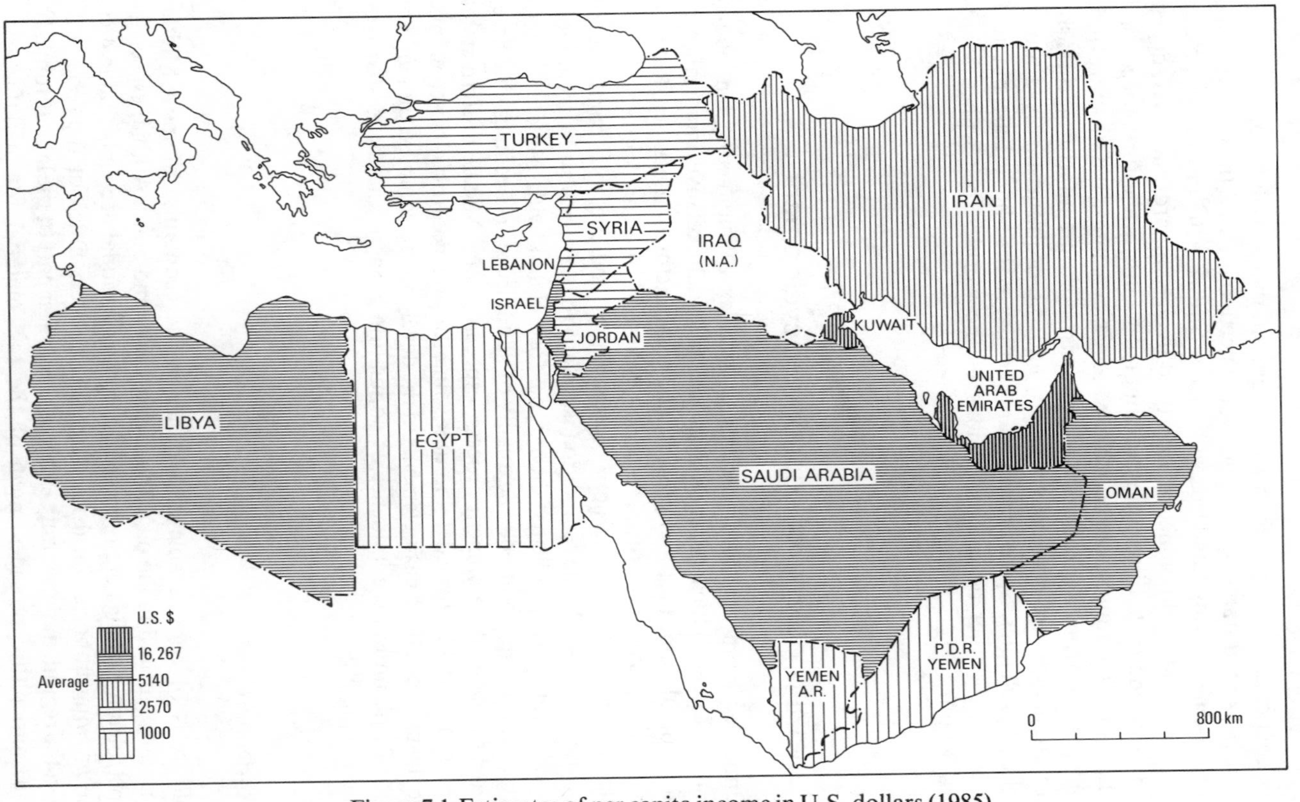

Figure 7.1 Estimates of per capita income in U.S. dollars (1985)

The very dominance and fundamental importance of agriculture create a number of problems for economic development. Most important is the consequent dependence of several states upon an unstable and uncertain economic base. Both the growing season and the range of crops are limited by the concentration of precipitation into one season and the excessive evapotranspiration characteristic of the summer. For environmental and technological reasons, yields are generally low (Chapter 4), but in the dry farming areas, which constitute the largest farming zone of the region, they also vary considerably from year to year according to the interannual variations in precipitation. Industrial supplies and exports are, therefore, uncertain and fluctuate greatly to the annoyance of the market. Since about 1950, the region's agriculture has not been able to meet demands made upon it for food, industrial raw material and foreign exchange. In fact, trade figures collected by the United Nations showed that Egypt, Iran, Jordan, Lebanon, Libya, Saudi Arabia (as well as the smaller Gulf states), Syria and PDR Yemen imported a considerably greater value of food and live animals during the late 1960s, 1970s and early 1980s than they exported. Over the same period, however, Turkey consistently exported more of these commodities than it imported, while the balance tended to oscillate for Israel. It is probably correct, though, to regard the whole of the Middle East as a food deficit region.[4] At the beginning of the 1980s, some countries, particularly in Arabia, became obsessed by food security and the risk of dependency on the major western exporters.

Several changes have brought the region to this situation. Population growth has been considerable (Chapter 5), and demand has consequently risen. At the same time, easily reclaimed land has virtually disappeared from the dry farming zone after more than a hundred years of spatial advance in the farming frontier. Reclamation of new land through irrigation has been limited, despite the publicity given to it, and expensive to affect. At the same time, land has gone out of agricultural use through the expansion of settlements (especially towns), salination and erosion. Land use has changed in favour of high value industrial crops, such as cotton, and fruit and vegetable production, with the result that proportionately less land is devoted to the staple grains. Government investment in agriculture and closely allied activities has continued to be small, despite the basic importance of the sector and low opportunity costs, which mean that relatively small investments can bring high returns.

Fundamentally, agriculture has failed to respond to the demands put upon it. This is partly the result of a lack of adequate government help for farming and partly the consequence of cheap food policies aimed at the volatile urban population. Despite these strictures, though, various measures have been taken to develop agriculture still further. Their appropriateness and effectiveness, however, are often debatable. Production might still be raised in the dry farming areas by ploughing up more land, but so little is now available that further horizontal expansion must largely depend, for example, upon the elimination of draft animals (and hence of pasture and land used for growing fodder) and greater use of tractors. The introduction of tractors has already meant that land is now in production which was once too distant from settlements to be cultivated.[5] Much of this is marginal in terms of precipitation, depth of soil and degree of slope. Any further expansion will be even more hazardous ecologically, and must depend, in any case, upon relatively high agricultural prices, if it is to be financially worthwhile. Cereal prices will inevitably remain above general world levels.

The introduction of crop rotation is another way of increasing production. By this means, wasteful fallow can be eliminated and soil fertility maintained. However, the rotation of crops requires a fundamental change in the entire system of land use. Boserup had argued[6] that land use is closely geared to population levels, and that a given system will change only when population growth reaches a level where there is no longer sufficient land available for the old system to provide basic subsistence at acceptable levels. According to this theory, crop rotation will expand greatly in the Middle East only when the existing system of fallowing and common grazing can no longer support the farming population and employ the labour available. As long as it is still possible to subsist using 'traditional' patterns of land use, there is no incentive to change. Indeed, Middle Eastern farmers often show considerable reluctance to adopt new systems.

Such reluctance has often been attributed to the inherent conservatism of 'traditional' socio-economic systems and often blamed specifically on Islam. This is based upon a number of misconceptions, largely stemming from a Euro-American viewpoint.[7] Too much emphasis has been given to the alien features of Islam and its allegedly all-embracing nature as a socio-economic system. Not enough allowance has been made for the elements of life which the Middle East shares with other developing regions. Most important has been a failure to appreciate the fundamental features of 'peasant economy' (Chapter 4). The peasant is not essentially a small-scale capitalist, and his activities are not geared to maximizing profit but to minimizing risk, so as to secure the subsistence needs of his family. He aims to employ all the labour at his disposal, rather than to maintain economic efficiency in a classic sense, while the pattern of work is closely related to the growing season and, accordingly, is characterized in the Middle East by short peaks of enormous activity and long troughs of unemployment. The peasant farmers will change their behaviour when its benefits are seen in terms of the peasant economy which still shapes their outlook and when the rewards for the increased drudgery associated with new methods and new crops become clear. Land reform may have had beneficial effects through allowing farmers to respond more readily to market demands, but the assessment of such programmes is difficult (Chapter 4). Higher prices, tighter but more sympathetic control of loans and credits, and better extension work would greatly assist the transitions now underway in the countryside.

Agricultural output can also be raised without much structural change in its socio-economic system. For example, the greater use of chemical fertilizers and the growing of new varieties of crops will improve yields. Both have happened in the region and productivity has certainly risen, dramatically in the case of the capital-surplus countries of Arabia where lavish investment has been possible. However, more drudgery is involved for the average small farmer since mechanization has not always proceeded as rapidly as it might, despite some real advances, and actual shortages of labour are being experienced in parts of the region as people leave the land. Using a ratio of agriculture's contribution to GDP to its share of the active population Weinbaum[8] showed that only slight improvements in labour productivity were achieved in Egypt and Israel and that declines were found elsewhere, notably in Iran and Iraq.

The extension of irrigation is a further solution to the problem of increasing agricultural production, though it is an expensive one. It also has the advantage of combating fluctuations in output due to interannual variations in precipitation. Cultivation can be carried on all year round, and a greater diversity of crops can be grown, as Egyptian experience with the extension of perennial irrigation over the last hundred years has shown (Chapter 19), while arid land can be brought into production. However, silting up of reservoirs and canals is often rapid, and evaporation is high. Management is generally poor, so that much water is lost in transport and innundation. The general neglect of drainage may mean that land is poisoned by salination. Water-carried disease usually increases, undermining the health of the farming population at the very time that the new system requires greater labour inputs to control water, clear irrigation channels and maintain production on a year-round basis. Finally, the introduction of irrigation is relatively expensive, even when simple diesel pumps are used, but especially when spectacular engineering works are necessary. The high cost of irrigation projects virtually dictates that high value crops, for export or local industrial use, are grown instead of cereals. Where cereals are produced, as in Saudi Arabia, it is with massive state support and largely for political reasons. The result is that the vicious circle is not always broken, but often perpetuated.

A fundamental solution to the problem of dependence upon agriculture is offered by a change in the structure of the economy. Manufacturing and service industries have increased their roles considerably since the Second World War,[9] and most states seek to promote industrialization still further. It is often forgotten, however, that industrial development in the Middle East is at least partially dependent upon the agricultural sector as a source of raw materials and investment, and that in numerical terms the market is still predominantly rural. Government stimulation of industrial development may even have detrimental affects upon agricultural progress in this situation, and consequently upon the pace of general economic development. Industrial development, and the work opportunities created, either directly or indirectly, are important factors in the drift of people from the countryside to the towns (Chapters 5 and 6). This has been regarded as perhaps inevitable, because of the differences in incomes and living standards, as well as what was thought until recently to be widespread underemployment in rural areas.[10] Removal of labour from the countryside, however, may have hitherto unsuspected results, and the dislocation of agriculture in Iraq following massive emigration to the towns, consequent upon land reform,[11] was simply an extreme case of a more widespread phenomenon. The new developments in farming outlined above, demand more labour, not less.

Labour shortages already mean that land is falling out of use in areas of high emigration. Mechanization is proceeding, but this too may require more human effort, because more land can now be cultivated, more attention may be necessary depending on the crop and a larger harvest has to be reaped, threshed and carried. It might even be argued that the drain of people from the countryside is reducing the pressure for structural change in land use, and thereby retarding the very processes of intensification which governments are seeking to promote. Population, in various ways, begins to look very much the key factor in economic development.

7.2.2 Population

Although the total population of the Middle East is still comparatively small (estimated at little more than 207 million in 1984) it has been growing at an average rate of 2.8 per cent per annum, which is amongst the highest rates in the world (Chapter 5). This rate of growth is fairly recent, and its economic significance is difficult to determine.[12]

Population increase appears to have been one of the engines of economic growth in the region.[13] There have been increasing numbers of people to feed, clothe and house, and a larger labour force available for employment. An expansion of the agricultural area was part of the response in the past, but recently mounting pressure on land in some districts has been instrumental in producing the kind of structural change in agriculture which the Boserup thesis would lead one to expect, though in others the necessary changes have failed to materialize. Migration contributes to the manpower needed for manufacturing and service industries; indeed, economic growth along the Gulf, for example, would have been difficult without it. Further, the youthfulness of the population, with its improving health and education, represents a considerable economic potential for the future. At the moment, though, a young population (40–50 per cent of the population under 15 years of age) means that a large proportion of the people is not economically active to any extent, even though child labour is still common, and, in a climate of modernization, requires considerable investment for its education. The high rate of population increase itself means that a considerable investment is necessary to maintain current living standards, let alone produce advances. However, structural changes in national economies may be expected to reduce crude birth rates, so that some of the problems associated with high rates of increase may be temporary.

Perhaps more serious is the rapid expansion of towns in this already comparatively urbanized region (Chapters 5 and 6). Much investment has to be put into maintaining the viability of towns – providing work, housing, transport, electricity, water and education, as well as maintaining public health and public order. Although this may promote general economic development in a state, the need to support the towns emphasizes two important development problems. One is the problem of feeding the urban population, already alluded to more than once in the preceding discussion. The other is the need for improving the distri-

Educational opportunities have improved greatly in recent years (top): a roadside scribe giving reading lessons in the Doha sūq, (The Press Association Ltd.); (bottom): Bahrain secondary school (Central Office of Information)

Middle East settlements have long suffered insecurity from natural and man-made causes (top): Nabatieh Camp for Palestine refugees, after Israeli attacks in May, 1974 (UNRWA photograph by F. Samia): (bottom): earthquake damage at Gediz in Turkey (Turkish Embassy, London)

bution of goods and services across national space. Improvements here are normally brought about by extending national road networks, but road building eats up capital in this region. Distances are often considerable, the terrain is difficult to surmount, even using modern engineering techniques, and large numbers of bridges and culverts are required to deal with seasonal floods, while many of the areas crossed generate little or no trade. But roads are not the only need. Marketing and storage facilities are often inadequate and need to be improved, though they receive proportionately less attention.

Despite these problems, some would see urban growth as forcing the economy along.[14] An opposite view is that rapid urban growth may be creating fundamental economic stresses, which may be more difficult to tackle than those so far discussed in this chapter. Immigration is an important element in urban growth, though its role can be exaggerated. Kinship networks appear to help the migratory and settling processes, but, because of the rate at which people are now moving to the larger towns, they may be weakening absorptive mechanisms found in 'traditional' urban communities. Rural attitudes are being maintained amongst the immigrants,[15] and the character of many larger towns appears to be changing. Dwyer argued that this trend, observable throughout the developing world, could destroy the innovating role of the city, thereby retarding economic development.[16] The braking effect appears to be reinforced by the strongly primate character of city-size distributions in the Middle East, and the gap between the largest city and the rest of the hierarchy may now be too large to be effectively bridged.[17] The problem is compounded by the 'traditional' attractions of urban life, and the expansion of the service character of most towns. These reinforce the tendency to neglect the economic development of the countryside, where agricultural change requires expert assistance. A further element is the possible retarding effects of releasing rural pressures by draining people to the towns.

In some cases, regression is apparent as land goes out of cultivation or is used to grow crops demanding least attention. Remittances may bring better housing and a range of consumer goods but they provide a precarious underpinning to the local economy since they are seldom put into productive investment.[18] If the long term effects of rapid urban growth are somewhat speculative, there is no doubt that, in general, the larger towns are increasing in population and economic activity more rapidly than the smaller ones. They lie at the heart of comparatively developed, growing regions and the rest of the country lags behind.

7.2.3 Regional inequalities

Little work has so far been done on regional inequalities in the Middle East, but, in general, the growth regions appear to be those which already have the advantages of standing astride major routes, and possess physical conditions which support a flourishing agriculture. On rather subjective grounds, examples of such core areas appear to include the northwestern region of Turkey,[19] the delta region

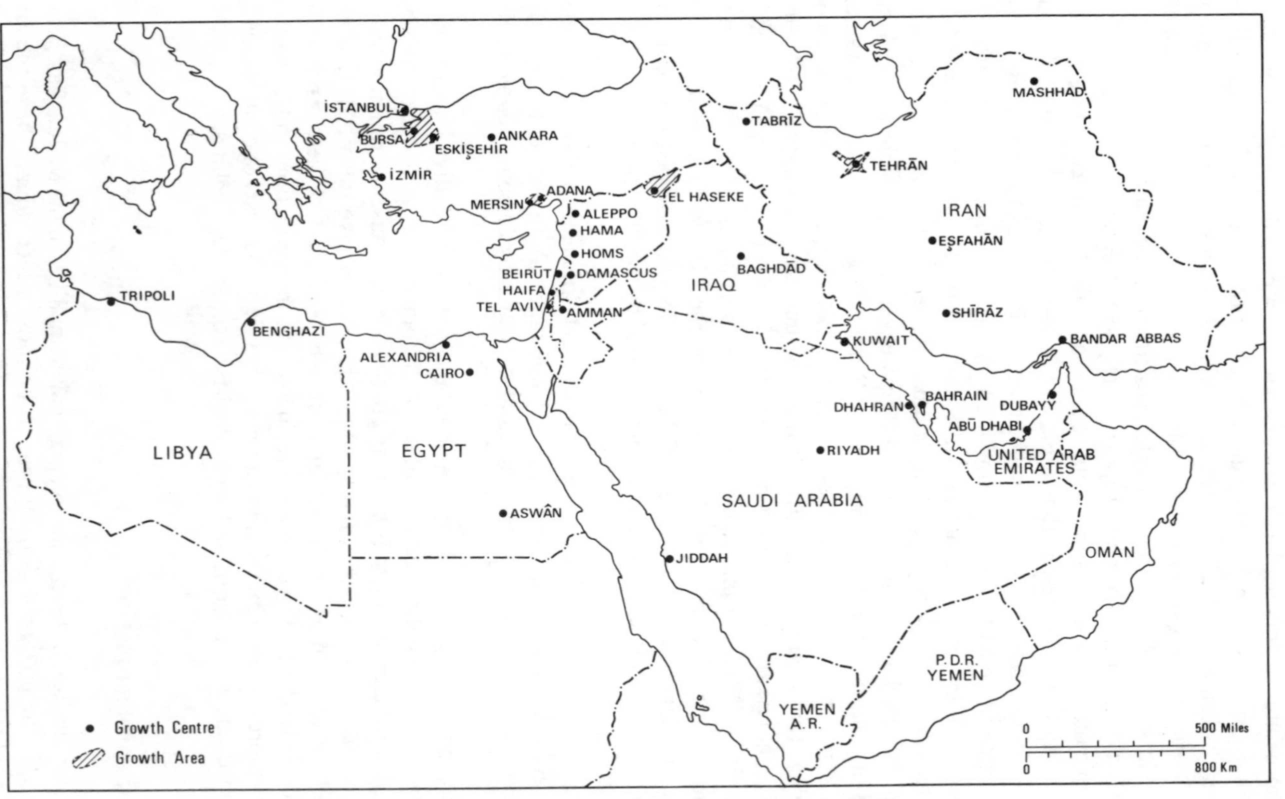

Figure 7.2 Growth centres and areas in about 1980

of Egypt, the Aleppo–Damascus axis in Syria,[20] and the capital city regions of practically all countries (Figure 7.2). The external economies achieved by continuing to locate new activity in these areas are very powerful, but the rest of the country appears to suffer when they are obeyed. Core areas not only draw upon the rest of the country for their food and raw materials, and thus assist regional growth; they also take a disproportionate share of infrastructural investment and drain the more backward regions of their manpower, especially its young, enterprising elements, as well as much of its wealth. Thus, the continued growth of the core may have a weakening effect upon the periphery, which seems to fall back, at least relatively.[21] Alonso argued that regional imbalance may be no more than an early phase in the development process, which will be followed by diffusion effects out towards the backward areas, leading to greater equalization of prosperity across national space.[22] In fact, though, the dominance of development cores seems to be strengthening by further concentration of people, economic activity and decision-making. Certainly, in the short term economic growth seems to be promoted at the expense of socio-economic equity. National aspirations may not be fully realized until greater equity has been achieved, in spatial as well as socio-political terms. Spread effects, so far, are weak. The structural changes necessary to national economic development are not being achieved. Indeed, various forms of regionally-based rivalries (linguistic, religious, economic) work actively against it for economic patronage has political ramifications. Progress in the peripheral regions requires substantial government intervention in the economic process, a large amount of goodwill on all sides and the injection of large amounts of capital if the disadvantages of distance, terrain, climate and resources are to be even partially overcome. National economic plans attempt to stimulate development away from the core areas, but the results are generally disappointing. Dams, factories and even new towns have been introduced into backward regions to promote development (Chapter 17), but their impact has been limited; much of the benefit has accrued to the cores, largely as a result of political centralization.

7.3 Specific problems of economic development

The development problems discussed so far are shared by most developing countries. Three specifically Middle Eastern problems must be raised now. They are the limited resource base of the region, the use of oil revenues, and the existence of conflict.

Mention has already been made of the belief that industrialization is the best means of promoting economic development. Ultimately, industrialization depends upon the availability of power and raw materials. With the exception of oil, which, it should be remembered, if not found in every Middle Eastern state in large quantities, the region generally lacks minerals and fossil fuels (Chapter 8). Fuel and raw materials, together with manufactured or semi-manufactured components, have to be imported, but the costs throw a heavy burden on the exports of primary products (including oil), which provide much of the necessary foreign exchange and, in the case of agricultural exports, distort the pattern of land use. Attention is being given to the search for commercial quantities of minerals and the exploitation of comparatively scarce commodities, such as chrome, for which there is a world demand. Meanwhile, much of the region's industry remains concerned with the processing of agricultural products into food, drink and clothes for largely local markets.

The development of more advanced types of industry, particularly in the essential facilitating sector of engineering, has been painfully slow and too much reliance has been put on imported technology without developing the necessary back-up in trained manpower. Paradoxically, then, despite all the efforts and some real gains in other fields, agricultural land remains the region's greatest industrial resource, though attempts to use it efficiently create the problems discussed earlier in the chapter.

Petroleum exploitation seems to offer a panacea for ancient ills to many countries in the region. The oil industry, however, is capital intensive and directly generates little employment. It is spatially confined, and, though small, highly developed areas are created, they remain islands in a sea of underdevelopment.

Revenues are certainly enormous (Chpater 9), but the best use still has to be made of them. Whilst capital-surplus countries have used their vast incomes to begin the diversification of their economies, much money has been wasted. Industrial and commercial facilities have been duplicated on the Arab side of the Gulf and are way in excess of the real needs of that sub-region. Free welfare provision on the Kuwaiti model has been lavish, whilst employment by the state has amounted to little more than an extensive series of sinecures. Consumerism has become rampant and draws in huge quantities of imports. Governments have largely failed to control the vast differences in wealth or to direct private investment into productive activities. The *nouveaux riches* have tended to become dependent upon their earnings from interest and dividends; middlemen and 'fixers' have proliferated. Economic structures on the Arab side of the Gulf are still precariously balanced on the export of petroleum (largely crude), imported goods (including food) and a disenfranchized immigrant labour force. The bubble may burst and there may be little to show in the long term for its expansion. The dangers are less serious elsewhere in the region, but the elites' capacity to consume, almost for the sake of it, and to demand luxuries and services on an extravagant scale is profoundly worrying to most observers. The attention of those with power and ability to improve the lot of their countrymen – the majority – is in danger of being diverted from the basic issues. These include real poverty and deprivation for many people, as well as the need to build sound economies with sustainable futures. The availability of oil and natural gas has encouraged the foundation of petrochemical industries, particularly along the Gulf, but their viability depends upon successful competition for markets in an era of increasing protectionism (section 8.5). In the 1970s, when new industries were being established and ambitious programmes of social welfare were being launched by the capital-surplus states, income was still available for investment abroad. Several states followed the example of Kuwait and created development agencies. Whilst their international record was good and capital scarce countries in the region such as Egypt and Jordan have benefited, falling oil prices in the early 1980s, accompanied by inflation at home and the effects of the world recession, encouraged retrenchment. Further investment of funds abroad is dependent not only on improvement in the economic conditions in the region and beyond, but also upon the political relationships between states.

A characteristic of the Middle East, particularly marked in recent years, though by no means new, has been conflict within and between states, notably the war between Iran and Iraq (1979–), the various conflicts in Lebanon since 1975, the ramifications of the Palestinian question, and the determination of the Kurds to win political recognition (Chapter 10). The existence of actual or potential conflict has a number of consequences for economic development. It is an important determinant of how resources, manpower and even national space are used. The geographical manifestations are particularly clear in the settlement and regional development schemes pursued in Israel, but they are also apparent elsewhere, for example in the continued existence of refugee camps, and the emphasis in Dhufar placed on forts and airstrips rather than on roads or agricultural improvement. On occasions, conflict is deliberately used by politicians to direct attention away from development issues and to explain why commitments have

TABLE 7.2
Defence expenditure as a percentage of GNP, 1965–84

Country	1965	1970	1975	1980	1984
Bahrain	na	na	na	na	6.3
Egypt	8.6	18.9	na	7.3*	9.6
Iran	4.4	7.2	17.4	3.6*	12.3
Iraq	10.2	7.7	na	na	51.1
Israel	11.7	19.9	35.9	23.2	24.4
Jordan	12.9	14.8	12.2	11.4*	13.4
Kuwait	na	na	na	4.3*	7.6
Lebanon	na	na	na	na	na
Libya	1.4	3.1	1.7	na	na
Oman	na	na	na	na	24.2
Qatar	na	na	na	13.6*	na
Saudi Arabia	8.6	7.0	18.0	20.5*	20.9
Syria	8.4	11.9	15.1	13.1	15.1
Turkey	4.3	3.5	9.0	4.2	4.4
UAE	na	na	na	na	8.3
Yemen AR	na	na	na	na	17.8
Yemen, PDR	na	na	na	na	16.3
Soviet Union	9.0	6.1	11.0	12.0	15.0
United Kingdom	6.3	4.8	4.9	5.1	5.5
United States	8.0	7.7	5.9	5.5	6.4

* estimates

Source: International Institute for Strategic Studies, *The Military Balance*, London, various dates.

not been fulfilled. The time lost, however, may be crucial, especially in dealing with such things as population growth.[23] Ordinary development capital, like tourists with their valuable foreign exchange, is discouraged. War itself has brought enormous physical devastation and massive losses of fixed capital in certain areas, notably Lebanon and parts of Iran and Iraq, as well as depriving countries of valuable assets such as the agricultural land and tourist resorts of the West Bank lost to Jordan as the result of Israeli occupation.[24] People have been driven from their homes, sometimes more than once, and become not only something of a political and economic liability to their hosts, but also, in the case of the Palestinians, the cause of further conflict. Seven years of war between Iran and Iraq (this is being written in 1987) not only affected the economies of the two protagonists, but led to the diversion of capital from the capital-surplus Arab states to bolster Iraq and to massive purchases of arms on their own account, in some cases for the first time.[25] Actual or projected conflict means that all Middle Eastern states maintain proportionately large defence forces and throughout the 1960s, 1970s and into the 1980s allocated substantial proportions of GNP to military purposes, probably more than indicated by the International Institute for Strategic Studies (Table 7.2) This drain on resources seems detrimental to general economic development. On the other hand, the existence of conflict may have distinct advantages. Unemployment is probably kept down to the currently estimated level of 10 to 20 per cent by this means, while military service provides

TABLE 7.3
Development Assistance, 1983

Receiving Region	Total ('000 US$)	Total per Head (US$)
Africa (– Egypt)	9,089.4	14.7
America	6,040.4	15.6
Asia(+ Egypt)	12,996.6	42.5
Bahrain	2.0	5.2
Egypt	1,619.6	35.3
Iran	– 1.8	0.0
Iraq	27.6	1.9
Israel	1,331.4	323.9
Jordan	124.8	38.4
Kuwait	5.2	3.1
Lebanon	123.7	46.9
Oman	11.7	10.4
Qatar	1.3	4.6
Saudi Arabia	43.3	4.2
Syria	147.6	15.4
UAE	2.7	2.2
Yemen AR	151.6	24.3
Middle East	3,592.5	36.8
Oceania	1,067.7	194.1
All Developing Countries	30,602.5	12.3

Source: United Nations, *Statistical Yearbook, 1983–84*, 1986

many men with modern technical skills otherwise difficult to acquire. Capital is attracted to the region on a surprising scale (Table 7.3) in an attempt to tip the political balance in favour of one bloc or the other, and this might not be available but for the possibilities inherent in a conflict situation. In a perverse sort of way, conflict even appears to facilitate economic development by providing a solvent, again through military service, for traditional prejudices and regional separatism, as well as acting as an acceptable explanation for otherwise unpalatable development schemes.[23] It is not without significance that socio-economic change in the region, for good or ill, has often been initiated by soldiers, whether Turkish mercenaries during the early Middle Ages or Egyptian colonels in the 1950s.

7.4 Conclusion

Military and political dimensions serve to underline the complexity of economic development and the interdependency of stimulating or retarding forces. Simplistic solutions to the problem of development are clearly inappropriate, especially when so little is really known about the existing socio-economic structures of the region. Change, if it is to be humane, requires greater knowledge, even of the constraints imposed by the physical environment. Development plans, which are important in ordering priorities and allocating investment, must be carefully conceived and flexibly implemented. In particular, the planners should be aware of the spatial dimension to economic development. Investment is applied in certain places, rather than others. The existence of growth areas is perhaps inevitable, but further growth in them increases regional imbalance. To some extent then, economic growth and equality of regional development are incompatible at least in the short run. Finally, attention should be given to indigenous forces of change and to domestic assets, such as population, rather than to the uncertain panaceas of imported solutions, technology and investment. Peoples who have already created some of the world's great civilizations have the capacity, and ought to possess the nerve, to build a reasonable economic future for themselves.

References

1. S. Kuznets, 'Problems in comparing recent growth rates for developed and less developed countries', *Economic Development and Cultural Change*, **20**, 185–209 (1971–72).
2. D. Forbes, *The Geography of Underdevelopment*, Croom Helm, London, 1984.
3. C. Issawi, 'Growth and structural change in the Middle East', *Middle East Journal*, **25**, 309–24 (1971).
4. (a) H. Bowen-Jones, 'Agriculture', in *The Middle East: A Handbook* (Ed M. Adams), Anthony Blond, London, 1971, 415–426.
   (b) K. McLachlan, 'The agricultural development of the Middle East: an overview', in *Agricultural Development in the Middle East* (Eds P. Beaumont and K. McLachlan), Wiley, Chichester, 1985, 27–50.

5. For example, W. Hütteroth, 'Getreidekonjunktur und jüngerer Siedlundsausbau im südlichen Inneranatolien', *Erdkunde*, **16**, 249–271 (1962).
6. E. Boserup, *The Conditions of Agricultural Growth: The Economics of Agrarian Change under Population Pressure*, George Allen and Unwin, London, 1965.
7. A. Abdel-Malek, 'Sociology and economic history: an essay on mediation', in *Studies in the Economic History of the Middle East*, (Ed M.A. Cook), Oxford University Press, London, 1970, 268–282.
8. M. G. Weinbaum, *Food Development and Politics in the Middle East*, Westview Press, Boulder; Croom Helm, London, 1982.
9. C. Issawi, 'Growth and structural change in the Middle East', *Middle East Journal*, **25**, 309–24 (1971).
10. C. A. Cooper and S. S. Alexander (Eds), *Economic Development and Population Growth in the Middle East*, American Elsevier Publishing Co, New York, 1972.
11. (a) F. Baali, 'Social factors in Iraqi rural-urban migration', *American Journal of Economics and Sociology*, **25**, 359–364 (1966).
(b) F. Baali, 'Agrarian reform in Iraq: some socio-economic aspects', *American Journal of Economics and Sociology*, **28**, 61–76 (1969).
12. C. A. Cooper and S. S. Alexander (Eds), *Economic Development and Population Growth in the Middle East*, American Elsevier Publishing Co, New York, 1972, 11.
13. C. Issawi, 'Growth and structural change in the Middle East', *Middle East Journal*, **25**, 309–24 (1971).
14. W. Alonso, 'Urban and regional imbalances in economic development', *Economic Development and Cultural Change*, **17**, 1–14 (1968–69).
15. P. J. Magnarella, 'From villagers to townsmen in Turkey', *Middle East Journal*, **24**, 229–40 (1970).
16. D. J. Dwyer, 'The city in the developing world and the example of South East Asia', *Geography*, **53**, 353–64 (1968).
17. (a) J. I. Clarke, 'Population in movement', in *Studies in Human Geography* (Eds M. Chisholm and B. Rogers), Heinemann, London, 1973, 84–124.
(b) J. I. Clarke and W. B. Fisher (Eds) *Populations of the Middle East and North Africa*, University of London Press, London.
18. J. C. Swanson, 'Some consequences of emigration for rural economic development in the Yemen Arab Republic', *Middle East Journal*, **33**, 34–43 (1979).
19. M. Albaum and C. S. Davies, 'The spatial structure of socio-economic attributes of Turkish provinces', *International Journal of Middle East Studies*, **4**, 288–310 (1973).
20. A. Drysdale, 'The regional equalisation of health care and education in Syria since the Ba'thi Revolution', *International Journal of Middle East Studies*, **13**, 93–111 (1981).
21. G. Myrdal, *Economic Theory and Underdeveloped Regions*, Duckworth, London, 1957.
22. W. Alonso, 'Urban growth and regional imbalances in economic developments, *Economic Development and Cultural Change*, **17**, 1–14 (1968–69).
23. C. A. Cooper and S. S. Alexander (Eds), *Economic Development and Population Growth in the Middle East*, American Elsevier Publishing Co, New York, 1972, 14.
24. E. Kanovsky, 'The economic aftermath of the Six Day War', *Middle East Journal*, **22**, 131–43, 278–96 (1968).
25. H. Maull, 'The arms trade with the Middle East and North Africa', *The Middle East and North Africa, 1987*, Europa Publications, London 1986, pp 148–53.

CHAPTER 8

Industry, Trade and Finance

8.1 Industrial resources

The resource base of a country or region largely conditions its patterns of industry and trade. Raw materials often control the form of industrial activity, while surpluses and deficiencies dictate the commodities of trade. The location, size and wealth of markets are also important, together with the qualities of the population, the availability of energy and the degree of transport development. On a world scale, and with the exclusion of petroleum, the Middle East is poorly endowed with resources, and this single fact goes far to explain the region's continuing economic backwardness, despite sustained growth during the period since the Second World War.

The region's greatest single resource, with the current exception of petroleum, is its agricultural land. Agriculture not only produces many of the leading exports of the Middle East, but also supplies raw materials to the region's manufacturing industry. Available statistics give an incomplete picture of the agricultural materials available for processing (Table 8.1), but they reveal the importance of cereals and cotton in terms of volume and regional availability. Sugar beet, olive oil, grapes and other crops are significant at a local level (in Aegean Turkey, for example), and often possess high value. The figures also show the concentration of production in the dry farming zone of the region, as well as the relatively diverse agricultural base available for industrial development in Iran, Syria and Turkey. The arid zone, however, is not without its importance, as the output of Kufra (in Libya) and other oases indicate. Despite the figures in Table 8.1, nomadic herdsmen throughout this section of the Middle East contribute important quantities of wool, hides and skins to their national economies and to international trade.

Although a wide range of minerals is found in the Middle East as a whole, including gold and silver, not all of them are now of commercial importance. Petroleum is undoubtedly the most important at the moment, and Chapter 9 describes its extraction and use. Clays and building stones, as well as limestones suitable for cement making, are widespread. Salt is also produced in many localities, by mining and evaporation (Table 8.1; Figure 8.1), but is important in the region mostly for seasoning food. Other minerals are more localized in occurrence. Phosphate rock is produced chiefly in the vicinity of the Dead Sea, in Israel and Jordan, though a large quantity is extracted in parts of upper Egypt accessible from the Red Sea or the Nile, as well as in Syria, while production has recently begun in northern Iraq.

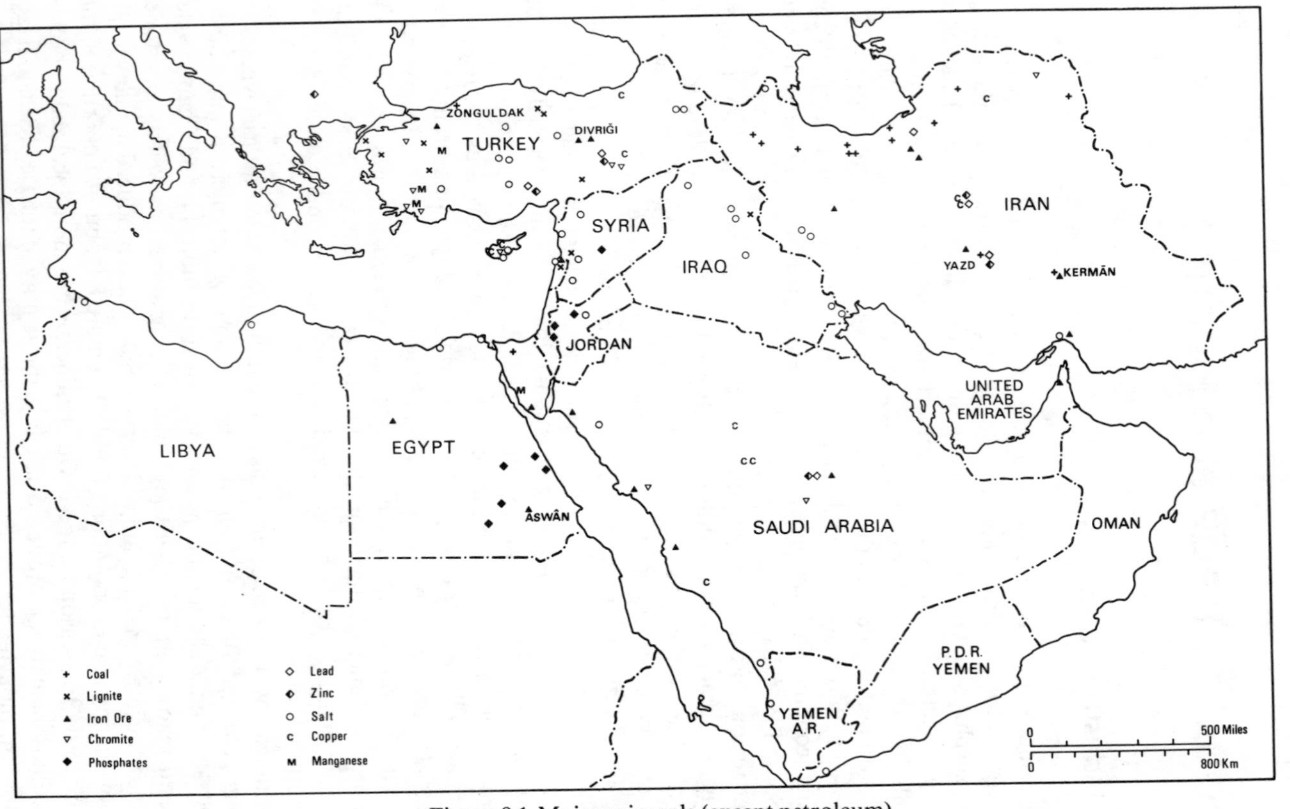

Figure 8.1 Major minerals (except petroleum)

TABLE 8.1
Production of selected raw materials (000 tonnes)

AGRICULTURAL PRODUCTS 1984

COUNTRY	Barley		Cotton Lint		Rice		Tobacco		Wheat		Wool	
	TOTAL (1)	% (2)	TOTAL (1)	% (2)	TOTAL (1)	% (2)	TOTAL (1)	% (2)	TOTAL (1)	% (2)	TOTAL (1)	% (2)
BAHRAIN												
EGYPT	145	1.6	390	29.2	2230	58.1			1815	6.7		
IRAN	1550	17.3	95	7.2	1230	32.1	25	9.0	5500	19.9	16	16.8
IRAQ	300	3.4	3	0.2	95	2.5	12.3	4.4	300	1.1	18	18.9
ISRAEL	9	0.1	94	7.0			0.5	0.2	130	0.5		
JORDAN	3						2.8	1	15			
KUWAIT												
LEBANON	5						4.0	1.4	18	0.1		
LIBYA	70	0.8							150	0.5		
OMAN							1.9	0.7				
P.D.R. YEMEN	2		5	0.4			0.8	0.3	15			
QATAR												
SAUDI ARABIA	12	0.1							1300	4.7		
SYRIA	302	3.4	160	12.0			14.6	5.2	1051	3.8		
TURKEY	6500	72.7	586	43.9	280	7.3	210	75.3	17235	62.5	61	64.3
U.A.E.												
YEMEN A.R.	50	0.5	2	0.1			6.9	2.5	50	0.2		
REGIONAL TOTAL	8948	100	1335	100	3835	100	279	100	27579	100	95	100

MINERALS 1983

COUNTRY	Coal		Iron Ore		Chromite		Copper Ore		Lead Ore		Manganese Ore		Zinc Ore		Salt		Phosphate Rock	
	TOTAL (1)	% (2)	TOTAL (1)	% (2)	TOTAL (1)	% (2)	TOTAL (1)	% (2)	TOTAL (1)	% (2)	TOTAL (1)	% (2)	TOTAL (1)	% (2)	TOTAL (1)	% (2)	TOTAL (1)	% (2)
BAHRAIN																		
EGYPT			1033	19.8											918	48.3	647	6.4
IRAN	800	6.8	450	8.6	17	3.2	48.5	100	26	100	3	48.4	39.9	100	753	39.6		
IRAQ																		
ISRAEL																	2065	20.6
JORDAN																	6119.6	60.8
KUWAIT																		
LEBANON																	5	
LIBYA																		
OMAN																		
P.D.R. YEMEN															73	3.8		
QATAR																		
SAUDI ARABIA																		
SYRIA															88	4.6	1231	12.2
TURKEY	11015 (3)	93.2	3739	71.6	507	96.8					3.2	51.6						
U.A.E.																		
YEMEN A.R.															70	3.7		
REGIONAL TOTAL	11815	100	5222	100	524	100	48.5	100	26	100	6.2	100	39.9	100	1902	100	10068	100

(1) Production of selected raw materials (000's tonnes)
(2) Percentages are those from the regional total
(3) Includes lignite (7934)

Source: United Nations, Statistical Yearbook, 1983–84, New York 1986

Metal ores, which are basic to a comprehensive industrial structure, are mainly associated with the fold mountains sweeping through Iran and Turkey, though small concentrations are also found in parts of Arabia and some sizeable deposits are situated in Libya and Egypt (Figure 8.1). Access is a great problem everywhere, and much of the mining is on a very small scale. Chromite is widespread in the northern part of the region, and production from Iran and Turkey together formed 6.0 per cent of the world's output in 1980, making chrome the region's second most important mineral in world terms after petroleum and gas. Deposits of other non-ferrous metals are more concentrated, with the Elâziğ (Turkey) and Yazd (Iran) districts as currently the most productive (Figure 8.1). Although iron ores are found in several parts of the Middle East, Egypt Iran and Turkey are the greatest producers (Table 8.1). Lignite is largely confined to Turkey, but poor quality coal is found not only there (in the Ereğeli–Zonguldak area) but also at various generally remote locations in Iran, and the deep mine at Maghara in Sinai has been redeveloped. Reserves are only about 0.03 per cent of the estimated total for the world. However, the variety of minerals and the scale of extraction mean that Iran and Turkey have the most diverse resource bases in the region. They are approached only by Egypt. The other countries of the region possess fewer minerals, and, with the exception of petroleum their industrial base lies in a range of agricultural produce; everything else required has to be imported.

Energy is important to modern industrial activity. Waterwheels and windmills (especially in eastern Iran) have been used for centuries to grind corn and lift water, but their power was little used for manufacturing purposes, mainly because of their dependence upon the seasons. When modern industrial development began, lack of coal was an important limitation on development in the region, and most of the fuel actually consumed down to the Second World War was imported. The situation has been transformed with the development of electricity generation. A relatively cheap and potentially ubiquitous form of power became available, and one which eased locational constraint for manufacturing industry, though the location of generators at first assisted an association with the larger towns. Electricity is now the most important source of power in the region. Production increased by 56 per cent between 1962 and 1970, and a staggering 355 per cent between 1970 and 1984 (Table 8.2). The amount of increase, however, has varied considerably across the region. With the exception of Yemen AR, the greatest increases took place in the oil-rich states of Kuwait, Libya, Oman, Qatar and Saudi Arabia, while the smallest changes occurred in those countries already comparatively well endowed with generators before 1962. This pattern emphasizes two facts. First, it accents the low level of development and the limited industrial potential of the desert states before the petroleum era. At the same time, the pattern of increase stresses the importance of regional oil production in the generation of electricity. Oil and gas provide the basic fuels in most countries and are vital commodities in intraregional trade. Regional variations in fuel availability, however, are equally reflected in the dominant role still played by low-grade coal and lignite in the production of Turkey's electricity. Hydroelectric power is well developed in Turkey (53 per cent of the total output for 1982) and

TABLE 8.2
Output of electricity, 1970–84
(million kWh)

Country	Output 1970	Output 1984	Increase 1970–84 (per cent)	Percentage of regional output, 1984
Bahrain	120	645	437.5	1.1
Egypt	4,357	5,409	124.1	9.0
Iran	2,197	13,025	492.9	21.7
Iraq	680	2,400	252.9	4.0
Israel	1,270	3,985	213.8	6.6
Jordan	80	730	812.5	1.2
Kuwait	779	5,230	571.4	8.7
Lebanon	421	668	58.7	1.1
Libya	160	1,300	712.5	2.2
Oman	33	664	1,912.1	1.1
Qatar	78	905	1,060.3	1.5
Saudi Arabia	316	12,200	3,760.8	20.3
Syria	301	1,823	505.6	3.0
Turkey	2,312	8,550	269.8	14.2
UAE	88	2,360	2,581.8	3.9
Yemen AR	6	115	1,816.7	0.2
Total	13,198	60,069	354.7	100.0

Source: United Nations, *Statistical Yearbook, 1983–84*, New York, 1986, Table 154

Egypt (48 per cent of the total output for 1982), where again there are distinct natural advantages, but many of the dams now being constructed throughout the region have electricity generation as one of their prime purposes. The greatest potential obviously lies where precipitation is high and where there is a reasonable flow all the year round, that is, chiefly in the northern countries of the region. At the moment, the greatest amounts of electricity potentially available to manufacturing industry are in Iran (21.7 per cent of the regional total in 1984), Saudi Arabia (20.3 per cent) and Turkey (14.2 per cent). Bahrain, Lebanon, Oman, Qatar and the two Yemens have the least electrical power available, though Oman and Qatar may be expected to improve over the next few years from a low base, just as Saudi Arabia did during the period 1970–84. Demand continues to grow as grid systems are extended and industrialization continues.[1]

A similar dichotomy to that in energy availability is found in manpower resources. Largely because of the origin of its people, Israel is fortunate in possessing a concentration of highly skilled people. Turkey, Egypt and Iran have the largest populations in the region (Chapter 5) and potentially the largest number of industrial workers. The other countries have considerably smaller populations and, to that extent, lower potential. Industrial development in Arabia, apart from the two Yemens, have been possible only through the immigration of labour from other parts of the region (especially, Egypt, Jordan, and Yemen AR), as well as from India and Pakistan. Illiteracy levels are high (over 60 per cent in Egypt, Iran and Libya), while technical skills are limited. Industrial employment is thus curtailed. At the same time, the types of industry which can be developed are restricted and the methods of production tend to be cheap and simple. The number of potential workers is reduced still further by the current youthfulness of the population (Chapter 5), 'traditional' social pressures against women working outside the home or the fields, the recent burgeoning of the service sector consequent upon the expansion of the socio-economic role of governments, and the continued dominance of agriculture, which can afford to shed labour only as farming structures change.

The relative poverty of the great mass of the Middle East's population also influences industrial development. Except where oil royalties are available, capital for investment in industry is relatively scarce. Since capital cannot readily be raised at home, it is sought abroad, either as foreign aid or by selling those very raw materials which Middle Eastern countries require for their own industries. Generally low purchasing power, even where national populations are large, tends to limit the division of labour characteristic of modern industrial economies, as well as effectively reducing the size of the market.

In fact, the number of workers in extractive and manufacturing industries has grown since 1970 to at least 33 per cent of the active population (Table 8.3). Turkey and, surprisingly, the UAE have larger proportions, though Israel, Lebanon, Egypt and Iraq are not far behind. Turkey, Iran and Egypt, however, have the largest shares of the region's industrial workers, a reflection of the relative diversification of their economies.

TABLE 8.3
Industrial employment

<table>
<tr><th>Country</th><th>Mining and Quarrying ('000)</th><th>Manufacturing ('000)</th><th>Percentage of active population in industry</th></tr>
<tr><td>Egypt</td><td>21.4</td><td>15.5</td><td>23.5</td></tr>
<tr><td>Iran (1976)</td><td>8.9</td><td>16.7</td><td>22.6</td></tr>
<tr><td>Iraq (1977)</td><td>3.7</td><td>2.4</td><td>20.7</td></tr>
<tr><td>Israel</td><td>4.4</td><td>2.9</td><td>31.0</td></tr>
<tr><td>Kuwait (1985)</td><td>7.0</td><td>0.5</td><td>19.6</td></tr>
<tr><td>Lebanon (1975e)</td><td>1.0</td><td>0.1</td><td>28.0</td></tr>
<tr><td>Libya (1978e)</td><td>20.4</td><td>0.5</td><td>21.3</td></tr>
<tr><td>Syria (1983)</td><td colspan="2">27.9</td><td>20.8</td></tr>
<tr><td>Turkey (1980)</td><td>13.2</td><td>19.8</td><td>30.1</td></tr>
<tr><td>UAE (1980)</td><td>11.8</td><td>0.3</td><td>44.4</td></tr>
<tr><td>Yemen AR</td><td>0.6</td><td>0.3</td><td>17.7</td></tr>
</table>

e = estimate
Source: United Nations, *Statistical Yearbook, 1983–84*, New York, 1986, Table 20

8.2 Manufacturing industry

Craft industries have made use of the region's resources for many centuries, but a diversity of trades[2] was particularly well developed in Turkey, Syria and Iran, where the resource endowment was relatively great and the population comparatively large. Traditional crafts still survive, especially in certain metal trades (coppersmithing, for example), while carpet-making in Iran and western Asia Minor has steadily increased following a revival of European interest in oriental carpets which began towards the end of the nineteenth century. Competition from European goods, which were not only cheaper but also given preferential treatment under capitulatory systems (see Section 17.2), brought about a change in fashion, then weakened and ultimately destroyed much of the Middle East's craft industry during the late eighteenth and early nineteenth centuries. Decline was never complete, while Persia seems to have fared better than the Ottoman Empire. Attempts were also made to resist the decay. In western Asia Minor and in Egypt new industries were established during the nineteenth century along western lines, using mechanical power and based upon machines collected into factories. The experiments were not particularly successful in the face of foreign competition and foreign control of the economy.[3] Sustained industrial development was initiated between the two World Wars, but the great advance has taken place only after the Second World War, and the breakthrough as late as the 1970s, even in the most industrialized countries.

A common temporal pattern of sectional development can be discerned within the general pattern of growth,[4] though the detailed experience of each country has varied according to its resource endowment and its political evolution. The first stage of industrialization involved the extraction of minerals and the processing of agricultural products, chiefly for export. It began in Egypt and other parts of the Ottoman Empire during the nineteenth century, but was not really experienced by Iran until the beginning of the twentieth century, and then only in petroleum. Important advances had to wait until the 1930s or even the 1950s and 1960s. Iraq, though part of the Ottoman Empire until the First World War, contained very few modern industrial plants by 1900, but experienced two forward surges, one in the 1930s and another in the 1950s. Both were associated with the exploitation of petroleum and stimulated by political changes in the country. Industrialization in Arabia was effectively delayed by the almost total lack of resources and investment capital until the 1960s, or even 1970s in the case of Oman though the conservative attitudes of the rulers were also important.

Amongst the earliest developments in the region were cotton ginning in Egypt (about 1821), the application of steampower to silk winding at Bursa (from 1838), the use of steam engines in milling and olive crushing in many places within reach of the sea, and the mining of chromite in Asia Minor. Expansion and spread of new techniques and modern machines went hand-in-hand with the expansion of cash cropping in many districts, so that, for example, mechanized cotton ginning was not taken up in Syria until the second cotton boom after the Second World War, though it was well-developed in western Asia Minor and the Çukorova before 1900 (Chapter 13). However, the most rapid, as well as the largest developments have taken place over the last 40 years with the expansion of petroleum exploitation. Although much of the region's oil is still shipped as crude, refineries and associated petrochemical industries have been established in many places, with a notable concentration on the Arabian side of the Gulf, where the oil fields lie (Chapter 9).

It is probably true to say that the industrial structure of all Middle Eastern countries has now moved beyond the first stage of development. The south Arabian states are perhaps still closest to it, since industrial development has been very limited (Chapter 11). Most states have reached at least the second stage of industrialization, while retaining aspects of the first phase which are basic to the viability of the entire industrial structure. The second stage may be described as one in which local raw materials are transformed into goods for the domestic market. It is associated with political independence and growth in national and per capita incomes, which produce an expanding domestic market. Manufacturing plants are normally small because risk capital is scarce, due to attractive alternatives in a developing agricultural sector and in construction, and it often has to be raised from the family or in the shape of short-term, high-interest bank loans. The state is not greatly involved at this stage, other than to provide tariff protection against imports, as for example in Egypt after 1930 when duties of up to 30 per cent ad valorem were applied to imported manufactured goods. Most states received some benefit from the self-sufficiency enforced by the Second World War and the availabilty of associated expertise. Local industry was virtually free from external competition, and use had to be made of local materials (the coordinating role of the Middle East Supply Centre was important here).[5] Allied forces not only had to be supplied, but also provided a large pool of additional purchasing power, especially in Egypt where, at its peak, Allied expenditure formed 25 per cent of the national income.[6]

Food processing, including flour milling, is characteristic of the second stage of development. It is largely free from foreign competition because of its perishable raw materials, while demand is strong and generally rising. The production of beverages, such as soft drinks and beer, is an associated development, together with the manufacture of the necessary bottles and glasses. Cigarette making and textile production are normally early developments, either because raw materials are locally available (cotton, for example) or are of such relatively high value that transport costs form a small proportion of the total costs, as in the case of the tobacco imported into Egypt for cigarette making before the First World War. Increasing populations mean that there is an expanding demand for clothing, while rising living standards, as well as Middle Eastern traditions of hospitality, are expressed in an increased consumption of cigarettes. In addition, textile and cigarette making involve technically simple operations, and are relatively easy to man with semi-skilled labour. Difficulties are experienced in all of these lines of development by several states in the Arabian peninsula where the volume and variety of agricultural production are limited by shortage of water. In consequence, imported raw materials are essential, as in the case of flour milling along the Gulf, and even simple industrial activity is possible only because of the large amounts of capital available.

Rising incomes, political independence and industrial and commercial development give rise to an expansion in construction. Transport costs relative to bulk and value make local manufacture of the necessary materials profitable, especially since limestones are available for making cement-based products, while clays can often be found which are suitable for fired bricks. Cement making and associated industries are widespread today. Nails, rivets and pipes are usually worth making in workshops dependent upon local scrap, imported ores, or, in favoured countries, indigenous ores. In this connection it is interesting to note that Egypt's iron and steel industry was founded in the late 1940s on scrap from the battlefields of the Western Desert, though it soon became more firmly based on local ores. The use of motor vehicles and diesel pumps has created an extensive and flourishing repair industry, while agricultural improvements have led to the manufacture of simple steel ploughs and other agricultural implements in many market towns and even to mass production. Ship repairing at Bahrain and Dubayy can be seen as lying in the same tradition since the principal customers were intended to be the supertankers and other bulk carriers serving the Gulf.

Thus, the second stage of industrial development is characterized by a diversity of light industries. This stage was possibly reached by Turkey before the outbreak of the First World War, and was certainly well established by 1939 (Chapter 17). Egypt had moved into the same position by the outbreak of the Second World War, and the existing patterns of industry were strengthened after the Revolution of 1952. The other states of the region, however, lagged behind. Iran, Iraq, Israel, Jordan and Syria may be said to have reached the second stage of development during the 1950s, while Libya and the oil-rich states of the Gulf attained it during the following decade and were consolidating their new position during the early 1980s.

Development into the third stage of industrialization has been dependent upon a definite and marked intervention by the state in economic affairs. State involvement is not new, but practically everywhere in the region its scale and nature are now massive. The aims are rapid, soundly-based industrialization, further diversification of the industrial sector and, by a restructuring of the economy, the improvement of living standards for the mass of the population. State intervention became necessary, regardless of ideology and political regime, because these desirable goals were not being achieved under a private entrepreneurial system. Only governments have the will and power to raise the necessary capital at home or abroad and to ensure the required patterns of investment. Protective measures are still continued, and imports of machinery and equipment are often subsidized, but some new devices have been adopted since the Second World War. Nationalization of much private and foreign industry has been carried out in the more socialist countries, beginning with Egypt from 1956, so that the benefits would accrue to the people. Various approaches have also been adopted for starting new and desired industrial activities. State companies have been set up in several countries, starting with republican Turkey during the difficult years of the

1930s (Chapter 4), but spreading to monarchical Saudi Arabia in 1962, when Petromin was established with the general responsibility for sponsoring industrial development (Chapter 11). Joint ventures with foreign firms have become increasingly common, since they appear to be successful in providing foreign expertise and technical training, as well as in establishing new lines of activity. But whatever the devices adopted, and regardless of the ideology of the political regime, practically all Middle Eastern states now make use of economic development plans to structure and allocate investment, largely in favour of industry.

State intervention has produced three lines of development.[7] Established food processing, textile manufacturing and construction goods industries have continued to be promoted everywhere. The degree of skill required is usually small, while markets and raw materials are available, often from local resources. Simple consumer goods, like soap, sugar and footwear, are also produced. At the same time, a second line of development has been followed, especially by the more industrialized countries, Egypt, Iran, Turkey and, increasingly, by Syria and Saudi Arabia. This has been the extension of import substitution to a range of high value consumer durables which are demanded by the growing numbers of affluent townspeople. The involvement of foreign companies is important and many of the components used to assemble such items as radios, television sets, refrigerators, washing machines and motor vehicles are imported, though local parts are increasingly being used. Pharmaceuticals are another large and expanding field where import substitution has advanced markedly, but where foreign involvement is again important.

The third line of development which is being actively pursued is the creation of basic heavy industry. Although plants producing industrial chemicals have been established, the production of iron and steel has been particularly pushed as providing the basis for a whole range of metal industries, where demand is expanding, and as being, it is thought, the surest way to transform the economy. Turkey was the leader in this field and began the establishment of the Karabük complex during the 1930s (Chapter 17), using her own coal and iron ore. No other country followed her lead until after the Second World War, partly because they lacked the essential raw materials and partly because the vast capital expenditure involved was prohibitive to private enterprise. Turkey has expanded its steel-producing capacity with plants at Ereğli (on the Black Sea), Iskenderun and most recently at Aliaga on the Aegean coast (1981) and Bursa (1983). The only other countries where a fully fledged iron and steel industry seemed possible in 'traditional' terms were Egypt and Iran, but the development of the direct reduction process made production possible in other countries. Thus, the conventional and recently expanded steel plants at Eşfahān (Iran) and Helwan (Egypt) have been joined by new ones at Khor az-Zubair in Iraq (1978), at Jiddah (1967) in Saudi Arabia, in Qatar (1978) and, if iron-pelletizing is included, in Bahrain too (1984). They face considerable problems, however, including world over-capacity and the need for ore, billets and scrap to be imported.

Other important developments which can be considered as part of the drive to create a sound industrial base are the construction of aluminium smelters and the development of petrochemical plants. Although an aluminium smelter is found at Nag Hammadi in Egypt, the principal plants are in Bahrain (1972, expanded late 1970s) and at Dubayy (1979). In both cases the availability of relatively cheap and abundant energy allowed the exploitation of global position between the sources of raw material (Australia) and the main markets (Europe). The move into petrochemicals was a logical extension of the development of refining in Arabia and was a decision simply to exploit the comparative advantage of cheap local feedstocks to meet a world demand for raw materials such as methanol, naphtha, ethylene and ammonia, as well as polymers and thermoplastics, which was rising rapidly in the 1960s and 1970s, notably in Japan and the emerging industrial countries of southeast Asia.[8]

Egypt, Israel, Jordan, Syria and Turkey dominate the pattern of manufacturing industry in the region. At least 10 per cent of their GDP was contributed by the sector in 1983 (Table 8.4), though there were declines in the case of Egypt and Syria since 1969 resulting, at least partly, from the expansion of service industries and a considerable strengthening of the 'black economy'. It is less easy to classify countries into distinct groups than it was a decade ago. The explanation for this is in two parts. Most important is the fact that the necessary statistical materials are no longer readily available on a comparative basis through the United Nations. The other part of the explanation is the greater complexity of manufacturing structures at the end of the 1980s compared with the 1970s. That said, it is possible to claim that Egypt, Iran, Israel, Turkey – and possibly Iran still – possess the most diversified manufacturing structures in the region. They combine a base in

TABLE 8.4
Contribution of manufacturing industry to GDP, 1983 (per cent)

	per cent
Egypt	12
Iran (1982)	8
Iraq	7
Israel	22+
Jordan	14
Kuwait	6
Lebanon	—
Libya	3
Oman	2
Qatar	6
Saudi Arabia	6
Syria (1981)	10
Turkey	25
UAE	9
Yemen, AR	6

+ = all industrial activity
Source: United Nations, *Statistical Yearbook, 1983–84*, New York 1986, Table 26

food processing, textile manufacture, cement-making and iron and steel production with diversification into fertilizer and chemical production, pharmaceuticals, electrical goods and a variety of assembly industries (including vehicles). They also possess their own armaments industries, particularly significant in the case of Israel where the industry can compete successfully on the world market. Iranian industry appears to be suffering from the effects of both the war with Iraq and the revolution against the Shah; it seems to be running at well below capacity in many branches.

Simple transformation still seems to be dominant in Iraq, Jordan and Syria, though each has a particular national emphasis. Food processing and textile production are important in Iraq, whilst construction materials (especially cement) and fertilizers are significant in Jordan. Manufacturing industry is more diversified in Syria and developing under quite firm protection, but textiles are particularly strong.

Saudi Arabia and the Gulf States have embarked upon ambitious industrialization schemes, but, apart from possessing a few large modern plants in aluminium, steel and petrochemicals, their industrial structures are essentially simple. They are dominated by food processing and the production of an array of construction materials, though a wide variety of light industry is being promoted both by local demand and government encouragement. Food is processed (including basic flour-milling) in the Yemens, whilst textiles and construction materials are manufactured, but the Aden refinery (built 1954) still dominates the economy of PDR Yemen.

Scarcity of capital and the low purchasing power of the local market largely explain the present size structure of manufacturing concerns in the region. Most manufacturing units are small, employing fewer than 50 people each, though a few really large plants generally account for most of the production. Size of employment varies according to the type of industry and the circumstances in which it was established. Sugar refining, petrochemicals and cement making are generally recent introductions and use modern techniques in large, purpose-built plants, but consequently, employ relatively few workers. Textile firms vary more widely in equipment and number of workers employed. Large state-sponsored units, as at Tanta in Egypt and Aleppo in Syria, have thousands of employees, but everywhere small and medium-sized units, founded by private enterprise and using second-hand machinery, are common. Light industry is generally dominated by workshop units in a spectrum merging into large assemby plants, at one end, and single-man booths in the bazaar, at the other. In all industries vertical integration is widespread, particularly in the larger firms, and all stages of production may be carried out on the same site. This is an indication of the low level of industrial development in the region. Competition for production factors, which are normally scarce, and for control of the generally limited home market have similarly ensured that horizontal linkage is well developed.

Horizontal linkage is part of the explanation for the concentration of industry in relatively few centres, chiefly the larger cities. Most of the industries also require locations near the markets, especially since transport is often inadequate.

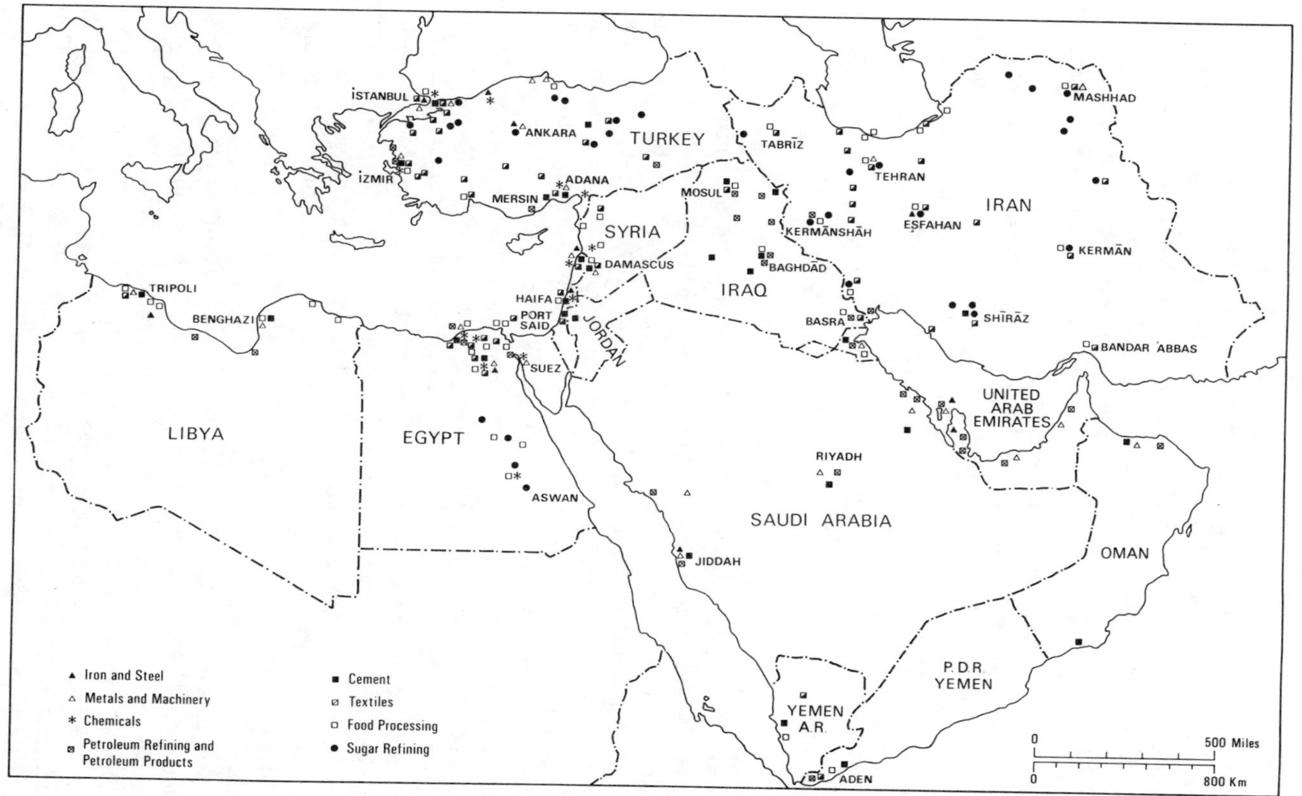

Figure 8.2 Major types of manufacturing industry in about 1985

In addition, factories require labour and access to such public utilities as clean water and electricity which, until recently, were confined to the larger centres of population. Major concentrations of industry on a regional scale are shown in Figure 8.2. They include the inland capitals of Tehrān, Baghdād, Damascus and Cairo, with their large and wealthy markets, as well as the major ports, such as Beirūt, Haifa, Alexandria and İzmir, where imports arrive and primary products are processed. Other concentrations exist in major provincial towns, like Aleppo, Eşfahān, Mosul and Eskişehir, which are situated in the more productive agricultural areas of the region and frequently close to important minerals. Small towns often contain a group of workshops, but, if their employment structure contains any significant degree of industrial work, it is usually provided by a single, comparatively large plant where local raw materials are processed or transformed. In such a case, the market seems to be essentially provincial, rather than national. It is the larger units, generally situated in the major cities, which appear to contribute most to internal, as well as international trade.

8.3 Foreign trade

'Trade is the acid test of industrialization.'[9] By this test, the Middle East emerges as divided into three camps in 1982 (Tables 8.5 and 8.6). The first consists of Israel and Turkey. Manufactured goods formed 65.2 and 60.5 per cent respectively of all their exports by value and covered a spread of goods from textiles, processed foods and chemicals to metal goods. Egypt, Jordan, Syria and Yemen AR form the second group. They exported a variety of manufactured goods, in which petrochemicals featured as a common element, as well as primary products. The third set of countries consists of the principal oil exporters. Unrefined petroleum formed more than 90 per cent of their exports by value (Iraq, Libya, Oman, Qatar, Saudi Arabia and the UAE) but petrochemical products came second, a long way behind. Bahrain may be linked to this group since over 80 per cent of its exports consisted of refined products. Agricultural products were still important exports for Turkey in 1982 (36.5 per cent), Egypt (19.7 per cent), Jordan (17.4 per cent) and Syria (16.7 per cent in 1979), and ranked second to petroleum and related goods for Iran. Finished and semi-finished goods actually declined as a proportion of total exports from Egypt and petroleum has become important since the 1960s.[10]

Primary products, in fact, were dominant exports from the region in 1982 as they had been in 1958, 1966 and 1970, and industrialization still has a long way to go. On the other hand, the structure of importation for 1982 indicates that industrialization was proceeding apace. Industrial supplies formed between 18.5 per cent (Bahrain) and 37.1 per cent (Egypt) of imports for all countries except Iran, Lebanon and PDR Yemen and for which there are data. Machinery imports ranged between 8.4 per cent (Bahrain) and 29.6 per cent (UAE). The structure of importation also reveals another important aspect of the Middle East's regional economy. Chapter 7 referred to the region as one of food deficit. This is clear in

TABLE 8.5
Selected exports, 1982

Country	Agricultural produce	Minerals	Manufactures
		(percentage of total)	
Bahrain	0.1	17.8	82.2
Egypt	19.4	55.7	24.9
Iran (ex. petroleum)[2]	90.2		7.5
Iraq[2]		98.6	
Israel	10.6	24.2	65.2
Jordan	17.4	31.9	50.7
Kuwait	0.5	41.9	57.7
Lebanon	—	—	—
Libya (1981)		99.6	
Oman	6.9	0.4	92.7
Qatar		93.9	3.9
Saudi Arabia	0.1	96.0	3.9
Syria (1979)	16.7	70.9	12.4
Turkey	36.5	3.0	60.5
UAE (1978)		98.6	
Yemen AR (1981)	3.7		96.3
Yemen, PDR	—	—	—

Sources: 1. United Nations, *International Trade Statistics Yearbook, 1984*, vol 1, New York 1986
2. *The Middle East and North Africa 1987*, 33rd edition, Europa Publications, London 1986

TABLE 8.6
Selected imports, 1982

Country	Foodstuffs	Industrial supplies	Fuel	Machinery
		(percentage of total)		
Bahrain	5.5	18.5	51.3	8.4
Egypt	24.8	37.1	4.0	17.0
Iran[2]	5.2	8.4	0.5	8.0
Iraq (1978)[2]	10.8	22.9	?	53.6
Israel	9.7	32.4	23.2	17.3
Jordan	16.0	24.0	21.2	12.7
Kuwait	12.2	24.8	0.6	19.8
Lebanon	—	—	—	—
Libya (1981)	13.3	35.5	0.2	24.8
Oman	11.6	23.1	10.1	21.3
Qatar	9.2	26.9	0.7	28.9
Saudi Arabia	11.2	28.3	0.4	23.4
Syria (1979)	11.8	32.9	24.7	14.9
Turkey	2.4	26.0	43.7	19.5
UAE (1978)	8.7	28.0	5.1	29.6
Yemen AR (1981)	29.9	25.9	7.7	11.9
Yemen, PDR	—	—	—	—

Sources: (1) United Nations, *International Trade Statistics Yearbook, 1984*, vol 1, New York 1986
(2) *The Middle East and North Africa 1987*, 33rd edition, Europa Publications, London 1986.

the value of food imports to the region. Whilst for Iran, Turkey and, unexpectedly, Bahrain and the UAE imports of agricultural products were less than 10 per cent by value, for the other countries of the region they were considerably greater. The major importers in terms of the percentage share of imports by value were not the oil-rich desert states, but Yemen AR (29.9 per cent) and Egypt (24.8 per cent). In the first case, high levels of food import reflect the importance of remittances to the local economy and the comparative neglect of agriculture. In the case of Egypt, though, the problem is the result of what must be recognized as serious over-population, on the one hand, and an emphasis within farming, on the other hand, on commercial crops (including onions, potatoes and citrus, as well as cotton) to sell abroad or, in the case of cotton, to supply the national textile industry which is largely orientated towards the domestic market. The improvement in the situation of Libya, Saudi Arabia and the other countries with respect to food imports since 1970 is partly a reflection of changed prices for food, partly the result of quite minor shifts in the composition of trade and partly the result of successful, if expensive, food security policies.

The pattern of trade for 1982 has been analysed by means of a matrix designed to show the extent of intraregional trade as well as the major trading partners (countries or blocs) outside the region (Table 8.7). The matrix was originally designed by Preston and Nashasnibi to examine trade patterns for 1958 and 1966.[11] As far as possible, the original form has been kept. The imports of a country are read vertically in the table and the exports horizontally. Thus, Turkey imported goods worth more than US$400 million from the UK (column 1, row 17) out of a total of US$8950 million (about 4 per cent), but exported goods worth only just over US$276 million (5 per cent) to the UK (column 17, row 1) out of a total worth US$5691 million.

Trade patterns in 1982 were dominated by the extraregional component, which comprised 88.4 per cent of all imports by value and 91.0 per cent of all exports. Whilst this was the situation also in 1958, 1966 and 1970, the sequence of figures shows a modest increase in the intraregional component over the quarter century. Although the pattern reflects the region's massive exports of petroleum, and is exemplified in Table 8.7 by Libya, Saudi Arabia and probably Iran, it also arises from the importance of all primary products in exports and the leading roles of industrial supplies and machinery in imports. The leading exporters in terms of value were Saudi Arabia, the Gulf States (Bahrain, Qatar and the UAE considered as a group) and Iran, a direct consequence of the importance of petroleum. Saudi Arabia was also the region's heaviest importer, followed by Iraq and the Gulf States. Oman and Yemen AR were insignificant contributors to the region's total trade, though commerce was clearly important to both countries. Egypt and Syria were both significant exporting countries in the past, but their position has deteriorated. The European Economic Community has been the region's major trading partner since it began in 1958. In 1982, the EEC accounted for 29.9 per cent of the region's exports as presented in Table 8.7 and 38.9 per cent of its imports. The exports are largely, but on a regional scale not exclusively, petroleum, as the Egyptian and Turkish figures demonstrate, for they

TABLE 8.7
Middle East trade matrix, 1982 (000 US dollars)

		1	2	3	4	5	6	7	8	9	10	11	12	13	14	15	
		TURKEY	EGYPT	JORDAN	SYRIA	LEBANON	YEMEN A.R.	P.D.R. YEMEN	IRAN	IRAQ	LIBYA	SAUDI ARABIA	KUWAIT	OMAN	GULF* STATES	ISRAEL	
1	TURKEY	–	126.2	87.4	52.9	111.6	–	–	791.1	610.4	271.5	334.1	93.4	7.5	43.2	11.0	
2	EGYPT	6.9	–	6.8	0.2	20.9	8.2	0.1	–	0.7	–	87.6	16.9	3.5	–	443.2	
3	JORDAN	6.1	0.4	–	27.6	6.9	1.2	5.6	–	225.6	10.9	121.3	44.1	1.0	27.4	–	
4	SYRIA	9.4	0.3	28.1	–	21.6	2.2	8.2	41.2	0.1	35.8	61.4	11.7	–	4.9	–	
5	LEBANON	–	35.6	57.6	46.3	–	3.7	–	–	–	8.3	313.8	77.5	2.3	73.6	–	
6	YEMEN A.R.	–	–	–	0.3	0.1	–	19.6	–	–	–	–	0.2	–	0.9	–	
7	P.D.R. YEMEN	–	–	–	–	–	20.7	–	–	–	–	–	–	–	0.1	–	
8	IRAN	747.7	–	–	788.3	–	–	–	–	–	–	–	14.4	–	13.6	–	
9	IRAQ	1417.6	–	4.2	295.4	–	1.6	–	–	–	–	–	–	–	0.1	–	
10	LIBYA	842.7	–	–	20.8	–	75.7	–	–	–	–	–	–	–	–	–	
11	SAUDI ARABIA	524.1	52.1	704.3	128.1	149.9	222.9	–	–	–	–	–	131.0	6.5	1082.9	–	
12	KUWAIT	82.0	12.9	26.1	7.3	16.2	51.6	6.3	67.6	891.1	–	302.0	–	23.9	304.2	–	
13	OMAN	–	0.9	2.7	0.1	0.3	8.6	–	10.9	11.1	–	90.9	8.1	–	141.5	–	
14	GULF* STATES	1.0	16.1	16.0	5.7	12.0	24.8	180.8	89.7	132.3	1.7	423.5	200.2	364.0	–	–	
15	ISRAEL	26.0	47.4	–	–	–	–	–	–	–	–	–	–	–	–	–	
16	USA	815.2	1922.2	409.5	164.7	–	30.7	–	–	845.7	301.0	7929.7	981.7	214.2	1669.8	1580.5	16864.9
17	UK	405.8	494.8	332.3	89.5	–	71.9	–	582.6	1525.8	513.8	2525.6	581.3	423.7	1631.4	528.3	9706.8
18	OTHER EEC	2024.9	3116.6	602.8	972.4	992.9	389.0	–	2982.6	7091.7	3810.8	11598.8	2265.2	456.6	2972.0	2422.1	41699.2
19	OTHER WEST EUROPE	704.7	716.5	169.1	266.8	177.0	39.3	–	714.7	1347.9	1013.5	2288.9	399.2	82.1	519.6	694.3	9133.6
20	JAPAN	355.4	536.5	247.9	180.9	–	202.6	–	934.9	2755.2	324.2	7199.9	1895.0	507.6	3252.7	186.3	17579.1
21	LATIN AMERICA	55.5	287.2	34.7	128.9	18.3	18.4	–	271.5	356.0	29.7	258.5	60.8	10.2	64.1	153.3	1747.1
22	AFRICA	102.3	55.1	22.9	12.1	59.6	35.4	–	24.0	64.6	102.0	369.6	–	25.6	35.3	190.0	1098.5
23	ASIA	76.0	222.8	86.2	74.3	42.0	163.3	–	456.9	207.6	58.9	2754.8	655.9	201.5	1113.0	94.1	6207.3
24	USSR	158.5	238.6	91.4	185.6	–	5.3	92.5	796.1	1345.8	164.6	26.2	12.0	–	4.3	–	3120.9
25	YUGO SLAVIA	22.9	155.8	32.2	18.0	20.9	–	–	201.5	755.4	165.4	46.2	24.8	5.6	34.9	27.0	1510.7
26	OTHER EAST EUROPE	294.2	510.7	137.0	385.0	50.7	16.9	–	363.0	687.7	361.3	187.7	92.1	3.6	28.2	63.9	3182.1
27	CHINA	–	56.4	42.1	74.4	–	62.1	–	–	–	5.5	234.1	134.0	18.0	153.6	–	780.2
28	OTHER	88.6	344.9	41.2	22.8	67.3	69.6	–	360.4	400.6	225.0	1125.7	349.4	74.9	552.5	165.1	3888.0
29	INTRA REGIONAL TRADE	3663.5	291.9	933.2	1373.0	339.5	421.2	220.6	1000.5	1871.3	328.2	1734.6	597.5	408.7	1692.4	454.2	15330.3
30	TOTAL IMPORTS	8767.5	8950.0	3182.3	3948.5	1768.2	1525.7	313.1	8688.7	19255.3	7403.9	38280.3	8048.9	2432.3	12724.8	6559.1	131848.7
31	29/30 AS %	41.8	3.3	29.3	34.8	19.2	27.6	70.4	11.5	9.7	4.4	4.5	7.4	16.8	13.3	6.9	11.6

	16	17	18	19	20	21	22	23	24	25	26	27	28	29	30	31
	USA	UK	OTHER EEC	OTHER WEST EUROPE	JAPAN	LATIN AMERICA	AFRICA	ASIA	USSR	YUGO SLAVIA	OTHER EAST EUROPE	CHINA	OTHER	INTRA REGIONAL TRADE	TOTAL EXPORTS	29/30 AS %
1 TURKEY	251.6	276.3	1479.1	491.9	43.2	8.8	168.3	66.7	127.6	20.4	187.7	17.5	11.6	2540.3	5691.0	44.6
2 EGYPT	147.5	399.4	960.8	130.6	122.6	17.1	106.2	33.4	275.7	52.5	256.6	35.4	6.9	595.0	3139.7	18.9
3 JORDAN	20.2	12.3	47.8	12.4	12.5	–	4.3	71.6	–	12.1	56.9	4.0	3.2	478.1	735.4	65.0
4 SYRIA	13.8	9.3	983.5	5.1	0.5	–	54.8	23.0	321.5	26.6	475.5	11.2	17.3	224.9	2167.0	10.4
5 LEBANON	–	–	–	33.5	–	–	28.6	–	–	–	–	–	2.4	618.7	683.2	90.5
6 YEMEN A.R.	–	1.8	5.7	–	–	–	0.2	1.2	–	–	–	0.1	–	21.1	30.1	70.0
7 P.D.R. YEMEN	–	–	–	–	–	–	–	–	–	–	–	–	–	20.8	20.8	100.0
8 IRAN	611.7	395.3	6508.9	1628.9	2566.9	513.9	55.4	2975.6	259.7	243.6	554.0	–	190.4	1564.0	18068.3	8.6
9 IRAQ	–	139.4	2398.7	1162.2	779.5	–	174.0	1062.9	25.1	364.7	266.1	–	131.7	1718.9	8223.2	20.9
10 LIBYA	7610.1	335.5	6837.7	1299.1	328.5	462.2	1.5	6.5	1549.7	359.5	685.8	–	121.5	939.2	20536.8	4.6
11 SAUDI ARABIA	7021.1	2396.7	20530.3	3824.4	19667.2	3564.1	1410.3	13875.8	–	81.1	116.7	–	1593.3	3001.8	77082.8	3.9
12 KUWAIT	36.3	139.7	1787.2	120.4	1585.4	303.1	201.7	2280.1	0.3	1.1	0.1	–	329.5	1791.2	8576.1	20.9
13 OMAN	4.4	15.0	8.6	1.2	851.8	–	1.4	24.2	–	–	–	–	6.8	275.1	1188.5	23.1
14 GULF* STATES	1430.3	433.9	3646.4	820.4	8976.7	1168.7	644.7	2310.5	–	–	–	11.9	462.0	1467.8	21373.3	6.9
15 ISRAEL	1181.0	449.8	1305.0	366.9	191.9	124.5	143.7	262.9	–	29.5	36.0	–	118.6	73.4	4283.2	1.7
	18328.0	5004.4	46499.7	9897.0	35126.7	6162.4	2995.1	22994.4	2559.6	1191.1	2635.4	80.1	2995.2	15330.3	171799.4	8.9

* Bahrain, Qatar and United Arab Emirates

NOTES

(1) Compiled from United Nations, International Trade Statistics Yearbook, 1984, Vol.1, New York. 1986

(2) Figures for Iran, Iraq and Lebanon were derived from the statistical series of their trading partners

(3) Export statistics for Libya and Qatar are for 1981

(4) Where imports and exports did not tally the figures have been adjusted by taking a mean of the two figures

(5) The imports of a country are read vertically in the table and the exports horizontally

(6) Import and export figures in million U.S. dollars

contain important quantities of agricultural produce and textiles. The various countries of the EEC have long been markets for many of the region's primary products. So important is the EEC to the Middle East countries that in 1977 Egypt, Jordan, Lebanon and Syria negotiated trade agreements with it, thereby joining Israel which negotiated similar arrangements in 1975. Yemen AR completed a five-year agreement on various aspects of cooperation, including trade, in 1984. Turkey went a stage further. It became an associate member of the Community in 1964, started the transition to a customs union in 1973 and, in April 1987, applied for full membership.

The region's second ranking trading partner is Japan. It took 20.4 per cent of the region's exports in 1982 (mainly petroleum and petrochemicals) and provided 13.3 per cent of its imports (including consumer goods). Japan's relations are particularly close with the countries of the Gulf. This is a recent development, resulting not only from Japan's need for imported energy and raw materials, but also its realization that the Gulf constituted a wealthy and potentially growing market. The rest of Asia occupied third place as the region's trading partner (13.3 per cent of exports; 4.7 per cent of imports). Whilst attributable to geographical location and spatial proximity, the relationship rests upon the demand for oil and petrochemicals from the industrializing countries of the south and southeast Asia, as well as their need for markets.

The USA and the USSR are relatively unimportant trading partners for the Middle East as a whole. The Soviet Union's position has declined since 1970, when it received 5.7 per cent of the region's exports and supplied 5.4 per cent of its imports. In 1982, it was the destination of only 1.7 per cent of the exports and provided 2.3 per cent of the imports. The United States has lost much more ground, largely as a result of competition from Japan and from an expanded EEC. Its share of the region's imports declined from 24.6 per cent in 1970 to 12.8 per cent in 1982. On the other hand, the Middle East's exports to the USA have more than doubled (from 4.4 to 10.9 per cent), largely as a result of the American need to import crude oil. As a consequence, the Gulf has become of vital strategic concern to the US administration.

Intraregional trade, though still a small part of the total in 1982 (10.0 per cent), grew during the late 1960s and the 1970s after a phase of decline or stagnation in the late 1950s, early 1960s. The main phase of growth seems to have been largely between 1973 and 1974, and is to be explained chiefly by trade in petroleum which increased in both volume and value.[12] The existence of the Arab Common Market since 1964 (initially Egypt, Jordan, Syria and Iraq) has not been of major importance, though many of the barriers to trade in agricultural products and raw materials have been reduced under its auspices.[13] Jordan and Syria benefited and their exports of fruit and vegetables have expanded. Further development of the Arab Common Market has been hampered in various ways. One has been political disagreements between the members, particularly after the Camp David accords between Egypt and Israel (1978) and the outbreak of the war between Iran and Iraq (1979), in which Syria supports Iran. Members have been unwilling to reduce tariffs on industrial goods since they wish to protect their own nascent industries, though it is arguable whether even Egypt, the most industrialized of the group, would be capable of taking advantage of the economies of scale associated with market expansion. Quotas remain in force, currencies are not convertible and foreign exchange controls are rigid, at least at the official level. To some extent, the economies of the members of the Arab Common Market are similar, rather than complementary, so that the trade potential of comparative advantage is reduced. In any case, established patterns militate against a rapid expansion of trade within the Arab Common Market, as they do against growth in intraregional trade as a whole. So long as primary products form such a large part of the exports and while imports of industrial supplies and machinery are important, then the extraregional component will dominate. Historic ties and consumer preference also support it.

The importance of intraregional trade varies from country to country (Table 8.7) and there has been comparatively little change over time. Saudi Arabia, Libya and Israel were the least involved in trade with the rest of the region in 1982, followed as far as imports are concerned, by Egypt and Kuwait, though both of the latter appear quite involved in trade with the region as far as exports are concerned. The export trade of Libya and Saudi Arabia is dominated by petroleum for which the major markets are outside the region, and most of their import needs cannot be met from within the Middle East. Kuwait is in a similar position, whilst Egypt imports largely from outside the region since it needs cereals and industrial goods but it has been able to sell manufactured goods (especially textiles) in the region. Political isolation, of course, has not helped its intraregional trade. Israel is also politically isolated in the region and her overt trade is confined to Egypt and Turkey, though covert trade with Iran has come to light. The states most heavily involved in intraregional trade in 1982 were Jordan, Turkey and Yemen AR, though Syria, Lebanon, Oman and the Gulf States (Bahrain, Qatar and the UAE) all showed a much higher than average orientation to the region. Jordan, Syria and Lebanon have benefited from the expansion of purchasing power in Saudi Arabia and along the Arab side of the Gulf. They supply fresh fruit and vegetables, as well as processed foods, to these countries but their trade statistics are inflated to some extent by goods in transit from outside the region. Yemen AR's inclusion with this group of countries is due to its export trade to PDR Yemen and to the fact that the intraregional component of its imports (33.5 per cent) is one of the highest and involves a spread of countries. PDR Yemen's own involvement in the intraregional trade is unclear from the available statistics, but its major asset, the Aden refinery, is dependent upon crude imported from Kuwait and the lower Gulf. The intraregional component of Turkey's trade developed during the 1970s and changed the country's virtual commercial isolation from the region quite dramatically. More than 40 per cent of its export trade went to the region in 1982 and about 42 per cent of its imports were drawn from it. The change is the result of various developments. First came the success of Turkish construction companies in gaining contracts in Libya and then in Saudi Arabia. These companies naturally drew their supplies from home, as far as possible. A second factor was Turkey's overproduction of sugar, for the surplus has been dumped, to some extent, on the Arab market where consumption has been rising. A third development was the realization by Turkish textile manufacturers that their products could compete successfully on price and quality in the Arab market. To some extent, though, this was a response to the problem of an almost satiated and contracting domestic market. The subsequent sales campaign was strongly supported by the government. Whilst Turkey's commercial involvement with the region might be regarded as something of a reversion to an historical pattern, the intraregional trade of Oman (23.1 per cent of exports; 16.8 per cent of imports) and the Gulf States (1.7 per cent of exports; 13.3 per cent of imports) is surprisingly low given the apparent continuance of long-established patterns in the Gulf and may be due to the disruption of the entrepôt trade caused by the Gulf War, as well as the slowing down in the growth of local economies. The various ports of the lower Gulf still compete with each other to become the distribution centre for goods imported from outside the region. Smuggling is probably significant, especially across the Gulf to Iran and outside it to India and Pakistan.

8.4 Transport and internal trade

Although the transport and internal trade systems of Middle Eastern countries are amongst the least studied aspects of the region's economic geography, sufficient information is available for a number of general observations to be made for further testing by research. Railways are still of considerable importance in the region, as statistics for passenger-km and net tonne-km show (Table 8.8). Another demonstration of their continuing importance was the construction and improvement of railway lines during the 1960s. The section of the Hejāz railway within Jordan was reconstructed as far south as Ma'ān, while a branch was started to Aqaba in 1972. The Baghdād to Basra line was converted to standard gauge by 1971, and an extension was constructed to the new port of Umm Qaṣr. Syria built a completely new line from Latakia, on the Mediterranean, to Aleppo, across to the Euphrates and then down the river to Deir ez Zōr, with the objects of servicing the Tabqa Dam project and tapping the agricultural wealth of the Jezira more effectively. In Saudi Arabia, the line from Dammān to Riyadh has been upgraded. Meanwhile, Iran and Turkey were joined by a railway line and ferry (on Lake Van) sponsored by CENTO and completed in 1970 (Figure 8.3). In effect, these developments continue policies of national integration developed between the two World Wars when Turkey's railway system was elaborated into a network across the whole country and Iran was traversed by axial routes from Gorgān to Bandar-e-Shāhpur and from Tabrīz to Mashhad. Other parts of the Middle East's railway network were constructed earlier to further imperialist ideals. This was the case especially with the extension of the Anatolian railway from Konya to Aleppo and Ras el'Ain (1918) as part of the Berlin to Baghdād railway project and with the building of the Hejāz railway from Damascus to Medina (1906). The earliest lines of all were built in the second

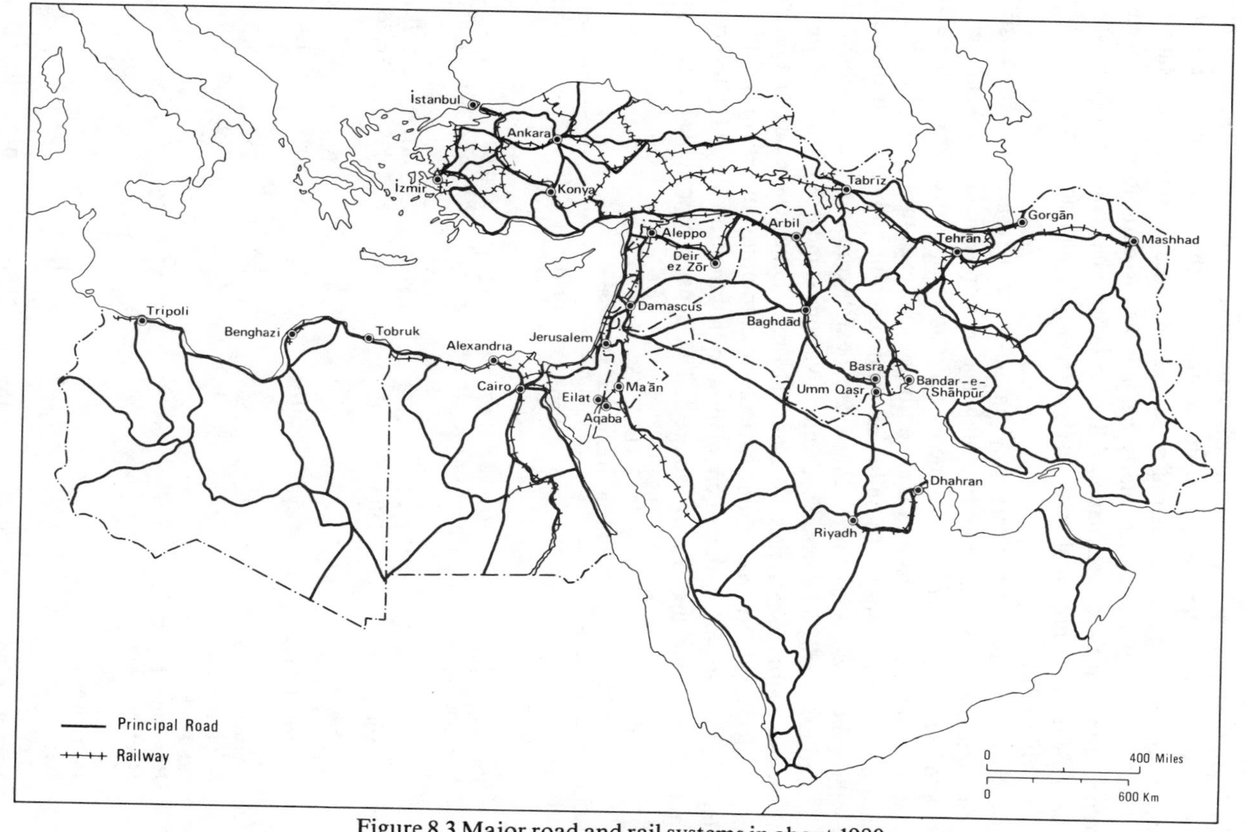

Figure 8.3 Major road and rail systems in about 1980

half of the nineteenth century to move goods from the interior to the Mediterranean coast for the benefit of western commercial interest, and began with the Smyrna to Aydin railway which was completed in 1867. Despite a long history, railways in the region are not particularly dense in terms of km/100 sq km, while their functions are increasingly being confined to the transport of cereals, minerals and some bulky industrial goods, chiefly because trains are relatively slow and the system is not particularly well-adapted to current economic needs. The net tonne-km ratio in Lebanon fell even before the civil war began, while passenger-km have grown slowly nearly everywhere and actually declined in Israel (Table 8.8).

Much of the traffic lost by the railways has passed to the roads, though statistics comparable with those presented in Table 8.8 are not available. The number of cars and commercial vehicles has increased enormously since 1953 (Table 8.9), and most states have embarked upon ambitious programmes of road building. Roads are not only cheaper to build and maintain than railways, but they also have the great advantage of providing a very flexible transport system which can reach into practically all parts of a country. Various standards of road now exist in the region, with the extremes represented by paved mule-tracks and multi-lane modern highways. Dirt roads are particularly common, but generally link the villages to the national system of circulation so that there are now very few rural communities which are wholly isolated. The density of metalled roads per

TABLE 8.8
Railway traffic, 1953 to 1984

Country	1953	1966	1970	1980	1984
Egypt					
Passenger-km	3,060	6,170	6,529	10,995	14,468+
Net tonne-km	1,215	3,387	3,333	2,174	2,303
Iran					
Passenger-km	377	1,161	1,800	2,704	5,784+
Net tonne-km	450	1,991	2,720	3,428	6,762
Iraq					
Passenger-km	525	431	467	1,215	1,375+
Net tonne-km	855	1,079	1,310	2,081	1,065+
Israel					
Passenger-km	169	368	358	252	215
Net tonne-km	97	318	468	819	970
Saudi Arabia					
Passenger-km	—	34	39	82	87+
Net tonne-km	—	52	34	261	194+
Syria					
Passenger-km	37	64	86	382	757
Net tonne-km	130	85	102	578	966
Turkey					
Passenger-km	3,016	4,189	5,561	6,012	6,277
Net tonne-km	3,071	5,494	5,618	5,028	7,532

+ = 1983
Source: United Nations, *Statistical Yearbooks.*

TABLE 8.9
Number of motor vehicles in use ('000)
1953 to 1983

Country		1953	1966	1970	1980	1983
Bahrain	Cars	—	—	23.2	47.0	60.1+
Commercial vehicles		—	—	9.4	18.4	20.0
Egypt	Cars	69.4	105.3	130.7	432.4	765.0
Commercial vehicles		19.5	27.5	35.4	130.0	231.0
Iran	Cars	22.4	142.4	278.2	95.8	90.0
Commercial vehicles		20.5	49.4	73.5	13.2	45.2
Iraq	Cars	13.0	58.2	67.4	237.1	345.9+
Commercial vehicles		11.5	37.0	42.0	145.2	206.3
Israel	Cars	14.0	91.5	151.2	409.5	571.5
Commercial vehicles		17.4	48.4	89.2	89.0	110.8
Jordan	Cars	3.4	13.6	15.4	89.7	137.7
Commercial vehicles		2.6	6.2	5.9	34.8	49.4
Kuwait	Cars	—	69.6	112.9	398.9	519.5
Commercial vehicles		—	25.3	36.8	144.0	178.7
Lebanon	Cars	16.6	105.4	136.0	—	—
Commercial vehicles		4.2	14.1	16.6	—	—
Libya	Cars	4.4	53.0	100.1	263.1	—
Commercial vehicles		2.8	25.2	45.4	131.3	—
Qatar	Cars	—	—	—	64.1	91.9+
Commercial vehicles		—	—	—	37.9	45.9
Saudi Arabia	Cars	—	37.5	64.9	—	—
Commercial vehicles		—	27.6	50.4	—	—
Syria	Cars	7.2	37.5	29.7	71.5	127.3
Commercial vehicles		6.0	27.6	18.3	88.3	108.7
Turkey	Cars	27.7	91.5	137.8	742.3	856.4
Commercial vehicles		33.5	102.3	159.8	304.2	342.5

Source: United Nations, *Statistical Yearbooks*.

100 sq km is still surprisingly small. Although this may genuinely reflect the degree of economic development in the region, it is also a function of the large amount of unproductive land within national territories. Both the road and rail networks, in fact, are densest in the most closely settled parts of the region (Figure 8.3). Maritime trade is even more limited in its location.

Merchant fleets grew considerably during the 1960s and 1970s (Table 8.10), but their importance lies in the international rather than the internal trade of the countries concerned. To accommodate increasing foreign trade, harbour works have been extended at ports like Beirūt and Jiddah, new general-purpose ports have been built at, for example, Ashdod in Israel and Umm Qaṣr in Iraq, while the hostility between Israel and her Arab neighbours has helped to revive the ancient ports of Aqaba and Eilat. While a rash of new ports appeared on the lower Gulf during the 1970s, existing facilities were improved elsewhere in the region. Container terminals were constructed and cargo handling has been much improved. The port congestion familiar in the Gulf and at Jiddah during the 1960s and early 1970s disappeared. Instead, there is overcapacity in the Gulf in the

Coastal shipping remains important, particularly in the Gulf. Loading a dhow at Sharjah. (Central Office of Information)

TABLE 8.10
Merchant fleet, 1970–83

Country	1970	1975	1980	1983
		('000 gross registered tonnes)		
Egypt	238	301	556	663
Iran	129	480	1,284	1,795
Iraq	37	311	1,466	1,561
Israel	714	451	450	690
Kuwait	592	991	2,529	2,548
Lebanon	182	167	268	459
Libya	4	242	890	904
Qatar	1	1	92	315
Saudi Arabia	49	180	1,590	5,297
Syria	1	8	39	48
Turkey	697	995	1,455	2,524
UAE	9	51	158	301

+ excludes oil tankers and ore bulk carriers
Source: United Nations, *Statistical Yearbook, 1983–84*, New York 1986, Table 173.

mid-1980s. This is due in part to the duplication of facilities in the 1970s and in part to the world recession, but it is also a consequence of the stage of development reached by the countries along the Gulf. In particular, they no longer need construction materials to the same extent as in the 1970s now that development projects have come to an end and their own industries can supply a large proportion of the basic requirements. Oil terminals increased in number in the Gulf and around the Gulf of Sirte in Libya, while national and international pipelines feeding to them have become numerous (Chapter 9). However, coastal shipping remains important to the Gulf States and in the internal trade of Turkey. Trade along the Black Sea coast is particularly well developed.[14] A small amount of traffic is found on the Caspian Sea. By contrast, inland waterways are of very little importance in the region. The Nile is still a major artery of Egypt, while the estuaries at the head of the Gulf were important to the international trade of Iran and Iraq before the war began in 1979.

The Iraq–Iran war has had a considerable impact on trade in the Gulf. Not only were insurance premiums increased substantially, but Iraq was forced to import goods by truck across neighbouring countries (particularly Jordan) and to export its vital oil through new pipelines across Saudi Arabia and southern Turkey. Iran reoriented its shipping network from the head of the Gulf and focused it on Bandar Abbas and a new port, Shahid Rajai, both near the mouth of the Gulf and upgraded road and rail links to this remote area. The expansion of other Iranian ports away from the war zone was under consideration in 1986–87. By contrast, the war has been a factor in scaling down the port expansion planned elsewhere in the Gulf, though recession in world trade was also involved. The civil war in Lebanon has virtually closed the port of Beirūt, but it has brought to life a number of smaller ports which service the various factions of that divided country.

Air transport is a comparatively recent development, though Turkey organized an internal airways system as early as 1933.[15] It is particularly important to the larger countries like Iran, Saudi Arabia and Turkey (Table 8.11), where surface distances are sufficiently great for air transport to become competitive. The urban nature of the nodes in the system is also particularly clear, and emphasizes the importance of large towns in generating traffic.

Small towns, however, play a more basic role in the internal trading patterns of Middle Eastern states. Agricultural goods flow into them for distribution up the urban hierarchy or export abroad, while the small towns distribute nationally manufactured goods and foreign imports to the countryside. Periodic markets and fairs are important, both in wholesaling and retailing, but they have been comparatively little studied.[16] Shops are found in the larger villages, though generally they are not specialized and sell goods in immediate demand (tea, sugar, cigarettes, plates, glasses) and in small quantities. The greatest variety of specialized shops, backed to some extent by specialist wholesalers, are found in the towns. A distinctive spatial arrangement appears to recur.[17] In the old core of the town, often near the citadel and the Friday Mosque, are the shops dealing with high-value non-perishable goods with a low turnover. Semi-perishable commodities and goods in everyday demand are sold in a zone outward from the first. Further away from the centre are shops specializing in cloth and clothing, but mixed in with them are often traders in perishable goods. These three zones, which are not rigidly defined, constitute the traditional bazaar of the typical Middle Eastern town where all the commodities essential to traditional lifestyles in the region may be found. Its physical characteristics are narrow alleyways, old property, small open-fronted shops and workshops and a degree of spatial con-

TABLE 8.11
Domestic civil aviation, 1984

Country	Passenger-km	Tonne-km
Bahrain	1,106	30.7
Egypt	1,459	20.7
Iran	4,089	90.3
Iraq	1,200	52.0
Israel	6,173	529.6
Jordan	3,610	139.9
Kuwait	3,749	171.0
Lebanon	831	487.7
Libya	1,607	8.1
Oman	1,107	30.7
Qatar	1,106	30.7
Saudi Arabia	15,507	458.8
Syria	923	11.3
Turkey	2,362	26.3
UAE	1,106	30.7
Yemen AR	584	7.8

Source: United Nations, *Statistical Yearbook, 1983–84*, New York 1986, Table 176.

centration in the various types of activity. A fourth shopping zone lies outside the traditional bazaar, along modern streets and at major intersections. This generally consists of western-style shops selling modern consumer goods, whether imported or manufactured locally. Large cities also contain suburban shopping centres, where the old and new types of shops are frequently mixed together. The residential areas of all towns are further supplied, especially with fresh fruit and vegetables, by peddlars.

Peddlars are still important in a different way to the remoter villages. Trading from pick-ups and small vans, they bring a selection of goods, like printed cloth and readymade clothes, which are normally only available in the towns.

Villagers often travel to the towns on foot or on the backs of their animals, but the village bus has become an important institution in most countries, especially for purchasers. Goods for sale are usually taken by truck or tractor trailer, and large items are brought home in the same way. The amount of freight moved between towns seems surprisingly small to the traveller used to western conditions. The flow seems reciprocal between the provincial towns, on the one hand, and the capital, chief port and major manufacturing centres, on the other. One would expect that distance, time and cost would be basic determinants of the type and volume of freight moved, but this remains to be investigated. Goods are often moved by individual merchants in their own trucks, but carrying firms run large trucks on something like regular weekly or even daily schedules, and appear to flourish.

8.5 Financial flows

Commerce is lubricated by financial flows between trading partners, whilst investment capital is essential to programmes of industrial and infrastructural development. Down to the oil price rises of 1973, the involvement of the Middle East in world financial patterns was comparatively small, though Beirūt emerged as an important regional financial centre during the 1950s and early 1960s. The rise in oil prices meant that some of the producer countries suddenly became extraordinarily rich in capital. This lead to the expansion of development programmes in Arabia and Iran and to the formulation of massive investment plans, but it also raised the problem of how to use the large surpluses whilst the absorptive capacity of the economies was growing. Financial services had to be improved in the oil rich states, not only to assist investment and development, but also to recycle funds abroad. New commercial banks were established, national banking systems were expanded and foreign finance houses were allowed to trade, especially in the states on the Arab side of the Gulf. Unofficial stock exchanges emerged in Saudi Arabia (Suq al-Asham) and in Kuwait, where the Suq al-Manakh collapsed in 1982 sending reverberations as far as Bahrain and the UAE. Offshore banking was promoted by Bahrain from 1975 in an attempt to exploit the island's position between time zones and major world financial centres via its telecommunications satellite. By the spring of 1977, 40 offshore banking units were in existence, much to the concern of Kuwait and Saudi Arabia which regarded 'suitcase banking' as undesirable, especially since their nationals generated much of the business. At the beginning of the 1980s, though, the OBUs found conditions less amenable and began to reduce the scale of their operations or to pull out altogether. Bahrain and Kuwait remain important financial centres for the region. Beirūt has been virtually eliminated by the Lebanese civil war, which not only destroyed offices but undermined confidence. By contrast with these capitalist developments in the Western mould, the region has been affected by the emergence of Islamic banking founded on the basis of risk and profit sharing, rather than interest payment.

Surplus capital and relatively small absorptive capacities allowed the creation of various development funds in the 1970s by Abū Dhabi, Iraq, Qatar and Saudi Arabia following the much earlier example of Kuwait (1962), whilst a multinational fund, the Arab Fund for Economic and Social Development began operations in 1974. The Kuwait Fund for Arab Development was originally created as a way of generating investment income for the state from development projects in the lower Gulf, where petroleum resources were still unexploited, and in advance of Kuwait's own major industrialization schemes coming to fruition. All the national funds have been significant in financing projects designed to assist development in poorer Arab countries (including those in Africa), but a large proportion of their investment has gone outside the Arab World. Qatar, Saudi Arabia and the UAE have regularly given more than 3 per cent of GNP to foreign aid, compared with the 0.7 per cent recommended by the United Nations. The Arab Fund aimed specifically at promoting the integration of the Arab economies and much of its investment has gone into improving transport and communications, though it did promote an ill-fated scheme to develop agriculture in Sudan as a way of supplying the food-deficit countries of the Arab world. More than 50 per cent of all Arab aid has gone to Arab countries. The main beneficiaries appear to have been Egypt, which received about half of all bilateral Arab aid between 1973 and 1978, followed by Syria and Jordan (front-line states in the struggle with Israel) and, lastly, Yemen AR, which has come back into the fold of political acceptability. Iran and Turkey have not derived much benefit from the Arab funds. They are eligible, however, for investment assistance from the Islamic Development Bank, re-established in 1964 by the Organisation of the Islamic Conference and largely capitalized by Arab countries.

Arab finance is now a force to be reckoned with. For example, the Kuwait Foreign Trading, Contracting and Investment Company has controlling interests in such major German companies as Daimler-Benz and Metallgesellschaft, a 25 per cent stake in the chemical and pharmaceutical firm Hoescht and 10 per cent of Volkswagen, as well as 27 per cent of the Savoy Hotel in London and control of Hays Wharf; it is also a major creditor of the cities of Bristol and Oslo. Declines in the price of oil during the first half of the 1980s threatened the financial flows from the oil-rich countries of Arabia, whilst the Iran–Iraq war absorbed large amounts of funds not only from the combatants themselves but also from the Arab states supporting Iraq. Even the smaller Gulf states have found themselves committed to large amounts of defence expenditure on their own accounts. The continued availability of investment capital from the Gulf is clearly dependent upon maintaining oil prices at a fair level and containing the inflation which surplus funds have stoked. It emphasizes yet again the importance of the region's petroleum resources, not only to the region but to the world.

References

1. C. H. Edwards, *Future Energy Prospects in the Middle East States*, Special Report on Energy, Economist Intelligence Unit, London, 1986.
2. H. E. Wulff, *The Traditional Crafts of Persia*, MIT Press, Cambridge, Mass and London, 1966.
3. (a) Z. Y. Hershlag, *Introduction to the Modern Economic History of the Middle East*, E. J. Brill, Leiden, 1964, 42–154.
   (b) C. Issawi (Ed), *The Economic History of the Middle East, 1800–1914*, University of Chicago Press, Chicago and London, 1966.
   (c) R. Owen, *The Middle East in the World Economy, 1800–1914*, Methuen, London and New York, 1981.
4. R. E. Mabro, 'Industrialisation', in *The Middle East: A Handbook* (Ed M. Adams), Anthony Blond, London, 1971, 442–449.
5. G. Hunter, 'The Middle East Supply Centre', in *The Middle East in the War* (Ed G. Kirk), Oxford University Press, London, 1952, 169–93.
6. C. Issawi, *Egypt in Revolution: An Economic Analysis*, Oxford University Press, London, 1963, 44–45.
7. R. E. Mabro, 'Industrialisation', in *The Middle East: A Handbook* (Ed M. Adams), Anthony Blond, London, 1971, 442–449.
8. H. G. Hamilton, 'The Saudi petrochemical industry: its rationale and effectiveness', in *State, Society and Economy in Saudi Arabia* (Ed T. Niblock), Croom Helm, London, 1982, 235–77.
9. R. E. Mabro, 'Industrialisation', in *The Middle East: A Handbook* (Ed M. Adams), Anthony Blond, London, 1971, 442–449.
10. R. Wilson, 'Egypt's exports: supply constraints and marketing problems', *Bulletin: British Society for Middle Eastern Studies*, **12**, 135–56 (1985).
11. L. E. Preston, *Trade Patterns in the Middle East*, American Enterprise Institute for Public Policy Research, Washington DC, 1970, Tables II–1 and II–2, 18 and 19.
12. B. W. Poulson and M. Wallace, 'Regional integration in the Middle East: the evidence for trade and capital flows', *Middle East Journal*, **33**, 467–78 (1979).
13. (a) H. Askari and J. T. Cummings, 'The future of economic integration within the Arab world', *International Journal of Middle East Studies*, **8**, 289–315 (1977).
    (b) A. Musrey, *An Arab Common Market*, Praeger, New York, 1969.
    (c) R. Wilson, *The Arab Common Market and Inter-Arab Trade*, Economic Research Paper 4, Centre for Middle Eastern and Islamic Studies, University of Durham, 1978.
14. J. C. Dewdney, *Turkey*, Chatto and Windus, London, 1971, Figure 40, p. 144.
15. V. Eldem, 'Turkey's transportation', *Middle Eastern Affairs*, **4**, 324–36 (1953).
16. H. Gaube, E. Grötzbach, E. Niewöhner-Eberhard, B. Oettinger and E. Wirth, 'Wochenmarkte, Marktorte und Marktzyklen in Vorderasein', *Erkunde*, **30**, 9–44 (1976).
17. (a) V. F. Costello, *Urbanization in the Middle East*, Cambridge University Press, 1977.
    (b) D. Potter, 'The bazaar merchant', in *Social Forces in the Middle East* (Ed S. N. Fisher), Cornell University Press, Ithaca, New York, 1955, 99–115.

CHAPTER 9

Petroleum

9.1 Introduction

One of the most spectacular economic achievements of the past 100 years has been the growth and development of the petroleum industry. In 1850 world oil production was virtually non-existent. Then, following discoveries in both the USA and Russia, total output began to rise rapidly after 1860 until by 1890 world production had reached 10 million tonnes per annum (Figure 9.1). From this date until the mid-1970s oil production continued to expand with world output doubling approximately every 10 years. 1975 recorded the first fall in oil production in over a century, but this quickly picked up to a new peak of 3225 million tonnes in 1979. Since then a downward trend has been registered.

At the beginning of the twentieth century oil and natural gas supplied less than 4 per cent of the total energy requirements of the Western world, while coal accounted for approximately 90 per cent. This picture quickly changed, however, with the share of oil and natural gas in total energy production rising to 12 per cent in 1920, 26 per cent in the mid-1930s, and 36 per cent following the Second World War in 1948. Since this time, the importance of petroleum in the world's economy continued to grow, so that by 1970 approximately 70 per cent of the total energy requirements of the West were being supplied by oil and natural gas. By 1985 although the relative importance of petroleum was declining, it still accounted for 63 per cent of the primary energy consumption of the OECD countries.[1]

The main users of petroleum and related products are the industrialized nations of Western Europe, North America, Australasia and Japan, which together consumed 56 per cent of the total production in 1985. These same countries, however, produce only 29 per cent of the world's oil and 50 per cent of the world's natural gas, and so have to satisfy their demands by imports. As a result a symbiotic relationship has grown up between the major producing and consuming countries.

9.2 The Middle East as an oil producing region

At the present time, the Middle East is a region where the output of petroleum is considerably in excess of demand. The oil production of this area, in terms of the total world output, has increased markedly during the twentieth century. In the 1930s and early 1940s it averaged between 2 and 5 per cent of total world production, but, following the rapid post-war revival and expansion of the industry

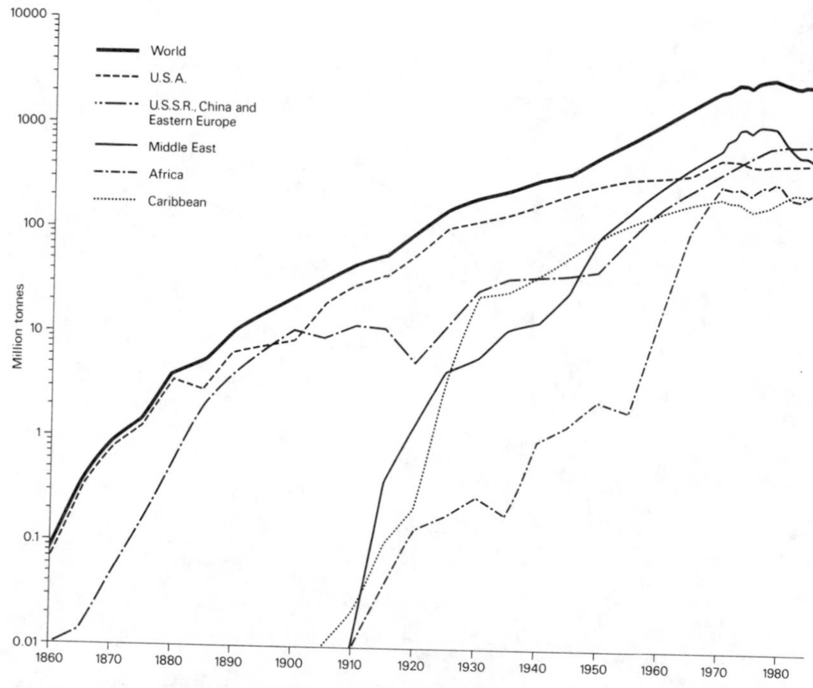

Figure 9.1 World oil production, 1860–1985

in the Gulf region, it had reached 15 per cent by 1950. Since then, the relative importance of Middle Eastern and North African oil in world production has continued to grow, passing the 25 per cent mark in 1960 and reaching 38 per cent in 1970. During the 1970s the relative importance of the region has declined so that by 1985 it only accounted for about 24 per cent of the world's total oil production. For natural gas the figure is even lower reaching only 5 per cent.

Although oil production in the Middle East first began in Iran in the early part of the twentieth century, it has only been since the Second World War that output begun to increase rapidly (Figure 9.2). During the early 1940s Iran was the largest oil producer in the Middle East, followed by Iraq and Saudi Arabia. After intensive oil exploration and development, petroleum output from both Kuwait and Saudi Arabia increased greatly until, by 1950, the production from the Saudi Arabian fields almost equalled that from Iran. In 1951 the oil industry in Iran was nationalized and as a result crude oil production slumped to less than two million tonnes in the following year, leaving Saudi Arabia and Kuwait as the two major Middle Eastern producers. During the decade 1955 to 1965 Kuwait became the largest oil producer followed by Saudi Arabia and Iran. A tremendous production increase in Iran occurred from 1954 onwards after a settlement of the dispute between the oil companies and the Iranian government. During the late 1960s output from both Iran and Saudi Arabia overtook that from Kuwait and these two countries vied for the leadership of Middle Eastern oil production. The

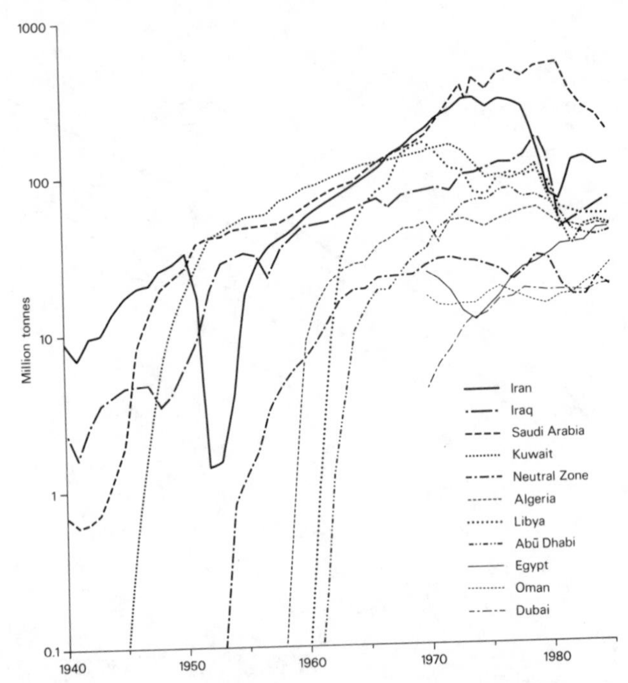

Figure 9.2 Oil production in the Middle East and North Africa

real success story in the 1960s belonged to Libya where production leapt from zero in 1960 to 150 million tonnes in 1969 when Libya ranked second, equal with Saudi Arabia. During the 1970s it was Saudi Arabia which came to dominate the Middle Eastern scene with an oil production which approached 500 million tonnes in 1980. Since then as part of a policy to try to keep oil prices high Saudi Arabia has substantially cut back its production so that by 1985 it was less than 200 million tonnes for the first time since 1970. Iran continued to be the second most important producer in the region until 1979 and the fall of the Shah, which led to a collapse of oil production. Subsequently the dislocation caused by the war with Iraq has meant that output has never regained the values of the mid-1970s. One of the most striking features of the 1970s and early 1980s has been the very significant decline in oil production in both Kuwait and Libya. Both were producing almost 150 million tonnes in 1970, yet by 1985 their output had dropped to round 50 million tonnes. Currently although Saudi Arabia is still the largest producer it does not dominate output in the way in which it did in the mid-1970s. (Table 9.1).

One of the major changes which occurred in the producing countries during the 1960s and 1970s was that they became significant users of their own petroleum and petrol products. This change occurred as local economies developed rapidly under the impact of growing oil revenues. In 1966 the total Middle East oil consumption was 39.3 million tonnes, but by 1985 it had increased to 98.8 million tonnes or 19 per cent of the region's production.

TABLE 9.1
Crude oil production in the Middle East and North Africa

	1976	1979	1982	1985
		(million tonnes)		
Abū Dhabi	76.8	70.3	42.5	41.5
Algeria	50.1	58.5	45.8	44.8
Dubayy	15.6	17.6	17.8	18.9
Egypt	16.4	26.5	34.9	43.4
Iran	295.0	158.1	119.8	110.5
Iraq	118.8	170.6	49.6	70.4
Kuwait	98.2	113.2	34.9	45.0
Libya	93.3	100.7	54.7	52.1
Neutral Zone	24.4	29.4	16.4	18.7
Oman	18.4	14.8	16.2	25.4
Qatar	23.9	24.7	16.3	14.5
Saudi Arabia	421.6	469.9	327.9	174.8

Source: British Petroleum Company plc, 1986, *BP Statistical Review of World Energy*, p 4.

9.3 Oil concessions

In the period prior to the Second World War the oil producing countries around the Gulf granted exclusive concessions for the exploration and development of their petroleum to Western oil firms. These concessions covered most or all of the countries concerned and were granted for long periods of 50 to 75 years.[2] At this time eight foreign parent companies, controlled by American, British, Dutch and French interests, produced all the oil output of the region. These companies were Standard Oil (New Jersey), Standard Oil (California), Mobil Oil, Texaco, Gulf Oil, British Petroleum, Royal Dutch Shell and the Compagnie Française des Petroles.[3] Although the situation has changed markedly since this time, even in the late 1960s these same eight companies accounted for more than 90 per cent of the petroleum production of the region.

The first major concession in the Middle East was granted in 1901 by Iran to the English entrepreneur, W. K. D'Arcy. This was acquired by the Anglo Persian Oil Company in 1909. It covered the whole country with the exception of the five northern provinces. Since 1954, following nationalization of the Iranian oil industry in 1951, production has been carried out by an international consortium with 14 participants, including the eight major companies already listed. In the agreement reached between the companies and the Iranian government, the concession area was reduced to 259,000 sq km. The negotiated contract was to run for 25 years with the consortium having an option on three further extensions of five years, each under stated conditions.[4]

In Iraq, the first concession was granted to an international consortium in 1925, the first to operate in the Middle East. This consortium known as the Iraq Petroleum Company, was controlled by British Petroleum, Shell, Compagnie Française des Petroles and the Near East Development Corporation (Standard

Oil of New Jersey and Mobil Oil). Its concession, following a number of extensions, covered the whole country, except for a narrow strip in the east close to the Iranian border. In 1961 more than 99 per cent of the original concession was re-appropriated by the unilateral action of the Iraqi government, leaving the Iraq Petroleum Company with only those regions actually in production. The dispute between the government and the Company which this action caused had not been settled prior to the nationalization of the company's assets in 1972.

The two other major producers of the region, Kuwait and Saudi Arabia, both granted concessions during the 1930s. The Kuwait concession, covering the whole country, was granted in 1934 to the Kuwait Oil Company, jointly owned by Gulf Oil and British Petroleum. In Saudi Arabia the original concession was awarded to Standard Oil of California in 1933 and later taken over by the Arabian American Oil Company (ARAMCO), controlled by four parent companies. In this case, the concessional area covered huge tracts in the central and eastern parts of the country.

Since the Second World War the concessions which have been granted have become much more beneficial to the host countries. This is particularly well illustrated in the case of Libya, where oil production first commenced in 1961. Here, a very different plan of exploration and development took place from that in the older producing countries. Following the enactment of the Petroleum Law in 1955, the oil companies were invited to bid for concessions. At first a number of attractive inducements were offered to encourage the international companies to participate. The concession areas were large, 30,000 to 100,000 sq km, and there was a depletion allowance of about 25 per cent besides other attractions.[5] As a result many companies representing American, British, French, Italian and German interests joined in the search for oil. Once it was obvious that the potential for petroleum was great, the Libyan government began to demand greater benefits. In 1960 the depletion allowance was abolished and the companies were required to reduce the size of their concessions to between a quarter and one third of their original size within a period of ten years. Libya joined the Organization of Petroleum Exporting Countries (OPEC) in 1962, two years after its formation, and was coerced into putting its petroleum tax system onto a similar footing to those of the other member states.

Pressure was brought to bear on the companies with older concessionary rights and eventually these were made to conform to the new proposals. Despite these stiffer terms the demand for concessions remained strong owing to the positional advantage possessed by Libyan oil compared with other Middle Eastern crudes, the low density and low sulphur of the oil and the fact that even the new concession terms were still less onerous than those in the Gulf region.

9.4 The oil fields

The distribution of oil in the Middle East is strictly controlled by geological conditions. Almost all the earlier finds were made in deep sedimentary basins along the margins of the Zagros Mountains and the Gulf. For many years it was believed that oil would not be discovered in commercial quantities outside this region. More recent work, however, has proved that thick sedimentary sequences with oil reserves are also found in other parts of the Middle East and North Africa, for example in Libya. Thus it is possible that new discoveries of oil may still be made. There can be little doubt, however, that none of these discoveries will match the huge oil fields already exploited in the Gulf region.

For oil to be found in commercial quantities certain geological conditions must be met. These are, first, a reservoir rock in which the oil can accumulate. This is usually a sandstone or limestone, with the oil collecting in the pore spaces or cavities in the rock. For the oil to accumulate there must be some structure to trap the oil. In most cases this is an anticline within the reservoir rock. To prevent the oil migrating to other rocks there must be a formation above the reservoir rock which is impervious to oil movement and so acts as a seal. This is commonly achieved by salt/evaporite deposits. Finally, there must be adjacent source rocks from which the oil has originally migrated.

In the Arabian–Iranian petroleum province there are five plays, or geological settings, which have produced suitable conditions for oil accumulation.[6] Each of these has a different georgraphical location, though in general all are located between the major folds of the Zagros Mountains and the Basement Complex of Arabia. The Upper Cretaceaous and Tertiary Play is situated mainly in the Zagros Fold Belt of Iran and Iraq. The structures which hold the oil were formed when the continental plates began to collide producing a series of NW–SE oriented anticlinal and synclinal structures. About three-quarters of the petroleum potential in this play is located in Iran, with the main reservoir rock being the Asmari Limestone. This is of Oligocene to early Miocene age and is between 30 and 500 metres in thickness. Cretaceous and Jurassic shales are believed to have been the source rocks for the petroleum. The Lower and Middle Cretaceous Play is located centrally in the Arabian–Iranian basin and includes the huge Burgan field. The reservoir rock is a quartz sandstone, from 60 metres to 350 metres in thickness and the source rocks are deltaic shales.

The Jurassic Play covers a large area of northeastern Saudi Arabia and extends northwards to Kuwait and eastwards to the Iranian border. The reservoir rocks are limey sandstones and these give rise to extremely large oil fields, including the Ghawār. The Hanifa and Sargelu formations of Middle and late Jurassic age are believed to be the chief source rocks. Finally the Permian Play is located in the southeast Arabian Gulf region and extends eastwards to the Zagros fold belt. The Khuff formation, consisting of carbonates with inter-bedded evaporites up to 650 metres in thickness, provides the reservoir rocks. Ordivician and Silurian shales, up to 1000 metres thickness, are the source rocks.

9.5 Early development of the oil fields

9.5.1 Libya

The first commercial discovery was made by Esso-Standard at Atshan in the Fezzan in 1958 (Chapter 20). Since that time several new fields have been proved, but nearly all of these were located inland from the Gulf of Sirte, far from the original find. The Zelten field, discovered in 1959, has been a major oil producer. It was connected by pipeline in 1961 to a tanker terminal at Marsa-el-Brega and has since been joined by a feeder system to adjacent fields. Large quantities of natural gas were also discovered here and export of gas began in 1969, following the construction of a gas pipeline to Marsa-el-Brega.

East of Zelten lies the large Intisar field from which export began in 1968, less than a year after the initial discovery. The crude oil from this field is transported by a large pipeline to a tanker terminal at Zuetina. This system also taps adjacent oil fields at Samah and Waha. Production from the Amal field is transported by pipeline to a terminal at Ras Lanuf, while output from the very isolated Sarir field is piped to the Marsa al Hariga terminal near Tobruk.

In 1969 the regime of King Idris was overthrown and a new republican government established. Difficulties arose between this government and the oil companies and, as a consequence, production dropped markedly. In 1970 a new state-owned firm, the Libyan National Oil Company, was formed and in 1972 the foreign oil interests in the country were nationalized.

9.5.2 Egypt

Drilling for oil in Egypt began in the late nineteenth century at Gemsa, on the western side of the Gulf of Suez, and a producing well was sunk in 1908. A second field was discovered nearby in 1913, followed by the proving of other small fields on the Gulf of Suez after the First World War. The discovery of petroleum in the Sinai peninsula was made at Sudr in 1946.

The Egyptian oil industry was nationalized in 1956 and after this grew rapidly. The first offshore field in the Gulf of Suez was discovered at Belayim Marine in 1961, followed during the mid-1960s by a series of finds in the western desert region of El 'Alamein, Umbaraka and Bîr Abu Gharâdiq. Following the 1967 Arab-Israeli War Egypt lost temporarily the crude oil production from the Sinai fields as a result of Israeli occupation, but these losses were made up by the development of the offshore El Morgan field in the Gulf of Suez.

9.5.3 Iraq

Oil was discovered by the Turkish Petroleum Company in Kirkūk in 1927. Two years later the name of the original company was changed to the Iraq Petroleum

Company. The company later gained further concessions west of the Tigris and in south Iraq, and established the Mosul Petroleum Company and the Basra Petroleum Company respectively to develop them. Production began from the Ayn Zāleh field of the Mosul Petroleum Company in 1952, and the Zubayr field of the Basra Petroleum Company in 1951. Subsequently new fields were brought into production at Butmah in the north and Rumaila in the south.

Owing to the isolated nature of the very large Kirkūk oilfield, output was unable to begin until pipelines were constructed to the Mediterranean coast at Haifa and Tripoli. Following the closure of the Haifa pipeline, with the establishment of the state of Israel, a new line was constructed to Tripoli and in 1952 a pipeline was completed to Baniyās. Output from the fields belonging to the Basra Petroleum Company was transported by pipeline initially to Al Fāw on the Gulf and later to a deep water terminal at Khor-al-Amaya.

In 1961 the Iraq government passed a law which expropriated all concession areas of companies not in production. This amounted to about 99.5 per cent of all concessions and was naturally resisted by the companies involved. The government-owned Iraq National Oil Company was formed and given the task of exploiting these expropriated areas. Discussion went on between the companies and the government for more than a decade without any resolution of the problems. Finally, in 1972 the Iraqi government nationalized all foreign oil interests within the country.

9.5.4 Saudi Arabia (Figure 9.3)

Oil was discovered in Saudi Arabia at Ad Dammām in 1938. Development was hindered by the Second World War, but from the mid-1940s onwards a series of further finds were made. Crude oil production was carried out by the Arabian American Oil Company (ARAMCO), originally jointly owned by Standard Oil of California (30 per cent interest), Standard Oil (New Jersey) (30 per cent), Texaco (30 per cent) and Mobil (10 per cent). By the early 1970s crude oil was being produced from nine onshore fields and four offshore fields. Most of the output was carried by pipeline to the large tanker terminal at Ras Tanūrah, but some was carried by a pipeline to a refinery in Bahrain. Part of the petroleum was transported by the Trans-Arabia Pipeline (TAPline) to the Mediterranean coast at Sidon.

9.5.5 Kuwait

Although petroleum was discovered in Kuwait in 1938 the first exports of crude oil did not begin until 1946. The initial discovery was made at Burgan, though subsequently several other major fields have been developed, of which the largest are at Rawdhatain, Ahmadī-Magwa and Minagish. The problems with Kuwait crude oil are a high sulphur content and a high specific gravity, which means that

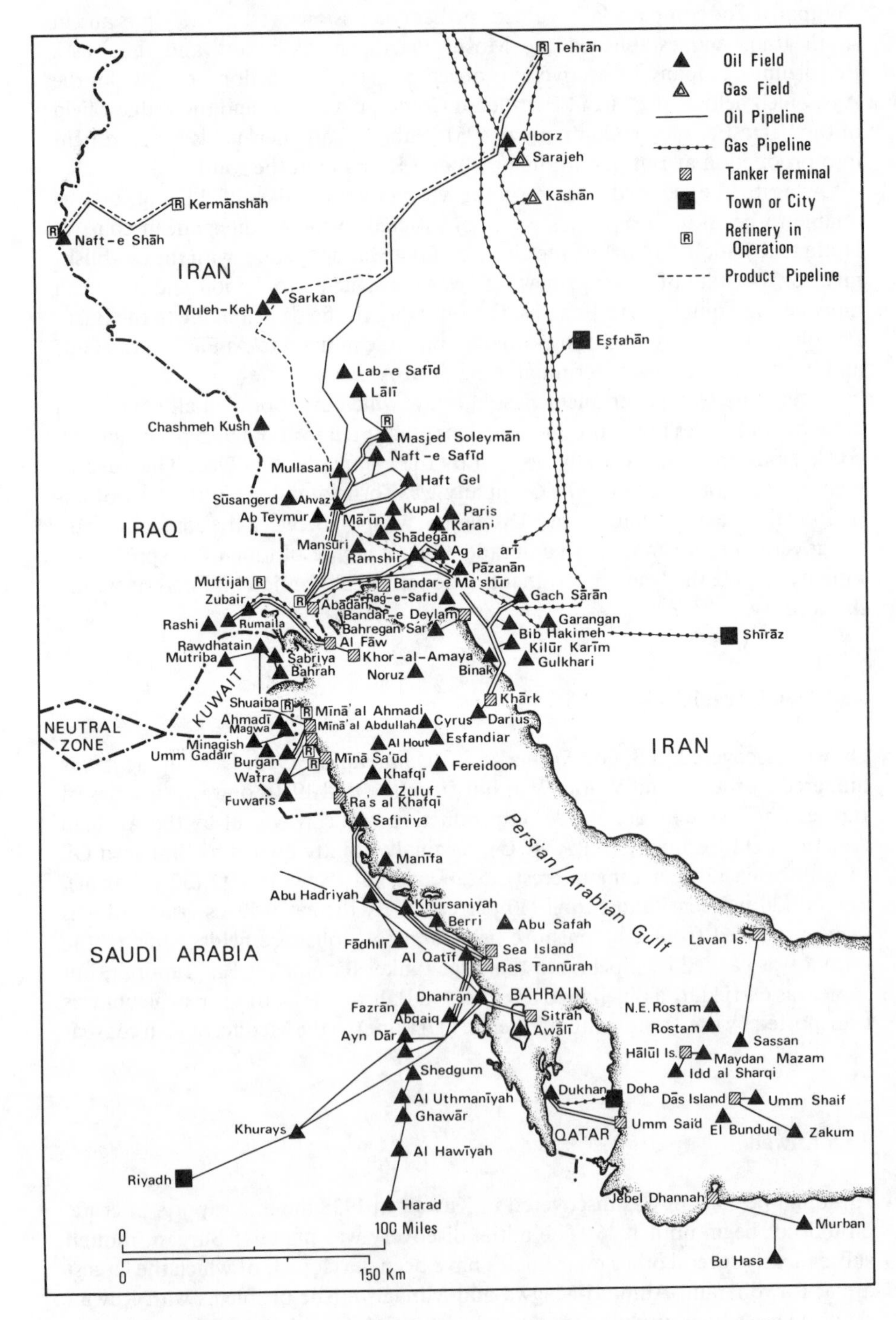

Figure 9.3 Oil and gas fields around the northern part of the Gulf

it is less acceptable in the Western European and Japanese markets than the lighter, low sulphur crudes. Crude oil is exported from the terminal of Mīnā al Ahmadī which is an artificial island 15 km offshore.

9.5.6 Saudi Arabia/Kuwait Neutral Zone

In the Neutral Zone, established in 1922, Saudi Arabia and Kuwait had equal undivided rights with regard to petroleum production. Oil was first discovered in this region in 1953 at the Wafra field following the granting of exploration concessions in 1948 and 1949. New fields have since been discovered at Fuwaris, producing since 1964, and at South Umm Gadair, an extension of one of the Kuwait fields. Most of the crude oil of the region has a high sulphur content. In 1960 an important offshore field was discovered at Khafqī by the Arabian Oil Company, which is mainly Japanese owned. Most of the crude oil exported from Wafra field went via the terminal at Mīnā al Abdullah in Kuwait. Later, another terminal was built at Mīnā Sa'ūd within the Neutral Zone. Output from the offshore Khafqī field is sent by pipeline to a terminal at Ra's al Khafqī.

9.5.7 Gulf States

The states towards the southern end of the Gulf produce sizeable amounts of petroleum.

Bahrain is the northernmost; here the oil find in 1932 was the prelude to extensive exploration and discoveries in the neighbourhood. The main onshore field, the Awālī, is operated by the Bahrain Petroleum Corporation. An offshore field, Abu Safah, also exists in an area shared equally between Saudi Arabia and Bahrain.

Oil in Qatar, although discovered in 1939, was not developed because of the Second World War, until 1949. The only onshore oilfield is at Dukhan on the extreme western side of the peninsula and is operated by the Qatar Petroleum Company. From here the crude oil is piped to a tanker terminal at Umm Sa'id on the east coast. In the early 1960s two large offshore oilfields, Idd al Sharqi and Maydan Mazam, were discovered by Shell off Qatar. Output was piped to a terminal on Hālūl Island.

Following extensive onshore and offshore exploration work in the 1950s Abū Dhabi became a major oil producer in the 1960s. Onshore, the Murban field was discoverd in 1960. Crude oil output began in 1963 and was transported by a pipeline to a terminal at Jebel Dhannah. Later discoveries were made at Bu Hasa and Abu Jidu. All of these fields are operated by the Abū Dhabi Petroleum Company. Offshore oil was discovered at Umm Shaif in 1958. A pipeline was quickly constructed to Dās Island, where a loading terminal was built, and production began in 1962. Two further offshore fields at Zakum and El Bunduq were also discovered at this time.

Crude oil production in Dubayy commenced from the offshore Fateh field in 1969. The operating company was Dubai Petroleum, owned by an international group of firms headed by Continental Oil.

In Oman a concession was granted in 1937 to an associate of the Iraq Petroleum Company to begin exploration. Drilling began in the 1950s, but following initial dry wells, the IPC partners, with the exception of Shell and Gulbenkien, decided to withdraw in 1960. These two interests, joined later by Compagnie Française des Petroles, continued exploration work and were rewarded in 1964 by the discovery of oil in the Natih/Fatud region to the west of the Jebel Akhdar. A terminal was built at Mīnā al Fahal and, following the construction of a pipeline, export of crude oil began in 1967.

9.5.8 Turkey

Exploration concessions for petroleum were first granted in Turkey in the late nineteenth century, but following the enactment of the Petroleum Law in 1926, foreign investors were effectively barred from exploration and development work. As a result a governmental agency carried out all petroleum exploration work until the law was amended in 1954. The first oil discovery was made at Raman Dağ in south-east Turkey in 1940 and nearly all subsequent discoveries have been made in the same area. Unfortunately the oil tends to be of the heavy variety with a high sulphur content. The principal producers are the government company TPAO (Türkiye Petrolleri Anomin Ortakligi), Shell and Mobil.

9.5.9 Syria

Oil was discovered at Karachuk, though only in small quantities, by a Syrian-American independent operator in 1956. In 1959 petroleum was proved at Es Suweidīya by the German Company, Deutsche Erdoel, but it was never granted a development concession. Subsequently, these concessions were expropriated and the oilfields developed by the General Petroleum Authority, a government agency, with the aid of Russian technical assistance. Other fields were also developed at Hanzan and Rumaidan. The crude oil is piped to the coast at Tartūs.

9.5.10. Israel

Israel has spent considerable effort in petroleum exploration since 1948, but so far with little success. Small oil fields were opened up at Helez-Bror-Koshav, to the southeast of Ashdod, and gas fields in the Dead Sea region. During its occupation of Sinai Israel augmented its oil supplies by production from the fields in that region. Crude oil pipelines link the port of Eilat on the Gulf of Aqaba with the Mediterranean coast.

9.5.11. Iran

The first discovery of oil in commercial quantities in Iran was made in 1908 at Masjed Soleymān on the western flanks of the Zagros Mountains. Subsequently, a pipeline was constructed from this field to Abādān, where a tanker terminal and a refinery were established. In the 1920s and 1930s considerable exploration for oil was carried out and a number of new fields at Haft Gel, Gach Sārān, Naft-e Safīd, Agha Jārī and Lālī were located.

During the Second World War, following the occupation of Iran by the Allied armies, the output from a number of fields was increased to supply the growing need for oil for military purposes. In 1944 the Agha Jārī field was connected by pipeline to the Abādān refinery, which was further expanded to a capacity of 500,000 barrels per day to become the largest oil refinery in the world at that time. By 1950 the oil producing regions of Iran could be divided into three major groupings. In the north was a single field at Naft-e Shāh, connected by pipeline to a refinery at Kermānshāh. Further south were the older Khuzestan fields, centred around the original discovery at Masjed Soleymān, and connected by pipeline to Abādān. Finally there were the two large fields of Gach Sārān and Agha Jārī which were linked to both the Abādān refinery and to a new tanker terminal at Bandar e Ma'shūr.

In 1951 Iran nationalized the oil industry and created the National Iranian Oil Company (NIOC). The Anglo Iranian Oil Company was forced to suspend its operations, but with the assistance of a number of Western countries, it managed to arrange a boycott of Iranian oil. Without any outlets for its oil Iran had to close down most of its fields and output slumped dramatically. In 1954 an agreement was reached between Iran and a consortium of Britsh, Dutch, American and French companies and, as a result, the boycott was lifted and production resumed. Under the new twenty-five year agreement the concession area of the Consortium was to be limited to only 259,000 sq km in southwest Iran, with oil rights for the rest of the country going to NIOC. Oil production picked up rapidly and by 1957 had passed the 1950 total of 30 million tonnes. In 1956 NIOC discovered oil on the central plateau at Qom, well away from the main centres of production but unfortunately the field proved difficult to develop. The Consortium enjoyed greater success when oil was discovered near Ahvāz in 1958. The field was developed rapidly and production, which was fed into the extensive pipeline system, helped to compensate for the declining production from the older fields at Masjed Soleymān, Naft-e Safīd and Lālī. Several other new fields were discovered by the Consortium in the early 1960s, including a number of offshore fields under the Gulf.

With the continued rise in oil production of the Consortium, the terminal at Bandar e Ma'shūr proved too small and so a new crude oil shipping point was constructed at Khārk Island on the Gulf. This was connected by pipeline with the Agha Jārī and Gach Sāran fields and the other pipeline systems. When the Khārk Island system was completed it became the sole crude oil exporting terminal for the Consortium and Bandar e Ma'shūr was converted to handle only refinery products from the Abādān refinery. In the 1960s NIOC encouraged oil exploration by companies which were not members of the Consortium. As a result of this work more than half a dozen new offshore fields beneath the Gulf have been discovered. To provide an outlet for these fields a terminal was built on Lavan Island and inaugurated in 1968.

A major problem in Iran associated with the exploitation of petroleum has been the wastage of natural gas. As late as 1960 only about 7 per cent of the natural gas was utilized and the rest flared. In 1966 a great step forward was made when an agreement was reached between Iran and the USSR. As part of the agreement Iran was to supply 6,200 million cubic metres of gas each year through a pipeline, which would be constructed, partly with Soviet money, from Khuzestan to Astara. In return the USSR agreed to build a steelworks in Eşfahān and to accept natural gas in payment for it. The pipeline, which was completed in 1970 also benefited Iran by supplying gas along small feeder mains to the towns en route.

Since 1964 NIOC has expanded its activities into international spheres, by providing technical assistance to other countries of the Middle East and Africa, and also by making bilateral trade agreements with East European and other countries, in which crude oil is sold in return for capital goods, industrial equipment or food products. In 1972 NIOC announced plans, in conjunction with BP, for oil exploration in the North Sea.

9.6 Petroleum reserves

One of the most important questions relating to the extraction of petroleum is that of the quantity of crude oil still remaining. Oil is not a renewable resource and once the reserves are exhausted man will have to turn to other energy sources to satisfy his requirements. Estimates of the world's oil resources are made at regular intervals by the petroleum industry. These figures are never constant owing to the dynamic nature of the industry. The total rises as new oil fields are discovered, but with every year of production the reserves are being depleted. On balance, however, the net total of known reserves has risen steadily since records were first made (Table 9.2). From 1970 the rate of new discoveries relative to oil consumption does appear to have slowed down considerably, though the trend still remains upwards.

The distribution of the estimated proven oil reserves throughout the world varies greatly from one region to another (Table 9.3). In particular the area of Southwest Asia alone contains 56.4 per cent of the world's reserves (1985). When North Africa reserves are added the figure rises to 61.2 per cent. Since 1970 the Middle East has retained its relative importance despite a marked increase in reserves in Western Europe and Latin America. In contrast, the Sino-Soviet area, Africa and North America have all recorded a decline in importance. Even within the Middle East and North Africa the reserves of individual countries exhibit

TABLE 9.2
Estimated proven oil reserves of petroleum – changes with time

Year	Million tonnes
1935[1]	3,300
1945[1]	9,300
1955[1]	26,400
1960[1]	41,400
1965[1]	48,300
1970[2]	84,600
1975[3]	90,400
1980[3]	88,900
1985[3]	95,000

Sources: 1. Petroleum Information Bureau 1967, *Oil – The World's Reserves*, p 1;
2. Institute of Petroleum Information Service, 1971, *Oil – World Statistics*, p 1;
3. British Petroleum Company Ltd, *BP Statistical Review of the Oil Industry 1975*;
British Petroleum Company Ltd, *BP Statistical Review of World Energy 1980*;
British Petroleum Company plc, 1986, *BP Statistical Review of World Energy*.

large variations (Table 9.4). For example, Saudi Arabia and Kuwait together account for 35.4 per cent of the worlds reserves in 1985.

What is interesting with the table is that despite high levels of production in the period between 1970 and 1985 many countries in the region now have larger known reserves than they possessed 15 years earlier. Indeed of the major producers only Iran and Libya show significant reductions in their proven reserves. Generally the countries with the largest oil production at the present have the largest reserves. However, the relative rates at which the reserves are being utilized varies greatly with a surprisingly large number of countries having supplies for only 40 years or less at 1985 extraction rates. In particular, the large North

TABLE 9.3
World petroleum reserves 1970 and 1985

	Million tonnes 1970	Per cent 1970	Million tonnes 1985	Per cent 1985
Middle East	47,818	56.5	54,000	56.4
Sino-Soviet area	13,699	16.2	11,000	11.5
Africa	9,621	11.4	7,400	7.7
North America	7,400	8.8	5,600	5.8
Latin America	3,577	4.2	11,900	12.4
Far East and Australia	1,946	2.3	2,500	2.6
Western Europe	508	0.6	3,400	3.5

Sources: 1. Institute of Petroleum Information Service 1971, *Oil – World Statistics*, p 1;
2. British Petroleum Company plc, 1986, *BP Statistical Review of World Energy*, p 2.

TABLE 9.4
Estimated proven reserves in the Middle East and North Africa 1970 and 1985

Country	Million tonnes 1970	Years to exhaustion of known 1970 reserves at 1970 production levels	Million tonnes 1985	Years to exhaustion of known 1985 reserves at 1985 production levels
Abū Dhabi	2,080	63	4,100	99
Dubayy	139	32	200	11
Iran	7,650	40	6,500	59
Iraq	3,950	52	5,900	84
Kuwait	9,860	72	12,400	276
Neutral Zone	810	68	800	43
Oman	416	24	500	20
Qatar	542	31	400	28
Saudi Arabia	19,000	107	23,000	132
Algeria	1,100	24	1,100	25
Egypt	416	25	500	12
Libya	4,160	26	2,800	54

Sources: 1. British Petroleum Company Ltd, 1970, *Our Industry Petroleum*, pp 431–53.
2. British Petroleum Company plc, 1986 *BP Statistical Review of World Energy*, pp 2–4.

African producers possess relatively small proven reserves. It is little wonder, therefore, that many nations are now considering a controlled and lower rate of annual exploitation of their existing resources. Even in those countries where large reserves are known, such as Saudi Arabia and Kuwait, there is a growing awareness that petroleum supplies will one day come to an end, and an increasing desire to ensure that the country's economy is on a sound footing before this happens. One of the main problems is that as oil is removed from a reservoir it becomes increasingly expensive to obtain what remains, until eventually the costs of extracting the oil can equal the revenues which can be obtained from it.

Any discussion of the problems within the oil industry nearly always returns to the question of the size of the world's reserves and a consideration of the reserves/production ratio. This term is defined as the ratio between the proved recoverable reserves within the ground and the annual production rate. It can be thought of as the number of years of production at the current rate of oil extraction. Since 1920 there has been a marked decline in the world R/P ratio to a figure of 36 in 1970 and 34.4 in 1985. Not all the oil which is present in an oil field can be extracted economically, and normally production has to cease when the R/P ratio falls to about 10:1.[7] For planning purposes on a world scale a more realistic figure is probably a R/P ratio of 15:1.

Besides the proven oil resources there is always interest for strategic planning purposes as to the amounts of oil which might still be discovered in a region in the future. The United States Geological Survey has estimated the probability of the amount of undiscovered oil in the Arabian–Iranian basin, which is the richest petroleum province in the world.[8] It claimed that there were likely to be between

TABLE 9.5
Assessment of undiscovered conventionally recoverable petroleum resources of the Arabian–Iranian basin by country.

	Low	High (billion tonnes)	Mean
Saudi Arabia	3.27	15.14	7.78
Iran	1.50	6.82	3.55
Iraq	4.37	20.46	10.64
UAE	0.27	1.77	0.95
Kuwait	0.27	1.23	0.55
Oman	0.14	0.55	0.27
Total	9.82	45.98	23.74

Note: The probability of more than the low value is 95 per cent.
Similarly the probability of more than the high value is 5 per cent.
Source: Masters C. D. *et al*, 1982.

9.86 billion tonnes and 45.98 billion tonnes, with a mean value of 23.74 billion tonnes (Table 9.5). Almost all of the undiscovered reserves are thought to be located in Saudi Arabia, Iran and Iraq, with Iraq likely to hold the greatest amount.

9.7 Transport of crude oil and petroleum products

In the Middle East one of the most important aspects of the petroleum industry is the transport of the product from the oil producing countries to the markets in Western Europe and Japan. By far the most important means of transport is the ocean going tanker. With the growth of tanker sizes after the Second World War, the patterns of movement between the Middle East and Western Europe began to change. During the 1950s and early 1960s, the main oil routeway between the Gulf fields and Europe was through the Suez Canal. With the advent of super tankers, however, this route was no longer usable for the larger vessels, forcing them to make the long detour around the Cape of Good Hope. Following the war of June 1967 the Suez Canal was closed until 1975 and during this time all traffic had to travel around Africa. This greatly increased journey times from forty to sixty days including loading and unloading.

The great heyday of tanker construction was the early 1970s with 109 million tonnes deadweight being constructed between 1971 and 1975. In the first five years of the 1980s tanker construction has been only about one-quarter that of a decade earlier. The main reason for this was the stagnation in the world economy following the oil price rise of 1974 and the consequent decline in oil consumption in the developed nations. Indeed a feature of the early 1980s has been the way in which tankers have been mothballed or scrapped. For example, between 1981 and 1985 102 million tonnes of tanker tonnage were eliminated from the world fleet owing to the great over-capacity which existed at this time.

In the period since 1970 there have been significant changes in the oil exports from the Middle East and North Africa in terms of both the amounts which are shipped as well as the actual destinations to which they go (Table 9.6). By far the most significant change is the huge decrease in oil exports to Western Europe with the 1985 figure for Middle East sources only 38 per cent of what it was in 1971. For North Africa the equivalent figure is 52 per cent. There are two reasons for this. The first and most important is the rise in oil production in the UK from zero in 1971 to almost 130 million tonnes in 1985. The second is greater energy efficiency in terms of usage and the rise of alternative fuels to oil throughout Europe. The result is that the Middle East and North Africa is much less vital for the survival of Western Europe than it was 15 years ago in terms of energy provision. A similar situation is seen with Japan, though to a lesser degree, which is now taking only 68 per cent of the oil it received from the Middle East in 1971. This shortfall is accounted for by oil from alternative sources and more efficient energy use. Of particular note is the fact that the USA imports very little oil from the Middle East and North Africa. Such oil only accounts for 12.5 per cent of the USA's imports, most of which is now obtained from Latin America.

Besides the ocean going tanker, the commonest means of bulk crude oil transport is by pipeline. When compared with tanker transport pipelines possess certain advantages and disadvantages. The capital costs of a pipeline are always high, while operating costs are low. A pipeline is also a static feature, and once constructed it is not easy to change the destination of the carried product. It is difficult to increase the capacity of a pipeline system and it is very vulnerable to attack. Over a given distance it is undoubtedly more economical to transport crude oil by large tanker than by pipeline, and at the same time there are benefits from greater flexibility: in being able to vary transported amounts in response to market and production fluctuations. Usually a pipeline can only compete with tankers if its route is shorter than the tanker route, or if sea transport is burdened

TABLE 9.6
Inter area movements of oil

	Middle East		North Africa	
	(million tonnes)			
Destination	1971	1985	1971	1985
USA	20.0	21.0	4.5	10.2
Canada	13.5	2.3	0.0	1.6
Latin America	22.5	30.4	4.3	4.7
Western Europe	378.8	144.2	158.3	82.6
Africa	25.3	16.7	0.3	1.0
Southeast Asia	48.5	66.4	0.0	0.0
Japan	191.8	129.6	1.0	2.6
Australasia	15.5	3.1	0.0	0.0
Rest of world	20.3	32.9	10.5	13.1
Unknown	19.5	0.0	7.0	0.0

Sources: 1. British Petroleum Company Ltd, *BP Statistical Review of the Oil Industry 1971*;
2. British Petroleum Company plc, 1986, *BP Statistical Review of World Energy*.

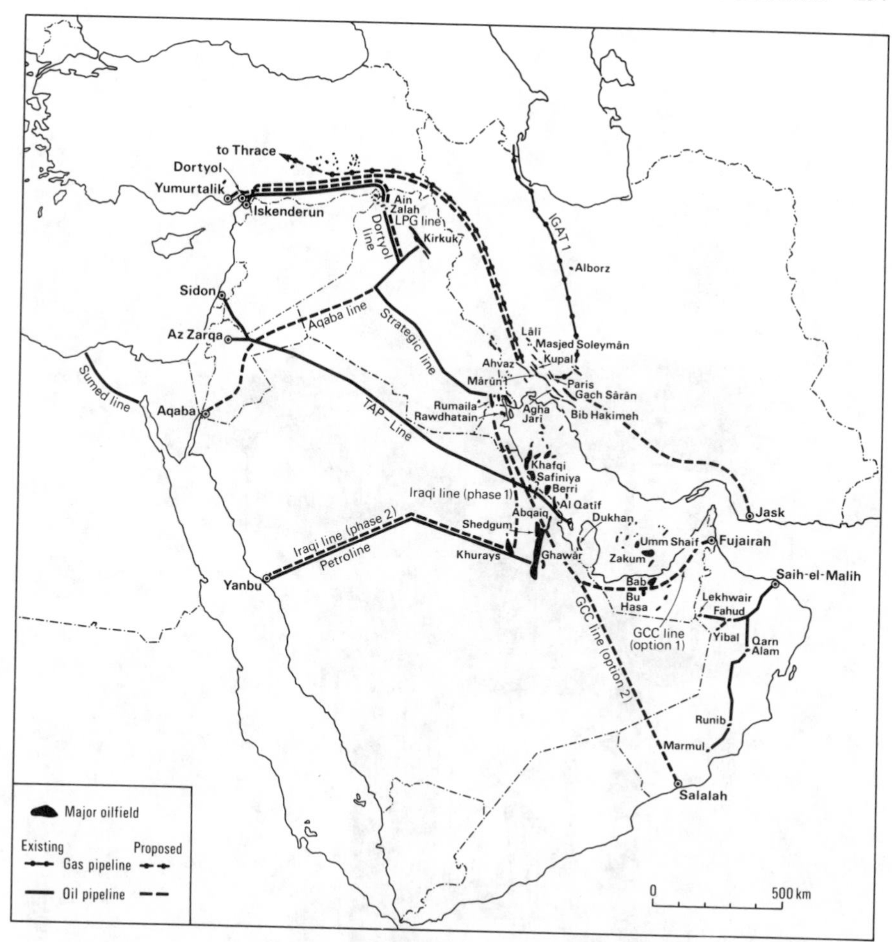

Figure 9.4 Pipelines in the Middle East

by heavy port charges, canal dues and high insurance costs. With the opening of oil fields in the interiors of the countries of the Middle East in the early part of the twentieth century, the only practical means of getting the crude oil to coastal locations was by pipeline. The first pipeline to be constructed in the Middle East was built by the Anglo-Persian Oil Company in 1911. It was 200 km in length and linked the newly discovered oil fields at Masjed Soleymān with Abādān on the Gulf.

The best known of the early pipelines in the Middle East is the Trans-Arabian pipeline (TAPline) connecting the Saudi Arabian fields with the Mediterranean Sea (Figure 9.4). At the time of its construction in 1950 this 1600 km long, 78.7/76.2 cm diameter pipeline with six pumping stations along its route, was the largest in the world and cost £80 million to build. The initial capacity was 15 million tonnes/annum, but this was subsequently increased to a throughput of 25 million tonnes/annum. Owing to the political problems of the Lebanon it is no

Increasing quantities of oil come from offshore fields. Wellheads in the Zakum field, Abu Dhabi marine areas (British Petroleum Co. Ltd.)

longer utilized, other than to send oil to Jordan. Other early large pipelines include the Iraq pipelines, connecting the Kirkūk oilfields with Tripoli and Baniyās on the Mediterranean coast, both of which are more than 800 km in length, and the Libyan pipeline from the Sarir oilfield to the tanker terminal at Tobruk. It is 510 km in length and has an initial design capacity of 15 million tonnes/annum.

An unusual crude oil pipeline was built in the 1970s in Israel between the ports of Eilat, on the Gulf of Aqaba, and Ashquelon, on the Mediterranean. The initial capacity of the line was about 12 million tonnes/annum, but it is considered that this could be raised to around 60 million tonnes/annum if necessary. One of the aims of the pipeline was to bypass the Suez Canal. It had been hoped that oil would be delivered from Iran, which under the Shah had been friendly towards Israel. However, with the overthrow of the Shah this was no longer a feasible proposition.

In recent years many new pipelines, including one across Saudi Arabia from the Ghawār oil field to Yanbu'al Baḫr, have been constructed (Figure 9.4). There were also proposals for new oil and gas pipelines from Iran, through Turkey, to the Mediterranean coast. However, in early 1987 it was decided to postpone construction of the oil pipeline, although work on the gas line was still under consideration.[9]

Another interesting proposed pipeline is from the Kirkūk oilfields to Aqaba, which would allow Iraq yet another outlet for its northern oilfields without crossing Syrian territory. A line to the Mediterranean coast through Turkey already exists. In Iran similar strategic lines are under construction or planned. One being built is from Ganāveh to 'Asalūyeh, which bypasses the Khārk Island terminal.[10] The proposed line is from the Gach Sārān oilfields to Jāsk, situated beyond the Straits of Hormuz. This will mean that tankers will be able to pick up Iranian oil without having to enter the Gulf, and so be less at risk from Iraqi air attack. All these provide examples of the growing strategic and political nature of pipeline construction, with the oil producing countries striving to maintain as flexible export policy as possible.

9.8 Revenues

With the earliest concessions the financial liability of the oil companies consisted of fixed royalty payments to the producing states. These arrangements proved extremely profitable to the oil companies and it was, therefore, almost inevitable that the governments in the producing countries would desire a larger share of the rewards. At first this was accomplished by royalty payments, but by the early 1950s the oil producing countries increased their demands for the revenues to be related to the profits of the oil companies. Following negotiations many governments agreed upon a tax formula which gave them 50 per cent of the net profits calculated on the basis of the posted prices.[11] This new arrangement greatly increased the revenues of the producing countries and, at the same time, produced no extra financial burden for the oil companies, which in most cases were able to offset the larger revenue payments against income tax liabilities in the country in which the company was registered.

In the late 1950s with increasing supplies of petroleum becoming available and increased competition between the producing companies, the posted prices of oil began to decline. This meant that the government oil revenues, which were related to the posted prices, also fell. The producing countries became increasingly worried over this trend and in 1960 a number of them joined to form OPEC, with the objective of protecting their interests against the actions of the international oil companies. In its first confrontation with the major companies, OPEC managed to stop the decline in 'posted prices'. From this time forward, 'posted prices' became, to all intents and purposes, nominal in character and of value only for calculating the revenues which the companies had to pay to the host countries. Petroleum taxes became, as far as the companies were concerned, a fixed payment on every barrel of oil which was produced.

During the 1960s the total revenues collected by the governments of the oil producing countries were further augmented by the introduction of a system known as the 'expensing of royalties'. With the old system, the royalty payment, usually of 12.5 per cent, was included with the 50 per cent of net profits going to the government. Under the new arrangement, the royalties were paid separately from the tax and were classified as a production expense to be deducted before net profits were calculated. The result was to give the governments yet further increased revenues. The implementation of this measure was naturally resisted by the oil companies, but eventually they were forced to accept the principal involved and, by the mid-1960s, most company/country agreements were employing this new method of payment.

Of the monies paid by the companies to the government of the oil producing nations, by far the largest amount comes from company taxes and royalty payments, mostly in the form of foreign exchange. Another source of income is from payments made locally during actual production for such items as labour, maintenance and related services. In the 1960s, such sources accounted for almost one fifth of the payments made to governments, but by 1970, with increasing production efficiency, this figure had fallen to 10 per cent or less of the total.[12]

Nowhere is this increased effiency witnessed better than in terms of local employment opportunities provided by the oil industry. In each of the producing countries local employment has been falling steadily since the beginning of the 1960s, despite continued increases in production. In Iran 37,000 people were employed by the Consortium in 1961. This had dropped to 26,000 by 1967, and to 17,800 by 1971.[13]

If the period up to 1970 was one in which the oil companies enjoyed a dominant role, the 1970s and early 1980s were a time when the oil producing countries seized the initiative in the struggle for power. This new phase can be said to have begun when the OPEC countries reached new agreements about tariffs in Tehrān in

TABLE 9.7
Middle East oil exports and revenues 1962–71
(Crude oil values in million tonnes; revenues in million US dollars)

	1962	1965	1968	1971	Percentage increase 1962–71
Kuwait					
Crude oil	91.7	108.5	121.8	145.0	58
Revenues	526.3	671.1	765.6	1395.3	165
Saudi Arabia					
Crude oil	76.0	101.2	140.9	222.0	193
Revenues	451.1	655.2	965.5	2159.6	338
Iran					
Crude oil	66.1	94.3	141.5	227.0	224
Revenues	333.8	522.4	817.1	1869.6	462
Iraq					
Crude oil	50.2	65.8	73.3	83.0	66
Revenues	266.6	374.9	476.2	840.0	214
Libya					
Crude oil	9.2	61.0	127.0	132.0	1370
Revenues	38.5	371.0	952.0	1766.6	4430

Source: Middle East Economic Digest, vol 16:43, 27 October 1972, p 1251.

January 1971, as a result of which the revenues paid by the oil companies to the governments increased dramatically (Table 9.7).

It was, however, the quadrupling of oil prices in 1973/74 following the October 1973 war between the Arabs and the Israelis which had the greatest effect on both the world economy and on the incomes of the oil producing countries. Immediately following this war oil supply was used by the Arabs as a political weapon. Crude oil supplies to the United States and the Netherlands were cut off completely for a period and reduced amounts sent to the rest of Western Europe and Japan.

Official crude oil prices are those charged by the oil producing countries to the major companies. They are often considerably different from the 'spot' market prices charged for oil in places like the Netherlands. In general 'spot' prices are higher than official prices in times of oil shortage and fall below the official prices in times of plenty. After the 1973 October War the official oil price jumped from under three dollars a barrel in mid-1973 to over ten dollars by early 1974 (Figure 9.5). During the mid-part of the 1970s the official price recorded a very slight increase, but then between 1979 and 1981 it more than doubled from 13.50 dollars per barrel to 34 dollars per barrel. As with the earlier price rise in 1974 this too had an enormous effect on the world's economy producing recession in the western countries and ever-growing debts in the countries of the Third World.

With the fall in industrial production the demand for oil dwindled in 1982 and so the official oil price began to decline, reaching 28 dollars a barrel by the end of 1985. It was, however, during 1986 that continued over-production by OPEC and falling demand led to a collapse in the oil price on the 'spot' markets, with the

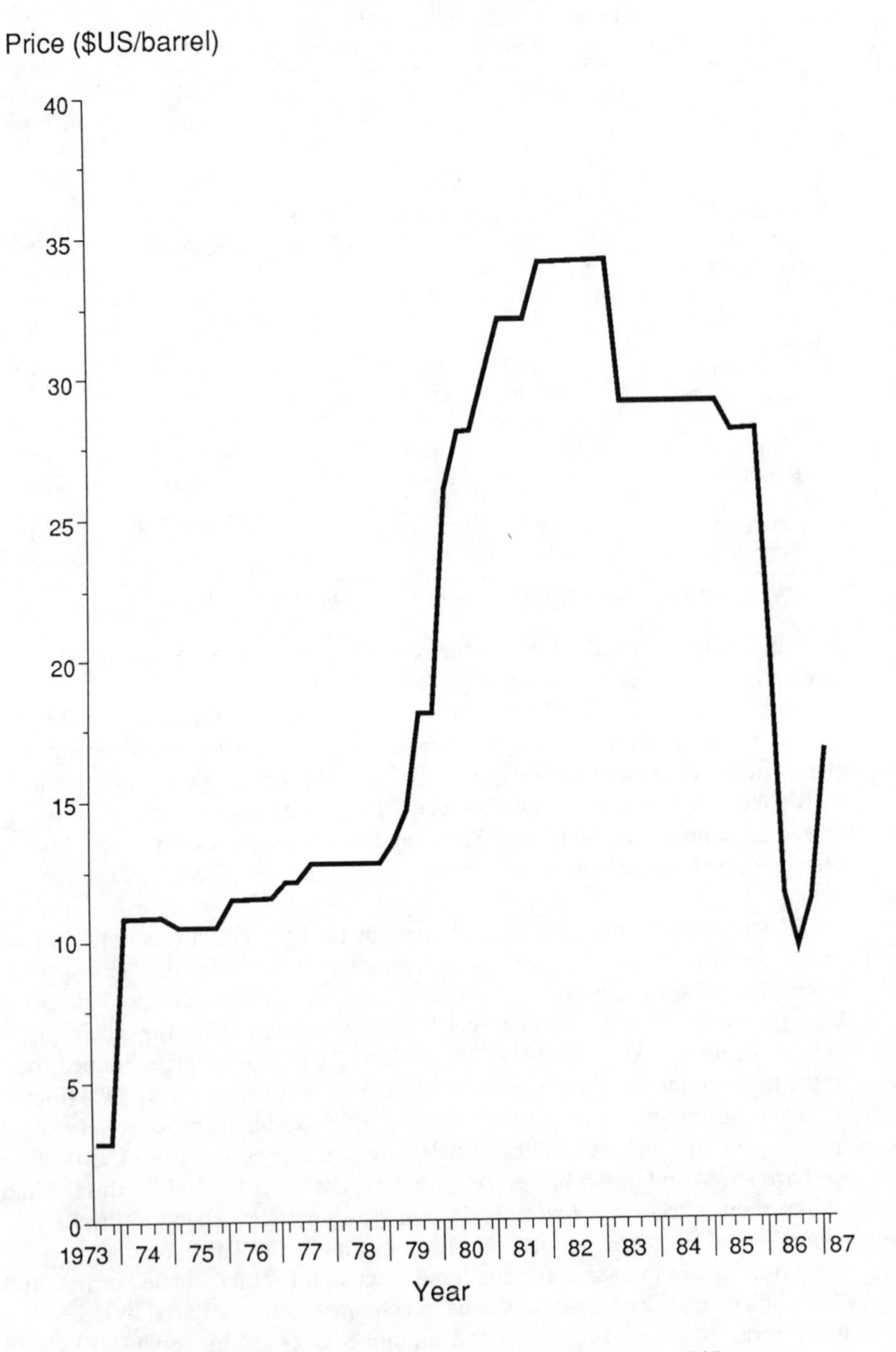

Figure 9.5 Official oil prices – Saudi Arabian Light – 1973–1987

result that prices in July 1986 fell to only 8 dollars a barrel. Since then prices have steadied and by early 1987 had risen to around 18 dollars a barrel.

As far as the individual countries are concerned the crucial figure is how much oil revenue the government receives. This is related to both the price of the oil and the volume produced. Over the last few years certain OPEC countries have attempted to keep their revenue totals high by increasing production when oil prices have been falling. The net effect of this has been to release more oil onto a market already suffering from over-production. The result has been to push prices even lower. Since the early 1980s OPEC has been aware of this growing problem and has attempted to enforce a system of production quotas on the member states, though with varying degrees of success. During this period it has only been the Saudi Arabian government's willingness to cut back its own production substantially which has prevented the price of oil from falling more rapidly than in fact occurred. Indeed, if it had not been for the fall of the Shah and the Gulf War between Iraq and Iran, which both disrupted oil production, then an oil glut would have depressed world prices even earlier.

In recent years the actual revenues received by the oil producing countries have reached staggering proportions, as is witnessed by the case of Saudi Arabia the largest producer (Figure 9.6). During the 1960s and early 1970s oil production in Saudi Arabia was rising rapidly, while oil revenues only increased slightly. Following the 1973 October War and the 1973/74 oil price increase, the revenues of Saudi Arabia jumped from less than 2 billion dollars in 1973 to more than 20 billion dollars in 1974. Then for the next few years while oil production varied somewhat, though at a high level, revenues increased gradually. The really crucial change occurred between 1978 and 1980 when the rapid increase in official oil prices shot revenues to over 100 billion dollars in 1981, despite only a modest increase in oil production since 1977. Following 1981 Saudi Arabian oil production plummeted and as a result revenues fell dramatically as well. From 1983 onwards the fall in revenues was intensified as official oil prices began to ease back as well. By 1985 the revenues received by Saudi Arabia were down to only 30 billion dollars.

This pattern of rapid rise and almost equally rapid fall in oil revenues has been experienced by all the other North African and Middle Eastern oil producers. All reveal a sharp increase in revenues in 1974 followed by a much greater increase in 1979, to produce peak revenues in the period 1980 to 1982. The actual year of greatest revenues depends largely on local production conditions. In future, although there are likely to be short periods of over-production of oil, when prices may fall, there seems little doubt that in the long term oil prices will continue to increase.

Over the years the importance of oil revenues has been in the development potential it has provided for the country which has received them. Since the Second World War these revenues have produced a period of considerable change. Although it is always difficult to generalize it can be said that the pace of development up to the late 1950s was relatively slow. In these early days it was oil companies themselves which were the major agents of change. They moved into

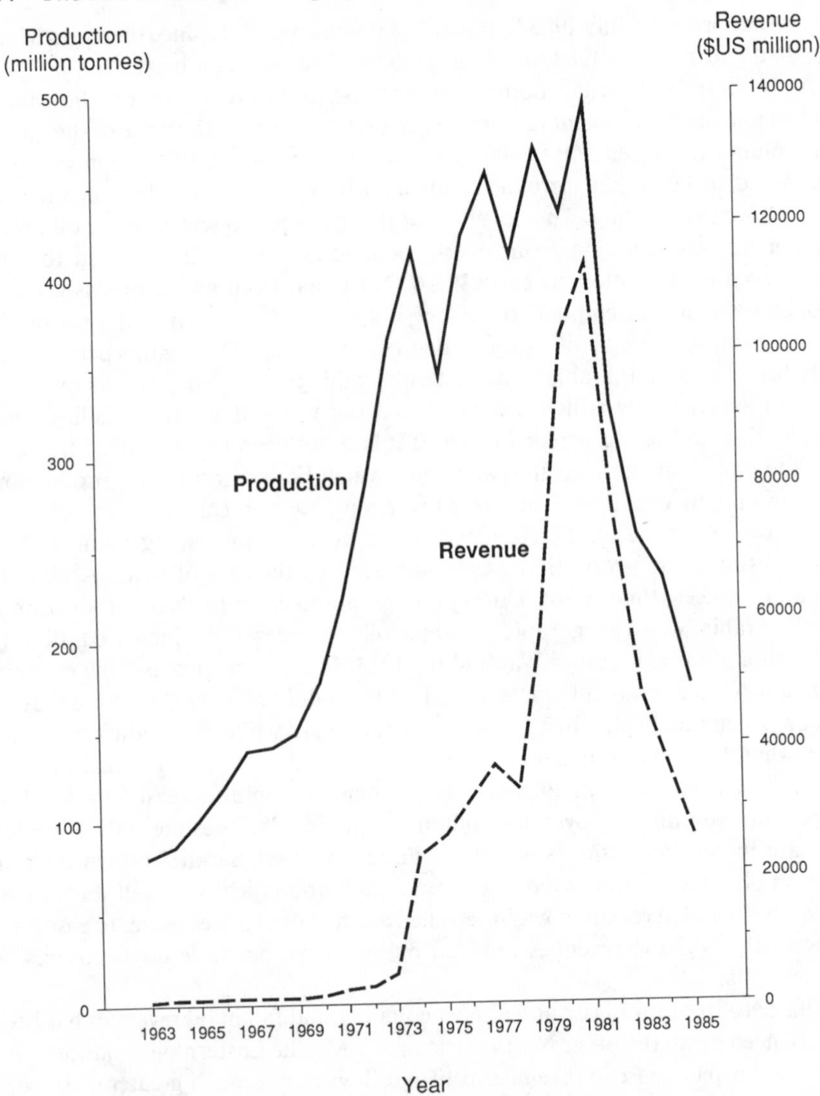

Figure 9.6 Oil revenues received by Saudi Arabia – 1963 to 1985

countries with traditional agricultural or trading economies totally unsuited to their high technology approach. As a result the oil companies had to construct their own infrastructures for producing the oil and shipping it. New roads and new ports were built and company towns established in which their overseas employees were housed. There is no doubt that at this time the oil companies were the major investors and consumers in most of the countries of the region and provided health, education and housing facilities which were often considerably superior to those offered by the state itself.

Zelten oilfield in Libya: pipelines converge on the gathering centre at Zelten (An Esso photograph)

In retrospect the 1960s were a time of transition. The growing oil revenues permitted the governments to undertake development planning on a scale much larger than anything which had previously occurred. Yet at the same time although the revenues were vital to the economies of the countries concerned, they had not yet come to dominate other aspects of the economy in the larger countries such as Iran and Iraq. In the smaller Gulf countries, with few natural resources, the economies were already being dominated by oil revenues. During the 1970s the increases in revenues meant that even the largest countries become dependent upon them to an extent that would have been difficult to predict a few years earlier. Much of the incoming money was used for infrastructure provision and for industrialization in an attempt to establish a more balanced economy. The size of the development plans was so large that it meant that any fluctuations in either oil price or oil output could seriously affect the planning programme. As prices fell in the early 1980s it resulted in many countries increasing production in an attempt to maintain their revenues. Unfortunately this pushed prices further down and so lessened the efficiency of OPEC as a guiding force in the oil industry.

9.9 The Future

The future of oil in the world economy is one of the most controversial topics in international affairs at the present time. With the different views held by the producing and consuming countries it seems that clashes of interest will occur, with perhaps even greater frequency in the future. There is no doubt that the Western governments were shocked with the use of oil as a political weapon by the Arabs in 1974 and also by the impact which this had on the world's economy. It caused all of them to re-evaluate their energy policies and in particular make greater use of alternative fuels and more efficient use of energy.

If changes in the world picture of primary energy use are studied over the period from 1972, that is before the energy crisis, to 1985, it is noted that oil consumption has only increased by 9 per cent (Table 9.8). All other fuels have increased by much larger percentages, though for water power and nuclear energy absolute usage still remains small. Of particular interest is the marked growth in the use of natural gas which before the early 1970s was flared without any beneficial use in many of the oil producing countries. A different pattern prevails in the USA over the same period. Here both the consumption of oil and natural gas have shown significant decreases of 4 and 20 per cent respectively. In the world context the USA is unusual in so far as it has been a major user of natural gas for many years now. Today petroleum products in the USA make up only 66 per cent of energy consumption, compared with over 80 per cent in 1972. This has been achieved largely by a considerable increase in the use of solid fuels and nuclear power. It is also worthy of note that total energy consumption in the USA has only grown by 8 per cent between 1972 and 1985, while world consumption has gone up by 37 per cent. This reveals the increased energy consciousness

TABLE 9.8
Primary energy consumption (million tonnes oil equivalent)

	1972		1985		Per cent change 1972–85
	Total	Per cent	Total	Per cent	
1. WORLD					
Oil	2,578.0	47.7	2,809.4	37.9	+9.0
Natural gas	1,005.7	18.6	1,491.4	20.1	+48.3
Solid fuel	1,709.3	31.6	2,277.8	30.7	+33.2
Water power	103.2	1.9	498.2	6.7	+382.8
Nuclear	12.1	0.2	337.1	4.5	+2,809.2
Total	5,408.7	100.0	7,414.3	100.0	+37.1
2. USA					
Oil	775.8	46.6	724.1	41.2	−4.3
Natural gas	559.4	33.6	444.5	24.7	−20.5
Solid fuel	301.5	18.1	443.3	24.6	+47.0
Water power	23.1	1.4	82.9	4.6	+258.9
Nuclear	4.8	0.3	104.6	5.8	+2,079.2
Total	1,664.6	100.0	1,799.4	100.0	+8.1

Source: 1. British Petroleum Company Ltd, *BP Statistical Review of the World Oil Industry 1973*, p 16;
2. British Petroleum Company plc, 1986 *BP Statistical Review of World Energy*, p 31.

within a country which still consumes 24 per cent of the world's energy, yet contains only about 5 per cent of the world's population.

At the present day the USA is almost self-sufficient in terms of natural gas, but it has to import about 45 per cent, or 224 million tonnes (1985) of its oil needs. In Western Europe the situation with regard to oil is even more critical as the consumption of 566.8 million tonnes (1985) is 198 per cent more than is produced. Even this seems insignificant when one considers Japan which produces only 0.5 million tonnes per year and yet needs 201.3 million tonnes (1985). These facts illustrate the strategic importance of oil movements to maintain the Western nations. Although the USA now obtains most of its oil needs from Latin America, Western Europe and Japan remain highly dependent on the Middle East and North Africa for their oil supplies both now and into the foreseeable future.

There can be no doubt, therefore, that the Middle East will continue to play a vital role during the remaining years of the twentieth century as a petroleum source for the developed and developing nations of the world. Equally, though it seems likely that oil and natural gas will continue to become less important as primary energy sources in the twenty-first century. Oil will have played an important role in the development of the world's economy, but before the bicentenary of its discovery in commercial quantities in 1858 is reached, its heyday will be well past.

References

1. Actual figures vary slightly depending upon the source of information
   (a) *Aramco handbook*, Dhahran, 1960, p 80.
   (b) Shell International Petroleum Company Ltd, *Oil in the World Economy*, London, no date, pp 33–8.
   (c) Institute of Petroleum Information Service, *Oil – World Statistics*, London, 1971, p 1.
   (d) British Petroleum Company plc, *BP Statistical Review of World Energy*, 1986, p 31
2. For a map of concession areas in the Middle East and North Africa in 1958, see Oxford Economic Atlas, *The Middle East and North Africa*, Oxford University Press, pp 46–7.
3. S. H. Schurr and P. T. Homan, *Middle Eastern Oil and the Western World*, American Elsevier Publishing Company, New York, 1971, p 111.
4. A. Melamid, 'Industrial activities'. In *The Cambridge History of Iran – The Land of Iran* (Ed W. B. Fisher) Vol. 1, Cambridge University Press, 1986, p 530.
5. S. H. Schurr and P. T. Homan, *Middle Eastern Oil and the Western World*, American Elsevier Publishing Company, New York, 1971, p 118.
6. C. D. Masters, H. D. Klemme, and A. B. Coury, *Assessment of undiscovered conventionally recoverable petroleum resources of the Arabian—Iranian basin*, United States Geological Survey Circular, No 881, 12 pages, 1982.
7. H. R. Warman, 'The future of oil', *Geogr. J.*, **38**, p 291, 1972.
8. C. D. Masters, H. D. Klemme, and A. B. Coury, *Assesment of undiscovered conventionally recoverable petroleum resources of the Arabian-Iranian basin*, United States Geological Survey Circular, No 881, 12 pages, 1982.
9. *Middle East Economic Digest*, **31**, No 15, p 11, 11 April 1987.
10. Iain Jenkins, 'Iran acts in pipeline war', *Middle East Economic Digest*, 16 November 1985, pp 4–6.
11. S. H. Schurr and P. T. Homan, *Middle Eastern Oil and the Western World*, American Elsevier Publishing Company, New York, 1971, p 120.
12. S. H. Schurr and P. T. Homan, *Middle Eastern Oil and the Western World*, American Elsevier Publishing Company, New York, 1971, p 101.
13. Iranian Oil Operating Companies, *Annual Review 1971*, Tehrān, 1971, p 23.

CHAPTER 10

The Political Map

10.1 Introduction

Those parts of Africa and Asia selected for detailed consideration in this volume comprise 16 states, together with the seven members of the United Arab Emirates. Some attention is also given to the three states of Morocco, Algeria and Tunisia as constituting the western limb of the Arab world – *Jezira al-Maghreb* to the Arab geographers, 'the island to the west'. The number of political units is not remarkable, but the political and economic interaction between them impinges on almost every aspect of the life of the region. Chapters 11 to 20 form a series of case studies chosen to elaborate some of the outstanding themes of the earlier systematic chapters. In all except Chapter 11 the state itself provides the unit of study, and the configuration, size, and location of the state will frequently be seen as of fundamental importance. It is therefore appropriate to consider the nature and origins of the political map in this chapter. No attempt is made to give a comprehensive account of the political geography of Southwest Asia and North Africa, although a number of key bibliographical references on geopolitical aspects are included.

10.2 Modern political history

The gradual decline of the Ottoman Empire in the nineteenth century and the growing ambitions and rivalries of the European powers, provide the background to the emergence of the modern political map. At the height of its power, the Ottoman Empire, like its predecessors the Roman and Arab Empires, embraced huge tracts of Southwest Asia and North Africa. Much of this territory was under Ottoman domination for three or four hundred years, although from the eighteenth century some parts acquired a measure of autonomy.

The broad outlines of the modern political map were established earlier in North Africa than in Sothwest Asia. The French conquest of Algeria began in 1830 and was effectively complete by 1845. Tunisia became a French protectorate in 1881, having also previous been under nominal Ottoman rule. Morocco, on the other hand, had managed to remain independent, but in 1912 its Sultan was forced to accept French protection, with Spanish protectorates being proclaimed in the north and south. An International Zone was also established around Tangier to ensure freedom of navigation through the Straits of Gibraltar.

The British occupied Egypt in 1882, although it remained nominally under

Ottoman rule until 1914, when a protectorate was formally declared. The chief reason for Britain's intervention was to ensure control of the Suez Canal. Egyptian independence was recognized in 1923, although until 1956 Britain retained the right to station troops in the Canal Zone. Libya became a distinct political unit only in 1912, when it was conquered by the Italians whose dream was to colonize the country and integrate it with the homeland in the way that the French had done in Algeria. During the Second World War the Italians were driven out of Libya, and after a brief period of Anglo-French rule, independence was achieved in 1951.

Thus between 1830 and 1914, European powers had gained control of the whole of North Africa, and although the boundaries of individual states were sometimes undefined, particularly in the Sahara, their future shape was already foreshadowed. But in Southwest Asia the political map in 1914 was quite similar to the pattern of previous centuries (Figure 10.1). Two important frontier zones marked the limit of Ottoman control. The frontier with Persia was generally defined according to the loyalties of the tribes, so that in some regions there was a zone some 100 km wide where neither Persia nor Turkey had much influence.[1] In Arabia, Ottoman control had never effectively extended to the interior beyond Al Hasa, Hejāz and Yemen. Most of Arabia's sparse population was ruled by local sheikhs, though in the course of the nineteenth century, two powerful Wahhābi empires were established. In the early years of this century the Wahhābis found a powerful leader in the person of Ibn Saud who had succeeded in imposing his sovereignty over much of Arabia by 1927, although Saudi Arabia was not so named until 1932.

During the latter half of the nineteenth century, new factors became evident in the politics of Southwest Asia. Britain, France, Germany and Russia looked with increasing covetousness at the weakening regimes in the Ottoman Empire and in Persia. British influence was growing as a result of treaties with the Trucial States (1820), the acquisition of Aden (1839) and Cyprus (1878) and the protection of Kuwait (1899). France had long-standing ambitions in Syria, while the Germans sought to promote their interests through railway building notably in Anatolia, Iraq, and the Hejāz. Russia also wanted to build a railway across Persia to the Persian Gulf, and in 1907 Britian and Russia actually agreed on their respective 'spheres' in Persia without reference to local governments (Figure 10.1). Britain's chief concern was still largely to do with imperial strategy, although oil was produced in Iran from 1908. In addition to this jockeying for position by the great powers, Zionist settlement in Palestine began in earnest in 1882, while among a few Arab intellectuals, Arab nationalism was developing as a new and powerful idea.[2] The period 1880 to 1914 in fact, might be described as the time of 'ideas' in Southwest Asia which, according to the theories of S. B. Jones, would lead to the emergence of 'political areas'.[3] The necessary catalyst was provided by the First World War from 1914 to 1918.

The post-war peace settlement created four new political units in Southwest Asia: Palestine, Transjordan and Iraq, to be administered by Britain under a mandate from the League of Nations, and Syria, mandated to France. This

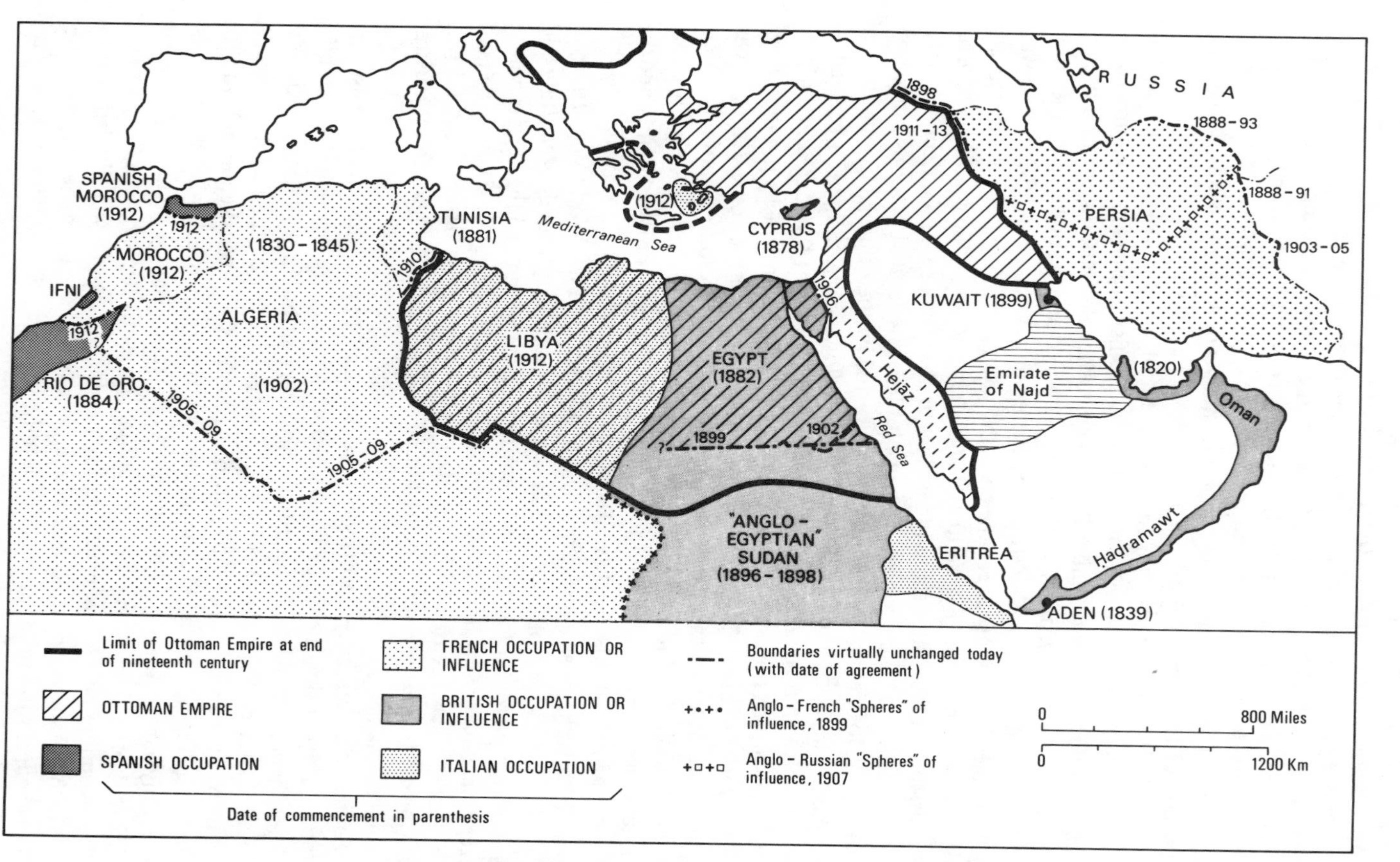

Figure 10.1 The political map on the eve of the First World War

arrangement closely resembled the secret wartime Sykes–Picot agreement between Britain and France for the dismemberment of the Ottoman Empire, and largely disregarded pledges concerning independence apparently made to Arab leaders in the McMahon correspondence.[4] Iraq became independent in 1932, and Transjordan in 1946. (Chapter 15). Lebanon was originally one of seven provinces of Syria intended to be predominantly Christian. After the fall of France in 1941, Lebanon and Syria declared their separate independence, which was recognized at the end of the Second World War.

Events in Palestine from 1918 and the advent of Israel in 1948 are beyond the scope of this book, although one aspect of Zionist expansion is examined in Chapter 16. From 2nd November 1917, when Arthur Balfour, the British Foreign Secretary, sent his declaration to Zionist leaders pledging support for the idea of a Jewish national home in Palestine, the eventual creation of a Jewish state was probably inevitable. Only the shape and timing were uncertain. Several proposals for partition were made (Figure 10.2), but understandably none was acceptable to the Palestine Arabs. These maps themselves, in the view of J. H. G. Lebon, amply demonstrated that the problem was insoluble.[5] The United Nations voted for partition in November 1947 along lines suggested by their Special Committee on Palestine, but during the fighting of the following year, Israel occupied a considerably enlarged area (about 77 per cent of the area of Mandated Palestine), which constituted the de facto state until June 1967. From Israel's point of view, the borders of this state raised many problems, rendering some settled areas highly vulnerable to attack,[6] and making rational use of resources, notably water, extremely difficult.[7] Although a relatively small addition to the political map, Israel is anathema to the Arabs, chiefly on account of injustices to the Palestinians, and the long-term implications of a Jewish state located at the heart of the Arab world.

Modern Turkey was also essentially a creation of the First World War. To begin with, the partitioning of Turkey itself seemed a possibility with Greece, Italy and France gaining footholds, but the deposition of the Sultan and the declaration of a republic under Mustafa Kemal ('Atatürk') in 1922 eventually secured the land for Turkey. Islam ceased to be the state religion, and the national capital was transferred from Constantinople (İstanbul) to the Anatolian plateau at Angora, renamed Anakara.

Several changes have occurred to the political map of the Middle East and North Africa during the past 60 years. Some changes were negotiated peacefully, while others occurred as a result of military intervention. In 1939 France ceded the Hatay region of Syria to Turkey partly in recognition of the region's Turkish majority, thus giving Turkey the port of Alexandretta (now Iskenderun). In 1964 Saudi Arabia and Jordan exchanged desert territory to give Jordan better access to its only part at Aqaba.[8] The Kuwait–Saudi Arabia Neutral Zone, established in 1922 to protect the grazing rights of nomads, was divided between the parties in 1969. A similar neutral zone between Iraq and Saudi Arabia was divided in 1975. In North Africa the most notable peaceful changes have been the disappearance of the Spanish and International Zones of Morocco in 1956, and the cession of the tiny Spanish Ifni enclave to Morocco in 1969.

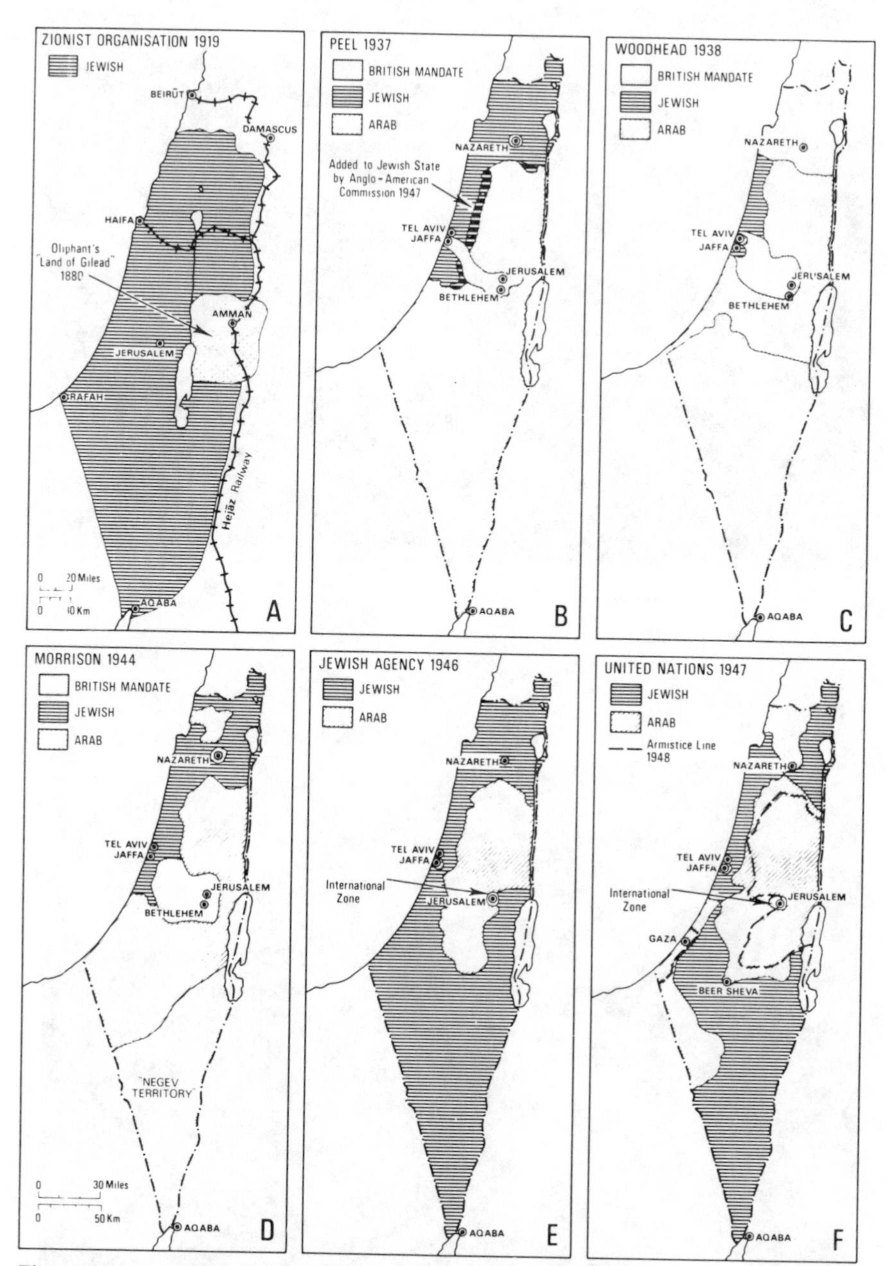

Figure 10.2 Proposals for a Jewish state in Palestine (reproduced by permission of Keter Publishing House Jerusalem Ltd)

Palestine refugee camps: Top: Dera'a Camp in Syria was established after the war of June, 1967 (UNRWA photograph by Sue Herrick-Cranmer) Bottom: Deir el Balah Camp is the smallest of the eight refugee camps in the Gaza strip (UNRWA photograph)

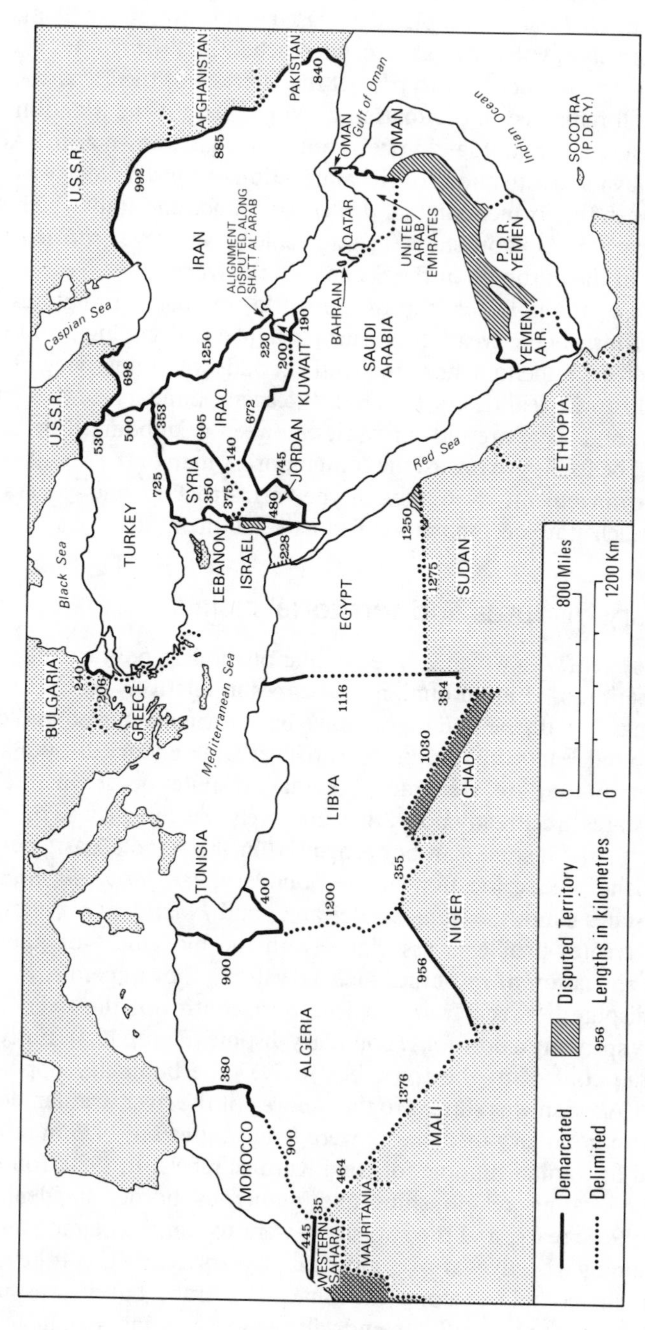

Figure 10.3 The political map in 1986

More spectacular changes have been brought about by force. By far the most important was the creation of Israel in 1948, and Israel's occupation of extensive Arab territories in 1967. Israel completed a phased withdrawal from Sinai in 1982 as part of an Israel–Egypt peace agreement reached in 1979. In 1974 Turkish forces invaded Cyprus, ostensibly to protect the interests of the Turkish Cypriot minority, and still remained in occupation of 36 per cent of the island in 1986.[9] Libya occupied a large area of land in neighbouring Chad known as the Aouzou strip in 1975. Libya's justification for the annexation of this territory was a 1935 agreement between the former colonial powers of France and Italy which would have given Libya the Aouzou strip. The agreement was never ratified by the parties because of the outbreak of the Second World War.[10]

Like so much of the political map of the world, the pattern of states in the Middle East was largely the creation of European interference (Figure 10.3). It is also a relatively recent phenomenon frequently at odds with underlying physical, cultural, and historic realities.[11] In several places boundary and territorial disputes persist, and there are bitter rivalries between neighbours. For all these reasons it would be foolish to regard the contemporary map of the Middle East as a permanent arrangement, even though the infrastructure and geographical pattern within each state become more distinctive all the time.

10.3 Boundary disputes and territorial claims

The boundary system with which we are familiar on modern political maps of the region was superimposed upon human landscapes in relatively recent times. To date, approximately half the boundaries have been formally agreed and demarcated on the ground.[12] It is thus perhaps surprising that boundary disputes are not more frequent than they are. Generally, boundary disputes occur when relationships between states are already poor; they are rarely the direct cause of conflict. Three major types of dispute can be recognized in the Middle East; *positional disputes*, over the precise location of the boundary; *territorial disputes* when neighbouring states claim the same border area; and *functional disputes*, when the boundary creates problems associated with the movement of goods and people, or the allocation of resources such as water, oil, or minerals.

The bitter dispute between Iran and Iraq over control of the river Shaṭṭ al 'Arab is strictly speaking a *positional* boundary dispute (Figure 10.4). Early agreements gave Iraq control of the river, but in 1937 the boundary opposite the Iranian port of Abadan was shifted to the *thalweg* of the river. Iranian demands for a mid-stream boundary throughout were finally agreed in 1975 in return for the withdrawal of Iranian support for Iraqi Kurdish rebels. In 1980, following a period of deteriorating relationships and numerous border incidents, Iraq abrogated the 1975 treaty, and the two states went to war.[13] The motives of the parties in sustaining a costly war over a period of years clearly go far beyond the question of their boundary along the Shaṭṭ al 'Arab, but it was used as justification for going to war in 1980, and will become a major issue in any peace negotiations.

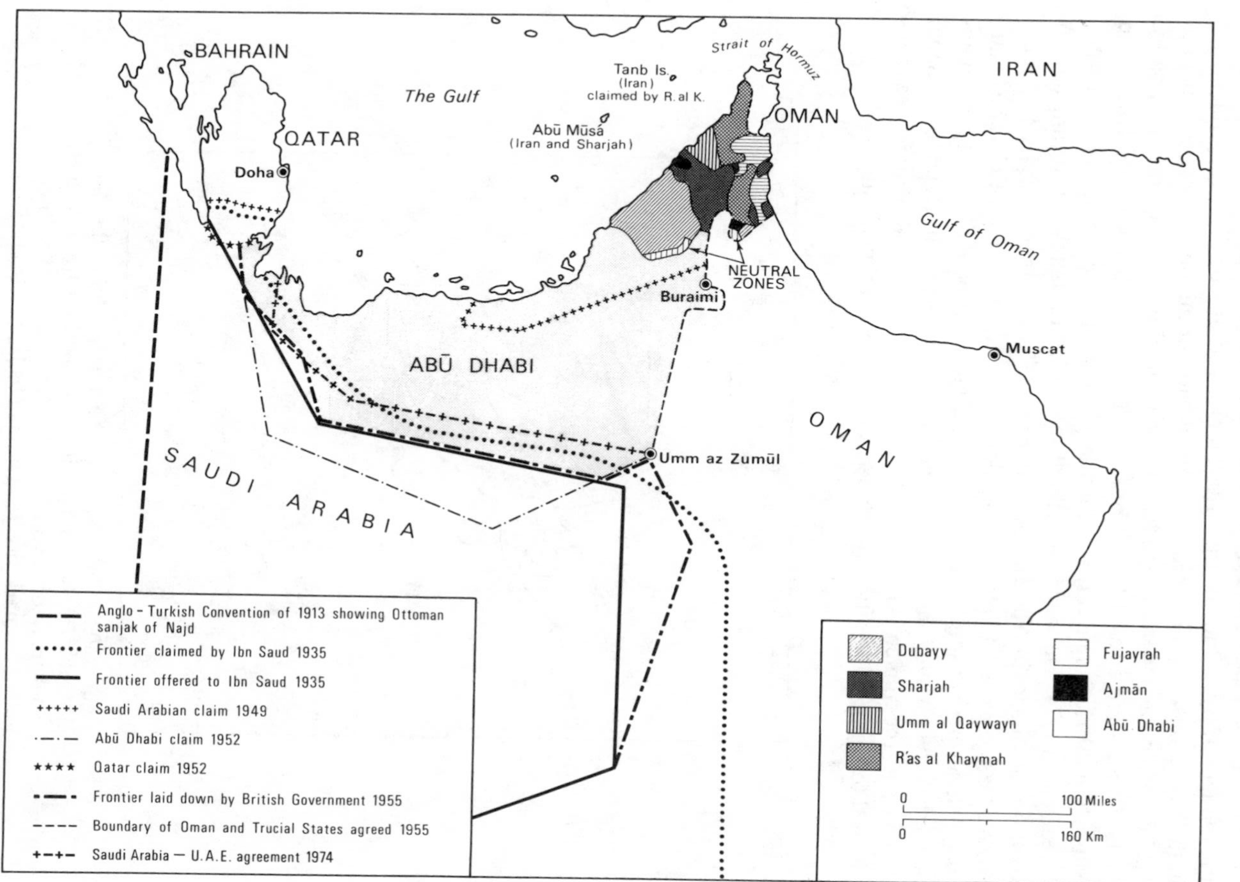

Figure 10.4 The political map of eastern Arabia

The difficulty of international boundary drawing is well illustrated by the uncertainty over delimitations in eastern and southern Saudi Arabia. As recently as the 1920s vast tracts of Arabia south and east of a line from Aden to Qatar were still thought of by Britain as 'the British hinterland of Aden'.[14] Territorial claims in the region have usually been maintained without military action, although in 1952 a small detatchment of Saudis occupied one of the Buraīmi group of oases. In 1955 they were ejected by a British-led force on behalf of Abū Dhabi and Muscat, and Britain laid down a boundary which became the basis for a formal agreement in 1974. This agreement finally ended Saudi claims to Buraīmi and left six Buraīmi villages in Abū Dhabi, and three in Oman. As part of the deal Saudi Arabia was conceded a valuable corridor to the sea south-east of Qatar. (Figure 10.5).[15] Elsewhere in southern Arabia long stretches of boundary still remain unaligned. Between 1969 and 1972 armed clashes occurred involving both Saudi Arabia and the Yemen AR with PDR Yemen whose radical Marxist regime is pledged to support liberation movements in the Arabian peninsula.

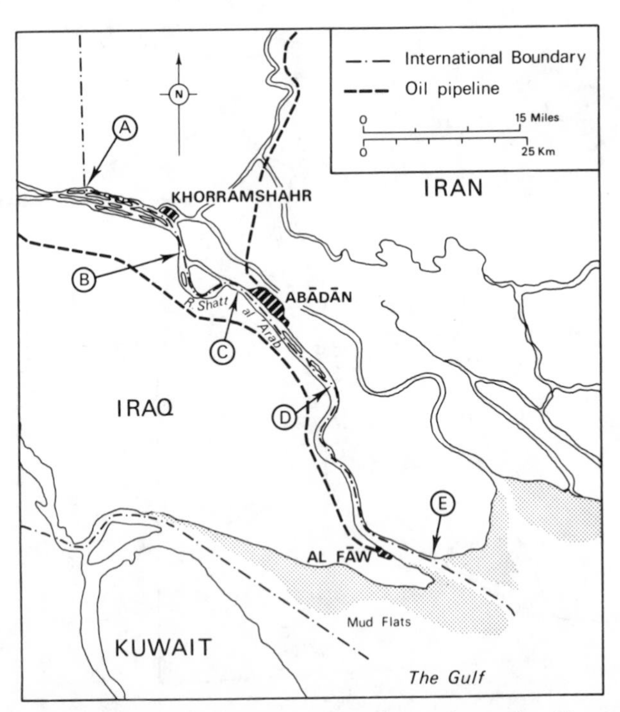

Figure 10.5 The Iran/Iraq boundary along the Shaṭṭ al'Arab. The boundary follows mid-stream from A to B, and from B to the sea. From B to C and D to E it follows the low water mark along the Iranian bank, and from C to D it follows the *thalweg* to give anchorage to Iran opposite Abadan (modified from C.A. Fisher, ed., *Essays in Political Geography*, Associated Book Publishers Ltd, 1968)

Functional disputes are well illustrated by problems created when international boundaries cross river basins. The best-documented of these is over the allocation of the waters of the River Jordan between Lebanon, Syria, Jordan and Israel. Although proposals for a regional water agreement have been made, it has never been implemented so that Israel and Jordan (see Chapters 15 and 16) have gone ahead with large schemes to draw water from the Jordan system. Israel's use of Jordan waters has been fiercely opposed by the Arabs, especially Syria, and as a result Israel may be very reluctant to hand back the headwaters of tributaries which rise in the occupied Golan Heights region.[16] An equally dangerous and far larger-scale river basin dispute remains unresolved between Turkey, Syria, and Iraq over the waters of the Euphrates. Iraq in particular is heavily dependent upon Euphrates water for irrigation and other purposes and argues that Syria and Turkey are taking more than their fair share. Syria's massive Euphrates dam was completed in 1973, and has apparently reduced the downstream flow of water into Iraq as irrigation is expanded. The new Atatürk dam in Turkey is likely to take even greater quantities of water from the Euphrates system, and both Syria and Iraq are extremely concerned about its effects on their own agricultural development plans. The Euphrates waters could become a major cause of international tension in the future.[17] The Nile waters could also create problems between Egypt, Sudan and possible Ethiopia in future (see Chapter 2). Although the allocation of Nile waters agreed by Egypt and Sudan in 1959 work well at present, rising demands for water in both states could result in serious quarrels before the end of the century.[18]

Talk of boundary disputes should not obscure the fact that many states have made considerable progress towards the elimination of boundary problems with their neighbours in recent decades. Several formal agreements have been followed up by careful surveying and demarcation of the boundary line on the ground, although in certain areas such as sandy deserts this may be impossible. By 1984 therefore more than half the region's international land boundaries had been agreed and at least partially demarcated (Table 10.1). It should be noted that great stretches of these boundaries traverse empty or sparsely populated regions,

TABLE 10.1
Land boundary status (1984)

Status	Number
Un-allocated	3
Allocated and delimited	13
Allocated, delimited, and demarcated	23
Allocated, delimited and partially demarcated	5
Disputed or partially disputed	5
Armistice lines	4
Total	53

Source: A.D. Drysdale and G.H. Blake, *The Middle East and North Africa: a Political Geography*, OUP New York, 1985, p 78.

where tribal groups are not divided as in tropical Africa. One notable exception is the Kurds who are divided between five states. The effect of boundaries in restricting the traditional movement of nomads has also tended to be overstressed. Apart from the USSR/Iran and Israel/Jordan boundaries migrations have not been greatly affected.

10.4 Maritime boundaries

While most of the international land boundaries of the Middle East and North Africa were drawn between about 1880 and 1930, offshore boundary delimitations only began in the 1960s and is still far from complete (Figure 10.6). The seas which fringe the Middle East and North Africa represent one of the world's highest concentrations of potential number of boundaries associated with Middle East and North African states is 62. Without the Maghreb States, the total is 54, though the figures do not include the complex federal boundaries of the United Arab Emirates. So far, approximately one fifth of the potential maritime boundaries associated with Middle Eastern and North African states have been formally agreed (Table 10.2). Progress towards offshore boundary delimitation has thus been slow, and the completion of the offshore map may take several decades. Among the chief difficulties are political differences between coastal states (in some cases over the location of their land boundary), disputed islands, and legal and technical arguments. In a number of cases no doubt, the parties are in no hurry to enter into potentially tricky and sometimes costly negotiations over offshore boundaries. On the other hand, the presence of offshore oil and gas fields may provide a powerful incentive to reach agreement, so that exploitation of the resources can proceed.

State jurisdiction is more complex at sea than on land because different types of state control are recognized – internal waters, territorial seas, contiguous zones, and continental shelves. In addition, many states claim exclusive fishing or exclusive economic zones. The highly important United Nations Convention on the Law of the Sea of 1982 entitles coastal states to 12 nautical miles of territorial sea and to a 200 nautical mile exclusive economic zone (or EEZ) in which all the resources of the seabed and waterbody are the property of the state. It seems likely that future maritime boundary agreements will delimit EEZs between adjacent and opposite states, whereas in the past, most maritime boundary agreements in the Middle East have divided continental shelf areas. Continental shelf rights give the coastal state exclusive control of the seabed resources of their own continental shelves, notably oil and natural gas. The new EEZ rights on the other hand also extend to living resources, and bestow certain other rights such as control of scientific research, and preservation of the marine environment.

The earliest maritime boundary agreements were in the Gulf where offshore deposits of oil and gas, in shallow waters were an attraction for the eight coastal states. The first agreement was a median line delimiting waters between Bahrain and Saudi Arabia in 1958. A controversial area of seabed (the Fasht Bu Safaa

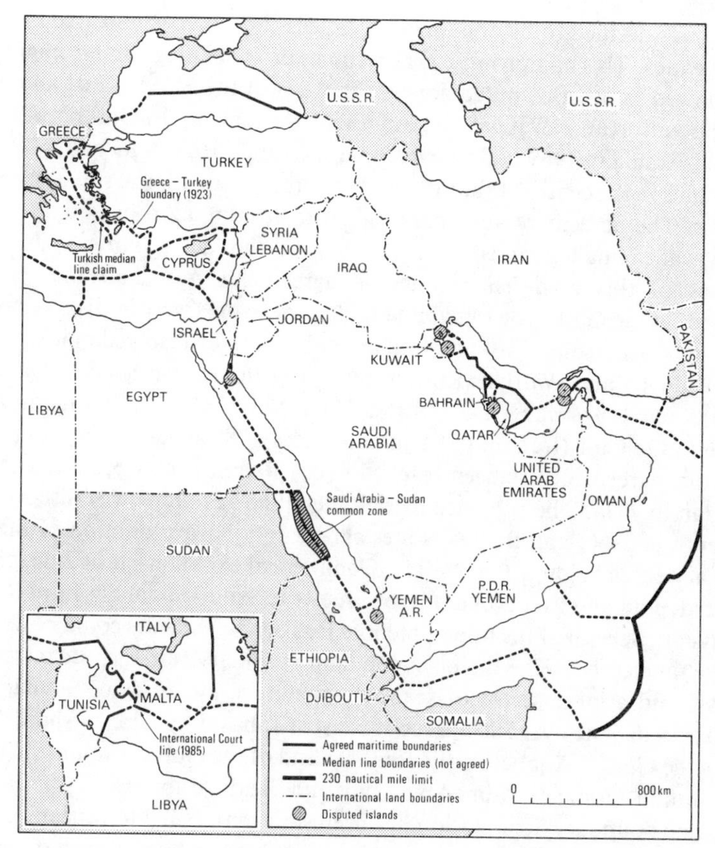

Figure 10.6 Maritime boundaries of the Middle East in 1986

TABLE 10.2
Maritime boundaries of the Middle East and North Africa

	Agreed	Not agreed	Total
Mediterranean Sea	3	20	23
Black Sea	1	1	2
The Gulf	7	7	14
Indian Ocean	0	5	5
Red Sea	1	14	15
Atlantic Ocean	0	3	3
Total	12	50	62

Source: G. H. Blake, 'Maritime boundaries of the Middle East and North Africa', in *Boundaries and State Territory in the Middle East and North Africa*, (Eds G. H. Blake and R. N. Schofield), Menas Press, 1987.

hexagon) was deemed to belong to Saudi Arabia, but the parties agreed to share its oil revenues. This compromise arrangement or something similar might have been followed in other disputed areas of the seas of the Middle East and North Africa, but unfortunately it has not. Another agreement in the Gulf between Iran and Saudi Arabia in 1969 had to overcome several difficulties including two disputed islands in the centre of the Gulf and the presence of several small islands off the coasts. The agreement gave Farsi island to Iran and Arabi island to Saudi Arabia, while small islands were ignored in the delimitation.[19] Judicious diplomacy of this kind has resulted in agreements over half the potential boundaries in the Gulf. The remaining boundaries are largely in dispute because of contested ownership of islands, and are unlikely to be resolved quickly. These include Bubiyan and Warbah (Kuwait and Iraq), the Huwar islands (Qatar and Bahrain), and the strategically located Abū Mūsa and Tunb islands at the entrance to the Gulf (Iran and Sharjah and Ra's al Khaymah).

Only one agreement has been reached in the Red Sea, between Saudi Arabia and Sudan in 1974. The presence of rich metalliferous muds was discovered at great depth in 1966, and the two states agreed to joint exploitation of these resources at depths below 1,000 metres. Commercial exploitation of zinc, copper, silver, lead, gold and iron is about to begin by the Saudi–Sudanese Joint Red Sea Commission. Several of the other potential Red Sea boundaries could be disputed when the time comes. The Egypt-Sudan land boundary is in dispute on the Red Sea coast and a large sector of seabed is therefore in question. Similarly, the Egypt–Israel dispute over 700 metres of coast at Taba is impeding offshore delimitation in the Gulf of Aqaba. At the southern end of the Gulf of Aqaba the islands of Sanafir and Tiran are claimed by Egypt and Saudi Arabia.[20]

Maritime boundary delimitation has been slow in the Mediterranean Sea for a number of reasons.[21] The prevalence of deep water has discouraged oil and gas exploration, and the presence of many islands and physically complex coasts have deterred boundary negotiators. The frequency of political disputes between 18 coastal states has also been a major obstacle, exacerbated by unresolved territorial questions such as Turkish-occupied northern Cyprus, the Gaza Strip, and Spain's possessions on the coast of Morocco. Thus Tunisia is the only Middle Eastern or North African state to have formally reached a boundary agreement in the Mediterranean. The Italy–Tunisian agreement of 1971 was one of a series negotiated by the Italians with their neighbours (Yugoslavia 1970, Greece 1980, and Spain 1978). With the judgements of the International Court of Justice in cases between Libya and Tunisia (1982) and Libya and Malta (1985), these states can also be expected to conclude formal boundary agreements in the near future. By far the most dangerous and the best-documented maritime boundary dispute in the Mediterranean world is between Greece and Turkey in the Aegean Sea.[22] Greece maintains, with some justification, that her ownership of almost all the islands in the Aegean gives her the right to the resources of the continental shelf around the islands. Turkey argues that the Aegean presents special geographical circumstances and that a median-line division of the seabed between mainlands (without prejudicing Greek sovereignty over the islands) would be more just and logical. Unfortunately the International Court has declined jurisdiction in this difficult case, because of the attitude of the parties. The Aegean dispute highlights the problems of maritime boundary delimitation of the region resulting from a combination of historic legacy and complex geography. As a result, the seas adjacent to the Middle East and North Africa could be left behind the rest of the world in establishing an agreed framework for offshore jurisdiction.

10.5 The political map: some implications

The contemporary political map of the Middle East comprises both macro and micro states. It includes four of the world's 16 largest states, while Israel, Kuwait, Lebanon, Qatar, Bahrain and six of the Untied Arab Emirates are among the world's 35 smallest states (Table 10.3). No state is landlocked, although the length of coastline (Table 10.3) and access to the sea vary greatly from the favourable location of Morocco, Egypt and Israel, to the geographically disadvantaged states of Jordan and Iraq. In geopolitical terms, the maritime orientation of all of the states in the region is of great significance, since land communications tend to converge on coastal locations in concordance with international boundaries, and not transverse to them. Political fragmentation and the interpenetration of the region by seas have given great strategic significance to four waterways; the

TABLE 10.3
State size and lengths of coastline

Rank		Area (sq km)[1]	Coastline (km)[2]
1	Algeria	2,381,741	1,038
2	Saudi Arabia	2,152,690 (approx)	2,474
3	Libya	1,759,540	1,685
4	Iran	1,648,000	1,833
5	Egypt	1,001,499	2,420
6	Turkey	780,576	3,558
7	Morocco	446,500	657
8	Iraq	437,924	19
9	PDR Yemen	287,683 (approx)	1,211
10	Oman	212,400 (approx)	1,861
11	Yemen Arab Republic	195,000 (approx)	452
12	Syria	185,180	152
13	Tunisia	164,150	1,027
14	Jordan	97,740	27
15	United Arab Emirates	78,000 (approx)	700 (approx)
16	Israel (pre-1967)	20,700	230
17	Kuwait	19,500	250
18	Lebanon	10,400	194
19	Qatar	10,360	378
20	Bahrain	598	126

Sources: 1. *The Middle East and North Africa 1971–72*, Europa, London, 1972, pp 117–757;
2. Geographic Bulletin No 3, *Sovereignty of the Sea*, US Department of State, Washington, 1969, pp 19–22.

Turkish Straits (minimum width, 1.8 km), the Red Sea (width of Bab al Mandab 26 km), the Gulf (width of Strait of Hormuz 39 km) and the Strait of Gibraltar (minimum width 15 km). The Red Sea also includes two sensitive waterways in the Gulf of Suez and the Gulf of Aqaba.

A further consequence of political compartmentalization is that it multiplies the opportunities for rival external powers to gain influence in the region. S. B. Cohen recognizes two world regions 'occupied by a number of conflicting states and caught between the conflicting interests of the Great Powers'[23] – Southeast Asia and the Middle East. These he calls 'shatterbelts'. His view of the strategic importance of the region has been shared by other political geographers, including N. J. Spykman and D. W. Meinig, for whom it constituted a critical sector of a 'Rimland' region where Soviet expansion could occur and must be contested.[24] This view is clearly oversimplified in an age of nuclear missiles and powerful naval forces, but it helps explain the quest for military bases in the region. The Soviet Union's first inroads into the Arab world were noted in 1952. Subsequently, it has acquired military interests from time to time in a number of states including notably Egypt, Syria, Libya, Iraq, and PDR Yemen. Britain retains bases in Cyprus and has military power in the region. The United States similarly has important client states in the Middle East to which arms and military assistance are available, especially Israel, Saudi Arabia, and Egypt. Turkey is a member of NATO and there are important United States bases located there. All this, together with the activities of Soviet and United States fleets in the Mediterranean is evidence of the strategic importance of the region on the southern flank of the Soviet Union, lying outside a 'ganglion' of land, sea, and air routes which are among the world's densest concentrations.[25] In addition, Southwest Asia possesses enormous reserves of oil and supplies a high proportion of western Europe's and Japan's energy requirements (Chapter 9). While North African reserves are smaller, and gas supplies from west of Suez are of particular value in Europe because of lower transportation costs.

The most remarkable state in the contemporary political mosaic is undoubtedly the United Arab Emirates, formed in 1971–72 out of seven Trucial Sheikdoms

TABLE 10.4
The United Arab Emirates

		1981 Population (estimated)	Area $(km)^2$ (approx)
1	Abū Dhabi	449,000	67,350
2	Ajmān	36,100	260
3	Dubayy	278,000	3,900
4	Fujayrah	32,300	1,170
5	Ra's al Khaymah	73,700	1,790
6	Umm al Qaywayn	12,300	780
7	Sharjah	159,000	2,600

Source: Middle East Review 1984, World of Information, Saffron Walden, 1984, 283–99.

with whom Britain had treaties of defence and friendship. The long-term viability of this federation seemed doubtful in a region where the tribes never achieved unity or much political stability.[26] As in Southwest Asia as a whole there are marked discontinuities of population and great differences in population size (Table 10.4) and resources. Abū Dhabi and Dubayy are the chief oil producers. There is modest production in Sharjah; the other states are non-oil producers. With a combined population of just over 1.6 million the United Arab Emirates has one of the smallest populations in the region. Moreover, the territories of individual member states are fragmented, and internal boundaries have rarely been agreed; yet somehow the federation has survived and achieved political credibility.

10.6 Future changes

It seems unlikely that the political map will change as much in the next 50 years as it did in the past 50 years. The most significant feature will be the partitioning of the adjacent seas between coastal states, and the gradual acceptance of an offshore area as an integral part of the state domain. On land, minor territorial adjustments will undoubtedly occur and boundaries will be more precisely delimited. The state system seems destined to persist, while the dream of a united Arab world will remain as elusive as ever. Five obstacles to Arab unity noted by Charles Issawi remain as valid as ever.[27]

First the great length of the Arab world, unmatched by corresponding depth; secondly, the discontinuity of settled population; thirdly, inadequate communications, exacerbated by the fact that most roads and railways were built by foreigners to facilitate overseas trade rather than link together the constituent parts of the Arab world; fourthly, the lack of economic integration both within and between countries despite various attempts to achieve it. Trade has always been limited because of the shortage of surplus produce for exchange. Several countries, notably in the Maghreb, would prefer closer association with the EEC. Fifthly, the lack of social integration, for example between town and country and nomad and peasant. A sixth problem at present is the diversity of political regimes within the Arab world. Indeed, it is arguable that the division of the region into nation states cannot be reversed.[28] Nevertheless, centripetal forces are also powerful, and with a common language and cultural heritage, these could eventually lead to unexpected levels of political co-operation. To begin with however, one can expect to see an increasing awareness of the advantages of close technical and economic co-operation between states. Theoretical studies of trade relations have shown that a good basis for a customs union exists between several states in Southwest Asia and Egypt.[29] Significantly, there has been an increasing interest in more overland pipelines in Southwest Asia in recent years, which could bring about further interdependence, while the completion of the improved road and rail links as between Turkey and Iran, and progress towards a North African highway are signs of developing integration between regions.[30] So far, probably the most successful example of regional cooperation has been the Arab Gulf Cooperation Council, an association of Gulf states which is partly defensive and partly economic.

One region where the political map may be expected to change is in the Arab territories occupied by Israel in June 1967 (Figure 10.7). Israeli forces' withdrawal from the vast Sinai peninsula (61,198 sq km) as part of a permanent peace treaty with Egypt was completed in 1982 entailing considerable economic and strategic sacrifices for Israel. Withdrawal from the other occupied territories of Golan (1,250 sq km) and the Jordan West Bank (5,900 sq km) is less likely to be achieved peacefully because of the crucial strategic importance of Golan for the defence of northern Israel, and the strong historic and emotional Israeli attachment to the West Bank (Chapter 16). Some territorial concessions by Israel are not inconceivable however in return for the prospect of lasting peace, and this might include some form of national homeland for the Palestinian people. The troubled state of Lebanon is already *de facto* partioned between a number of rival factions, and some formalization of this arrangement could occur in future.[31] Lebanon was once widely regarded as a progressive and stable Arab state with many geographical assets, which may illustrate the folly of political prediction in the Middle East.

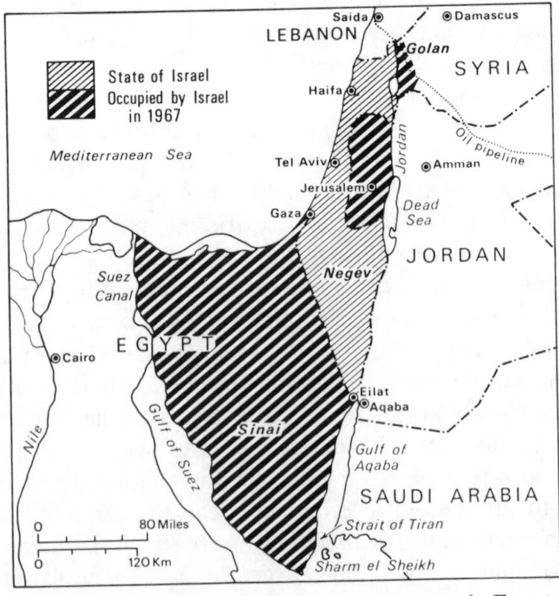

Figure 10.7 Areas of Syria, Jordan and Egypt occupied by Israel in June 1967

References

1. A. Melamid, 'The Shaṭṭ al'Arab boundary dispute, *Middle East Journal*, **22**, 350–7 (1968).
2. G. E. Kirk, *A Short History of the Middle East*, University Paperbacks, London, 1964, Chapter 5.
3. S. B. Jones, 'A unified field theory of political geography', *Ann. Ass. Amer. Geogr.*, **44.** 111–23 (1954).
4. W. Laqueur (Ed), *The Israel–Arab Reader*, Weidenfield, London, 1969, 12–18.
5. J. H. G. Lebon, 'South-West Asia and Egypt'. In *The Changing Map of Asia*, 5th edn, (Eds W. G. East, O. H. K. Spate, and C. A. Fisher), Methuen, London, 1971, 120.
6. J. S. Haupert, 'Political geography of the Israel–Syrian boundary dispute 1949–67', *Prof. Geogr.*, **21,** 163–71 (1969).
7. (a) M. Brawer, 'The geographical background of the Jordan water dispute'. In *Essays in Political Geography*, (Ed C. A. Fisher), Methuen, London 1968, 225–242.
   (b) C. G. Smith, 'The disputed waters of the Jordan', *Trans. Inst. Brit Geogr.*, **40,** 111–28 (1966).
8. Document: 'Agreement for the delimitation of boundaries between Jordan and Saudi Arabia', *Middle East Journal*, **22**, 346–48 (1968).
9. M. P. Drury, 'The political geography of Cyprus'. In *Change and Development in the Middle East*, (Eds J. I. Clarke and H. Bowen-Jones), Methuen, London, 1981, 289–304.
10. M. Muller, 'Frontiers: an imported concept: an historical review of the creation and consequences of Libya's frontiers'. In *Libya Since Independence* (Ed J. A. Allan), Croom-Helm, London, 1982 165–180.
11. E. H. G. Joffe, 'Frontiers in North Africa'. In *Boundaries and State Territory in the Middle East and North Africa* (Eds G. H. Blake and R. N. Schofield), Menas Press, Cambridge, 1987, pp 25–43.
12. A. D. Drysdale and G. H. Blake, *The Middle East and North Africa: a Political Geography*, Oxford University Press, New York, 1985, 75–104.
13. R. N. Schofield, *The Evolution of the Shatt al-Arab Boundary Dispute*, Menas Press, Cambridge, 1986.
14. Anon. 'The boundries of the Nejd: a note on the special conditions.' *Geogrl. Rev.*, **17,** 128–134 (1927).
15. The details of this agreement have not been published and Figure 10.4 is speculative in respect of the corridor.
16. A. D. Drysdale and G. H. Blake (1985) *The Middle East and North Africa: a Political Geography*, Oxford University Press, New York, pp. 97–99.
17. P. Beaumont, 'The Euphrates river – an international problem of water resource development,' *Environmental Conservation*, **5 (1),** 1978, 35–43.
18. J. Waterbury, *Hydropolitics of the Nile Valley*, Syracuse University Press, Syracuse, 1979, 239.
19. A. A. El-Hakim, *The Middle Eastern States and the Law of the Sea*, Manchester University Press, 1979.
20. A. A. El-Hakim, *The Middle Eastern States and the Law of the Sea*, Manchester University Press, 1979, p. 137.
21. F. Bastianelli, 'Boundary delimitation in the Mediterranean Sea', *Marine Policy Reports*, **5 (4),** University of Delaware College of Marine Studies, (1983).
22. C. Rozakis, *The Greek-Turkish Dispute over the Aegean continental shelf*, Law of the Sea Institute, Kingston, Rhode Island, 1975.
23. S. B. Cohen, *Geography and Politics in a World Divided*, Oxford University Press, New York, 2nd edn 1973, p 251.

24. N. J. G. Pounds, *Political Geography*, McGraw-Hill, New York, 2nd edn, 1972, pp. 429–434.
25. J. H. G. Lebon, 'South-West Asia and Egypt'. In *The Changing Map of Asia* (Eds W. G, East, O. H. K. Spate and C. A. Fisher), Methuen, London, 5th edn 1971, 53–126.
26. D. Hawley, *The Trucial States*, George Allen and Unwin, London, 1970.
27. C. Issawi, Political disunity in the Arab World, in *Readings in Arab Middle East Societies and Cultures* (Eds A. M. Lutfiyya and C. W. Churchill), Mouton, The Hague, 1970, 278–84.
28. A. H. Hourani, 'Race, religion and nation-state in the Near East', in *Arab Middle East Societies and Cultures*, (Eds A. M. Lutfiyya and C. W. Churchill), Mouton, The Hague, 1970, 1–19.
29. J. E. McConnell, 'The Middle East: competitive or complementary?' *Tijdschr. econ. Geogr.*, **2**, 82–93 (1962).
30. G. H. Blake, J. C. Dewdney, J. K. M. Mitchell, *Cambridge Atlas of the Middle East and North Africa*, Cambridge University Press, 1987, Maps 42–45.
31. N. Kliot, 'Lebanon: a geography of hostages', *Political Geog. Quarterly*, **5 (3)**, 1986, 199–220.

CHAPTER 11

Tradition and Change in Arabia

11.1 Introduction

Arabia may be regarded as the heartland of the region studied in this book. Not only does it merge northwards into the upland at the eastern end of the Mediterranean Sea, the foreland of the high Armenian mountains and the lowlands of the Tigris and Euphrates valleys, which form the modern states of Iraq, Syria and Jordan (Chapters 12, 13 and 15) and lead on to Lebanon, Israel, Iran, Turkey and Egypt (Chapters 14, 16 and 17 to 19), but Arabia was also the birthplace of two forces which have produced a significant degree of cultural integration in the region, Islam and Arabic (Introduction). The present chapter focuses upon the southern part of the peninsula – a region which may be conveniently defined in terms of its present political units. The Kingdom of Saudi Arabia is the largest state and borders upon all the others. In the southwest lie the Arab Republic of Yemen and its neighbour, the People's Democratic Republic of Yemen. The southeast corner is occupied by the Sultanate of Oman. On the east is the United Arab Amirates (the Trucial States before 1971) of Abū Dhabi, Ajmān, Al Fujayrah, Dubayy, Ra's al Khaymah, Shārjah and Umm al Qay-wayn. North of these lie the peninsular state of Qatar, the small archipelago of Bahrain, and Kuwait.

Within this immense region, a variety of economics evolved over the centuries in response partly to differences in physical conditions and partly to the degree of isolation from the outside world. An attempt will be made to outline the main features of these 'traditional' economies as they existed around the beginning of the twentieth century, though they are often inadequately documented, despite the vast literature generated by the fascination long exerted by Arabia on the imaginations of north Europeans. The second half of the chapter will be devoted to an account of the modifications and changes brought about during this century. Petroleum exploitation is usually singled out as the sole agent of change. There is no doubt of its importance, but only comparatively recently has petroleum radically altered the patterns of life followed by the majority of people in the peninsula, while the direct impact of the oil industry has been localized. Other forces of change have been at work in Arabia, some of them indigenous, but others emanating from outside, and it is these – perhaps arguably – which have frequently had the most profound effects.

11.2 'Traditional' economies

The physical environments of Arabia are generally harsh and uncompromising. Daytime temperatures are high, but there is often a severe diurnal change. Skies are generally clear so that the sun beats down relentlessly, while heat and glare are flung back from bare rock and sand surfaces. The light itself is pure and searing. Scorching blasts of air, dust devils and mirages are frequent during the day. Nights are cool, often cold, and the intensity of starlight and moonshine make them particularly attractive, while the relative comfort turned them into a preferred time for travel. Precipitation is low practically everywhere over the peninsula, and not only extremely variable from one year to the next, but also apparently subject to cycles of drought which might last from 10 to 30 years. However, even a slight fall of rain is sufficient to send fierce and destructive torrents (*seils*) down the wadis and cast a green flush over the desert.

Despite these similarities across the region, there are a number of distinct climates. The coasts of the Red Sea and the Gulf have a broadly tropical climate characterized by stiflingly high humidity (60 to 80 per cent) resulting from onshore winds and precipitation as low as 8 mm per annum. More temperate climates are found in the mountains which form the western, southwestern and southeastern sections of the peninsula. Frosts and snow are frequent in the western mountains, while precipitation ranges from about 500 mm to over 750 mm. Yemen receives most of its rain between July and September, but dew and mist are significant throughout the year. Precipitation is lower and has a late winter maximum over the southern mountains. It declines in amount from west to east, though an accident of exposure raises totals again in Dhufar. Interior Arabia has an mean annual rainfall of between 25 and 150 mm, with a winter maximum. Winter nights here are bitterly cold, due to elevation, low humidity, rapid diurnal cooling and exposure to icy blasts from further north.

Lack of water, steep mountain slopes, lava fields, sand seas and salt marsh (Figure 11.1) have made large parts of Arabia either totally unsuitable for human use or capable of exploitation only at infrequent intervals of time. The scale varies from the local to the regional. Local tracts of uninhabitable country are represented by the lava fields on the inner side of the western mountains, the steep slopes separating the Yemeni plateaus or the Oman valleys, and the salt marshes which impinge upon the east coast. The Rub'al Khālī (The Abode of Emptiness), on the other hand, is a vast, largely uninhabited region, the fringes of which are penetrated only when the rains give an flush of grazing or, mainly in the past, when persecution and banditry made its empty spaces attractive to certain groups. Such desolate tracts have divided the inhabited areas from one another, and enforced a relative isolation which, even with the development of modern means of communication, is still important. In the not too distant past, links were provided only by caravans of dromedaries, with their capacity to cross almost waterless and sandy country. In certain areas, coastal shipping was important.

The inhabited areas were those where some water was available, in the form of precipitation or surface and subterranean flow, and where physical conditions in

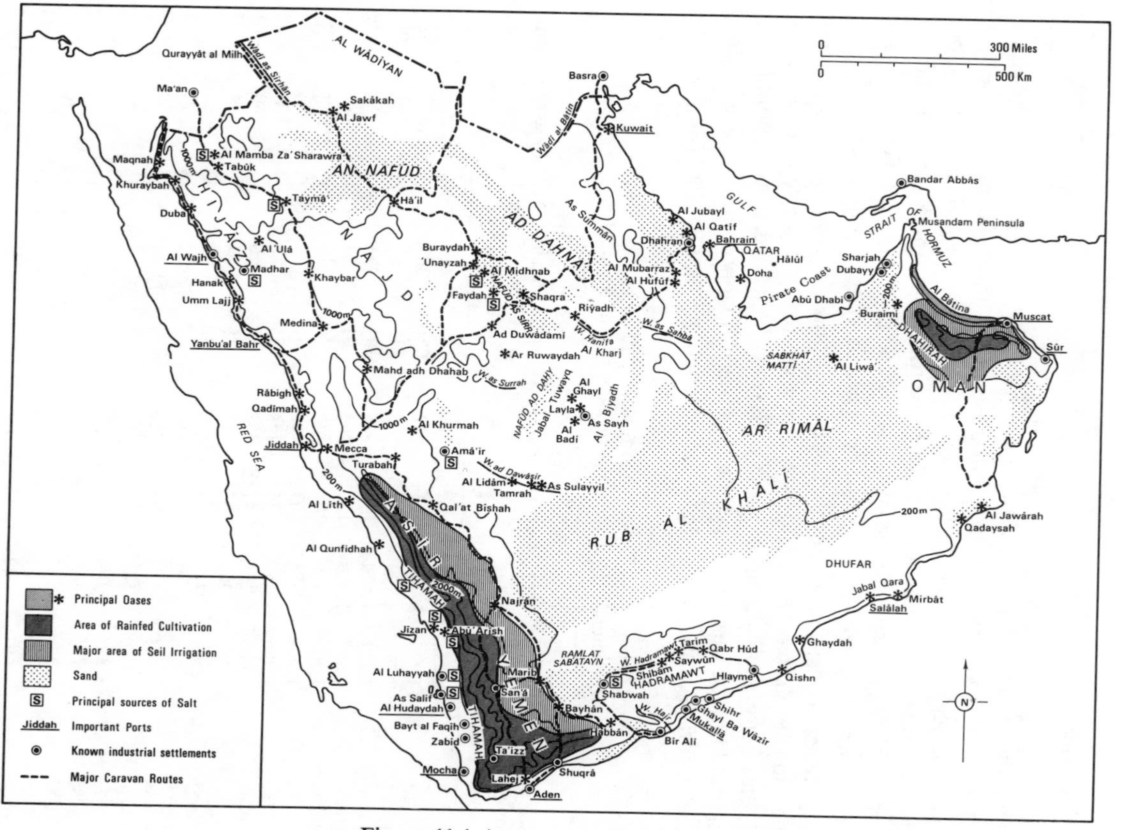

Figure 11.1 Aspects of 'traditional' Arabia

general combined to produce a resource potential in the shape of, for example, periodic grazing, cultivable land or shoals of fish. It is to the 'traditional' economies based on these and other resources that attention is first given, beginning with the most basic, cultivation.

11.2.1 Cultivation

Although the cultivated area of Arabia was probabaly unable to feed the entire population at the beginning of the twentieth century, arable farming has been of basic importance for many centuries. Not only were cultivators in a majority, but by sale, exchange and tribute they also helped to support the pastoral and other economies in the peninsula. Some land was worked by owner occupiers and some was in communal ownership, but most belonged to sharecropping estates. Size of holding varied considerably throughout the peninsula, as figures for Saudi Arabia indicate (Table 11.1), largely in response to the relative abundance of water, the intensity of cultivation which this allows, and the density of population which could thus be carried. Fragmentation was probably commonplace because of the division of property between all male heirs, but its intensity varied regionally.[1]

TABLE 11.1
Distribution of land holdings in Saudi Arabia in about 1965

Size Group (donums)	Total	South	North	Central	Regions Medina	Jiddah and Mecca	At Ta'if	Eastern	Qasim
0– 5	43,444	19,461	6,186	1,715	5,280	827	4,749	4,201	6,186
5–10	17,236	9,254	1,313	932	1,021	309	1,761	2,232	1,313
10–15	7,515	4,170	306	686	255	93	860	798	306
15–20	3,479	1,670	148	408	144	44	320	538	148
20–25	2,866	1,328	122	445	93	24	323	291	122
25–30	1,308	446	68	275	46	16	120	146	68
30–35	1,204	386	63	265	45	25	99	195	53
35–40	994	539	34	134	25	3	42	109	34
40–45	1,089	199	41	331	20	110	110	60	41
45–50	779	373	35	165	19	4	58	84	35
50–55	565	117	24	176	14	3	49	39	24
55–60	319	72	17	93	—	2	28	21	17
60–65	695	218	24	218	14	16	58	69	34
65–70	289	48	19	89	4	1	31	28	19
70–75	332	108	15	106	3	1	18	15	15
75–80	226	50	10	75	6	—	11	29	10
80–85	454	53	11	152	2	9	53	10	11
85–90	199	31	9	64	2	—	24	9	9
90–95	393	42	6	154	9	11	32	33	6
95–100	215	38	9	89	5	—	20	12	9
100+	3,510	303	174	1,448	29	104	310	179	174
Total	87,111	38,806	8,624	8,020	7,036	1,501	9,076	9,098	8,624

Source: Kingdom of Saudi Arabia, Ministry of Finance and National Economy, Central Department of Statistics, *Statistical Yearbook, 1387 AH* (AD 1967–68), p 138.

Permanent rain-fed cultivation was concentrated where average annual precipitation exceeds 240 mm, that is in the mountains of Asīr, the plateaus of Yemen and the higher valleys of Oman (Figure 11.1). Durah, millet, wheat, barley, legumes and fodder were the most extensive crops grown in wadi-floor fields or stone-built terraces, often laboriously maintained with soil collected from elsewhere and fertilized from middens in the settlements. Cash crops were grown mainly in the southwest at the beginning of the twentieth century. Cotton, indigo and later tobacco were important within Arabia, but coffee and qāt (*Catha edulis*) were pre-eminent in Yemen from perhaps the sixteenth century, succeeding to that position long after the collapse of the ancient trade in incense and myrrh from Ḥaḍramawt and Dhufar.[2]

Irrigated cultivation was found throughout Arabia (Figure 11.1), but with particular concentrations in the mountains and along their fringes, as well as in the valleys of the scarp-and-vale country of Najd. A variety of water sources was used, but the greater certainty of water created an altogether more productive form of agriculture than dry farming. Wheat and barley, maize, millet and durah were grown, together with a variety of fruit and vegetables, but the most important crop, at altitudes below about 1300 m, was dates. More than 70 varieties of date are known.[3] The palm trees, of course, provided vital shade to other crops, useful timber, and leaves for thatching, mat making and basketry. They have the great advantages of tolerating moderately saline water, which is all that is available for irrigation in some areas, of providing more food per unit area than any other crop grown in Arabia, and of being susceptible to several preparations for eating. Mature dates, which are about 58 per cent sugar, are easily preserved, so that they were favoured as an item of commerce and preferred by many nomads as an easily packed food.

11.2.2 Other 'traditional' economies

Around the beginning of the twentieth century, food produced by the cultivators helped to support a number of other economic activities in Arabia. These included fishing, pearling and trading, but the best known – and most typically associated with the interior of Arabia – was nomadic pastoralism.[4] Although animals were kept by cultivators, at least for draught purposes, large herds could only be maintained in the arid environments of the peninsula by a nomadic existence which made fullest use of the scattered and uncertain nature of grazing and of the limited supply of drinking water in seasonal pools and permanent wells. The broad patterns of nomadic pastoralism have been described in Chapter 4 and need not be repeated here. Although sheep and goats were kept to produce butter, cheese, wool and hides as well as fat for cooking, dromedary rearing was the most prestigious of a group of activities, including limited horse breeding, which provided the nomads of Arabia with an income to buy food and even tent cloth. Dromedaries were reared not so much for their own products, though this could be important at times, as for sale to local sedentary communities or, through the

The city of Mecca in the early nineteenth century (Radio Times Hulton Picture Library)

great camel fairs, to Iraq, Syria and Egypt where they were used for all manner of draught purposes. Circular and elliptical movements (Chapter 4) were characteristic within tribal territories in northern Arabia and the adjacent parts of Iraq, Syria and Jordan, while constricted-oscillatory movements seem to have been more common in what is now PDR Yemen.

The variety of fish found around the coasts of Arabia supported scattered fishing communities, particularly along the Gulf, where ample fish food was provided through a mixing of local and oceanic waters as well as the discharge from the Shaṭṭ al'Arab.[5] Pearling was also a major activity along the Arabian shore of the Gulf, particularly in water up to 20 fathoms deep in a zone stretching from Ra's al Khaymah to just south of Kuwait. Bahrain was the main centre for this activity.[6] As well as providing bases for fishing and pearling vessels, the coves and creeks of Arabia were also the homes of the trading ships. There was always some overseas trade wherever the more fertile parts of the interior impinged directly on the coast or lay close behind it. Trade was facilitated in the Gulf and the Red Sea by the almost parallel nature of the coasts and the relatively short distances between them. But the narrow seas led into the Indian Ocean, which associated Arabia, Egypt, Persia and Iraq with India and East Africa in a single trading system which was worked by various types of lateen rigged ships[7] and powered by the alternation of the monsoons. By the early twentieth century, the African trade was probably the most important, and local commodities like dates, dried fish, salt and coffee were traded, while return cargoes included cloves and ivory but above all mangrove poles which were widely used on the largely treeless coasts of Arabia.[8] Where the ships went, there also the pirate lurked and the migrant travelled. Short-term migration to East Africa or the East Indies was marked in some parts of Arabia in the early twentieth century,[9] but piracy had been suppressed by the British, and communities along the western approaches to the Straits of Hormuz in the lower Gulf were no longer able to exploit the natural advantages of a major trade route passing near the hazards of reefs and shoals along a lee shore.[10]

While cultivation supported all these activities to some extent, it also provided raw materials for local craft industries and some of the commodities for internal, as well as external trade. Pottery making was limited by the lack of suitable clays, but salt production, metal working and textile manufacture were widespread, but with sufficient local specialisms, like the *beedis* (blanket-like coats) of Ta'izz and the shawls of Zabīd or the indigo dyeing of al Hudaydah or Mukallā (Figure 11.1), to generate a degree of trade above that in local agricultural produce, imported cereals and exotic luxuries such as tea, coffee (often from Brazil or the East Indies), sugar and tobacco. Internal trade was more important than is often appreciated. Weekly markets and annual fairs provided the basic mechanisms, though simple shops were found in the larger centres and peddlars moved amongst the smaller settlements and nomadic camps. The caravan, generally composed of dromedaries, each capable of carrying up to 180 kg, was the main form of transport. In Yemen and Ḥaḍramawt, caravans were organized by nomads and, changing animals at each stage, followed major routes from the coast to the interior along wadi systems, and then along the interior of the Yemeni or Omani mountains towards the north. Caravans were organized by urban-based merchants in northern Arabia, and draught animals were collected at the various urban termini for use along the entire length of the route. They crossed the territories of the various nomadic tribes safely only when protection money was paid and local escorts hired. The principal routes in the centre and north coincided with the major pilgrim roads throughout Islamic times, and followed lines of wells along the main wadis. They linked the coasts of the Red Sea and the Gulf, and tied the Najd district to Iraq and Syria, but the road system as a whole focused upon Mecca, the pilgrims' goal.

At the beginning of the twentieth century, local pilgrimages to the tombs of saints and prophets were widespread in south Arabia, which was largely unaffected by the Wahhābi 'reformation' of the late eighteenth and early nineteenth centuries or its subsequent revival. Much more important in the whole peninsula, however, was the Great Pilgrimage (*Hadj*) to the Holy Places of Mecca and its vicinity.[11] Mecca is set in a particularly sterile part of Arabia, and for centuries it was virtually dependent upon the Pilgrimage for its livelihood. The other caravan towns also benefited, but it was clear by the early twentieth century that competition from steamships on the sea routes had reduced the size of the Pilgrimage caravans and created something of a crisis for the interior of Arabia.[12] Nevertheless, considerable revenues were collected on goods imported to Jiddah and Mecca for the benefit of pilgrims, and in 1930, before the discovery of oil in what was then Saudi Arabia, the Pilgrimage was the major source of income to Abdul Aziz Ibn Saud. About this time, though, change was already apparent in Arabia, and change is the theme for the second part of the chapter.

11.3 Modification and change

A generalization is often made to the effect that the socio-economic patterns of Arabia changed little between the mission of the Prophet in the seventh century and the discovery of oil in the twentieth. Socio-economic patterns, however, have never been completely fossilized, but have often been much more dynamic than is generally appreciated. This chapter concentrates upon recent changes which appear likely to have lasting and far-reaching consequences. The exploitation of oil has undoubtedly been important in shaping the socio-economic patterns of Arabia in recent years, largely through the vast incomes which it has provided to individual states. Change, however, began before the first oil discoveries were made, stimulated partly by local vision and pressing need, and partly by changes taking place beyond Arabia.

11.3.1 Cultivation

The age-old system of agricultural life and the use of 'traditional' techniques continues in many parts of Arabia today. Change has largely taken the form of an expansion in the cultivated area, almost exclusively by the use of irrigation, and a diversification of cropping patterns, though both have been accompanied by a degree of mechanization. Development started first in Saudi Arabia, but in the last twenty years it has spread throughout the peninsula.

Change in Saudi Arabia was initiated by the father of the state, Abdul Aziz Ibn Saud, himself. The death in 1908 of his great rival, Abdul Aziz Ibn Rashid, allowed him to consolidate his control over the Najd. Subsequent expansion involved in the formation of a confraternity of warriors – former bedouin – who became known as *Ikhwān* (Brethren), and were bound together by their devotion to the Wahhābi concept of a purified Islam.[13] The *Ikhwān* were established in permanent military colonies (*hijras*), containing between 2,000 and 10,000 inhabitants, grouped around a water source used to irrigate cultivated land. Beginning in 1912, with the foundation of al Artāwīyah, over two hundred such settlements were established before 1928. Although the *Ikhwān* rebelled against Ibn Saud in 1929 and were destroyed as a fighting force, their settlements were important in reviving settled ways of life and extending cultivation. Agricultural development was taken a stage further in 1937 when Ibn Saud's finance minister, Sheikh Abdullah Suleiman, installed a number of pumps, an Iraqi farmer and a Palestinian vegetable grower in the oasis of al Kharj, some 88 km southeast of Riyadh.[14] From small beginnings, the project blossomed into a large-scale experimental farm with the assistance of ARAMCO and successive American agricultural missions. The area has remained in the forefront of agricultural innovation ever since, with dairy farming the most recent development (1980).

The original Al Kharj project was followed in the late 1940s by the establishment of experimental farms in different parts of the country, and an attack was launched on the problems of Saudi Arabia's largest oasis, al Hasa. Al Hasa had been suffering from severe economic depression. Population pressure had become extreme and was increasing, while the cultivated area was continually being diminished by drifting sand and salinity. Date palms, the main crop, were low yielding, partly because of their age, and partly because of a high water table. Prices for dates had slumped as a result of Ibn Saud's ban on the export of dates to the Gulf coast, in order to cheapen this staple for the nomads, and they remained depressed. Water, though abundant, was inefficiently used. Attempts to improve the situation began in 1949. Crop diversification was introduced, improved plant strains were employed and the use of water improved. Pumps allowed the extension of irrigation, while the construction of the Ad Dammān to Riyadh railway (completed 1951), and the subsequent completion of modern roads, has solved the marketing problem by allowing export to the capital, on one hand, and to the thriving oil settlements, on the other.[15] Further developments took place in the 1970s.

Building on the initial successes, and with capital readily available from oil royalties, Saudi Arabia has initiated a number of other agricultural schemes. A series of storage dams has been built in the wetter southwestern parts of the country, for example near Jīzān (1971), Abhā (1974) and Najrān (1980). Land has been distributed to would-be farmers in former tribal territories, deep wells have been sunk, and an elaborate system of subsidies, loans and grants has been created. As a result, the cultivated area has increased and cropping patterns have changed. Subsidies on wheat brought about a dramatic increase in production over the decade 1974–84 (132.9 thousand tonnes to 1,300 thousand tonnes), whilst the amount of durah grown declined. Vegetable-growing also expanded and new lines such as artichokes, carrots and potatoes were introduced in response to rising urban demand.[16]

Despite all the activity and huge investments, the value added by Saudi agriculture has been small (3.5 per cent at standard prices in 1973–78), its contribution to GNP less than 3 per cent (1984–85) and the proportion of the land surface cultivated no more than 1 per cent.[17] Some of the problems which beset Saudi agriculture are physical and fundamental; others are social and intractable. Irrigation water is essential but scarce, and much of it is saline. Continuous pumping has produced falling water tables. Cultivable land is often confined to discontinuous tracts in the bottoms of wadis where there is a constant threat of *seils* destroying irrigation works and washing out planted areas. The greatest potential lies in the Qasim, Qunayfidah and al-Hasa areas. The isolation imposed by distance and terrain is still restrictive, despite road-building. Local labour is scarce, partly because some Saudi citizens find certain kinds of agricultureal work demeaning (*'aib*), and partly because the towns offer more attractions than the villages. Foreign labour makes up the deficit, especially in areas where agriculture has only a short recent history. Many landowners are now urban-based businessmen, and farming is perhaps more of a hobby to them than a serious economic proposition. At the same time, some small farmers in the old-established farming areas treat government loans and grants as income and so feel that they can neglect their land.

Similar constraints on agricultural development are found elsewhere in the peninsula. In Kuwait and Qatar precipitation and surface water are so scarce that agriculture is virtually the creation of oil wealth, and the economics of producing fruit and vegetables have not been seriously calculated. By contrast, in Bahrain, where there has long been an agricultural sector to the economy, many of the springs have dried up from over-exploitation for irrigation and other purposes, and perhaps as much as 75 per cent of the cultivable land has simply been abandoned as unfeasible to work for economic or physical reasons.[18] The resources in land and water are greater in the United Arab Emirates, especially on either side of the Hajar Mountains, and the cultivated area doubled during the 1970s. This was achieved, though, at great financial cost. It also initiated both a serious fall in the watertable and also increased salinity.[19] Developments have been more cautious and limited in Oman, but oil wealth has allowed the rejuvenation of *falaj* (or *qanāt*) systems, the construction of dams and the spread of irrigation pumps. Much of the labour on the new schemes, though, is of foreign origin. Both Yemeni states have initiated modern agricultural development schemes which make use of pumped irrigation water, rather than *seil*-irrigation techniques. They have encouraged diversification in cash-cropping. During the 1970s, though, massive labour emigration from Yemen AR and the availability of remittances on a larger scale than before brought contrasting developments in agriculture. On the one hand, the system as a whole appears to have stagnated. About 7.5 per cent of the cultivated area actually went out of use in 1975–80, whilst terraces have been allowed to collapse, thus threatening the low-lying land where most modern developments have taken place, with devastating floods. On the other hand, truck farming has developed to meet rising consumer demand. Melons and vegetables are traded from the coastal plain (Tihama) into the mountains.[20] In the mountains *qāt* production has expanded. This has been in response to demand from a market enlarged by road building and cash availability, as well as the extremely high yields per unit of land which the crop allows.[21]

11.3.2 Pastoralism[22]

While cultivation is diversifying throughout Arabia and both expansion and contraction are apparent on the local scale, nomadic pastoralism has completely changed its character, particularly in northern Arabia. On the one hand, sedentarization has proceeded rapidly.[23] Adbul Aziz Ibn Saud encouraged it in his kingdom for political and religious reasons, and his successors have continued the same policy, though in different forms (e.g. grants and deep wells). They have been helped by two other developments. The first was the steady fall in demand for dromedaries from the beginning of the century as pumps replaced ancient lifting devices and motor trucks supplanted caravans. The second was drought (for instance, between 1955 and 1963) in northern Arabia. This reduced both the number and the ranging possibilities of livestock. Incomes diminished to a level where many nomads were only too glad to accept government assistance, build semi-permanent villages of tin huts and breeze-block houses and take up farming,[24] though some preferred to find work in the towns, with oil companies or in the Saudi National Guard. Pastoralism, however, has continued in many districts. As ecological conditions improved in the 1970s there was some reversion by groups which had settled to cultivation a decade earlier. Only in Jīzān emirate in Saudi Arabia was the proportion of pastoralists at the 1974 census recorded as less than 10 per cent, but the total number in the country was put at no more than half a million people. Where it survives, pastoralism is no longer entirely traditional in either its stock mix or its methods.[25] Despite a continuing attachment to them, even classic dromedary rearers such as the Rwala have increased the proportion of sheep in their herds. Sheep cannot range as far from water as dromedaries and this, together with the provision of secure water points, has reduced ranging and concentrated the herds and flocks to the extent that over-grazing is severe in some areas. It has been increased by the expansion of cultivation, a rise in the number of animals kept and the use of trucks for transporting stock and tankers for carrying water.[26] In contrast to this type of motorized nomadism, herds of dairy cows are now being kept in various parts of Arabia using modern feeding and managerial techniques to produce fresh milk for the urban market.

11.3.3 Other activities

Maritime activities have probably experienced an even greater change than pastoralism. Seaborne trade is more important to the region than ever before, but few sailing vessels make the voyage to East Africa and oil tankers, large freighters and container ships dominate the traffic. To accommodate them, ambitious new facilities have been constructed at ports along the Gulf and at Jiddah and Yanbu'al Baḥr on the Red Sea.

Pearling virtually disappeared in the 1920s and 1930s with the perfection of the culturing process by the Japanese, but recently other types of fishing have been transformed. Kuwait and Oman led the way and now modern ships and equipment are used at sea, whilst deep-freezing facilities have been provided on shore. Shrimp production for export has been particularly successful but spillage from offshore oil rigs and pollution from land-fill operations are proving to be major hazards.

Internal movement has been transformed. The motor vehicle, especially with four-wheel drive and sand tyres, has opened up virtually the whole of Arabia, while important networks of roads have been constructed (Figure 11.2). Movement is much more frequent than in the past, some periodic markets have become permanent and the market areas of the towns have considerably expanded. Partly in consequence, the populations of inland towns appear to be growing, though figures are difficult to obtain. Intraregional trade is increasing, especially within the diverse territory of Saudi Arabia, where over 26,000 km of surfaced road have been constructed. Crops produced by dry farming in Asīr now find a ready market in the east of Saudi Arabia, while Ra's al Khaymah trucks fruit and vegetables to Dubayy and even Abū Dhabi. Arabia's only working railway (recently upgraded) has been important in movements between the Najd and the Gulf coast. The development of air transport has been very significant, especially in spanning the vast distances of Saudi Arabia, as well as in integrating the Gulf states and the different parts of PDR Yemen.

Ease of communication has had two further results of note. It has allowed the more efficient distribution of imported and locally manufactured goods throughout Arabia, and reduced the severe effects of local crop failures which took a heavy toll of life in the past. Improved transportation has been responsible, secondly, for a remarkable revival in the Pilgrimage. In consequence, Jiddah has expanded enormously in both population and areal extent, while Mecca itself has been reshaped. The Great Mosque has been extended, and tall apartment blocks and tree-planted squares have replaced much of the old squalor. An express highway now links the two Holy Cities.

Artisans continue to meet local and tourist demands for traditional products, but modern industries have been introduced. Oil revenues are the reason, directly or indirectly. Although increased per capita income has stimulated some development through private initiative, the really large and important introductions have been government sponsored. Oil revenues provide a rich income and have raised living standards for many, but the reserves are finite; they will eventually run dry –

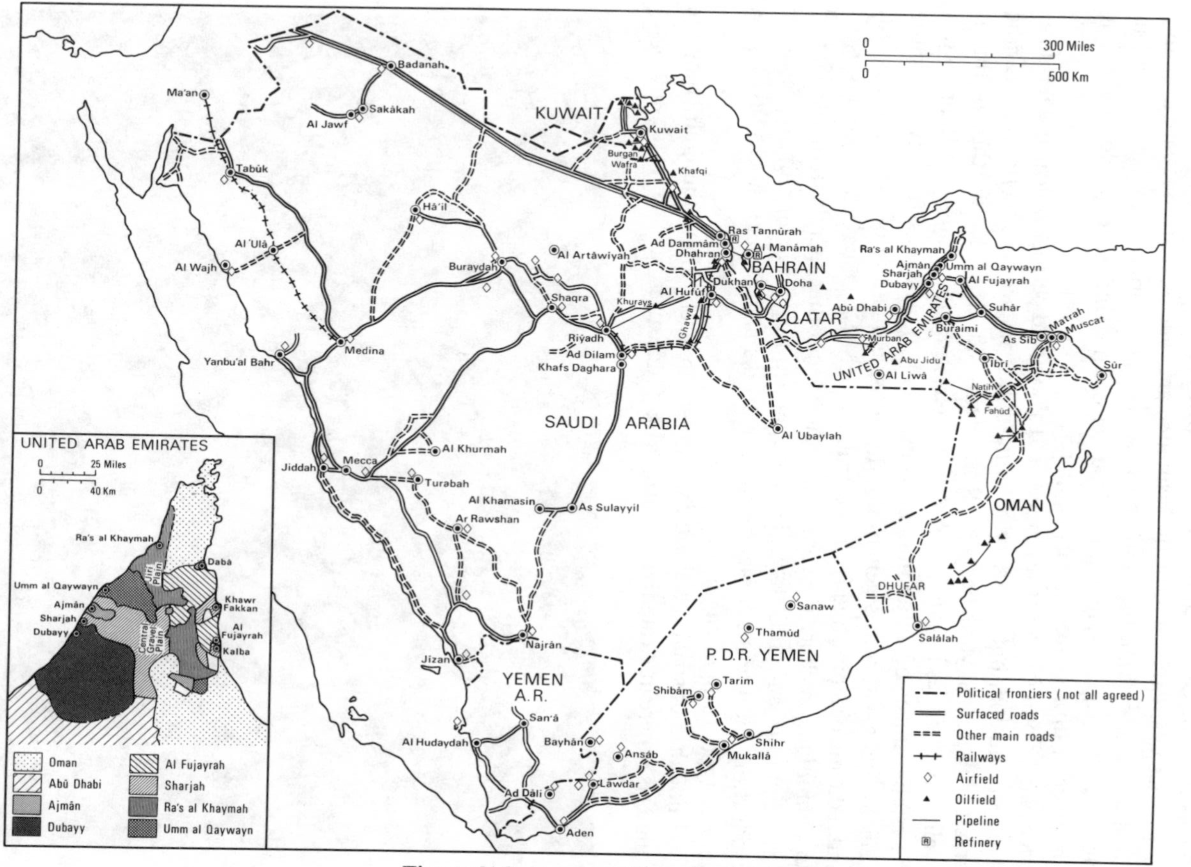

Figure 11.2 Aspects of 'modern' Arabia

in Bahrain, before the end of the century. Diversification away from dependence upon the sale of crude oil is thus seen as an absolute necessity by all the oil-rich states in order to safeguard their future and avoid a worse disaster than that which struck the Gulf with the decline of pearling. Most of the development has taken place where the oil industry is situated and where oil revenues are available – that is, largely along the Gulf coast. A common pattern may be detected, in which building supplies, light and consumer-oriented industries predominate, but a number of divergences are also apparent.

Kuwait is dominated by the refinery and associated petrochemical and fertilizer plants at Shuaiba, but a variety of other industries are found. Bahrain has the largest ship repair yard between Rotterdam and Hong Kong, though this is rivalled by facilities in Dubayy. Another interesting development in Bahrain was the establishment of an aluminium smelter (1972) using gas from two offshore wells to fuel gas turbines generating the massive quantity of electricity needed, and reducing ore imported from Australia. Since 1979 Dubayy has also possessed an aluminium smelter. By contrast, industry in other members of the United Arab Emirates and in Qatar is largely confined to petroleum refining, fertilizer production and the manufacture of soft drinks and cement blocks, as well as various types of light engineering.

Although industrial developments are so apparent along the Gulf, they are found elsewhere in Arabia, too. Various light industries, such as textile and cigar-

The coming of the tarmac road to Hili oasis in Saudi Arabia brought changes in agricultural practice. Date gardens formed the core of the oasis, but small holdings growing vegetables and fodder were found on the periphery (Hunting Surveys Ltd.)

Dubayy Creek, a finger-like inlet of the Gulf, separates the old city, with its be-flagged fortress and traditional wind-towers, from Deira, the high-rise banking and commercial centre. Note the dhows which are still important trading craft (photograph ARAMCO, 1986)

ette manufacturing, were established in Aden under the British, who also established the refinery. Several of the larger towns of Saudi Arabia contain small-scale modern industrial plants producing building materials, plastics, furniture and soft drinks or engaged in light engineering (Figure 11.2). Jiddah, however, emerged as the country's leading industrial centre. A cement plant and a refinery were in production by 1968, while the first stage of an iron and steel complex, a 180 m long steel-rolling mill, was in operation by 1970. The mill uses billets imported from West Germany and local electricity, but plans have been laid for the smelting of ore from the mountains between Mecca and Yanbu'al Baḥr. During the late 1970s and early 1980s, however, Yanbu'al Baḥr and Al Jubal (on the Gulf) emerged as Saudi Arabia's industrial growth centres. Huge revenues were available from the rise in oil prices (1973, 1979), and the government took the decision to attempt a fundamental transformation of the economy based upon the petrochemical industry.

Several problems, however, are involved in industrializing Saudi Arabia and her neighbours still further. The most basic of these is a general lack of raw materials, apart from petroleum and agricultural or pastoral products. Although a number of surveys have shown that Saudi Arabia, for example, possesses a variety of non-ferrous metals, these are situated in remote and difficult country. The result so far has been the importation of most raw materials, thus adding extra costs to manufacturing. The provision of adequate supplies of water is fast emerging as another major constraint, especially since water is also required for agriculture and an expanding urban population.[27] In addition, the infrastructure in terms of roads and power is still inadequate, though the position is being remedied by sensible overall development plans. The small size and comparatively low purchasing power of local markets is a great problem, and is intensified by the almost parallel development taking place in each state. Traditional markets in Pakistan, India and East Africa may offer some scope for industrial exports as these countries advance their economies still further, while Japan and the emergent economies of Taiwan and South Korea have a growing demand for processed and semi-processed petrochemicals. Lack of manpower, particularly skilled industrial and managerial labour, is a major problem, at present uneasily resolved by immigration. Technical education is being pushed ahead, but many local people show a reluctance to take on anything but a desk job, while poor skills, high urban rents and various forms of subsidy are powerful disincentives.

11.3.4 Petroleum

Petroleum is so important to modern Arabia that some specific comments on its effects are in order here, though the general situation is surveyed in Chapter 9. The direct effects of petroleum exploitation are confined to a number of small, but drastically changed localities in the eastern parts of the peninsula, where oil is extracted and exported (Figure 11.2). Production sites are characterized by derricks, pipes, tanks and often flares burning off waste gases, while the specially constructed harbour facilities are marked by more storage tanks, and long piers running out into deep water. Between the two areas run pipelines. Refineries are found along the Gulf coast, with outliers at Riyadh, Jiddah and Aden. Petrochemical plants have been added as governments became aware of the high values added by downstream operations (Figure 11.2). Settlements nearby were specially built to accommodate oil industry personnel and company officials. They are marked by regular, grid-iron street plans and air conditioned bungalows surrounded by jasmine and oleander. The prototypes were the ARAMCO settlements in Abqaiq, Dhahran and Ra's Tannūrah in Saudi Arabia, but similar developments may now be seen at Al Ahmadī in Kuwait and on Dās Island, the exporting centre for Abū Dhabi's oil.

The indirect effects of oil exploitation are much more widespread. Some of them have been touched upon already. Exploration for oil, taking place since the 1920s, opened up large parts of Arabia by necessitating at least the grading of tracks, some of them completely new, to supply the drilling sites. The imminence of discovery, and later the finding of oil, led to the abandonment of neutral zones in the United Arab Emirates and between Kuwait and Saudi Arabia, to the allocation of territory definitely to one state or another and to general agreement about boundaries, though with insufficient precision about their lines to be a fruitful source of wrangling. In the same way, offshore drilling has required the delimitation of claims to territorial waters and areas of seabed.[28] Revenues from the exploitation of oil have led to many changes. Immediate steps have generally involved the improvement of administration, the extension of education at all levels, the establishment of hospitals, clinics and free medical services, and the gradual introduction of adequate water supplies, especially to the coastal towns. A considerable amount of money, however, is being spent on imported consumer and capital goods, as well as on construction. For example, within ten years of the first shipment of crude from Abū Dhabi (1962), the island town was transformed into an agglomeration of concrete buildings of standard, international design which line wide, surfaced roads. Kuwait underwent a similar transformation at an earlier date, while new towns are being built in association with the new economic facilities throughout the peninsula.

There is no doubt that wealth from oil has been used to bolster up the 'traditional' patriarchal oligarchies of Arabia and, though there have been changes in the exercise of power, political structures remain much the same. It is not by chance that the only revolutionary regimes are found in the two Yemeni republics, where oil has only recently been discovered (near Mārib, Yemen AR, in 1984). Where the proportion of immigrant workers is high (as in the Gulf principalities), the indigenous population feels threatened, while the newcomers themselves are excluded from power, as well as from most social benefits in the affluent societies which they have helped to build. At the same time, the Yemen AR has had up to 25 per cent of its population abroad in any one year and, since most of these are males, emigration has weakened local and family leadership. It is doubtful, though, whether modernization and traditionalism are as much in conflict in most of Arabia as has sometimes been supposed. Modern technology is being used to support traditional power structures, while kinship and clientalism often ease the processes of change. Most of the friction which surfaces from time to time comes from two other directions. One is the feeling of relative exclusion from modern developments and the 'easy' life experienced by much of the rural population of Arabia, despite the improvements which have been made in the countryside. The other threat is from religious fundamentalism in Saudi Arabia and the Gulf states. Whilst Wahhābi doctrines have fostered a pragmatic accommodation to events which is shared by many of nomadic background, Shiite 'enthusiasm' has begun to feed on feelings of relative deprivation and exclusion found in the Eastern Province of Saudi Arabia, where Shia Muslims form a local majority in al-Hasa, and amongst the descendants of Iranian settlers in the Gulf ports. One expression of Muslim fundamentalism is a rejection of the westernization of attitudes and behaviour often associated with recent economic and technical developments. Its Shiite manifestations have been encouraged by the success of the fundamentalist revolution in Iran, while the ruling Sunni families of Arabia have become nervous of Iranian promises to extend both the revolution and the war with Iraq. Wahhābi and Khomeni fundamentalism may yet clash in Arabia with fearful consequences.

References

1. H. Dequin, *Die Landwirtschaft Saudisch-Arabiens und ihre Entwicklungmöglichkeiten*, DLG-Verlag-GMBH, Frankfurt am Main, 1963, 48–87.
2. R. L. Bowen and F. P. Albright (Eds), *Archaeological Discoveries in South Arabia*, American Foundation for the Study of Man, John Hopkins, Baltimore, 1958.
3. V. H. W. Dowson, 'The date and the Arab', *Jl. R. cent. Asian Soc.*, **36**, 34–41 (1949).
4. (a) J. L. Burckhardt, *Notes on the Bedouins and Wahabys*, Colburn and Bentley, London, 1831.
   (b) A. Musil, *The Manners and Customs of the Rwala Bedouins*, American Geographical Society, New York, 1928.
   (c) P. G. N. Peppelenbosch, 'Nomadism in the Arabian peninsula: a general appraisal', *Tijdschr. econ. soc. Geogr.*, **59**, 335–46 (1968).
   (d) L. E. Sweet, 'Camel pastoralism in north Arabia and the minimal camping unit'. In *Man, Culture and Animals: The Role of Animals in Human Ecological Adjustments*, (Eds A. Leeds and A. P. Vayda), American Association for the Advancement of Science, Washington DC, 1965, 129–52.
5. (a) R. Serjeant, 'Fisher folk and fish-traps in al-Bahrain', *Bulletin of the School of Oriental and African Studies*, **31**, 486–514 (1968).
   (b) I. F. Wallen, 'Non-oil trade and resources', in *The Princeton University Conference and Twentieth Annual Near Eastern Conference on Middle East Focus: The Persian Gulf*, (Ed T. C. Young), Princeton University Conference, Princeton, NJ, 1969, 107–110.
6. (a) C. D. Belgrave, 'Pearl diving in Bahrain', *Jl. R. cent. Asian Soc.*, **21**, 450–452 (1934).
   (b) R. L. Bowen, 'The pearl fisheries of the Persian Gulf', *Middle East Journal*, **5**, 161–80 (1951).
7. J. Hornell, 'A tentative classification of Arab seacraft', *Mariners' Mirror*, **28**, 11–40 (1942).
8. D. N. McMaster, 'The ocean-going dhow trade to East Africa', *East African Geographical Review*, **4**, 13–24 (1966).
9. (a) M. S. El Attar, *Le Sous-développement économique et social du Yémén. Perspectives de la révolution yéménite*, Editions Tiers Monde, Algiers, 1964, 65–67.
   (b) H. Ingrams, *Arabia and the Isles*, John Murray, London, 1966, 337.
   (c) R. Levy, *The Social Structure of Islam*, Cambridge University Press, London, 1957, 43.

10. (a) L. E. Sweet, 'Pirates or politics? Arab societies of the Persian or Arabian Gulf, eighteenth century', *Ethnohistory*, **11**, 262–80 (1964).
(b) R. G. Lander, 'The modernisation of the Persian Gulf: the period of British dominance', in *The Princeton University Conference and Twentieth Annual Near Eastern Conference on Middle East Focus: The Persian Gulf*, (Ed T. C. Young), Princeton University Conference, Princeton, NJ, 1969, 1–29.
11. (a) Article on the *Hadj* in *The Encyclopaedia of Islam* (Eds B. Lewis, C. Pellart and J. Schaht), **2**, new edn, E. J. Brill, Leiden, and Luzac and Co., London, 1965, 31–38.
(b) R. King, 'The Pilgrimage to Mecca: some geographical and historical aspects', *Erdkunde*, **26**, 61–72 (1972).
12. D. G. Hogarth, *The Nearer East*, William Heinemann, London, 1902, 221–4.
13. J. S. Habib, *Ibn Sa'ud's Warriors of Islam*, Social, Economic and Political Studies of the Middle East 27, Brill, Leiden, 1978.
14. (a) D. D. Crary, 'Recent agricultural developments in Saudi Arabia, *Geogr. Rev.*, **41**, 366–83 (1951).
(b) R. H. Sanger, *The Arabian Peninsula*, Cornell University Press, New York, 1954, 58–72.
15. F. S. Vidal, *The Oasis of Al-Hasa*, Aramco, New York, 1955.
16. (a) K. Abdulfattah, *Mountain Farmer and Fellah in 'Asir, South-west Saudi Arabia*, Erlangen Geographische Arbeiten 12, Erlangen.
(b) E. G. H. Joffe, 'Agricultural development in Saudi Arabia: the problematic path to self-sufficiency', in *Agricultural Development in the Middle East*, (Eds P. Beaumont and K. McLachlan), Wiley, Chichester, 1985, pp 209–693
17. R. El Mallakh, *Saudi Arabia: Rush to Development*, Croom Helm, London, 1982, 77–95.
18. R. Dutton, 'Agricultural policy and development: Oman, Bahrain, Qatar and the United Arab Emirates'. In *Agricultural Development in the Middle East*, (Eds P. Beaumont and K. McLachlan), Wiley, Chichester, 1985, pp 227–40.
19. (a) R. Dutton, 'Agricultural policy and development: Oman, Bahrain, Qatar and the United Arab Emirates'. In *Agricultural Development in the Middle East*, (Eds P. Beaumont and K. McLachlan), Wiley, Chichester, 1985, pp 227–40.
(b) J. H. Stevens, 'Arid zone agricultural development in the Trucial States', *J. Soil Wat. Conserv.*, **24**, 181–3 (1969).
20. (a) S. Carapico, 'Yemeni agriculture in transition'. In *Agricultural Development in the Middle East*, (Eds P. Beaumont and K. McLachlan), Wiley, Chichester, 1985, pp 241–54.
(b) H. Kopp, 'Land usage and its implications for Yemeni agriculture'. In *Economy, Society and Culture in Contemporary Yemen* (Ed B. R. Pridham), Croom Helm, London, 1985, pp 41–50.
(c) M. Mundy, 'Agricultural development in the Yemeni Tihama: the past ten years'. In *Economy, Society and Culture in Contemporary Yemen*, (Ed B. R. Pridham), Croom Helm, 1985, pp 22–40.
21. S. Weir, 'Economic aspects of the qat industry in north-west Yemen'. In *Economy, Society and Culture in Contemporary Yemen* (Ed B. R. Pridham), Croom Helm, London, 1985, pp 64–82.
22. (a) A. S. Helaissi, 'The bedouins and tribal life in Saudi Arabia', *International Social Science Journal*, **11**, 532–8 (1959).
(b) P. G. N. Peppelenbosch, 'Nomadism in the Arabian Peninsula: a general appraisal', *Tijdsch. econ. soc. Geogr.*, **59**, 335–46 (1968).
(c) D. P. Cole, *Nomads of the Nomads: The Al Murrah Bedouin of the Empty Quarter*, AHM Publishing Corporation, Arlington Heights, Illinois, 1975.
23. S. S. Abu'Adhirah, 'Sedentarisation and settlement of the bedouin', *Arabian Studies*, **4**, 1–5 (1978).

24. (a) M. Katakura, *Bedouin Village: A Study of a Saudi Arabian People in Transition*, University of Tokyo 1977.
(b) A. A. Shamekh, *Spatial Patterns of Bedouin Settlement in Al-Qasim Region, Saudi Arabia*, Department of Geography, University of Kentucky, Lexington, 1975.
25. (a) J. S. Birks, 'Development or decline of pastoralists', *Arabian Studies*, **4**, 7–20 (1978).
(b) D. P. Cole, 'The enmeshment of nomads in Sa'udi Arabian society: the case of Āl Murrah'. In *The Desert and the Sown: Nomads in the Wider Society* (Ed C. Nelson), Institute of International Studies, Research Series No 21, University of California, Berkeley, 1973, pp 113–28.
26. H. F. Hendy, 'Ecological consequences of bedouin settlement in Saudi Arabia', in *The Careless Technology*, (Eds M. T. Farvar and J. P. Milton), Natural History Press, Garden City, New York, 1972, pp 683–93.
27. P. Beaumont, 'Water and development in Saudi Arabia', *Geogrl. J.*, **143**, 42–60 (1977).
28. A. Drysdale and G. H. Blake, *The Middle East and North Africa: A Political Geography*, Oxford University Press, New York and Oxford, 1985, pp 75–145.

CHAPTER 12

Iraq – A Study of Man, Land and Water in an Alluvial Environment

12.1 The economy

With an area of 434,000 sq km and a population of 15.5 million in 1985, Iraq is one of the least densely populated countries of the Middle East. Yet, in terms of the availability of flat lands and water, it has perhaps the greatest agricultural potential of any of the countries within the region. At the present time the economy rests on oil production and to a lesser extent agriculture. Over the last two decades, although the population has increased markedly from only 10.4 million in 1972 to its present figure, agricultural production has stagnated or even declined in some sectors. As a result, a larger proportion of the oil revenues, needed so urgently for development projects, has had to be diverted to pay for the import of foodstuffs. For example, in 1980 although Iraq produced 2.22 million tonnes of cereals, it still had to import 2.48 million tonnes.[1] This situation is in marked contrast to the period before 1955, when Iraq was largely self-sufficient in staple crops and animal produce. Even more startling is the way in which the cereal production per capita has declined in Iraq from 268 kg/capita in 1950 to only 170 kg/capita in 1980. To comprehend the present position it is necessary to outline the environmental characteristics of the nation, as well as the sequence of land use changes which have occurred during the historical period.

Four major regions can be identified in Iraq. In the extreme northeast are the high ranges of the Zagros Mountains. This is a barren and harsh environment, with little human activity except for pastoral nomadism in the broad upland basins during summer. At lower altitudes, the foothills of the Zagros provide a much more hospitable landscape for human settlement. This region, the home of the Kurds, once had a natural vegetation cover of forest. Today it is covered by scrub, as the result of centuries of deforestation and overgrazing. Precipitation ranges from 400 mm to 500 mm per annum, so that sufficient moisture is available for rainfed agriculture, at least in the more northerly and easterly parts. This area, part of the famed 'Fertile Crescent', is one in which most of the country's cereal crops of wheat and barley are grown. Between the rivers Tigris and Euphrates is the Jezira, a barren zone frequented today by nomads, but with considerable agricultural potential if irrigation water can be supplied to it.

Central and southern Iraq is an alluvial plain formed by the deposition of sediment from the rivers Tigris and Euphrates. This is the main agricultural region of the country, and the site of some of the earliest civilizations in the world.

Dates, wheat and barley, rice, cotton and vegetables are the principal crops. At the southern end of the alluvial plain lie extensive marshlands, while along the Shaṭṭ al'Arab the richest date groves in Iraq are found. It is in this area where most of the fighting in the Iraq–Iran war has taken place since 1980. To the west of the alluvial plain, covering more than half the total area of the country, is the desert stretching away towards Syria, Jordan and Saudi Arabia. In this region, land use is limited to nomadic pastoralism.

The wealth of Iraq rests upon its oil production and the revenues derived from it. Like all other oil producers Iraq has benefited greatly from the oil price increases of the 1970s (Figure 12.1). Between 1970 and 1979 oil production rose by 2.2 times, yet revenues increased 41 times. Unlike most other countries in the region Iraq shows a collapse in both oil production and revenues since 1980 with the beginning of its war with Iran. Although oil production is now increasing slightly, income is still less than half of what it was at the beginning of the 1980s.

Most of the oil comes from the old Kirkūk fields in the north, but other deposits have been discovered at Ayn Zāleh near the Turkish border and also close to the head of the Gulf at Ramailah. Although in the early days of the industry most of the oil was exported by pipeline through Syria to Baniyās on the Mediterranean coast, in the period prior to the start of the Iraq–Iran war, a growing proportion was being sent by pipeline to the new terminal at Al Fāw near Basra.

Besides the oil industry, very little large scale industrialization exists in Iraq, and what little there is is concentrated in the capital. Even in the late 1960s of the

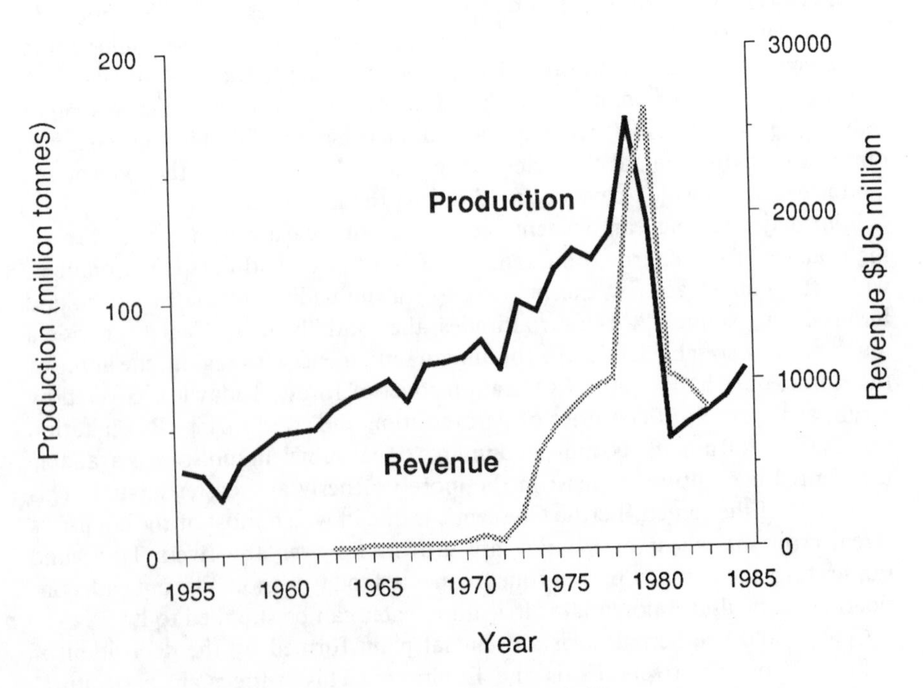

Figure 12.1 Oil production and oil revenues

1,200 units employing more than ten or more people, more than half were located in Baghdād.[2] At this time food processing and light industries predominated, though textiles were found at Mosul and at other centres. In the last National Economic Plan (1976–80) before the beginning of the Iraq–Iran war, the highest priority was given to the development of industry. A similar situation was found with the 1981–85 Plan, though it has not been possible to implement much of this Plan because of the war with Iran.

A characteristic feature of the population during the twentieth century has been the drift to the towns.[3] After the Second World War approximately 65 per

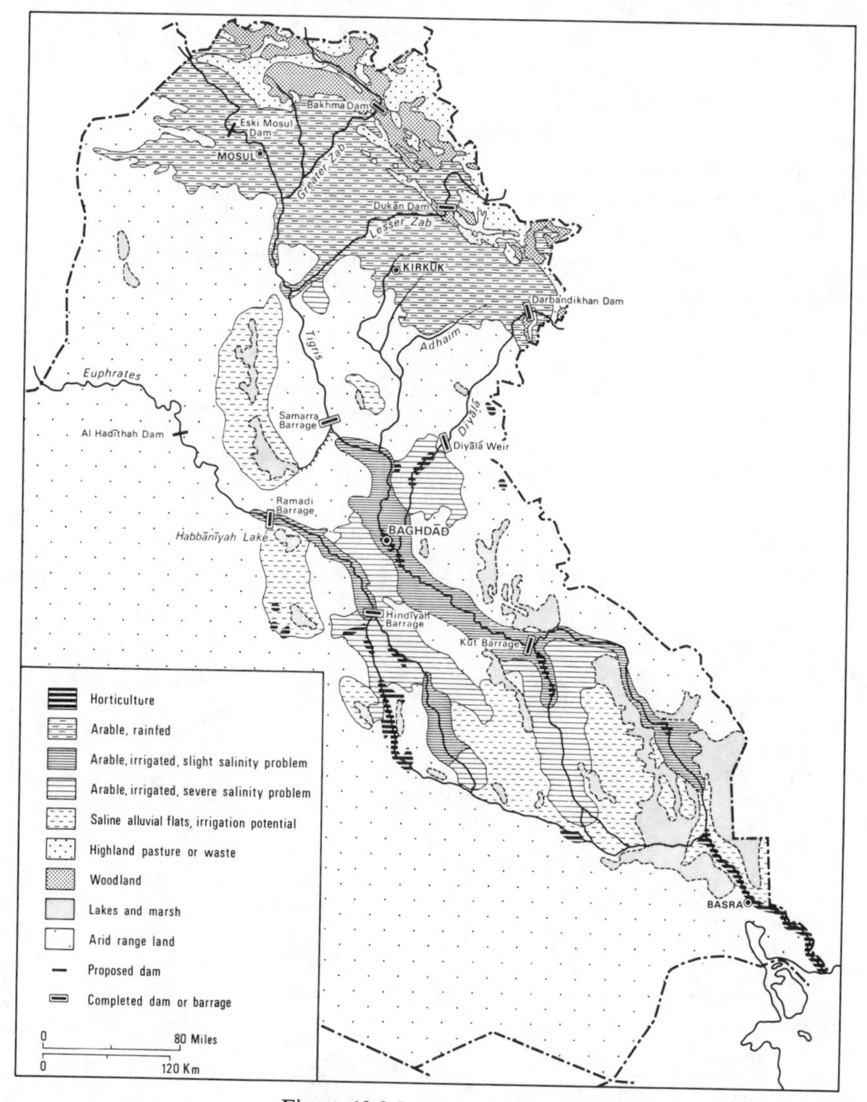

Figure 12.2 Land use in Iraq

cent of the population lived and worked in rural areas. By the early 1970s half of the population was concentrated in urban centres and by 1985 the figure had risen even higher to 68 per cent. About one fifth of the total population of the country is now located in the capital, Baghād. In 1985 the total population of Iraq was 15.5 million and the natural rate of increase was 3.3 per cent.[4] By the year 2000 numbers are expected to grow to 24.9 million and to 39.3 million by 2020.

The total area of cultivated land in Iraq is 5.45 million hectares[5] (Figure 12.2). An unknown area remains fallow each year. Wheat and barley are by far the most important crops and make up well over half of the cropped area in any one year (Table 12.1). Production of cereals is concentrated in the northern, wetter parts of the country and in the Tigris–Euphrates lowlands (Figure 12.3). In the uplands wheat and barley are grown by dry farming methods, while on the lowlands irrigation has to be employed. Actual production totals of wheat and barley are highly susceptible to changes in precipitation. Over the last twenty years wheat has shown production totals from as low as 471,000 tonnes in 1983 to as much as 2,625,000 tonnes in 1972 (Figure 12.4). Similarly barley totals have ranged from 462,000 tonnes in 1973 to 992,000 tonnes in 1968. Yields of both crops remain low and variable. Rice is the only other cereal crop of major significance, with almost all production concentrated in the Muhafadhas of Qadissiya, Maysan, Kerbela

TABLE 12.1
Major crops – areas under cultivation in Iraq

	1976	1977	1978 (100 mesharra)	1979
Cereals				
Wheat	59,972	34,304	59,826	43,112
Barley	23,028	21,435	28,573	30,412
Rice	2,096	2,539	2,189	2,348
Industrial crops				
Cotton	1,013	837	685	610
Tobacco	339	500	469	388
Linseed	36	39	35	114
Sesame	480	367	723	533
Sugar cane	120	130	169	169
Sunflower	320	327	180	513
Vegetables				
Dry onion	389	421	409	328
Dry broad beans	633	582	588	324
Chick peas	547	598	575	725
Okra	399	461	498	548
Tomato	1,825	1,625	1,613	1,145
Broad beans, green	723	695	560	535
Water melons	1,791	1,826	1,782	1,372
Cucumber	577	790	799	980

(Note: 1 mesharra equals 0.25 ha)
Source: Central Statistical Organization, Annual Abstract of Statistics 1979, Baghdad.

and Thi-Qar. The highest production totals for rice were attained in the late 1960s and the early 1970s, when over 300,000 tonnes were produced from over 100,000 ha. In the 1980s the area of rice production has fallen markedly to around 50,000 ha, partly as a result of the Gulf War, but mainly because it has proved difficult to maintain the very high water demands for the crop at a time when increasing pressures are being put on the available water resources of Iraq. An impressive increase in rice yields took place in the mid and late 1960s as the result of the introduction of new varieties. These high yields, of between 2500 and 3000 kg/ha, have been maintained, although recently the effects of the Iran–Iraq War have been making the cultivation of rice increasingly difficult.

The other major crops are vegetables, fruit and dates. Vegetables are grown, with the aid of irrigation, around all the villages and towns. In recent years there

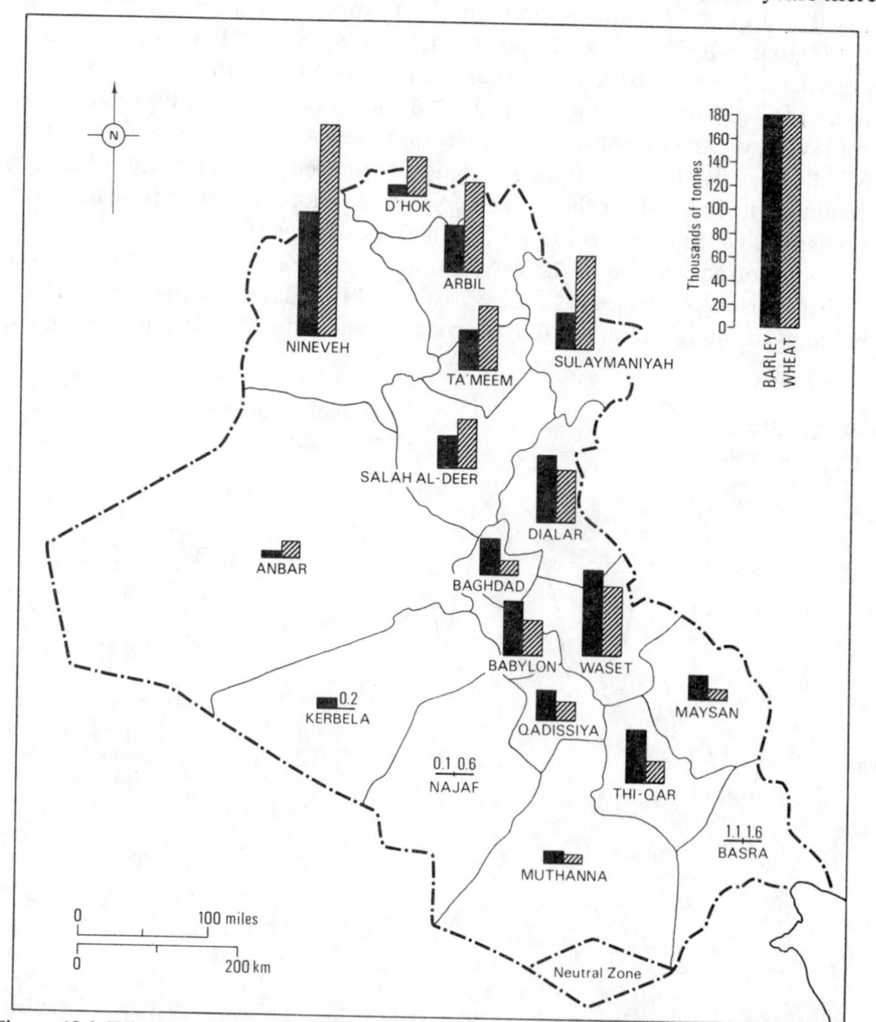

Figure 12.3 Wheat and barley production in the major administrative units of Iraq in 1979

has been increasing production of vegetables and citrus fruit for both local use and to supply the lucrative markets in the Gulf states. Dates are grown along the lower Tigris and in particular in the Basra region. Production has suffered greatly as a result of the Iran–Iraq war.

Throughout the remoter parts of the country traditional methods of agricultural production still prevail, though mechanization has been proceeding steadily in areas close to the larger urban centres. In the late 1960s about 10,000 tractors were in use. This number had risen to around 20,000 in the mid-1970s and by 1983 to approximately 30,000. About 2750 combine harvesters were also in use in the early 1980s.

The use of chemical fertilizers has increased, particularly for production of rice, cotton and vegetables. There is still very little employed for the cultivation of wheat and barley. Total consumption of all types of fertilizer has grown from 17,000 tonnes in 1969/71 to 77,000 tonnes in 1981/82.[6] In terms of application rates per unit area, fertilizer usage has risen from 0.5 kg/ha for all cultivated land in Iraq in 1961–65, to 6.9 kg/ha in 1974–76 and to 14.1 kg/ha in 1981.[7] However, Iraq still possesses one of the lowest fertilizer use rates of all the countries of the Middle East. With the continued depletion of nutrients from the soil, which has continued unchecked for thousands of years, it is not really surprising that crop yields are low. Little use is made of herbicides and pesticides.

An important change in the agricultural structure of the country took place with the passing of the Land Reform Law in 1959.[8] This limited the size of private holdings to 1000 donums (250 ha) of irrigated land, and 2000 donums (500 ha) of

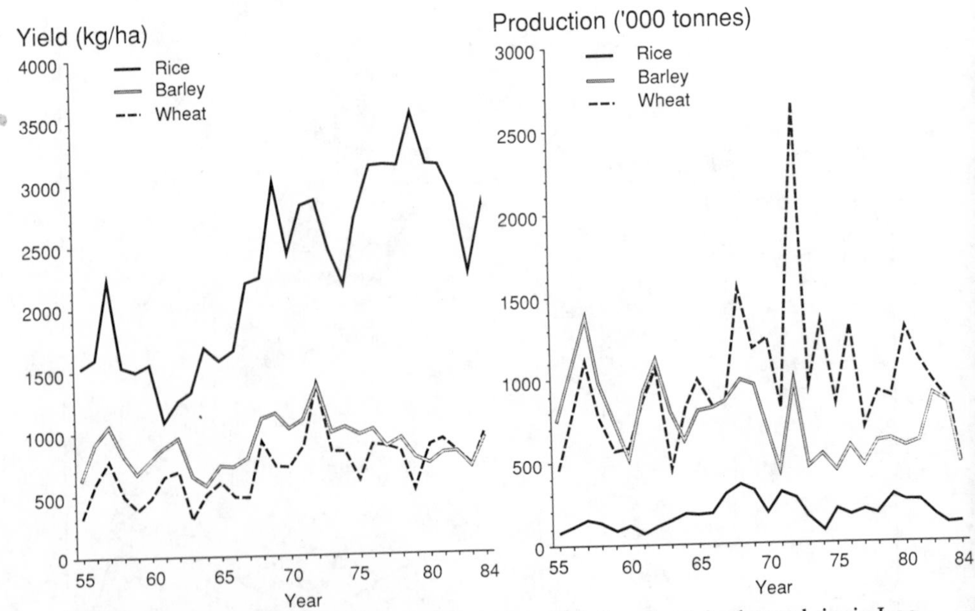

Figure 12.4 Variations in annual production and yields for wheat, barley and rice in Iraq, 1955–1984

rainfed land. As a result large areas have been redistributed among landless labourers, but administrative and political difficulties, especially the lack of managerial experience, limited the overall success of the scheme in terms of agricultural production.[9] In 1970 a further land reform programme was introduced, which further reduced the maximum size of individual holdings. Estimates suggest that under the 1959 law 4.2 million ha were expropriated and 6.0 million ha with the 1970 legislation.[10] Currently the agricultural system of Iraq is dominated by state farms and cooperative villages. However, in the last few years the government has relaxed its policies and it is no longer essential for farmers to belong to cooperatives in order to obtain credit, supplies and equipment. Whether this will generate a strong private sector within agriculture still remains to be seen.

12.2 Agriculture and settlement in the alluvial lowlands

Lowland Iraq provides one of the harshest environments for man in the Middle East. Despite this fact, the riverine lands gave rise to one of the world's oldest civilizations, dating back to at least the fourth millenium BC, when the settlements of Akkad and Sumer were established in the northern and southern parts respectively of the Tigris–Euphrates delta (Chapter 6). Then, as now, agriculture in lowland Iraq was only possible when an assured supply of irrigation water was available, and the growth of these civilizations was entirely dependent upon the life-giving waters of the rivers Tigris and Euphrates. The story of the interrelationships between man and his environment in lowland Iraq throughout history provides one of the most fascinating studies which can be found anywhere in the world. It illustrates the dominant control which the natural environment has imposed on man at certain times and in certain areas, and also shows how human organization and determination have been able to overcome, or at least modify, the effects of some of these controls.

The stage on which this human drama has been enacted is situated in the lower reaches of the drainage basin of the rivers Tigris and Euphrates. Together these catchments cover an area of 785,000 sq km and are located at the present day within the boundaries of Iraq, Iran, Saudi Arabia, Syria and Turkey. Only about 46 per cent of the total area of the two catchments lies in Iraq (Figure 12.5).

Most of the precipitation within the basin falls in the belts of rugged fold mountains, commonly reaching elevations of between 1,500 m and 3,000 m, in Turkey and Iran. In these regions annual totals of more than 1,000 mm are often recorded. Very little of this precipitation occurs during the summer months and much of it falls as snow.

In Iraq, the alluvial flood plain region, characterized by poor surface drainage and swamps, is found around Baghdād and to the south. Rainfall totals throughout this area are less than 300 mm, and often below 150 mm per annum. Almost all of the precipitation is concentrated in the period from November to April. Summer temperatures commonly rise to more than 40°C and frosts are rare.

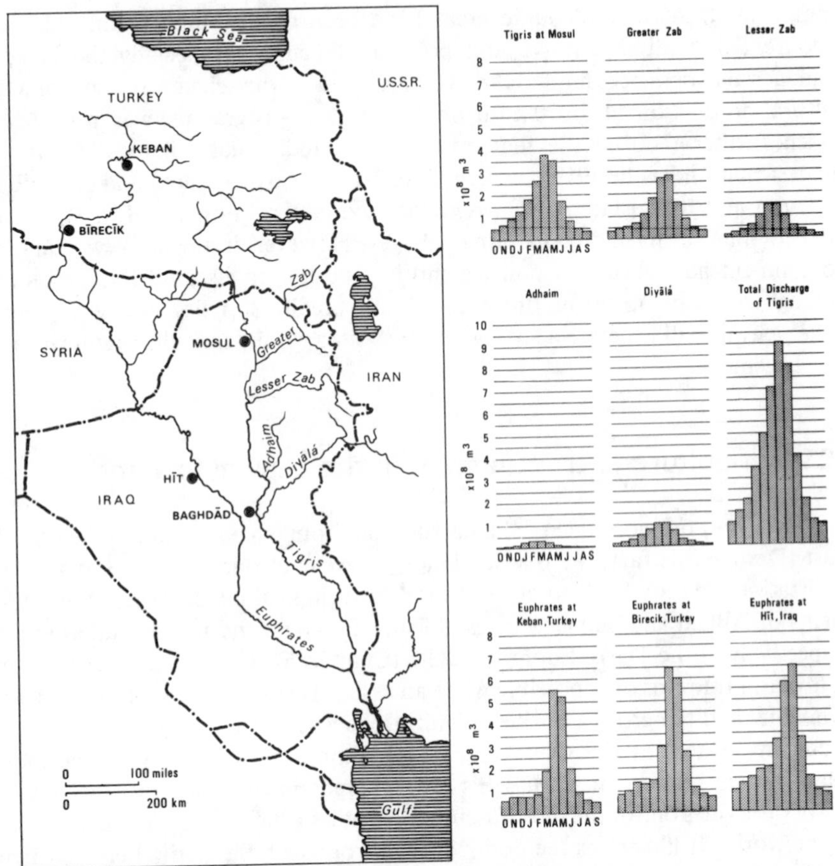

Figure 12.5 The drainage basins of the rivers Tigris and Euphrates, together with selected regime hydrographs

Xerophytic vegetation predominates throughout the zone. It is on this alluvial floodplain that human activity has especially been concentrated, although there have been long periods when the lowlands have been virtually uninhabited.

Throughout history, certain related natural hazards have limited man's ability to live successfully in the Tigris–Euphrates lowlands. Four problems stand out as of particular importance. These are floods, often of a sudden and devastating nature; drought, with its slow and relentless pressure on man and his resources; salination, a process which with time renders crop growth impossible; and siltation, causing a progressive deterioration in the efficiency of the man-made irrigation systems.

Major floods and droughts in a river system are extreme events of relatively infrequent occurrence. To fully comprehend their environmental significance in Iraq, it is necessary to consider the hydrology and water balance of the whole of the Tigris–Euphrates basin. In this catchment, an annual water surplus occurs only in the highland region to the north and east, while large water deficits are found elsewhere. This has meant that arable farming is only possible in the lowlands of Iraq if water, which has fallen elsewhere as precipitation, can be imported into the region, either by natural river flow or by irrigation canals. In the case of Iraq the two largest importing conduits are the rivers Tigris and Euphrates, which transport water surpluses from the highland zone in the northern and eastern parts of the basin to the alluvial lowlands, where water is desperately needed for agricultural activity.

Both the Tigris and Euphrates rivers rise in the mountains of southern Turkey and flow southeastwards into Iraq. The Euphrates possesses only one large tributary, the Khābūr, which joins the mainstrean in Syria, while in contrast, the Tigris has four main tributaries, all of which unite with the mainstream in Iraq. The largest of these, the Greater Zab, has its source in Turkey, while the Lesser Zab and the Diyālá rise in Iran. All of the catchment of the Adhaim, the smallest stream, is situated in Iraq. In southern Iraq the Tigris and the Euphrates unite to form the Shaṭṭ 'al Arab, which in turn flows into the Gulf.

Much of the discharge of the Tigris results from the melting of snow accumulated during the winter season in Turkey. However, winter rains, which are common in late winter and early spring, falling on a ripe snowpack in the highlands, can greatly augment the flow of the main stream and its tributaries, giving rise to the violent floods for which the river Tigris is notorious. The period of greatest discharge on the Tigris system as a whole occurs during March, April and May, and accounts for 53 per cent of the mean annual flow. The highest mean monthly discharge takes place during April. Minimum flow conditions are experienced in August, September and October and make up seven per cent of the annual total discharge.

The total flow of the Euphrates is not as great as that of the Tigris, although the river regimes are similar. It, too, rises in the highlands of Turkey and is fed by melting snows, to an even greater extent than the Tigris, but lacks the major tributaries which the former possesses. In Iraq, the period of maximum flow on the Euphrates is shorter and later than that of the Tigris, and is usually confined to the months of April and May. Discharge during these two months accounts for 42 per cent of the annual total. Minimum flows occur in August, September and October and contribute only 8.5 per cent of the total discharge.

These mean values, however, conceal the fluctuations in discharge which can occur from year to year, for it must be remembered that floods, as well as droughts, are themselves of variable magnitude. Both floods and droughts can have serious effects on the environment, which ancient man, with his limited technology, was unable to control. During the spring floods, although water levels varied from year to year, the effects on the landscape were similar. Natural levees were overtopped, large areas of low-lying ground were flooded and irrigation control works were often seriously damaged by the swirling river waters.

Drought, or water shortage, is caused by long periods of dry weather. Such conditions are, of course, commonplace in lowland Iraq, where annual rainfall totals are so low, although their importance in terms of the water budget of the region is not great. Of far greater importance to the alluvial lowlands is the occurrence of droughts in the upper part of the Tigris–Euphrates basin. In such a case, their effects are always experienced elsewhere, through the transporting agency of the river, and are registered downstream by low spring water levels and insufficient quantities of water for adequate irrigation.

Salinity problems are common in arid and semi-arid regions, especially where soils have high silt and clay contents, and natural drainage is poor.[11] If the water table is shallow, water is drawn upwards through the soil profile by capillary action. As this water evaporates from the ground surface or is transpired by plants, the less soluble calcium and magnesium salts, already concentrated in the groundwater owing to high evaporation, will be precipitated as sulphates and carbonates in the soil profile. Sodium ions are left in the soil solution for the longest period before they too are deposited. It is these sodium ions which have the greatest deleterious effect on the soil, for by deflocculating the clay-sized particles, they cause the structure of the soil to be destroyed. As a result, drainage is further impeded and the ability of the soil to grow crops successfully is seriously reduced. Unfortunately, the process of salination is both progressive and cumulative, and, consequently, the problem worsens from year to year, unless remedial measures are introduced. In the Tigris–Euphrates lowlands, salinity has been introduced into the region largely as the result of human activity. The use of excessive amounts of irrigation water, much of which percolates into the ground, has caused the water table to rise, and this, coupled with the lack of adequate drainage facilities, partly due to the low natural gradients, has greatly speeded up the concentration of salts within the soils.[12] If irrigation is stopped for a long period and the land withdrawn for cultivation, then the water table may fall sufficiently by natural processes to permit cultivation to be resumed, following an initial washing of the soil to remove the more soluble sodium salts. Salination will, of course, occur once again if the environmental controls remain unchanged.

The deposition of silt is a naturally occurring phenomenon in the lowland reaches of most large river systems. Rates of deposition are usually not uniform and tend to be greatest along the margins of the major channels, where levees are built up. The formation and growth of such levees in ancient Iraq often meant that large areas of land had their natural drainage lines to the river blocked, resulting in the development of swamp conditions. Swamps such as these seem to have been transient features in lowland Iraq, with some being drained and others created as the rivers changed their courses relatively frequently on the level alluvial plains.

In the irrigation canals, sedimentation from the turbid waters over the years greatly reduced the carrying capacity of the system, making periodic cleaning operations essential for the preservation of the irrigation network. Even after passing through the canals, the irrigation water which was led into the fields often still carried large quantities of sediment in suspension. Here the sediment would be deposited under tranquil conditions, causing a progressive rise in the level of the fields annually, which made irrigation increasingly difficult.

The soils of the lowlands are formed entirely from alluvial sediments and, as a consequence of their youth, do not reveal any marked horizon differentiation. They do not possess a well-developed structure and, thus tend to break down to individual particle sizes when wetted. To some extent this absence of a well-defined structure is a reflection of the low organic matter content of the soil. This is due to the general absence of vegetation and the high summer temperatures which promote rapid decomposition of organic material. Throughout the lowlands, calcareous alluvial soils cover by far the largest area. They are often well drained, but, when drainage is poor, more saline alluvial soils predominate. In the larger depressions, which may be continually wet for part of the year, Solonchak soils are found.

12.3 History of land use

Our knowledge of the ancient history of settlement and land use in the Tigris–Euphrates lowlands is still fragmentary. However, following detailed archaeological work, a reasonably clear picture of the changing patterns of agriculture, irrigation and settlement can be outlined in certain areas.

Research on the Diyālá Plains, near Baghdād, has revealed successive and distinctive patterns of water use and settlement.[13] From the beginnings of cultivation and settlement in the Tigris–Euphrates lowlands during the late fifth millenium BC, the Diyālá region was characterized by a linear pattern of dispersed settlements along the major water courses and by irrigation works which had little impact on the natural environment. During this period, irrigation seems to have been achieved by breaching the natural levees of the braided channel system of the lower Diyālá river or by the construction of small canals. One of the chief characteristics at this time appears to have been the durability of the irrigation network, for the same system of canals and distribution channels was in existence at both the beginning and end of the period. To some extent, the longevity of the system can be explained by the ease and simplicity with which the maintenance of the levees and canals could be accomplished by the villagers themselves. The settlements which evolved with this pattern of agricultural activity were relatively small, with the majority of the inhabitants apparently engaged in food production for their own consumption. Regional integration of agriculture does not seem to have occurred, and there is little evidence for the existence of coordinated networks of irrigation canals (Figure 12.6).

A second phase of land use, which lasted more than a thousand years, was initiated towards the close of the first millenium BC, following a long period of agricultural decline, and perhaps even the abandonment of land throughout the region. During this phase, the growth of urban centres commenced and became quite rapid, especially in the period following the conquests of Alexander the Great, and the spread of Greek influence. Later, under Parthian and Sassanian rule, the political capital was established at Ctesiphon, and as a result the Diyālá Plains became the centre of a large and important empire. So great was the expansion of the cultivable area at this time, in an attempt to feed the growing

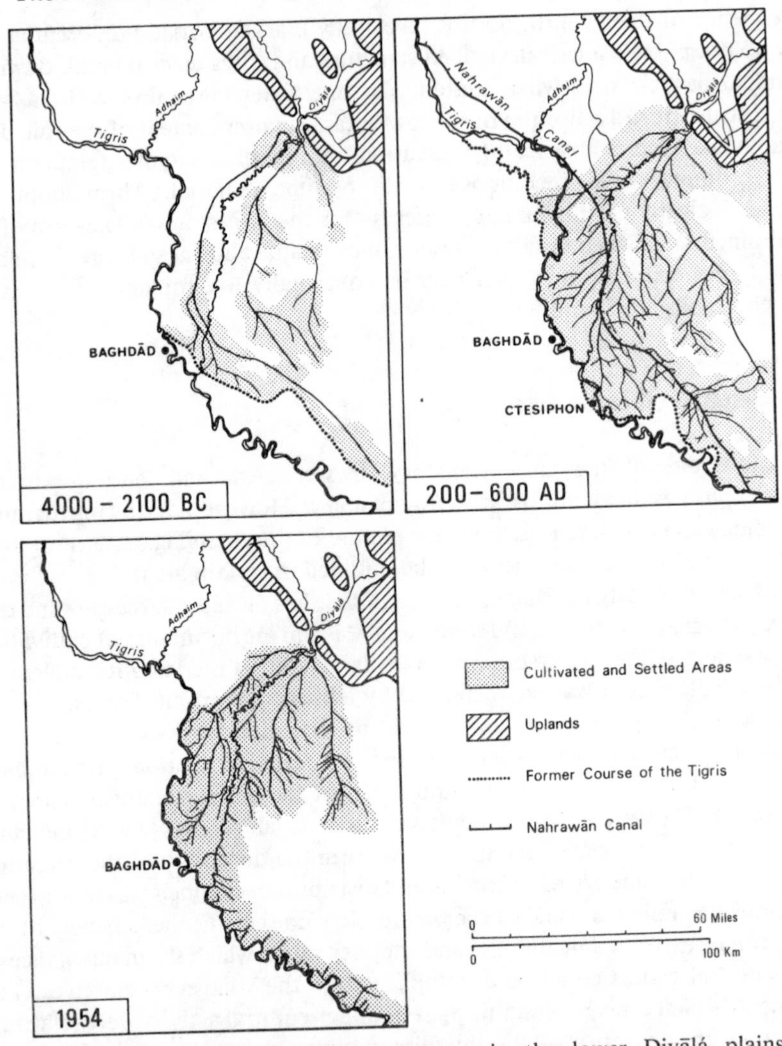

Figure 12.6 Cultivated and settled areas in the lower Diyālá plains (reproduced by permission of University of Chicago Press)

population, that land, as well as water, seems to have been in short supply. With the consequently increased demands for water, the irrigation system had to be enlarged. To ensure that these needs could be met, the central government became increasingly involved in the creation and operation of large integrated canal networks. The supreme example of such a construction was the Nahrawān canal, built during the sixth century AD. This canal, which was 300 km in length and in places more than 50 m in width, abstracted water from the river Tigris near Sāmarrā and transported it southeast to the lower plains of the river Diyālá, where it was used for irrigation purposes. Its water permitted the almost continuous cultivation of the region.

One of the major reasons for the success of this complex irrigation network was the establishment of an efficient system of drainage, which prevented water-logging of the soil and consequent salination of the land. Throughout the lowland as a whole, this drainage was achieved by supplying irrigation water from the Euphrates in the west, and the Nahrawān canal in the east. This permitted the river Tigris, which was situated between the two, to function as a drain, and to collect water from the adjacent agricultural lands. So efficient was this system, that it supported widespread cultivation of the land in the region for many years without a serious decline in land quality.

The crucial disadvantage of this land use system was that it depended upon a strong central government to ensure that maintenance operations were carried out. On the other hand, the high productivity of the irrigated land enabled the rulers to accumulate large stores of agricultural produce, chiefly of cereals, in the form of taxes from the peasant farmers. With these supplies, they were able to maintain and feed large bodies of soldiers and labourers, who were able to protect and construct new irrigation facilities.

The term 'hydraulic civilization' has been used to describe societies similar to those which existed in the alluvial lowlands of Iraq, and which required large scale management of water supplies by the bureaucracies of central governments for widespread agriculture to be feasible.[14] Such societies based on government control of complex hydraulic structures and intensive irrigated farming by peasants, seems to have come into existence through an organizational, rather than a technical revolution, which necessitated the establishment of a new system for the division of labour to ensure that the irrigation network functioned efficiently. The great advantage of the system was that it allowed the production of large quantities of food from a given acreage and so permitted the support of high population densities in what were extremely harsh natural environments.

The maximum limits of agricultural expansion in the Diyālá Plains seem to have been attained during the Sassanian period (AD 226–637) (Figure 12.6).[15] With the collapse of Sassanian rule, a marked deterioration in agricultural conditions occurred, which continued almost unchecked for centuries. One of the characteristic and perhaps surprising features of this period was that, though the rural agricultural economy was in a state of progressive decline, urban growth continued at certain sites. New centres at Baghdād and Sāmarrā quickly outgrew earlier settlements, even though the provincial towns seemed to be stagnating and declining. At its zenith, the imperial capital at Baghdād is claimed to have possessed a population of at least several hundred thousand and perhaps even a million inhabitants, with the majority engaged in non-agricultural pursuits of administration, commerce and service activities.[16] With this marked concentration of power in the capital, little interest was shown in the rural regions which were mainly regarded as the source of tax revenues and food.

The reasons for the agricultural decline are complex, but the major one was probably the decreasing effectiveness of the central government, which meant that the necessary reconstruction and maintenance of the irrigation networks tended to lapse. Progressive siltation of the major canals occurred, reducing the efficiency of water transmission, and the irrigation control works fell into disrepair. These reasons, together with the artificial nature of the complex irrigation system built up during the Sassanian and earlier periods, meant that single destructive acts, or local deterioration at key points in the system, could quickly bring about the complete breakdown of water supply and distribution over wide areas downstream from the affected point. Such events, sometimes of quite small magnitude, were often sufficient to precipitate crises which the local population did not have the resources to remedy without the aid of the central government. The result was that, by the middle of the twelfth century AD, following brief revivals under 'Abbasid rule, large areas of the alluvial lowlands could not be used for arable farming, and the land reverted to poor pasture and range land, which was only capable of supporting a few nomadic peoples. By the time of the Mongul invasions of the twelfth and thirteenth centuries AD, the abandonment of the once fertile land was almost complete. While it is true to say that the Monguls did destroy a number of canal headworks and massacred local populations, their total effect upon the decline of the irrigation system of the alluvial lowlands appears to have been minimal. The irrigation complex, to all intents and purposes, had been destroyed by neglect before they arrived.[17]

From the end of the 'Abbasid Caliphate until the later part of the nineteenth century, when some of the old canals were cleaned of silt, irrigation within the Tigris–Euphrates lowlands was carried out only locally on a non-coordinated basis, and agricultural activity was at a particular low level. Wherever possible, the ancient canals were used to supply water and were even in places repaired, but nowhere was the work undertaken as part of a comprehensive scheme on the scale witnessed in Sassanian or even 'Abbasid times. In summary, for the 600 years prior to the beginning of the twentieth century, it would seem that the physical environment of Iraq dominated the agricultural response of man more strongly than at any time since the first millenium BC (Figure 12.6).

Although the agricultural recovery of the Tigris–Euphrates lowlands began during the late nineteenth century, with the cleaning of a number of the ancient canals, it was not until the early part of the twentieth century that the first modern river control work, the Al Hindīyah barrage (1909–13) was constructed on the river Euphrates.[18] Its original function was to divert water into the Al Hillah channel, which was running dry, but later, following reconstruction in the 1920s, it was also used to supply other canals. Between the two World Wars, considerable attention was given to the Euphrates canal system, and many new channels were constructed and new control works established. On the river Tigris, development work tended to come later. The building of the Al Kūt barrage commenced in 1934, but was not completed until 1943, while on the Diyālá, a tributary of the Tigris, a weir was constructed in 1927–28 to replace a temporary earthen dam which had to be rebuilt each year following the winter flood. The weir allowed six canals to be supplied with water throughout the year.

Following the Second World War, river control schemes tended to concentrate on the problems of flood control. Two of the earliest projects, completed in the mid-1950s, were situated towards the upper part of the alluvial valley. The

Sāmarrā barrage was constructed on the Tigris river with the objective of diverting flood waters into the Tharthar depression to provide a storage capacity of 30,000 million cu m. A similar scheme was also built on the river Euphrates, where the Ar Ramādī barrage diverted floodwaters into the Habbānīyah reservoir and the Abu Dibis depression. It had been hoped that the stored water from these two projects might be used for irrigation during the summer months, but it was discovered that the very large evaporation losses, together with the dissolution of salts from the soils of the depressions, seriously diminished water quality and rendered it unsuitable for irrigation purposes. In conjunction with the barrages on the mainstreams themselves, two major dams were constructed on tributaries of the Tigris. The Dukān dam, with a reservoir storage capacity of 6,300 million cu m, was completed on the Lesser Zab river in 1959, while further south, on the Diyālá river, the Darbandikhan dam, with 3,250 million cu m of storage was opened in 1961.[19]

12.4 Contemporary environmental problems and their solution

One of the major problems facing Iraq today is the need to feed the growing population, which seems set to attain 25 million people by the end of the century. Unfortunately the deterioration of land in Iraq over the last few centuries has been so severe that carefully planned large-scale schemes will be necessary to ensure any major improvement. The extensive and wasteful use of land for cereal production in the alluvial lowlands has been necessitated by the high salt content of the soils and the lack of field drainage. Over the years, a system of agriculture had evolved which at least makes cultivation possible, but which provides no lasting solution to the salt problem. Following the cereal harvest, the land is allowed to remain fallow for at least a year so that it might dry out and so that the water table, which has risen as the result of percolating irrigation water, may decline. During the hot summer season, the soils dry out to depths of approximately one metre, and deep cracks often develop. Just before the preparation of the ground for the next season's crop, the land is flooded to a depth of a few centimetres in an attempt to dissolve the soluble salts occurring close to the ground surface. The salts are then carried under the influence of gravity to the deeper layers of the soil profile where they enter into the groundwater system. The soil, following preparation, is planted with cereals. Wheat is the preferred crop, but barley is more salt tolerant and is now widely grown throughout the alluvial lowlands. During the early stages of growth, the crop thrives in the salt-free upper layers of the soil, provided that sufficient water is available through irrigation. Unfortunately, as the season progresses, water rises in the soil and is lost through evapotranspiration from the ground surface. The result is a progressive increase in the salt content of the surface layers of the soil. By harvest time, salt levels are often quite high, causing the crop yield to be low.

In recent years, agricultural water use in the lowlands has averaged

13,300 cu m/ha or 1330 mm/annum in the irrigated regions.[20] This figure is considered to be greater than the amount necessary for plant growth and for the leaching of salt from soil horizons. As a result of this excessive water use, about 60 per cent of the irrigated lands of Iraq now suffer from some form of salinity. Indeed, is has been claimed that since the beginning of modern large scale irrigation, 20 to 30 per cent of the country's total cultivated land has been abandoned because of this problem.[21]

Despite the serious difficulties which agriculture in Iraq is facing, it is reassuring to know that some of them are capable of solution, at least at a technical level. For example, soil salinity can be alleviated by surface leaching and an efficient drainage system. Once this is achieved, and provided that adequate water for continued leaching of the soil is available, then crop yields are likely to increase.

Drainage waters with very high salt contents will cause further environmental problems if they are allowed to drain directly into the major rivers. In such cases, the quality of river water can be so seriously reduced that it might become totally unsuitable for irrigation further downstream. A temporary solution to the problem would be to allow the saline waters to drain into enclosed desert basins and there evaporate. Any long term solution, however, must include the construction of a separate drainage network to ensure that the saline waters are carried directly to the sea without risk of contaminating river waters. At the present time, the waters of the Tigris and Euphrates rivers at the head of the alluvial plain have total dissolved salt contents of between 200 and 400 parts per million (ppm) and, therefore, as yet pose little danger to irrigated farming.

Although soil salinity usually receives most attention in discussions of agriculture, the really crucial problem for the alluvial lowlands of Iraq is the availability of water for irrigation. Without additional water supplies, expansion of arable farming in the region will be impossible. The total amount of water available for the Tigris–Euphrates catchment is still a matter of some dispute, with estimates ranging from 73,000 million cu m/annum to 84,485 million cu m/annum.[22] Of these the highest figure is probably the most accurate as it is based on the 40 year period 1931 to 1969. Of the total flow of 84,485 million cu m, 37.7 per cent is from the Euphrates, 27.5 per cent from the Tigris above Mosul and 34.8 per cent from major tributaries of the Tigris.

To utilize the available water fully, Iraq desperately needs a comprehensive irrigation and water resource development programme which considers the country as a single unit. Any integrated scheme in Iraq would have to establish efficient methods of water distribution and control at both the local and regional level, bearing in mind the finite amount of irrigation water which is available and ensuring that sufficient water is at hand for leaching the soil and thus preventing salinity build-up. Better still would be to plan for basin-wide development. In the 1970s an agreement was reached between Turkey, Syria and Iraq for the rational development of the waters of the Euphrates. However, the growing political differences between Syria and Iraq in the late 1970s and early 1980s has meant that any plans have never been fully implemented.

Today, irrigation water in Iraq is almost exclusively obtained from surface water sources. Since the Second World War, the average withdrawal of water from the Tigris–Euphrates system has increased markedly, although the area of arable lands has only shown a moderate expansion. The available data on water use suggest a withdrawal of about 19,000 million cu m/annum between 1940 and 1949; about 28,000 million cu m/annum between 1950 and 1959; and a very rapid rise in the 1960s to an average value of around 49,000 million cu m/annum.[23] A detailed breakdown of the figures in terms of the individual drainage basins is shown in Table 12.2. The average water which returns to the rivers as the result of drainage in unknown. The actual water demands for crops grown in central and southern Iraq are high. For example, cereals such as wheat and barley require 4,300 and 5,200 cu m/ha, while cotton and vegetables need from 10,000 to 12,400 cu m/ha.[24] The largest water users are lucerne (alfalfa) and rice which consume between 18,000 and 21,450 cu m/ha. What Iraq has to decide upon is whether it can go on with the cultivation of crops with very high irrigation water demands or whether it would be better to switch the water to other uses.

These water withdrawals from the two rivers have grown to such a scale that the future development of agriculture in Iraq is likely to become increasingly difficult, unless rigorous and strictly enforced controls of water use are introduced. Even by the early 1970s the rate of water abstraction from the Tigris–Euphrates system was approximately 60 per cent of the average flow of the major rivers. Since this time water consumption is believed to have increased further, though it is difficult to obtain accurate current data.

In low flow years, the discharge can fall to 25,000 million cu m/annum, a value which is only about one half of annual water consumption.[25] This clearly demonstrates the serious nature of the problem facing Iraq in terms of its water resource development. To protect the natural ecosystem of the rivers, as well as to transport agricultural and industrial effluents, it seems necessary to maintain a flow of at least 10,000 miliion cu m/annum. Subtracting this value from the mean annual value of 84,000 million cu m/annum suggests that up to 74,000 million cu m/annum of water is available for beneficial uses over a long period for the basin as a whole. However, to make full use of this maximum figure, it is essential that storage capacity is available, so that water in years of high discharge can be stored and utilized during periods of drought.

TABLE 12.2
Volume of irrigation water abstracted from the major rivers in Iraq

River	million cu m/annum
Euphrates (Hit to Hindīyah)	17,213
Tigris and tributaries (Mosul to Fathah)	4,190
Tigris (Fathah to Baghdād)	14,052
Diyālá	5,139
Tigris (Baghdād to Kūt)	8,614
Total	49,208

Source: K. Ubell, 'Iraq's water resources', *Nature and Resources*, 7, p 4, (1971), by permission of UNESCO, ©UNESCO 1971.

In contrast to earlier times, when the provision of irrigation water was the sole concern of any hydraulic control works, river engineering in the twentieth century has had three major objectives in Iraq. These are the provision of water for irrigation and the reclamation of neglected agricultural land; the prevention of floods; and, wherever possible, the generation of power. All three objectives have necessitated the creation of reservoirs capable of providing sufficient storage capacity to hold the flood water of one year and so allow it to be used in subsequent years. The storage of water within the Tigris-Euphrates system is most efficiently achieved in the region outside the alluvial plain proper, where dam sites are more plentiful and potential evapotranspiration losses lower. Ideally, the location of such sites are to be found in the upland margins of the basin. On the Euphrates, most of the better sites lie within Turkey or Syria, while for the Tigris, Iraq is more fortunate in that a number of suitable locations lie within her territory, along the tributaries of the Greater Zab and the Diyālá. A number of these latter sites have already been developed. The Dukān dam on the Lesser Zab and the Darbandikhan on the Diyālá were primarily planned as flood control structures, but naturally fulfil other needs as well. For example, it is hoped to divert waters stored in the Dukān dam southwards to the river Tigris. The Bakhma dam on the Greater Zab could impound flood waters which might be used to supplement the flow of the Tigris, and so compensate for any water which is diverted from the Lesser Zab at the Dukān dam. Any surplus water from the Bakhma dam could also be utilized for irrigation on the lowland around Arbīl. A new dam on the Tigris at Mosul will hold back a reservoir with 10.7 million cu m of water storage. Other potential reservoir sites exist at Eski Mosul and Al Fathah. These could supply varying amounts of irrigation water, flood protection and power.

Almost 90 per cent of the discharge of the Euphrates originates in Turkey. This has led to a considerable dispute arising between Turkey, Syria and Iraq as to how the water resources should be divided. In Turkey, the Keban dam has been built, and plans made for considerable use of irrigation water in the southeastern parts of the country, where it is believed that up to 800,000 ha can be irrigated from the Euphrates. In Syria, the Tabqa dam has also been constructed and large quantities of water are being used for both urban/industrial and irrigation uses. Iraq too has also built a large dam at Al Hadīthah, with a reservoir capacity of 6.4 million cu m, to control the flow of the river and to provide a limited amount of storage. At the present time it is difficult to obtain reliable information on the actual amounts of water which are being used in Turkey and Syria. As yet little seems to be being abstracted in Turkey, though in Syria substantial quantities are being used. If all the water resource development schemes of the three countries are carried out as planned, there can be little doubt that the total amount of water necessary will be well above the mean annual river flow[26] (Table 12.3). From this, it is obvious that some form of binding agreement between the three countries must be reached as soon as possible so that scarce capital resources are not squandered on projects which can never produce the designed amounts of water. However, given the political differences between Iraq and Syria any agreement would seem to have little chance of implementation.

TABLE 12.3
Potential water use along the Euphrates River (annual)

	million cu m/annum
Mean total discharge at Hit (Iraq)	31,820
Turkey	
Evaporation from reservoir above Keban Dam	476 (max)
Evaporation from reservoirs above Karakaya, Gölköy, and Karabaka Dams	607 (max)
Potential water withdrawal for irrigation	3,500– 7,000
Syria	
Evaporation from reservoir above Tabqa Dam	630 (max)
Potential water withdrawal for irrigation	5,000–10,000
Iraq	
Current water use (1960–69)	17,213
Potential evaporation from reservoir above Al Hadīthah Dam	602 (max)
Predicted total water use as a result of schemes actually built, under consideration, or planned	Minimum – 28,028 Maximum – 36,538

Source: P. Beaumont, 'The Euphrates River – an international problem of water resources development, *Environmental Conservation*, **5**, No 1, pp 35–43, (1978).

In Iraq, the future development of the irrigation system depends basically upon a more efficient use of water. The irrigated area is unlikely to increase significantly, but total agricultural production would rise dramatically if two crops per year were to be cultivated. Emphasis, of course, would be placed on producing more summer crops. Such increased cultivation would require an assured water supply, which in turn means the use of large-scale water storage schemes on the major rivers. At the local level, besides the provision of efficient drains, an important need is the levelling of the fields to ensure adequate and equal irrigation dosages. The results would be the use of smaller amounts of water to cultivate a larger area. Yields would tend to be higher as well, and the amount of labour input would decrease. Water distribution networks can also be improved by providing adequate linings to the larger canals to ensure that water transmission losses are reduced. To a large degree, such losses have been significantly reduced since the Second World War with the introduction of pumps which lift water from the canals and rivers directly to the fields, and so do away with the need for intermediate links in the system.

Even more fundamental to the future development of agriculture in Iraq is the emphasis which is to be placed on rainfed systems. To date most of the investment has gone into irrigated agriculture infrastructures in the Tigris–Euphrates lowlands, at the expense of new developments in dry farming regions. However, if the water available to lowland Iraq does decline and the population continues to grow at its present rate, then there can be little doubt that Iraq will have to obtain a greater proportion of its agricultural produce from rainfed systems. Unlike many other countries in the region Iraq is fortunate to possess large areas of land in the north which might be successfully developed for more intensive agricultural production.

12.5 Conclusion

Irrigated agriculture in Iraq now seems to be entering yet another crucial stage in its development. Growing populations in all the countries within the Tigris–Euphrates basin are causing increasing pressure on available resources. At the present time, a number of water development projects are being constructed to meet these needs, but little or no coordination of effort is taking place between the various countries involved. Already it would appear that the completion of certain projects on the Euphrates will mean that other schemes downstream will never be able to reach their design potential. Just as the large irrigation networks in ancient Iraq were dependent for successful operation upon a strong central government, so today the increasing size and complexity of environmental problems necessitates the creation of a centralized water authority with complete control of development projects throughout the whole of the Tigris–Euphrates basin. In this context at least, the existing states of the region are of insufficient size to be able to ensure efficient water resource development. In the future some international authority, with responsibility for the whole of the basin, must come into being to coordinate irrigation, drainage, power generation and land reclamation schemes. Unfortunately the creation of such an agency did not seem imminent even before the Iran–Iraq war. Today other matters are of greater concern to the Iraqi government.

References

1. P. Beaumont and K. S. McLachlan, *Agricultural Development in the Middle East*, Wiley, London, 349 pages, p 315, (1985)
2. R. A. Fernea and E. W. Fernea, 'Iraq', *Focus*, New York, **20**, p 5, (1969).
3. R. I. Lawless, 'Iraq – changing population patterns'. In *Populations of the Middle East and North Africa,* (Eds J. I. Clarke and W. B. Fisher), University of London Press, Chapter 4, pp 97–129, (1972).
4. Population Reference Bureau, *Population Data Sheet 1985*, Washington, DC.
5. *FAO Production Yearbook 1984*, FAO, Rome, p 53, 1985.
6. *FAO Fertilizer Yearbook 1982*, FAO, Rome, p 121, 1983.
7. P. Beaumont, 'The agricultural environment: an overview'. In *Agricultural Development in the Middle East*, (Eds P. Beaumont and K. S. McLachlan) Wiley, Chichester, p 23, 1985.

8. D. Warriner, 'Revolutions in Iraq'. In *Land Reform in Principle and Practice*, Oxford University Press, London, Chapter 4, pp 78–108, 1969.
9. J.L. Simmons, 'Agricultural development in Iraq: planning and management failures', *Middle East Journal*, **19**, pp 129–40, 1965.
10. H.W. Ockerman and S.G. Samano, 'The agricultural development of Iraq', in *The Agricultural Development of the Middle East*, (Eds P. Beaumont and K.S. McLachlan), Wiley, Chichester, p 192, 1985.
11. M.M. Elgabaly, 'Water in arid agriculture: salinity and waterlogging in the Near East region', *Ambio*, **6**, pp 36–9, 1977.
12. V.R. Aart, 'Drainage and land reclamation in the Lower Mesopotamian Plain', *Nature and Resources*, **10** (2), pp 11–17, 1974.
13. R.McC. Adams, *The Land behind Baghdād*, The University of Chicago Press, 1965, 187 pages.
14. K.A. Wittfogel, 'The hydraulic civilisations'. In *Man's Role in Changing the Face of the Earth*, (Ed W.L. Thomas, Jnr). University of Chicago Press, pp 152–164, (1965).
15. T. Jacobsen and R.McC. Adams, 'Sand and silt in ancient Mesopotamian agriculture', *Science, N.Y.*, **128**, p 1256, (1958).
16. R.McC. Adams, *The Land behind Baghdad*, University of Chicago Press, p 116, (1965).
17. T. Jacobsen and R.McC. Adams, 'Sand and silt in ancient Mesopotamian agriculture', *Science N.Y.*, **128**, p 1257, (1958).
18. Naval Intelligence Division (Great Britain), *Iraq and the Persian Gulf*, Geographical Handbook Series, p 438, (1944).
19. (a) C.W. Mitchell, 'Investigations into the soils and agriculture of the Lower Diyala area of eastern Iraq', *Geogrl. J.*, **125**, pp 390–7, (1959).
    (b) C.W. Mitchell and P.E. Naylor, 'Investigations into the soils and agriculture of the middle Diyala region of eastern Iraq', *Geogrl. J.*, **126**, pp 469–75, (1960).
20. K. Ubell, 'Iraq's water resources', *Nature and Resources*, **7 (9)**, (1971).
21. W.C. Brice, *South-West Asia*, University of London Press, 1966, p 379.
22. (a) 73,000 million cubic metres/annum: W.H. Al-Khashab, *The Water Budget of the Tigris–Euphrates Basin*, Department of Geography, University of Chicago, Research Paper No 54, 1958, p 39.
    (b) 79,000 million cubic metres/annum: M. Clawson, H.H. Landsberg and L.T. Alexander, *The Agricultural Potential of the Middle East*, Elsevier, New York, 1971, pp 202–5.
    (c) 85,000 million cubic metres/annum: K. Ubell, 'Iraq's water resources', *Nature and Resources*, **7**, p 3, (1971).
23. K. Ubell, 'Iraq's water resources', Nature and Resources, **7**, p 4, (1971).
24. N.S. Kharrufa, G.M. Al-Kaway and H.N. Ismail, *Studies on Crop Consumption Use of Water in Iraq*, Pergamon Press, Oxford, 1980.
25. K. Ubell, 'Iraq's water resources', *Nature and Resources*, **7**, p 9, (1971).
26. P. Beaumont, 'The Euphrates River – an international problem of water resources development', *Environmental Conservation*, **5 (1)**, pp 35–44, 1978.

CHAPTER 13

Agricultural Expansion in Syria

13.1 Introduction

Modern Syria (Figure 13.1) was carved out of the Ottoman vilayets of Aleppo, Beirūt, Damascus, and Zōr-and-Jezira in the aftermath of the First World War. Since foundation, its existence has been characterized by political turmoil, though different factors have combined to produce this situation, including the imposition of a French mandate (1920 to 1945), the struggle for independence, and the Arab–Israeli conflict. Stress was increased by the socialist measures taken during Syria's union with Egypt in the United Arab Republic (1958–61) and under the Ba'ath (Arab Resurrection) Party which came to power in 1963. Rapid

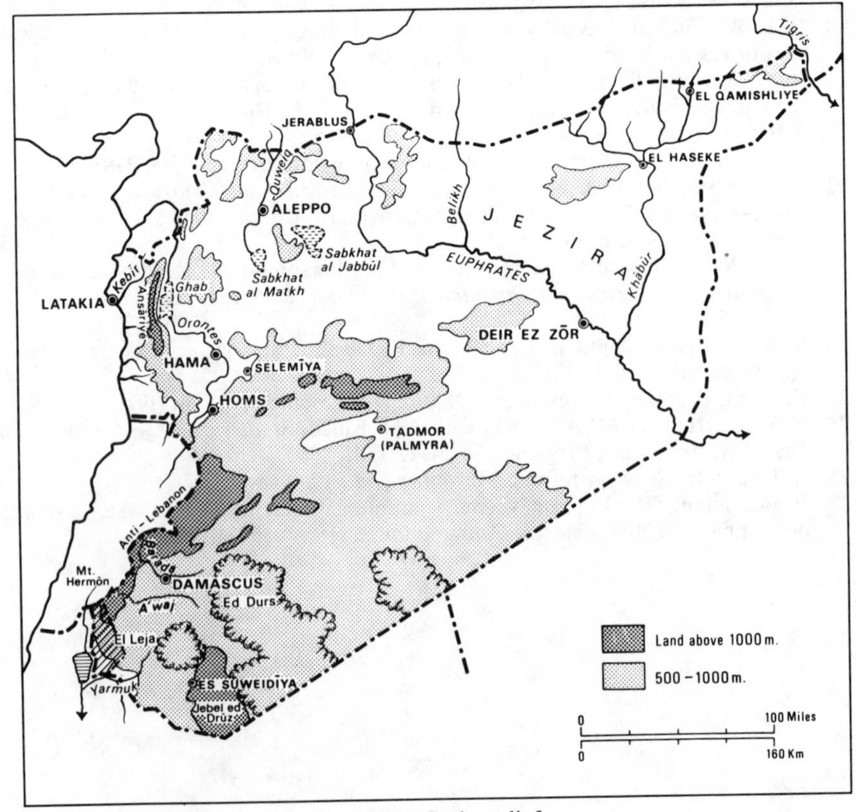

Figure 13.1 Syria: relief

TABLE 13.1
Percentage growth in Syria's GDP, 1953–83

Years	Annual Rate of growth
1953–57	5.5
1957–64	4.6
1964–69	4.6
1969–74	5.5
1974–79	7.1
1979–83	8.2

Sources: Kanovsky 1977, 9; Firro 1986, 43

economic development during the 1950s and 1960s (Table 13.1) also caused strains. A degree of political stability followed the seizure of power by Lieutenant-General Hafiz al-Assad (August 1970), but the ascendancy of fellow Alawites in the state provoked trouble, notably the Muslim Brotherhood. Continued economic growth during the 1970s (Table 13.1) helped to reduce tension, as did conscious attempts to integrate the state and to ensure that the provinces, as well as the major cities, benefited from economic and social improvement.

Until the nationalization of the oil industry in 1964 and the boom in oil revenues after 1973, the comparatively high rates of growth enjoyed by Syria were based largely on increases in agricultural production. Although the contribution of agriculture to GDP has fallen substantially since then (32 per cent in 1962, 25.6 per cent in 1972, 18.5 per cent in 1982), the value added by the agricultural sector increased at a mean annual rate of 3.5 per cent between 1963 and 1978.[1] Under the Third (1971–75), Fourth (1976–80) and Fifth (1981–85) Five Year Development Plans agriculture received considerable amounts of investment, even if these were not as large as planned. The aims were to achieve greater self-sufficiency in food, reduce imports, provide more raw materials to industry and to improve the nutrition of the population.[2] This degree of attention recognises the basic importance of agriculture to Syria. Agricultural development is the theme of the present chapter.

A large proportion of the increased output from agriculture during the 1950s and 1960s resulted from the steady advance of cultivation into areas long unused for arable farming. It was the continuation of a process set in train during the early nineteenth century, but the pattern of advance was closely influenced by physical conditions, especially the characteristics of precipitation (Figure 13.2).

For several centuries, dry farming was associated with the wetter areas of the west. The Ansāriye and Anti-Lebanon mountains, which close off much of Syria from Mediterranean influences, constituted one important, though fragmented zone. Steep slopes provided relatively secure situations against the marauders who overran so much of the country, while high ridges, exposed to the rain-bearing winds, raised precipitation to a reliable 750 to 1,000 mm/annum. A more continuous belt of dry farming stretched across rolling upland between the

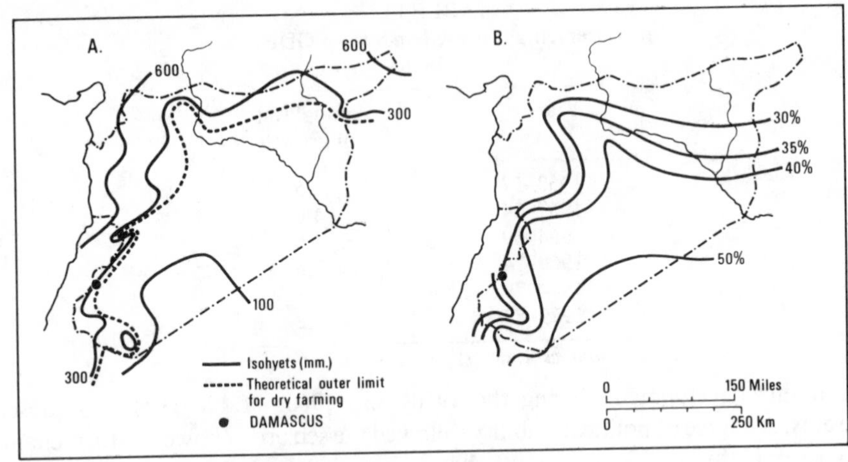

Figure 13.2 Precipitation in Syria
(a) annual precipitation
(b) relative interannual reliability
(after de Brichambaut and Wallén 1963)

Orontes depression, at the foot of the mountains, and Aleppo, and was made possible by an average precipitation of over 300 mm/annum with a 65 per cent reliability. Smaller areas lay scattered amongst the volcanic mounds and basaltic tracts of the Hauran and in the valleys and hollows of the neighbouring Jebel ed Drūz, where average annual precipitation is over 250 mm. Wheat and barley are basic field crops in all the traditional dry farming districts, while figs, olives and vines often form extensive plantations.

Opportunities for dry farming existed northeastwards of the Euphrates, in the rolling country of the Jezira, the 'island' between the Euphrates and the Tigris. Here average precipitation totals reach 200 to 400 mm/annum, with interannual variabilities of 20 to 30 per cent. The rivers Belikh and Khābur, as well as the Euphrates, offer scope for irrigation. The potential, however, could not be realized until after about 1930, when the region's nomads had been pacified.[3]

Much of Syria receives less than the critical average of 240 mm of precipitation and interannual variabilities are greater than 37 per cent (Figure 13.2). The transition zone runs across rolling country in the north, and permanent rainfed cultivation becomes impossible with distance southeastwards. Precipitation, however, is usually sufficient to produce extensive but poor quality seasonal grazing in various parts of the Syrian desert. The grazing was used by thousands of sheep and camel-rearing nomads until recently. But expansion of cultivation and changes in cropping practice in the crucial base-areas on the margins of the dry farming zone, or in the oases, combined with government sponsored sedentarization schemes to reduce the number of nomads to fewer than 70,000.

Outside the zone of permanent dry farming, cultivation takes place only where irrigation is possible, as at the famous oasis of Tadmor (ancient Palmyra). The main areas of traditional irrigated farming, however, lie further north and west, either within the dry farming zone itself or on its margins, since these are the situations where most water is available. An important transitional belt stretched along the Orontes between Homs and Hama, though the widest part of the rift valley, called the Ghab, was little more than a malarial swamp for centuries. Wheat, barley and beans were the main crops here. Further south, the rivers Barada and A'waj flow down from the Anti-Lebanon range to nourish intensive, garden-like cultivation in the oasis of Damascus. Some irrigated farming has always been found along the Euphrates, though the incised nature of the river in many reaches restricted development before the era of the large dam. Technical advances, however, have allowed the expansion of agriculture to break away from the age-old constraints represented by critical isohyets and interannual variabilities.

13.2 Expansion of cultivation

Western commentators in the late eighteenth and early nineteenth century constantly drew attention to the number of ruined villages to be seen in what is now Syria. They were found in every region, but appeared most numerous east of the Aleppo–Damascus road and in the Hauran. Inhabited villages and cultivation survived mainly to the west of this axial road, a region corresponding roughly with a narrow tract in which permanent rain fed cultivation was possible and the risks of dearth comparatively small.

The sorry state of affairs over much of Syria was the result of a long phase of land abandonment caused by a complex of interacting factors. Capricious taxation, rather than environmental factors, was probably the most important of these. By the second half of the eighteenth century, the raising of taxes had degenerated into an almost *ad hoc* system based upon farming and greatly influenced by the needs of new governors to recoup the price paid for their position, as well as to meet various emergencies, real or imagined. The peasant's ability to pay fluctuated with the harvest, which in turn depended upon rainfall conditions. In areas where rainfall variability exceeded 30 per cent a year and rents were accordingly below average, cultivation ceased to yield the landlords any profit. When this happened, landlords seem to have concentrated their tenants in areas where cultivation remained possible and the 'marginal' land was simply abandoned. Frequently, it was taken over by nomads, some of whom may have been the descendents of cultivators who abandoned agriculture because of the lighter taxes levied on livestock. Nomads certainly ranged far and wide across the country during the eighteenth and early nineteenth centuries.

Raiding and demands for protection money increased after the end of the seventeenth century, when indigenous Arab tribes were forced westwards by the northern movement of the Shammar and Anizeh confederations. Many peasants found their burdens excessive and were only too pleased to remove into large villages further west, where the harvest was more certain and agglomerated population offered a degree of protection.

The situation began to improve during the brief period of Egyptian rule, 1831 to 1840. Over 150 villages were reoccupied in the Aleppo district and a similar number in the Hauran. However, some abandonment was reported by British consuls in the 1850s and 1860s. Ruined villages remained common as late as 1920, but the process of reoccupation pushed ahead sporadically and locally throughout the second half of the nineteenth century. By 1860 the frontier of settlement and cultivation had moved to a point about halfway between Aleppo and the Euphrates. At the start of the First World War it had reached the river.

Several parallel developments assisted the expansion. Nomads, defeated in the tribal wars of the eighteenth and early nineteenth centuries, gradually gave up their old habits and became permanently settled in the fringes of the cultivated area. This was especially marked around Aleppo. Here, as elsewhere, little colonies of cultivators also gradually established themselves away from large parent villages. They began to reclaim land, either on their own account or for some city magnate or nomadic chief. During the 1870s, numbers of Circassian refugees were deliberately settled on the cultivation frontier.

A classic example of the reoccupation process is afforded by the Selemīya district southeast of Hama[4] (Figure 13.1). Despite the fame of its water in medieval times, when *qanāts* were used for irrigation, the whole district lay abandoned at the beginning of the nineteenth century. In 1854, a group of Ismā'īlīs (a Shī'a sect, some of whom were known as 'The Assassins') moved into it from the Ansāriye mountains and settled on the site of the old town. They partially reactivated the old irrigation system by cleaning out the *qanāts* so that within twenty years the cultivated areas stretched outward for about 13 km from Selemīya. Two daughter settlements were founded on old sites in the 1860s and a further seven in the 1870s. Circassians joined the original colonists in 1885, adding a further three villages of their own. Other colonists moved to the district subsequently. After the First World War, semi-nomads, originally driven eastwards by the expansion of cultivation around Selemīya, began to make cultivation a major activity and built a number of permanent settlements.

The recolonization of the Jebel ed Drūz was somewhat different. During the eighteenth century much of the Jebel Hauran, as it was known, was exposed to nomadic raiding and many villages had been abandoned. The situation, however, had already begun to change. From about 1711, small groups of Druzes, defeated in clan wars in the Lebanon, began to make their way eastwards and to establish themselves, against weaker opposition, in the vicinity of Es Suweidīya (Figure 13.1). In the early nineteenth century they were joined by larger groups driven from the Lebanon when Emir Bashīr (1789–1840) broke the power of their chiefs. Fighting between Christians and Druzes in the 1840s and early 1860s added more colonists. These tended to settle away from Es Suweidīya and gradually brought the whole Jebel under control. About 120 villages are currently inhabited.

Reclamation and expansion were facilitated by economic and political developments both within and beyond the Turkish Empire during the second half of the nineteenth century. The most important single factor was a rising demand for agricultural produce, especially for cereals. This seems to have begun during the

Anglo-French wars of 1793 to 1815, when large numbers of foreign, principally British, troops were garrisoned around the Mediterranean and required supplies. Despite an Ottoman ban on cereal exports, Greek merchants were able to buy grain from landowners in the coastal provinces of the Turkish Empire and, with the connivance of local governors, ship it to the foreign garrisons. Coastal Syria was, no doubt, tapped. Cereal production was further stimulated by rising demand from the increasing population in northwestern Europe and by evolving regional specialization within the Turkish Empire. Although Britain imported much of her grain from the Baltic and the Ukraine until the 1870s, a small amount was also taken from Syria, which enjoyed the advantage of a harvest at least a month in advance of European suppliers. Syrian grain even replaced that from the Ukraine during the Crimean War. During the nineteenth century, some of the provinces of the Turkish Empire began to develop agricultural specialities. Mount Lebanon, for example, specialized increasingly in the production of silk, and the planting of mulberry trees reduced the area available for cereals. Since Lebanese harvests were already inadequate to feed the growing population, grain had to be imported and Syria was close at hand and able to meet the demand.

The cereals exported principally represented the landlords' share of the normal crop, although it was marketed by merchants. Long-established landlords and credit-giving speculators were able to acquire extensive properties by manipulation of the Ottoman Land Law of 1858. Designed to secure the peasant's title to land, the law resulted in property being registered under the names of powerful individuals. They realized that profits could be made from cereal farming, and, accordingly, were prepared to introduce colonists to their newly acquired land. In some cases, the chiefs of semi-nomadic tribes, whose collective property was registered in one name, were able to follow suit and persuade their tribesmen to expand cultivation. Somewhat later on, the Sultan himself encouraged the settlement of imperial domain lying on the then existing frontier of cultivation.

Colonization was assisted by a general tightening-up of Turkish local government. Central direction became more effective as the nineteenth century advanced and distant provinces were brought within the range of the telegraph and railway. Active steps were taken to contain and pacify nomads. Regular, well-armed troops were sent on expeditions against them, garrisons were maintained at strategic points, such as Tadmor and Deir ez Zōr, and a network of police posts was established.

The First World War may have stimulated further expansion of the cultivated area to meet the needs of Turkish armies campaigning in Mesopotamia and Palestine, though the subject has not been investigated.

Sporadic expansion of the cropped area certainly continued in some districts between the World Wars, while infilling with daughter settlements was more characteristic in others. Altogether some 7,000 sq km were added to the cultivated area of Syria between about 1850 and 1940, and about 2,000 villages were created.

The outbreak of the Second World War meant that the countries of the Middle East were forced to provide all their own food. At the same time, large numbers of British and French troops were garrisoned in the region or were operating in adjacent parts of North Africa and had to be fed. Syrian response to the situation was to initiate reclamation in the recently pacified Jezira. Economic exploitation was facilitated by a government decision to grant immense areas of fertile but uncultivated state land to tribal sheikhs. Sheikhs leased out their rainfed land to town-based entrepreneurs for 10 to 15 per cent of the crop, but sold their irrigable property. The entrepreneurs often came from the western towns, chiefly Aleppo. Low population forced them to adopt mechanized methods, as is shown by a rise in the number of tractors from about 30 in 1942 to 500 in 1950 and in the number of harvesting and threshing machines from about 20 to 430.* The area of cultivated land rose from about 216,000 ha in 1942 to 302,200 in 1945.[5]

Capital accumulation during the War, a rising national population and increasing urbanization and industrialization allowed developments in the Jezira to continue. By 1951, about 500,000 ha were cultivated. The total reached 1,400 sq km in 1960, representing an increase of nearly 500 per cent since the War. In 1961 the cultivated area of the whole country was estimated at about six million ha, though at least 40 per cent of the total probably lay fallow. Nevertheless, the figure represents the attainment of an optimum situation and the physical limits were reached, or even exceeded, apparently even in the Jezira, where attempts to extend rainfed cultivation south of El Haseke proved disastrous. Recent studies suggest, in fact, that nearly 30 per cent of the dry farmed land is really unsuitable for its present use.[6]

Expansion in the Jezira is worth further consideration because of its results. Speculative mechanized farming in the region resulted in the monoculture of cereals or cotton, or an alternation of the two. Fertility accumulated through centuries of disuse has been mined, and declining yields were commented upon by the International Bank Mission as early as 1955,[7] though they became more noticeable under drought conditions at the end of the decade. Extensive mechanization also increased the effects of wind erosion, so that parts at least of the Jezira are in danger of resembling an American dust bowl.

Ploughing reduced the grazing available to nomads. Some moved southwards into Iraq, but others have attempted to maintain their old way of life, though on herds reduced to sizes incompatible with a satisfactory standard of living and at the cost of more extensive wandering. Many tribesmen, however, became cultivators for their sheikh. Others found work as labourers in the expanding towns and villages of the region or as tractor and lorry drivers.

At the end of the First World War, the population of the Jezira was low and the settled element minimal. Armenian and Kurdish refugees from Turkey and Iraq in the 1920s increased the population, as did the arrival of some 9,000 Assyrian Christians from Iraq in 1933. By 1938, the settled population was about 103,500. This was too small to meet the needs of the 'merchant-tractorists', as the entrepreneurs have been called.[8] Labour had to be imported from the western parts of the country. Unlike earlier colonists in other districts, these were paid

*Figures refer to Haseke province, which covers much of the *high* Jezira.

regular wages, though attempts to introduce a modified form of share cropping were reported by Warriner.[9] Colonization and the settlement of nomads brought the established population of the Jezira to about 340,000 in 1960. There was little spontaneous migration to the region from the congested areas of the west; emigrants from the backward districts of the Ansāriye mountains and the Jebel ed Drūz preferred to settle in neighbouring large towns.

Population growth in the Jezira has been accompanied by an increase in the number of settlements and the development of towns. Relatively few permanent settlements existed east of the Euphrates at the end of the First World War, but within ten years there were about 280 and by 1960 there were at least 2,000. Some villages were built by the 'merchant-tractorists' to house their workers. Others crystallized around fillng stations and cafés established on routes leading westwards out of the Jezira. The town of El Haseke was little more than a military post in the 1920s, whilst El Qamishliye was a completely new town established to replace Nusaybin, the original regional centre, which was allocated to Turkey. Both grew on the influx of refugees.[10] They subsequently enjoyed boom conditions as centres for mechanized farming and marketing. An atmosphere of frontier rawness persisted into the 1960s, despite the appearance of apartment blocks and modern shops. By 1981, El Haseke had a population of 72,589 and El Qamishliye one of 93,385.[11]

Despite the dominance of large estates, the Jezira appears to have been affected comparatively little by the troubled history of land reform which had such a disruptive impact on the agricultural sector of Syria's economy during the 1960s. This may be due to two special reasons. Much of the region was not worked by share-cropping tenants, like the traditional dry farming areas, since labour was in short supply, and successive governments may have been reluctant to assist the numerically dominant community in the Jezira, namely the Kurds. In any case, a lot of appropriated land remained in state hands.

As elsewhere in the Middle East, the major aim of land reform was to break the power of the great landlords and redistribute confiscated property to landless families. In 1958, when reform was initiated, about 0.6 per cent of the rural population was estimated to own about 35 per cent of the cultivated land, while some 240,000 rural families, at a minimum, were said to be in need of land. Although the scope of the programme remained comparatively modest throughout the vicissitudes of the reform (officially complete by 1970), considerable resistance was encountered from the landlords, many of whom resorted to violence and refused to cultivate their surviving properties. Not only were they aggrieved at the actual loss of their estates, where they did in fact lose them, but they felt that the proposed reforms were unrealistic in that they fixed the maximum holdings at levels which were too low (initially 80 ha of irrigated and 300 ha of unirrigated land) for Syrian conditions, where extensive dry farming is characteristic and not intensive irrigated farming, as in Egypt, whence many of the original reformist ideas were derived. There were further problems attendant on land reform. Title to land became insecure for many families at all levels of the rural social hierarchy (despite the wishes of the reformers), while credit and extension services were inadequate to the tasks required of them in the new circumstances. Moreover, reform was launched at an unfortunate time, during a period of severe drought, when repeated crop failure drove many dispirited people from the land. Altogether, then, it is not surprising that even in 1970 as much as 62 per cent of the normally cultivated land lay idle, though it must be remembered that this figure still included a large amount of fallow. By the same date, the contribution of agriculture to GDP fell by about 12 per cent since 1962. The decline was due not only to precipitation conditions, but arguably also to the direct effects of the land reform programme. Structural changes in the economy of the country were of little consequence at that time. Some recovery took place after 1970, when rainfall conditions again improved and, more particularly, when the landlords began to receive more sympathetic attention from the government. The agricultural contribution to GDP rose to about 26 per cent in 1972. Its subsequent decline owes more to the advance of other sectors of the economy than to anything else. Nonetheless, the total cultivated area of the country apparently decreased for a decade from the mid-1960s (1963–66: 6.5 million ha; 1975: 5.5 million ha). Whilst the decline owes something to modifications in the official definitions, it may also reflect a withdrawal from some of the more marginal land brought into cultivation during the 1950s.[12]

13.3 Cropping (Figure 13.3)

The expansion of cultivation in the 1950s increased the quantity of cereals available. Recent figures indicate the trend with respect to wheat and barley, the leading cereals, and show quite clearly the decline associated with land reform in the early 1960s (Table 13.2). Wheat and barley between them occupied about 75

TABLE 13.2
Areas devoted to major crops

Year	Wheat ha	Barley ha	Cotton ha
1938[1]	538,000	300,000	37,000
1945[2]	751,000	348,000	17,500
1950	992,000	416,000	78,000
1955	1,463,000	614,000	228,000
1960	1,549,000	742,000	260,000
1965[3]	1,213,000	682,000	295,411
1970[3]	1,340,000	1,126,000	299,072
1975[4]	1,590,000	1,172,000	169,000[5]
1980[4]	1,449,000	1,210,000	139,000[5]

Source: 1. Naval Intelligence Division, Geographical Handbook Series *Syria*, London, 1943, p. 259.
2. Figures for 1945–1960 from M. Rosciezewski, Méditerranée **6**, 1965, 179–180.
3. L'Office Arabe de Presse et de Documentation, *Rapport 1971–1972, sur l'Economie Syrienne*, Damascus, n.d., A33, B61.
4. FAO, *Production Yearbooks*.
5. K. Firro 1986, 48.

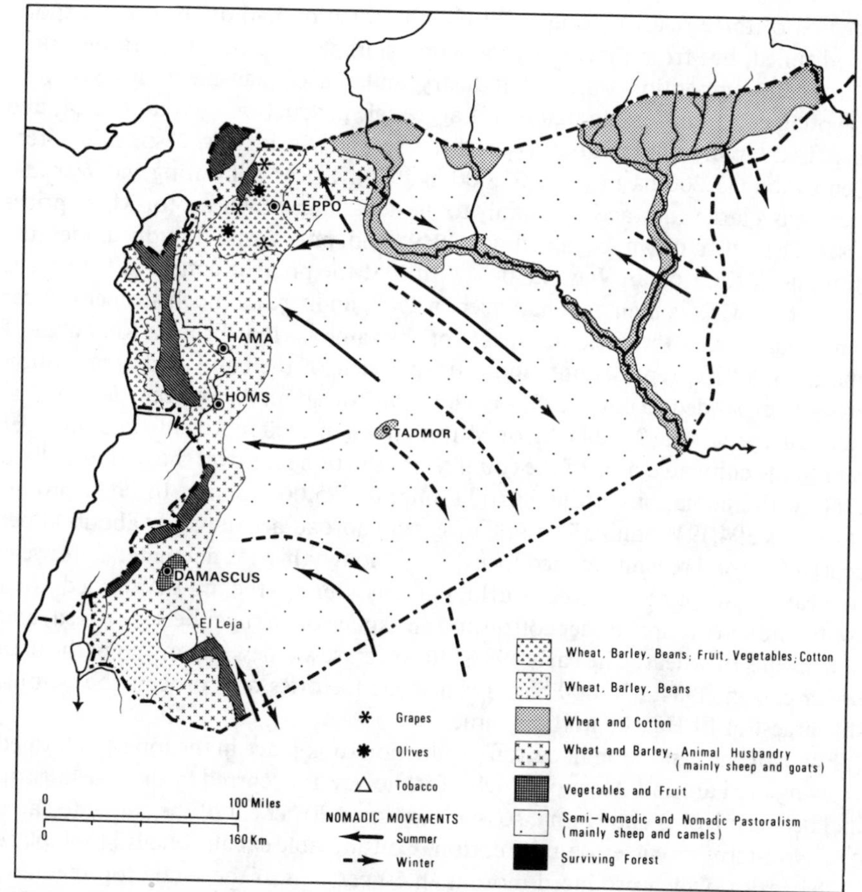

Figure 13.3 Generalised land use in Syria (updated from Hudson, *Focus* 18, 1963, published by the American Geographical Society)

per cent of the cultivated area of the country down to 1970 but during the 1970s the proportion declined to about 70 per cent. The precise proportion of cultivated land devoted to the two crops varies regionally. It is smaller (up to 60 per cent) in the western provinces, where polyculture of olives, vines, fruit trees and cotton is important, and much greater (up to 90 per cent and over) in the Jezira where monoculture of cereals and cotton is frequent. Legumes, vegetables and fodder crops take up about 10 per cent of the cultivated area nationally, but are rather more important in the west (Figure 13.3). Since 1976 considerable efforts have been made to increase sugar production and the area under beet more than doubled by 1982 (26,000 ha).

In terms of area and economic importance, cotton was the second rank crop in Syria until well into the 1970s. Its fortunes, however, fluctuated over the previous 140–150 years. Some cotton has been grown in Syria since medieval times. In the eighteenth century, it was used in the textile industries of Aleppo and Damascus. During the brief period of Egyptian rule in the early nineteenth century, it enjoyed the first of three relative booms. By 1840, production had doubled and exports quadrupled, but from that year production slumped as a result of the effects of European competition on local industry and the displacement of Syrian by Egyptian cotton in foreign markets. Large scale production revived in 1924, and increased rapidly until the outbreak of the Second World War, despite the Great Depression (13,200 ha in 1934; 37,000 in 1938). Prices fell during the War and there was a large contraction, mainly to the advantage of wheat, for which prices rose. The third boom began about 1949, and prices rose rapidly under the stimulus of the Korean War (1950–53). In 1951 the price of cotton at Damascus was about twice as high as it had been in 1949, an increase much steeper (28 per cent) than that in the price of wheat for the same period. From a total area of about 25,000 ha, representing about 0.6 per cent of the cultivated area, cotton growing expanded to cover 228,000 ha in 1955 or about 4.5 per cent of the total cultivated area. The 260,000 ha of cotton grown in 1960 represented about 7 per cent of the cultivated area of the country. Production rose over the period 1949 to 1960, with fluctuations, from 38,100 tonnes to 295,000 tonnes. In 1968, production was 394,193 tonnes. The sale of cotton abroad accounts for about 40 per cent of the total revenue earned by exports, though this has fluctuated from year to year. Low prices on the world markets during the mid-1970s led to a contraction in the area under cotton and an expansion in sugar beet (as noted) and new strains of wheat. The early 1980s, however, saw a new expansion in the area under cotton (1984; 170,000 ha). Twenty-one factories now process the output, the largest at El Haseke in the Jezira.

Some of the expansion in cotton production took place in the long-established growing area around Aleppo. Much of it, however, occurred in the coastal strip and between Homs and Hama, so that now up to 20 per cnt of the cultivated area of the western provinces is under cotton. Considerable expansion also took place in the Jezira, which produced more than 60 per cent of the total crop regularly during the 1960s. Here, as in other areas, both dry and irrigated methods are used, depending on water availability, though the use of irrigation is expanding. Excessive use of water was reported from the central hill and steppe region, where monoculture of cotton is carried out on some estates. Not only did salination develop in the soil, but the water-table also dropped locally so that some villages were reported to be short of drinking water.

13.4 Expansion of irrigation

Much of the expansion of cultivation up to the Second World War was achieved in the rainfed zone using the long-established methods of dry farming. Since the Second World War, irrigation has become increasingly important. Reserves of cultivable rainfed land have been used up and the need to increase yield and to control its fluctuations has also been realized. Experience of severe drought in 1958 to 1961 may have increased the sense of urgency.

Until the Second World War, irrigated farming was found in four distinct areas. It was especially characteristic of the Damascus Oasis, where river-fed canals were used. On the Orontes at Homs and Hama, the current turned enormous wooden wheels which lifted water into high-level canals. Similar devices were also found along the Euphrates. The Aleppo district was a fourth irrigated area. Here a variety of sources was tapped – the Orontes west of the town, the Quweiq in its immediate vicinity, the Euphrates to the east and a number of scattered wells. Small and dispersed patches of irrigation were found elsewhere in the country, in the coastal strip, for example, and at oases such as Tadmor. *Qanāts* and springs were tapped, often by systems of great antiquity, in favourable districts, while elsewhere water was lifted from wells by wheels similar in design to those on the Orontes and Euphrates but worked by animal power (*challuf* or *doulab*). These ancient lifting devices were steadily being replaced by pumps during the interwar period, though the process was to gather momentum after the Second World War. Pumps were also used to expand the strips of cultivated land along the Euphrates and Khābūr.

Despite local expansion, less than 10 per cent of the cultivated area of the country was irrigated in 1947. Between 1947 and 1959 an increase of over 97 per cent was achieved to bring the irrigated area to a total of 584,000 ha. By 1967, some 640,000 ha were irrigated, although the proportional increase which this represents is minimal. Much of the expansion took place in the western provinces (the amount of irrigated land in Hama and Homs provinces increased by about 300 per cent from 1947 to 1959). As in the interwar period, the greater part of it was the result of small-scale developments. Improvements in pumping techniques, the reduction of seepage by lining canals and an increase in the number of pumps all made more water available.

A number of government schemes also helped (Figure 13.4). The earliest scheme was actually begun in 1930, under the Mandate. It involved raising the level of the ancient dam retaining Lake Homs and the cutting of a canal for some 70 km along the western side of the Orontes. This was completed in 1944 and an additional 21,600 ha were irrigated. Six other schemes were completed before 1967. The copious and perennial Sinn river, which enters the sea between Latakia and Baniyās, was dammed to irrigate a small area of coastal plain, primarily for the benefit of Palestinian refugees. Further along the coast, the northernmost of the two Kebīr rivers was regulated by a dam to prevent flooding in the coastal marshes south of Latakia and irrigation works are proceeding. On the Mzerib, a tributary of the Yarmuk, a small dam diverted water into canals along both banks to irrigate some 3,460 ha. At Matkh, south of Aleppo, flood water is diverted from January to April to irrigate some 14,400 ha of good valley land, whilst farther west, at Roudi, mountain streams were regulated and a swamp area drained to provide some 3,840 ha of irrigated land. Work also began on a scheme to make fuller use of the Khābūr. The head spring at Ras el'Ain is used to irrigate about 10,000 ha, though there are plans to irrigate up to 35,000 ha using storage reservoirs on the river.

Down to 1968, the most ambitious scheme completed involved making greater use of the Orontes and reclaiming the Ghab. Work began in 1954.

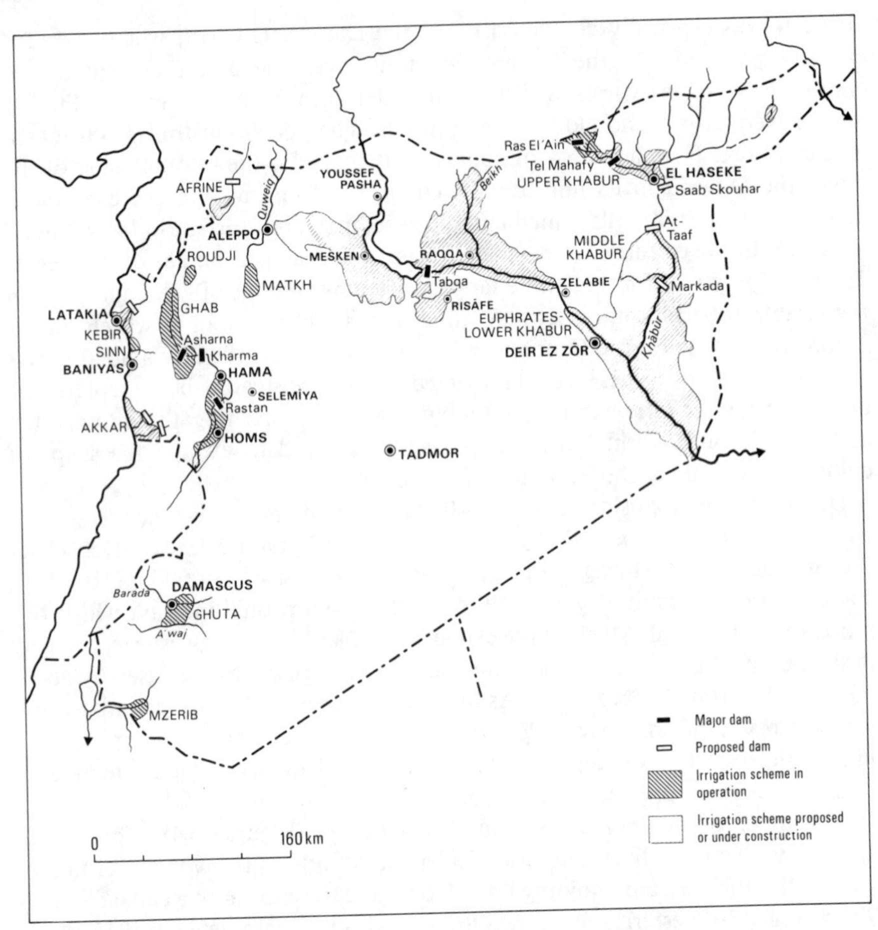

Figure 13.4 Syrian irrigation schemes in about 1980

Flow was improved by deepening and widening the river bed at the point where it cuts through the basalt sill at the north end of the Ghab (Kārkūr). Drainage was further assisted by cutting a new and short course for the river through low ground to the west of the swamp. A large concrete dam at the divergence near Asharna directs water to two main irrigation canals. The river has been regulated in the gorges above the Ghab by the construction of dams at Karma and Rastan, incidentally rendering the flow too low to turn the traditional water wheels. By 1968, when the original project was virtually complete, some 43,200 ha of swamp had been reclaimed and irrigation extended to a total of 68,000 ha. These developments were accompanied by the construction of new roads and settlements. The range of crops was extended by adding rice, cotton, and sugar beet to the long-established winter combination of wheat and barley alternating with vetch or clover.

In 1968 work began on a long-discussed, but much delayed scheme to exploit the Euphrates still further. The key is a huge, earth-filled dam at Tabqa. Completed in 1973, it is 4.5 km long and 60 m high. Lake Assad will eventually extend for about 80 km upstream and cover an area of some 640 sq km. About 25,000 ha of irrigated land producing cotton and 43 villages will be flooded. In return, 1600 million kW/h of electricity are being produced, 15 new villages have been built, a new town (Al-Thawra: 'The Revolution') has been created with a population in excess of 40,000, and it is hoped to irrigate 240,000 ha of land in the Euphrates valley itself and a further 400,000 ha in neighbouring areas. The initial plan was to bring the new land into production within 20 years, but this is proving wildly optimistic. By 1978, only 7,400 ha of land was actually being irrigated. The delay is due to inadequate preparatory work, in particular superficial appraisal of land capability.[13] Yet the project has been enormously expensive. For example, it absorbed nearly a quarter of all public investment in Syria during the Third Five Year Plan (1971–75).

Cost is a major problem facing any major extension of irrigation in Syria. Further expansion can be achieved, in large measure, only by constructing more dams (on the Orontes and its tributary, the Afrin; on the Arouse river in the Akkar plain; and on the Khābūr). The net increase in irrigated land would be about 30 per cent on the 1980 total of about 539,000 ha, if all the current plans were implemented and the estimates fully realized. In fact, the estimates appear to be optimistic, and the difficulties with the Euphrates scheme encourage scepticism about their realization. Several schemes have been launched, nonetheless, with much of the finance coming from abroad. Meanwhile, the number of motor pumps more than doubled during the 1960s and 1970s so that about 44 per cent of all irrigated land is serviced by them. Overpumping of groundwater resources is now a serious problem in the Orontes basin and in the neighbourhood of Aleppo and Damascus. Official statistics even suggest that there was a net loss of about 20 per cent in the amount of land irrigated between 1963 and 1980. If the decline is not a statistical aberration, its scale is worrying and suggests that more attention must be given to such side-effects of irrigation as water-logging and salination.

A further constraint on the expansion of irrigation, at least along the Euphrates, may be international agreements. The Euphrates is an international river. Iraq has ambitious plans to extend irrigation in its territory. Both Iraq and Syria suffer from the reduced flow and poorer quality of the water resulting from the construction of the Keban Dam in Turkey (1974). Matters will get worse as the Turks complete the Karakaya Dam, the first in a projected series. International agreement is clearly called for, but it will limit what the riparian governments will be able to undertake in their own territories.

13.5 Communications and industry

Part of Syria's increasing agricultural production has been consumed directly by her rising rural population (4.8 million in 1981), but some has been exported. Agricultural products formed about 17 per cent of Syria's exports by value in

1982, compared with 60 per cent a decade earlier. The principal export is cotton and, in the early 1980s, the major buyer was Algeria. The rest of Syria's agricultural production goes to the towns, to feed the urban population (48 per cent of the total of 9.2 million in 1981) or for use as industrial raw material.

The leading towns are located in the west, the traditional core of cultivation in the country. Food, raw materials and important exports are drawn from this region but, until recently, flows were hampered by poor communications and correspondingly high transport costs. Polarization of production, resulting from developments in the Jezira, exacerbated the situation, for the distances involved are considerable and communications were scarcely adequate to the need. Much has been done over the last fifteen years to improve the situation.

A new railway was built from Aleppo to Tabqa for the construction work on the dam and extended down to Deir-ez-Zōr. It has since been extended across the Jezira to El Haseke and El Qamishliye. At the latter it crosses the frontier to join the old Baghdād Railway which runs just inside Turkey. The new line has helped to consolidate developments in Syria's northeast and to reduce the region's peripheral status. Aleppo is linked to the western end of the Baghdād Railway which also has a branch running southwards to Homs and Hama. A direct extension to Damascus, with a branch to Tadmor, was completed in 1982. South of Damascus, the Hejāz Railway has been rebuilt after partial destruction during the First World War, but it is largely used by Jordan. The spurs from the main north–south axis to Beirūt and Tripoli became less important for freight with the eruption of fighting in Lebanon (1975), but that from just south of Aleppo through to Latakia has become of increasing significance following the development of the Orontes Ghab and the coastal plain, as well as the expansion of the port itself. A line from Latakia to Tartūs, where the harbour was deepened and improved in the 1970s, is under consideration. Even without this link, the whole railway system is now much more integrated than it was a decade ago. Bottlenecks in the movement of goods have been eased, with the result that the amount of freight carried had increased nearly six times by 1982.

Roads have borne much of the burden of Syria's recent economic development. They were instrumental in opening up the Jezira and they integrated the western provinces through a particularly dense network on either side of the axial route between Der'a in Hauran and Aleppo. Pipe lines became important when oil production started at Es Selemīye (1968) and in the far northeast, at Karachuk (1969). They fed the crude to the refinery at Homs and to a new terminal at Baniyās. Oil production reached a peak of 10 million tonnes in 1974, but was running at below 8.5 million tonnes per annum during the early 1980s. Despite initial difficulties in marketing her low-quality crude with a high sulphate content, Syria secured 51.3 per cent of her export earnings from it in 1982. Her major customers are Greece, France, Italy and the USSR. Even so, since 1981, Syria has been a net oil importer.

As well as containing the largest urban populations in the country, the western towns are the principal centres of manufacturing industry. Although inertia has played a part in this, for the tradition of handicraft industry is very ancient here, it is largely a result of the consumer orientation of most modern industry. Some development took place between the wars, but the main stimulus came with the Second World War. Imported manufactured goods became scarce and prices rose, with the result that local industry enjoyed a unique advantage. Import substitution largely came to an end about 1950, though tariffs and embargoes fostered local enterprise. Capital was diverted into construction and agriculture, and little new industrial development took place. In fact, the profitability of agricultural development acted as something of a brake on the rate of industrial development in the 1950s. Nationalization in 1964 and 1965 caused considerable dislocation in industry and a serious flight of capital abroad, though, according to official indices, output continued to rise from 100 in 1963 to 129 in 1966 and 238 in 1970. The rising trend continued through the 1970s (1970 = 100, 1977 = 191, 1981 = 271).[14]

In the early 1980s industry still contributed 20–25 per cent of GDP and employed just over 30 per cent of the workforce. Manufacturing is still based squarely on agriculture, though much less so than a decade ago. Food processing (olive oil processing and refining, sugar and tobacco processing) contributed about 7 per cent of industrial output in 1982 compared with about 33 per cent in 1971. Textiles contributed about 25 per cent in 1982, instead of about 32 per cent. Food processing is widely distributed, but the major concentrations of plant are found in the west. Sugar refining has been boosted over the last decade, mainly in the hopes of processing beet grown in the Euphrates basin. Cotton textiles are made chiefly in Aleppo and Damascus, though printing and dyeing now take place on a significant scale at Homs and Hama as well. The woollen industry consists largely of carpet making at Aleppo and Damascus, though knitwear is also made, chiefly in the capital. Syria imports the wool used, despite being an exporter. Similarly, the long-established silk industry, mostly found in Aleppo, uses imported yarn, though a small amount is still produced locally.

The decline in the relative importance of the long-established industries is due principally to the increased output from comparatively new activities. One of these is cement. In addition to major plants at Damascus, Aleppo and Homs, new units have been built at Adra, Aleppo, Hamra, Musulmiya and Tartūs; output exceeded 5 million tonnes in 1985. Iron and steel production started at Hama in 1975 and reached 2.85 million tonnes in 1983, though this is well below the planned target. Other industries which have grown include furniture making, glass, matches, plastics, soap and rubber, as well as the assembly of tractors, refrigerators and television sets.

Plans for a car assembly plant, however, were postponed following financial problems in 1976. Salt is still made in a number of places, especially in the salt marshes south of Aleppo. Natural asphalt is exploited in the Ansāriye Mountains near Latakia, whilst phosphates have been extracted in the Tadmor area since 1972. Superphosphates are now made at two plants at Homs. Oil production has already been noted. With these developments, and the related expansion of commerce and services, Syria is clearly moving away from being an agricultural country. Nonetheless, the basic importance of farming remains. It provides much of the population's food and a renewable resource for industrial transformation at home and export abroad.

References

1. USAID (United States Agency for International Development), *Syria: Agricultural Sector Assessment*, Washington DC, 1980.
2. USAID (United States Agency for International Development), *Syria: Agricultural Sector Assessment*, Washington DC, 1980.
3. E. de Vaumas, 'La Djézireh', *Annls. Géogr.* **65**, 64–80 (1956).
4. N.E. Lewis, 'The frontier of settlement in Syria; 1800–1950', *International Affairs*, **31**, 48–60 (1955).
5. E. de Vaumas, 'La Djézireh', *Annls, Géogr.* **65**, 64–80 (1956).
6. USAID (United States Agency for International Development), *Syria: Agricultural Sector Assessment*, Washington DC, 1980, vol 1.
7. International Bank for Reconstruction and Development, *The Economic Development of Syria*, Johns Hopkins, Baltimore, 1955, 300.
8. D. Warriner, *Land Reform and Development in the Middle East. A Study of Egypt, Syria and Iraq*, Oxford University Press, London, 2nd ed, 1962, 89.
9. D. Warriner, *Land Reform and Development in the Middle East. A Study of Egypt, Syria and Iraq*, Oxford University Press, London, 2nd ed, 1962, 91–92.
10. E. de Vaumas, 'La Djézireh', *Annls. Géogr.* **65**, 64–80 (1956).
11. M.L. Samman, 'Le recensement syrien de 1981', *Population*, **38**, 184–88 (1983).
12. I.R. Manners and T. Sagafi-Nejad, 'Agricultural development in Syria', in *Agricultural Development in the Middle East*, (Eds P. Beaumont and K. McLachlan) John Wiley, Chichester, 1985, 255–78.
13. USAID (United States Agency for International Development), *Syria: Agricultural Sector Assessment*, Washington DC, 1980, vol. 2.
14. UN, *Statistical Yearbooks*.

CHAPTER 14

Religion, Community and Conflict in Lebanon

14.1 Introduction

Many states are composed of groups of people recognized as distinctive on grounds other than those of class or colour. Middle Eastern states are no exception. What is perhaps different about the region is the habitual classification of its communities on religious grounds. Ever since the Muslim Arab conquests the gross divisions of Middle Eastern society have been recognized as lying between the Muslims (those who claim that they submit to the full revelation of God through the Qur'ān) and the 'People of the Book' (that is, in Muslim eyes, those who have accepted part of God's self-revelation but corrupted it in one way or another - the Jews by confining God's love and mercy to themselves; the Christians by behaving towards Jesus, son of Mary, as if he were Son of God). Further subdivision is based upon specific patterns of belief and practice. Nowhere in the Middle East is the fragmentation of society into communities more developed than in Lebanon where 17 groups are officially recognized. Friction between some of these communities has fed a ten-year conflagration which has killed at least 70,000 people (2 per cent of the population) to date. This chapter seeks to sketch the origins of the community divisions of Lebanon and to outline some of the consequences.

Lebanon is not a natural unit in any physical sense. It consists basically of two roughly parallel mountain ranges, Mount Lebanon (Arabic *Lubnān*) and Jebel esh-Sharqi (Anti-Lebanon with Mount Hermon), separated by a broad upland valley known as El Beq'a (Biqa or Bekaa). Mount Lebanon itself is essentially a limestone plateau, crowned by a number of peaks (up to 3083 m) on which snow lies for much of the year. Above about 1,600 m, the so-called *jurd* contains typical karst scenery and looks distinctly bare and bleak, except where a few tattered remnants of the famous cedar forest have survived. The vegetation now is largely spiky shrub, grazed in summer by herds from El Beq'a and the hill country of 'Akkar in the northwest. The mountain slopes above about 1,600 m are characterized by huge cirque-like features lying at the head of deep valleys and ravines which cut to the coast and offer considerable obstruction to north–south communication. These are separated by crags, great tabular blocks and truncated pyramids, all of which combine to give the western slopes an irregular stepped appearance. The middle section of the eastern slopes is more precipitous, as a result of faulting, though the lower slopes are fairly gentle. Middle slopes throughout the mountain, the so-called *wusut*, are fortunate in containing a narrow out-crop of clays, marls and sandstones where water, absorbed by the higher limestones from a winter precipitation of 700 to 1500 mm and considerable snow melt, is forced to the surface in a number of often copious springs. The outcrop also provides relatively fertile soils, where cereals and fruit of various kinds (vines, olives, apples and some mulberries) are grown. Terracing has been necessary to exploit these conditions on the western side of the mountain and spectacular flights of terraces may be seen, though their higher levels have often fallen out of use. On the eastern side, slopes are either too precipitous even for terracing or, at lower levels, are sufficiently gentle for it to be unnecessary. Woodland survives in a ravaged condition in many places, but is most extensive on northeastern slopes. Loosely clustered, stone-built villages are numerous on defensive sites at an average height of 1400 to 1500 m and close to springs. Population densities are fairly high (about 177 person/sq km, on the western side), especially in the central area near Beirūt.

The mountain falls steeply to the coast for much of its length, making access to the inhabited zone fairly difficult, but in the north and south it descends through a series of ridges to merge with undulating plateau areas – the ‘Akkar plateau in the north, which affords communication between the coast and the plain of Homs in Syria, and Upper Galilee in the south. Extensive tracts of chalky marl occur in Upper Galilee and form such a hard crust that cultivation is difficult. Along the coast, mountain spurs separate small plains where low cliffs and sand dunes alternate with areas of intensive cultivation. Bananas, citrus and olives are grown, while market gardening is important around the country’s principal towns of Beirūt, Tripoli, Saida (Sidon) and Sūr (Tyre) which are linked by a coastal road making use of old wave-cut terraces to move from plain to plain. The towns were once fortified against both land and sea attack, and threat of the latter perhaps explains the clustered form of old rural settlements and their preference for cliff-top sites. Dispersion is characteristic of comparatively new settlement in the plains.

Anti-Lebanon consists of a series of mountain ranges which form a considerable barrier to east–west communication for much of their length. The main ridge is a broad, barren limestone upland with an average height of 2100 m. It falls steeply into El Beq’a on the west and is scored by wild and rocky ravines. Although average annual precipitation over the whole mountain is 400–600 mm, the northern parts are so sheltered by the highest peaks of Mount Lebanon that they receive much less. Springs are scarce and there are few villages and little cultivation, though some transhumance takes place.

South of the Zebdani depression rises the whaleback of Mount Hermon. Its oval-shaped summit is topped by three small peaks (2814 m), which are snow covered except at midsummer. The western slopes are exposed to rain-bearing winds, and their lower levels are composed of sandstones and basalts with the result that copious springs are found, together with vineyards, orchards and patches of deciduous oak woodland. Villages are relatively numerous in a zone up to about 1000 to 1100 m, and a few larger agglomerations occur.

El Beq'a is a high valley some 112 km long and up to 26 km wide. It is divided into northern and southern sections by the watershed (1002 m) between the Āsi (Orontes) and the Līṭāni in the vicinity of Ba'albek. The north is crossed by low ridges strewn with loose stone, though patches of productive land exist near springs along the foot of the Anti-Lebanon range and, to a lesser extent, at the contact of the plain with the foothills of Mount Lebanon. Mount Lebanon shuts out the westerly winds, reducing average annual precipitation to an unreliable 300 mm. South of Ba'albek the valley floor is undulating and its red alluvial soils prove more fertile under a much higher rainfall. Wheat is a traditional crop but maize and, increasingly, cotton and apples, are also grown under irrigation. Large, highly compact villages are fairly numerous, both along the Līṭāni itself, where they occupy elevated sites to avoid flooding, and along spring lines at the foot of the mountains to west and east.

The main valley is closed to the south by the limestone ridge of Jebel al-Arbi. However, the Līṭāni continues southwards through a series of deep gorges along the western side of the mountain before finally turning sharply westwards to reach the sea eventually near Sūr. The line of El Beq'a itself is continued southward by the Wādī el-Taym. Despite being drained by the Hasbani, one of the head streams of the Jordan, this is a remote and largely inaccessible area with a low density of population. Its villages are poor, since possession of elevated, defensible sites has deprived them of adequate water and cultivable land.

These different physical units were put together by France after she received the mandate for Syria at the San Remo Conference of 1920. In retrospect, French action has been seen as the cynical application of a classic policy of divide and rule. The core of the new state in a political sense was Mt Lebanon itself. Here an autonomous polity of Druzes and Maronite Christians emerged within the Ottoman Empire in the sixteenth and seventeenth centuries and survived, in various guises, down to the eve of the First World War. To this core the French added the coastal fringes which had formed the Ottoman vilayet of Beirūt, and various parts of the vilayet of Damascus – the lower lying hill country south of the Mountain in Upper Galilee, the upland valley of El Beq'a, and the eastern mountain ranges of Anti-Lebanon and Hermon. These areas were inhabited largely by Muslims: Sunni in the north and the coastal towns, Shī'a in the south, and a mixture in El Beq'a.

14.2 The origins of the principal communities (Figures 14.1a)

The mosaic of communities in the new state of Lebanon was the product of a specific historical evolution in which the security potential of the mountains featured prominently. At the time of the First Crusade (1096–99), most of the territory of the future state seems to have been only sparsely inhabited. Many of the people appear to have belonged to the Shī'a branch of Islam; they were known later in Lebanon as Mitwālis. The Shī'ites in general originated as the political and religious supporters of 'Ali, the Prophet's nephew and son-in-law,

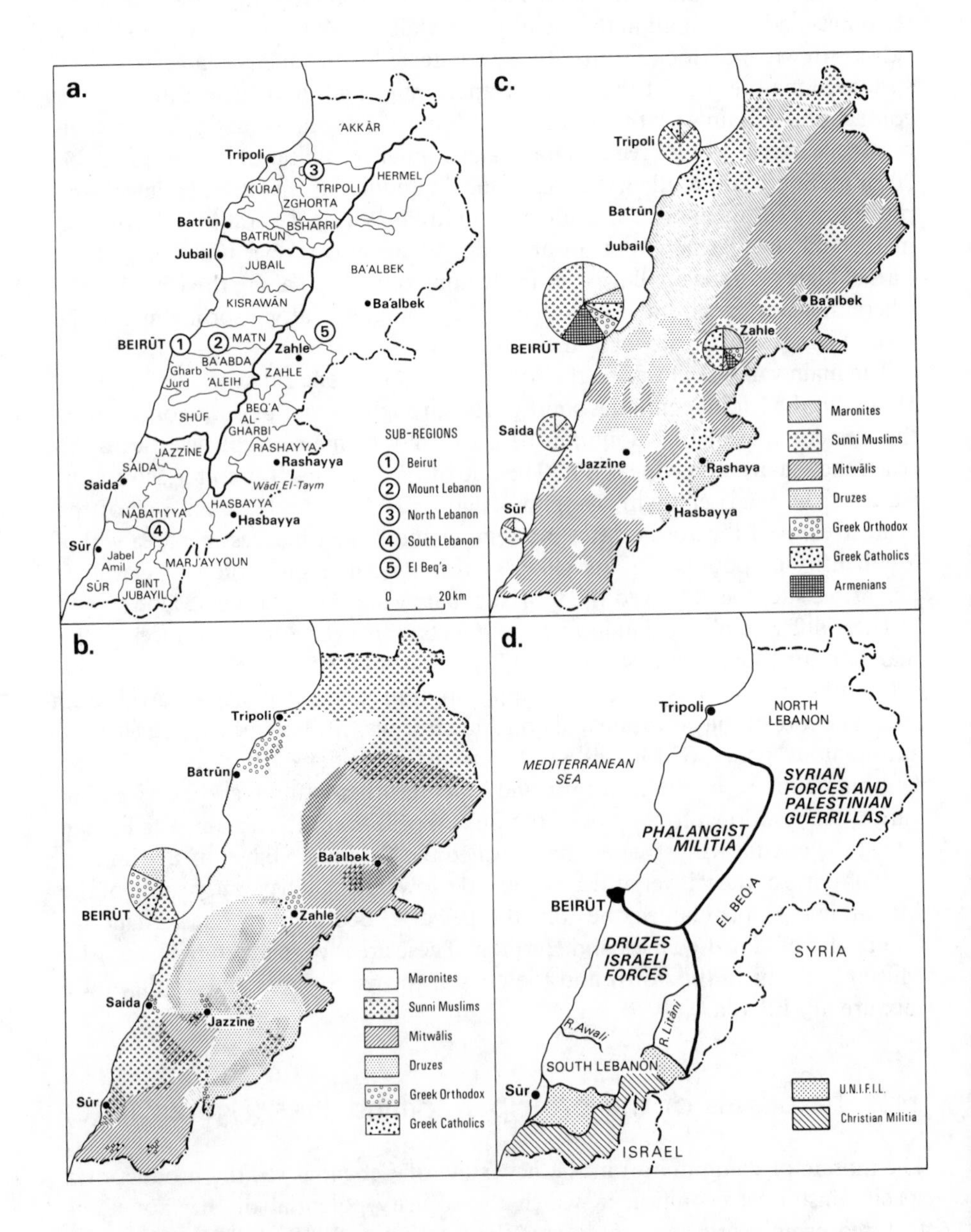

Figure 14.1 Changing community distributions in Lebanon, c. 1840–1982
(a) Administrative divisions used in text
(b) Community distribution, c.1840
(c) Community distribution, 1953
(d) Division of Lebanon, December 1982

first in his claims to be the Caliph or Successor to Muhammad as the leader of the Muslim community and its recently created state, and, subsequently, in his efforts to maintain his position as the Fourth Caliph. 'Ali's murder in AD 661 transformed him into a martyred saint. The partisans of 'Ali then declared his eldest son, Hasan to be the Imam (prayer leader, but to the Shī'ites a divinely appointed leader) and when he abdicated appointed the second son in his place. Husain was killed at Karbela in Iraq on his way to join his sponsors (AD 680) and, like his father, was immediately recognised as a martyr. His tomb became a place of pilgrimage. The Shī'ites soon established the custom of marking the anniversary of Husain's death with ten days of mourning, involving processions and flagellation, and a play depicting the saint's suffering and sacrificial death. A theology of suffering and martyrdom emerged which readily appealed to the poor and oppressed of the region, though over time it became also the ideology of ruling groups in some parts of the Middle East, notably Iran, Oman and Yemen. The teaching embraced the idea that the divine was made manifest in a person, the Imam. Some Shī'ites went on to believe that when a particular Imam died, the divine element passed to his successor, though it might also be temporarily withdrawn from mankind altogether. The Mitwālis of Lebanon, like many of the Shī'ites across the Middle East, recognise a line of twelve Imams in direct succession from and including 'Ali, the last of whom went into concealment in 800. Closely identified with these ideas is that of the Madhī ('The Guided-One') who will appear in the fullness of time to restore justice and righteousness to a corrupt world.

The reasons for the crystallization of Shī'ite ideas amongst people in the mountain ranges at the eastern end of the Mediterranean are obscure. Life was hard and basic. The possibilities of finding secure refuge certainly helped passive resistance to persecution organized through the towns, whether this came from Orthodox Christians in the days of the Byzantine Empire or from Sunni Muslims when these opponents of 'Ali triumphed. At the same time, the difficulties of communication within mountainous terrain not only reduced interaction and the spread of ideas, but allowed heterodox views to flourish unchecked. Other factors were probably a strong sense of local identity and opposition to ruling elites from outside. Both expressed themselves through religious dissent.

Whilst all Muslims accept the centrality of the Qur'ān in their lives and beliefs, Sunnis insist upon strict adherence to its teachings and the careful performance of all the rituals and obligations which flow from it into the Sharia or Holy Law. Legalism, rather than redemption through suffering, has been their response to the need for a system of sustaining beliefs. Sunnis are also more inclined to accept the *de facto* descent of political power and religious authority away from 'Ali and his line after the death of the Prophet.

Sunni presence within the boundaries of Lebanon appears to have developed relatively late. The Sunnis came as administrators and soldiers with the establishment of Mamluk supremacy over the area during the thirteenth century from their base in Egypt. Originally confined to the coastal towns, they were able to persuade and coerce Mitwālis in the northern part of the country and in El

Beq'a to convert, whilst settling Sunni Türkman and Kurdish groups in their midst during the fourteenth century. Ottoman expansion in the early sixteenth century added to the numbers in the coastal towns, whilst the economic prosperity of the hinterland from the seventeenth century onwards encouraged Sunni merchants to settle. In El Beq'a the Sunni element was reinforced by the voluntary sedentarization of some bedouin groups.

Mitwāli spatial and social dominance was further eroded by the expansion of two other communities, Maronite Christians and Druzes. Druzes took the Shī'ite concept of the Imam to its logical conclusion by recognising the highly eccentric sixth Fatimid Caliph of Cairo, al-Hākim ibn-'Armri'llah (996–1021) as the One and Unique embodiment of the Divine. He went into concealment to test the commitment of his followers. Although they read the Qur'ān, and also honour Muhammad as the Prophet of God, Druzes do not use mosques, do not make the Pilgrimage to Mecca and do not fast during Ramadan. In fact, for many, it is difficult to accept that they belong to the House of Islam. Their own customs and rules reinforce community solidarity and exclude strangers. How they came to be established in Mt Lebanon is something of a mystery, but no doubt the antinomian characteristics of Druze belief appealed to a beleaguered and exploited peasantry. Under persecution, the Druzes evolved a particularly tough and self-reliant society with a reputation for fierce fighting. It was clearly stratified with a mass of peasants leasing land from relatively few landowners who frequently feuded with each other. In the later sixteenth and early seventeenth centuries one of the leading landed families from the Shūf district of the southern part of the Lebanon range, the Ma'an, was able to secure control over Mt Lebanon and some neighbouring areas through a series of confused and repetitive wars. Their power subsequently collapsed. A new autonomous state was built in the early eighteenth century by the Shihābs, a related family originally from the Wādī el-Taym some of whose leaders eventually became Christians. Factional struggle gradually weakened their power and many Druzes fled southeastwards to safety in the Jebel Hauran, subsequently renamed Jebel ed Drūz.

Maronites appeared in Mt Lebanon much earlier. Their origins, too are somewhat obscure. They trace their history back to the split in the church over the question of whether Christ had a single nature or two, human and divine. A compromise theology emerged – that Christ had two natures but a single will – but this was accepted only by an Aramaic-speaking community distributed across a strip of territory between Antakya and Urfa. The Byzantine state turned against them under Justinian II (685–95, 705–11). Their principal monastery near Antakya (Antioch-on-Orontes) was destroyed. Some of the community were transported to Thrace. The survivors fled southwards and established themselves in the high fastness of Mt Lebanon, chiefly along the upper reaches of the Qadisha valley and its tributaries. The earliest known Maronite church in Lebanon was built at Ehden in 749. The Maronite community survived. It established contacts with Europe during the Crusades. Further rapprochement with the West was fostered by political and commercial contacts with France and Italy, as well as through clerics tempted to study abroad. Reunion with Rome took place as a result of synods held in Lebanon at Qannūbin and Luwayza in 1596 and 1736 respectively. Contacts with the West and the educational activities of Roman Catholic missionaries in Lebanon itself gradually built up an affinity with Christian Europe, especially France, amongst the Maronite leaders. France extended her protection to them, particularly during the nineteenth century, and the Maronites generally acquiesed in the French mandate.

Maronite settlement spread in the area of initial colonization, now regarded as something of a holy land. There were few indigenous inhabitants and, as well as defensibility, the area possessed cultivable soils and numerous springs.

The availability of similar, poorly exploited resources and the remnants of the forests, together with a decimated Mitwāli population, allowed Maronite farmers to spread southwards into Kirawān during the early fourteenth century. Further expansion of this energetic peasantry took place throughout Mt Lebanon during the seventeenth and eighteenth centuries, when the community enjoyed the protection of the Druze Ma'an and Shibāb emirs. They were favoured not only because their higher standard of education made some of them useful to the emirs in their administration, but also because of their proven ability to turn waste land to profit by building terraces, cultivating cereals and producing silk, for which there was a rising demand abroad. At the same time, the Maronites were helped by their own social customs. Their system of land holding was looser than that prevailing amongst other groups, especially the Druzes, whilst inheritance within the nuclear family, rather than the wider kin, and a ban on marriage within the patrilinear group meant that they were not tied to one locality, like Muslim share-croppers. Maronites also spread into the coastal plains, El Beq'a and Wādī el-Taym – a process of colonization perhaps growing from the mountaineer's habit of finding seasonal work on the large estates of the lower land.

However, Maronite expansion into traditional Druze districts was one of a number of factors which finally changed the centuries-old balance of power between the two communities. It was further undermined by the Egyptian occupation (1831–40) and the rise of a Christian middle class. Egyptian rule effectively undermined the bases of Druze power and, when the occupation ended under pressure from Britain and Austria-Hungary, Mt Lebanon was divided into two separate autonomous areas under Ottoman officials. The arrangement is often described as the Dual Kaymakamate (1842–60). The Maronites controlled the area to the north of the Beirut–Damascus road and the Druze controlled the area to the south. The new Christian middle class grew with the increasing role of Beirūt as a port and the expansion of silk production throughout the Mountain, especially during the 1850s. It gradually came to control and capitalize agriculture throughout Mt Lebanon, as well as to handle much of the export trade. Class conflict merged into violence, with Maronites fighting each other, as well as the Druzes, in a series of revolts and local wars during the 1840s. In 1858–60, encouraged by Ottoman officials and anti-Christian demonstrations in Damascus, the Druzes started to massacre the Christians in their midst. As many as 11,000 Christians were killed, 4000 starved to death and 100,000 fled.[1] When order was restored under Great Power pressure, Mt Lebanon was given special status under a Christian governor but within the Ottoman Empire. This was the United Mutesarrifate (1861–1909). Maronite expansion, however, had stopped and their distribution contracted.

Apart from the Maronites, other Christian communities are found in modern Lebanon. The Greek Orthodox are the most numerous. Although a degree of continuity can be claimed with the dominant Christian group of the Byzantine Empire, at least in beliefs and Liturgy, perhaps the majority of the Greek Orthodox Christians in the country are descended from refugees. The first group sought security in Mt Lebanon from persecutions in Palestine and Syria following the expulsion of the Crusaders, finally accomplished in 1291. Other refugees arrived in the eighteenth and early nineteenth centuries as persecution broke out in the cities of Syria and Wahhābi* influence grew. Many of the Greek Orthodox community moved into commerce as this increased from the seventeenth century onwards and came to enjoy a profitable, almost symbiotic relationship with the expansionist Maronites. Greek migrants were important to the rapid rise of Zahle during the nineteenth century when the town developed as a collecting point for grain from El Beq'a destined for the mountain villages where silk production had expanded at the cost of basic subsistence crops, a market for bedouin from further east, and a staging post on the carriage road from Damascus to Beirūt opened in 1858. Jesuit activity during the seventeenth century split the Greek church, and some of its congregations reunited with Rome in 1727. They are usually known as Greek Catholics, and are largely peasant farmers.

Armenian Christians differ from the rest of the population in speaking a language other than Arabic, a fact which has helped to make them unpopular with the local population. Most Armenians belong to the Gregorian church, one of the monophysite groups which separated from the universal church over the doctrine of Christ's nature in the fifth and sixth centuries, and whose Patriarch lives in the Soviet Union, but a few of them are attached to the Armenian Catholic church which separated from the Gregorian and re-established communion with Rome in 1740. Although some Armenians were known in Lebanon from at least the eighteenth century, the majority came as refugees from south-eastern and Aegean Turkey in the aftermath of the First World War, following repressive measures and a series of 'massacres' (1890–93, 1915 and 1922). Their numbers were increased in 1939, when the sanjak of Alexandretta was finally ceded to Turkey, and again in 1946–47, as a result of Russian action in the Armenian Soviet Socialist Republic. The refugees settled mainly on the eastern edge of Beirūt. Their original settlements were shanty towns bearing names evocative of their homes in Turkey, but these were gradually transformed as the Armenians improved their position in commerce and the professions. Smaller Armenian groups were found in the Zahle and Sūr districts.

The remaining Christian communities are also small in size, but some of them have played a significant part in fostering the friction which now exists between Christian and Muslim in Lebanon. Two of the Christian communities are relics of

*A reformist, almost fundamentalist, Sunni Muslim sect originating in central Arabia.

ancient and once extensive churches. The Syrian Orthodox (Jacobites), a monophysite church, uses Syriac in the liturgy and retains a few Syriac speaking villages in the northern part of Mt Lebanon. Some of their number returned to Rome in 1662. The Chaldean or Assyrian church as been in communion with Rome since 1552, though originally a Nestorian church which believed in the separate personalities of Christ, human and divine. Chaldeans form a very poor group at the bottom of the urban social hierarchy. Many of its adherents originally came to Lebanon to avoid persecution in the new state of Iraq. The other Christian communities are introductions from the West. The Latin or Roman Catholic church is composed of western teachers and missionaries resident in the country, and has a history going back to the seventeenth century at least. Although separate from the Uniate churches,* the community has had a profound effect upon them through its educational activities. Although some of Lebanon's Protestants are westerners, many are converts, won as a result of western, largely American, missionary activity since 1823. Like the Latins, the Protestants have exerted a great influence on the country through their schools and colleges, principally the American University of Beirūt, which was established in 1866 as the Syrian Protestant College.

14.3 Some implications of community identity

A background of persecution and refuge seeking is likely to have had certain consequences for the people concerned. They include, first, a strong awareness of group identity, particularly as defined against, or in hostility to, other groups; second, a determination to survive, come what may; third, an identification with territory where the community has a majority and to which it feels a particular attachment. In Lebanon, these characteristics were added to community solidarity built upon kinship and religious affinity. Each community is thus an integrated socio-economic and political unit (a kind of embryonic polity) whose norms constrain individual behaviour and condition interaction with outsiders.

Another possible consequence of persecution is introversion. This has tended to mark the Druzes. They have been unable to find support from similar religious communities; they do not exist. The Mitwālis have been in a similar position within a predominantly Sunni framework. Of course, if isolation is one reaction to persecution and insecurity, another is to seek friends beyond the group's own territory. The Mitwālis have been offered help from Iran in the 1980s. The Maronites, though, were able to find powerful outside support at a much earlier date. Through commercial and religious contacts from the sixteenth century onwards, they came to know the West and to be known there. By contrast, the Sunni Muslims brought into the state after 1920 had wide contacts in the Muslim world, shared the rich cosmopolitan culture of the dominant community of the

*Indigenous churches in communion with Rome but retaining separate hierarchies and customs.

Figure 14.2 Palestinian refugee camps in Lebanon

Ottoman Empire, and identified themselves with their Arab brethren in neighbouring countries. But, as well as being a consolation and a source of strength, outside support could also be a weakness. At the very least, the opportunity was created for interference.

Against the background outlined above, it is clear that any polity embracing Mt Lebanon, let alone adjacent territories as well, would have to take community affiliation into account. Equally, though, any configuration of power adapted to the balance of community interest at a particular moment of time would be at risk from a variety of directions. Shifts in the size of different communities, resulting from differential rates of natural increase and migration, could produce strains. Conflict might also be generated by variation in economic success. A new configuration would have to be produced to accomodate these, as in the Dual Kaymakamate and the United Mutesarrifate.

14.4 Demographic and political developments, 1846–1941 (Figure 14.1b)

The Dual Kaymakamate was an attempt to divide Mt Lebanon politically according to the geographical distribution of the major communities. Reliable statistics, however, were not available and the division was made on the basis of assumptions. The nature of these is revealed by two contemporary sources. Guys (1846) produced figures which suggested that, out of a total population in the Mountain of 300,919, 83 per cent were Christians (62 per cent Maronite) and 16 per cent Muslims (including 32,493 Druzes).[2] Figures given by the Russian consul at Beirūt, Konstantin Basili, suggest that Maronites predominated in the northern section of the Mountain but that in the south Druzes were numerically, as well as politically, dominant, though about 43 per cent of the total number of families were probably Maronite.[3] As already indicated, this mixing of communities added fuel to the conflicts of 1858–60.

The *Règlement Organique* of 1861 established a unified polity in the Mountain under Ottoman Suzerainty and international guarantee. It appears to have recognized the numerical superiority of the Christians (89 per cent; Maronites 73 per cent) in a total population estimated at 296,000 by General Ducros.[4] The basis of these figures is unclear. In the light of later official Ottoman statistics, however, they look suspect. In fact, the solution to the community problem in Mt Lebanon imposed by the Great Powers seems like a particularly good example of the effectiveness of the special pleading which Christian communities in the Ottoman Empire were able to make in a context of western concern for the non-Muslim minorities under what Europeans characterized as an oppressive and corrupt regime.

By the end of the nineteenth century, official Ottoman population statistics become available. Karpat has argued that these are as accurate as anything produced in the West at the time and certainly more reliable for the Empire than the figures published by contemporary western experts and local community leaders.[5] The general census of 1881/82–93 gave an estimate for the *Cebel-i Lübnan* of about 100,000 but provided detailed figures which allow the calculation of the total population of an area which approximates that of the future state. It was 339,505. These figures are well below the totals given by Cuinet (1895) (Mountain 399,953; Greater Lebanon 806,777).[6] If Cuinet's statistics for the total population can be so different from the official Ottoman figures, there is no reason to believe those that he gave for the sizes of the different communities. The Ottoman census does not provide a community breakdown for the Mountain, but it does for the areas which were added to it to make the independent state. These reveal that Christians formed only 28.6 per cent of the population of these areas compared with the 47.8 per cent given by Cuinet. It is possible that Christians, and especially Maronites, were demographically dominant in the Mountain itself. However, their position is likely to have been weakened by large scale emigration. In the period 1860–1914, at least 600,000 people emigrated from Syria and Mt Lebanon to north and south America, 80–85 per cent of them Christians.[7]

By 1914, the total population of the area of the future state of Lebanon was about 511,580 according to official Ottoman sources. Again, this is considerably lower than a western estimate published in 1920 (838,705).[8] The community composition of the Mountain cannot be compared between the two sources. For the added territories, though, comparison reveals considerable discrepancies.

Samné showed Muslims to form 59.4 per cent of the population and Christians 39.7 per cent. The Ottoman sources give proportions of 69.3 per cent and 32.7 per cent respectively. Similar distortion is expected for the Mountain, where an apparently official but otherwise unknown source quoted by Samné gives the breakdown as 77–79 per cent Christian and 20–22 per cent Muslim.

This discussion of the population figures for Lebanon suggests that the size of the Muslim population (including Druzes) was consistently underestimated by influential western sources during the nineteenth century. So effective may these have been in the contemporary information battle that Muslims themselves may have come to believe the picture which they seemed to reveal. The only official census taken by the new state in 1932 produced a total population of 793,426 (including residents and people only temporarily absent).[9] In community terms, the population was shown to be about equally divided, 50 per cent Christian (28 per cent Maronite) and 48 per cent Muslim (including Druzes, 6 per cent). Whilst the community balance may have been fairly close to reality, examination of the nineteenth century data leads one to suspect bias. For total population, the difference beween the estimate of 1914, based on official Ottoman sources, and the census total of 1932 is 281,846. An annual increase of 3.06 per cent over 18 years is not absolutely impossible, but it is doubtful, especially in view of the deprivation and starvation experienced in the Mountain during the First World War when the flow of remittances stopped. Apart from the possibility of error in the Ottoman statistics, the discrepancy can perhaps be explained most charitably by the inclusion of a large number of long-term emigrants in the 1932 figures. Such a device would not only inflate the total population of Lebanon, but it would also exaggerate the proportion of the Christians. Nonetheless, the numerical superiority of the Christians was accepted at the time. It provided some justification for two other developments. One was the power and influence which Maronites enjoyed under the French mandate. The other was the power-sharing agreement gradually worked out in advance of the end of French rule and finally ratified in the unofficial National Pact of 1941. The arrangements made then survived until the 1970s.

14.5 Demographic and political developments, 1941–76 (Figure 14.1c)

The starting point for the new arrangements was the balance of communities in Lebanon revealed in the 1932 census. The president of the Lebanese Republic was to be a Christian, and presumably a Maronite. His prime minister was to be a Sunni Muslim, and the president of the Chamber of Deputies (later the National Assembly) a Mitwāli. Seats in the Chamber were allocated in the proportions of six Christian to five Muslim, but were filled by a complex system of list-voting from constituencies which often contained a mixture of communities. Despite shifts in the distribution of population, and especially the massive losses from the countryside during the 1950s and 1960s, constituencies were not changed. Voters were required to return home for the elections. There they were exposed to a range of narrow political rivalries between the leading families, but cast within a community context.[10] In effect, communities put up separate lists of candidates and these were then voted on by the whole electorate in the district. Cooperation was thus essential. The candidates were local community leaders (the *zu'ama*), or their nominees. The *zu'ama* were often landlords or wealthy merchants; sometimes they were both. They had their clients, their 'minders' and their strong-arm boys. Political parties were usually personal followings and community based, as with the National Liberal Party of Maronite, Camille Chamoun, and the Progressive Socialist Party which is entirely Druze and composed of the followers of the Jumblatt family. Parties without a clientalist basis, like the Communists, were weak and ineffective. There was one notable exception. This was the *Kata'ib* or *Phalange*, founded as a sort of Maronite fascist party by Pierre Gemayel in 1936 to protect what was seen as the Christian homeland against the Muslim enemy recently brought into the state and to support the romanticized virtues of Maronite village life in the Mountain against the inroads of commercialism already becoming manifest in, and close to, Beirūt.

The overriding importance of community affinities and the nature of the National Pact meant that the central government formed after independence was rather feeble, though its administrative arm created a useful pool of jobs. As in other facets of Lebanese public life, government jobs were allocated on the basis of patronage, in a sort of spoils system, loosely regulated by the 6:5 convention. An exception to this pattern was the army. Although officered largely by Christians and manned mainly by Muslims, it was originally something like a genuine national institution. Otherwise, Lebanon as such had little charisma and inspired few loyalties.

The system created in the early 1940s could survive only so long as certain conditions were fulfilled. These were:

- conformity to the conventions governing the political game;
- restraint on the part of the President in pushing his interests and those of his community;
- and, critically, stability in the perceived demographic balance of the communities.

Infringement of the first two conditions in 1952 and 1958 strained the political system and brought short outbreaks of inter-community violence. Although things seemed to get back to normal, in fact the way was being prepared for the collapse which came in 1975. When it did come, the collapse was fundamentally the result of shifts in the community balance.

No census was taken after 1932. Instead, population figures were adjusted on the basis of registered births and deaths, as well as official statistics on emigration. This means that the population figures quoted in various discussions over the forty-three years, 1932–75, cannot be entirely trusted. Demographic trends have to be established by indirect means. Emigration continued to drain the number of Maronites after independence, though the statistics may have been massaged to disguise its effects. Natural increase appears to have continued to be high amongst

Muslims, especially in the poorest and least educated community, the Mitwāli. The combined result seems to have been to tilt the community balance in favour of the Muslims. By 1977, the best estimates suggest that the Muslims (including Druzes) constituted 60 per cent of the total population of 3.056 million, with the Mitwālis predominant (27 per cent). The Christian population had sunk to about 40 per cent.[11]

Once the demographic trend was grasped, the political cauldron began to seethe. It was stirred by three further developments in the socio-political structure of the country. These were the differential impact of economic growth during the 1950s and 1960s, the increasing dominance of Beirūt, and the emergence of the Palestinians as a political force.

Although the Lebanese economy as a whole expanded by an average 7 per cent a year, much of the growth was in services, particularly commerce, tourism and banking.[12] Lebanon exported mainly fruit and vegetables in the period 1951–65, but changed over to chemicals and construction materials as 'lead' exports in 1965–70 and then to consumer goods in 1970–73. By 1974, 66 per cent of Lebanon's merchandise exports went to neighbouring Arab countries.[13] Much of the country's trade, however, was goods in transit: grain and flour, hides, cotton and wool from Syria, Iraq and Jordan; vehicles, machines and manufactured goods of all kinds from outside the region. Whilst the expansion of the transit trade reflected an ability to exploit location effectively, it was also assisted by the continuance of the Arab–Israeli conflict which prevented competition from ports further south. The closure of the Suez Canal (1967–75) was also beneficial since it diverted goods to the land routes. Tourism similarly capitalized upon location, but in addition made good use of the country's environment. Beaches and snowfields are in close proximity, as are ancient ruins and the cosmopolitan life of Beirūt. The Mountain offers a cool and green setting which contrasts starkly with the heat and aridity of neighbouring lands. Access was good.

Lebanon's development as a major financial centre began after the end of the Second World War and was confined almost exclusively to Beirūt. In 1945, only nine banks operated in the city. By 1966, at the height of the financial boom, there were 85, 68 of them Lebanese-owned. In addition, numerous money-changers and finance houses operated in the city. Beirūt's rise to prominence depended upon a number of factors. The most basic of these included the legal confidentiality of accounts, free convertibility, the free import of gold and location on the edge of the Arab Middle East. But these would have made little difference without the flood of wealth coming from the Gulf and Saudi Arabia after the exploitation of oil resources there during the 1950s and 1960s. Political instability in other Arab countries was important, for wealthy individuals and the new political leaders often had so little trust in their own people that they thought it wise to move their assets abroad. Cairo might have been preferred, but Egypt's growing socialism diverted the funds to Beirūt. Lebanon's own political stability was thus crucial.

Whilst services expanded, agriculture largely stagnated and industrial development was limited. Agriculture suffered from a number of classic problems. Holdings were often too small to support a family at reasonable levels of subsistence, despite a remarkable degree of intensification which produced such relatively valuable items as fresh fruit and vegetables, wine, olives and olive oil. Marketing was poorly integrated. Opportunities for extending irrigation were not taken and, though a dam was built on the Līṭāni river, the associated major irrigation scheme was aborted, mainly because of community rivalry. Labour was leaving the land rapidly, partly because of inadequate holdings but partly because of economic opportunities elsewhere.

Although Lebanon was one of the most industrialized countries in the Middle East in terms of the proportion of its active population employed in manufacturing (18 per cent),[14] it has few resources, a limited home market and enjoyed no protection from foreign competition. The refineries at Tripoli and Saida were the largest industrial employers. Other concerns were small, family concerns and probably undercapitalized. They were found in the food processing, textiles and light engineering sectors. Most of the industrial plants were concentrated in, or close to, Beirūt. A few were found in Tripoli, but they were generally lacking in the rest of the country.

The prosperity generated by economic development gave individuals a deceptively high per capita share of GNP. Christians – and to a much lesser extent some groups of Sunni Muslims – were the chief beneficiaries. This was partly because of the better education of most Christians and a long experience of doing business with the West. Remittances from abroad were available for investment purposes. Most of the local banks were owned by Greek Orthodox Christians. Location within Lebanon was also important. Much of the expanding economic activity was concentrated in Beirūt.

Although a Muslim city in terms of community structure even in 1956 (54 per cent Sunni, 12 per cent Mitwāli; 34 per cent Christian)[15] Beirūt is in fact close to some of the more densely settled Christian districts of the country. Workers were able to commute in from their home villages which, in any case, controlled some of the more fertile land in the country. By contrast, the Muslim areas of north and south Lebanon were relatively undeveloped and poverty was actually widespread, especially amongst the Mitwālis of the south. Government attempts to encourage investment in these areas were thwarted by powerful, community-based interests and little was done.

Poverty and hopes of betterment encouraged people to migrate to Beirūt from the disadvantaged areas. The growth of the city's population was phenomenal – from about 700,000 in the 1950s to over a million twenty years later (equivalent to about half of the country's entire population). All these people were crammed into a relatively restricted peninsular site. Whilst modern apartment blocks characterized the inner city areas, a 'belt of misery' developed around them. This was where the poor from the rural areas lived, in shanties and off the crumbs which fell from richmen's tables. Palestinian refugee camps were found in the same zone. The emergence of the Palestinian factor is the third strand in the drift towards civil war.

The Palestinian factor in Lebanese affairs began to emerge with the very first

Arab–Israeli war of 1948.[16] As a direct result of the war some 180,000 refugees arrived in Lebanon, mainly from Lower Galilee but in smaller numbers also from Jaffa and Haifa. They were joined after the 1967 war by people fleeing from the West Bank and in 1970 by those who left Jordan after King Hussein decided to deal forcibly with the Palestinian guerrillas based there. Altogether, there were about 350,000 Palestinians in Lebanon by 1975. Perhaps a third of them still lived in camps. These were concentrated near Beirūt, Saida and Sūr (Figure 14.2). Over the years the camps took on a degree of permanency as shacks replaced tents and were themselves replaced by brick-built houses. They were overcrowded. Village communities from Palestine reconstituted themselves socially and to some extent politically. More than 64 per cent of the population in the 1970s was under twenty years of age. These youngsters were often bored, and engaged in endless political discussions.[17] Unlike the Armenian refugees of a generation or so before, the Palestinians were not given citizenship rights automatically, though perhaps as many as a third of the total did eventually become registered as Lebanese citizens. The remainder were 'non-persons', dependent initially upon world charity channelled through the United Nations Relief and Works Agency (UNRWA).

From the first, the Maronites regarded the Palestinians with suspicion as likely to destabilize the country. This view intensified as the Palestinian presence became more obvious. As already noted, their numbers more than doubled by the 1970s. During the 1960s south Lebanon became the launching pad for numerous attacks into Israel with consequent Israeli retaliation. The Palestinian cause became ever more popular with Muslims. After 1968, an embryonic Palestinian state emerged within Lebanon. It had its own social and commercial institutions, its organs of government dealing with essential services like garbage collection and sewerage, and its own centrally directed economic projects.[18] For Muslims, the Palestinian Resistance became the focus of pan-Arab sentiment, whilst its socialist rhetoric attracted radical groups alienated in one way or another from the Lebanese system. The Palestinians themselves actively sought alliances with local political groups as a means of self-preservation following the lessons of the debâcle in Jordan. Meanwhile, the ranks of the poor increased around Beirūt as people forced from their towns and villages by Israeli military action in the south sought security and support. Discontent festered in Beirūt's 'belt of misery' even more when the world economy went into recession after 1973 and workers started to come home from working in the Gulf region. Amongst the Mitwālis the 'Movement of the Deprived' emerged; by 1974 it had its own military arm, *Amal*. As this was happening, the Maronites were already arming and organising, especially under the banner of *Kata'ib*.

14.6 Social war and invasion (Figure 14.1d)

The fighting which ensued had a complicated pattern. Numerous incidents built up to the pitched battles of 1975–76. These were fought out across towns and villages, particularly in and around Beirūt. They were not just contests between clearly defined Christian and Muslim groups. Alliances ran along the lines of clientship, as well as immediate community; they changed and shifted. Endless truces were arranged and rapidly broken. In December, 1975 the Palestinians joined in, somewhat reluctantly. About the same time, a ferocious struggle for territory began. Israel started to support *Kata'ib*. Syria entered the fray in June, 1976, though it had been diplomatically active for some months before.[19] Its objective seems to have been to restore political stability in Lebanon. Thereby Syria would be able to maintain its prestige in the Arab world and deflect attention from some of its own internal problems. By this stage, the Maronites appeared to be losing. Syrian intervention, arguably, saved them and prevented the partition of the country. The fighting ended in October, 1976. A Syrian army, however, stayed (in the north and El Beq'a, even for a time in Beirūt), a complicating factor in the Lebanese balance of power.

The social war caused a considerable amount of damage (especially in Beirūt) and distrupted the economy, particularly its financial sector. Between 25 and 50,000 people may have died. Population was displaced on a large scale, with perhaps as many as a third of the people forced to migrate. Christians were forced out of Tripoli and the 'Akkar district in the north, as well as from settlements in the northern part of El Beq'a valley. Muslims were ejected from east Beirūt and Christians cleared from the western part. The city was divided along community line. South of Beirūt, the Maronite town of Damour was emptied and its Christian population replaced by Palestinians driven from the camp at Tel al-Zaater near Beirūt. These movements altered the distribution of communities in Lebanon and, with the failure of the national government to reassert control, they produced a set of virtually autonomous polities within the national territory. The new structure was consolidated by subsequent events which brought Israel into the country.

Israel was already deeply involved in the affairs of Lebanon.[20] She had watched the build up of the Palestinian Resistance Movement with unease and attacked it in a series of retaliatory actions. In May, 1970 and again in February, 1972 major operations were carried out against Palestinian positions on the slopes of Mt Hermon ('Fatahland'). September, 1972 saw a short-lived Israeli penetration as far as Bint Jbeil. Before and during the Lebanese social war, Israel developed strong ties with Maronite Christian groups, providing arms and military training. After the war, Israel supported various Christian militia forces which emerged to protect their villages in the south from Palestinian depredations. This had the effect of creating a *cordon sanitaire* for northern Israel. By 1978, however, it was clear in Israel that its policy was neither weakening the Palestinians nor limiting their ability to strike. Accordingly, in March 1978, Israeli forces occupied southern Lebanon as far as the Lītāni River. The objectives were to remove the Palestinian Resistance Movement to a safe distance, and possibly defeat it outright, as well as to destabilize the political situation in which Palestinians, Lebanese and Syrians were becoming reconciled. The damage was enormous. An estimated 2000 civilians died, 2500 houses were demolished and a further 5200 partially destroyed.[21] Israel was induced to withdraw when a small (9000) United

Nations Force (United Nations Interim Force in Lebanon – UNIFIL) was inserted between the Christian-controlled border zone and the Līṭāni River to act as a buffer. Nonetheless, fighting continued in southern Lebanon between Palestinians and Israelis or their proxies. An estimated 200–300,000 people, chiefly Mitwālis, fled and most of them ended up on the edge of Beirūt.[22]

A final solution to the Palestinian problem in Lebanon was attempted in 1982. After two days of air strikes, Israel launched a well-prepared and full-scale invasion on 6 June. Although represented to the world as a limited retaliation, the aim of the attack was to use technological superiority and rapid movement in a strike at Beirūt which would destroy the Palestinian Resistance once and for all. By 11 June, Syrian forces were driven out of the southern part of the Mountain and the southern Beq'a. Beirūt was entered on 24 June, though the Palestianians held out until 21 August when their evacuation began. Within about a week 12,000 had left. As different Lebanese groups continued to fight each other, the United States persuaded Britain, France and Italy to join in providing a multinational peacekeeping force. Its role was ill-thought out and, after some disasterous experiences, it was withdrawn. Under mounting international as well as domestic pressure, Israel began a phased withdrawal of its troops. Harassment and reprisals accompanied it. The process was completed in July, 1985, though some special units remained behind in south Lebanon.

14.7 New patterns

The Israeli invasion and occupation were disasterous for Lebanon. Field walls were deliberately destroyed. Orchards and olive groves were uprooted; wheat was burnt. At least 130 industrial plants, including the oil refinery at Saida, were wrecked. Physical damage to property in 32 villages was estimated to be at least $2 billion. Israeli goods and financial services penetrated southern Lebanon.[23] Plans for diverting the waters of the Līṭāni River into Lake Tiberias for the benefit of Israel were again openly discussed. Thousands of men, often of military age, were taken away and interned for months in an interment camp at Ansar; more than 700 were taken south and imprisoned at Athlit in Israel. The Israeli umbrella provided the opportunity and the means for *Kata'ib* to assert itself in Beirūt where its militia massacred 1000–2000 inhabitants of two Palestinian camps (Shabra and Chatila); in the Shūf where it tried to expand its territorial control; and in the south where it and other Maronite groups harassed Mitwālis and Palestinians. By the end of 1983, Druze forces had won in the Shūf with the result that 25–30,000 Christians left, as usual mainly for Beirūt, though some went to Saida and Jezzin. About 17,000 refugees were allowed to return in the following spring. Lebanon was destabilized. A new pattern of community distributions emerged.

Northern Lebanon and the northern part of El Beq'a are under Syrian control at the time of writing (January, 1986). Since many of the Christians in 'Akkar and Tripoli districts fled during the social war, the population is almost entirely

Muslim. Sunnis predominate in the west, though some Alawite (Shī'ite) migration from neighbouring parts of Syria has been reported and produced some degree of friction. The Sunnis, though, are divided politically and have fought each other for dominance. The islands of Christian population in El Beq'a vanished during the social war, and the thinly scattered Muslim settlements are predominantly Mitwāli. They play host, however, not only to Syrian forces but also to perhaps 2000 Palestinian Resistance fighters. A further 4000 Palestinian guerillas were forced out of the town of Tripoli in December, 1983, as a result of internecine strife in which a Syrian-backed anti-PLO group was successful.

Most of the northern section of Mt Lebanon is Christian and predominantly Maronite. It is not politically homogeneous, however. Christian control extended to Zahle, but this outlier in a pool of Muslim settlements has been lost. East Beirūt is in Christian hands. Beirūt, though, is a divided city now, but with a Mitwāli predominance in terms of overall population (perhaps a half of the total) and an effective power base in the southern suburbs. Sunni Muslims (perhaps a quarter of the population) live in the single district of Barta, in west Beirūt.

The situation in the southern part of Mt Lebanon and throughout the south generally displays a more complicated pattern of communities. Christians and Druzes still live side by side in the districts of Shūf, Aley, Ba'abola and Matn in the southern part of Mt Lebanon, as well as on the slopes of Mt Hermon (Rashayyha, Hasbayya). The Jumblatt and Arslan families are dominant.

Mitwālis occupy the southern parts of El Beq'a and most of the rest of southern Lebanon, but they are divided politically. There are inliers of Christians, strengthened by Palestinian attacks on Christian villages near Saida in April, 1985 (e.g. Jezzine) and Sunnis (e.g. Sūr). A surprising number of Palestinians are still living in the south, mainly in camps near Sūr. Not surprisingly in view of its recent history, south Lebanon as a whole remains politically unstable. Sporadic outbreaks of fighting occur. Together with Israeli harassment, these incidents are continuing to sift out the population and separate the different communities into defensible concentrations.

At the time of writing efforts are continuing under Syrian patronage to reconcile the different leaders and their communities in Lebanon. Reconciliation, however, requires some give and take.[24] The Mitwālis will have to have a larger share in government and one which recognizes their numbers. Some accommodation will have to be reached with the remaining Palestinians. The historical role of the Druzes in creating an independent Lebanon needs to be acknowledged with a greater share in national power. The Maronites will have to give way on the 6:5 ratio which has been to their advantage for so long.

There is some ground for hoping that these changes may be possible. The traditional leaders of Lebanon's socio-religious communities share common political and economic interests, whilst the leaders of the various militia groups may not be as representative of their communities as they like to think. Nonetheless, the relocations of population which have taken place since 1975 are unlikely to be reversed, certainly in the short term. The principal communities in Lebanon are now more spatially concentrated than they have ever been. Whilst inter-communal violence will be difficult to root out, and a slow process, the new community patterns may provide the basis for some sort of a cantonal political system involving a degree of autonomy. Communities have long run their own affairs; they are now responsible for their own security and taxation. Separation but co-operation may be the way forward. Much will depend, however, upon the attitudes of Syria and Israel. Syria appears to want a stable and friendly Lebanon. Although some Zionists entertain hopes of annexing southern Lebanon as historically part of The Land to which they lay claim through God's gift, most Israeli politicians might settle for a secure northern frontier. The recognition of a new community balance in Lebanon would be to everyone's advantage.

References

1. K. Salibi, *The Modern History of Lebanon*, Caravan Books: New York 1965, 106.
2. E. de Vaumas, 'La répartition confessionelle au Liban et l'equilibre de l'état libanais', *Revue Géogr. alp.*, **43,** 511-604 (1955).
3. K. Basili, *Memoires from Lebanon, 1839–47* (Hebrew), Yad Ben Zvi, Jerusalem 1984.
4. E. de Vaumas, 'La répartition confessionelle au Liban et l'equilibre de l'état libanais', *Revue Géogr. alp.*, **43,** 511-604 (1955).
5. K.H. Karpat, *Ottoman Population, 1830–1914: Demographic and Social Characteristics*, University of Wisconsin: Madison and London 1984.
6. E. de Vaumas, 'La répartition confessionelle au Liban et l'equilibre et l'état libanais', *Revue Geógr. alp.*, **43,** 511-604 (1955).
7. K. H. Karpat, 'The Ottoman emigration to America', *Int. J. Middle East Stud.*, **17,** 175–209 (1985).
8. E. de Vaumas, 'La répartition confessionelle au Liban et l'equilibre de l'état libanais', *Revue Géogr. alp.*, **43,** 511-604 (1955).
9. E. de Vaumas, 'La répartition confessionelle au Liban et l'equilibre et l'état libanais', *Revue Géorgr. alp.*, **43,** 511–604 (1955).
10. (a) I. Harik, 'Voting participation and political integration in Lebanon 1943–1974', *Middle Eastern Stud.*, **16,** 27–48 (1980);
    (b) J. G. and N. W. Jabbra, 'Local political dynamics in Lebanon: the case of 'Ain al-Qasis', *Anthropological Qtly.*, **51,** 137–51 (1978).
11. D.C. Gordon, *The Republic of Lebanon: Nation in Jeopardy*, Westview Press: Boulder; Croom Helm, London 1983, 11.
12. S.A. Makolisi, 'An appraisal of Lebanon's postwar economic development and a look at the future', *Middle East J.*, **31,** 267–80 (1977).
13. A. E. Chaib, 'Analysis of Lebanon's merchandise exports, 1951–1974', *Middle East J.*, **34,** 438–55 (1980).
14. 1975 official estimates, *The Middle East and North Africa 1986*, 32nd ed, Europa: London 1985, 558.
15. D.C. Gordon, *The Republic of Lebanon: Nation in Jeopardy*, Westview Press: Boulder; Croom Helm: London 1983, 11.
16. M. C. Hudson, 'The Palestinian factor in the Lebanese civil war', *Middle East J.*, **32,** 261–78 (1978).
17. B. Shirhan, 'Palestinian refugee camp life in Lebanon', *Journal of Palestine Studies*, **4,** 91–107 (1975).
18. R. Khalidi, 'The Palestinians in Lebanon: social repercussions of Israel's invasion', *Middle East J.*, **38,** 255–66 (1984).

19. (a) A. I. Dawisha, *Syria and the Lebanese Crisis*, Macmillan: London 1980;
(b) P. B. Heller, The Syrian factor in the Lebanese civil war', *J. of South Asia and Middle East Studies*, **4**, 56–76 (1980).
20. (a) S. Ryan, 'Israel's invasion of Lebanon: background to the crisis', *J. of Palestine Studies* **11,12**, 23–37 (1982);
(b) Y. Sayigi, 'Israel's military peformance in Lebanon, 1982', *J. of Palestine Studies*, **13**, 24–65 (1983).
21. D. Gilmour, *Lebanon: The Fractured Country*, Sphere Books: London, rev ed 1984, 149–50.
22. D. Gilmour, *Lebanon: The Fractured Country*, Sphere Books: London, rev ed 1984, 149–50.
23. W. Khalidi, *Conflict and Violence in Lebanon*, Harvard University Press: Cambridge, Mass. 1979, 116–20.
24. M. K. Deeb, 'Lebanon: prospects for national reconciliation in the mid-1980s', *Middle East J.*, **38**, 267–83 (1984).

CHAPTER 15

Jordan – The Struggle for Economic Survival

15.1 Introduction

From the sixteenth century to the end of the First World War, the territory occupied by Jordan formed part of the Ottoman Empire. Towards the end of these times, the Damascus vilayet, with its western boundary on the Jordan, extended from Ḥama to Aqaba, while the country west of the Jordan, Palestine, was divided between the independent sanjak of Jerusalem in the south, and the vilayet of Beirūt in the north. Following its capture by British and Arab forces, and the defeat of Turkey in 1918, Britain became responsible for the region under a Mandate from the League of Nations. Under British administration the area was divided in 1923 into the territories of Palestine, west of the Jordan river, and Transjordan to the east under the rule of the Emir Abdullah (Chapter 10). After a relatively uneventful period of indirect British rule in Transjordan, the Mandate ended on 22 May 1946, and in 1948 the country became known as the Hashemite Kingdom of Transjordan, under King Abdullah. When the British Mandate of Palestine ended on 14 May 1948, the Jews in Palestine proclaimed the establishment of the state of Israel, and widespread fighting broke out between Jews and Arabs. Eventually a ceasefire was established, and an armistice on 3 April 1949 demarcated *de facto* borders for the new state of Israel. As a result of this armistice, and, to acknowledge this fact, the country was renamed in 1950 as the Hashemite Kingdom of Jordan.

15.2 Transjordan

The main theme of this chapter is the economy of the state of Jordan. The economic position of Transjordan in the 1940s will first be examined and then the changes since this date. Transjordan covered an area of 89,000 sq km. Although no census of the population was taken under the British Mandate, it was estimated to be between 300,000 and 350,000 in the early 1940s.[1] The population was overwhelmingly Arab, although small and distinct groups of Circassians, Chechens, Turkomans and Armenians were present. The vast majority of the people were Muslim. The Arabs were descendants of nomadic tribes which had moved northwards in a slow migration from the Arabian peninsula. In their movements to the north, some of the groups settled when they reached favourable agricultural areas on the southern fringe of the Fertile Crescent in northern Transjordan. Others took up a semi-nomadic type of existence.

The population clearly showed this threefold division into sedentary, semi-nomadic and fully nomadic groups. The settled inhabitants, including all the non-Arabs, lived in towns and villages concentrated in the north-west of the country with main centres at Irbid, Ajlūn and Jarash. In the districts to the south, centred around the towns of Amman and Salt, approximately half of the population could be classified as settled. Elsewhere in Transjordan, sedentary populations were only found near Kerak, Ma'ān and Aqaba. The semi-nomadic population, numbering 130,000 to 150,000, a figure similar to that of the sedentary population, was organized in tribal groupings and lived in tents. These people, besides their pastoral activities, cultivated the land and occupied the highland area around Amman and southwards towards Ma'an. The bedouin, numbering between 40,000 and 50,000, were scattered throughout the rest of the country. These people were entirely dependent on their flocks of sheep and goats and herds of camels, and cultivated no land whatsoever. Living on the fringe of the desert these tribes undertook seasonal migrations in the search for pastures. During the wet winter months they moved eastwards into the desert, or downwards to the Jordan valley, while in summer movement was westwards and upwards into the better watered highlands.

During the period of the British mandate there was little market activity within Transjordan with the majority of the population dependent upon subsistence arable farming or nomadic pastoralism. The total cultivated area was estimated at 445,000 ha and was concentrated in the highlands of the northwestern part of the country. Of the cultivated area, only some 91,000 ha received an annual rainfall of more than 500 mm. Much of the land which was cultivated was really unsuited for this type of activity. Soil erosion was widespread and a large area of formerly arable land had lost its top soil cover. Irrigated land comprised about 7 per cent of the total cultivated area and was found mainly in the Jordan valley. Even in the 1940s it was estimated that most of the easily available water in the valley was already being used, although the distributions systems were often wasteful. In the highland areas, little irrigation was practised except to supply the villages with fruit and vegetables. On the plateau and uplands of the north and west, wheat and barley were the main winter cereals, often grown in conjunction with leguminous crops, such as vetches, kersennah peas, and lentils. The chief summer crop was durah, a form of millet, but melons and cucumbers were also grown. On the slopes overlooking the Jordan valley, fruit cultivation was more important with vineyards being common around Salt and Ajlūn. Finally, in the Jordan valley, besides the production of wheat and barley, bananas and other tropical fruits were grown under irrigation. Throughout the country, livestock, particularly flocks of sheep and goats, formed an essential part of the agricultural economy.

With regard to the mineral resources of the country, no mines were worked at this period, and apart from the huge salt resources of the Dead Sea, no minerals other than phosphates, had been proved in large quantities. Rock with a rich phosphate content was known to occur at Er Rusaifa, to the northeast of Amman, and in 1936 a company was formed to exploit these reserves, with the result that by the early 1940s the quarried phosphatic rock was being used for the manufacture of fertilizers in Palestine. The only other mineral activity was the local quarrying of limestones, around Jarash and Amman for building purposes, and the sporadic digging of the salt deposits at El Azraq.

Industrial activity, in the western sense, was practically non-existent and was confined to two small tobacco factories at Amman, and three small distilleries near Salt. In the larger towns, tailoring and dyeing were of some importance but were concentrated in extremely small units. Some processing of agricultural products did take place. *Samne*, a clarified butter, was manufactured, while rugs and sacks were made from locally produced wool and hair when prices were low.

With the low level of economic activity in Transjordan, commercial activity was very restricted and little trade passed over the country's frontiers. That which did occur was mostly confined to the neighbouring states of Palestine, Syria, Iraq and Saudi Arabia, but no accurate statistics of the amount of trade existed. Exports were largely agricultural products such as wheat, barley, sheep and fresh fruit, while imports were dominated by products which could not be produced in Transjordan. Chief amongst such items were textiles, sugar, petroleum, rice and kerosene. From this short description it is clear that Transjordan in the early 1940s was a very poor country, with very few natural resources, a low level of economic activity, and already, evidence of population pressure.

15.3 Jordan (Figure 15.1)

With the *de facto* partition of Palestine following the establishment of the state of Israel, Jordan acquired the West Bank territory, which increased its area from 89,300 sq km to 96,600 sq km. At the same time, the state received a massive influx of population. Some 460,000 people were residents of the West Bank when Transjordan was enlarged to become the Hashemitite Kingdom of Jordan, but besides these people, a further 350,000 to 500,000 refugees from Palestine had already fled into the West Bank area and Transjordan, while the fighting was going on. When the armistice was signed in 1949 these refugees were never allowed to return to their lands in Israel. So by 1952, while the area of Transjordan had been enlarged by about 10 per cent to form the new state of Jordan, the population had risen 130 per cent from 375,000 in 1946 to at least 1.1 million.[2]

Since 1950 the population of Jordan, including the refugees, has risen steadily at the rate of about 3 per cent per annum. As a result the population had grown to approximately 2.1 million in 1967 prior to the June War. Of this total 720,000 were registered as refugees with the United Nations Relief and Works Agency.[3] The 1961 census revealed that 45 per cent of the refugees were less than 15 years of age. The population density of the country was 21 persons/sq km. Unfortunately, owing to the harshness of the environment the population is markedly concentrated in a very small portion of the total area and in this zone population densities were more than 190 people/sq km.

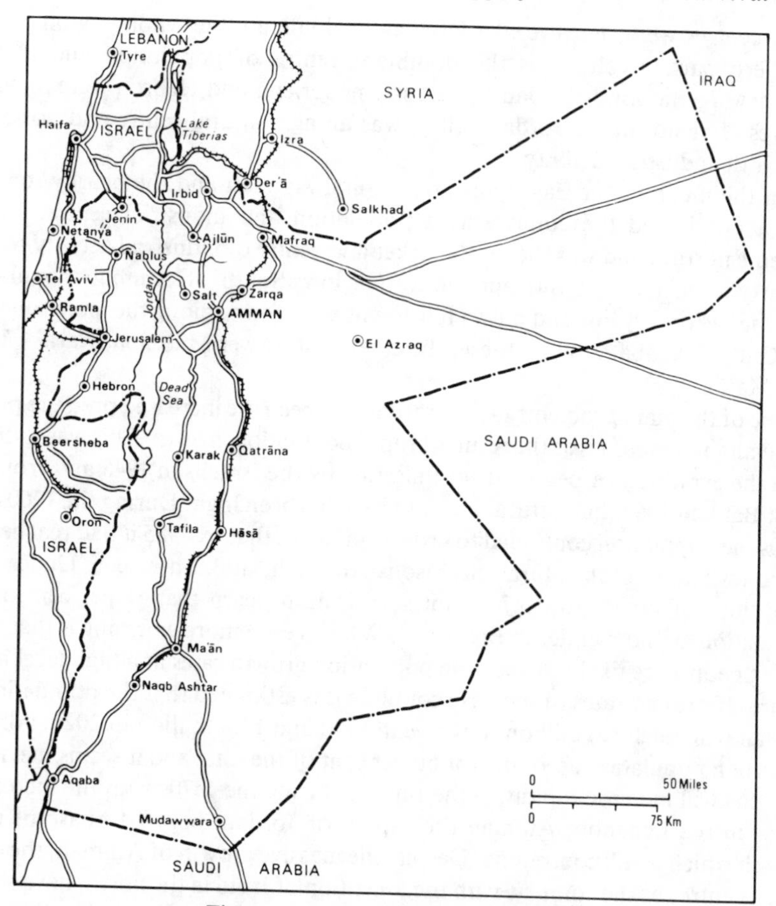

Figure 15.1 The Kingdom of Jordan

From the time of the formation of the state of Jordan, there was a steady drift of population from the West Bank to the East Bank, so that by the early 1960s a slight majority of Jordan's population were residents of the East Bank. Many reasons help to account for this population drift, but the main one would seem to have been the concentration of industrial development in the larger towns of the East Bank, especially Amman and Zarqa, and it was here that opportunities for employment were greatest. In all rural areas, population pressure on the land increased markedly, in itself giving rise to a drift of population to the urban centres. Unfortunately, however, unemployment rates in cities were high and underemployment of labour was everywhere common. Following the war of June 1967, when Israel took over Jordan's West Bank region, another 200,000 refugees fled from the area to the East Bank and so made an already bad situation even worse.

On the West Bank of the Jordan, occupied by Israel in 1967, a number of local urban centres were found around which population was concentrated. None of these centres were, however, so large as to dominate the whole region. In the northern part, Nablus was the dominant centre of population and also of industry. To the south Jordanian Jerusalem, Jericho and Hebron were the main centres. Jericho, in the Jordan valley, was an agricultural centre and possessed very little industrial activity.

On the much larger East Bank, the capital Amman and the nearby town of Zarqa dominated the country as a population and industrial centre. In the extreme north, Irbid was the local market and centre of industrial activity for the country's most productive agricultural region. South of Amman population densities were very low and only a few towns were found along the main routes to the Gulf of Aqaba. Among these, the chief centres wer Mārdaba, Karak, Tafila and Ma'ān.

One of the main problems facing Jordan has been the increased pressure put on the available resources as the result of rapid population growth. This was initiated with the expulsion of people from Palestine by the Israelis in 1948 and from the West Bank in 1967, but natural increase has also been high. During the 1960s and 1970s the population continued to grow rapidly, so that by 1985 it had reached 3.6 million.[4] The rate of natural increase is still high, at 3.8 per cent. Under such conditions of rapid growth it is not surprising to learn that 51 per cent of the population are now under the age of 15. What is even more worrying is that these young people are likely to keep the population growth rates at a high level in the future. If present rates of increase continue it is estimated that the population of Jordan will reach 6.6 million in the year 2000 and 12.3 million in 2020. Already the urban population accounts for 60 per cent of the total and it seems inevitable that this will increase rapidly in the future. During the 1970s with the increasing chaos in the Lebanon, Amman the capital of Jordan, began a phase of rapid growth which is still continuing. Despite the massive growth of Amman, the other urban centres of the country, with the exception of Irbid in the north, have shown little evidence of development.

15.3.1 Agriculture

Of all the problems facing Jordan in the early 1950s, none was more immediate than the problem of feeding its new citizens and of placing its agricultural production on a sound basis. If one was willing to accept a risk, the former of the two problems could at least be eased by the cultivation of those areas which were not really suited for the task, due either to the uncertainty of the precipitation or the instability or insufficiency of the soil. In many cases this happened, but the result of such a practice has been to further worsen the overall situation by substantially increasing the processes of land degradation and soil erosion. Therefore, any food increase resulting from this practice has been bought at the price of potentially less food production in the future. Although the new Jordanian government realized the problem at the time it was unable to initiate conservation measures owing to lack of funds.

Utilization of the waters of the Jordan valley has been hindered since 1948 by the political difficulties of the area.[5] Four states, Lebanon, Syria, Jordan and Israel, control the Jordan catchment, and although a number of schemes for the rational utilization of the water resources of the region have been proposed, none have been implemented. Instead, each state has gone ahead with its own plans, often to the detriment of other nations. In Jordan, the major irrigation scheme which has been implemented is known as the East Ghor project.[6] This consists of the diversion of the waters of the Yarmuk, a tributary of the Jordan, to irrigate land in the northern part of the Ghor (Figure 15.2).

The Ghor is a terrace feature which occurs some 50 m above the present river, and forms the main area of level land in the Jordan valley. It is found on both sides of the flood plain of the Jordan and at its outer margin abruptly meets the fault escarpments which delimit the valley. It is well developed over the 120 km between Lake Tiberias and the Dead Sea. Composed largely of alluvial material, the soils developed upon it are well suited to agricultural production. Unfortunately, precipitation in the Jordan valley is low and unreliable. At the northern

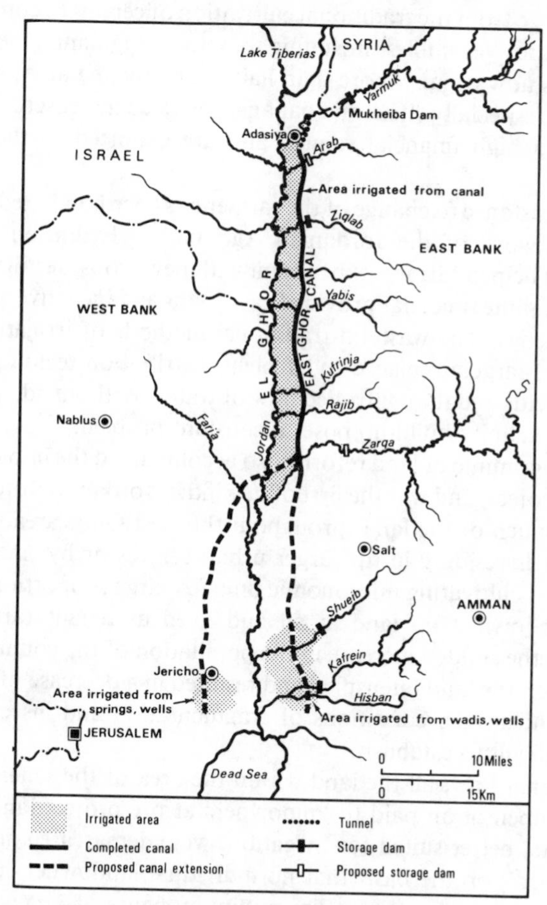

Figure 15.2 The East Ghor irrigation project

end of the valley near Lake Tiberias, rain-fed agriculture can occur, but further south towards the Dead Sea, where precipitation totals are less than 100 mm, arable farming cannot be undertaken without irrigation. As a result, pastoral farming was the primary economic activity. Very little of the Ghor was brought under cultivation for the first time by the East Ghor Project, as irrigation using the waters of the streams flowing down into the rift valley had been practised over a long period. Unfortunately, such water sources have always proved unreliable, especially in spring and summer during the time of maximum irrigation demand.

Work on the East Ghor project was initiated in August 1958, and by the summer of 1966 water was being delivered through the system to most of the land supplied from the main canal.[7] The canal taps the water of the Yarmuk at a point approximately 8 km above its confluence with the Jordan and conveys it through a concrete-lined canal parallel to the Jordan river for a distance of some 70 km. Lateral canals then distribute the water by gravity flow to fields situated between the canal and the Jordan river.

The scheme has had a tremendous effect on farming practices, and has resulted in a land use change from the traditional cultivation of cereals to commercially orientated agriculture. Vegetables, including tomatoes, eggplants, melons, cucumbers and peppers now occupy more than half the cultivated area, while the area devoted to fruits, especially citrus and bananas, has also increased. Initially, all of these produced a high financial return, and are exported to the Gulf are in particular.

The natural resistance to change of the farmer has been combated in this region by the valuable work of the Jordanian Agricultural Extension Service. This organization has helped farmers experiment with new crops, fertilizers and pesticides, and, at the same time, has provided cash loans and incentive grants through cooperative societies. The wasteful traditional methods of irrigation have been discouraged, and largely replaced by efficient distribution techniques involving the levelling of fields and the controlled flow of water. Without adequate drainage in this area, soil salinity build-up poses a constant problem.

A detailed programme of land reform also accompanied the implementation of the East Ghor project, and was the first of its kind in Jordan. Before the initiation of the project, much of the land throughout the East Ghor area was owned by absentee landlords residing in the larger urban centres or by smaller landlords often owning and cultivating uneconomic units. A large proportion of the population, however, owned no land at all and lived as tenant farmers or farm labourers. With the sudden increase in the population of the country in the early 1950s, pressure on the land intensified and resulted in a decrease of farm size and higher rents. Under such conditions of fragmentation and insecurity, efficient farming was difficult to establish.

Under the Canal Law, all the land within the area of the scheme was appropriated and compensation paid to landowners at pre-project land values. New rectangular units, better suited for irrigation, were demarcated and realloted to the former owners, on the basis that no individual landowner could now own more than 20 ha or less than 3 ha.[8] The minimum figure of 3 ha was regarded as the smallest unit capable of providing a family with a reasonable standard of living and at the same time of being capable of repaying the construction costs. This ruling allowed a redistribution of property with the virtual abolition of both very large and very small farms. The overall effect of this policy was a 20 per cent reduction in the total number of landlords. As a result, the scheme did not provide any opportunity for the tenant farmers or farm labourers to obtain land of their own, and indeed, many were worse off than before, owing to the lower labour requirements.

Associated with the East Ghor scheme was a programme of soil conservation and farming modernization in the Ajlūn and Amman Highlands.[9] These highlands form a dissected upland area between Amman and Irbid, bounded on the west by the Rift valley of the Jordan, and on the east by the Syrian desert. The two highland masses, each rising to more than 1000 m, are separated by the valley of the Wadi Zarqa. The eastern part of the area is a plateau sloping gently eastwards, while the western part has been deeply dissected by a series of steep-sided wadis which flow down into the Jordan. The rocks of these highlands are mainly terrestrial and marine sediments dating from the Jurassic to Tertiary times.

With an altitudinal range of more than 1500 m, temperature variations within the area can be considerable. From the data available it would seem that frosts occur on between five and fifteen days/annum on the plateau around Amman, while they are only rarely recorded in the Jordan valley. Everywhere summers are hot. Almost the whole region receives more than 300 mm of precipitation, with the two areas of highest elevation recording totals of more than 600 mm. A feature of the precipitation of the region is its annual variability, with a number of stations recording maximum annual values four times greater than the minimum ones. Approximately 95 per cent of the winter precipitation falls in the winter season from November to March, and within this period the three months of December to February account for about 70 per cent of the total. As a result, the number of rain days is surprisingly small, averaging between 30 and 50 over most of the higher ground.

Soils tend to correlate closely with geological outcrop. Terra Rossa soils cover the greatest area in the Ajlūn and Amman Highlands, and occur almost exclusively on hard crystalline limestones, while Rendzinas occur on the chalky or marly formations. Slope and alluvial soils are common and a significant proportion of the land area is bare rock. The natural vegetation of the region was forest, but at the present time, only stunted remnants of the original cover are found.[10]

The most important problem of this region is undoubtedly that of soil erosion, which although caused largely by human activity, has certainly been intensified by natural conditions. The soils, owing to a lack of structural stability, tend to break down to individual particles when subjected to wetting. This means that the pores in the upper layers become clogged by the particles and so an impermeable surface crust is formed. In turn, this reduces infiltration rates, and promotes increased runoff and water erosion. Another important environmental feature is the steepness of the slopes, with much of the region possessing gradients steeper than 15 per cent (Figure 15.3).[11] On such slopes runoff is both rapid and erosive.

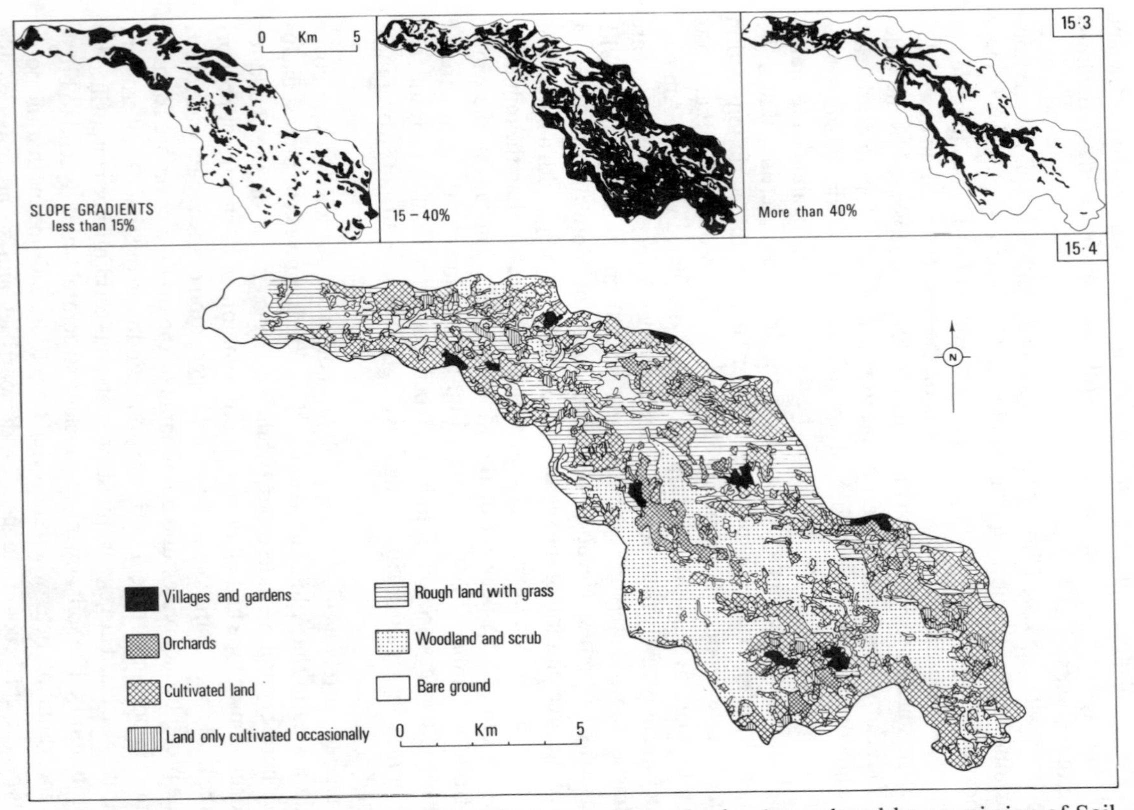

Figure 15.3 Steepness of slopes in the Wadi Ziqlab, northern Jordan (reproduced by permission of Soil Conservation Society of America)

Figure 15.4 Land use in the Wadi Ziqlab, northern Jordan

Over the centuries deforestation has been practised to provide additional agricultural land, as well as timber and fuel supplies. This activity is particularly difficult to correct, as forest growth is such a long term process. Deforestation continues, but fortunately at a much reduced rate, while the replanting of forest areas is now official government policy. The overgrazing of pasture land dates back to Biblical days and earlier. In theory, this is a problem which can easily be solved by legislation, and in some areas goat husbandry has been severely limited. Unfortunately, in a country where wealth is often measured by the number of animals owned, such a decree has proved difficult to enforce. As a result, overgrazing continues in many regions. Throughout the Ajlūn and Amman Highlands, between one half and three-quarters of the total cultivated area is devoted to the cultivation of cereals, especially wheat and barley (Figure 15.4). The most serious effect of this cropping pattern is that the soil surface is bare of vegetation during the winter months, when precipitation totals are highest. This, coupled with the lack of soil stability when wet, means that sheet erosion of soil is extremely widespread throughout the region.

As all of the westerly drainage of the northern highlands flows into the Jordan valley in a series of short wadis, flooding and sedimentation present serious problems to the success of the east Ghor canal scheme. To minimize these problems a series of dams at the mouths of the Wadis Ziqlab, Shueib and Kafrein were constructed in the 1960s to regulate the discharges. These dams are, of course, susceptible to sedimentation and consequent reduction in capacity owing to the severe soil erosion within the watersheds behind them. It was, therefore, essential that a conservation programme was initiated in this area.

Conservation measures were initiated within the Ajlūn and Amman Highlands as part of a large scale project in the early 1960s. The prime aim of these measures was to reduce the rate of soil removal by introducing physical barriers to downslope soil movement and, at the same time, increasing the density of the vegetation cover. Contour walling, gradoni terraces and gully plugs were constructed in selected areas in association with an afforestation policy. While the measures were clearly successful in reducing the rate of soil erosion, the lack of funds meant that the programme could not be extended throughout the whole area. So far, relatively little attention has been paid to the improvement of agricultural practices, which are, after all, one of the main causes of the erosion problem. In the future, it is hoped that the emphasis will change from the direct government implementation of conservation projects to schemes of village and area rehabilitation carried out by the villagers themselves, under the guidance of trained conservationists. At the same time, new cropping practices, which will increase the proportion of tree crops, such as olives, vines and citrus, will help to ensure a more stable agricultural pattern in the area. In the mid 1980s the Jordanian government announced that it was undertaking a Zarqa River Project, which would focus on integrated rural development, soil conservation and land use programmes for the 5,000 farms in the Zarqa basin facing severe soil erosion problems. This scheme represents a continuation of the work done in the Northern Highlands in the 1960s.

On the West Bank, the agricultural system was similar to that of the East Bank, and was dominated by the cultivation of wheat and barley, and the grazing of sheep and goats. However, this region has always had a much higher proportion of crops such as vines and olives than the East Bank, and many areas have been terraced. As a general rule, the West Bank always gave the impression of being more prosperous and the soil better conserved than on the East Bank. However, the possibilities of increasing agricultural production within this area are not high.

At the margin of the desert on the East Bank, considerable areas of land were brought under cultivation for cereal production in the 1960s and 1970s. Unfortunately, most of this land is at best marginal for any type of cultivation and crop yields have been very dependent on rainfall totals. Indeed, in some years crop failures have been extensive. Throughout Jordan though it must be admitted that wheat production in particular has always been subject to vast annual variations in production as a result of precipitation fluctuations. Coupled with the uncertainty of production is the added hazard of soil erosion by strong winds blowing across the desert. Following ploughing, the soil is extremely friable and particularly susceptible to wind blow. Given a reliable source of water and the provision of windbreaks, however, this land could be very productive. Further south, in the region around Qatrāna, a number of projects were initiated to explore the possibility of building small check dams to impound the winter runoff. Precipitation is low, and so no large scale increases in the cultivated area can occur, but it may be sufficient to ensure a more reliable agricultural yield for the local population.

Considerable groundwater supplies also appear to exist in eastern Jordan. Seven major aquifers have been recognized in rocks of Mesozoic or younger age, with the Amman-Wadi Sir aquifer system being the most important.[12] This outcrops in the high rainfall zone of the northern highlands and is also present beneath younger sediments on the east Jordanian plateau. Water recharge to the system is believed to average about 336 million cu m/annum, with most of this generated from local precipitation and runoff. About two-thirds of the total is thought to be discharged from springs and seepage points along the rift escarpment, with the rest moving eastwards as groundwater flow.[13] With careful management this groundwater should permit a moderate development of the cultivated area.

During the 1970s and the early 1980s the most significant agricultural scheme was the development of commercial irrigated agriculture in the Dead Sea lowlands, using water from the River Yarmuk, delivered through the East Ghor canal. The first stage of the project, which permitted the irrigation of 22,000 ha was completed in 1979. This was dependent on water control and storage facilities provided by the King Tahal dam and irrigation networks along the Ghor itself. It also involved an 18 km extension to the East Ghor canal, the first stages of which had been built in the 1960s. During the second phase of the project in the early 1980s, the East Ghor canal was to be extended a further 14 km and an irrigation scheme developed in the Wādī ‘Araba. This would add a further 7000 ha of irrigated land. Another phase involves the irrigation of lands to the south and east of the Dead Sea, where 4,600 ha are to be developed.

These irrigation projects along the Ghor permitted the introduction of large scale commercial agriculture to Jordan. Vegetables have been the chief crops produced, especially tomatoes, cucumbers and melons, and these are often grown in long polythene greenhouses to control environmental conditions. Initially, production was destined for the local Jordanian market and the rich urban markets of the Gulf states. The Gulf markets proved highly lucrative at first, though recently increasing competition from home produced vegetables in the Gulf states has meant that Jordan has had to develop new markets in countries such as Turkey. With mild winter temperatures the Ghor region does possess marked advantages with regard to the production of early crops.

The Jordan Valley Authority, the organization which oversees the development of the Ghor project, has instituted a policy of piping irrigation water to farms wherever possible. From here the farmers will connect either sprinkler or drip irrigation systems. With such procedures the JVA estimates that in the long term water usage might be halved. The current use of the Jordan valley irrigated lands has been largely for the production of vegetables, citrus fruit and bananas, dependent on what the farmers felt was likely to be most profitable. However, this will change in future as farmers will have to comply with a Ministry of Agriculture cropping pattern, which will limit the production of tomatoes and cucumbers. In their places the Ministry will encourage the cultivation of wheat, onions and potatoes.

The major crop in Jordan is wheat. It covered an area of 943,000 donums (94,300 ha) and in 1985 its production was 62,800 tonnes (Table 15.1). As with

TABLE 15.1
Agricultural production – main crops 1985

	Production (tonnes)	Area (donums)
Field crops		
Wheat	62,827	943,556
Barley	19,681	399,202
Lentils	4,063	57,847
Vetch	2,309	35,727
Chick peas	1,589	28,900
Tobacco	2,692	56,500
Tree crops		
Olives	22,610	297,749
Grapes	52,637	126,156
Citrus	158,230	53,031
Apples	3,035	11,816
Bananas	30,804	10,668
Figs	2,698	10,955
Guava	5,282	3,995

Source: Statistical Yearbook 1985, Amman, Jordan.

barley, this crop is mostly grown without the aid of irrigation, and consequently, total production varies from year to year in response to the amount of precipitation which had fallen in the preceding winter period (Figure 15.5). As a result yields reveal large variations from between 200 and 1000 kg/ha. Of the tree crops olives, grapes and citrus fruits are the most widely grown.

The most important vegetable crops are tomatoes, water melons and melons, and cucumbers (Table 15.2). The increasing importance of the Ghor region as a producer of tomatoes and cucumbers is clearly seen. These crops are often grown with the aid of irrigation and, as a result, production does not reveal marked annual fluctuations, as do cereals. Sheep and goats are the most numerous livestock. In 1985 sheep numbered about 960,000 and goats 400,000. Over recent years the numbers of both animals have been declining. They are generally reared on poor marginal pastures and their numbers fluctuate annually in response to precipitation conditions.

TABLE 15.2
Vegetable production 1985

Crop	East Bank		Ghor	
	Production (tonnes)	Area (donums)	Production (tonnes)	Area (donums)
Tomato	118,779	49,296	293,489	87,771
Summer squash	12,652	12,344	36,873	24,715
Cucumber	25,628	11,782	179,120	53,035
Cauliflower	27,003	14,251	8,405	2,404
Cabbage	12,067	6,637	16,847	4,845
Water melon	61,347	35,779	3,664	2,232
Sweet melon	30,755	21,545	20,886	13,693
Eggplant	8,499	4,500	67,773	22,662
String bean	2,233	2,582	18,261	16,364
Potato	9,614	4,448	16,585	11,056

Source: Statistical Yearbook 1985, Amman, Jordan

TABLE 15.3
Imports of wheat and wheat flour compared with domestic production (000 tonnes)

Year	Wheat flour imported	Wheat imported	Domestic wheat production
1977	107.6	139.0	92.5
1978	85.1	173.3	53.3
1979	78.5	201.3	16.5
1980	118.1	162.9	133.5
1981	64.4	348.1	50.6
1982	138.5	209.2	52.2
1983	19.8	318.7	116.0
1984	12.2	450.5	49.7
1985	8.8	376.9	62.8

Source: Statistical Yearbook 1985, Amman, Jordan.

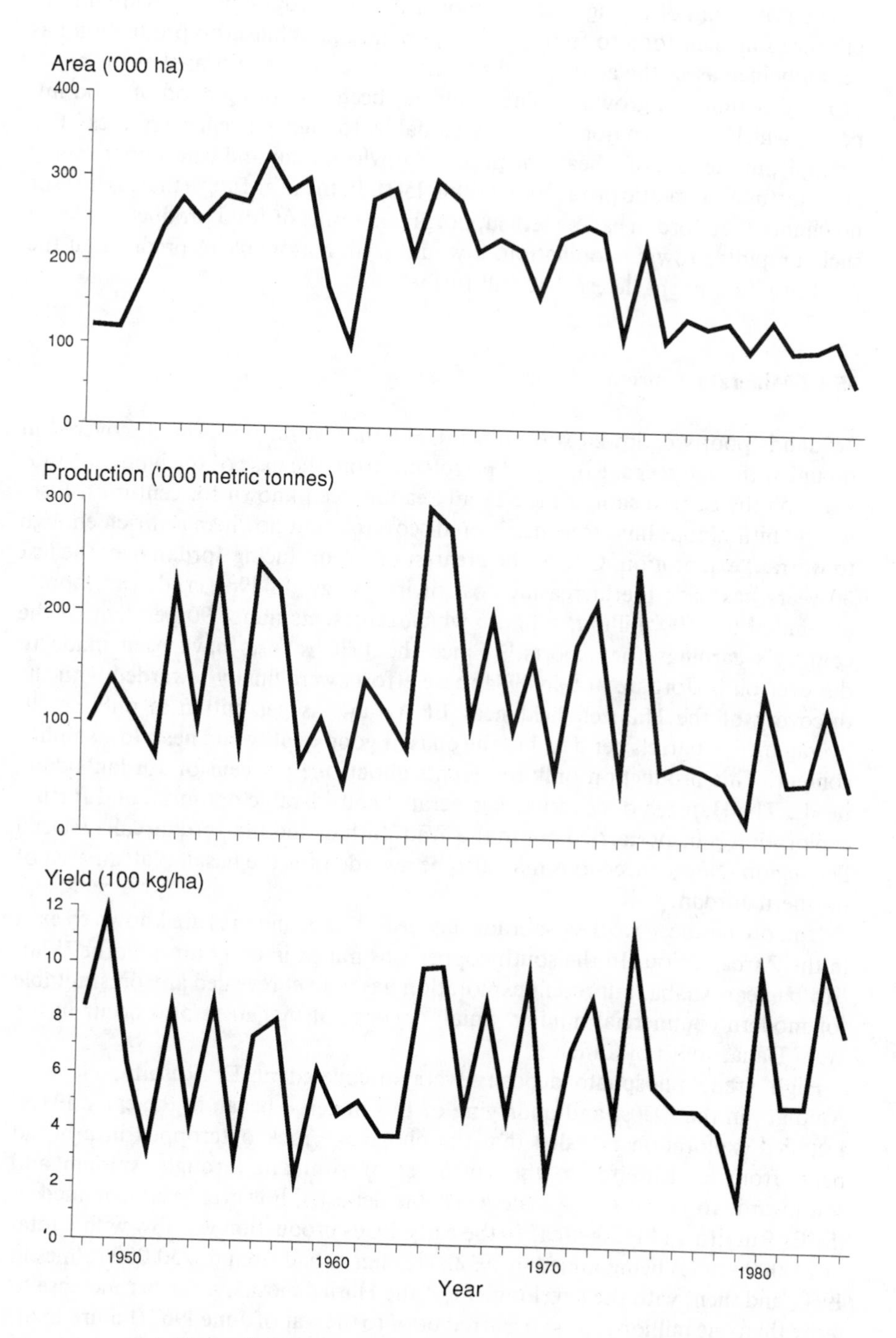

Figure 15.5 Variations in wheat production in Jordan

The prime aim of the agricultural modernization programme is to attempt to produce sufficient food to feed the local population. While crop production has certainly increased, the actual food situation has often deteriorated as the result of rapid population growth.[14] The result has been a growing flood of food imports, which are a major drain on valuable foreign currency reserves. For example, in the case of wheat, the imports of wheat flour and wheat are running at many times domestic production (Table 15.3). In the near future there is little or no chance that Jordan can be self supporting in terms of food production. With such a rapidly growing population, it would seem reasonable to predict that the food situation might deteriorate still further.

15.3.2 Mineral resources

Jordan is poorly endowed with minerals. Oil has so far not been discovered in quantity, though seepage of liquid petroleum from the base of the Nubian Sandstone on the eastern shore of the Dead Sea has been known for centuries. Elsewhere, bituminous limestone has been discovered, but nowhere is it rich enough to warrant exploration. One of the greatest problems facing Jordan over the last 30 years has been the increasing cost of its energy. In 1986 crude oil imports amounted to $600 million, a figure which represents about 90 per cent of the country's earnings for exports.[15] Since the 1930s efforts have been made to discover oil in Jordan, and in 1984 these efforts were finally rewarded with the discovery of the Hamzeh field near El Azraq. As yet output is only small, averaging 600 barrels per day, but the oil is of good quality and has a low sulphur content. This production only represents about one per cent of Jordan's daily needs. The Hamzeh discovery has generated considerable optimism and further exploration is being carried out in the Wādī Sirhān and the Jordan Valley/Dead Sea region. New concessions may also be awarded for the basalt plateau area of northern Jordan.

Iron ore has been worked sporadically near Ajlūn, and ores are known to exist in the Zarqa region. In the south copper was mined in early times in the Wādī 'Araba near Aqaba, but recent exploration has not yet revealed any ores suitable for modern commercial mining. Small deposits of maganese also occur in the Wadi Dana, south of Tafila.

High grade phosphate deposits were discovered at Er Rusaifa, close to Amman, in the 1930s, and quarrying of this material began in the early 1940s. Detailed exploration revealed that the phosphate rock outcropped in a broad band from Er Ruseifa, to the northeast of Amman, through Amman and southwards to Hāsā. Large scale workable deposits, however, were confined to the Er Rusaifa and Hāsā area. In the early 1950s production was low with a total of 25,000 tonnes being mined in 1952. This figure had risen to 250,000 tonnes in 1960, and then, with the development of the Hāsā deposits, a further increase to more than one million tonnes occurred prior to the war of June 1967 (Figure 15.6). Following this war and the internal troubles of the country in 1970, production

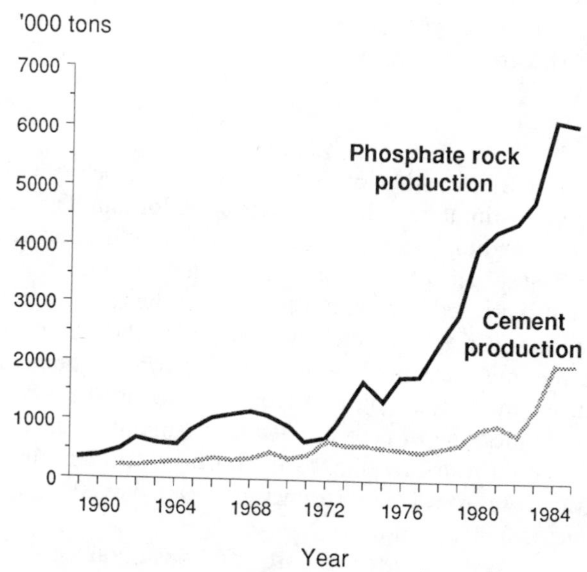

Figure 15.6 Phosphate and cement production in Jordan

dropped to below 700,000 tonnes in 1971. Since then production has continued to rise, particularly in the 1980s so that by 1985 production was 6.1 million tonnes per annum.

The deposits at Hāsā are suitable for mining by opencast methods and, therefore, production costs are considerably cheaper than at the older Er Ruseifa works. This fact, linked with its proximity to Aqaba, means that Hāsā production is relatively easily moved to the coast for shipping. Aqaba now possesses modern storage and handling facilities for the phosphate rock. Owing to unfavourable market conditions the Er Rusaifa mine closed in 1985.

The salt resources of the Dead Sea, in particular potash, were exploited commercially by the Palestine Potash Company between 1929 and 1948 at its two plants at the northern and southern ends of the Dead Sea. As a result of fighting in 1948, the northern plant at Kallia (on the West Bank) was destroyed. A new plant was built at Safi on the southeast shore of the Dead Sea and this is now operated by the Arab Potash Company. It was connected to Aqaba by a newly constructed road link. In 1986 production of potash reached 1.1 million tonnes. Plans are now being considered to increase production at the plant to an eventual 2 million tonnes, with a figure of 1.4 million tonnes being planned for the end of the decade.

The only other kind of natural resources exploited are high quality limestones, chiefly in the Amman and Jarash districts. Such material is widely used for building in the larger urban centres, and it seems likely that demand in the future will increase. The export potential of this product, owing to the high transport costs, is not likely to be great. In the Amman area, limestone is also quarried for use as the raw material in cement manufacture. With the rapid growth of Amman in the 1970s there has been a massive increase in production. In 1982 production was

793,000 tonnes, but three years later in 1985, this had risen to slightly over two million tonnes (Figure 15.6).

15.3.3 Industry

The establishment of the state of Israel, and the subsequent closing of the frontiers, helped to stimulate industrial activity in Jordan. The demands of the domestic market for building material, basic household goods and foodstuffs could no longer be supplied from what was formerly Palestine, and, as a result, local production increased. The opportunities for large scale industrialization within Jordan, however, were severely limited, owing the restricted range of raw materials and the small size and low purchasing power of the domestic market. Even today heavy industry is still poorly represented in the country. A cement plant using local limestone was constructed near Amman in the mid-1950s and a larger plant was built near Suweilih, to the north of Amman, in the late 1960s. Total production rose from 110,000 tonnes in 1959 to 480,000 tonnes in 1969 and 2 million tonnes in 1985.

An oil refinery was constructed at Zarqa and this obtained its crude oil from the Trans-Arabian pipeline (TAPline) which crossed the country. In the early days it was very small, with a capacity of about 7,500 barrels per day, but was capable of supplying the needs of the country. By the mid-1980s the capacity of the refinery has been increased to approximately 60,000 barrels per day, with the oil still being delivered along the TAPline.

In the early days of the state the other main industries were concerned with the processing of agricultural produce. Olive oil, largely pressed in small local mills, was of great importance on the West Bank, although it tended to be of a poor quality with a high acid content. Only one modern vegetable oil plant was operating prior to the June War of 1967 to the south of Nablus. Using olive oil as a raw material there was a small but important soap industry centred in the Nablus region which used to supply most of Jordan's requirements. The milling of flour was another important industrial activity based on agricultural production, and was centred around Irbid, Nablus and Jerusalem.

A number of small clothing factories existed, especially in the main urban areas such as Amman. These manufactured goods of a high quality, but in very small quantities. At many places throughout the Kingdom, tanning of goat and sheep skins took place under primitive conditions. A dairy industry existed to supply the urban centres, but again on a very small scale. Associated with all the larger urban centres were a whole range of service industries, including automobile repairing, building and contracting and similar activities. In many centres, but particularly close to the main tourist sites, the production of handicrafts for sale to overseas visitors was an important part of the local economy. Most of the work was done in small individual units, still largely by hand labour, but the quality of the product was often high. Jerusalem pottery, mother-of-pearl brooches from Bethlehem, olive wood carvings, and embroideries were known and appreciated by tourists from all over the western world.

During the 1970s and early 1980s considerable diversification of industry has taken place, though most of it is small in scale. In particular, a number of food processing industries have developed associated with the establishment of commercial agriculture in the East Ghor project. These has included the preparation and packing of tomato paste, the extraction of fruit juices and the making of soft drinks. With the growth of Amman into a major urban centre the production of dairy products has also become more important. Almost all this new industry has been located in the Amman/Zarqa conurbation. It is diverse in nature and includes plastics, textiles and light engineering. In Irbid, the only other centre with a major manufacturing base, a new industrial estate is currently planned and it is here also that a new iron foundry is to be located. Equally important has been the development of Amman as a commercial and service centre since the mid-1970s. Banking and other financial activities have become much more important and in the late 1970s a stock market was opened. The growth of such financial organizations greatly helped industrial development in the country by providing quick access to venture capital. Many other services, including retailing and catering, have expanded greatly so that Amman now has many of the characteristics of a large western city.

When Jordan became independent in 1950 her economic development depended largely on the provision of better communication and transport facilities. One of the main needs was to reduce the dependence of the country on the port of Beirūt, because of the high costs of crossing the Syrian and Lebanese frontiers, now that the outlets of Haifa, Jaffa and Gaza had been lost. This has been achieved by the development of Jordan's only port at Aqaba into an important shipping, commercial and tourist centre. As late as 1950, Aqaba was only a small fishing village and yet, by 1962, it was handling more than 500,000 tonnes and more than a million tonnes by 1966. In the first three months of 1986 imports were 833,000 tonnes, exports 2.9 million tonnes, and transit trade 2.6 million tonnes.[16]

In the 1950s and 1960s land communications were greatly improved. A new 'Desert Highway' was built between Aqaba and Amman, and modern motor roads were constructed to the West Bank and Jerusalem and to Irbid in the north. Even at the village level, a network of all weather roads, often with an asphalt topping, was built in the more densely populated regions. The Hejāz railway provided a link northwards to Damascus from Amman and southwards to Naqb Ashtar, south of Ma'ān. Subsequently it was extended southwards to Aqaba. The airports at Amman and Jerusalem (before 1967) had been developed and were capable of handling large commercial aircraft. Jordan also possessed its own airline – Alia. More recently a new international airport has been built at a distance of about 25 km out of Amman.

Perhaps the greatest resource of all which Jordan possesses, is the unique collection of historical remains stretching back over more than 10,000 years. Unfortunately many of these, including the great Christian shrines, were situated on the occupied West Bank, and so after the Israeli invasion of 1967 revenues from these sites were lost. During the 1950s, the tourist industry, although important, did not play a crucial role in the economy of the country. In the early 1960s, however, the number of tourists increased by almost 100,000 a year from 210,000 in 1962 to a record 616,000 in 1966, and revenues from tourism increased proportionately in the same period from US$ 5.05 million to US$ 31.6 million. Tourists from North America and Western Europe numbered abut 150,000 per annum in the mid-1960s (Figure 15.7). New hotels, catering for the wealthy Western tourist were opened during this period and the number of smal businesses dealing with the tourist trade greatly increased. The tourist industry appeared to be entering a boom period, and it was hoped, according to the Seven Year Plan of 1966, that the annual number of tourists would reach more than a million by the early 1970s. Following the 1967 hostilities, the sector of the Jordanian economy which was hit most was tourism, with numbers dropping to as low as 257,000 in 1971. This was due partly to the physical loss of many of the historical sites on the West Bank, but perhaps more importantly, owing to the political uncertainty and instability which existed within the country. As a consequence, Western tourists, who provided most of the tourist revenue, dwindled to only 20,000 in the late 1960s compared to more than 150,000 in 1966.

The late 1960s and the early 1970s were a time of confusion and conflict in the Middle East, which disrupted economic growth, investment and the tourist industry in Jordan as in other countries of the region. It was also a period of political volatility when long-term decision making was not easy to carry out and as far as Jordan was concerned was a time of stagnation and uncertainty. Calmer conditions prevailed in the late 1970s, but in the early 1980s, with Israel's invasion of the Lebanon the tourist trade dwindled once again (Figure 15.7). Between 1981 and 1985 the number of visitors entering Jordan has varied from 2.3 to 2.7 million each year. Of this total visitors from North America and Western Europe have averaged between 160,000 and 200,000 each year.

From the mid-1970s onwards Jordan entered a phase of consolidation, followed by one of remarkable growth. One of the main reasons for this was the breakdown of political authority in the Lebanon. At this time Jordan appeared a centre of relative calm and, as a result, many organizations with offices in Beirūt relocated in Amman. Cairo was the other centre to which these organizations could have moved, but its infrastructure was in a sorry state and not capable of providing the services needed by modern commercial firms. During the late 1970s and early 1980s Amman registered an extremely rapid growth rate, in terms of both population numbers and economic activity, leading to an unprecedented boom in land prices. Almost all the development which occurred at this time was located in the Amman–Zarqa region. The only other centres to grow significantly were the port of Aqaba and the northern city of Irbid. The whole country did, however, benefit from improvements to infrastructures, such as roads, electricity, water supplies and sewerage. Examples of this include the new thermal power station at Aqaba, the first phase of which was opened in 1986 and the rebuilding of the Hejāz railway down to Aqaba. This growth was also encouraged by huge investments made in Jordan by the oil countries of the Gulf, which in the late 1970s and early 1980s possessed enormous sums of money obtained from oil revenues.

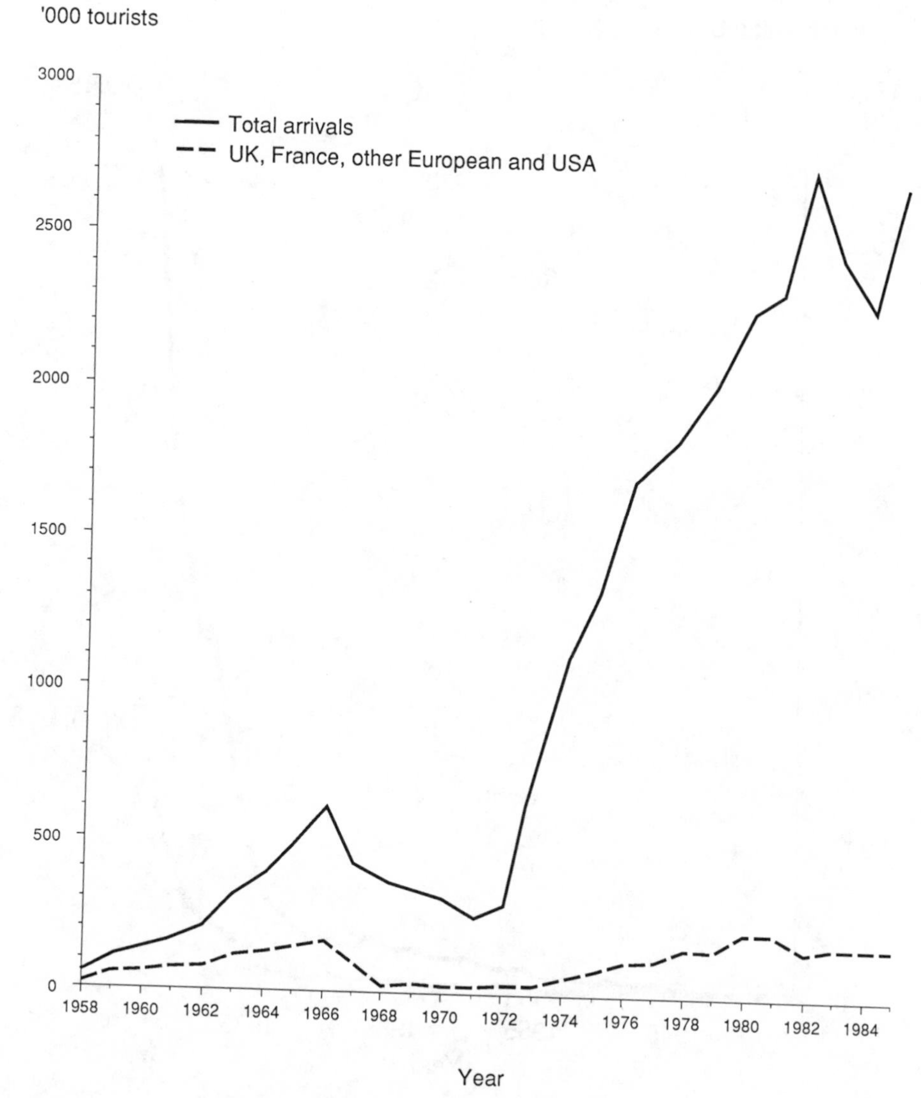

Figure 15.7 Number of visitors to Jordan

15.4 Conclusion

A useful index of the viability of a state is its balance of trade. Unfortunately this has always been heavily weighed against Jordan since the country's formation (Figure 15.8). As late as the end of the 1950s, it was thought that Jordan had little economic future owing to the paucity of the natural resources of the country and the political difficulties of the King. Then during the 1960s, prior to the June war, there occurred a period of rapid economic growth, made possible by the wise use

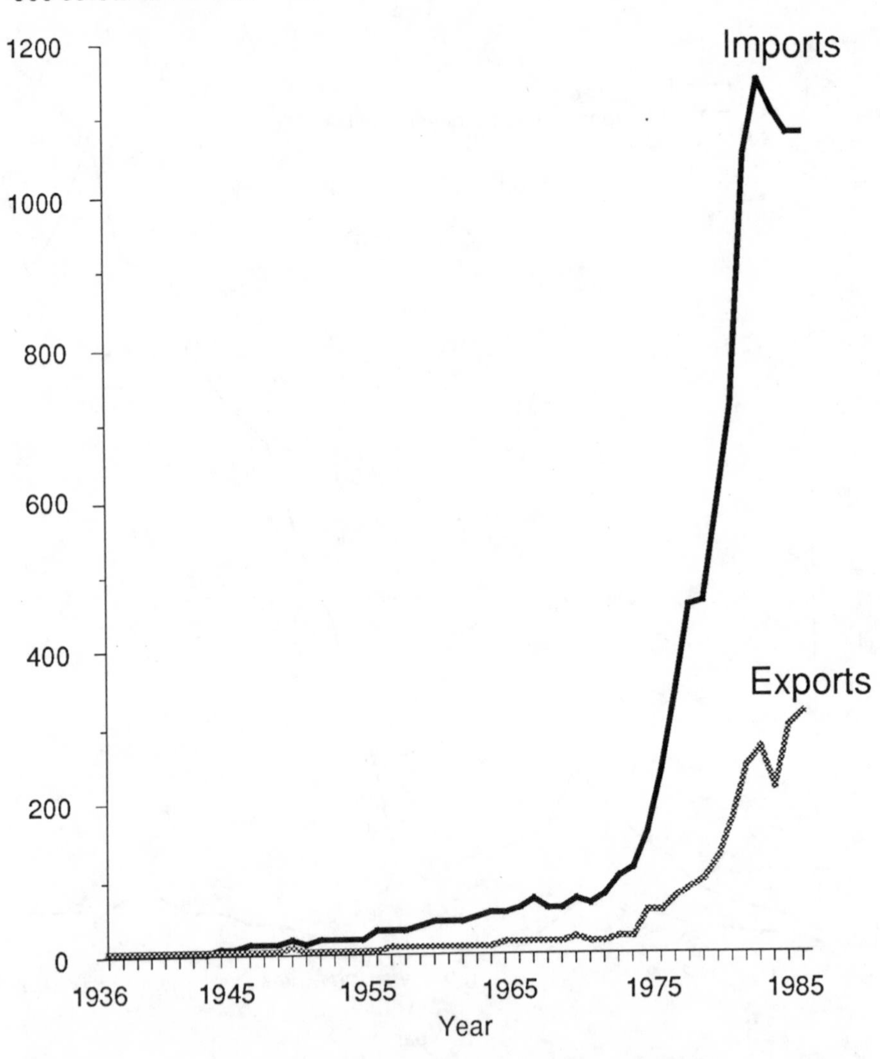

Figure 15.8 Jordan's foreign trade—exports and imports

of foreign economic assistance, mainly from the United States and Great Britain. As a result the GNP rose from US$140 million in 1954 to US$575 million in 1967, a rate of increase of about 9 per cent per annum. This period of rapid economic growth had given rise to the hope that Jordan would reach the economic 'take-off' point by the mid-1970s, but hopes of achieving this were smashed by the effects of the war of June 1967, the internal strife of 1970, and the war of 1973. During the 1960s the level of imports were low and only varied form about JD42 million to JD68 million. However, the ratio of imports to exports was of the order of at least five to one for most of the decade.

Although the ratio of the deficit was large, it is seen that the absolute cash amounts involved were, by Western standards, relatively small. For example, in 1972 exports from Jordan, made up largely of fruit, vegetables and phosphates, brought in a revenue of JD17 million. The main destination of these goods were Saudi Arabia 17 per cent; Kuwait 12 per cent; Lebanon 12 per cent; Iraq 12 per cent and India 11 per cent. On the other hand, the cost of imports, mainly petrol, textiles, capital goods, motor vehicles and foodstuffs, amounted to JD95 million. Here the major supplying countries were USA 17 per cent; West Germany 9 per cent; United Kingdom 8 per cent; Lebanon 5 per cent and Japan 4.5 per cent. Revenue from the tourist industry did, of course, help to reduce the balance of payments deficit.

Since 1973 there has been a steady growth in the economy. This has been the result of generous foreign aid, expatriate remittances and profitable regional markets elsewhere in the Middle East. Of particular significance was the fact that the oil rich Gulf states had plenty of money at this time for investment capital and for financing government borrowing. In the 1970s the level of imports began to rise very rapidly, increasing in money terms almost tenfold during the decade. Exports too revealed a rapid growth in the 1970s, but the ratio between imports and exports still remained at around 5 to 1.

As far as Jordan was concerned this growth in the economy was halted in 1982/3 by the oil price slump, which greatly reduced investment. Throughout the early 1980s the trade balance of the country remained high with a deficit of more than US$2,100 million in 1982 and 1983. As a result the government became concerned and has now introduced import controls in an attempt to keep the annual trade deficit below US$2,000 million. This action appears to have been successful and the rate of increase of imports is now much less. Between 1980 and 1985 it was exports which revealed the greater relative increase and this has now cut down the ratio of the deficit to around three to one. In absolute terms though the deficit remains huge with a monetary value in 1985 of about JD760 million. In 1985 imports totalled JD1,074 million, while exports were running at JD311 million. GDP has also continued to grow steadily with an average annual rise of 9.4 per cent in the period between 1981 and 1985. In 1984 tourism and services contributed 18 per cent to the GDP; manufacturing 15 per cent; transport 12 per cent; banking 12 per cent and agriculture 8.5 per cent.

In terms of the direction of trade in 1985, Saudi Arabia is the largest source country by value (Table 15.4). The USA is second largest, followed by Italy, Iraq, Japan, West Germany, the UK and Switzerland. Together these countries account for 64 per cent of the total imports to Jordan. On the other hand one hundred countries are involved in trade with Jordan. With exports Iraq, India and Saudi Arabia are the main receiving countries, and these alone account for 59 per cent of the total trade.

Of the domestic exports amounting to JD255 million in 1985, the largest single item was crude materials (inedible, excluding fuels), which totalled JD98 million. Of this JD66 million was accounted for by phosphates. Chemicals provided JD50.9 million and cement a further JD39.7 million. These three, namely, phos-

TABLE 15.4
External trade 1985

		Amount (million JD)
From: 1. Imports		
Saudi Arabia		159.1
USA		128.0
Italy		73.4
Iraq		72.6
Japan		67.8
West Germany		65.6
UK		63.3
Switzerland		58.9
	Sub-total	688.7
	Total imports	1074.4
To: 2. Exports		
Iraq		65.9
India		45.3
Saudi Arabia		39.1
Rumania		10.0
Indonesia		9.1
Kuwait		7.7
	Sub-total	177.1
	Total exports	255.3

Source: Statistical Yearbook of Jordan 1985, Amman, Jordan

phates, chemicals and cement, make up 61 per cent of all exports. The only other major item was food and live animals recording JD43.6 million. Of this 52 per cent was made up by the export of vegetables and tomatoes. Overall it is basic materials which dominate the export trade with only a small input from manufacturing.

With imports, totalling JD 1,074.4 million in 1985, four main areas stand out. These are mineral fuels (JD223.3 million), manufacturing and transport equipment (JD205.6 million), food and live animals (JD174.9 million), and manufacturing goods (JD169.2 million). Together these four account for 72.5 per cent of all imports. What is particularly worrying for the Jordanian economy is that food and live animals form so large an element of all imports.

The next few years are covered by a Five Year Plan 1986–90, during which time it is estimated that a total of JD3,115 million (US$9,727 million) will be spent. The prime objective of the plan is to stimulate economic growth and produce a rise in the Gross Domestic Product of 5 per cent per annum, a figure considerably less than that achieved in the early 1980s. Considerable emphasis is placed on job creation and it is hoped that 200,000 new jobs will be generated. It is felt that these will be necessary as increasing numbers of expatriate workers will be returning from areas such as the Gulf in the late 1980s. Of the total investment 52 per cent will be allocated to the public sector for water, irrigation, roads, communications, education and other social service projects. Per capita income is expected to rise from US$2,170 in 1985 to US$2,307 in 1990. As part of the Five Year Plan an investment of JD461.5 million is proposed for the occupied territories, with 40 per cent of this to be spent on housing and 17 per cent on education.

Despite these objectives it seems likely that there will be relatively little overall change in the economic viability of the country. Although it is true to say that the ratio between exports and imports had fallen since the late 1960s and early 1970s, the trade deficit remains large and can only be reduced by aid and investment from elsewhere. Even with such financial assistance Jordan was still having current account deficits of between US$200 and 500 million in the early 1980s. In the future there seems little likelihood that the situation will change, unless substantial quantities of oil are discovered. At the moment the major natural resources continue to be phosphates and potash, and these like all minerals are subjects to considerable variations in market demand and price. Although it had once been hoped that irrigated agriculture would make a significant contribution to the economy, this seems less likely as competition increases elsewhere. Similarly tourism, with visitors from Western Europe and North America, was once felt to be possible major source of income, but owing to the general instability of the Middle East this no longer appears to be realistic, at least for the next few years. Jordan will, therefore, have to accept that its long term economic position remains insecure and that it will have to continue to depend on the goodwill of other more propsperous neighbours and allies to assist its economy.

References

1. Naval Intelligence Division (Great Britain), *Palestine and Transjordan*, Geographical Handbook Series, BR 514, 1943, p 465.
2. P.G. Phillips, *The Hashemite Kingdom of Jordan: Prolegomena to a Technical Assistance Program*, Department of Geography, University of Chicago, Research Paper, No 34, 1954, p 70.
3. For a detailed study of the population of Jordan up to 1970 see W. B. Fisher, 'Jordan – a demographic shatterbelt', in *Populations of the Middle East and North Africa* (Eds J.I. Clarke and W.B. Fisher), University of London Press, 1972, Chapter 9, pp 202–219.
4. Population Reference Bureau, *World Population Data Sheet – 1985*, Washington, (1986).
5. C.G. Smith, 'The disputed water of the Jordan', *Trans. Inst. Br. Geogr.*, **40**, pp 111–128, (1966).
6. (a) J.L. Dees, 'Jordan's East Ghor canal project', *Middle East J.*, **13**, pp 357–372, (1959).

   (b) R.A. Smith and B.P. Birch, 'The East Ghor irrigation project in the Jordan valley', *Geography*, **48**, pp 407–409, (1963).
7. I.R. Manners, 'The East Ghor irrigation project', *Focus, N. Y.*, **20**, pp 8–11, (1970).
8. (a) I.R. Manners, 'The East Ghor irrigation project', *Focus, N.Y.*, **20**, p 11, (1970).
   (b) J.S. Haupert, 'Recent progress on the Jordan's East Ghor canal', *Prof. Geogr.*, **18**, pp 9–13, (1966).

9. K. Atkinson, and P. Beaumont, 'Watershed management in northern Jordan', *Wld Crops*, **19**, pp 62–65, (1967).
10. K. Atkinson and P. Beaumont, 'The forests of Jordan', *Econ. Bot.*, **25**, pp 305–311, 1971.
11. P. Beaumont and K. Atkinson, 'Soil erosion and conservation in northern Jordan', *J. Soil Wat. Conserv.*, **24**, pp 144–147, (1969).
12. J.E. Mitchell, 'Planning of water development in the Hashemite Kingdom of Jordan', in *Water for Peace - Vol 7 - Planning and Developing Water Programs*, Washington, DC, US Gov. Printing Office, pp 282–296, (1967).
13. UNDP/FAO, *Investigation of the Sandstone Aquifers of East Jordan - Jordan - The Hydrogeology of the Mesozoic-Cainozoic Aquifers of the Western Highlands and Plateau of East Jordan*, Rome, (1970).
14. P. Beaumont, 'Trends in Middle Eastern agriculture', in *Agricultural development in the Middle East*, (Eds P. Beaumont and K. S. McLachlan) Wiley, London 349 pages, pp 305–322, (1985).
15. M. Kielmas, 'Jordan: oozing confidence about oil', *Middle East Economic Digest*, 14 February 1987, p 16.
16. *Middle East Economic Digest*, 1 Nov, 1986, p 17.

CHAPTER 16

Israel and the Occupied Areas: Jewish colonization

16.1 Introduction

Although one of the smallest states in Southwest Asia or North Africa, Israel has been the subject of a torrent of literature on every conceivable aspect of its social, political and economic life. Geography is no exception, with a considerable quantity of high quality geographical literature being produced by Israeli scholars.[1] Thus a whole range of possible themes suggested themselves for this chapter, but the process of Jewish colonization stood out as being of particular geographical interest, and of cruicial political significance. Pre-1948 Jewish colonization in Palestine laid the foundations for the emergence of Israel in 1948. The configuration of Israel's pre-1967 boundaries were largely determined by the location of Jewish rural settlements, and the distribution of the Jewish population was the basis of UN partition proposals. National institutions and types of rural settlement were evolved which survive today, while the nation's future leaders acquired an understanding of the strategic importance of land settlement. Jewish colonization of the Arab territories of Sinai (with the Gaza Strip), the Golan, and the West Bank began after the Six Day War of June 1967. Here again, the geographical dimension of the settlement strategy is of great interest, but the real significance of the occupied area settlements is that they may hold the key to the future of the state of Israel.

Ottoman rule in Palestine lasted from 1517 until British troops captured Jerusalem from the Turkish army in 1917. After the First World War the Mandate for Palestine was entrusted to Britain and continued until the eve of the birth of Israel in 1948. The boundaries of Palestine in the east followed the natural divide of the Jordan valley and the Araba depression. In the southwest the 1906 boundary between the Ottoman Empire and Egypt remained, from the Gulf of Aqaba to the Mediterranean Sea south of Gaza. This border, with the Gaza Strip enclave on the coast, was Israel's border with Egypt from 1948 to 1967, and it was re-established following Israel's withdrawal from Sinai in 1979. The northern and northeastern boundaries of Palestine were established by Anglo-French agreements in 1920 and 1923, and remained unchanged until Israel seized the Golan Heights in June 1967. Although substantially larger than the area popularly thought of as Biblical Palestine, the Mandated territory was one of the smallest political units in Southwest Asia. Its total population was only 752,000 at the 1922 census. Yet throughout history, control of this part of the earth's surface has been the ambition of succeeding powers, not only because of its unrivalled strategic significance at the crossroads of Asia, Africa and Europe, and between the Mediterranean and Red Sea, but also because Palestine is revered by millions of Muslims, Christians, and Jews, for its religious associations.

16.2 The personality of Palestine

Palestine is part of a physiographic region consisting of marine sedimentary formations lying along the western margins of the ancient Arabian landmass, which were folded in Miocene and early Pliocene times to form a long anticline running roughly parallel with the Mediterranean coast. Through most of Palestine the axis lies almost north to south, curving southwest towards the Sinai frontier in southern Palestine. Heights range from 600 m to 1,000 m. Strong vertical movements have occurred, resulting in the formation of the deepest inland depression on earth, containing the Jordan valley, the Dead Sea depression (over 790 m below the level of the Mediterranean Sea in some places), the Araba, and the Gulf of Aqaba. Other minor transverse faults have created a number of smaller depressions, including notably the 'Emek Yizre'el in the north. Volcanicity was often associated with these vertical movements, and basalt outcrops widely in eastern Galilee. The almost straight, harbourless shoreline may also be attributable to faulting, but uplift of the land and regression of the sea have also been suggested as possible causes.[2] Surface drainage is largely by a few east and west flowing streams, some of which have cut deeply into the highlands with their numerous headstreams. Throughout the southern half of Palestine streams are ephemeral. The largest river, the Jordan, is entirely an inland river terminating in the Dead Sea.

The structure and topography of Palestine are thus complex and varied. At least three dozen subregions have been recognized by geographers, but the following fourfold division should serve to illustrate the personality of the country.

16.2.1 The coastal plains and the 'Emek Yizre'el

The great spur of Mount Carmel interrupts the coastal plain just south of the Bay of Haifa. To the north lies the small plain of 'Akko; to the south are the wider plains of Sharon and the southern coastal plains. They nowhere exceed 20 km in breadth. The coast itself as far north as Caesarea is fringed by sand dunes, in places more than 5 km inland, covering good agricultural soils. The soils of the coastal plains are formed largely from alluvial deposits brought down from the highlands. The fertile Red Mediterranean Soils (Arabic *'Hamra'*) of the western parts are particularly suitable for citrus cultivation. Irrigation is commonly practised on these soils, and erosion has been severe in the past. In recent geological times they were enriched and darkened by swamp vegetation associated with streams blocked at their mouth by sand dunes. Soils of the eastern plains and the plain of 'Akko are heavier, formed by deposits of Terra Rossa from the eastern highlands; these soils have also been partially darkened by swamp vegetation in Pleistocene times. Certain crops can be cultivated on these heavier soils without irrigation. The coastal plain is rich in water resources from wells and springs. Up to 700 mm of precipitation may be received over the hills and permeable strata dipping westward yield valuable groundwater for the coastal plains. This gave the early Jewish settlers a wide choice of sites for their villages, while today it is one reason why Israel wishes to retain control of the West Bank (see below). The aquifer supplying the majority of settlements lies at depths of 18 to 120 m.

The 'Emek Yizre'el is a tectonic trough formed by subsidence, floored by a thick layer of alluvium weathered from the limestone and basalt dykes of the surrounding hills. The soils are dark, heavy, and rich in organic matter derived from the thick swamp vegetation which flourished localy until Jewish colonization in the 1920s.

16.2.2 The mountains and hills

This division includes the historic areas of Upper and Lower Galilee, Samaria, and Judaea. Upper Galilee is structurally part of the mountains of Lebanon – a picturesque limestone plateau dominated by Mount Hermon (2814 m). Lower Galilee to the south, is broken into many smaller hills of lower altitude with gentler slopes. Galilee as a whole is the wettest region of Palestine, receiving over 1000 mm of precipitation annually in places, and both springs and streams are more numerous than in Judaea.

The hills of Samaria roughly correspond with the heartland of the ancient Kingdom of Israel, between the 'Emek Yizre'el and the plateau of Judaea. Samaria is lower in elevation than Galilee or Judaea, and the climate and vegetation is transitional. Although rainfall reaches 630 to 750 mm, surface water is not plentiful. Both Galilee and Samaria were well populated with Arab Muslim villages before 1948, and Galilee also supported a number of Christian Arab and Druze communities.

The boundary between Samaria and Judaea is not physically well defined, but may be thought of as passing some 15 km north of Jerusalem. Samaria is dissected into hills and valleys, whereas Judaea is more like a high plateau between 450 m and 900 m high. The Judaean landscape is bleaker and bare rocks and loose stones dominate the scenery. Apart from a few valley floors, the possibilities for cultivation are limited. Three parallel subregions of Judaea can be identified: the foothills in the west which receive up to 500 mm of precipitation, and contain several wide fertile valleys; the Judaean hills themselves in the centre rising to nearly 1000 m, with precipitation of up to 700 mm; and, to the east, the dry virtually uninhabited Wilderness of Judaea, with under 300 mm of precipitation.

The commonest soil type throughout these mountain regions is the reddish brown Terra Rossa, enriched in parts of Galilee by the weathering products from the basalt dykes. Where it has sufficient depth, Terra Rossa is excellent for cultivation, but highly susceptible to erosion so that large areas have lost nearly all their top soil. Nevertheless, the mountains and hills contained the majority of the Arab rural population of Palestine. Good defensive sites were plentiful and cisterns for water storage could easily be hewn out of the soft limestone rocks to supplement springs and wells. In the lowlands on the other hand, where defensive sites are scarce, Arab settlement was sparse until improved security during the nineteenth century encouraged the founding of a number of new villages.[3] Even so, the prevalence of malaria continued to be a strong disincentive to settlement.

16.2.3 The rift valley

By far the most impressive physical feature of Palestine, the rift valley, extends the entire length of the country from beyond the Lebanon border to the Gulf of Aqaba. Its width varies from three to 25 km. The lowest point in the Dead Sea depression is some 790 m below sea level. The water level, which fluctuates, is about 390 m below sea level. Although clearly belonging to a single structural region bounded by steep faulted sides, the rift has several subregions. In the north is the Hula basin, about one third of which was formerly occupied by the shallow Lake Hula where papyrus and other reeds formed malarial swamps. It is now an area of fertile farmland. Both Lake Hula and Lake Tiberias to the south were originally formed by basalt flows across the rift valley. Beyond Lake Tiberias, the Jordan river meanders some 100 km to the Dead Sea through the Ghor. The flood plain of the Jordan was often inundated and covered in parts by thick vegetation, though much has now been cleared. Although rainfall is generally under 300 mm, irrigation sustains successful agriculture in parts of the Jordan valley. The Dead Sea region and the Araba further south have never been important for settlement, though some experimental *kibbutzim* have been established there by the Israelis. The chief importance of the Dead Sea is its vast mineral wealth including large quantities of potash, common salt, bromide, magnesium chloride and calcium chloride.

16.2.4 The Negev

This large triangular desert region constituted about half the area of Palestine and over six-tenths of Israel before June 1967. Its limits in the north are somewhat indeterminate, but the traditional margin between the closely settled lowlands to the north and the semi-desert steppe to the south is an accepted guide, corresponding roughly with the 300 mm isohyet.

The only possibility for extensive agricultural settlement in the Negev is in the northwest, adjacent to the congested Gaza strip. It is a depression covered with a thick deposit of loess soil and considerable areas of desert sand. The loess is potentialy fertile but, being liable to serious erosion, sometimes degenerating into 'badlands', it requires skilful management. The northwest also receives a fair but unreliable rainfall; the rest of the Negev receives from 200 mm to less than 50 mm of rainfall annually. A ridge of mountains and hills runs across the central Negev at heights between 500 m and 600 m, rising towards the Sinai frontier to above 900 m in places. The Eilat hills in the extreme south expose the ancient basement rocks of the Arabian block – a variety of crystalline and metamorphic rocks which have been sculpted into an impressive arid landscape.

The Negev was never thickly populated. Even in Nabatean times, the population was probably no more than 50,000 (Chapter 5). The present Jewish population is under 400,000, most of whom are urban dwellers. There are 20,000 or so largely sedentary Arab bedouin in the Negev. Its economic significance however is considerable, since almost all Israel's important minerals such as copper, phosphates, natural gas, and glass sand are found there.

Climatically, Palestine lies in a transitional zone between the Mediterranean and the deserts of Asia and Africa. The chief features are the rainly and dry seasons, and the regional contrasts in climate due in part to the variety of topography outlined above.

The rainy season begins in October or November and ends in April; for the rest of the year there is very little rain, three or four summer months being completely rainless. Thus rain is received during the cool winter months when evaporation is low. The amount varies, decreasing generally from north to south (Table 16.1). Westerly winds bring rain to the western slopes of the highlands, but in the Negev they carry little moisture, having passed over the deserts of North Africa and

TABLE 16.1
Israel: selected climatic data

Region and station	Altitude	Average annual rainfall (1931–1969)	Monthly mean of daily minimum & maximum temperatures (°C)				Mean relative humidity (per cent)	
			January Min	January Max	August Min	August Max	January	August
Coastal plain:								
Tel Aviv	20 m	564 mm	8	18	22	31(1949–58)	74	73
Hill:								
Har Kenaan	934 m	718 mm	4	10	18	29(1940–49)	—	—
Jerusalem	810 m	486 mm	6	13	20	30(1951–60)	65	54
Rift valley:								
Sedom	−390 m	47 mm	12	21	29	39(1961–70)	56	38
Deganya	−200 m	384 mm	9	18	24	37(1949–58)	70	54
Negev:								
Beersheba	280 m	204 mm	7	17	19	33(1961–70)	65	58
Eilat	12 m	25 mm	10	21	26	40(1956–65)	46	28

Source: State of Israel, *Statistical Abstract of Israel*, **23**, Jerusalem, 1972, pp 8–10.

Sinai. Annual precipitation totals also fluctuate considerably; in wet years the total may be more than twice that of a year of low rainfall. Some rain comes in the form of violent storms and cloud bursts, in the course of which up to 100 mm may fall in 24 hours. Snow is quite common in the highlands in winter.

Temperatures also show great regional contrasts, the chief determinants being distance from the sea and altitude. On the whole, conditions are pleasant and healthy (Table 16.1). Fresh westerly breezes off the Mediterranean moderate temperatures by day, and at night breezes blow from the great deserts to the east and south towards the sea. Winters are cool on the coast but are colder inland. Occasional frosts occur in winter, though rarely at low altitudes. While citrus trees are in no danger from frosts, deciduous fruits and vines on the hills experience moderate cold spells from which they actually benefit. Daily ranges of temperature are greatest in the rift valley and in the southern and central Negev, with the greatest ranges recorded during summer. For most of the year relative humidity is high in the coastal plains, averaging from 65 to 70 per cent. In the rift valley, it is sufficiently high to add to the discomfort of heat in summer, but the air in the central and southern Negev is agreeably dry.

Perhaps the outstanding feature of the geographical personality of Palestine is the rich variety of its landscapes and physical environments. Within the compass of an area measuring some 420 km by 100 km, seas, lakes, mountains, valleys, lowland and desert are all found. The climates range from Subalpine through Temperate to Mediterranean and Tropical, with corresponding contrasts in vegetation and agricultural potential. The effect of these contrasts on the historic events narrated in the Bible for purposes of moral instruction and communicating the divine revelation are remarkable.[4] Palestine has many beautiful landscapes and a certain quality of light and air which engender deep and lasting affection among its inhabitants. The nature of relief, soils, and climate offer man an environment capable of high productivity, but liable to serious ecological deterioration if neglected or mismanged. In recent centuries the land had indeed become pitifully impoverished. The hillsides which may have seen the invention of the art of terracing, aided by the natural structure of their step-like slopes,[5] became stony and barren. It has even been suggested that over one metre of soil has been carried away since the breakdown of terrace agriculture from the seventh century AD.[6] Parts of the central coastal plain and the northern Negev were also severely eroded in places. Forests of oak and Aleppo pine were once extensive, but by 1918 following a great wave of destruction by the Turks which was the climax to centuries of cutting and burning, they had almost entirely disappeared. On the coastal plains between Haifa and Tel Aviv many west-flowing streams were choked with sand, forming swamps in their lower courses. Parts of ’Emek Yizre’el had degenerated into pestilential swamps, though according to several European travellers in the late nineteenth century other parts were under regular cultivation. The Beyt Shean and central Jordan valleys, once extensively cultivated, were the domain of pastoral nomads.[7] Altogether, the land bore of the scars of centuries of neglect and insecurity, and of the incursions of nomadic groups from the east.

16.3 Jewish rural settlement, 1882–1948

During the latter half of the nineteenth century more than half the Jews in the world lived in Eastern Europe and Tsarist Russia, where their conditions were as miserable as they had been anywhere in Europe for several centuries. In the 1880s their distress was added to by a series of anti-Jewish riots in southern Russia which resulted in large scale emigration. Among these was the first wave of Zionists to reach Palestine in modern times. It was their conviction that the only possible solution to the plight of Jewish communities in the east, and the threat of assimilation in the west, was the creation of a Jewish state. By 1903 altogether 20,000 to 30,000 immigrants had arrived, chiefly from Russia, Rumania and Poland.[8] In 1882 there were already 24,000 Jews in Palestine out of a total population of 450,000,[9] but these were chiefly devout urban dwellers in the towns of Jerusalem, Safad and Tiberias who had been in the land for many centuries. The newcomers, on the other hand, were determined to establish rural settlements in Palestine, as being the only practical way of laying the foundations of a Jewish society free from interference, and of restoring the land itself.

16.3.1 1882 to 1903 (Figure 16.2)

The first period of immigration lasted from 1882 until 1903, during which time 20 Jewish villages were successfully founded in Palestine, though not without several disasters. These early Zionist settlements (‘*moshava*’, plural ‘*moshavot*’) strongly resembled the grain-growing villages of eastern Europe in their layout. Farmsteads were arranged along both sides of a broad village street. A proportion of each holding was attached to the farmyard at right angles to the street. The rest of the village farmland was divided into several large blocks on the basis of their suitability for particular crops, each farmer being allocated a plot in each, an imitation of the practice in Palestine Arab villages. Gedera, founded on the southern coastal plain in 1884, provides a good example (Figure 16.1). The form of tenure was always private ownership and partnerships were very rare. The number of holdings in the early *moshavot* varied with the location and in accordance with the type of farming practised. Zikron Ya’akov (1882) and Petah Tiqva (1883) were exceptionally large with 150 to 200 holdings by the turn of the century, but 30 to 40 units was much more typical.[10] The small number of holdings was partly due to the difficulties of purchasing large blocks of continuous land from the Arabs, and the fact that the type of farming adopted required large holdings. As far as possible, farm units were of comparable size in individual villages, but varied considerably between villages from less than 10 ha to 30 ha each. Central services were not well developed in most *moshavot*, though Gedera had a flour mill, olive press and school, as well as a pharmacy and synagogue.

For a number of reasons, including the inexperience of the settlers, the *moshavot* were not economically a great success, and but for financial assistance

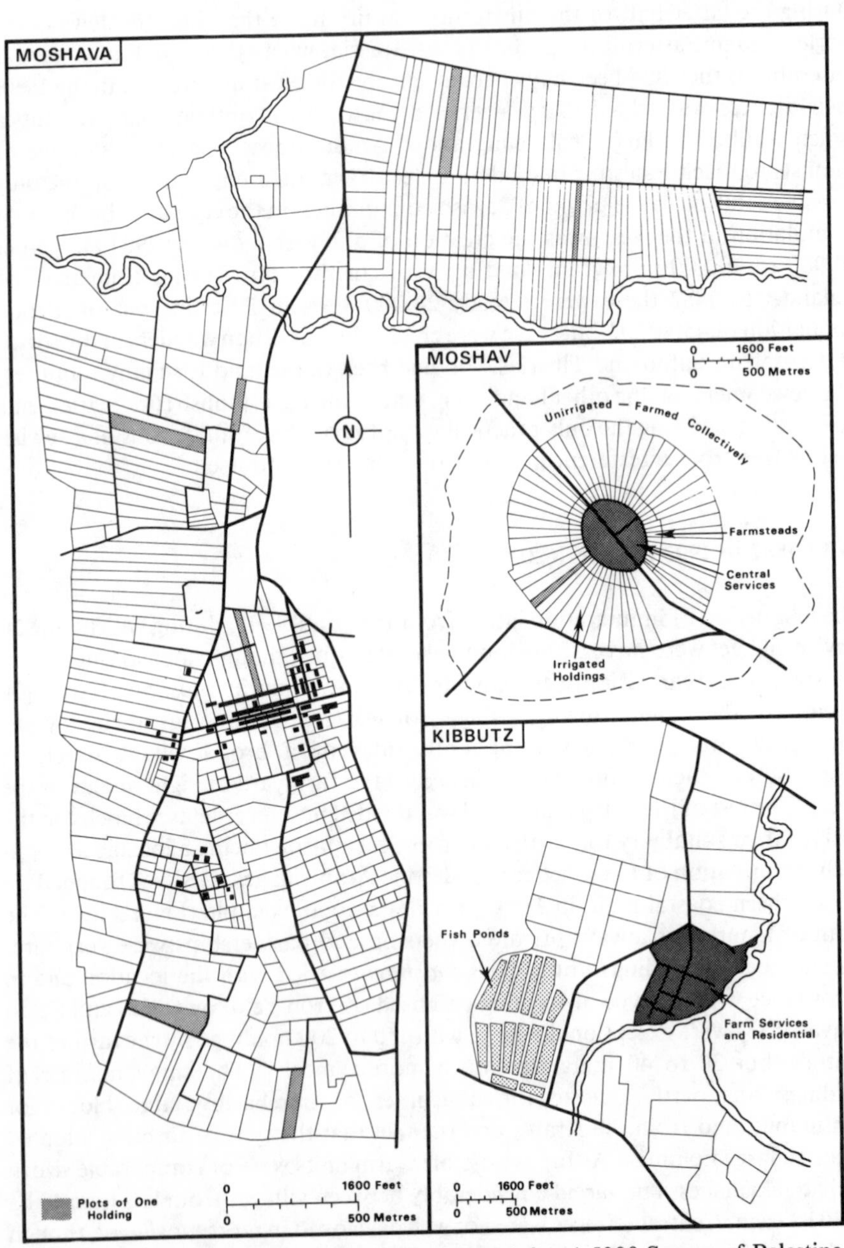

Figure 16.1 Gedera, a *moshava* founded in 1884 (after 1:5000 Survey of Palestine, 1943). **Inset**: layout of typical *kibbutz* and *moshav*

from the Jewish millionaire Baron Edmond de Rothschild, many would have collapsed altogether. He was responsible for introducing vines and citrus fruit, which meant that few families could now manage without hired Arab labour, particularly at harvest time. In due course, many became dependent on Arab labour and began to assume the habits of gentlemen-farmers.

In 1904, a second wave of immigrants began to arrive in Palestine from Russia. These people regarded the *moshavot* as wholly incompatible with their ideals, and their search for something better resulted in the emergence of both the *kibbutz* (plural, *kibbutzim*) and later the *moshav* (plural, *moshavim*) as alternative forms of settlement. The second period of immigration, which ended in 1914 brought another 35,000 to 40,000 Jews to Palestine. Unlike their predecessors, a high proportion of these immigrants were young manual workers and students, often penniless, but with a passionate concern for social justice derived, it seems, from the teachings of Karl Marx and the Hebrew prophets. Two of their ideas are particularly important for understanding the *kibbutz* movement. First, they intended to revitalize the Jewish personality by creating a class of farmers to work the soil for themselves. Secondly, they would create a just society, unlike other nations, which could be a model for the rest of the world. After competing unsuccessfully with Arab labour for work in the *moshavot*, and following the failure of a number of experiments with large commercial farms, a small group of these idealists succeeded in founding the first fully collective farm at Deganya in 1910.[11] The ideology and spirit of this first *kibbutz**, 'from each according to his ability, to each according to his need' remains essentially the ideal of the *kibbutz* movement today. Six important principles are generally recognized: (a) no wages are paid; (b) everything is held in common; (c) farming and other forms of production are fully collective; (d) government is by consent of the majority; (e) children live and are educated collectively; (f) there is restricted use of outside labour. The *kibbutz* had an immediate appeal, nine being established before the end of the First World War.

Figure 16.2 shows the extent of Jewish colonization in Palestine by 1918. The scattered distribution of Zionist villages in these early years was largely a reflection of the availability of land for puchase from Arabs, frequently on or near malarial swamps. In fact the first settlers, accustomed to the black earths of Russia, considered the dark swampy soils to be ideally suited to grain growing. Most villages founded in the first decade were within 35 km of the only port at Jaffa, but villages also appeared in the North Sharon plain and in eastern Galilee. Groundwater supplies were generally plentiful in the lowland areas colonized at first. Yesud Hama'ala was founded on the shores of Lake Hula as early as 1892, and Metullah appeared in the extreme north in 1896. Penetration of these remote districts was not as yet politically motivated, but it had political results. For example, the extreme north of the Hula basin was included within Palestine by an Anglo-French agreement in 1923 (Figure 16.3). Although the *kibbutzim* were at an early stage of development in 1918, their suitability for isolated and difficult

*The earliest collectives were called *Kvutsot*, but they were in all essential aspects, small *kibbutzim*.

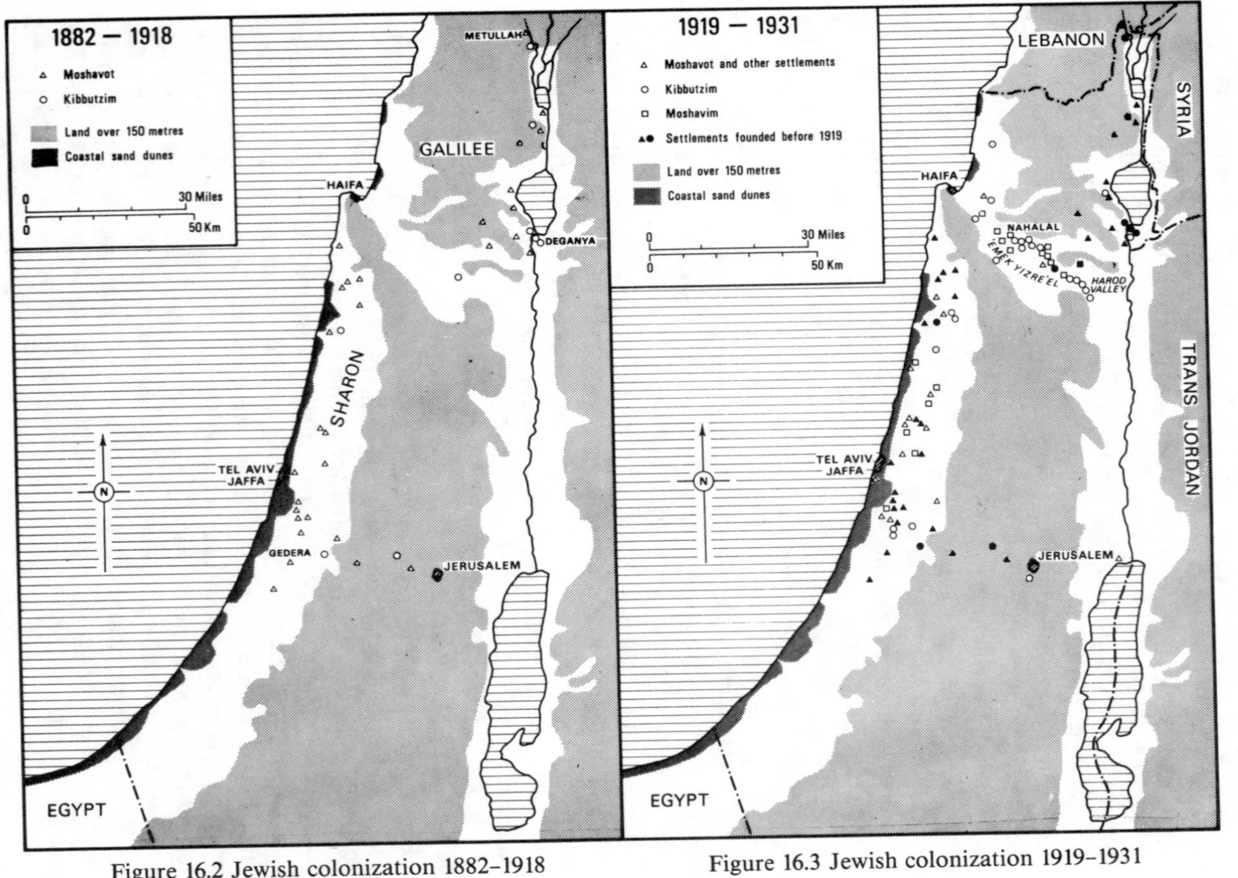

Figure 16.2 Jewish colonization 1882–1918

Figure 16.3 Jewish colonization 1919–1931

locations was already being demonstrated. In time, this unique form of rural settlement was to become the spearhead of Zionist colonization, and its ability to survive in pioneer situations was a big factor in extending Jewish influence.

16.3.2 1919 to 1931 (Figure 16.3)

Two further waves of immigration reached Palestine during this period. From 1919 until 1923, following the optimism created among Jews by the Balfour Declaration and the commencement of the British Mandate, some 35,000 immigrants arrived, chiefly pioneer elements from several European countries, many of them already prepared by Zionist organizations to undertake any tasks required of them. A second wave began to arrive in 1924, chiefly from Poland, where Jews were now being effectively excluded from certain trades. Many were middle class immigrants chiefly interested in settling in the towns. By 1931, they numbered some 82,000. Thus the 1931 census of Palestine revealed 175,000 Jews in a total population of 1,036,000, some 17 per cent.[12] This was a period of great significance for Jewish rural settlement since is saw the emergence of the *moshav*, an entirely new type of village designed to provide an alternative to the *kibbutzim* and *moshavot*. More than 70 *moshavim* were founded before 1948, compared with twice as many *kibbutzim*, but after 1948 the *moshav* became numerically more important.[13]

Even during the earliest years of the *kibbutz* movement there were those who wanted more individual initiative than collective farming would allow, and others who favoured more normal family life. To meet these serious objections and yet preserve something of the spirit and security of the *kibbutz*, Eliezer Yaffe proposed the creation of *moshavim* in 1919. The *moshav* would be governed by four fundamental principles: (a) national ownership of land, with inheritable leases; (b) self-labour on family farms; (c) mutual aid and responsibility among members; (d) cooperative purchasing and marketing. A fifth unwritten principle was equality of opportunity in the size and quality of holdings. Two developments had taken place during the previous few years to make these proposals feasible. The Jewish National Fund had begun work in Palestine, purchasing large blocks of land, not for resale to private individuals but for letting at low rents and held in trust on behalf of the Jewish people. The shortages of the First World War had forced many Jewish settlements to take to mixed farming, adding dairying, poultry and vegetables to the previous main branches of production – citrus, cereals, and vines. The new type of production was common in Jewish villages all over Palestine, and persisted almost unchallenged until the mid-1950s. Its importance lay in providing a relatively even labour schedule throughout the year, intensive production leading to a reasonably high standard of living, and self-sufficiency. At the same time, a rapidly growing Jewish urban community in the 1930s created a demand for dairy products and vegetables.

Mixed farming, with dairying as the mainstay, became the economic basis of the new *moshavim*. The first *moshav*, Nahalal, was founded in the 'Emek Yizre'el in 1921 and was quickly followed by several more. The number of holdings was generally between 70 and 100, small enough to permit a reasonably compact groundplan yet large enough to support central services such as school, dairy, cold store, tractor station, granary and so on. The size of farm units was small by comparison with the *moshavot*, but much of the income was derived from irrigated crops and dairy products. In Nahalal for example, holdings comprised 10 ha of irrigated land, together with a share of the income from non-irrigated land farmed collectively for cereals. In all pre-1948 *moshavim*, the main irrigated plots were attached to the farmsteads. Since cows were stall-fed with irrigated fodder cut daily, this was clearly the most efficient arrangement. The typical *moshav* plan, with holdings grouped around a closed village street containing services at the centre, provided a great sense of community and a measure of security, but it left no way of increasing the number of holdings for a second generation.

The period 1919 to 1931 also saw changes in the *kibbutz* movement. Many new *kibbutzim* were established, sometimes as large settlements from the beginning, whereas previously they had tended to acquire land and members gradually. The use of machinery was greatly increased. Mixed farming became predominant, so that by the end of the period the agricultural income of all *kibbutzim* was divided between dry farming (40 per cent), animal husbandry (28 per cent), poultry (16 per cent) and vegetables and outside employment on neighbouring farms (15 per cent).[14]

The activities of the Jewish National Fund led to the purchase, reclamation and colonization of large areas by almost continuous chains of Jewish villages. The most notable developments before 1931 were the draining and reclamation of the 'Emek Yizre'el and Harod valleys, which now came to have the greatest concentration of *kibbutzim* and *moshavim* (Figure 16.3). Most were highly successful settlements. A few more privately owned villages were also founded, chiefly by middle class immigrants for the purpose of citrus production for export. However, Jewish rural settlement was still largely confined to the lowlands of central and northern Palestine.

16.3.3 1932 to 1939 (Figure 16.4)

The fifth and final pre-1948 wave of immigration began to reach Palestine in 1931 as a result of the rise of the Nazis in Germany. By the end of 1939, 230,000 immigrants had entered Palestine, chiefly from Germany and Austria, with smaller numbers from Poland and Rumania where anti-Semitic policies were also being followed. The estimated population of Palestine in 1940 was 1,530,000 of which 457,000 (22 per cent) were Jews.[15] Once again, many immigrants in this period were middle class Jews, with capital and skill, who preferred life in Tel Aviv, Haifa or Jerusalem to pioneering in rural areas. Nevertheless, a sufficient number, particularly the young, took to agriculture, making this a period of rapid increase in the number of all kinds of rural settlements. More than 50 *kibbutzim* were founded, many of them between 1937 and 1939. Almost as many *moshavim* were established, the largest number between 1932 and 1935.

From the beginning of the British Mandate in Palestine, the Arab population had understandably become anxious about the political future of the country. They feared that their interests and aspirations were secondary to those of the Jewish minority and that they had been cheated by the great powers, but most of all they deplored the growing number of Jewish immigrants from 1919. After 1936, when the numbers increased markedly, their exasperation, hitherto expressed by rioting and isolated acts of terrorism, broke out into full-scale rebellion against the British authorities. A Royal commissiòn under Lord Peel was sent to Palestine to investigate the causes of Arab unrest; this was the fourth official inquiry since 1919.[16] Lord Peel's Commission recommended the partitioning of Palestine into Jewish and Arab states under British sovereignty, with certain regions retained under direct British control.[17] Figure 16.4 shows the relationship of the proposed arrangements with existing Jewish settlements; no Zionist village was to be in the Arab State, though large numbers of Arabs would be in the proposed Jewish State. The Arabs rejected the Peel Commission, and violence was renewed. The idea of partition was examined more closely by the Woodhead Commission which reported in 1938.[18] Although regarding partition as unworkable, the Commission reluctantly drew up sevral alternative plans, all of which assigned less of northern Palestine to the Jewish state, which was now reduced to a narrow coastal strip. The whole of the south would be either part of an Arab state, as in the Peel proposals, or would remain under British administration. The British government however also rejected partition, at least for the time being, but announced new policies for Palestine in a White Paper published in 1939.[19] Palestine was to become an independent Arab–Jewish state within ten years. Jewish immigration would be limited to 75,000 over the next five years, enough to bring the Jewish population to one third of the total. Any subsequent immigration would require the consent of the Arabs. Finally, the purchase of land by Jews was restricted to a small part of western Palestine.

Jewish settlement in the period 1932 to 1939 took place against this disturbed and tragic political background, and many of the developments shown in Figure 16.4 were politically motivated. The most striking example was the acquisition and settlement of the Beyt Shean valley, settled by *kibbutzim* after the 1937 Peel proposals had excluded it from the Jewish state. Attempts were also made to consolidate settlement of the Hula basin by further land purchases and colonization, while more settlements appeared in western Galilee. *Kibbutzim* were founded in the Menashe Hills to connect Jewish centres in the Sharon, generally designated as the core of any future Jewish state, with 'Emek Yizre'el, whose future seemed more ambiguous. Another important development during the period was the puchase of the Hefer valley, which now became closely colonized by *kibbutzim* and *moshavim*, linking Jewish settlement in the northern and southern parts of the Sharon plain. There were also signs of activity in the south of Palestine with the appearance of a *kibbutz* at Negba in 1933, and the

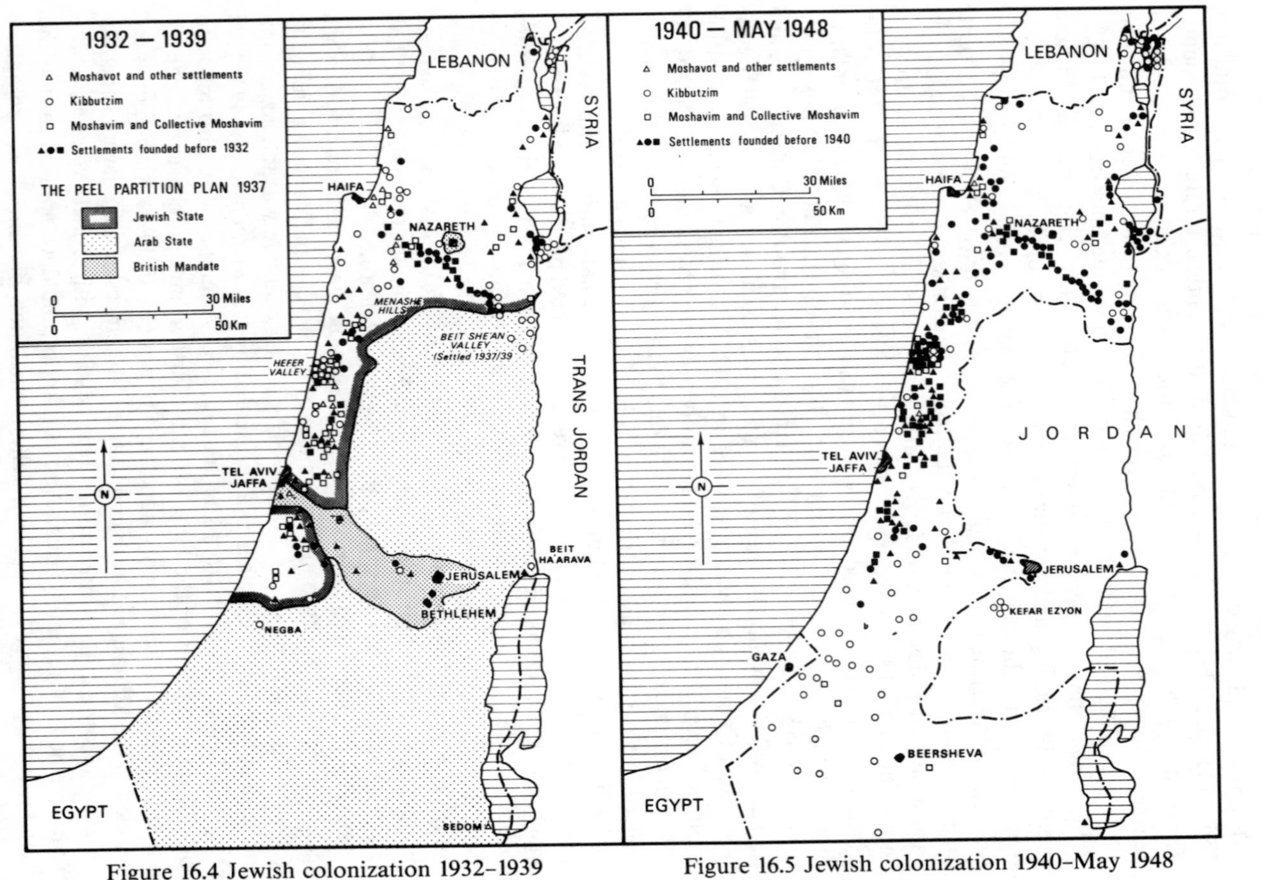

Figure 16.4 Jewish colonization 1932–1939

Figure 16.5 Jewish colonization 1940–May 1948

commencement of operations at Sedom by the Palestine Potash Company, which already had installations at the northern end of the Dead Sea, at Kallia.

In many ways the immediate pre-war period was the golden age of the *kibbutz*. Recruits were plentiful, many of them already prepared for the life by pioneer youth movements in east and central Europe. The *kibbutz* population nearly quadrupled between 1931 and 1936. The number of *kibbutizim* rose from three dozen to over 50 by 1939, and most of these were established more cheaply and more quickly than in previous years, often in difficult situations. While production became more intensive with the expansion of vegetable growing and dairying, the economic base of most *kibbutzim* was broadened to include light industrial activities, which earned nearly one fifth of their total income by 1936.[20] Some *kibbutzim* in the central Jordan valley were now producing dates, pomegranates and bananas, and the first fish ponds were established there. Experiments with new crops and techniques were being undertaken. At *kibbutz* Beit Ha'arava north of the Dead Sea, for example, record yields of winter vegetables and fruit were obtained from highly saline soils after treatment with large quantities of Jordan water. A countrywide marketing organization, TNUVA, was set up, which greatly improved access to the growing Jewish urban markets. This was extremely important since the Arabs could produce more cheap food for the towns, but Jewish urban dwellers purchased Jewish products for nationalistic reasons and as a result of a 'campaign of persuasion'.[21] Indeed, many *kibbutzim* and *moshavim* at this time were non-economic, in that many of their products were not strictly competitive and the real cost of settlement was borne by the Jewish National Fund and other Zionist agencies content to charge non-economic rents. Some settlements established for political reasons were never intended to pay their way; others, chiefly engaged in the export of citrus fruit, had quickly become self-supporting.

An interesting new form of settlement to emerge at this time was the collective *moshav* or *moshav shitufi*, an attempt to combine the best principles of the *kibbutz* and *moshav*. While family life was preserved in individual households, colletive production was undertaken on the farm. The first collective *moshavim* were founded by German and Bulgarian Jewish immigrants in 1936, but in spite of their theoretical merits, only seven such villages appeared before 1948 and only a further 15 between 1948 and 1970. Precisely why the *moshav shitufi* has not proved more popular remains obscure. It is a sound proposition economically, combining the social advantages of the *moshav* with the economic advantages of the *kibbutz*. The usual explanations that it combines the disadvantages of both, or that existing farms provide sufficient choice, seem somehow unconvincing.[22]

16.3.4 1940 to May 1948 (Figure 16.5)

Between 1940 and May 1948 Jewish immigration into Palestine was officially restricted, but in practice more than 110,000 people managed to enter the country, the vast majority of whom were refugees from Nazi Europe. Although the British also attempted to enforce the land puchase regulations, the Jewish National Fund acquired more land, though some holdings purchased before 1939 had not yet been colonized. Some 60 *kibbutzim* were established during this period, and about 20 *moshavim*. Many of the new *kibbutzim* were strategically located in the upper Hula basin and throughout the north and centre of Palestine, but above all in the south. This was the period of 'tower and stockade' settlement in which *kibbutzim* (and one or two *moshavim*) were established literally in a day, at first with crude wooden defences, and later with more permanent structures. The objective was frankly political and strategic, to establish a Jewish presence on land already in Jewish ownership. The most extraordinary instance of the technique was in 1946 when 11 settlements were founded in the northern Negev in the course of one night. The region is near the southern extremity of a Pleistocene aquifer which supplied all the pre-1948 settlements with water delivered in a network of six-inch pipes. Three Jewish observation posts had been conducting experiments in the region since 1943, and their conclusion was that irrigated agriculture in the Negev could provide a sound economic basis for permanent settlement. Their views have proved economically justified since 1948, but in the period immediately preceding the birth of Israel, they also proved politically justified from a Zionist point of view. When the United Nations resolved to partition the country in November 1947, the whole of the northwestern Negev, excluding Beersheba, was included within the Jewish state, together with the southern Negev. Apart from the Haifa Bay district, the whole of western Galilee was to become part of an Arab state.

The British Mandate in Palestine terminated on May 14, 1948, and the State of Israel was proclaimed by the Jews on the same day. Several months of desperate fighting between Israel and surrounding Arab states followed. In 1949 armistice lines were agreed with Egypt, Jordan, Syria and Lebanon, roughly reflecting the front lines when military operations ceased. The area contained within Israel, which remained unaltered until June 1967, included all but six Jewish settlements established before 1948 (Figure 16.5). Its total area was 20,700 sq km, about one fifth larger than that proposed by the United Nations. The only radical difference between the new armistice lines and the old frontiers of Palestine were the Gaza strip and the West Bank territory of Jordan. The Gaza region was dealt a serious blow since its population was now swollen by many thousands of refugees, and its economic links with the rest of Palestine were severed. The West Bank of Jordan fared little better. Jerusalem was a divided city. Many Arab villages found themselves cut off from their lands, and the region was overwhelmed by refugees. Transjordan now no longer had access to the ports of Jaffa and Haifa through which to pursue a modest trade (Chapter 15).

From Israel's point of view, the armistice lines constituted more of a security risk than an economic problem. Two particularly vulnerable corridors had been created in the south between Gaza and the West Bank of Jordan, and in central Israel, where Jordanian territory came within 17 km of the Mediterranean sea. Elsewhere, Israeli settlements were overlooked by Arab highlands, notably from the Golan Heights. For nearly 70 years geographical factors seem to have invited

Jewish settlement of lowland areas, leaving the highlands to their Arab inhabitants. In this sense, geography can be said to have played a part in the Palestine tragedy, as always seeming to hold out the prospect of partition or physical separation of Arabs and Jews, whereas greater integration in town and country might conceivably have forced some kind of enduring co-operation.

The population of Palestine would have reached about 2,065,000 by May 1948, including 650,000 Jews, about 31 per cent. Just under six per cent of the total land area, or 15 per cent of the cultivable land, was owned by Jews.[23] In Novermber 1948, following the flight and expulsion of approximately 650,000 to 700,000 Arabs from within the borders of Israel, and the immigration of over 100,000 Jews, the population of the young state was 873,000, 82 per cent of whom were Jews.

16.4 The Jewish urban population, 1882 to 1948

In spite of their immense political significance in laying the foundations of the Jewish state, rural settlers were generally no more than a small proportion of the total Jewish population in Palestine. The reasons for this are fairly obvious. In the first place, few Jews in the dispersion were farmers, and the decision to take to the soil was based upon an ideological desire to build a Jewish peasant class, in order to recreate the Jewish personality corrupted by centuries of oppression in Europe, and to found a Jewish community independent of interference by other people. During the 1920s, no more than 4 per cent of the economically active Jews in Europe were farmers; in 1907, the proportion for Germany was 1.3 per cent[24] and for Russia in 1897 it was less than 4 per cent.[25] Moreover, a high proportion were engaged in occupations least calculated to prepare them for arduous pioneer agriculture, in the professions, commerce, and industry. Agricultural colonization in Palestine offered a life of unremitting toil, and in some cases ill-health and death. Nor was land available in unlimited quantities; it had to be purchased piecemeal, chiefly by Zionist agencies, and the number of suitable settlers usually exceeded the land available. Finally, it should be remembered that the Jewish urban population grew most rapidly with the advent of Polish, German, and Austrian refugees, the first large-scale influx occasioned more by 'push' factors than 'pull' factors, and for whom Tel Aviv, Haifa or Jerusalem provided attractive and welcome refuge.

The level of the Jewish urban population in Palestine from 1900 to 1948 appears to have been between 10 and 16 per cent. Overall, the proportion of rural dwellers has remained surprisingly low. At the beginning of Zionist activity in 1882, Jewish communities were confined to Jerusalem, Safad, Tiberias and Hebron. These were chiefly devout Orthodox Jews, descendents of ancient communities, or a few who had migrated to the Holy Land to pray and die there. Most were supported by charity. An attempt to found an agricultural village had been made in 1870, but in 1882 Jewish rural settlement was confined to the small

Mikwe Israel farm school. The Jewish population was therefore effectively 100 per cent urban in the early 1880s.

Jerusalem, Tel Aviv and Haifa were responsible for attracting the great majority of urban immigrants (Table 16.2). Together they contain 68 per cent of the Jewish population of Palestine, and 56 per cent of the total population in 1948. Although the historic and religious associations of Jerusalem had an emotional appeal, and the city functioned both as administrative capital and the headquarters of a number of Jewish organizations, it grew relatively slowly. Tel Aviv and Haifa on the other hand grew rapidly. Tel Aviv, founded in 1909, was the only purely Jewish city in the world, and rapidly assumed the role of cultural and commercial capital of Palestine. Many small industries grew up, and a jetty and lighter harbour began operations in 1936, following disturbances in the neighbouring Arab port of Jaffa. About 24,000 Jews lived in Jaffa itself, on the border with Tel Aviv. Haifa enjoyed two advantages: a beautiful setting, and a good harbour which stimulated its industrial expansion. Economically, Haifa was the leading city of Palestine, with an oil refinery, railway workshops, foundries, cement works, and a variety of other industrial activities after the end of the First World War.

Approximately 15 per cent of the Jews in Palestine lived in eight small towns in 1948, the largest of which, Petah Tiqva, had scarcely more than 20,000 inhabitants. Five others possessed over 10,000 inhabitants and two 9,000 inhabitants. Three of these towns had been founded during the 1920s as satellite townlets for Tel Aviv, but engaging in some agriculture. The remainder were *moshavot* which had become increasingly urbanized since the late 1920s, so that by the early 1940s aproximately four-fifths of their active populations were engaged in non-agricultural activities,[26] chiefly light industries such as textiles, food processing, wine making and diamond polishing. The process of urbanization was made possible in the *moshavot* by private ownership of land, and was stimulated by their favourable location within the populous coastal zone near Tel Aviv or Haifa.

TABLE 16.2
Jewish populations of Jerusalem, Tel Aviv–Jaffa, and Haifa, 1914 to 1948

	Jerusalem	Tel Aviv–Jaffa	Haifa
1914	45,000	1,400	?
1922	?	22,000	6,200
1931	53,000	46,000	16,000
1935	70,000	135,000	50,000
1941	85,700	180,000	57,100
1948 (November)	84,000*	244,300	95,400

(*Israeli Jerusalem only)

Sources: (a) A. Bein, *The Five Aliyot and Their Achievements*, South African Zionist Federation, Johannesburg, 1943–44, pp 17–19.
(b) Naval Intelligence Division (Great Britain), *Palestine and Trans-jordan*, Geographical Handbook Series, BR 514, 1943, pp 184–186.
(c) State of Israel, *Statistical Abstract of Israel,* **12**, Jerusalem, 1961, pp 36–37.

Apart from these eight centres, there were no towns of any size in Palestine, except the Arab town of Nazareth with nearly 17,000 inhabitants.[27] The population of Palestine, particularly the Jews, was thus polarized between large cities on the one hand and the small agricultural villages on the other. Centres of intermediate size were few, and by no means all of these functioned as intermediaries in the provision of services. In 1925, 'Afula was founded as a regional town for the 'Emek Yizre'el, but proved a great disappointment. Its population in 1948 was only 2,500. The failure of 'Afula was largely due to the nature of the *kibbutz* and *moshav*, which were almost self-sufficient for most services, and depended largely on nationwide Jewish marketing organizations to sell their products in the towns. Indeed, the absence of medium-sized towns in the pre-state era presented no real problem, but it clearly had important implications for the planners of Israel's future settlement pattern. It is interesting to note that the British drew up a proposed hierarchy of settlement for Samaria in 1942 (revised in 1946), including rural centres and regional centres along the classic lines later adopted in Israel.[28]

16.5 The occupied areas

Jewish rural settlements were founded in Israel at a greatly accelerated rate following the establishment of the state in 1948. The prime objectives were to secure border areas against invasion and infiltration, and to promote the spread of Jewish population to the sparsely inhabited regions away from the centre of the country. Secondary objectives included boosting agricultural production, and assisting with the absorption and integration of large numbers of immigrants. The two forms of rural settlement evolved during pre-state colonization, the *kibbutz* and the *moshav* proved to be complementary in these new tasks. The *kibbutz* was ideal in hazardous border locations or where soils and climatic conditions were harsh, while the *moshav* proved most effective for the production of vegetables and industrial crops which superseded mixed farming. The *moshavim* were more easily integrated into schemes of regional settlement and were successful in the absorption of non-farming immigrants. By the end of 1953 more than 300 new settlements had been established in Israel, largely peopled by tens of thousands of new immigrants. Many of these villages were founded in northern Israel where there was a substantial Arab majority, and in the Negev which had suported only about one per cent of the Jewish population in 1948. As part of the national settlement strategy a network of new towns was established to serve the rural communities, while some were planned as part of integrated regional models as in the Lachish region.

During the early 1960s relatively few new settlements were founded in Israel, the emphasis being on consolidation of existing *kibbutzim* and *moshavim*. The Six Day War of 1967 however, stimulated Jewish settlement activity considerably, particularly in the Arab territories captured by Israel in the fighting. Once again the motives involved national security, economic opportunity, and idealism. A new wave of immigrants as well as the second generation of settlers from *kibbutzim* and *moshavim* founded in the 1950s, were looking for land on which to settle, and they turned to the occupied areas, and also to the hilly areas of Galilee. Good arable land is scarce both in the West Bank region and in Galilee, while both have large populations of Arabs. As a result, an entirely new form of Jewish rural settlement has evolved known as the *yishuv kehillati* (or community settlement) for use in these regions.[29] From an Israeli viewpoint it has been extremely successful in creating a Jewish presence in key areas.

16.5.1 Sinai and the Gaza Strip

Israel occupied Sinai and the Gaza Strip (61,558 sq km) in June 1967. Israeli forces finally withdrew from Sinai as part of the peace agreement with Egypt in April 1982, leaving the Gaza Strip under occupation. Although Sinai was extremely valuable to the Israelis as a strategic buffer-zone, and for its oilfields, the opportunities for rural settlement were limited by aridity, and remoteness. Altogether some 22 Jewish settlements were established in Sinai supporting approximately 4,500 Israelis. The bulk of these were *moshavim* established near the Gaza Strip in a region known as the Rafah Salient, lying astride the major invasion route from Egypt to Israel. The other Sinai settlements were largely tourist resorts on the Gulf of Aqaba, with a small town at Ofira near the southern end of the Sinai

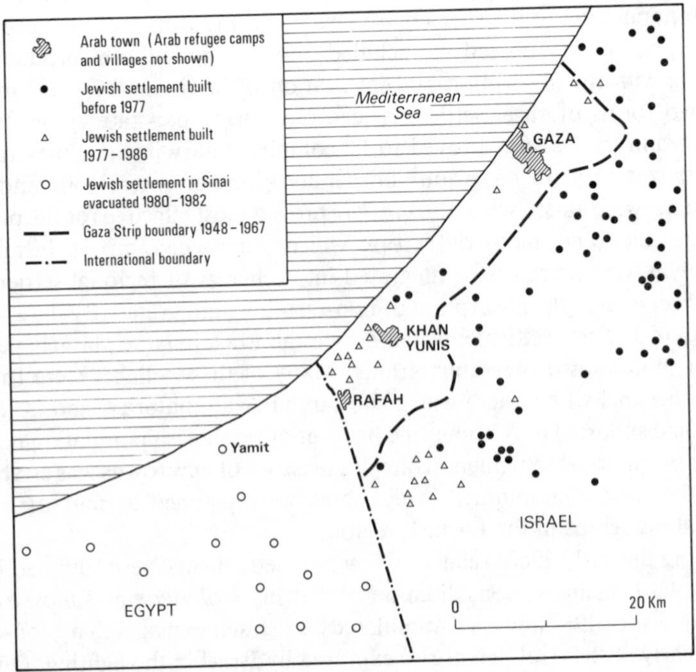

Figure 16.6 Jewish settlements in and around the Gaza Strip in 1986

peninsula. All Israeli settlements in Sinai were progressively abandoned and mostly demolished as the region was handed back to the Egyptians between 1979 and 1982, but not without fierce opposition from Jewish settlers. The political significance of the surrender of these settlements, which included the developing towns of Ofira and Yamit may be noted; Israel may be prepared to pay a high price for peace.[30]

Israeli settlement in the Gaza Strip has been encouraged by the government since Sinai was returned to Egyptian control. In spite of a dense Arab population of more than half a million which is predicted to reach 865,000 by the year 2,000, 19 Israeli settlements have been planted in Gaza, many of them since 1980. Their economies are largely dependent upon hothouse agriculture and some tourism. Their aim is presumably to separate Palestinian population concentrations, and create security outposts in a politically volatile area.[31] Although they appear impressive on the map (Figure 16.6) their total civilian population is probably not much more than 2,000, while they all depend heavily upon military personnel for protection. On the other hand, possibly as much as one third of the land in Gaza is under Israeli control.

16.5.2 The Golan

The Golan region of Syria, often referred to as the Golan Heights, was probably physically the most attractive of the occupied areas from the viewpoint of Jewish rural settlement. Apart from 13,000 Druzes who remained, the Arab population of 95,000 had fled during the fighting of June 1967. Although small in area (1250 sq km) the opportunities for agricultural production in the Golan are varied, including rearing dairy and beef cattle, cereal cultivation, and tree crops such as apples and plums. Thus the region was the first of the occupied areas to be settled after the June War in 1967 with nine *kibbutzim* and *moshavim* during the first year. The urgency with which settlement went ahead is partly explained by the crucial strategic importance of the Golan in controlling Syrian access to northern Israel. The effectiveness of settlements in securing territory was put to the test in October 1973. Although four settlements were over-run by Syrian tanks, and the remaining ten had to be evacuated, Israel remains wedded to the idea of territorial security by settlement.[32] It seems very doubtful whether Israel will ever seriously consider the peaceful surrender of the Golan, particularly the western region, which dominates the Huleh valley and Tiberias Basin. Another powerful reason for retaining control of the Golan is that approximately 75 per cent of the headwaters of the Jordan river originate there.[33]

There were approximately 40 Israeli settlements in the Golan in 1987, with a total Jewish population of about 10,000 (Figure 16.7). These included *kibbutzim, moshavim* and a number of *moshav shitufim*. Before 1973 most settlements were designed to depend upon agriculture for a livelihood, but after 1973 greater diversity was introduced, and a number are now engaged in industrial activity. One objective was to facilitate colonization of the central Golan whose

Figure 16.7 Jewish and Druze settlements in the Golan region in 1986

agricultural and touristic prospects are limited. Qazrin has been established in this area as a regional town with a target population of 45,000. In general Jewish population on the Golan has not grown at the rate the government of Israel would have wished. It has to be remembered that colossal spending on infrastructure (roads, water, electricity, sewerage) has been carried out to undergird the settlement effort, and to date the cost per capita would be very high indeed.

16.5.3 The West Bank

In many ways the West Bank is by far the most important of the remaining occupied areas, being not only the largest (5,900 sq km) and most populous, but the natural focus of Palestinian political aspirations. The 1984 Arab population, including those temporarily abroad, was over 900,000, an increase of approximately one third since Israeli occupation began in 1967. The West Bank is also the focus of intensive Jewish settlement effort, spurred on by a variety of factors. First there is a historic and emotional attachment to Judea and Samaria which roughly coincide with the West Bank. Secondly, it lies in close proximity to Israel's major urban centres in Jerusalem, Tel Aviv and Haifa, and thirdly, it is regarded as strategically important by Israel's military leaders. Indeed, Israel's determination to control and colonize the West Bank has resulted in land acquisition on a massive scale, by means which have earned international condemnation.[34] It is worth noting at the same time that the establishment of permanent civilian settlements in the occupied areas is contrary to international law.[35]

From 1967 to 1977 under Labour governments, settlement strategy in the West Bank followed the Allon Plan based primarily on military considerations. This called for a line of *kibbutzim* and *moshavim* overlooking the Jordan valley away from the hills to the west, with a large Arab population. Extremist groups however pressed for more widespread Jewish settlement in the historic heartlands of Judea and Samari, spearheaded by the religious Gush Emunim movement. From 1974, several unofficial settlements sprang up to which the authorities turned a blind eye. After the 1977 elections the Likud coalition approved these settlements, and encouraged the founding of a large number of settlements on the central West Bank (Figure 16.8). Private land purchases were also permitted which further accelerated transfers to Jewish hands. By 1985 there were 52,000 Jewish settlers in 115 villages, although 72 per cent of these were confined to 15 large settlements;[36] Israeli authorities had control of 52 per cent of the land of the West Bank.[37] Plans for future Jewish settlement envisage six towns with one million inhabitants and a network of Jewish villages. Most of the rural settlements are likely to be *yishuv kehillati* based upon light industries, some agriculture and animal husbandry, but above all commuting to work in Israel's major urban centres.

Plans for future settlement activity on the West Bank foresee the intensification of effort in the central region with existing types of village clustered around regional centres, and continued attention to the eastern one third of the region. The emergence of large suburban developments within the spheres of Tel Aviv and Jerusalem are foreseen, in response to demands for housing in semi-rural surroundings. Even so it is doubtful whether the projected Jewish population of 100,000 will be reached by 1990.[38] In all these ambitious plans, the interests of an indigenous Arab population on the West Bank are largely ignored. The infrastructure being painstakingly constructed by the Israelis does nothing to integrate Jews and Arabs of the region.

Israeli stake in the West Bank is now so great that voluntary withdrawal seems inconceivable. Apart from the heavy financial and political investment in the new settlements discussed above, Israel is now dependent upon the West Bank for some 475 million cu m, of water out of a total of 1900 million cu m, a quarter of the annual water potential.[39] This would be less critical if Israel was not already over-exploiting its water potential while facing increasing demands for water which could result in a deficit of 230 to 340 million cu m/annum by 1990.[40] Since 1982 Israel's national water company Mekorot has been integrating the West Bank supplies into the Israeli network. It seems clear that control of these sources will not be surrendered. The West Bank Arab population meanwhile feels badly

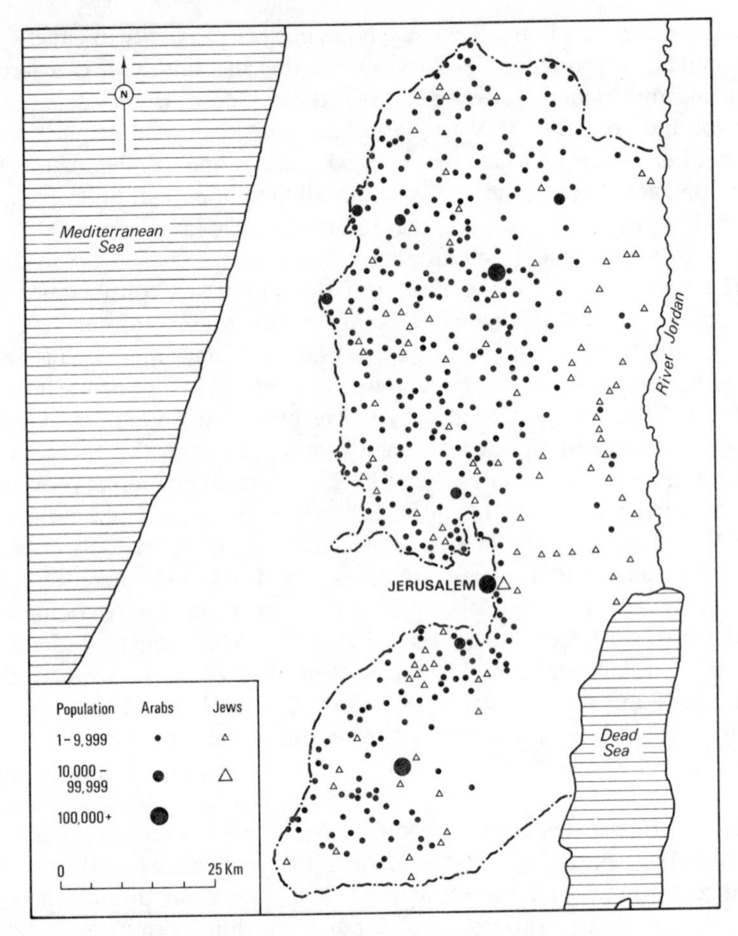

Figure 16.8 Arab and Jewish settlements in the West Bank in 1986

cheated; Israeli West Bank settlements plus water diverted into Israel probably account for 70 per cent of West Bank water. Water available for Arab agriculture is frozen at 90 to 100 million cu m by high prices and rationing, while water available to Jewish West Bank settlements will increase by more than 100 per cent in the 1980s.[41]

Israeli rural settlement on the West Bank must be put into the context of what is happening around Jerusalem. Israel annexed East Jerusalem after the June War in 1967, incorporating a large surrounding area of the West Bank, drawing the municipal boundaries to exclude certain Arab villages. A series of large suburban settlements have since been built on the hills surrounding Jerusalem, mostly in former Jordanian territory. These are massively constructed and already provide homes for over 100,000 Jews. In the early 1970s a second outer ring of small industrial satellite towns were founded, some of them beyond the municipal boundary. By 1987 the Jerusalem metropolitan area within 15 km of the city, had some 600,000 inhabitants. Thus more than twice as many Jewish settlers are to be found in the Jerusalem areas as in the whole of the rest of the West Bank. The Israelis do not regard Jerusalem as negotiable.

16.6 Conclusion

Zionist villages staked out the Jewish claim to a national home in Palestine in the pre-state period before 1948. When fighting broke out between Arabs and Jews in 1948 Jewish rural settlements proved the decisive factor in Israel's victory. The area and shape of Israel until June 1967 was thus largely a reflection of the course of pre-1948 colonization. Previous proposals for the partitioning of Palestine had similarly reflected this pattern. What appears to have been critical was not population size, which was modest, but the physical presence of a large number of Jewish settlements on the ground.

In certain respects, the colonization of areas occupied by Israel in 1967 is a repetition of the experience of the pre-state years. Jewish settlements create at least some potential hindrance to military action by the Arabs to regain lost lands. They also present a powerful bargaining counter in any peace negotiations, as well as a very influential lobby inside Israel which cannot easily be ignored. This is especially true of some of the settler organizations on the West Bank, who firmly believe in their religious duty to settle what they see as the historic homeland of Judaism.[42] All this means that the Israelis will be very difficult to dislodge. In Gaza, there is some chance that the settlements would be forfeited as they were in Sinai. In Golan some, but not all, may be negotiable in return for peace with Syria. The West Bank on the other hand is unlikely to be returned to Jordan, and the pattern of Israeli settlement makes any kind of political partition extremely difficult to implement.

References

1. S. Waterman, 'Israeli human geography since the Six Day War' *Progress in Human Geography*, **9,** (2) June 1985, 194–234.
2. E. Orni and E. Efrat, *Geography of Israel*, Israel Universities Press, Jerusalem, 1971, 37.
3. A. Granott, *The Land System in Palestine*, Eyre and Spottiswoode, London, 1951. 35–36.
4. (a) G. Adam-Smith, *Historical Geography of the Holy Land*, Hodder and Stoughton, London, 7th edn, 1900, 713 pages.
   (b) D. Baly, *The Geography of the Bible*, Lutterworth, London, 1957, 303 pages.
5. E. Orni and E. Efrat, *Geography of Israel*, Israel Program for Scientific Translations, Jerusalem, 1964, 2.
6. W.C. Lowdermilk, *Palestine Land of Promise*, Victor Gollancz, London, 1946, 14.
7. (a) D. Nir, *La vallée de Beth-Chéane*, Librarie Armand Colin, Paris, 1968, 176 pages.
   (b) Y. Ben-Arieh, 'The changing landscape of the central Jordan valley', *Scripta Hierosolymitana*, Jerusalem, **15,** Pamphlet 3, 1–131 (1968).

8. S.N. Eisenstadt, 'Israel', in *The Institutions of Advanced Societies* (Ed A.M. Rose). University of Minnesota Press, Minneapolis, 1958, 385–386.
9. E. Orni and E. Efrat, *Geography of Israel*, Israel Program for Scientific Translations, Jerusalem, 1964, 157.
10. A. Granott, *The Land System in Palestine*, Eyre and Spottiswoode, London, 1951, 259.
11. J. Baratz, *A Village by the Jordan*, Ichud Habonim, Tel Aviv, 1960, 174 pages.
12. Naval Intelligence Division (Great Britain), *Palestine and Transjordan*, Geographical Handbook Series, BR 514, 1943, 172.
13. State of Israel, *Statistical Abstract of Israel*, **23**, Jerusalem, 1972, 31.
14. U. Paran, 'Kibbutzim in Israel: their development and distribution', *Jerusalem Studies in Geography*, **1**, Hebrew University, Jerusalem, 1970, 1–36.
15. Naval Intelligence Division (Great Britain), *Palestine and Transjordan*, Geographical Handbook Series, BR 514, 1943, 172–181.
16. Cmd 1540: *Disturbances in May 1921: Reports of the Commission of Inquiry*, HMSO London, 1930 (The 'Shaw Commission').
    Cmd. 3686, and Cmd. 3687 (Maps): *Report on Immigration, Land Settlement and Development in Palestine*, HMSO, London 1930 (The 'Hope-Simpson Commission').
17. Cmd 5479: *Palestine Royal Commission Report*, HMSO, London, 1937 (The 'Peel Commission'), 380–395.
18. Cmd 5854: *Palestine Partition Commission Report*, HMSO, London, 1938 (The 'Woodhead Commission').
19. Cmd 6019: *Palestine–Statement of Policy*, HMSO, London, 1939 (The 'Macdonald White Paper').
20. U. Paran, 'Kibutzim in Israel: their development and distribution', *Jerusalem Studies in Geography*, **1**, Hebrew University, Jerusalem, 1970, 13.
21. A. Rubner, *The Economy of Israel*, Frank Cass, London, 1960, 99.
22. Klatzmann, *Les enseignements de l'expérience Israélienne*, Presses Universitaires de France, Paris, 1963, 163–177.
23. Y.A. Sayigh, *Palestine in Focus*, Palestine Research Center, Beirūt, 1968, 27–28.
24. A. Granott, *Agrarian Reform and the Record of Israel*, Eyre and Spottiswoode, London, 1956, 19–20.
25. *Jewish Encyclopedia*, **1**, Funk and Wagnalls, New York and London, 1901, 246–252.
26. H. Halperin, *Changing Patterns in Israel Agriculture*, Routledge and Kegan Paul, London, 1957, 75–76.
27. For populations of settlements in 1948 see: State of Israel, *Statistical Abstract of Israel* **12**, Jerusalem, 1961, Tables 9 and 10.
28. H. Kendall and K.H. Baruth, *Village Development in Palestine During the British Mandate*, Crown Agent, London, 1949, 18.
29. D. Newman, 'Ideological and political influences on Israel's rurban colonization: the West Bank and Galilee mountains', *Canadian Geographer*, **28**, (2), 1984, 142–155.
30. The traumas of settlement removal are discussed in: N. Kliot, 'Here and there - the phenomenology of settlement removal from northern Sinai', *J. Applied Behavioural Science*, No.1 (1987), pp 35–51.
31. R. Locke and A. Stewart, *Bantustan Gaza*, Zed Books and Council for the Advancement of Arab-British Understanding, London, 1985, 63–66.
32. (a) M. Gilbert, *The Arab-Israeli Conflict: Its History in Maps*, Weidenfeld and Nicolson, London, 3rd edn 1979, p 94.
    (b) W.W. Harris, *Taking Root: Israeli Settlement in the West Bank, The Golan and Gaza-Sinai 1967–1980*, Research Studies Press, John Wiley, Chichester, 1980.
33. A.M. Farid and H. Sirriyeh (Eds), *Israel and Arab Water*, Arab Research Centre, London, and Ithaca Press, London, 1985.
34. G. Rowley, *Israel into Palestine*, Mansell Publishing, London, 1984, 84–89.

35. H.J. Hansell, Letter from State Department Legal Adviser Concerning Legality of Israeli Settlements in the Occupied Territories, April 21, *Americans for Middle East Understanding*, New York, 1978, 10–12.
36. M. Benvenisti, Z. Abu-Zayed and D. Rubinstein *The West Bank Handbook: a Political Lexicon*, The Jerusalem Post, Jerusalem, 1986, 49–50.
37. M. Benvenisti, *1986 Report: Demographic, Economic, Legal, Social and Political Developments in the West Bank*, West Bank Data Base Project, Jerusalem, 1986, p 29.
38. M. Benvenisti, *West Bank Data Project: a Survey of Israel's Policies*, The Jerusalem Post, Jerusalem, 1984, 60.
39. M. Benvenisti, Z. Abu-Zayed and D. Rubinstein, *The West Bank Handbook: a Political Lexicon*, The Jerusalem Post, Jerusalem, 1986, 223–24.
40. V. Davis, 'Arab water resources and Israel's water policies', *Israel and Arab Water*, (Eds A.M. Farid and H. Sirriyeh), Arab Research Centre and Ithaca Press, London, (1985) 16–24.
41. M. Benvenisti, Z. Abu-Zayed and D. Rubinstein, *op.cit*, 1986, 224.
42. D. Newman (Ed), *The Impact of Gush Emunim: Politics and Settlement in the West Bank*, Croom Helm, London 1985.

CHAPTER 17

The Industrialization of Turkey

17.1 Introduction

All the countries of the Middle East are attempting to industrialize (Chapter 8). Turkey was one of the first to begin the process, and by 1980 manufacturing industry employed about 54 per cent of the active population of 22,284,000, contributed 22 per cent of GDP and accounted for more than 40 per cent of total exports. Thus, Turkey can fairly be described as an industrialized country. The point is strengthened by looking back at the statistics for 1970 and by making a comparison with Japan. In 1970, Turkish manufacturing industry employed about 12 per cent of the active population, contributed 14 per cent of GDP and made little show in exports. In Japan, which began to industrialize about the same time as Turkey but became a successful industrial power much earlier,[1] 25 per cent of the active population of 55,810,000 worked in manufacturing industry in 1980. Manufacturing contributed 30 per cent of Japan's GDP but produced over 90 per cent of the country's exports.

This chapter outlines the spatial aspects of Turkey's struggle to industrialize and distinguishes four phases of development, two of them virtually new beginnings. Each phase was characterized by different combinations of productive factors and marked by particular socio-political circumstances.[2] Some elements, however, remained more or less permanent and exerted profound influences upon industrial development.

The chapter is concerned with the Asiatic section of the Turkish Republic, though it is impossible to leave aside the İstanbul conurbation which covers a considerable area on both sides of the Bosphorus. Asiatic Turkey will frequently be called by its traditional name, Anatolia (Turkish *Anadolu*), partly to avoid confusion when discussing industrialization under the Ottoman Turkish Empire and partly to retain the subregional terminology currently in use. This region has presented formidable barriers to national economic development by its size, shape and physical variety.

Anatolia is a rectangular peninsula stretching more than 1,600 km from the Aegean coast to the eastern frontier, but generally less than 800 km between the Black Sea and the Mediterranean, distances which in 1915 would have taken about 13 and seven days of continuous motoring to traverse respectively.[3] The area of 755,681 sq km is framed on the north and south by chains of fold mountains which merge in the east to form a tangled knot (Figure 17.1) through which communications are difficult. Between the great mountain chains is a belt of terrain within which movement is comparatively easy. Its northwestern corner

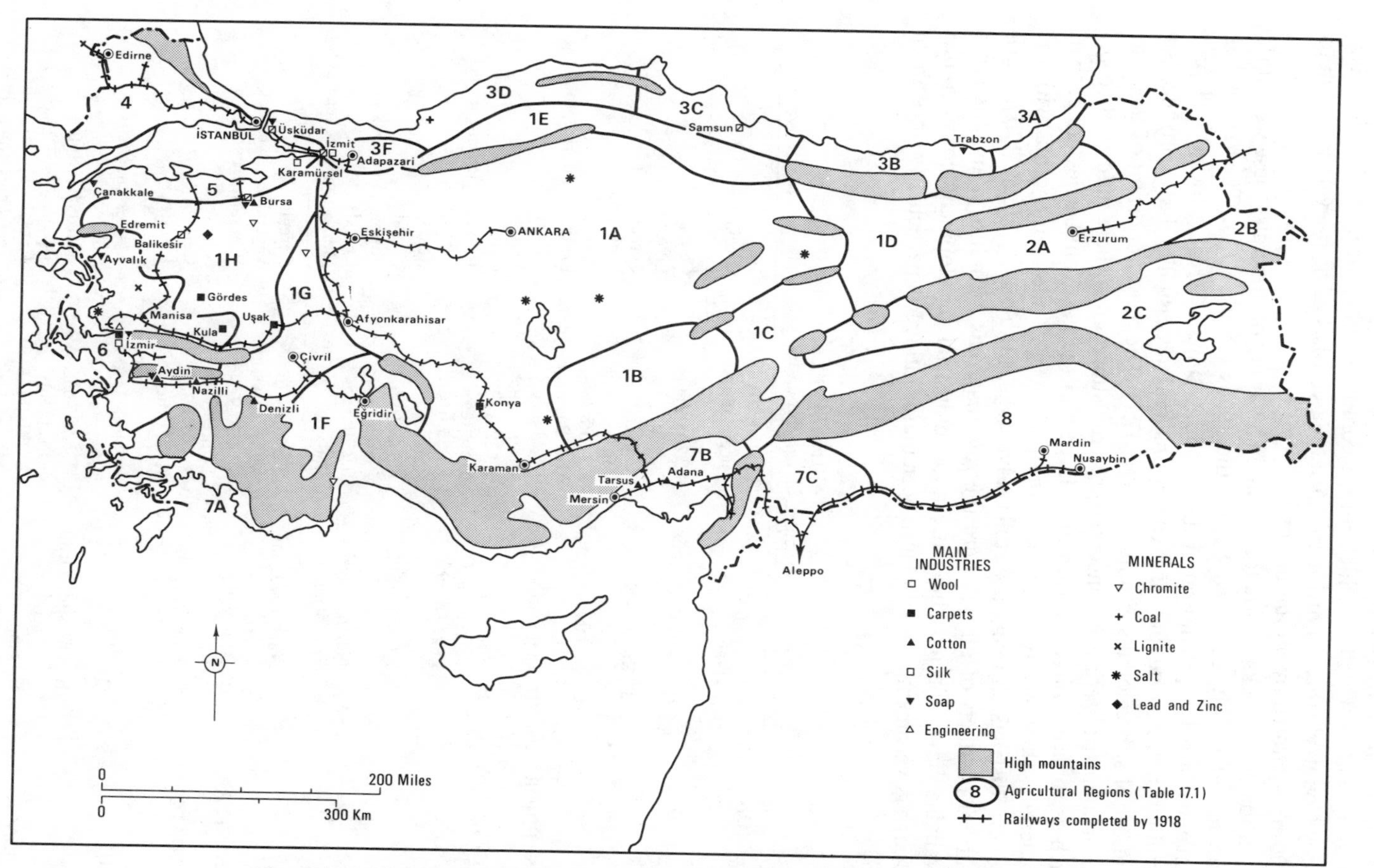

Figure 17.1 Distribution of factory industry and mineral production in Turkey, 1900–18

is an upland mass fringed by plains. To the south lies a series of horsts separated by long, wide rift valleys which afford communication with the interior. The 'Grey Country' (*Bozkır*) of Interior Anatolia consists of high-level plains (up to 1000 m) separated by mountain ranges.

Although the rocks of Anatolia contain a variety of minerals, these were little known and poorly exploited before the 1930s (Figure 17.1). The region has remained largely agricultural from Neolithic times. Cereal growing still predominates, but local physical conditions and accessibility have produced a number of subregional specialities (Table 17.1). The most important of these, as far as industrialization is concerned, are the subregions where industrial crops, such as tobacco, olives, cotton and, largely in the past, mulberries are grown. Generally, these are found around the periphery of the country, but with the western districts enjoying a distinct advantage, (Table 17.1; Figure 17.1). Agricultural products of all types are the raw materials of industry, and the types available have affected the pattern of industrial development considerably, though government action has also been major influence, particularly since about 1934.

TABLE 17.1
Agricultural regions of Turkey

REGION		SUB REGION		
		Name	Physical conditions	Specialities
I	A	Central Anatolia	– 0.2 20.0 360 Plateau	Cereals Livestock
	B	Kayseri–Niğde	0.3 22.7 357 Volcanic soils	Cereals (rye) Fruit Vines
	C	Malatya–Elâzığ	– 1.2 26.8 368 Mts, Basins, Plateaus	Cereals Fruit, cotton, rice, tobacco
INTERIOR	D	Erzincan	– 3.6 23.8 365 Mountains	Cereals Vines
ANATOLIA	E	Northern Transitional	– 1.3 20.3 438 Mountains	Cereals Tobacco, rice, sugar beet
	F	Lake District	1.7 23.0 615 Karst	Cereals Roses
	G	Afyonkarahisar	0.3 22.0 461 Plateau	Cereals Poppies, livestock
	H	Northwestern Transitional	0.2 20.5 552 Hilly	Cereals Maize, tobacco, vegetables
II	A	Kars–Erzurum	– 8.6 15.0 476 Mts, depressions	Cereals Vegetables, livestock
EASTERN	B	Aras Valley	– 10.1 20.9 546 Depressions	Vines, rice, cotton
ANATOLIA	C	Van–Tunceli	– 3.4 22.1 383 Mts, valleys	Cereals Livestock
III	A	Rize	6.9 19.8 2440 Mts, plains	Maize, tea, citrus Tobacco
	B	Giresum–Ordu	7.2 22.7 836 Mts, plains	Maize, hazelnuts Beans, tobacco
BLACK	C	Samsun	6.9 20.0 731 Mts, plains	Cereals Vegetables, tobacco

TABLE 17.1 (Contd.)

REGION	SUB REGION Name	Physical conditions	Specialities
SEA	D Kastamonu–Kocaeli	6.0 19.2 1245 Mts, basins	Cereals Maize
	E Istranca	5.5 22.3 735 Mountains	Cereals Maize
	F Düzce–Adapazari	6.6 23.1 774 Basins, plains	Cereals, maize, potatoes Fruits, sugar beet, tobacco
IV INTERIOR THRACE		2.0 21.9 609 Plateaus, basins	Cereals, maize, hemp Sugar beet, tobacco, vines
V MARMARA		5.4 21.6 740 Plateaus, basins	Fruit, vegetables, cereals Tobacco, olives, vines
VI AEGEAN		8.6 24.8 693 Mountains, valleys	Cereals, tobacco, cotton Vines, figs, olives
VII MEDITERRANEAN	A Muğla–Mersin	10.0 25.0 1030 Mountains, plains	Cereals Cotton, flax, sesame, citrus
	B Seyhan–Ceyhan (Çukorova)	9.1 25.0 611 Plains	Cereals Cotton, early fruit, citrus
	C Hatay–Gaziantep	8.0 26.9 1141 Mountains, valleys	Cereals Vines, olives, pistachios
VIII SOUTH EAST		5.0 27.7 452 Plateaus, valleys	Cereals, livestock Rice, vegetables, vines, fruit

Key to physical conditions
Jan. mean temp. (°C), July mean temp. (°C); Average annual precipitation (mm) for representative stations.

Sources: Devlet Meteoroloji İşteri Genel Mürdülüğü, *Ortalama ve Ekstrem Kiymetler*, İstanbul, 1962; S. Erinç and N. Tunçdilek, 'The agricultural regions of Turkey', *Georgl. Rev.*, **42** 189–203 (1952).

17.2 Phase I: the beginnings, about 1800 to 1914

The Ottoman Empire may have appeared to be the 'Sick Man of Europe' during the nineteenth century, but her vast territories contained valuable minerals, which could be exploited for the benefit of European and American industry, and produced a diversity of crops, some of which could be transformed in the place of origin either for export or for sale in a market of over 20 million people,[4] even if many of them were impoverished.

In Anatolia itself, mining was carried on in an haphazard and sporadic way by a number of foreign companies, mainly in the west and comparatively near the coast (Figure 17.1).[5] Handicraft industry had been badly affected by imports during the first half of the nineteenth century, but not as badly as used to be thought.[6] About 1000 workshops were listed in the Turkish Trade Annual for 1900,[7] and this number is probably a vast underestimate of the true position, for in 1921 a survey of industrial activity in those parts of Anatolia controlled by the

Nationalist Government listed over 33,000 establishments.[8] Local demand for traditional products remained strong, while distance and poor communications blunted the competitive edge of imports in the interior of the empire.

Power-driven factory industry forms the subject of this chapter, and at the end of the nineteenth century it was much more restricted than handicrafts in both scale and location (Figure 17.1).

Industrial censuses were carried out principally for the İstanbul and İzmir regions in 1913 and 1915. They covered plants employing motive power of at least 5 hp, paying not less than 750 day-wages per year and having capital assets of £T1000 or more. The 1915 census gives a total of 282 establishments. Of these, 78 processed food and 78 manufactured textiles. Whilst the censuses emphasize the importance of the İstanbul region (55 per cent of the plants in 1915) and İzmir (22 per cent), it is also clear that some modern industrial firms were located elsewhere in Anatolia, with a notable outlier of modern textile factories in the cotton-growing region of the Çukorova.[9]

The concentration in western Anatolia was probably real, as well as apparent from the incomplete source material. The region produced a range of industrial crops (though mainly for export) and possessed a relatively dense and wealthy population. Western capitalists had provided it with railways (Figure 17.1) which had reorientated local communications networks and intensified land use.[10] A well-developed system of sea communications was focused on İzmir and İstanbul and allowed the ready export of goods in demand in Europe (silk thread and carpets, for example) and the import of coal from the Zonguldak area, though western Anatolia itself possessed reserves of lignite which were already being exploited at the end of the century (Figure 17.1). In addition, both great ports were long-standing centres of western commercial activity.

Foreign enterprise, in fact, laid the foundations of mechanical industry in Turkey. The Capitulations, a series of agreements granting specific privileges in the empire, had been extended to such an extent after the Anglo-Turkish Commercial Convention (1838) that foreigners enjoyed almost unrestricted freedom of movement and activity, as well as preferential tariff rates. Not only did this allow an increase in imports, but it also allowed foreign capital to penetrate deeply into the economy. Foreign industrial entrepreneurs were attracted by the availability of raw materials, low wages, some skilled labour, a potential market of about 11–12 million people in Anatolia alone, and savings in transport costs. Their agents were often Armenian and Greek Christians to whom the privileges of the Capitulations were extended, and these 'Europe Merchants' (*Avrupa Tüccari*) soon became industrialists in their own right.

A comparatively small part was played in industrialization by the state during the nineteenth century, though much attention has been given to it by commentators.[11] Efforts to establish modern factory industry effectively began during the early years of the *Tanzimat* ('Reordering') period (1839–76) when influential elements in the Ottoman ruling class rejected traditional rescripts for reform and turned to new institutions to revivify the Empire.[12] The specific aims of the industrialization policy were to supply the large imperial army with clothing and equipment. With the exception of a copper refinery at Tokat and an unsuccessful papermill at İzmir, the new state factories produced textiles and, like the armaments' industry created by Selim III (1789–1807), were located in, or comparatively close to, the centre of political power. Very few plants were established by the state and those which survived to the end of the century had been revived under Abdülhamid II (1876–1909).[13]

The reasons for failure in the state sector, as of Ottoman industrialization in general, are not far to seek.[14] There was virtually no protection of nascent local industries against cheaper foreign imports which, in any case, enjoyed a certain prestige. Production costs were high, whilst relatively poor communications in the interior of Anatolia created severe bottlenecks. Many of the managers and skilled workers, like the machines themselves, came from abroad and had little incentive to train local labour. The state itself did not give industry a high place in its list of priorities, at least not until almost on the eve of the First World War. Then some modest increases were made to import duties and, in 1909, a *Law for the Encouragement of Industry* was introduced. As revised in 1913, the Law gave privileges to would-be industrialists by providing free sites, giving tax and customs concessions and furnishing easier credit.[15] Some success was achieved and, as indicated above, a modest degree of industrialization had been effected by the time the First World War broke out.

17.3 Phase II: a new state, about 1920 to 1940

The First World War shattered the Ottoman Empire but, together with the War of Liberation (1919 to 1923), fought against a Greek invasion in the west of Anatolia and a French occupation of Cilicia in the south, it produced the Turkish Republic. The first task of the new government in Ankara was to rebuild a country torn by more than a decade of war which had decimated the population; virtually ruined the economy; devastated İzmir and several small towns and strained the communications system. The long term aims were the transformation of the suppressed and exploited people of Anatolia into an independent, western, industrial nation.

Much attention had to be given to reconstruction and to laying the legal foundations of the new order, but industrialization was encouraged in various ways which began to bear fruit in the 1930s. Communications were gradually restored and improved, chiefly by building railways (Figure 17.2). The silk industry was saved by government action. Sugar production was started when the government introduced beet and built refineries at Apullu (in Thrace) and Uşak (in western Anatolia). Two other government measures were important. The İş Bankasi was established in 1924 to finance private business enterprises, including industrial ventures, and a new *Law for the Encouragement of Industry* (1927) expanded the scope of the Ottoman legislation. Industrialization in the 1920s, though, was beset by considerable difficulties.[16] The country had lost its important Greek and Armenian entrepreneurs. Capital was in short supply. An obligation to repay the

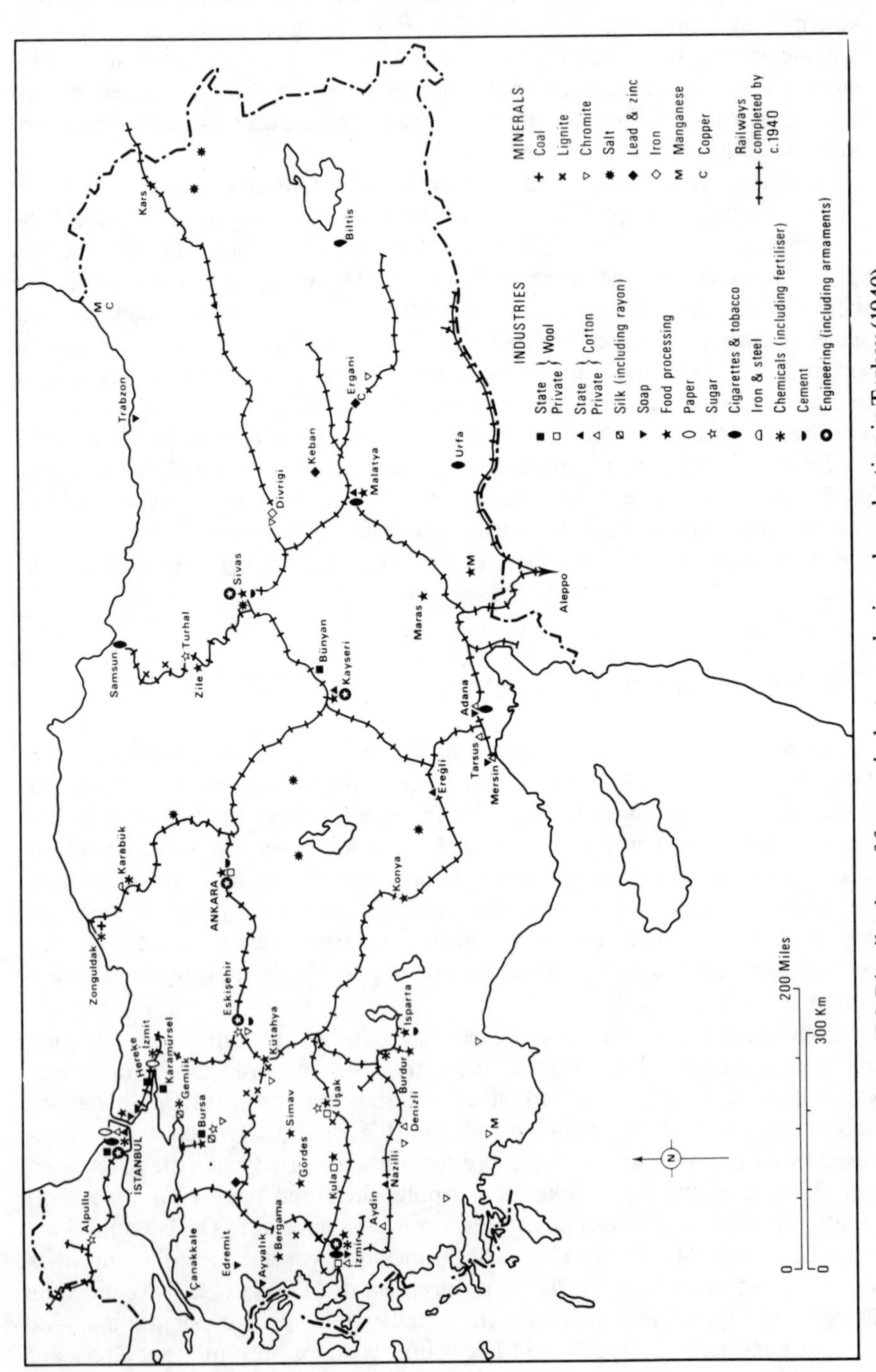

Figure 17.2 Distribution of factory industry and mineral production in Turkey (1940)

Ottoman Debt was shouldered, whilst customs duties were reduced to the levels of 1916 and the economic concessions in existence before October, 1914 were retained. The removal of the Capitulations undermined foreign confidence, already weakened by the Nationalists' military and political success. Despite the difficulties, an industrial census taken in 1927 revealed that there were at least 64,743 manufacturing enterprises in the country and that they employed about 231,000 workers. Whilst most of the entreprises were very small (a mean of 3.6 workers), 2822 (4.4 per cent) were using inanimate power, 54 per cent of them electricity. Food processing, metal working and textiles dominated the industrial structure. Most plants were in the west of the country, particularly in and around İstanbul and İzmir, whilst central and eastern Anatolia emerged from the statistics as containing comparatively little manufacturing industry.[17] In 1927, industry as a whole contributed about 14 per cent of GNP. By 1932, 1473 new concerns had been created under the provisions of the *Law for the Encouragement of Industry.*[18]

Figures for 1939 show a decline in the number of relatively large modern plants covered by the *Law for the Encouragement of Industry* to 1144, but the contribution of industry to GNP had increased to 18.0 per cent. Food processing (468 establishments) and textile industries (249 establishments) remained dominant, but several new industries had appeared. The most important were artificial silk (at Gemlik and Bursa), cement (in which Turkey soon became more than self-sufficient), paper, chemicals and steel (at Karabük). Mining, which had been stagnant during the 1920s, received substantial encouragement (Figure 17.2).[19]

A marked feature of the 1930s was the spread of modern industry to the Central Anatolia and Kayseri–Niğde subregions of Interior Anatolia. Single modern plants were established near railways on the outskirts of several towns with populations of 10,000 or more at the census of 1927. Small concentrations of industrial activity emerged in major provincial cities such as Eskişehir, Kayseri and Sivas, which were important railway nodes, whilst new industrial regions began to emerge along the Gulf of İzmit and around the ends of the Karabük–Zonguldak axis (Figure 17.2). The general effect, though, was to disperse industry in a way which was socially and politically justifiable, but which was often economically inefficient and could not prevent the Aegean, Marmara and Northwestern Transitional regions from retaining their advantage. Only Ankara could compete as an attractive industrial location with İstanbul and İzmir. The town's increasing and wealthy population provided a valuable market for industrial goods, while developing centrality gave access to the national market. A vast building programme stimulated the local production of construction materials, though the state took a hand by its investments in infrastructure and the making of cement and armaments.

The government, in fact, was responsible for all the new industrial developments away from the western parts of the country, apart from some at Ankara. The 'Brasilia approach' could be used only once, and the main impetus came from direct action. The new approach, usually called *Etatism* (Turkish *Devletçilik*), was never adequately defined, but the aim was clearly 'to initiate and develop projects in fields which were of vital concern to the strength and well-being of the nation'.[20] Various influences produced this policy. Despite the encouragements, private investment had not gone into industry to any great extent during the 1920s, while the world depression emphasized the lack of industrial development in Turkey and her dependence on exports of primary products. The ending of tariff restrictions in 1929 allowed the government to implement a national economic policy, while an example of what state action could achieve was available in the first Soviet Five Year Plan, launched in 1927. Finally, the Ottoman legacy was strong, for the Republic inherited not only several state enterprises but also a long tradition of state intervention in economic affairs.[21]

Investment was structured by two Five Year Plans (1934–39, 1938–42). The first aimed at import substitution by using local raw materials to establish consumer industries, while the second emphasized energy provision as well as producer and capital goods. Both plans were very unsophisticated by modern standards, and were really little more than 'listings of industry, mines and infra-structure which the government considered desirable'.[22] Only state industry was covered. The previous arrangements were continued for the private sector, but agriculture, despite its basic relationships to industry, was almost completely neglected. Funds were channelled through two, originally three, development corporations.* The Sümer Bankası was established in 1933 with the major responsibilities of operating the existing state concerns, as well as planning and ultimately running new industrial enterprises. The Eti Bankasi was founded in 1935 to develop mining in accordance with a special Five Year Plan launched in 1936. Capital continued to be raised from government monopolies on tobacco, spirits and salt, as well as from confiscated *evakf* (*wakf*, mortmain) properties, but new sources were also tapped. High taxes were applied internally, customs duties were raised, prices were fixed at high levels and loans were contracted from Britain and Russia.

Industrial development was further assisted by the type of activity promoted by the government and, to some extent, by a favourable combination of socio-economic conditions within the country. Several of the new activities, like iron making and construction, had important linkages, while an expansion in textile production was well adapted to national circumstances, particularly the income elasticity of demand, the availability of raw material and the great mobility of the products. A population increase of 1.8 per cent per annum and a slight improvement in agricultural incomes over the period 1927 to 1940 increased the national market, especially for processed foods, cigarettes and textiles, while small-scale and often seasonal migration provided the labour force for the emerging industrial towns. Migration was to become one of the main socio-political problems of the 1950s, and brought a renewed government commitment to industrialize the less-developed provinces.

*The Deniz Bankası did not operate effectively and its responsibilities were transferred to the Ministry of Communications.

17.4 Phase III: experiment, 1950 to 1960

Although Turkey was not involved in the Second World War, the state of emergency in the country produced a decade of virtual stagnation, particularly in industry and its contribution to GNP dropped to 13 per cent. The post-war period, particularly the years of Democratic Party rule, 1950 to 1960, was marked by four changes which had considerable effect upon the industrialization of Turkey and helped to raise the contribution of industry to GNP to about 16 per cent. These were the encouragement given to private enterprise, the availability of large amounts of foreign aid, the relative prosperity of agriculture and a rapid rise in population.

Private enterprise had not been neglected under the Etatist system, but its scope had certainly been restricted by the priorities and privileges given to state enterprises (SEEs). Post-war criticism of the state sector produced a series of measures designed to encourage private enterprise in industry.[23] The main instrument of government assistance was the *Türkiye Sinai Kalkınma Bankasi* (Industrial Development Bank of Turkey), established in May, 1950 by a consortium including the Central Bank, the International Bank for Reconstruction and Development, which had suggested the idea, and private business.[24]

The availability of large amounts of capital through the International Bank and direct from the United States (after 1947) allowed Turkey to embark upon a vast investment programme. Industry was helped by direct investment and loans, but it was also assisted by expenditure on infrastructure. The road network was greatly extended and opened up the country as never before (Figure 17.3),[25] while new power stations were built to realize some of the country's considerable thermal and hydro potential. In the west, an electricity grid was completed. Agriculture prospered during the 1950s, particularly in the wet years 1951 to 1953, but prosperity was achieved mainly by ploughing up marginal land in Interior Anatolia with the help of tractors. Mechanization integrated agriculture more firmly into the industrial market economy, while improved rural incomes produced something of a boom in consumer goods and encouraged some landlords to invest in industry, especially in food processing.[26] Population increase also appeared to assist industry by expanding the potential market. However, an increase of more than 3 per cent per annum had the effect of reducing the arable land available to the average rural family. Increased pressure on the land, together with mechanization, which tended to end sharecropping arrangements and create unemployment,[27] promoted migration. Migration increased to a level equivalent to about 30 per cent of the estimated rural increase, compared with an estimated 10 per cent over the previous 23 years.[28] Towns expanded rapidly, especially Ankara, İzmir and İstanbul where *gecekondular* (shanties) proliferated alarmingly. Eventually, the influx of people from the countryside brought a modification in government industrial policy.

The encouragement of private industry helped to achieve an average growth rate in industrial output of 8 per cent. It also maintained textiles and food processing as the main lines of activity, though 40 per cent of the Industrial Develop-

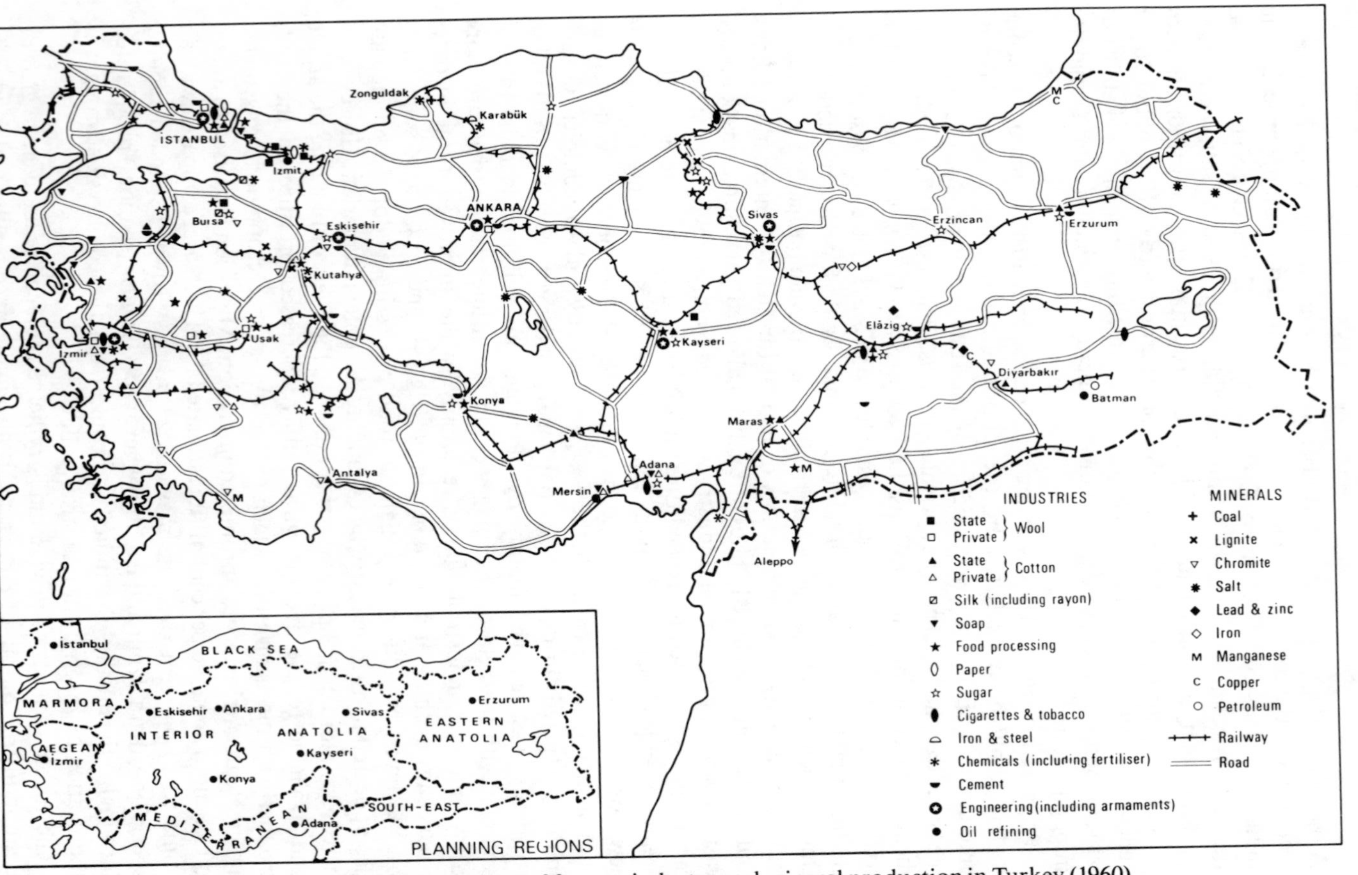

Figure 17.3 Distribution of factory industry and mineral production in Turkey (1960)

Rugged terrain requires spectacular engineering on many of the regions major roads. The Sha'ar descent, 'Asir, Saudi Arabia (photograph ARAMACO, 1986)

ment Bank's investments were in engineering, chemical and metallurgical industries. At the same time, the encouragement of private enterprise concentrated activity in the already industrialized regions. More than 1,000 of the 5,000 private enterprises using more than 10 HP and/or employing more than 10 workers in 1957 were situated in İstanbul vilâyet (province) and about 500 were found in İzmir, about 350 in Bursa and 234 in Ankara vilâyets. Seventy-five per cent of the plants sponsored by the Industrial Development Bank were similarly located in the Aegean, Marmara and Northwestern Transitional regions.[29] Concentration was due to the locational advantages already enjoyed by these regions. İstanbul and İzmir offered not only the largest and wealthiest markets in the country but also the ones in which purchasing power was increasing most rapidly. The government's promotion of the three leading cities as show places stimulated construction, thus expanding the market for building materials of all kinds. The second advantage enjoyed by the growth regions was in transport. İstanbul and

The Bosporus bridge, opened in 1973 (Freeman Fox and Partners)

İzmir remained the most important ports and their harbours were improved. The road building programme emphasized their nodality, together with that of Ankara, easing the flow of goods between them as well as to and from the provinces. At the same time, the improvement of communications in western Turkey reduced any need for manuacturers to locate their factories away from existing industrial regions, while the relative paucity of transport links in the eastern regions, as well as their remoteness and general backwardness, were positive disincentives to anyone seeking to set up a factory there. Capital availability was the third advantage possessed by the already industrialized regions. Private wealth, whether generated by trade or drained from the land, was concentrated in the three main cities. Ankara, İstanbul and İzmir were the main banking centres, and entrepreneurs in the provinces had relative difficulty in securing access to investment capital. Finally, the leading industrial regions possessed the advantages of external economies because so many industries already existed there and the necessary infrastructure had been developed.

Direct government intervention in industrial development was revived largely by the migration of large numbers of people to the towns which had resulted from the continued regional imbalance.[30] All but 12 of the 40 new SEEs were located outside Aegean, Marmara and Northwestern Transitional regions in towns with populations of 10,000 or more. The pattern was dispersed, as during the 1930s, but even less clearly related to local resources or markets, while lack of co-ordination with other development activities meant that plants were incapable of acting as the foci for industrial concentrations. In general, the counter-magnets to Ankara, İstanbul and İzmir, which these *ad hoc* developments might have produced, failed to emerge. Interior Anatolia benefited to some extent from governmental industrial activity, but Eastern Anatolia, the Mediterranean coastlands between Muğla and Mersin and much of the Black Sea region were still neglected.

17.5 Phase IV: the return to planning, 1960 to 1980

Extensive government investments designed to develop the peripheral regions of the country were instrumental in causing the Turkish economy to overheat during the second half of the 1950s. Mounting inflation and a serious trade deficit resulted. Rigid government controls were introduced in 1958 in an attempt to curb inflation and restore international confidence, but growing economic dislocation and increasing social frustration (especially amongst career officers and civil servants) produced a political upheaval culminating in a military *coup d'état* on 27 May, 1960. Civilian rule was soon restored, but from then on the Army has loomed large in the background of Turkish politics as the self-appointed but widely acknowledged guarantor of democracy and socio-economic justice.

The late 1960s saw divisions open and widen in the ruling Democratic Party, whilst further industrialization and continuing inflation brought the expansion of trade unions and an increased radicalization of their leaders. Political violence emerged on university campuses in 1968 but soon spread beyond them. The military intervened again in March, 1971 in an attempt to bring in a strong, credible government. During the 1970s, though, neither of the main political parties could command an adequate majority in the National Assembly and had to form coalitions with the smaller parties which began to proliferate. Governments turned to irresponsible free spending to buy support and to the development of what can only be described as a 'spoils system' which reached far into the civil service and police. New balance of payments difficulties emerged, partly as a result of the rises in oil prices in 1973 and 1979 and which were particularly serious for Turkey because it had become recklessly dependent upon oil imports. Turkey borrowed heavily abroad, but was forced to turn for help to the IMF in 1977–79 with the consequence that unpopular, deflationary and monetarist measures were introduced. Left wing policies and revivalist Islam gained increased support as the financial crisis threatened and unemployment increased. Political violence grew on both the left and the right, whilst the question of Kurdish autonomy emerged in violent form, as in the 1920s. On 12 September, 1980, the Army again stepped in. The country was rapidly brought to order. Measures were introduced to stabilize the economy, liberate 'free' market forces and then to encourage the export of manufactured goods. Civilian rule was allowed to return in the autumn of 1983 when elections were again held, though the old political parties had been banned and new ones hastily formed and martial law continued in some provinces.[31]

During these twenty turbulent years, economic policy moved from an emphasis on centralist planning directed through state agencies to the reassertion of the virtues of private enterprise, albeit state-guided in something like the Japanese manner.[32] The principle of planning was enshrined in the Constitution of 1961 and the necessary machinery created. A series of five-year plans was implemented (1963–67, 1968–72, 1973–77, 1978–82) within a perspective of 15-year rolling planning horizons. They were based on a relatively simple macro-economic growth model relating 'income levels to gross investment with relevant capital-output ratios'[33] and, consequently, they were more comprehensive than the industrial development plans of the 1930s (section 17.3). The aims of all four plans were essentially the same: to achieve high rates of growth in GDP (7 per cent per annum); to eliminate balance of payments difficulties; to create additional employment for a fast expanding population; to achieve greater social justice for all sections of the community; and to reduce disparities in the levels of socio-economic development between the various provinces.[34] Manufacturing industry was given a major role, both in raising national income and also in developing the more backward regions.

The well-publicized difficulties which faced the Turkish economy in the late 1970s should not detract from the considerable achievements of the period being surveyed in this section. GDP grew by an average of 5.8 per cent per annum between 1960 and 1969 and by 6.0 per cent during the years 1970–79.[35] Industry's contribution to GNP rose from 15.9 per cent in 1960 to 25.5 per cent in 1978, whilst its share of exports reached 43 per cent by 1980 compared with 20.7 per cent in 1963–67. The policy of import substitution paid off, with Turkey reaching, or close to self-sufficiency in most basic intermediate and technical goods. New steel works were built at Ereğ (near Zonguldak) and at Iskenderum. Oil refineries came into production at İzmir, İzmit and Mersin, and subsequently became the centres for a developing petrochemical industry. Overall, consumer goods industries continued to dominate manufacturing (62.3 per cent in 1962; 49.0 per cent in 1977).[36] The production of textiles and clothing remained important with about 20 per cent of total manufacturing output, a significant (but fluctuating) part in exports and about a third of the employment in the manufacturing sector. By 1971, though, the motor industry had become the second largest employer in manufacturing and contributed about 5 per cent of total manufacturing output.

The spatial distribution of manufacturing plants in 1977 is shown in Figure 17.4. Of the 5972 enterprises employing at least 10 workers, nearly 46 per cent were concentrated in İstanbul and a further 14 per cent in the adjacent Marmora region.[37] The Aegean region ranked second in terms of the number of enterprises (15.6 per cent), with İzmir the focus, whilst concentration of about 8 per cent of the enterprises in Ankara boosted Central Anatolia to third rank. Eastern and Southeastern Anatolia came out worst, with 1.5 and 2.3 per cent respectively of the establishments covered. In terms of industrial structure, enterprises making consumer goods were numerically dominant in each of the seven planning regions but of greatest significance in Eastern (63.2 per cent) and Southeastern Anatolia (61.1 per cent), an indication of the relative under-development of these regions. Intermediate goods were most important in the Black Sea (33.3 per cent) and Southeastern Anatolia regions (largely because of the steel and associated plants), though they were also numerous in the country's largest cities, İstanbul (29.0 per cent) and Ankara (27.8 per cent). Investment goods ranked second to consumer goods in the number of manufacturing enterprises in the Aegean region (37.1 per cent), where İzmir's industrial structure is dominated by them.

The predominance of the west and northwest in the distribution of manufacturing industry in the late 1970s is identical with the concentration of the urban population (Table 17.2). In reciprocal fashion, both are the product of past processes. The west and northwest were probably the most economically advanced regions of Anatolia when the industrialization process began in the nineteenth century and they have continued to receive disproportionately high levels of investment from the state, from Turkish private industrialists and holding companies and from foreign corporations. Financial services are located in the three principal cities and these are the locations preferred for the offices of major holding companies,[38] whilst the SEEs are, of course, directed from Ankara. Economies of scale derive from locating near activities of similar type. Personal contacts can be developed and exploited. Galaxies of suppliers develop. The physical infrastructure is comparatively good in the west and northwest, in part as a result of the foundations laid by foreign capital in the nineteenth century and again in the 1950s. The exploitation and transportation of raw materials is better than in the eastern parts of the country. Other potential sources of materials and parts are relatively close at hand in Europe, as are Turkey's prefer-

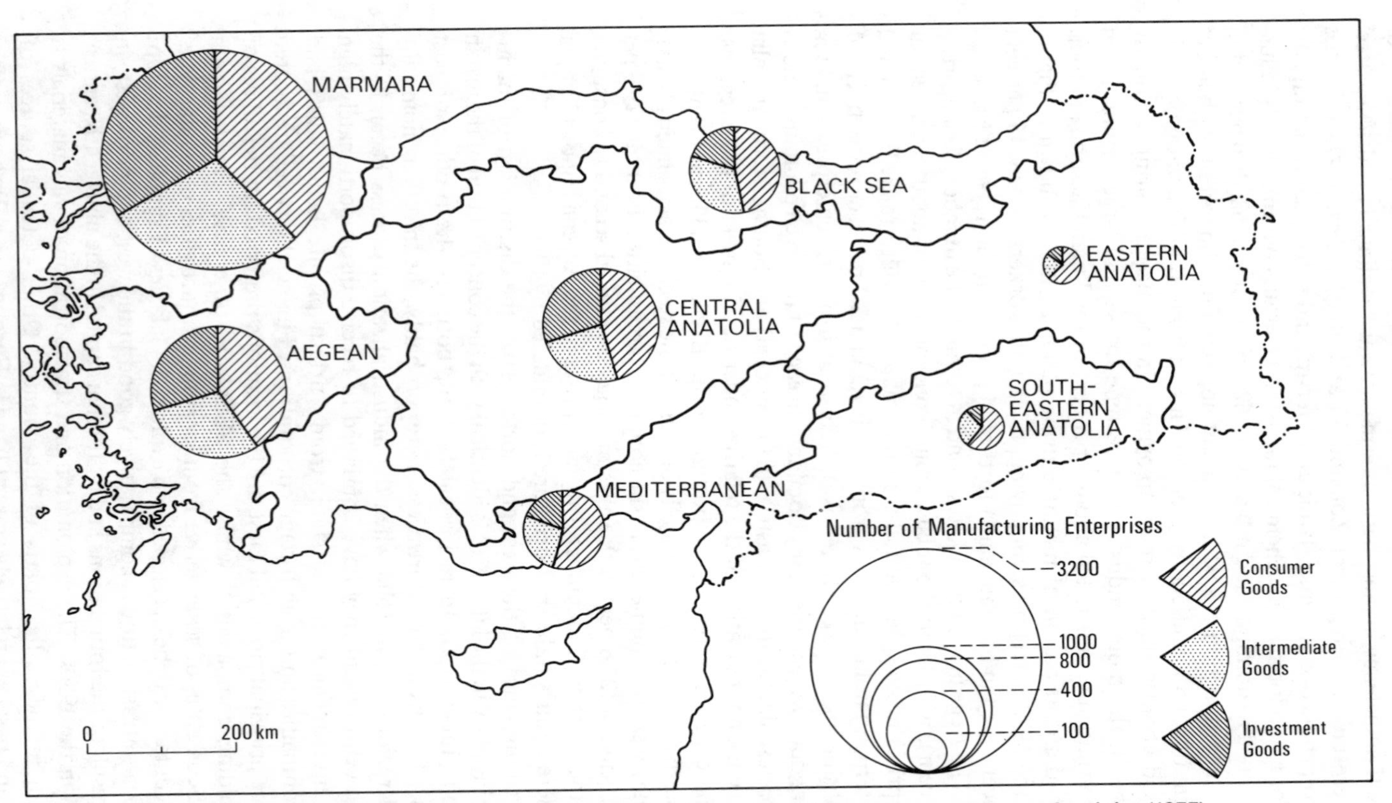

Figure 17.4 Distribution of manufacturing industry: number of enterprises and types of activity (1977).

TABLE 17.2
Urban population (places with 10,000 or more people) by planning region, 1960–80

Planning Region	1960	(percentage of population) 1970	1980
Marmara	42.4	50.4	68.7
Aegean	29.5	34.2	48.6
Central Anatolia	25.0	36.5	47.4
Eastern Anatolia	13.0	20.3	27.2
Southeastern Anatolia	16.0	25.8	36.5
Mediterranean	33.0	39.7	48.8
Black Sea	12.0	17.8	24.0

Source: Başbakanlık Develet İstatistick Enstitüsü, *Genel Nüfus Sayımı*, 1960, 1970, 1980.

ential markets. İstanbul and İzmir are the country's leading ports. Distribution of finished products is also easier to the domestic market, a very large proportion of which is concentrated in these favoured regions.[39] The eastern and southeastern provinces remain relatively unindustrialized, as do a number of scattered provinces in the centre and the northeast. Partly as a result, these areas emerge as underdeveloped on most of the usual indices like per capita incomes, literacy, and welfare provision. This is striking because successive regimes in Ankara have believed firmly that investment in industry would raise incomes and improve the quality of life in precisely these areas, whilst at the same time reducing the emigration which has caused worrying socio-economic problems to emerge in the cities over the last twenty years.[40] The First Five Year Plan (1963–67) spoke of the elimination of regional imbalances and the acceleration of development in the backward regions. Under the Second Plan (1968–72) priority investment was to be put on regionalized sectoral investment around selected growth centres. The Third Plan (1973–77) created a special authority within the State Planning Organisation to make recommendations to reduce regional imbalances, whilst various incentives were devised to encourage private companies willing to invest in the designated priority regions. These were continued under the Fourth Plan (1979–83). In addition, considerable state investment has gone into the Southeast Anatolia Project which has as one of its major objectives the generation of some 762.2 MW (26774.7 GWh) of HEP from a series of dams on the Tigris and especially the Euphrates.[41]

The failure of manufacturing industry to develop the backward provinces has several inter-related causes. Development has been hampered by limited resources, difficult terrain, poor infrastructure and remoteness from decision-making centres. Investment has consistently avoided the least productive locations. The continuance of the Kurdish problem in the southeast has acted as a disincentive to private investors since the late 1960s, whilst the predominantly conservative and influential landowners of the region have tended to discourage innovation. At the same time, much of the blame for the failed hopes for the backward provinces can be attributed to inconsistencies in government policy, apparent at least from 1960 but also present under the early Republic. The twin planning goals of overall growth in national income and the development of the backward regions are probably incompatible. It is difficult, if not impossible to achieve both simultaneously. Under the pressure of balance of payment problems and the need for loans from abroad, successive governments have probably been forced to give priority to income growth. Despite the plans, this has meant that most investment has been placed where it would be most productive, that is in the west and northwest or near Ankara. The use of single massive plants, such as cement works, to bring industry, and thus 'development', to an area proved mistaken. They were expensive to set up, mainly because practically all of the infrastructure had to be provided, and they were incapable of generating multiplier effects in the local economy or much lasting employment. The use of the SEEs to boost regional development is also incompatible with a requirement to create profits for reinvestment and to be judged on purely commercial criteria. Their borrowing requirements to cover the almost inevitable deficits have been a major contributor to the government's financial problems. It is thus difficult to escape from an economic whirlpool which perpetuates existing spatial patterns.

References

1. R.E. Wary and D.A. Rustow (Eds), *Political Modernization in Japan and Turkey*, Princeton University Press, Princeton, 1964.
2. R. Stewig, 'Die industrialisierung in der Türkei', *Erde*, **103**, 21–47 (1972).
3. B. Darkot, *Türkiye Iktisadî Coğrafasi*, Bermet, İstanbul, 1958, 165.
4. K.H. Karpat, *Ottoman Population, 1830–1914*, University of Wisconsin Press, Madison, 1985, Appendix 1.18.
5. G.P. Meriam, 'The regional geography of Anatolia', *Econ. Geogr.*, **2**, 86–107 (1926).
6. (a) O.C. Sarç, 'Tanzimat ve sanayimiz', in *Tanzimat*, İstanbul, 1941, 423–440, translated as 'The Tanzimat and our industry' in *The Economic History of the Middle East, 1800–1914* (Ed C. Issawi), Chicago University Press, Chicago and London, 1966, 48–59.
   (b) V.Cuinet, *La Turquie d'Asie, Géographie Administrative, Déscriptive et Raisonée de Chaque Province de l'Asie Mineure*, Ernest Leroux, Paris, **1**, 1890.
7. Z.Y. Hershlag, *Introduction to the Modern Economic History of the Middle East*, E.J. Brill, Leiden, 1964, 70.
8. R. Owen, *The Middle East and the World Economy, 1800–1914*, Methuen, London, 1981, 210–11.
9. (a) G. Ökçün, *Osmanlı Sanayii: 1913–1915 İstatistikleri*, partial revision, Hil Yayın, İstanbul, 1984.
   (b) V. Cuinet, *La Turquie d'Asie*, Ernest Leroux, Paris, T. 3, 1893, 406–12; **4**, 1894, 619f.
10. (a) O. Kurmuş, 'The Role of British Capital in the Development of Western Anatolia, 1850–1913', (unpublished PhD thesis, University of London, 1974).
    (b) D. Quartet, 'Limited revolution: the impact of the Anatolian Railway on Turkish transportation and the provisioning of İstanbul, 1890–1908', *Business History Review*, **51**, 139–60 (1977).
    (c) D. Quartet, 'The commercialisation of agriculture in Ottoman Turkey, 1800–1914', *International Journal of Turkish Studies*, **1**, 38–55 (1980).

11. O. C. Sarç, 'Tanzimat ve sanayimiz', in *Tanzimat*, İstanbul, 1941, 423–440, translated as 'The Tanzimat and our industry' in *The Economic History of the Middle East, 1800–1914* (Ed C. Issawi), Chicago University Press, Chicago and London, 1966, 48–59.
12. B. Lewis, *The Emergence of Modern Turkey*, 2nd ed, Oxford University Press, London, Oxford and New York, 1968, 107–08.
13. E. C. Clark, 'The Ottoman industrial revolution', *International Journal of Middle East Studies*, **5**, 65–76 (1974).
14. E. C. Clark, 'The Ottoman industrial revolution', *International Journal of Middle East Studies*, **5**, 65–76 (1974).
15. K. Göymen, 'The Nature and Practice of Turkish Etatism', (unpublished PhD Thesis, University of Leeds, 1973), 70–77.
16. Z. Y. Hershlag, *Turkey: The Challenge of Growth*, E. J. Brill, Leiden, 1968, 16–27.
17. Başvekâlet İstatistik Umum Mudurlüyü, *İstatistik Yıllığı*, 1930/31, Ankara, 1931.
18. İktisat Vekâleti, *İkmci Beş Yıllık Sanayı Plânı*, Ankara, 1936.
19. K. Göymen, 'The Nature and Practice of Turkish Etatism', (unpublished PhD thesis, University of Leeds, 1973).
(b) W. Hale, *The Political and Economic Development of Modern Turkey*, Croom Helm, London, 1981.
(c) Başbakanlık Devlet İstatistik Enstitüsü, *Census of Manufacturing and Business*, Ankara, 1964, 299–325.
20. B. Lewis, *The Emergence of Modern Turkey*, 2nd ed, Oxford University Press, London, Oxford and New York, 1968, 286.
21. (a) Z. Y. Hershlag, *Turkey: The Challenge of Growth*, E. J. Brill, Leiden, 1968, 72.
(b) O. Okyas, 'The concept of Etatism', *Economic Journal*, **75**, 98–111 (1965).
22. M. D. Rivkin, *Area Development for National Growth. The Turkish Precedent*, Praeger, New York and London, 1965, 68.
23. R. W. Kerwin, 'Private enterprise in Turkish industrial development', *Middle East Journal*, **5**, 21–38 (1951).
24. W. Diamond, 'The Industrial Development Bank of Turkey', *Middle East Journal*, **4**, 349–351 (1950).
25. R. W. Kerwin, 'The Turkish roads programme', *Middle East Journal*, **4**, 196–208 (1950).
26. A. P. Alexander, 'Industrial entrepreneurship in Turkey: origins and growth', *Economic Development and Cultural Change*, **8**, 349–365 (1960).
27. R. D. Robinson, 'Turkey's agrarian revolution and the problem of urbanisation', *Public Opinion Quarterly*, **22**, (1958), quoted by M. D. Rivkin, *Area Development for National Growth, The Turkish Precedent*, Praeger, New York and London, 1965, 104.
28. M. D. Rivkin, *Area Development for National Growth, The Turkish Precedent*, Praeger, New York and London, 1965, 98–100.
29. (a) J. Hiltner, 'The distribution of Turkish manufacturing', *J. Geogr.*, **61**, 251–258 (1962).
(b) M. D. Rivkin, *Area Development for National Growth. The Turkish Precedent*, Praeger, New York and London, 1965, 115.
30. G. Ritter, 'Landflucht und Städtewachstum in der Türkei, *Erdkunde*, **26**, 177–96 (1972).
31. (a) C. H. Dodd, *Politics and Government in Turkey*, Manchester University Press, 1969.
(b) C. H. Dodd, *Democracy and Development in Turkey*, Eothen Press, Beverley, 1979.
(c) C. H. Dodd, *The Crisis of Turkish Democracy*, Eothen Press, Beverley, 1983.
32. D. Barchard, *Turkey and the West*, Chatham House Papers 27, Royal Institute of International Affairs; Routledge and Kegan Paul, London, 1985, 34.

33. (a) I. I. Poroy, 'Planning with a large public sector: Turkey (1963–1967)', *International Journal of Middle East Studies*, **3**, 348–360 (1972).
(b) W. W. Snyder, 'Turkish economic development: the First Five Year Plan, 1963–67', *Journal of Development Studies*, **6**, 58–71 (1969).
34. W. Hale, *The Political and Economic Development of Modern Turkey*, Croom Helm, London, 1981, 142–44.
35. W. Hale, *The Political and Economic Development of Modern Turkey*, Croom Helm, London, 1981, Tables 7.2 and 7.3, pp 130–31.
36. Başbakanlık Devlet Planlama Teşkilâti, *Dördüncü Beş Yıllık Kalkinma Planı, 1979–1983*, Ankara, 1979.
37. Başbakanlık Devlet Planlama Teşkilâti, *Dördüncü Beş Yıllık Kalkinma Planı, 1979–1983*, Ankara, 1979.
38. I. Tekeli and G. Menteş, 'Development of "holdings" in Turkey and their organisation in space', *Planning Industrial Development* (Ed D. F. Walker), John Wiley, Chichester, 1980, 149–74.
39. B. W. Beeley, *Migration – The Turkish Case*, Open University, Milton Keynes, 1983, Map 1.
40. (a) B. W. Beeley, 'Investment and the spatial pattern of development in Turkey' (unpublished paper, 1975).
(b) B. W. Beeley, *Migration – The Turkish Case*, Open University, Milton Keynes, 1983, Map 2.
(c) N. Levine, 'Anti-urbanisation: an implicit development policy in Turkey', *Journal of Developing Areas*, **14**, 513–37 (1980).
41. J. K. M. Mitchell, *The South East Anatolia Project – Commercial Opportunities: Preliminary Report*, (mimeographed, Ankara, 1986).

CHAPTER 18

Iran – Agriculture and Its Modernization

18.1 Introduction

Until recently agriculture has provided the major source of national income in all the countries of the Middle East. Indeed, it is only in the last 40 years, with the growth of oil revenues and industrialization, that the relative importance of agriculture in the economies of a few favoured countries has begun to decline. A characteristic feature of all countries of the region has been that agricultural activity has changed little over the past 1000 years until the time of the Second World War. In Iran, agriculture has been dominated by the production of cereal crops for small scale local economies. This has been characterized by a high labour input, low levels of mechanization, little seed selection and only minor use of fertilizers. Under these conditions, agricultural productivity has been largely dependent upon the vagaries of local weather, in particular the availability of water, rather than on the labour input of the individual farmer. In an attempt to lessen the degree of environmental control of farming activity, Iran has initiated programmes of agricultural modernization, financed by oil revenues which have produced a large and assured source of national income since the late 1950s.

In addition to the environmental controls on agricultural development, there are also those imposed by human or cultural conditions. The old system of land tenure, with farming activity strictly controlled by landlords, stifled innovation. The low level of income of the rural peasant meant that, until recently, farming implements had to remain simple, cheap and capable of being fabricated locally, Similarly, the only sources of power in the fields were provided by man and animals. Rural education in any formal sense was almost totally lacking, and this, coupled with the relative isolation of most communities owing to poor communications, meant that new ideas and beneficial farming practices diffused slowly, or not at all.

Since the Second World War Iranian agriculture has been subjected to a number of pressures which have influenced its development. Firstly there has been the 'westernization' of agriculture as indicated by the growth in the use of commercial fertilizers and the massive increase in mechanization. Effectively what has occurred is that Iran has changed from agriculture based on solar energy to one in which fossil fuel subsidies are now essential. On the social side there has been a major land reform programme which has disrupted the traditional fabric of the countryside. Researchers differ on the effectiveness of the land reform programme, but all acknowledge that the impact in rural areas has been considerable. In 1979 a revolution overthrew the Shah and brought about massive changes in the way in which the country was governed. Finally, since 1980 Iran has been at war with Iraq in what has turned out to be a very bloody and debilitating struggle.

One of the most pressing problems facing Iran has been the need to establish a sound and productive modern agricultural system which is capable of feeding the country's growing population. From a total of around 10 million in 1900 the population of the country rose slowly to a total of 14.5 million in 1940.[1] Since then it has grown rapidly to 27 million in 1966 and to a reported figure of 50 million in 1986. The current growth rate is claimed to be close to 4 per cent. If such rates are maintained the country will have to support a population in excess of 70 million people by the early years of the twenty-first century. Associated with this rapid population growth has been a marked change in the balance between urban and rural dwellers. At the beginning of the century the urban population was probably only about 20 per cent of the total. By 1966 this had doubled to 40 per cent and by 1985 has reached 50 per cent. However, although the relative proportions of urban and rural populations have changed since the beginning of the century, it is important the realize that the absolute numbers of people living in rural areas have risen sharply. For example, in 1900 it is estimated that about 7.8 million people were classed as rural dwellers, while by 1966 this figure had more than doubled to 16.5 million.[2] By 1986 the total had risen still further to almost 25 million. These people live scattered throughout the country in more than 65,000 villages. Until recently incomes were low and illiteracy high. One of the factors which has caused the rapid population growth has been the introduction of medical care which has meant that the impact of disease and disaster is no longer as great as it was in the past. In particular fewer children die young, with the consequence that today 43 per cent of the population of Iran are under the age of 15 years.[3]

Agriculture has always been important to the Iranian economy in terms of GNP, although since the 1960s it has declined relatively as the industrial and oil sectors have expanded. At the beginning of the twentieth century, agricultural production appears to have made up at least 80 per cent of the GNP.[4] By the early 1960s it had fallen to about 28 per cent and to 22 per cent in 1968. By 1985 it had decreased still further to less than 10 per cent.

Of the total area of Iran only about 10 per cent is cultivated at any one time (Table 18.1). The fundamental basis of Iranian agriculture is cereal production aimed at supplying local needs with wheat being by far the most important crop, though barley is important locally in the less favoured areas. Rice is the only other significant cereal and the growth of this crop is mostly confined to the Caspian Sea lowlands. Together these crops occupy between one half and three-quarters of the total cultivated area in any year. A wide range of vegetables are also cultivated. The other mainstay of the peasant diet is food obtained from animals, particularly whey, milk and cheese, and to a much lesser extent, meat. Sheep and goats are the most important animals and are raised on mainly poor quality pastures around the villages. Cattle and donkeys used to be the traditional draught and transport animals.

TABLE 18.1
Land use in Iran
1974–76

Type of land use	Area ('000 hectares)	Percentage of total land area of country
Arable land	15,850	9.6
Permanent crops	600	0.4
Permanent pasture	44,000	26.7
Forest and woodland	18,000	10.9
Other land	85,150	51.7
Total area (including water)	164,800	

Source: FAO Production Yearbook 1984, FAO, Rome 1985, p 53.

Agricultural land in Iran has always tended to belong to urban residents.[5] Unlike western Europe in medieval times, no hereditary landed aristocracy ever developed, owing mainly to frequent conquests and the Islamic laws of multiple-inheritance. As a result, there were frequent changes in the families composing the land owning class, although the privileges and functions of this class as a whole remained unchanged. The dominant type of land ownership has always been absentee. Many of the large estates had become fragmented and it was not uncommon for an owner to possess parts of different villages in widely separated areas. At the same time, aggregation of holdings did take place as a man or family gained more power. The relationship between the peasant working in the fields and the landlord was, in most cases, in the form of a sharecropping agreement which involved the payment of 'rents' in kind. Five elements – land, water, seeds, draught animals and labour – often formed the basis of fixing the shares between landlord and peasant. Frequently, the landowner provided the first two of these, so that the peasant received only three-fifths of the crop. Locally, however, other practices were found which were influenced by the nature of farming and the types of crops grown. Under a sharecropping agreement, the peasant possessed no permanent right to the land he farmed, though customary and hereditary rights were often recognized. Redistribution of the land amongst the peasants was practised by the landlords from time to time. This was done partly to stop the peasant acquiring any real interest in the land he worked, but also to adjust the amount of land to the size of the peasant's family. Under such conditions, there was little security of tenure for the peasant, and little incentive to improve the land.

18.2 Factors influencing agricultural activity

The distribution of agricultural activity in Iran reflects the availability of certain natural resources. Amongst the most important are the presence of a long enough growing season for a particular crop, good soil, flat land and, above all, water.

Only rarely in Iran do these four factors combine to produce optimum conditions for agricultural production, and in most regions one or more of these environmental parameters fall well below optimum development. The result is that farming activity reveals a very patchy distribution within the framework of the country as a whole (Figure 18.1). Indeed, on the plateau and in the highlands, the pattern of settlement and agriculture is one of regional oases separated by barren wildernesses. Only in one part of the country, the Caspian lowlands, does agricultural land use provide an almost unbroken mosaic over a very large area.

Water, or more accurately, its absence, is the overriding control of agricultural activity throughout most of Iran. Successful agriculture without irrigation requires ar least 240 mm annual precipitation with an interannual variability of 37 per cent. Unfortunately, about half of the country receives annual totals of less than this amount[6] (Figure 18.2). Those regions where annual precipitation totals are greatest are generally the highland areas, where farming is made difficult by shallow soils, the scarcity of flat land and a restricted growing season (Figure 18.3). A compromise, therefore, has to be reached in which water is transported from the uplands of water surplus to the dry fringing basins and alluvial plains,

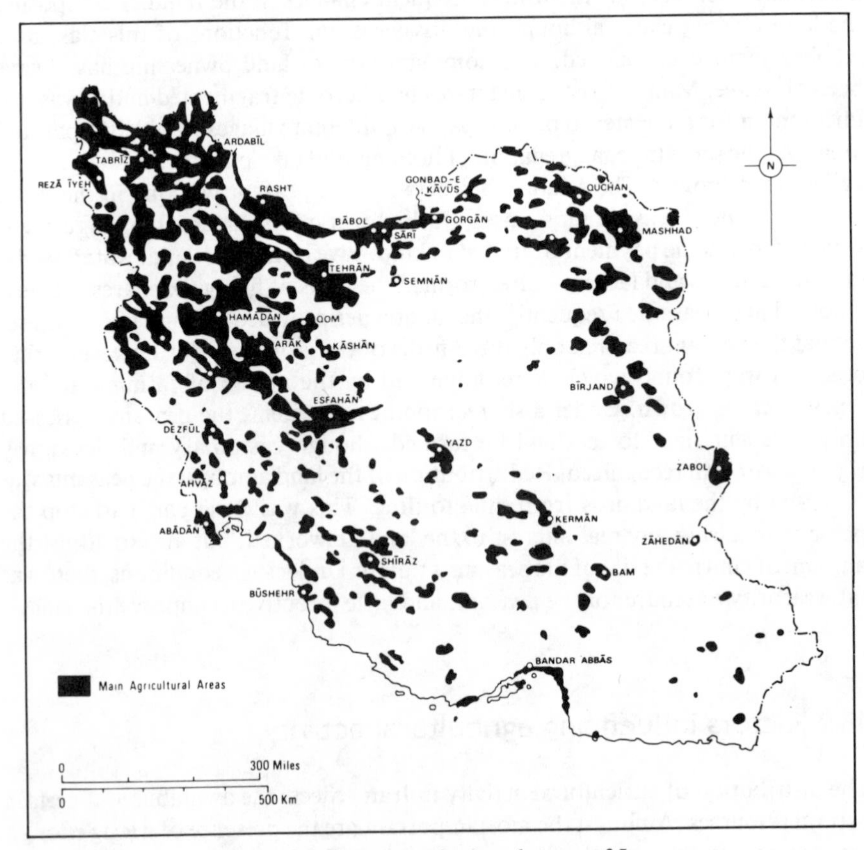

Figure 18.1 Main agricultural areas of Iran

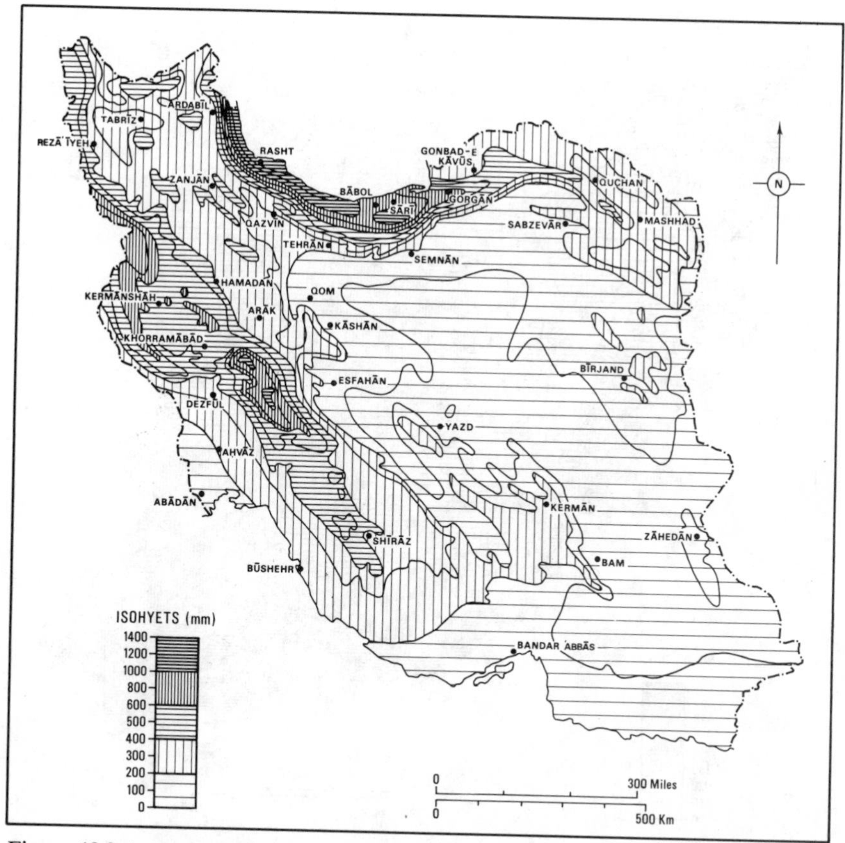

Figure 18.2 Annual precipitation totals in Iran (after *Climatic Atlas of Iran*, 1965, Plan Organization, Tehrān)

which possess a longer growing season and more fertile soils. In the natural environment, such water movement is accomplished by rivers which convey water surpluses from one region to another under the influence of gravity. Once the water reaches a zone amenable to agricultural activity, it is immediately utilized by the rural farming community. This is achieved by the construction of small, hand-dug canals which lead the water to the fields in which the crops are to be planted. Such canals, from 0.5 m to more than 5 m in width, are generally unlined, and as a result, percolation losses through the bed and banks can be up to one half or more of the total intake volume. They usually commence at a point where the river leaves its upland course and radiate like the veins of a leaf to all parts of the cultivated area.[7] In the individual fields breaches are made in the canal banks to permit flood or furrow irrigation.

Rivers flowing from highland regions are also the major sources of groundwater replenishment in the adjacent plains and basins. This recharge of underground water reservoirs takes place mainly where rivers form large alluvial spreads and fans at the margin of the upland zone. Use of groundwater for irri-

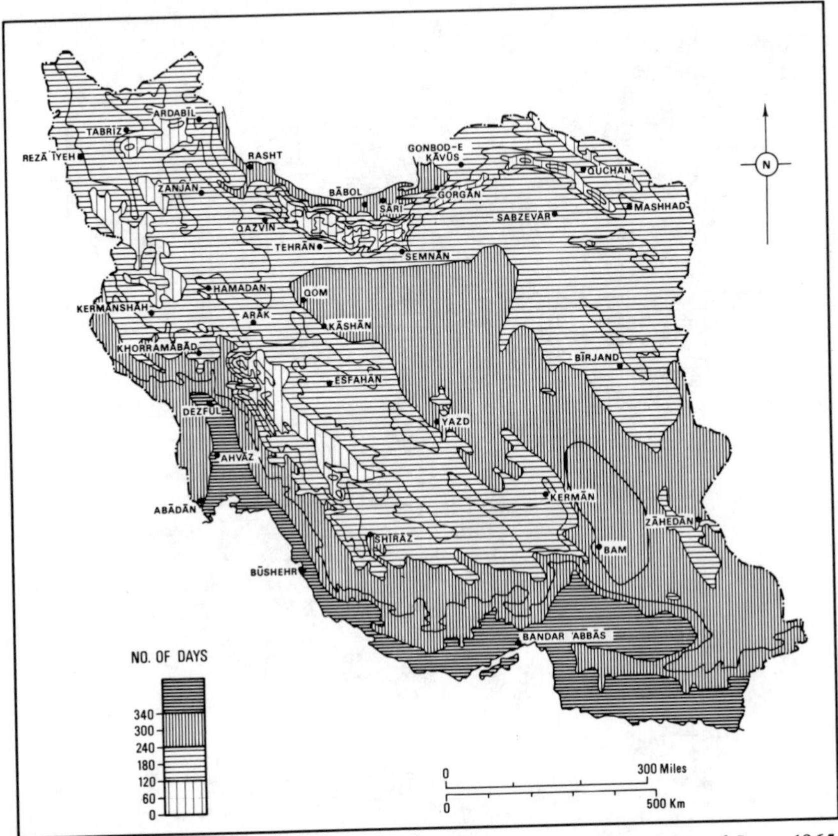

Figure 18.3 Length of growing season in Iran (after *Climatic Atlas of Iran*, 1965, Plan Organization, Tehrān)

gation is common throughout the plateau region of central Iran and in the Zagros Mountains. Unlike many other parts of the Middle East, these water resources have not until recently been exploited by pumped wells, but rather by highly individual engineering constructions known as *qanāts*[8] (Chapter 2). Groundwater possesses the advantage over surface water supplies in that it is commonly available throughout the year.

In one respect, Iran is fortunate in possessing a crescent of high mountain ranges, since most of the precipitation occurring during the period from October to March is in the form of snow. Owing to the altitude, the snow does not melt immediately, but rather remains as deep snowpacks. Snowmelt begins in the spring, swelling the streams and rivers to provide maximum water discharges during April and May, at a time which coincides with the beginning of the growing season.[9] Water supply is, therefore, not as much out of phase with the season of maximum crop growth as precipitation data alone would at first suggest. Once the early summer discharge peak has passed, the availability of water from surface sources diminishes rapidly. Unfortunately, this is at a time when water demand for the growing crops is reaching a maximum. Where water cannot be supplied from alternative sources during this period arable farming is impossible.

18.3 Agricultural production

The major changes in agricultural production in Iran during the twentieth century have largely been due to the increased and changing demands of an expanding population. Table 18.2 indicates the changes in production of selected major crops over the last 50 years, and reveals marked increases for most of the crops. At the time of the revolution in 1979 there seems to have been considerable disruption to crop production and in particular to those crops which required processing. For example, both sugar beet and cotton production dropped drastically between the mid-1970s and 1980, but then recovered appreciably. Of all the crops it is the sugar beet and tea which have registered the most remarkable increases from the 1930s to the late 1970s. Sugar beet production grew by almost 300 times, while tea registered a 60-fold increase. The extremely rapid rise in output of these latter crops has been the result of government policy since the 1920s in an attempt to cut down the high level of imports.

Data on the number of livestock in Iran are particularly unreliable. In general, it seems that the numbers of cattle, sheep and goats have increased markedly since the Second World War, while the number of camels appears to be decreasing, owing to the increasing use of mechanized transport throughout the country and the sedentarization of nomads. Estimates for the late 1960s reveal 32 million sheep, 15 million goats, 6.2 million cattle and 2.2 million asses. The number of poultry was thought to be about 48 million. By 1984, despite the large increases in population, animal numbers have changed surprisingly little. Sheep are estimated at 34 million; goats 13.6 million; cattle 8.2 million and asses 1.8 million. What has revealed a big increase is the numbers of poultry which are now believed to be around 74 million.

TABLE 18.2
Agricultural production in Iran

'000 tonnes	1934–38	1950	1960	1965–6	1970	1974–6	1980	1982	1984
Wheat	1,869	2,263	2,590	3,648	4,262	5,438	5,700	6,600	5,500
Barley	638	875	684	935	1,083	1,263	1,100	1,903	1,550
Rice	423	450	651	681	1,350	1,436	1,212	1,605	1,230
Sugar beet	17	62	588	1,411	3,800	4,665	1,500	4,146	3,740
Cotton[1]	na	28	99	120	160	175	60	103	75
Tea	1	7	9	11	20	22	19	36	45
Tobacco	15	15	11	24	17	16	23	21	25

Sources: 1. Iran Almanacs, Echo of Iran, various years.
2. *The Middle East and North Africa*, Europa Publications, various years.
3. Bharier, J., *Economic Development in Iran 1900–1970*, Oxford University Press, 1971.
4. *FAO Production Yearbooks*, Rome, various years.

The only comprehensive survey of what might be termed the traditional agricultural system in Iran was made in October 1960, when the first national census of agriculture was undertaken. This census preceded the land reform measures of the 1960s, which resulted in a reallocation of land amongst rural dwellers. The census revealed a land tenure system which had prevailed with little change for centuries. The sizes of the individual holdings was small, with four-fifths of the total number composed of less than 10 ha and with more than half less than 3 ha. By way of contrast, two-thirds of the total agricultural area was included in holdings of between 5 and 50 ha in size and less than 10 per cent in holdings smaller than 3 ha. The type of power used for agricultural production on these holdings was mostly animal, with mechanical power being of importance only in those areas close to the urban centres or adjacent to the major routeways. The use of human labour as the only power source reached its highest values in the agricultural regions of central and southern Iran.

The distribution of agricultural activity within Iran, as revealed by the 1960 census, showed considerable variation between the different administrative units. Of the total area of agricultural land, by far the largest proportion was concentrated within the wetter northern and western parts of the country, with only minor amounts in the central drier regions. Temporary crops made up from 5 to 60 per cent of this total area in the different census units, while permanent crops contributed only 0.2 to 14 per cent. In the arid central parts of Iran, temporary fallow land was from a fifth to a half of the total agricultural area, with particularly low values of less than 10 per cent concentrated along the Caspian lowlands. Permanent pastures, accounting for 2 to 60 per cent of the total area, showed more varied distributions. The major areas of land classified as potentially cultivable following improvements were found in the eastern regions, bordering the central deserts.

The proportion of temporary crops which were irrigated varied widely, dependent upon climatic conditions together with the types of crop cultivated. In the wetter parts of the north and west, between a fifth and a quarter of the temporary crops were irrigated, while towards the east and south the figure rose to between a half and four-fifths of the total area. Interestingly, in the wettest part of Iran, the western margin of the Caspian Sea lowlands, irrigation of temporary crops accounted for three-quarters of the total area. Here rice, with its high water needs, and not wheat and barley, was the major cereal produced. Irrigation of permanent crops is of crucial importance to cultivation in almost every part of the country, with the sole exception of the Caspian Sea lowlands.

Amongst the temporary crops, wheat and barley dominated the picture in terms of the total area under production, and revealed a distribution which closely paralleled that of population (Figure 18.4). Taken together, these two crops account for between 50 and 94 per cent of the area devoted to temporary crops in all the administrative units with the exception of the western part of the Caspian Sea lowlands, where rice production was particularly important (Figure 18.4). Rice was also cultivated outside the Caspian Sea lowlands in places where water supplies are available. In particular, the Khuzestan lowlands, the Gulf

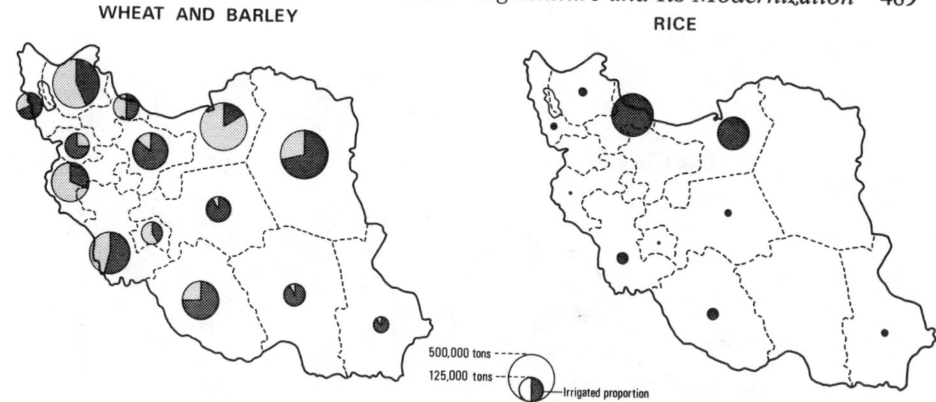

Figure 18.4 Wheat and barley production in Iran (left); rice production in Iran (right)

coastal region, parts of Sistan-Baluchistan and the interior desert oases all produced significant, though small amounts. Cotton, one of the major cash crops of Iran, although requiring an assured water supply, demands much smaller quantities than rice. As a consequence, its production was chiefly concentrated in the drier eastern parts of the Caspian Sea lowlands, where it was often grown without irrigation (Figure 18.5a). Locally important centres of cultivation of cotton were also found in Khurasan, the plateau region around Tehrān and in the southern parts of the Zagros Mountains.

The distribution of permanent crops exhibited much greater regional variations than temporary crops. This is largely because of a more pronounced susceptibility to environmental conditions shown by many of these crops. A variety of fruits form an important sector of agricultural production in Iran, and one that appears destined to grow even larger as standards of living increase. The growth of some of these crops is closely controlled by environmental conditions. Date cultivation, which requires a long and hot growing season, was confined almost solely to the dry southerly *Ostans* (administrative regions), especially the Khuzestan lowlands and Sistan-Baluchistan (Figures 18.3 and 18.5b). Citrus fruits, sensitive to frosts, were concentrated along the Caspian Sea littoral and in the southern *Ostans* bordering the Gulf. In contrast, the vine is somewhat less demanding of its environment, although it, too, is sensitive to late frosts in spring and early summer. As a result, the distribution of the vine was more evenly spread through the country than either the date or citrus fruit. Apples, pears, quinces and the stone fruits are more temperate crops and their production was concentrated in the more northerly Ostans of Azerbaijan and Khurasan (Figure 18.5c). Tea is almost exclusively confined to the wetter western parts of the Caspian Sea lowlands (Figure 18.5c). Other tree crops, almond, hazel, pistachio and walnut, showed a concentration in the Caspian Sea lowlands, Azerbaijan, and parts of central and southern Iran. This pattern of crop production has changed little to the present day, though the areas under certain crops have increased markedly in some regions.

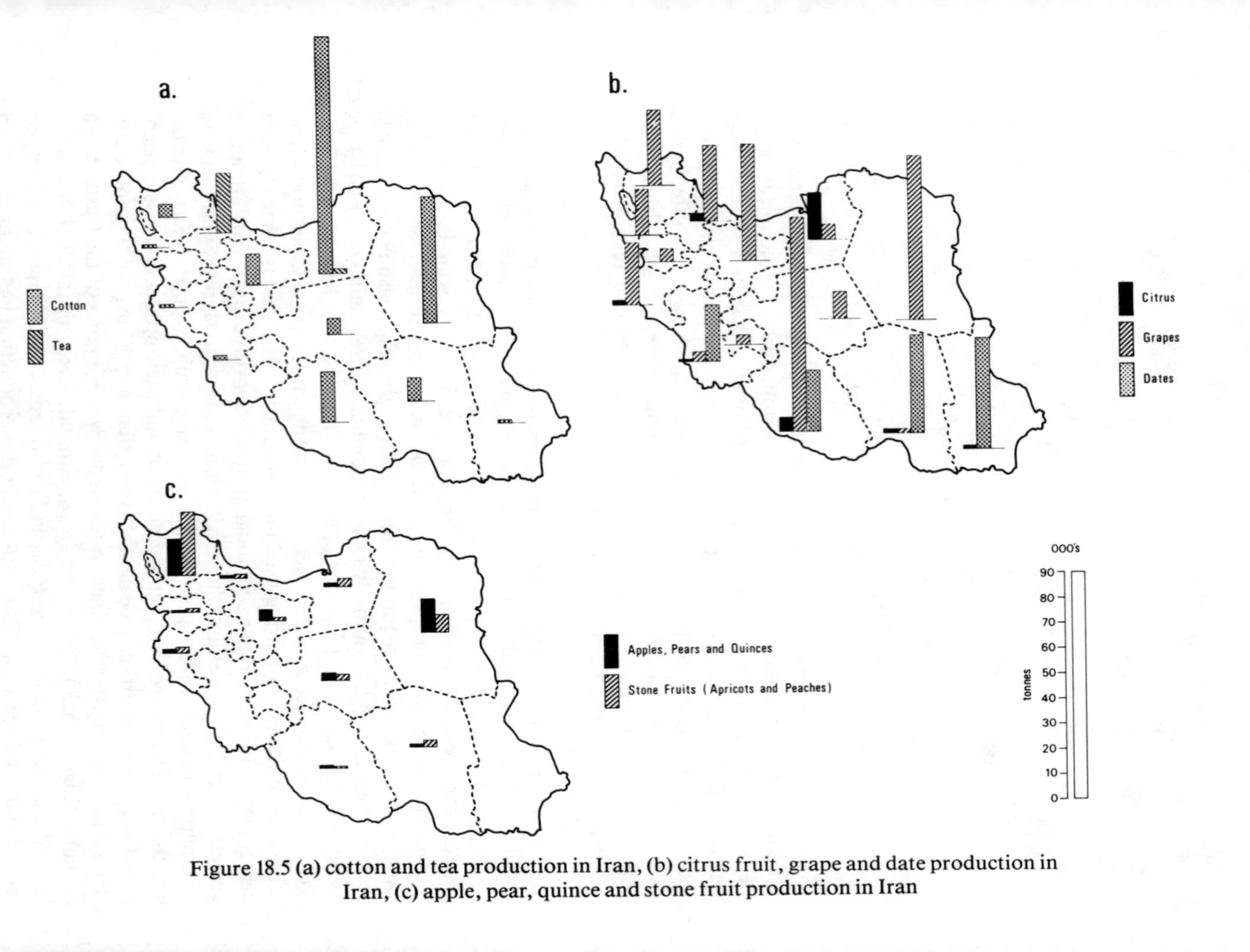

Figure 18.5 (a) cotton and tea production in Iran, (b) citrus fruit, grape and date production in Iran, (c) apple, pear, quince and stone fruit production in Iran

18.4 Modernization of agriculture

18.4.1 Economic planning

Significant changes in the traditional agricultural system of Iran have occurred only since the end of the Second World War, and particularly from the late 1950s. Resistance to any change has often been great, both from landlords and peasants, though for very different reasons. The greatest changes have been brought about by government action through a programme of economic planning and land reform, financed largely by oil revenues.[10]

The first Iranian economic plan was introduced more than 35 years ago, with one of its mains aims being the 'westernization' of methods of agricultural production. This First Seven Year Plan (1949–56), originally proposed to devote about 25 per cent of its income to agriculture, 24 per cent to transport, and 19 per cent to industry. Administration and implementation of the schemes and objectives was to be carried out by Plan Organization, a newly established government agency. Unfortunately, owing to Iranian expropriation of the Anglo-Iranian oil company's installations in 1953, and the subsequent world boycott of Iranian oil, government revenues dropped sharply. As a consequence, many of the objectives of this first Plan had to be changed or at least postponed, and the money that was available concentrated in small capital projects.

The initiation of the Second Plan (1956–62) began with the resumption of oil revenues in 1954, and it is this Plan which laid the base for the rapid economic growth of the country. Lack of experience and shortage of accurate statistics created considerable difficulties during the course of the Plan, which was not really working satisfactorily until 1957. Some of the greatest achievements were the construction of a number of large multi-purpose dam schemes aimed at providing irrigation and flood control, domestic and industrial water production and the generation of electricity. Four large projects on the Dez river, the Safīd river, the Karaj river and one near Hamadan – were all completed during the plan period. At the same time a number of smaller projects were built, and studies begun for the construction of other large dams throughout the country. Attempts were also made to improve the low yields of all crops grown in Iran through the greater use of fertilizers. The consumption of chemical fertilizer, which was first used on a large scale in Iran in 1956, when 3000 tonnes were imported, increased markedly to more than 25,000 tonnes by 1960, as a result of this stimulus. A fertilizer plant was set up near Shīrāz during this plan. Funds were set aside in the field of animal husbandry to reduce disease by large scale vaccination programmes and efforts were made to increase the amount and quality of meat production.

The Third Plan (1962–68) was concerned with more specific objectives, building on the foundations established by the Second Plan. This Plan coincided with the new land reform laws and, therefore, its aims in the agricultural sector concentrated on the development of regionalism the formation of cooperatives and rural development. Numerous structural projects, such as public baths, schools and water wells, together with sanitation and water improvement projects, were carried out in the village communities, while almost 9,000 km of feeder roads were built to improve rural communications. Finally, to provide education for the villages, a number of what became known as 'revolutionary corps' were set up: Literacy Corps; Health Corps; Extension and Development Corps. During the Third Plan the Government decided upon the nationalization of forest and water resources in order to permit more rational and efficient use. Within this context, the supply of irrigation water became one of the main aims of the Plan, and this was implemented by the drilling of deep wells, the repair and improvement of *qanāts*, the construction of irrigation networks and the completion of projects began under the Second Plan. New construction also began on three major dams. With the completion of these multi-purpose water resource projects it had been hoped that a 4 per cent growth rate of agricultural production could be achieved during the Third Plan. However, owing to adverse weather conditions and the dislocation of land ownership through land reform, less capital investment in farming took place than had been envisaged by the Plan, and the average rate of growth was actually only 2.6 per cent.

The Fourth Development Plan (1968–73) had, as its main objective, the raising of the economic growth of the country to 9 per cent per annum.[11] This, it was hoped, would be achieved by an annual growth rate of 5 per cent in the agricultural sector and of 13 per cent in the industrial sector. Once again, a large proportion of the funds of this Plan were devoted to rural development. In the agricultural sector, the Plan placed special emphasis on the use of chemical fertilizers, mechanization, the improvement of seeds and plants, better crop protection and further agricultural research.

Chemical fertilizer consumption had already risen from 40,000 tonnes in 1962 to 250,000 tonnes in 1969, and was expected to rise to 350,000 tonnes by the end of the Fourth Plan. A new tractor plant was built at Tabrīz, and the number of tractors in use increased from about 17,500 at the beginning of the plan to almost 23,000 in 1973. At Atrek, a heavy engineering works began production of agricultural implements, such as ploughs, disc harrows, and seed drills. The Ministry of Agriculture also began developing better wheat and rice seeds at its research stations. Largely as the result of the introduction of new rice strains, production rose to a record of 1.2 million tonnes in 1969, while a recently developed strain of Mexican wheat proved highly suitable for Iranian conditions. The cultivation of a wide variety of oil seeds was encouraged, and the area of land under oil seed production rose from 16,000 ha in 1968 to 68,000 ha in 1969. At the same time, storage facilities were being developed to handle agricultural produce, as at this time it was estimated that approximately 40 per cent of perishable goods were being lost because of inadequate storage.

The Fourth Plan aimed to increase the availability of irrigation water by 15 per cent from the 1969 volume of about 29,000 million cu m/annum to 33,000 million cu m/annum in 1973, through further water resource development schemes. With this water 400,000 ha of new land would be brought under cultivation and the irrigation of a further 500,000 ha improved. Several major multi-purpose reservoir dams were either completed or initiated during the Fourth Plan and these provided greater volumes of water for urban and industrial uses, as well as electricity for a rural electrification programme.

Finally, in an attempt to utilize Iran's agricultural potential more efficiently on a large scale the government established a number of large agro-business concerns to develop the irrigated lands below large dams. Such schemes, financed in part by foreign capital, were planned to operate on 8,000 ha of land, mainly in Khuzestan. Unfortunately, almost all of these have proved to be both environmental and economic failures.

The Fifth Development Plan (1973–78) was drawn up before the massive increase in oil revenues which occurred in 1973. As a result the Plan had to be revised to allow for an estimated investment total almost double that which had been estimated earlier (Table 18.3).

Of the investment from public funds only 8.4 per cent was to go into agriculture. Communications received the largest allocation amounting to 14.2 per cent, followed by 12.4 for industry and 11.7 per cent for oil. At this time the economy was growing rapidly and emphasis was placed on the development of heavy and basic industries, including steel and non-ferrous metals, chemicals and petrochemicals, mechanical and electrical industries and vehicle construction. The agricultural sector was very much a poor relation at this time. The rapid growth in oil revenues also caused the oil sector to play an ever increasing part in terms of its contribution to GNP. At the beginning of the Fifth Plan the oil sector contributed about 20 per cent to GNP, while towards the end it was around 50 per cent. As a consequence the relative importance of agriculture declined from just over 18 per cent in 1973 to less than 10 per cent of GNP in 1978. Interestingly, during this Plan Iran became more and not less dependent on oil; a result at odds with the objective of economic planning in the country.

It was undoubtedly the rapidity of economic growth during this Plan which was one of the major indirect factors leading to the overthrow of the Shah. In 1973–74 GNP growth reached a staggering 50 per cent, and throughout the plan period it averaged around 25 per cent per annum. Under the stress of such growth rates it was inevitable that ordered development could not be sustained. In Tehrān, the capital, the inflation rate soared, while at the same time, consumer goods such as cars and other expensive items poured into the country. As a result traffic congestion and air pollution in Tehrān became serious problems.[12] The increasing affluence meant that imports were sucked into a country which did not

TABLE 18.3
Development plans – Iran

	period	Total investment (million US dollars)
First Plan	1949–56	68
Second Plan	1956–62	979
Third Plan	1962–68	2728
Fourth Plan	1968–73	10853
Fifth Plan	1973–78	68600

possess the infrastructure to cope with them. At the port of Khorramshahr in 1975 the waiting period for ships to be unloaded reached five months, while on the Turkish–Iranian overland road, transport suffered from long delays.

In the cities shortage of manpower for the numerous building projects meant that wage rates rose rapidly, causing male workers to flock to the towns from the impoverished rural areas. During 1975/76 alone average wage rates rose by 35 per cent. However, frequent shortages of cement, bricks and steel necessitated laying workers off for long periods, and so gave rise to the first stirrings of unrest. When the downturn in the world economy occurred in 1977 oil revenues dropped and many projects had to be curtailed. Hundreds of thousands of rural workers who had moved into Tehrān and other large cities were laid off to become the dissatisfied masses out of which the revolution was born.

In retrospect, the Fifth Plan was one which, with the fevered pace of urban/industrial development, completely overlooked the importance of rural life and the agricultural sector. This was not the result of a lack of understanding as to the importance of the agricultural sector in the development of the country's economy, but rather that the various government departments were completely overwhelmed by the sheer pace of change. So much investment was taking place that it proved impossible for government departments to evaluate and coordinate activity, and in some cases near chaos prevailed. Indeed, from 1975 onwards the government clearly realized the dangers of 'supergrowth' and in the preparatory work for the Sixth Development Plan (1979–83) envisaged a much more controlled expansion of the economy. However, the Sixth Plan was never implemented owing to the overthrow of the Shah in 1979. The timing of his overthrow is somewhat ironic as in 1979 there began to occur another huge increase in revenues of the oil producing states, which in the case of Iran might well have been sufficient to revitalize the flagging economy and so provide work for the large dissatisfied urban labour force.

With the coming to power of the Khomeini government there occurred a three-year period of adjustment during which little planning of any kind took place. As far as agriculture was concerned subsistence agriculture continued with little change. However, the transport and marketing of vegetables, fruits and cash crops was seriously disrupted, though the lack of statistical information for this period makes it impossible to be certain as to what actually occurred. Before a new organizational and planning framework could be established the country had to placed on a war footing as a result of invasion of its territories by Iraq. This did have certain advantages for the Khomeini government in that it meant that the energies of the growing numbers of unemployed urban workers could be focused on an external enemy. The long continual nature of the Gulf War and the consequent disruption of oil exports from Iran has meant that the country has been unable to invest much in new projects. As far as the countryside is concerned there seems to have been a return to more traditional methods of cultivation as fertilizers became scarcer and spare parts for machinery more difficult to obtain. The lack of detailed statistics, however, makes it difficult to be precise as to what is actually going on.

Despite the Gulf War the Khomeini government did commence a programme of economic planning with the preparation of the First Development Plan of the Islamic Republic of Iran (1983–88). In this Plan agriculture was viewed as the basis for all economic development in the country and was accorded the necessary priority. This view was in contrast with all previous economic planning in Iran since the 1920s which had concentrated on industrialization as the key factor in the development process. A key objective of the new plan was the provision of food and clothing for Iranians through indigenous agricultural production.

> 'To attain self-sufficiency in agricultural products, while considering an appropriate pattern of nutrition and hygiene based on the genuine Iranian and Islamic traditions and customs, is among the most important steps to be taken to reach economic independence'.[13]

As far as arable farming was concerned it was hoped to be able to increase productivity per unit area, as well as the amount of land actually under cultivation. Attention was also to be focused on improving the efficiency of both dry farming and irrigated agriculture. The modern methods of production were not to be overlooked, and an adequate supply of commercial fertilizer was to be made available. Even more important was the promise of the provision of effective credit facilities for the farmers at reasonable rates. Lack of credit had always been one of the most serious problems holding back rural development in Iran. With pastoral farming it was hoped to revive traditional animal husbandry practices and at the same time get maximum production from the many industrial animal husbandry units which had been established. Increases in foodstuff production were also regarded as a priority area.

In the first year of the Plan it was estimated that the agricultural sector would receive 13.8 per cent of total plan investment, rising to 16.7 per cent at the end of the plan. Total investment in the Plan was expected to be 169 billion dollars from government sources. Most of the investment, however, it was hoped, would be generated from the private and co-operative sectors, with the government providing funds for technical assistance, development work and for infrastructure provision. The Plan clearly recognizes the importance of establishing and protecting reasonable prices for farm products, which all encourage higher levels of investment and production within the agricultural sector, and at the same time, help to reduce the flow of rural dwellers moving to the cities. Unlike any previous development plans the agricultural sector is seen as the driving force for the rest of the economy. It is, therefore, essential that growth in this sector is achieved, so that vital raw materials, such as cotton, sugar beet, sugar cane and oil seeds, can be fed into the industrial and processing sectors of the economy. Whether these new approaches will prove any more successful than earlier ones will have to wait until probably well into the time of the Second Plan of the Islamic Republic.

18.4.2 Land reform

One of the most important aspects of the agricultural revolution in Iran has been the government-instigated programme of land reform which resulted in the transfer of land from wealthy land-owning families to small peasant farmers. Iranian land reform was very definitely a product of the 1960s and it is important to see it in that context. The 1966 census, taken in the middle of the land reform programme, revealed a total population of 27 million. In the rural areas about 16 million people lived in around 67,000 villages, defined as having 5000 people or less. The location of these villages was determined by the presence of agricultural land, and, in the drier areas, water availability. In general, most of the villages were concentrated in the wetter northern and western parts of the country. The vast majority of the villages were small, with 49,000, or three-quarters of the total numbers, having less than 250 inhabitants each. However, villages of this size range accounted for only 26 per cent of the total rural population. In contrast, the larger villages of more than 500 people made up only 11.4 per cent of total village numbers, but comprised 50 per cent of the people living in rural areas.[14]

At the time when the land reform programme was initiated in Iran rural Iran was largely controlled by absentee landlords who managed their lands through agents. These agents made contracts with the villagers who then undertook the cultivation of the fields for a specified share of the crop. The share was often so small that the villagers were not able to survive without accumulating considerable debts. By the late 1950s large landowners, making up only 2 per cent of all owners, controlled about 55 per cent of the cultivated land, and, as might be expected, possessed very great political power.[15] It was not uncommon for these large landowners to possess more than 20 villages, and being absentee, they left the detailed administration to their agents, who in turn exercised considerable influence. The agents were paid in cash or kind in relation to the volume of the crops produced. It was in their interest, therefore, to extract maximum work from the peasants and, as a result, they were always unpopular. Occasionally they also acted as the headman of a village, or more usually were involved in his appointment. Other lands were owned by the government or were *vaqf* lands, which were endowed in perpetuity for religious or charitable purposes.

Throughout Iran there were many small landowners as well. Some of these were non-farmers, such as merchants, professionals or craftsmen, who had obtained land as an investment. Many of these too were absentee landlords and often employed agents. The majority of the small owners were peasants, though these accounted for only 5 per cent of all peasants in rural areas. They were often very poor and owned impoverished land in isolated villages which was not wanted by other more properous owners.

Throughout rural Iran the crucial division amongst the rural dwellers was the possession or otherwise of a '*nasaq*'. This was the customary right to cultivate land and to receive a portion of the harvest. In 1960 it was estimated that about 60 per cent of the households of rural Iran were *nasaq* holders. The *nasaq* conveyed the right to cultivate a portion of an owner's land and to use village water to irrigate the plot. It was gained by working as a sharecropper for at least two seasons. The cultivation rights of the *nasaq* were not strictly inheritable, though most landlords usually permitted the rights to be transferred to a cultivator's eldest son. It is important to realise that the *nasaq* did not convey the right to cultivate a specific piece of land, but merely the right to cultivate some land belonging to the owner. In effect, landowners reallocated land amongst the *nasaq* holders often on an annual basis to prevent the peasants becoming attached to any piece of land and from acquiring rights to it through the cultivation of tree crops.

The third group in rural Iran, besides the landowners and the peasants, were the *kwushnishins*. These were a heterogeneous group with the most affluent being middle men acting as traders, merchants and bankers, who made up about one twentieth of all *kwushnishins*. Others provided services or made things. These were the blacksmiths, the carpenters and the owners of draught animals. However, the vast majority of the *kwushnishins*, certainly well over 80 per cent of the total, were agricultural labourers who depended largely on seasonal farm work for their livelihood. As a group these agricultural labourers were always poor and their numbers seem to have grown substantially in the twentieth century in line with the rapid population growth. By 1960 they accounted for at least a quarter, and often close to a half, of all the people in individual villages. As a group they could almost be regarded as people surplus to the labour needs of the rural environment. It was rare for many of them to work for more than 100 days per year, and as a result many migrated to the towns in search of jobs at times when rural labour requirements were low.

The land reform movement in Iran, like so many other countries, was generated amongst middle class intellectuals in urban areas. Since the 1920s there existed a wish to modernize the country, though the method by which this should be achieved was not always well articulated. Reza Shah, who rose to power at this time, believed that industrial development was the key to prosperity and so under his rule the agricultural sector was neglected. In the post Second World War period, following the abdication of Reza Shah in 1941, his son, Mohammed Reza Shah (known in the West as the Shah of Iran), continued his father's policy of economic development through industrialization as the best way forward for the country.

During the early 1950s the most important event was the coming to power of the Musaddiq government and the nationalization of the assets of the Anglo-Iranian Oil Company. Although the Musaddiq government is known mainly for oil nationalization schemes, it did also propose in 1952 a series of rural reforms, which were never implemented owing to the fall of the government in 1953. Of these perhaps the most far reaching was the proposal that following the division of the crop 20 per cent of the landlord's share had to be returned to the village as a kind of tax. Following Musaddiq's fall the Shah regained full control of the country, but was obviously shaken by what had happened. What he was quite clear about was the need to broaden his power base. He decided that the best way to achieve this was to initiate a land reform programme. This he hoped would get him popular support and at the same time strengthen the power of the central government by weakening the hold of the landowners.

In 1959 the Shah introduced a land reform bill, which was resisted by the landowners in the *Majlis*, the Iranian parliament, but eventually passed into law in

1960. The bill limited individual holdings to 400 ha of irrigated land and 800 ha of dry farmed land. All land in excess of these holdings had to be sold to the peasants. The great problem was that since areal measures had been chosen, maps of all the villages would be required before redistribution of land could begin. The landowners knew that this would take many years to accomplish.

At this stage it is important to realize that the lands associated with each village was traditionally divided into six equal parts or '*dang*'. Each *dang* consisted of a number of agricultural units measured in terms of the area which could be cultivated or irrigated in a given time period. As this depended on many local factors the size of the *dang* varied quite considerably from one place to another.

With a change in government in 1961, Hassan Arsanjani, who had always been a strong believer in land reform, became Minister of Agriculture. Arsanjani felt that the 1960 law was unworkable given the lack of adequate maps and so he proposed amendments to the legislation. The key proposal was that maximum land ownership should be limited to one six-*dang* village. This required no surveys as no absolute areas were specified, and so could be implemented immediately. All landlords with more than one village had to sell the land to the govenment at a price determined by the tax payable on the land. This too was a brilliant stroke as most landlords had undervalued their land for tax purposes, but, of course, dare not now admit it.

The land acquired by the govenment was then to be sold to the people who held cultivation rights. The basic idea being to create a class of peasant landowners who would move away from subsistence agriculture and embrace a production system based on market forces. The problem of obtaining credit in rural areas still remained, but this was to be overcome by the creation of Rural Co-operative Societies, which would be able to provide low interest loans from government funds.

In January 1962 the new regulation was passed and the landlords had to sell excess villages to the government. They could choose which they kept and if they so wished could keep one *dang* in each of six villages. Unfortunately, more than half the villages in the country were exempt from the law. These exceptions included orchards, land which was cultivated by mechanical methods and land governed by religious endowments. The land obtained by the government was sold to the peasants at 10 per cent above the purchase price, so that administrative costs could be recovered. The right to obtain the redistributed land was granted by the possession of a title to cultivate the land. The scheme for the sale to the peasant consisted of a one-year grace period followed by fifteen equal instalments. The implementation of the 1962 land reform law is known in the literature as the 'First Phase' of the Land Reform programme. Originally it had been hoped to complete this stage in one year under the supervision of the Land Reform Organization, but the lack of adequate trained manpower meant that this was impossible.

The land reform programme did, however, raise the aspirations of the rural dwellers and when changes did not take place as quickly as expected pockets of unrest broke out. As a result the Shah felt that the situation needed to be stabilized and so the pace of change was slowed still further. The landlords realized that the slowing of the programme presented them with the opportunities to escape some of the more severe consequences. Although each individual could only own one village, this was easily got round by giving villages to wives, children and close relatives. Similarly, the fact that mechanized land was exempt from the laws meant that many unscrupulous landowners purchased tractors and so avoided the need to dispose of land. Despite all these actions the results of the First Phase are impressive, for by September 1963, the land of 8042 villages had been redistributed and 271,000 peasants had benefited.

In 1963 five Additional Articles to the Land Reform law were proposed dealing with villages not covered by the original legislation. In the parliament the Additional Articles were modified considerably. The Articles were only to apply to sharecropped land and the landlords could chose in which of five possible ways the land was to be treated. These were a written rental agreement between the landowner and the peasants; the division of land between the owners and the peasants; the sale of land to the peasants; the formation of owner-cultivator joint stock farm corporations or the purchase by the landlord of the peasant's cultivation rights. The passing of these Additional Articles is usually taken as marking the beginning of the Second Phase of the programme, though for two years little was achieved. At this time the Shah became increasingly worried about the growing independence of the peasants, though at the same time he realized that they could be a strong force of support for his throne. His feeling of caution prevailed and so the land reform programme was suitably slowed down.

The implementation of the Additional Articles did not take place until February 1965, but during the Second Phase it was estimated that 54,000 villages and 1.7 million sharecroppers were affected.[16] However, it would seem that over 80 per cent of the peasants had their sharecropping agreements merely exchanged for tenancies with little overall benefit. Only 1.5 per cent of the sharecroppers obtained land by sales from landowners, while another 4.5 per cent acquired some land by division of the land between owner and peasant. In general this stage of the land reform was unpopular in the rural areas as the landlords were virtually in control of all that was happening. The cash rents in particular were greatly disliked as it was often impossible to attain them in years of water shortage, and so the peasants were drawn into even deeper debt. At least with the old sharecropping system the share of the crop either increased or decreased dependent on how much was actually produced.

The growing dissatisfaction in rural areas in 1967 and 1968 led many government planners to feel that agricultural production could only be increased if the smaller and uneconomic tenancies were abolished. The peasants farming such lands had very little commitment to them and output was low. As a result the land reform law was amended in 1969 so that peasants holding land on 30 year leases were able to purchase their holdings at a cost of twelve times the rent. The implementation of this law was not carried out as efficiently as it might have been, but nevertheless the Land Reform Organization did claim that 60 per cent of the tenants with 30 year leases did become owners. In many areas though landowners merely ignored the provisions of the law.

The late 1960s were also a time when the government believed that the only way to increase agricultural output was by large scale investment and the modernization of agricultural production. One of the main aims of the modernization programme associated with the Third Phase of land reform was to promote the establishment of rural cooperatives. The idea behind the cooperative was to form groupings of peasant families into production and consumption units, in order to give them more control over their livelihood. At the same time it was realized that some organization had to be set up to take over the administrative role which had been performed by the landlords. The number of cooperatives grew steadily and by the end of 1966 there were more than 8,600. Cooperatives had the immediate advantage of increasing the purchasing power of the farmers, and also provided them with access to information on modern farming techniques. To gain even greater economies, a grouping of the cooperatives took place in some areas to form regional cooperative unions. By 1971 a hundred of these had come into operation. Over a longer period of time it was hoped that the cooperatives would increase the political maturity of the rural areas and so allow a greater delegation of responsibility from the central government.

Associated with the Third Phase of land reform was a scheme by which the government planned to encourage large scale farming methods. This was achieved through a new law passed in 1968 which encouraged the establishment of agribusinesses on the land downstream from the major dams. These agribusinesses were partly financed by government funds and partly by foreign capital. The chief area selected for this approach was the land below the Dez dam in Khuzestan. Here 58 villages were taken over, covering 68,000 ha, and 55,000 people were forcibly relocated. The water from the lake impounded by the dam was thought capable of irrigating 125,000 ha of land in the Khuzestan lowlands as well as producing 520,000 kW of electricity. Before the construction of the dam, about 91,000 ha had been irrigated within the region and a further 33,000 ha dry farmed. Wheat and barley formed the major non-irrigated crops, and wheat, barley, beans, rice and sesame the irrigated ones. Yields of all crops were poor and rural standards of living were generally low.

The implementation of the Khuzestan lowlands project was not without its problems. The distribution of water and power from the Dez dam was controlled by the Khuzestan Water and Electricity Authority, and it was this authority which also supervised land reform within the region. As agricultural practices were also supervised by the same organization many felt that with land reform they had merely exchanged one landlord for another.[17] Unfortunately, despite that fact that careful planning based on sound economic criteria had gone into the agricultural organization, the human element had tended to be neglected. In many cases there was little contact or understanding between the officials and the farmers. The problem was compounded by the fact that a number of the officials were sons of former landowners. Many of these had been trained abroad and so were even more out of touch with the feelings of local farmers.

The oldest agribusiness in Khuzestan was the government's sugar cane plantation at Haft Tappeh, which had begun production during the Third Development Plan. The rest of the 68,000 ha of the Dez irrigation project were divided amongst a number of concerns. These were Iran California with 10,000 ha; Hashim Narraghi Agro-Industries of Iran and America with 20,000 ha; Dezkar with 5,000 ha; Iran Shellcott with 15,000 ha; the Ahvāz sugar refinery with 1,000 ha and a corporation for the Dez farmers with 17,000 ha.[18] For their right to farm the land, the companies had to pay a high water and ground rent and were also compelled to make a fixed investment per hectare. Government organizations were responsible for the supply of water through large scale canals to the project areas, but the individual concerns had to arrange and provide the field distribution and irrigation systems at their own expense.

In Khuzestan most of these agribusinesses proved to be failures for a variety of reasons. One of the main ones was that the units, which were largely experimental in terms of Iranian conditions, were too large to be managed efficiently. Costs were high and labour both unskilled and uncommitted. A lack of knowledge of local conditions meant that serious environmental mistakes were made, including burial of top soil in land levelling operations. The result was that in a few years most of the agribusinesses were bankrupt, despite large cash injections from the government. Elsewhere throughout Iran other, often smaller, agribusinesses were set up, so that by 1978 thirty-eight were in operation farming an area of 200,000 ha. Some of these proved more successful than their Khuzestan counterparts.

At the local level farm corporations were set up to modernize farming techniques and to encourage large scale reclamation schemes.[19] These corporations acted as a means of concentrating small capital sums and allowing the economic administration of many small farms. In a region where a corporation was planned, the landowners were faced with the choice of joining or not. If the majority elected to form a corporation, the others had to join, sell or rent the land to others who were willing to do so. Once the corporation had been established, the peasants transferred their rights to the land to the corporation in perpetuity.

The peasants then obtained shares in proportion to the value of their land. At the end of the year peasants received profits relative to the number of shares they held, although if they so wished, they could also work for a daily wage. In the late 1960s, some 20 corporations were in existence cultivating more than 15,000 ha. By 1978 there were 98 farm corporations in existence. These covered 400,000 ha and contained more than 300,000 people in 850 villages. Despite considerable injections of capital from the government the results were disappointing, with productivity remaining low. One of the main problems was lack of commitment amongst the peasants who felt that their status had declined by having to become agricultural labourers. They felt no tie to any piece of land and, therefore, their efforts to improve the land were minimal.

An overall assessment of land reform in Iran is extremely difficult as accurate data on what actually happened are very scarce. The official results of land reform are impressive. Under the First Phase of legislation up to July 1969, 15,710 villages had been purchased by the government and redistributed to 730,000 farmers. In all it is claimed that 3.6 million people benefited from these changes.

Under the Second Phase up to the same date, it is stated that 2,457,982 farmers benefited, making with their families a total of 12 million people. Some scholars, however, have claimed that the official statistics have been in some cases too optimistic. In an agricultural sample survey in 1960, undertaken by the Iranian government and FAO workers, it was found that 14.4 per cent of the employed rural population were wage labourers and that 33.1 per cent were family workers. Thus about 47.5 per cent of the rural employed population received no land in either the First or Second Phase of land reform. In summarizing the available data, both from official and unofficial sources, it has been estimated that about 8 per cent of Iran's farmers obtained land during the First Phase of land reform, while during the Second Phase another 6 to 7 per cent of peasants received some land, making a total of 14 to 15 per cent of Iran's cultivators as new landowners,[20] However, A. K. S. Lambton did note that farmers in many areas had become politically more mature, with a growing belief in themselves and of their importance in the community.[21]

Although it was true that the large landowners had been done away with, the power and influence of these very rich people still remained as a bulwark against rural change. The problem of absentee landlords also still remained. By the late 1970s it was thought that at least half of all the agricultural land was still owned by 200,000 landowners, most of whom were absentee. A new type of absentee landlord also appeared, who bought land around the large towns and cities as a speculation, hoping to sell at higher price for industrial or related activities. Such owners often kept the land fallow, so that it could be disposed of quickly. Naturally this caused much resentment amongst the peasants.

The land reform programme applied to about 16 million ha of land which was being cultivated in the 1960s by over two million peasants. Even if all this land had been divided equally amongst all the peasants it would have made a plot size of only 8 ha, which is close to the minimum required to sustain a family under Iranian conditions. As has been shown earlier, not all the land was in fact redistributed and in many cases peasants got much less than 8 ha. Indeed, estimates in 1976 showed that about 73 per cent of all peasant owners, totalling 1.7 million, had holdings of less than 6 ha.[22] The land they received was often highly fragmented making efficient utilization almost impossible. These small land holdings meant that most of the peasants could not support their families adequately through agricultural activities. As a consequence many were forced to take up work in the urban area on a daily or seasonal basis. By so doing they helped to swell the growing mass of dissatisfied people which formed the basis for the revolution.

It was, however, amongst the *kwushnishins* that the greatest problems occurred, and in particular, with the agricultural labourers. In the early 1970s there were at least 1 million *kwushnishins*, who, not possessing a *nasaq*, could not benefit from the land reform. They also formed part of a group which was rapidly growing in numbers. With land redistribution there was much less casual labour needed as the small farmers used family labour to a greater extent. Given the inadequately sized holdings, which even *nasaq* holders obtained, many small owners were often in competition with the *kwushnishins* for paid agricultural work. At the same time the increasing use of mechanization meant that the demand for manual labour was actually falling quite considerably.

The net result was that with not enough land and not enough work in rural areas, there was a widespread drift to the cities and towns to obtain work. In the period from 1971 to 1976 the demand for labour was high, and this in fact had accelerated the movement, for even people in the remotest villages heard tales of the high wages being paid in urban areas. Then, in 1977 and 1978, with growing world recession, the Iranian economy faltered, and under and unemployment became for the first time a problem even in the urban centres.

Under these conditions the country became ripe for revolution. Dissatisfaction grew and outbreaks of violence began to occur in 1978. This was put down savagely by government forces. However, with a stagnant economy entrepreneurs were not willing to invest in the future and decision making in government departments almost ceased. By late 1978 there was to all intents and purposes a breakdown in law and order in the poorer urban areas where the male workers were concentrated. This situation was in marked contrast with conditions forty or more years earlier, when it had been the more peripheral rural areas which had been most resistant to government control. This time most of the rural areas remained peaceful. By late 1978 the government had ceased functioning and in early 1979 the Shah was deposed. In summary, there can be no doubt that land reform, instead of bolstering the power of the Shah as had been hoped, was indeed one of the main factors which contributed to his overthrow.

18.4.3 Developments in crop production

One of the most spectacular changes in Iranian agriculture which has taken place in the post-Second World War period has been that of mechanization. The earliest mechanization took place on dry-farmed land in Gorgān in the 1950s sponsored by wealthy Turkoman merchants and later by entrepreneurs from Tehrān.[23] These farms, which were often large, were also unusual in that they used hired farm labourers and the owners managed the farms themselves. The process of mechanization was further encouraged in the 1950s by the Crown disposing of large areas of uncultivated, but often potentially fertile land in Gorgān. This permitted many entrepreneurs to buy large areas suitable for mechanized agriculture. The main crop which was grown initially was wheat, but conditions also proved favourable for cotton cultivation. By the early 1960s Gorgān was the most mechanized area of Iran with almost one third of the tractors in use.

Accurate details on the number of tractors in Iran are difficult to obtain, but FAO estimates suggest that numbers grew from only 6000 in 1963–64 to about 75,000 by the early 1980s (Table 18.4).

The use of agricultural machinery was facilitated by the establishment of the Agricultural Machinery Development Bureau in 1952.[24] Loans were made available by this organization so that the farmer only required to pay a deposit of 20

TABLE 18.4
Tractors in Iran

Year	Numbers
1963–64	6,000
1974–75	43,900
1979	57,000
1983	75,000

Source: FAO Production Yearbooks, Rome, various years.

per cent of the purchase price of the machinery. The rest was paid back at low interest rates over a four year period. A government requirement was that service facilities had to be provided for every type of machinery which was on sale. Despite this obtaining spare parts did remain one of the major difficulties with the mechanization programme.

Gilan and Mazanderan, other areas like Gorgān close to the Caspian Sea, also experienced early mechanization of their agriculture. The chief form of production here was paddy rice. Mechanized tillers from Japan were introduced in the late 1970s and proved ideal for local conditions and were cheap to purchase. By 1965 over 10,000 were in use and by 1980 the figure had risen to almost 75,000 (Figure 18.6). Estimates show that by the early 1970s tillers were being used in about 70 per cent of all paddy fields in the region.[25] The rapid sale of tillers appears to have been the result of aggressive marketing policies by the Japanese manufacturers. These companies provided attractive financial terms for buyers and also ensured adequate service facilities.

The great advantage of mechanization was that it allowed the more rapid cultivation of the fields and also permitted much larger areas to be cultivated. It did, however, also mean that agricultural labour requirements were considerably reduced. In the 1960s the mechanization programme was greatly facilitated by the land reform legislation, which excluded mechanized land from the provisions of the new laws. As a result many landowners rapidly went in for machinery purchase to keep their land outside the land reform measures.

Another facet of mechanization in agriculture was the use of pumped wells. The first drilled wells in Iran had been sunk by the Allied armies during the Second World War, but it was not until the 1950s that their use became widespread. They were primarily used by entrepreneurs opening up new lands. Initially they were very successful and permitted high yielding crops to be grown in areas where crop production without irrigation was not feasible. By the early 1960s, however, it had become obvious that throughout Iran the overpumping of aquifers was leading to falls in the local water table, which often caused the traditional *qanāt* systems to decline in output or even dry up completely. In 1968 the government acknowledged the problem by nationalizing water resources. A feature of this law was that any new well which was proposed required a government permit before construction could begin. The idea was that such permits would only be given if it was felt that the effects on local groundwater resources

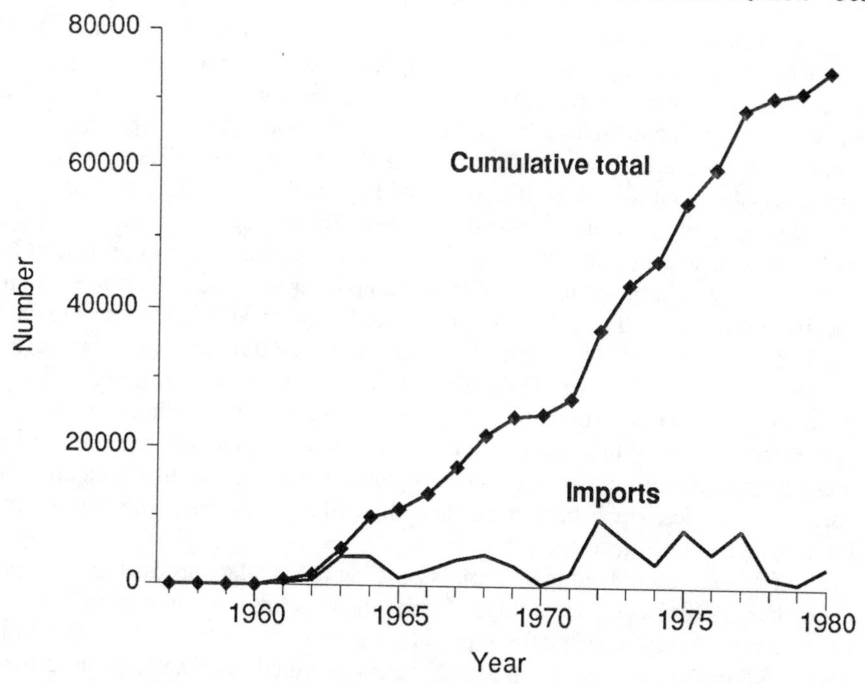

Figure 18.6 Imports of tillers and two-wheeled tractors from Japan and total numbers in use in Iran

would not be harmful. Unfortunately, the law came too late to protect *qanāt* systems in many parts of the country, and as a result many villages suffered considerable hardship as the volume of water delivered from their *qanāts* declined.

Modernization has also been associated with the growing use of commercial fertilizers. In the traditional agricultural system of Iran only organic fertilizers were available, consisting mainly of crop residues and animal dung. In the Eşfahān oasis pigeon towers were used to collect bird droppings, which could then be used as a fertilizer on high value crops. A feature of the post Second World War period has been the growing use of commercial fertilizers. In the early 1960s consumption of commercial fertilizers was about 25,000 tonnes per annum. During the 1960s there was a rapid increase in use so that by 1970 it had reached 120,000 tonnes per annum. The same trend continued through the 1970s and by 1981/82 consumption was 676,000 tonnes.[26] So great was this growth in absolute use of fertilizer that it was able to keep ahead of the increase in the cultivated area. As a consequence the relative use of fertilizer on cultivated land had increased from 1.6 kg/ha in 1961–65, to 22.1 kg/ha in 1974–76 and to 42.3 kg/ha in 1981.[27] Even so these consumption levels remain well below the fertilizer usage of Egypt at 247.5 kg/ha and Israel at 199.6 kg/ha in 1981.

All these changes have permitted an extension of agriculture onto lands often unsuited for cultivation, but overall owing to the high fossil fuel subsidies average yields have been maintained. Average yields are, of course, notoriously bad indicators of what is happening in particular regions. In Iran the yields of many crops in the more favoured agricultural areas appear to have risen markedly, but in the poorer areas yields still remain low. The growing use of commercial fertilizers has made one significant change in Iranian agriculture and that is the much less frequent use of land fallowing. The addition of commercial fertilizers now means that land can be cultivated continuously without a significant fall in yields.

The impact of growing population has been to produce, especially since the early 1960s, a situation in which Iran has been unable to feed its population by crops grown within the country. This has been particularly the case with regard to cereals. Between 1950 and 1980 the cultivated area of all cereals has shown a marked increase (Table 18.5). Wheat has almost tripled its area, while both barley and rice have doubled theirs. The maize area has grown more than sevenfold, but from a very small initial area. A study of yields reveals that for wheat, barley and maize, there has been little change over the past few decades. This means that for these crops increases in production totals parallel increases in cultivated area. In contrast, with rice, yields have risen more than 50 per cent, with a corresponding increase in total production.

If the cereals are bulked together it is possible to calculate production on a per capita basis. This has the advantage of showing what has been happening in terms of food consumption. In 1950 the population of Iran is estimated to have been about 17.5 million people. With a total cereal production of 2.9 million tonnes,

TABLE 18.5
Cereal production in Iran

Year	Wheat	Barley	Maize	Rice
1. Cultivated area of cereals – 000 hectares				
1948/49–1952/53	2,080	934	6	220
1969–71	5,370	1,532	25	362
1974–76	5,839	1,432	31	452
1982	6,192	1,841	45	483
1983	6,042	2,007	45	429
1984	5,800	1,800	45	420
2. Cereal production – 000 tonnes				
1948/49–1952/53	1,872	636	6	425
1969–71	3,946	1,042	35	1,041
1974–76	5,438	1,263	52	1,436
1982	6,660	1,903	60	1,605
1983	5,956	2,034	50	1,216
1984	5,500	1,550	50	1,230
3. Yields of cereals – kg/ha				
1948/49–1952/53	900	1,010	1,030	1,930
1969–71	735	81	1,400	2,875
1974–76	931	882	1,685	3,175
1982	1,076	1,034	1,333	3,322
1983	986	1,014	1,111	2,831
1984	948	861	1,111	2,929

Source: FAO Production Yearbooks, Rome, various years.

Circular earth mounds around the ventilations shafts mark the lines of qanāts to a village near Mashhad, Iran (National Cartographic Centre, Iran)

this means that the per capita cereal consumption from home grown crops was 168 kg/capita/annum. By 1984 the population had risen to 43.8 milion and production per capita was 232 kg/capita/annum. This reveals quite conclusively that cereal production had more than kept up with population growth. Yet despite this level of production, cereal imports have shown a sharply rising trend so that by 1980, while total cereal production was around 8.1 million tonnes, cereal imports were 3.4 million tonnes. In per capita terms home production of cereals was 213 kg/capita/annum, and imports 90 kg/capita/annum. This suggests that total consumption was of the order of 303 kg/capita/annum in 1980.

From these figure there can be little doubt that the per capita consumption of cereals has revealed a very marked increase since the 1950s. Part of this is undoubtedly the result that today people in Iran are better fed than they were 40 years ago. However, it must also reflect a trend whereby the cereals are being processed more prior to consumption. Such processing is associated with much higher wastage rates than in a traditional society. Just how great these losses are remains unknown. What is clearly seen, at least up to the mid-1980s, is that Iranian agriculture had been able to maintain a level of cereal production on a per capita basis which exceeds the per capita production values of the 1950s. What it has not been able to do though is to keep up with the increased level of per capita consumption which is a feature of modern Iran.

With the current rapid growth in population and the predicted rise to a total of 70 million in the early years of the twenty-first century, it is interesting to speculate as to what is likely to happen with regard to food production. If it is assumed that indigenous cereal production is maintained at about the 1982 level of around 10 million tonnes, then the cereal provision for 70 million people will be 142 kg/capita/annum. If the 1980 total consumption level of about 300 kg/capita/annum is maintained, then cereal imports would have to rise to 11 million tonnes per annum.

On the other hand, if cereal production in Iran attempted to maintain the per capita values of the early 1980s (c 230 kg/capita/annum), output would have to rise to 16.1 million tonnes to support 70 million people. To achieve this would need an approximately 60 per cent increase in either the cultivated area or in crop yields.

Finally, it should be remembered that the figures above only maintain the relative balance between domestic output and imports of the early 1980s, when imports were 42 per cent of total use. To be self-sufficient in cereals at early 1980s consumption rates, that is around 300 kg/capita, would require a production of 21 million tonnes to feed 70 million people. These figures while not providing any answers to the food problem of Iran do reveal just how big a problem it is going to be to feed so many people. The solution which has been chosen by the Islamic Republic is one of national self-sufficiency in terms of food and basic agricultural raw materials. Whether this can be achieved or not before the end of the twentieth century still remains to be seen.

References

1. B. D. Clark, 'Iran – changing population patterns', in *Populations of the Middle East and North Africa*, (Eds J. I. Clarke and W. B. Fisher), University of London Press, 1972, p 79.
2. B. D. Clark, 'Iran – changing population patterns', in *Populations of the Middle East and North Africa*, (Eds J. I. Clarke and W. B. Fisher), University of London Press, 1972, p 79.
3. Population Reference Bureau, *World Population Data Sheet 1985*, Washington, DC.
4. J. Bharier, *Economic Development in Iran 1900–1970*, Oxford University Press, London, 1971, p 131.
5. For a detailed study of traditional rural life in Iran see A. K. S. Lambton, *Landlord and Peasant in Persia*, Oxford University Press, 1953, 459 pages.
6. For a detailed study of climate in Iran see M. H. Ganji, 'Climate', in *The Land of Iran* (Ed W, B. Fisher). The Cambridge History of Iran, **1**, Cambridge University Press, 1968, pp 684–713.
7. P. Beaumont, 'Water resource development in Iran', *Geographical Journal*, **140**, pp 418–431, 1974.
8. P. Beaumont, 'Qanat systems in Iran', *Bull. int. Ass. scient. Hydrol*, **16**, pp 39–50, 1971.
9. P. Beaumont, *River Regimes in Iran*, Occasional Publications (New Series) No. 1, Department of Geography, University of Durham, 1973, 29 pages.
10. (a) K. S. McLachlan, 'Land reform in Iran', in *The Land of Iran* (Ed W. B. Fisher), the Cambridge History of Iran, 1, Cambridge University Press, 1968, pp 684–713.
    (b) A. K. S. Lambton, *The Persian Land Reform*, Clarendon Press, Oxford, 1969, pp 1–28.
11. Plan Organisation, *Fourth National Development Plan, 1968–1972*, Plan Organisation, The Imperial Government of Iran, Tehran, 1968, 355 pages.
12. P. Beaumont, 'Environmental management problems – the Middle East', *Built Environment Quarterly*, **2**, pp 104–112, 1976.
13. Plan and Budget Organisation, *The First Five Year Economic, Social and Cultural Macro-Development Plan of the Islamic Republic of Iran 1362–1366 (1983/84 – 1987/88)*, Tehran, p 30, (1982).
14. For a study of changes in Iranian villages see: E. Ehlers, 'The Iranian village: a socio-economic microcosm', in *Agricultural development in the Middle East*, (Eds P. Beaumont and K. S. McLachlan), Wiley, Chichester, pp 151–170. (1985).
15. E. J. Hoogland, *Land and Revolution in Iran, 1960–1980*, University of Texas Press, Austin, p 12, (1982).
16. E. J. Hoogland, *Land and Revolution in Iran, 1960–1980*, University of Texas Press, Austin, p 64, (1982).
17. A. K. S. Lambton, *The Persian Land Reform*, Clarendon Press, Oxford, 1969, p 280.
18. M. Field, 'Agro-business and agricultural planning in Iran', *World Crops*, **24**, pp 68–72, (1972).
19. Ministry of Information. *Farm Corporations in Iran*, Ministry of Information, Tehran, 1970, 10 pages.
20. N.K. Keddie, 'The Iranian village before and after land reform', *Journal of Contemporary History*, **3**, p 87, (1968).
21. A. K. S. Lambton, *The Persian Land Reform*, Clarendon Press, Oxford, 1969, 386 pages.
22. E. J. Hoogland. *Land and Revolution in Iran, 1960–1980*, University of Texas Press, Austin, p 91, (1982).
23. S. Okazaki, 'Agricultural mechanisation in Iran', in *Agricultural development in the Middle East*, (Eds P. Beaumont and K. S. McLachlan), Wiley, Chichester, pp 172–73, (1985).

24. S. Okazaki, ‘Agricultural mechanisation in Iran’, in *Agricultural development in the Middle East*, (Eds P. Beaumont and K.S. McLachlan), Wiley, Chichester, p 176. (1985).
25. O. Aresvik, *The agricultural development of Iran*, Praeger, New York, pp. 160–63. (1976).
26. *FAO Fertilizer Yearbook 1982*, vol 32, FAO, Rome, p 117, (1983).
27. P. Beaumont, ‘The agricultural environment: an overview’, in *Agricultural development in the Middle East*, (Eds P. Beaumont and K. S. McLachlan), Wiley, Chichester, p 23, (1985).

CHAPTER 19

Egypt: Population Growth and Agricultural Development

19.1 Introduction

The Arab Republic of Egypt is one of the best documented countries in the Arab world, and perhaps one of the least understood. With a highly distinctive geographical personality, the country's complex social, economic and political problems have been the subject of much oversimplification, often resulting in great concern about the future. The aim of this chapter is to give some perspective to one of the most geographical aspects of Egypt's development problems – the expansion of agricultural production, seen against the background of rapid population growth, a vacillating economy and an increasingly uncertain political relationship with its neighbours.

There is still much truth in the old saying of Herodotus that Egypt is 'The Gift of the Nile'. Although located in the great Saharan-Arabian desert belt, the Nile Valley and delta support over 95 per cent of Egypt's population. Within this settled region communications are easy and centralized political control is relatively simple to impose, which partly accounts for the many centuries during which Egypt was successfully dominated by external powers. It also helps explain the relative cultural homogeneity of the country. The physical environment of the valley and delta is also highly conducive to intensive agriculture and settlement. The nature of the valley itself, enclosed by scarps sometimes rising to over 400 m above the valley floor, enables the river to flow without serious losses by seepage and evaporation. The alluvial soils of the valley and delta are also favourable to agriculture (Chapter 2). Crops can be grown all the year round in three main growing seasons – winter (*shitwi*), summer (*seifi*) and autumn (*nili*) – because of the continuous warmth and high levels of insolation. The temperature regime throughout is ideal for crop development, without extremes of hot or cold. In the past the annual flood added a valuable layer of fertile silt to the land. For a significant proportion of the population the physical environment of the valley and delta remains of paramount importance, but the economic life of the country is also vastly more complicated than in the time of Herodotus.

Apart from the period between 1966 and 1973, when it was deprived of the Sinai peninsula, Egypt has the unique distinction of being an African state with a foot in Asia. One outcome of this has been Egypt's role at the forefront of conflict and diplomacy with Israel. Prior to the agreement at Camp David in 1979, Egypt was Israel's most powerful and aggressive neighbour. Since the peace initiatives of President Sadat however, Egypt has reached a formal peace agreement and achieved some co-operation with Israel, placing the country in an awkward position with its Arab neighbours. The consequential loss of investments from countries like Kuwait, the United Arab Emirates and Saudi Arabia has been partly responsible for some of Egypt's economic problems throughout the 1980s[1]. With the small exception of the Taba area, the Egypt–Israel boundary is now formally agreed. Egypt's boundaries with Sudan and Libya were largely products of European influence; the Sudan boundary was fixed in 1899, and the Libyan boundary in 1925. Its present area covers 997,738 sq km, although only about five per cent of this is inhabited or cultivated. Until the last three or four decades, Egypt's territories outside the Nile Valley and delta were largely neglected, but the discovery of a range of important raw materials, including oil and gas, iron ore, manganese and phosphates has changed this attitude, while the possibility of desert reclamation and settlement has been a subject for considerable debate.

19.2 Population growth 1800 to 1986

In 1800 the population of Egypt was probably about two and a half million.[2] The first population census was in 1882, but it is not regarded as reliable. Subsequently, nine censuses have been conducted, so that Egyptian population statistics are more plentiful than for most developing countries. Table 19.1 summarizes the results.

The accelerating rate of population growth is entirely the result of natural increase since immigration is negligible in Egypt. If sustained, the present rate of increase would lead to a population of about 70 million by the end of the century.[3] The remarkable decline in death rates has been chiefly responsible for this increase. (Figure 19.1).

TABLE 19.1
Populations of Egypt 1897 to 1986

Census Year	Population	Rate of Increase Per cent per annum
1897	9,714,500	—
1907	11,190,000	—
1917	12,718,300	1.3
1927	14,177,900	1.1
1937	15,920,700	1.2
1947	19,038,500	1.8
1960	26,085,300	2.4
1966	30,083,400	2.5
1976	36,626,204	2.5
1986*	50,200,000	2.7
2000**	70,000,000	—

Source: United Nations, *Demographic Yearbook*, New York, various years.
* Estimate ** Projected

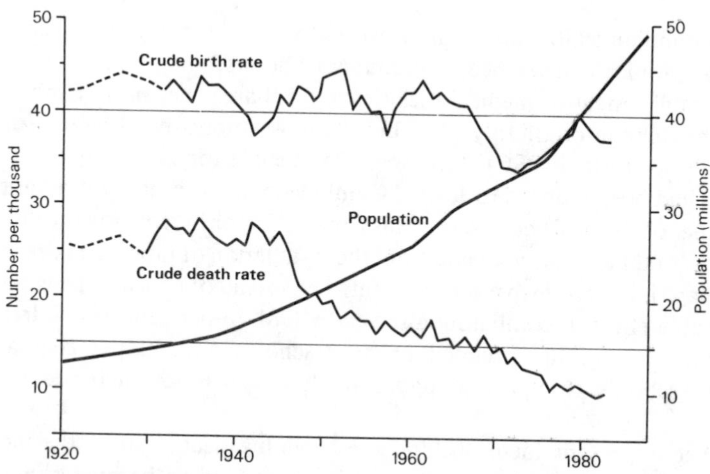

Figure 19.1 Crude birth rates and death rates for Egypt, 1920–1986

Before the Second World War, Egyptian death rates (25 to 29 per thousand) were among the highest in the world as a result of several related factors. To begin with, the peasants and their families suffered appallingly poor health. Many endemic diseases in Egypt are waterborne and affect a high proportion of the rural population because of the dense network of irrigation canals and ditches. Waterborne diseases increased with the spread of perennial irrigation. Several common diseases such as bilharziasis and malaria, were a constant source of misery and reduced efficiency. Anykylostomiasis (hookworm) and trachoma, which results in total or partial blindness were also very widespread. The chief killers however, were bronchitis, heart diseases, tuberculosis and all kinds of liver diseases. Cholera epidemics spread through Egypt several times from 1916, the final epidemic being in 1947 when over 20,000 people died.[4] Resistance to disease in rural areas was also low because of widespread undernourishment; in the 1950s it was estimated that as many as 75 to 80 per cent of the population received an inadequate diet.[5] The intake of animal protein and protective foods, such as calcium and vitamins, was found to be extremely low, particularly in some of the poorer rural areas, although this has since been partly offset by an increase in vegetable intake. Cheese, dates, melons, and more recently, meat and fish have also become more readily available. During the past two decades, most of the required calories have been derived from wheat products, and Egyptians are now the world's greatest per capita consumers of wheat.[6] Indeed, by 1981 Egyptians consumed 116 per cent of their daily required calorie intake.[7] Another cause of so much disease is poor housing and inadequate sanitation. The low houses, built of sun-dried Nile mud bricks, generally consist of one or two dark rooms which the family sometimes share with their animals. Settlements consist of closely-knit groups of houses with twisting narrow lanes; many villages have less than 5000 inhabitants. While a considerable effort has been made to improve sanitation conditions in Egypt, especially in the towns as exemplified by the US$55 million sewerage scheme in greater Cairo, the smaller villages are still amongst the most insanitary and unhealthy places in the world to live. Everyday life in such villages has been graphically described in a number of books.[8]

The establishment of medical facilities in rural areas began on a modest scale before the Revolution of July 1952, but the most impressive developments have occurred since then. In 1960 there were 2550 people for every physician, but by 1980 this had been reduced to 1,970.[9] Combined with a similarly dramatic rise in the number of medical centres, this was a major factor in reducing the death rate, especially infant mortality. In addition, the installation of piped drinking water to every village is an impressive achievement. The spread of education has also had a considerable effect in combating disease. In 1981 76 per cent of children in the appropriate age group attended primary schools and 52 per cent attended secondary schools. Comparative figures for 1960 were 60 and 16 per cent respectively.[10]

Another important factor behind the declining death rate is the increasing proportion living in urban areas where incomes tend to be higher, and social services superior to those in rural areas. In 1937, 25 per cent of the population of Egypt were classed as urban compared with 43 per cent in 1970. During the 1960s, when natural increase was about 2.5 per cent, the average annual growth of the urban population was 4.6 per cent. Urban growth was then maintained at about 3.8 per cent until the mid 1980s, when it began to rise once more. The slight decrease in the 1970s is partly explained by the large number of migrants who left Egypt to work in more prosperous petroleum producing countries, many of whom have returned since the oil price crisis of the mid-1980s. Even allowing for higher natural increase in the towns, rural-urban migrants were an important component in the urban growth. Their contribution was greatest in the large cities of Cairo and Alexandria, and to a lesser extent in other towns with over 100,000 inhabitants. Many small towns have recorded modest growth rates during the past two or three decades. Thus it is in the largest towns that the greatest concentrations of rural migrants occur, and where problems of unemployment are most severe. Despite rapid industrialization, far too few jobs have been created, and urban slum conditions are multiplying on the outskirts of many large urban centres.

The current crude birth rate in Egypt lies between 35 and 38 per thousand, compared with 40 to 44 in the pre-Second World War period (Figure 19.1), indicating modest success for family planning measures which have existed for two decades. Between 1960 and 1982 there was a 45 per cent decrease in the crude birth rate.[11] However, success in reducing the birth rate has been confined mainly to urban areas and high birth rates remain in rural areas. This is due partly to strong economic incentives to have large families where children can work in the fields and provide security for ageing parents, the practice of young marriages, and very high divorce rates.

The social and economic implications of rapid population growth in any developing country are well known. With some 45 per cent of the population of Egypt under the age of 15, a high proportion are in the 'dependent' age group. Every year there are more than one million more people to feed, resulting in heavy expenditure on imported food, approximately half on cereals and milling products; in 1984 Egypt spent over US$4000 million dollars on importing wheat. The government also has to build more schools, hospitals and houses to keep pace with the rising population, while every year thousands of school leavers seek jobs in the towns.

19.3 Man and the land

The rate of population growth in Egypt is not exceptionally high compared with several other Middle East countries (Table 5.2), but against a background of severely limited resources of cultivable land and water, the growing population has serious implications. Less than 5 per cent of the country is settled and cultivated, with the Nile valley and the delta supporting the bulk of the population, three-fifths of them more or less directly by agriculture. Rural population densities are already high, averaging 1381 persons per sq km in 1985,[12] an average which is much higher in parts of the delta. In some regions, agricultural holdings are already too small to yield a reasonable standard of living and present levels of production could be generally achieved by a greatly reduced labour force. Moreover, productivity is already high. Each cultivated feddan (1.038 acres or 0.42 hectares) produces an average of 2.0 crops a year, and over 99.5 per cent of farmland is irrigated. More fertilizer is used in Egypt than in the whole of the rest of North Africa, and levels of application are higher than in Britain and the United States.

Table 19.2 shows that both cultivated area per capita and crop area per capita are declining. An unnecessarily gloomy picture is sometimes given by quoting the figures in column (v) whereas column (viii) is a more realistic indicator. Thus, while the *cultivated* area per capita has fallen by about two-thirds since 1897, the *cropped* area per capita of the *rural* population has only declined by just over one third. Certainly there is no room for complacency; these figures do not reveal that a high proportion of proprietors own less than one feddan and that output per capita has declined markedly. Hired labour is also in less demand than a few years ago partly as a result of mechanization on the large estates. During the next two or three decades the proportion of Egypt's population living in rural areas will continue to decline, while the overall rate of population increase could be slowed down by a successful family planning programme. Nevertheless, by the end of this century Egypt's non-urban areas might still have to support at least another 15 million people. The challenge of absorbing such numbers in the rural areas, while raising the living standards of the rural population as a whole, is a formidable one.

The following sections show how this problem has been dealt with since the Revolution of 1952, and how present and future development must be oriented to meet the increasing land, food and employment demands of the Egyptian people. Four principal approaches to agricultural planning can be identified: agrarian reform which dominated the period to 1972; land reclamation and intensification

TABLE 19.2
Man and the land in Egypt: some statistics

Year	(i) Total popn (millions)	(ii) Urban popn (percent)	(iii) Cultivated area (million feddans	(iv) Cropped area (million feddans)	(v) Feddans/capita of total popn	(vi)	(vii) Feddans/capita of rural popn	(viii)
					Cultivated	Cropped	Cultivated	Cropped
1820	2.5	12	3.0	3.0	1.20	1.20	1.40	1.40
1897	9.7	20	5.0	6.8	0.51	0.70	0.64	0.87
1907	11.2	19	5.4	7.7	0.48	0.68	0.60	0.85
1917	12.7	21	5.3	7.7	0.41	0.60	0.53	0.77
1927	14.2	23	5.5	8.7	0.39	0.61	0.50	0.79
1937	15.9	25	5.3	8.4	0.33	0.53	0.44	0.70
1947	19.0	31	5.8	9.2	0.31	0.48	0.44	0.70
1960	26.0	38	6.1	10.3	0.30	0.38	0.49	0.64
1966	30.0	40	6.5	10.5	0.27	0.35	0.36	0.58
1970	33.9	42	7.4	10.7	0.22	0.32	0.37	0.54
1982	44.7	50	8.3	14.0	0.18	0.31	0.36	0.61

Sources: E. Garzouzi, *Old Ills and New Remedies in Egypt*, Dar al Ma'aref, Cairo, 1958, 15;
P. O'Brien, 'The long-term growth of agricultural production in Egypt: 1821–1962', in *Political and Social Change in Modern Egypt*, (Ed P. M. Holt), Oxford University Press, London 1968, 162–95.
Daily Telegraph, Report, 15 January 1971.
Statistical Handbook of the U.A.R. 1952–1966, Cairo, 1967, 22–25.
Statistical Abstract of the U.A.R., Cairo, 1971, 26–28.

which were features of the 1970s; and a general restructuring of the economy which is likely to dominate agricultural policy to the end of the century.

19.3.1 Agrarian reform

Agrarian reform dominated agricultural planning in Egypt for two decades (1952–72), and its legacy is sufficiently important to warrant some discussion of its successes and failures.

Agrarian reform bills had been introduced in 1945 and 1950, but were decisively rejected by the landlord-dominated parliament. In September 1952, scarcely six weeks after the Revolution, the Young Officers implemented the First Agrarian Reform. It had four main objectives: first, to break the power of the landlords; secondly, to improve the living conditions for the rural population; thirdly, to divert capital from agriculture to industry, and finally to raise agricultural output. In 1952, out of a total cultivated area of 5.9 million feddans, 40 per cent was held by less than one per cent of owners, while 72 per cent of owners together held no more than 13 per cent of the cultivated area, and the average size of their holdings was less than half a feddan (Table 19.3).

Since two feddans was generally regarded as the minimum required to yield a reasonable standard of living in Egypt, many *fellaheen* were clearly in great poverty, although some were able to supplement their incomes through hired employment on estates or by selling craft work. The average size of holdings had shown a steady decline since the nineteenth century; before 1900, there were still no owners with less than one feddan in their possession, but by 1913 there were nearly one million, farming over seven per cent of the cultivated area of Egypt.[13] The chief reasons for this progressive decline were the division of property between all the male heirs, and the growing importance of large estates. There were also large numbers of tenant farmers and landless wage earners in Egypt, probably as many as 1.5 million families.[14] Generally, rents were exorbitant, wages extremely meagre, and the very high price of land precluded its purchase by any but a few. Between the Second World War and 1952, land prices rose fourfold. The eagerness of capitalists to purchase agricultural land was largely due to the promise of a guaranteed income, and the absence of attractive alternatives.

The implications of the maldistribution of land in 1952 are obvious enough. Tenants and wage earners were readily exploited by landowners or their middlemen. On the other hand, many small landowners experienced a decline in real income as their holdings diminished in size. The First Agrarian Reform of 1952 was a cautious attempt to deal with these problems. The maximum size of individual holdings was reduced to 200 feddans, though landowners were permitted to transfer up to 100 feddans to their children. Expropriated land was normally given to landless labourers and to owners with less than five feddans. As a condition of this land grant, each new peasant-proprietor was required to join one of the newly-formed land reform co-operatives. Maximum rents were fixed for agricultural land, reducing them in effect by about 40 to 50 per cent. Written

TABLE 19.3
Egypt: landownership in 1952 and 1965

Size (Feddans)	Before Land Reform (1965)				After Land Reform (1965)			
	Owners	Area	Per cent of all		Owners	Area	Per cent of all	
	(000s)		Owners	Area	(000s)		Owners	Area
<1	2,018	788	72.0	13.0	3,033	3,693	94.5	57.1
1–4	624	1,344	22.2	22.5				
5–9	79	526	2.8	8.8	78	614	2.4	9.5
10–19	47	638	1.8	10.7	61	527	1.9	8.2
20–29	13	309	0.5	5.0	29	815	0.9	12.6
30–49	9	344	0.3	5.7				
50–99	6	429	0.2	7.2	6	392	0.2	6.1
100–199	3	437	0.1	7.3	4	421	0.1	6.5
>200	2	1,177	0.1	19.8	0	0	0.0	0.0
TOTAL	2,802	5,982	100.0	100.0	3,211	6,462	100.0	100.0

Sources: Garzouzi, *Old Ills and New Remedies in Egypt*, Dar al Ma'aref, Cairo, 1958, 79.
Statistical Handbook of the U.A.R. 1952–1969, Cairo, 1970, 54–57.

tenancy agreements, with a minimum tenure of three years became compulsory, and a minimum agricultural wage was fixed. Although wages for landless labourers rose, the improvement was not as great as originally planned. Agricultural workers were authorized to form trade unions, and fragmentation of holdings into lots of less than five feddans was henceforth forbidden.

Under the law of 1952, about one tenth of the cultivated area was eligible for redistribution. This was completed by 1956. The area involved was not great, but the social and economic implications of reform, the first in the Arab world, were colossal. The transfer of land occurred without bloodshed, and the legal status and human rights of the *fellaheen* was proclaimed. A Second Reform Law in 1961 reduced the maximum size of holdings to 100 feddans per person, although the ceiling on family owned land remained at 300 feddans. Compensation was paid in government bonds, but these were cancelled in 1964. In 1962, all foreign-owned land in Egypt was expropriated for redistribution, while in 1969 the maximum size of holding per individual was reduced to 50 feddans, and family holdings to 100 feddans, affecting about 13 per cent of the cultivated land. By about 1972, land reform had almost run its course with only 32,000 feddans being redistributed throughout the 1970s. Nevertheless, it is worth noting that the fundamental character of land ownership in Egypt remains unchanged, with the vast majority consisting of small properties. By the mid-1980s the average size of holding per peasant family seemed to have stabilized at 0.9 feddans.

One of the most important achievements of the Land Reform Authority was the introduction of multi-purpose cooperatives. About 1700 cooperatives already existed in 1952, largely for the provision of credit, and the wholesale purchase of seeds and fertilizers, but their membership was limited and they did not undertake marketing. The new-style cooperatives offered additional services such as technical and practical advice, arrangements for the marketing of produce, and the general supervision of farm production through a resident manager. Within a decade of the introduction of agrarian reform the number of cooperatives had almost tripled and nearly one million *fellaheen* could secure loans from them. The most striking feature, however, was the introduction of a consolidated triennial crop rotation. Under the traditional system in Egypt, farms typically consisted of several strips of land scattered throughout three or more large blocks of land. Within each block, individual farmers could grow any crop on their own strips, resulting in inefficient application of irrigation water and waste of land in paths, boundaries, and ditches. Some crops were watered insufficiently, others too much, for example when cotton was planted next to rice. Insects and plant diseases spread easily from crop to crop, and the application of fertilizers and pesticides was difficult. The traditional system also relied on biennial crop rotation which had an adverse effect on soil fertility since the land lay fallow for only two months every two years. Clover, which is essential for restoring nitrogen to the soil, could only be grown once in every two years.

Under the new system the lands of the village were divided into three large consolidated blocks each being assigned annually to one of the three main groups of crops. The blocks could then be ploughed, watered, drained, sprayed and harvested as a unit. Sowing, weeding and tending crops remained individual activities which the *fellaheen* continued to conduct on their own plots as before. Farming thus became a mixture of collective and individual enterprise. Crop consolidation permitted the adoption of a triennial crop rotation which enabled soil to rest under clover cultivation twice every three years. It was hoped that yields would increase by up to 20 per cent as a result. In some villages, land consolidation as well as crop consolidation took place, with plots belonging to individual *fellaheen* being consolidated on an exchange basis. Under both systems land remained in private hands, and no attempt was made to standardize the size of ownership. After consolidation the farmer's share of the crop was calculated in proportion to his holding after deductions for production costs.

Participation in the scheme became obligatory for all recipients of redistributed land, and attempts were made to introduce it more generally in other villages. In association with all-purpose co-operatives the new triennial rotation stimulated new advances in Egyptian agriculture. In one village, a 58 per cent increase in cotton yields occurred after consolidation and all crop yields were at least 20 per cent higher than in neighbouring villages.[15] Reasons for higher yields included more efficient farming techniques, land saved from ditches and boundaries, a scientific crop rotation, and the fact that family workers tended farm plots more intensively than hired labourers on the old estates.

Most small landowners received a modest improvement in their average net incomes through the greatly increased yields. The beneficiaries of land distribution, for example, more than doubled their net income *per feddan* between 1953 and 1965.[16] Tenant cultivators were also better off due to lower rents and higher yields. The most unfortunate group remained the landless labourers. The statutory minimum wage proved unenforceable because of the large surplus of unemployed. In some regions, there was hardship because the break-up of large estates reduced the demand for casual labour.

It would be a mistake to assess Egypt's attempts at agrarian reform simply in the light of crop yields and incomes. Changes in tenure and the introduction of cooperatives, for example, restricted much of the individual's choice of crop and many efficient medium-sized farming units were broken up. By the mid 1970s, and especially after the law 43 'Open Door' policies introduced by Sadat in 1974, many cooperatives found that dealing with the private sector offered more advantages than dealing with the state one. Elsewhere, some large cooperatives were under-financed and achieved only low productivity of land and capital. One particularly large cooperative on the Tahrir Province reclamation project was abandoned due to technical and managerial difficulties.[17] Nevertheless, taken as a whole, the changes linked to agrarian reform secured significant financial and agricultural advantages. Hand in hand with the economic and technical programme, a substantial social programme was designed to improve the quality of life in rural areas. The most outstanding aspect of this was the construction of over 800 combined rural centres throughout the country, each providing comprehensive social services catering for about 15,000 people. Typically, these centres included a health unit, school, and training centre for rural crafts, a social unit with library and assembly hall, cooperative society offices, and housing for the professional employees of the centre. In selecting villages for the new centres, communications and centrality were key considerations. It has been suggested that, if urban industry cannot absorb Egypt's surplus rural population, handicrafts and light industries, such as food processing, might be widely established outside the towns in some of these rural centres.[18]

Overall, it can be seen that agrarian reform achieved considerable positive changes in the agricultural sector of the Egyptian economy, but placed against the rapid increase in population the policy could not come anywhere near to solving the country's long term food needs. The limitations of reform were summed up succinctly by one writer, 'The fundamental land tenure problem is not so much one of distribution but of an overall scarcity.[19] Partly stimulated by the completion of the High Dam at Aswan in 1970, agricultural development in the 1970s was marked by new attempts to extend land by reclamation combined with greater attempts to intensify land use.

19.3.2 Expansion of cultivated areas

Until the construction of the High Dam (1958-70) the water from the Nile could irrigate little more than six million feddans. Apart from some minor cultivated areas in the oases and along the northwest coast, this figure was approximately equivalent to the cultivated area of Egypt in the early 1960s. The additional water made available by the Aswan High Dam made a potential extra 1.3 million feddans of cultivable land available in the Nile valley and the delta and enabled the conversion of 900,000 feddans in Upper Egypt from basin to perennial irrigation (Chapter 2). By 1972, 800,000 feddans had already been reclaimed using the extra Nile water to produce crops, and much of the remainder was taken up by 1980.

Figure 19.2 shows the chief features of the expansion of cultivated land in Egypt from 1960 to 1980. Six major projects are shown in and around the delta and western Sinai:[20]

(1) *Tahrir ('Liberation') Province*, one of the earliest large scale settlement schemes in Egypt was begun in 1957. Nearly 200,000 feddans were reclaimed and 25,000 people were established in seven villages.[21] Water is obtained from the Nile by canal, but in the southern sector it is also supplemented by local groundwater.

(2) *The Maryut region*. This scheme, covering some 100,000 feddans included the draining and reclamation of Lake Maryut and adjacent land near Alexandria. A feature of this project was the use of processed sewage water to supplement water from the Nile.

(3) *The Nubariya desert scheme* was undertaken using Nile waters on some 217,000 feddans southwest of Alexandria.

(4) *The northern delta*, south of Lakes Idku and Burullus. A vast region of

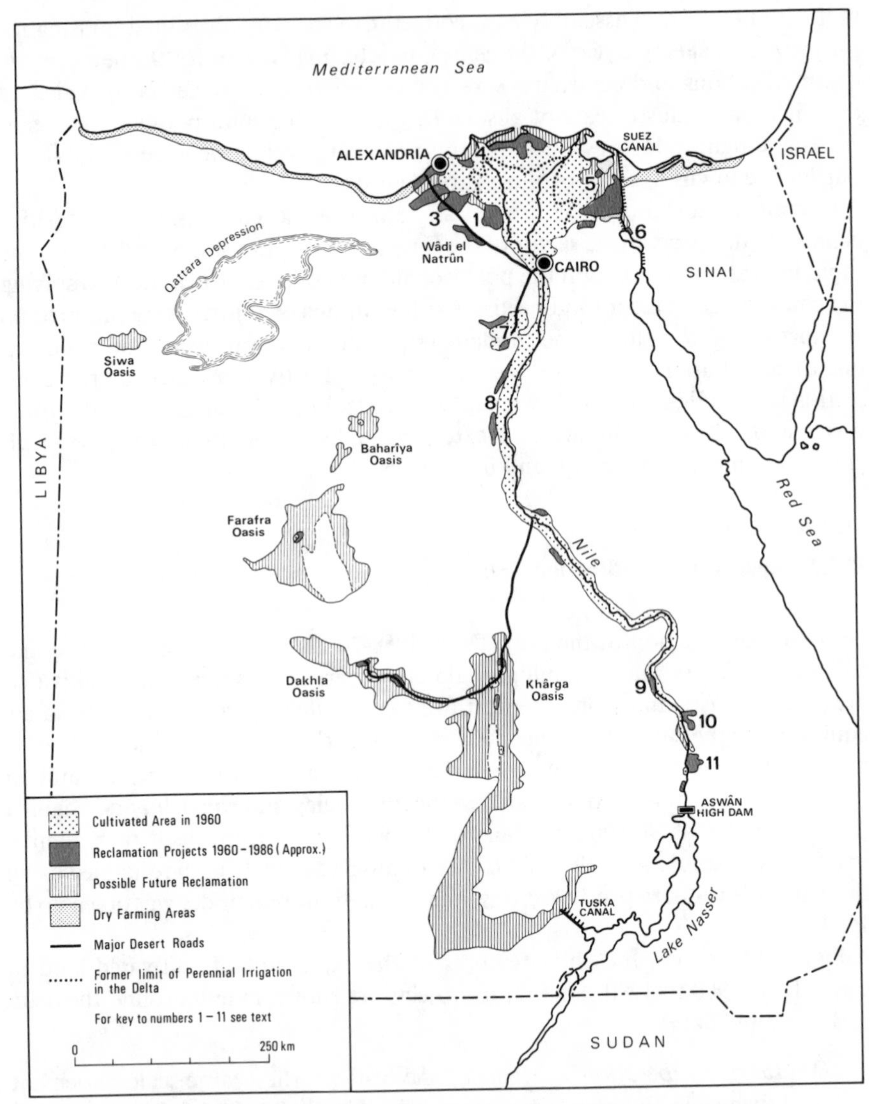

Figure 19.2 Expansion of the cultivated area of Egypt, 1960–1986

lakes, swamps and lagoons known locally as *Barari* was drained and the saline soils reclaimed at great cost, bringing about 120,000 feddans into cultivation.

(5) *The region south of Lake Manzala* was drained and irrigated, adding about 80,000 feddans for agricultural development.

(6) *The western Sinai scheme*. Over 2000 feddans out of a projected 20,000 was reclaimed before 1967. The rest of the scheme was delayed until the Salem Canal began in 1979, which pumped water underneath the Suez Canal from the eastern Nile.

These six projects were supplemented by five smaller schemes along the Nile valley utilizing Nile water, which added some 142,000 feddans by 1972. These were:

(7)	The Faiyum depression	– 9,500 feddans
(8)	El Minya region	– 58,500 feddans
(9)	Kena Province	– 17,000 feddans
(10)	Radesia and Wadi Abbady	– 13,000 feddans
(11)	Kom Ombo	– 44,000 feddans

The Kom Ombo scheme was the first phase of a resettlement project for Nubians of Upper Egypt dispossessed of their lands by Lake Nasser.

Between 1952 and 1980 a total of 1,072,000 feddans of new land was reclaimed from the Nile valley and the delta of which less than 15 per cent was added after 1976. A further 880,000 feddans of basin land was also reclaimed in Upper Egypt.[22] However, the actual amount of land available for cultivation was much less than the figures quoted as considerable areas of existing land were lost; in 1986 President Mubarak claimed that in real terms only about one million feddans had been added to the cultivated areas since 1909 although the introduction of multiple cropping following completion of the High Dam did mean that the actual cropped area was somewhat greater. Estimates suggested that by 1980 as much as 20,000 feddans of land were being lost each year for building development and for brick making.[23] In 1985 a new act was introduced prohibiting the manufacture of bricks from the fertile Nile clays. Further land was also being lost due to problems of waterlogging and salinity, both of which were exacerbated by the completion of the High Dam as more intensive cropping and less concern with drainage resulted in a general rise of the water table.

The combined effects of rapid population growth and the unexpected loss of cultivable lands in traditional farming areas forced the Egyptian government to initiate a new phase of land reclamation. Plans were announced in 1981 to reclaim an additional 1.2 million feddans of new land by the end of the century. The majority of this new land was centred on three large projects:

(1) *The Nubariya Desert* scheme is expected to reclaim another 300,000 feddans. This project is supported by US$80 million worth of loans from the International Development Association to establish specialist sugar beet production facilities.
(2) *Tahrir Province*. A new project concentrating on the Mediterranean coastal belt in the middle delta could add up to 200,000 feddans.
(3) *South of Port Said*. The reinitiation of a scheme first proposed in the 1960s to irrigate 200,000 feddans in a desert area south of Port Said. This scheme will form part of a new plan to construct up to 1000 new settlements to help relieve pressure from urban areas in the lower Nile Valley.

One of the most controversial aspects of Egypt's land reclamation policies has been the development of desert lands, especially in the so-called New Valley area of the Western Desert which links a number of oasis depressions, most notably those at Bahriya, Farafra, Abu Munqar, Dakhla and Kharga. Plans have envisaged water supplies coming from a variety of sources such as from deep artesian wells, desalinated sea water and, most recently, from water draining from the Nile via a canal into a new artificial lake in the Tuska Depression (Figure 19.2). This canal is partly built[24] and has the extra role of flood protection by draining Lake Nasser if its level gets too high.

The principal criticism of the desert reclamation schemes proposed by the national agencies, General Organization for Desert Rehabilitation (GODR) and its predecessor, the Executive Agency for Desert Projects (EADP), has been the vast discrepancies between the sizes of the proposed areas for reclamation and those which have actually materialized. At various times since 1960 up to three million feddans have been proposed for reclamation from the desert, but by 1985 the actual amount reclaimed was only about 200,000 feddans, less that 7 per cent of the maximum claimed potential. The principal deterrent has been the high cost, especially when war with Israel was placing great stress on the national economy. In 1972 Government finances forced a temporary suspension of all desert reclamation projects linked to the completion of the High Dam. To date the most successful developments have been at the southern end of the New Valley, especially in the Khârga and Dakhla oasis areas. Over 50,000 feddans were reclaimed from these desert areas during the 1960s and 1970s, but only about half of this was still being cultivated in 1978, mainly due to unforseen problems of over-hasty planning and implementation of the land reclamation, inadequacies in drainage and water management, reduced flow of water from the deep wells, and the loss of cultivated land caused by shifting dunes.[25] By 1980 it was clear that agricultural yields hardly justified the sums invested, but there were a number of other achievements, such as the improvement of community facilities in new or modernized settlements, which helped to stem high levels of out-migration, and important lessons were learnt about the costs and technologies required for successful development in other desert areas. The potential for long term exploitation of the desert areas remains obvious, especially by utilizing drip and pivot irrigation for high quality produce, but the comparative costs of developing such areas needs to be considered in the broader context of the hydropolitics of the Nile valley (see below). Oil and gas extraction also seems to offer significant potential in these western areas. It is clear that any further large scale development of the Western Desert will have to be part of general regional, economic and resettlement strategies where the agricultural costs will be evaluated in a more general economic and social perspective. This can be demonstrated in the Sinai peninsula where, since Egypt regained a considerable amount of land from Israel following the Camp David Agreement, it has been seen to be of great political advantage to resettle the arid lands of this region, almost regardless of cost. Plans exist to resettle more than one million Egyptians in the Sinai by the end of the century, although heavy dependence upon agricultural development has been partly offset by considerable offshore oil discoveries in the Gulf of Suez.[26]

Finally, in connection with extending the cultivated area, projects to develop rainfed cultivation deserve mention. In the northwestern coastal zone from Alexandria to the Libyan frontier at Salum, 15,000 feddans were prepared for the permanent settlement of bedouins. Drought-tolerant barley species were introduced for mechanized cultivation.[27] Wind pumps were also erected and small dams built to allow some irrigation from an extension of the Maryut region project.[28] The success of this project has yet to be evaluated.

19.3.3 Intensification of agricultural production

Whilst the completion of the High Dam was a major stimulus to the horizontal expansion of agriculture through land reclamation, it also facilitated vertical expansion through modified agricultural techniques. This was achieved by improving yields through more efficient farming and by introducing new crops.

One of the most important consequences of the completion of the High Dam at Aswan was the elimination of the traditional Nile floods, which were replaced by consistent controlled levels of the Nile waters. This led to a rapid transformation of agricultural techniques which had been practised for centuries. Multiple cropping was introduced, giving a cropped area today in excess of 11 million feddans from only 6.3 million feddans of actual land. Crops could also be produced according to strict irrigation timetables. The regularity and certainty of supplies not only allowed more crops to be grown during the year, but also offered greater opportunity for introducing new crops and varieties. This of course was also facilitated by the introduction of the triennial crop rotation systems and the proliferation of cooperatives which had been a integral aspect of agrarian reform. Fertilizers were also applied in vast quantities, to offset the loss of fertile Nile silts which no longer enriched Egypt's soils, but instead remained in Lake Nasser causing long term physical problems of lake infill. The changes introduced altered the traditional balance of the area devoted to cash crops, staple crops and vegetables and fruits. The most dramatic increases in yields in the 1970s were for maize, wheat and cotton (Table 19.4).

The principal explanation for the cereals was the successful production of summer varieties under controlled conditions, which replaced the traditional natural flood (*nili*) varieties. Between 1952 and 1984 wheat yields increased from

TABLE 19.4
Yields of three main crops 1952 to 1984

Crop	Yields (tonnes/feddan)				
	1952	1970/4	1975/8	1979	1984
Wheat	0.78	1.34	1.44	1.35	1.30
Maize	0.89	1.57	1.60	1.71	1.98
Cotton	0.70	0.89	0.93	1.10	1.14

Sources: Adapted from C. Daniels, *Egypt in the 1980s: The Challenge*, Economist Intelligence Unit Special Report No 158, London, 1983.
Economist Intelligence Unit, *Country Profile – Egypt 1986*, London.

0.78 to 1.3 tonnes/feddan, whilst maize increased from 0.89 to 1.98 tonnes/feddan. The increased yields for cereals were matched by an increase in the cultivated area, but this pattern was not repeated for cotton. Between 1952 and 1981 cotton yields increased from 0.7 to 1.14 tonnes/feddan, but whereas about 20 per cent of the cropped area was devoted to cotton in 1952, this had declined to less than 10 per cent by the mid 1980s. Of greater significance was the introduction of long varieties of cotton in preference to the more traditional medium and extra long varieties, which declined to less than 10 per cent of the production by 1983. The removal of the medium variety, which gives significantly lower yields than the other two, was one of the main reasons behind the boost in yields. The cash returns however did not match the scale of the increased yields, as quality was compromised and the area allocated to extra long cotton increased again from 1984.[29] Another important agricultural change introduced in the 1970s was a significant boost in the area devoted to lucerne (alfalfa), the principal livestock fodder and an important nitrogen fixer for the soil,[30] so that by 1984 it accounted for about 30 per cent of all arable land. High value crops such as fruit and vegetables were not greatly affected by agricultural changes although they did increase in importance during the 1980s.

Among the attempts to improve irrigation efficiency was the conversion of open drainage systems to a system of underground pipes. This reduced evaporation losses and allowed water to be concentrated more easily where it was required. More than 2.5 million feddans were converted in this way in Lower Egypt during the 1970s. Another advantage of this new system was expected to be a better organized system of drainage to help reduce problems of salinity and waterlogging, especially in the delta. However an unforseen consequence was that many farmers became more carefree in their use of water and the problem actually worsened in many areas. Other changes helped to improve the efficiency of irrigation, most notably the introduction of sprinkler irrigation in parts of the delta, and small scale drip irrigation projects at various locations along the Nile valley; both systems have now been widely introduced in the Sinai peninsula. The question of drainage has remained an important aspect of agricultural planning, mainly due to a rapid rise of the water table which resulted from the more intensive irrigation. More than three million feddans of irrigation land are expected to have tile drainage systems installed by the end of 1987, which is a considerable achievement.

There is no question that agricultural production was boosted by the combined effects of agrarian reform, the reclamation of new land and by more efficient and intensive production methods. Despite these modifications however, the population was continuing to expand faster than indigenous food production and between 1970 and 1976 food imports more than doubled by weight. By the late 1970s the conclusion of armed conflict with Israel, the development of oil, gas and other minerals, and the large amounts of remittances being sent to Egypt by migrants who had departed to the petroleum states of the Gulf, allowed the Egyptian government to adopt a new approach which has since dominated agricultural policy.

19.3.4 General restructuring of the economy

One of the principal features of agricultural policy during the 1980s has been the reduction in government investments. Between 1978 and 1982 agriculture received only 8 per cent of the national budget, although additional sums were allocated specifically for land reclamation, irrigation and drainage. In the 1982–87 plan these elements together still only received about 40 per cent of the budget, whilst 25 per cent was invested in industry and 22 per cent in transport and general infrastructure.

Partly to offset the reduction in investment there has been greater concentration on the production of high value crops, for export, to pay for more basic food imports. This has led to increased production of fruits, vegetables, sugar, and livestock produce, much of this on recently reclaimed land. Egypt is currently self-sufficient in fruit and vegetables and has begun to earn considerable sums from exports. In 1984 183 tonnes of citrus fruit were exported, enough to finance the import of 1.5 million tonnes of subsidized wheat from the EEC and North America.[31] Not only have fruits and vegetables increased in volume, but there has been a growing emphasis on high value products such as bananas, which more than doubled their cultivated area during the past two decades, and strawberries which are believed to offer considerable earning potential. More than 37,000 tonnes of strawberries were exported in the 1985/86 season, generating US$110 million.[32] Important consequences of this market-based strategy are that Egypt is vulnerable to fluctuating world prices while in Egypt merchants wield far more effective power than the *fellaheen*.

The principal objective of the export oriented agricultural policy is to support the import of staple foods. Egypt produces only about 35 per cent of its wheat needs, the most important staple crop, and 80 per cent of its maize. By 1987 Egypt was importing more than half of its food requirements and this figure is increasing rapidly. In only two years (1982 to 1984) Egypt's food bill increased from US$3500 million to US$4100. In the mid 1980s, estimates suggest that food production is increasing at a rate of 3.5 per cent, whilst demand is increasing at nearly 5 per cent, a rate greater than can be expected from population increase alone. This can be explained by a controversial policy of subsidies. Food subsidies, mainly for wheat, were intoduced to offer consumer protection against fluctuating world prices, but they have been set too high resulting in considerable wastage of wheat products. In 1960 Egyptians consumed only 80kg of wheat per capita, but by 1984 this had risen to 184 kg,[33] making Egyptians the world's largest consumers per head of population. Wheat prices have been so low that some farmers have found it profitable to feed bread to livestock. The obvious solution of reducing food subsidies has so far proved impossible, as whenever the government has tried to reduce the subsidy, riots have occurred, sometimes resulting in bloodshed.

Although the subsidies issue remains, there has been a slight shift in agricultural policy. Apart from the new land reclamation schemes already described, there have been more cash incentives for farmers to grow rice, wheat and maize, and renewed attempts to reduce the problems of inadequate drainage.

Some concern has also been expressed at the high costs of keeping livestock and, in common with some other Islamic countries, the government has declared meatless days but the problem of excessive number of draught animals remains unsolved.

The continued rapid population growth, the need to resettle people in desert areas to relieve some of the desperate problems of urbanization, and the utilization of most of the extra water generated by the building of the High Dam has made a review of the potential utilization of the Nile waters necessary once more. It is clear that Egypt has little more opportunity for increasing water supplies within its own boundaries and this has resulted in new debate about the international hydropolitics of the whole Nile basin.

19.4 Hydropolitical issues in the Nile valley

There have been a number of major studies of the agricultural potential of the Nile valley.[34] A number of important areas have been identified for irrigation development, for example in the delta, the basins of Upper Egypt and northern Sudan, the vast fertile clay plains of the Gezira between the Blue and the White Niles, and huge irrigation schemes using water previously lost to evaporation in places like the *sudd* marshes of southern Sudan. The exploitation of some of these potential resources is clearly dependent upon international cooperation, heavy capital investment, the availability of skilled manpower and appropriate agricultural methods. Of all the limitations to the development of these Nile lands the most important has been the political division between Egypt and Sudan. Since Classical times Egyptians have been aware that the Nile waters originate in territories beyond their immediate control, and the search for the source of the Nile stimulated many expeditions. The construction of the High Dam was partly an attempt to bring security to Egypt by keeping a large reservoir of water within its own territory, even if it was located in one of the driest areas of the world with massive evaporation losses. Nevertheless, the construction of the dam seemed to have been justified in 1984 and 1985 when the Nile floods were amongst the lowest ever recorded. Vast irrigated areas in the Sudan suffered from major water deficiencies whilst those in Egypt were hardly affected.[35]

The question still remains however as to how much mutual cooperation could exist with Sudan. Egypt has demonstrated its willingness to cooperate on many occasions. Formal agreements on the allocation of Nile waters were accepted in 1929 and 1959. In 1937 the Jebel Aulia dam was completed south of Khartoum to hold back the waters of the White Nile so that they could be released for Egyptian use during the period of minimal flow – the so-called 'timely season'. More recently the Egyptians agreed to help finance the Jonglei Canal in southern Sudan, the main objective of which was to save up to 4,000 million cu m of water currently lost to evaporation and which could be shared between the two states. The canal was two-thirds completed when civil war in Sudan forced its abandonment, an act which clearly demonstrated Egypt's vulnerability to political events beyond its domain.

It is hardly surprising that Egypt has tried to concentrate its agricultural expansion projects within its own boundaries, but the time has come when limits are beginning to be reached. All new land reclaimed in Egypt is now more marginal and costs more to irrigate; for example in the desert reclamation schemes in the New Valley, irrigation costs can be up to four times greater than areas supplied directly from the Nile. Discussion continues in Egypt as to whether a better policy would be to utilize some of its finances and manpower to help expand irrigation on some of the extensive fertile clay lands in Sudan under some form of cooperation.[36] Economists and hydrologists have demonstrated the sense behind such an arrangement. The principal problems revolve around issues such as Sudanese immigration policy, improved communications between the two states and the willingness of Egypt to invest in territories beyond its jurisdiction. A key factor will be the political relationship between the two states. Until 1984 Egypt and Sudan had maintained harmonious relations for several decades, but the overthrow of President Nimeri in Khartoum followed by new cooperation between Sudan and Libya soured this relationship. According to some authorities, Egypt and Sudan could also be competing for a greater share of Nile waters by the 1990s.

It appears that cooperation must eventually succeed between Egypt and Sudan over the sharing of water, land, manpower and investment, but until long term political confidence is restored few significant developments are likely to occur in this direction.

19.5 Conclusion

It seems reasonable to conclude that increased agricultural production in Egypt should be sufficient to help maintain present rural standards of living, at least until the end of the century. Nevertheless, Egypt's long term prosperity will depend upon the successful development of all sectors of the national economy. It has been shown that agricultural policies have been integrated more fully into general economic strategies and this is likely to continue, especially as further agricultural expansion will prove more costly unless firm and lasting agreements can be formed with Sudan about the use of Nile waters. Further land reclamation will continue, but water availability, especially from the Nile, will limit its extent. Egyptians lament the loss of water from the Jonglei Canal project, but it is possible to question how valuable that water might have been. The project would have secured an extra 2 billion cu m/annum of water for Egypt, yet in 1985 the Under Secretary at the Ministry of Agriculture in Cairo admitted that Egypt wasted three times this volume every year, mainly from unlined and poorly maintained canals, and because of lack of commitment to reusing irrigation water for second or third irrigation. This inefficiency does not take account of the vast evaporation losses from Lake Nasser. Concentrated efforts on trying to reduce this wastage could still offer Egypt considerable possibilities for land reclamation using Nile waters.

Water availability alone is, of course, not the only limitation to an extension of agricultural land. Rapid population growth and the desire to live in urban areas is eating up considerable amounts of prime agricultural land. Settlements expand most readily along the banks of the river, forcing agricultural land away from the river into areas where the soil is normally less fertile, where irrigation costs are greater, and where there are increased opportunities for water wastage. It is hardly surprising that resettlement is an integral part of population planning, and new towns of varying sizes have been constructed in the last decade, normally away from the immediate vicinity of the Nile. Some, such as Sadat City, King Khaled City and Tenth of Ramadan City,[37] are already housing several hundred thousand people, offering some relief to the older settlements. Other smaller settlements are being constructed in areas with industrial or agricultural potential away from the Nile valley and delta, although some manufacturing industries have found that relocation incentives to move to such areas have been outweighed by considerable amounts of unnecessary bureaucracy. One leading catalyst in opening up new settlements has been the exploitation of minerals, most of which are found in the more remote parts of Egypt. Oil, for example, has been found in large quantities in the sea off northern Sinai[38] and the north of the Qattara Depression,[39] although its real value to Egypt is at the mercy of international oil pricing agreements. However, it will help to offset some of Egypt's internal energy supply problems. Electricity produced from turbines at the High Dam is no longer sufficient to cope with internal demand and considerable effort has gone into identifying potential alternative energy sources from coal, oil, gas and nuclear power stations,[40] some of which are already being exploited.

Egypt's economy is relying more and more on foreign trading partners and this is proving to be its Achilles heel. In 1984 Egypt's four principal earners of foreign currency were petroleum exports, remittances from workers in other states, especially in the Gulf and Iraq, revenue generated by the Suez Canal, and tourism. In recent years all of these have suffered setbacks. The fall in world oil prices from 1985 has had a direct impact of reducing revenue, but it has also meant that many Egyptians are no longer able to secure high paying jobs overseas, leading to a fall in remittances and the return of many migrants. Suez Canal dues and revenues from the SUMED pipeline have fallen as world oil shipments have declined. Tourism has also suffered setbacks since 1986. Despite considerable investments in the industry by private and government sources, European and American tourists have been dissuaded from visiting the country due to the Egyptian reaction to the hijacking of an Italian cruise liner, major riots in one of the main tourist areas of Cairo, and perceived dangers of visiting 'Arab' countries following the American bombing of Libya.

Considering all of Egypt's economic and social problems however, one factor above all others is vital for securing the future prosperity of the country. Its population growth will continue to stifle economic developments if growth rates continue at their present levels. The main efforts must continue to focus on reducing birth rates, a remedy which appears to have had limited successes over the past two decades, especially in urban areas.

References

1. *Middle East and North Africa Yearbook: 1986*, Europa, 1986, 361.
2. P. O'Brien, 'The long-term growth of agricultural production in Egypt: 1821–1962', in *Political and Social Change in Modern Egypt* (Ed P.M. Holt), Oxford University Press, London, 1968, 174.
3. Economist Intelligence Unit, *Country Profile - Egypt: 1986*, London, 1986, 7.
4. L.D. Stamp, *The Geography of Life and Death*, Fontana, London, 1964, 35.
5. J.M. May and I.S. Jarcho, *The Ecology of Malnutrition in the Far and Near East*, Hafner, New York, 1961, 651.
6. *Middle East and North Africa Yearbook: 1986*, Europa, 1986, 363–64.
7. *World Development Report 1985*, Oxford University Press, London, 1985, Table 24.
8. (a) H.M. Ammar, *Growing up in an Egyptian Village*, Routledge and Kegan Paul, London, 1954.
   (b) H.H. Ayrout, *The Egyptian Peasant*, Beacon Press, Boston, 1963.
   (c) W. S. Blackman, *The Fellahin of Upper Egypt: their Religious, Social and Industrial Life*, New Impression, Frank Cass, 1971 (first published 1927).
   (d) S. Radwan and E. Lee, *Agrarian Change in Egypt: An Anatomy of Rural Poverty*, Croom Helm, London, 1986.
9. *World Development Report 1985*, Oxford University Press, London, 1985, Table 24.
10. *Ibid*, Table 25.
11. *Ibid*, Table 20.
12. Economist Intelligence Unit, *Country Profile - Egypt: 1986*, London, 1986, 7.
13. A. Granott, *Agrarian Reform and the Record of Israel*, Eyre and Spottiswoode, London, 1956, 208.
14. E. Eshag and M. A. Kamal, 'Agrarian reform in the United Arab Republic', *Bull. Oxf. Univ. Inst. Statist.* **30**, (1968), 81.
15. G.S. Saab, *The Egyptian Agrarian Reform 1952-62*, Oxford University Press, London, 1967, 192.
16. E.Eshag and M. A. Kamal, 'Agrarian reform in the United Arab Republic', *Bull. Oxf. Univ. Inst. Statist.* **30**, (1968), 87–93.
17. R. I. Lawless, 'The Agricultural Sector in development policy', in *Agricultural Development in the Middle East* (Eds P. Beaumont and K. McLachlan), Wiley, Chichester, 1985, 118.
18. M. Adamowicz, 'Transformation of agricultural structure in the United Arab Republic', *Africana Bulletin*, **43**, University of Warsaw, (1970), 76.
19. *Middle East and North Africa Yearbook: 1986*, Europa, 1986, 365.
20. Compiled from the following:
    (a) *The Guardian*, Manchester, 14 April 1966, 14–15.
    (b) H. Hopkins, *Egypt, The Crucible*, Secker and Warburg, London, 1969, 129–139.
    (c) Ministry of Land Reclamation: *Land Reclamation Development in the Arab Republic of Egypt*, Cairo, 1972, 18–36.
    (d) R. R. Platt and M. B. Hefny, *Egypt: a Compendium*, American Geographical Society, New York, 1958, 17, 61–78.
    (e) Supplement on Egypt, *The Times*, London, 24 July 1969, i–xx.
    (f) Economist Intelligence Unit, *Annual Supplement on Egypt*, London, 1970, 9.
    (g) Economist Intelligence Unit, *Annual Supplement on Egypt*, London, 1978, 9.
    (h) Economist Intelligence Unit, *Annual Supplement on Egypt*, London, 1982, 9.
    (i) Ministry of Information: *Agriculture in the Arab Republic of Egypt*, Cairo, 1983, 3–4.

21. A. B. Mountjoy, 'Egypt cultivates her deserts', *Geogr Mag.* **44**, 241–50 (1972).
22. Economist Intelligence Unit, *Annual Supplement on Egypt*, London, 1982, 10.
23. C. Daniels, *Egypt in the 1980s: the Challenge*, Economist Intelligence Unit Special Report No 158, London, 1983, 127.
24. Ministry of Information: *Agriculture in the Arab Republic of Egypt*, Cairo, 1983, 23.
25. G. Meyer, 'Effects of the "New Valley" project upon the development of the Egyptian oases', *Applied Geography and Development*, **15**, (1980), 323.
26. *Financial Times*, London, 20 March 1986.
27. H. Hopkins, *Egypt, the Crucible*, Secker and Warburg, London, 1969, 323.
28. Ministry of Information: *Agriculture in the Arab Republic of Egypt*, Cairo, 1983, 4.
29. Economist Intelligence Unit, *Country Profile - Egypt: 1986*, London, 1986, 19.
30. D.N. Wilber (Ed), *The United Arab Republic, its People, its Society, its Culture*, Human Relations Area Files, New Haven, 1969, 310.
31. *Financial Times*, London, 5 June 1985, vi.
32. Economist Intelligence Unit, *Country Report - Egypt: 1986* London, 1986, 21.
33. *Financial Times*, London, 5 June 1985, vi.
34. (a) J. Waterbury, *The Hydropolitics of the Nile Valley*, Syracuse University Press, New York, 1979.
    (b) D. Whittington and K. E. Haynes, 'Nile water for whom? Emerging conflicts in water allocation for agricultural expansion in Egypt and Sudan', in *Agricultural Development in the Middle East* (Eds P. Beaumont and K. McLachlan), Wiley, Chichester, 1985, 125–49.
35. *Financial Times*, London, 5 June 1985, vi.
36. D. Whittington and K.E. Haynes, 'Nile Water for Whom? Emerging Conflicts in Water Allocation for Agricultural Expansion in Egypt and Sudan', in *Agricultural Development in the Middle East* (Eds P. Beaumont and K. McLachlan), Wiley, Chichester, 1985, 142–48.
37. A. D. C. Hyland, A. G. Tipple and N. Wilkinson (Eds.), *Housing in Egypt*, University of Newcastle upon Tyne, School of Architecture House Course Working Paper Series No 1, Newcastle upon Tyne, 1984.
38. *Financial Times*, London, 20 March, 1986.
39. *Ibid*, 25 March, 1986.
40. *Ibid*, 22 August, 1986.

CHAPTER 20

Libya: Oil Revenues and Revolution

20.1 Introduction

With a total area of 1,759,500 sq km Libya is among the world's 15 largest states, yet its population (3,637,488 at the 1984 census) is among the smallest. The underlying cause of this sparse population is aridity. Nearly 95 per cent of the country receives less than 100 mm of rainfall per annum, and even in the well watered areas such as the Gebel al Akhdar in Cyrenaica and the Gefara plains in Tripolitania, where rainfall may be as high as 350 to 500 mm per annum, serious drought can occur. Throughout the south practically no rain falls and there are no perennial surface streams. Groundwater however is relatively plentiful in Libya and soils are potentially fertile over large areas including some parts of the desert. Altogether probably no more than 5 to 10 per cent of the land can be put to economic use.[1] and less than 2 per cent is suitable for rainfed cultivation[2].

Population distribution reflects the scarcity of cultivable land in Libya and the rather limited range of alternative economic opportunities. Most of the people are concentrated in a northern coastal belt, 92 per cent of them in coastal areas around Tripoli and Benghazi on less than 10 per cent of the land. By contrast, the oases of the Saharan zone support no more than 150,000 inhabitants at average densities below one person per sq km.[3] The proportion of nomads and semi-nomads among the rural population has diminished in recent years as a result of employment in the cities and oil industry and government efforts to bring about sedentarization. In 1964 17 per cent of the population was classified as nomadic or semi-nomadic, but it is doubtful whether the proportion now exceeds one per cent.

Libya was one of the poorest countries in the world when granted independence in 1951, with an average per capita income of less than 50 US dollars a year and a total budget of about 25 million US dollars.[4] Over 70 per cent of the indigenous population were farmers or herders, and in Cyrenaica half were nomadic. Most were illiterate. With such a poor, scattered internal market, local manufacturing industry had hardly developed at all. There were almost no known mineral resources. The chief exports were esparto grass, olive oil and scrap metal collected from the desert battlefields of the Second World War. The common view that Libya was 'chained by the harsh environment to never-ending poverty'.[5] Economic problems were compounded by internal political rivalries between the three federal provinces of Tripolitania, Cyrenaica and Fezzan, and a top-heavy administration. The Libyan economy was not only an extreme example of economic backwardness, but an example of economic exhaustion made worse by destruction and disruption during the Second World War. Not only was property and infrastructure damaged, but millions of mines had been planted by both sides which hindered agricultural activity and transportation, and endangered livestock and people.[6]

Libya's economic problems in 1951 were also in part the result of the Italian occupation. The Italian conquest began in 1912, but Libyan resistance was fierce and it was not until 1929 that the whole country was under Italian control. Almost half the indigenous population had been killed or fled the country.[7] During the short period of colonial rule some superficially impressive developments were begun including road building, town planning and land reclamation. Altogether they invested the equivalent of 150 million pre-war US dollars in Libya.[8] Unfortunately their efforts were largely for the benefit of Italian colonists, who numbered 110,000 by 1940, while the welfare of the Libyans was neglected.[9] The Italians were strongly influenced by the fact that in Roman times the coastal areas of Libya apparently supported two or three times more people than in the 1930s and magnificent cities such as Leptis Magna and Sabratha flourished. The exodus of many Italian settlers and Jews after the Second World War left the country with a critical shortage of personnel in key positions.

Until 1959 Libya received massive international aid, partly in return for the use of military bases by Britain and the United States. Possibly the only optimists in Libya in the 1950s were the oil companies. Intensive exploration began in 1955 after the passing of the Libyan Petroleum Law, whose generous terms appealed to the oil companies. The first oil strike occurred in 1958 at Atshan in the Fezzan, but the first major discovery was at Zelten in the Sirte Basin in 1959. Libyan oil was first exported in September 1961, and a favourable balance of payments was registered for the first time in 1963. Few areas have ever been explored so intensively; by 1961 the whole northern half of Libya and much of the coastal waters were under concession. Companies were obliged to relinquish one quarter of their blocks after five years and another quarter after eight. These blocks were awarded again and success by one company sometimes followed failure by another. Thus exploration was rapid, and a mosaic of concessions emerged quite unlike the huge blocks familiar in Southwest Asia. Many difficulties were encountered in exploration; thousands of mines had to be cleared, distances were great, and vast expanses of sand and rocky desert made movement of heavy equipment difficult. Nor was there any guarantee of success. The Shell Company for example drilled 70 wells over a period of 20 years at a cost of over US$100 million without success.[10] But there were powerful incentives for the oil companies in their search for Libyan oil, the most important being proximity to the world's fastest growing oilmarket in Western Europe. By the early 1970s half of all international oil imports were to Western Europe, and Libya's freight advantage to these markets had greatly increased after the closure of the Suez Canal in June 1967, which led to Libyan oil production being boosted by 50 per cent. This freight advantage is illustrated by the 1971 cost of shipping one barrel of crude oil from the Gulf to Rotterdam (US$9.18) compared with the cost from Libya (US$2.95). Libyan crude also proved attractive to the customers of Western Europe since it has low

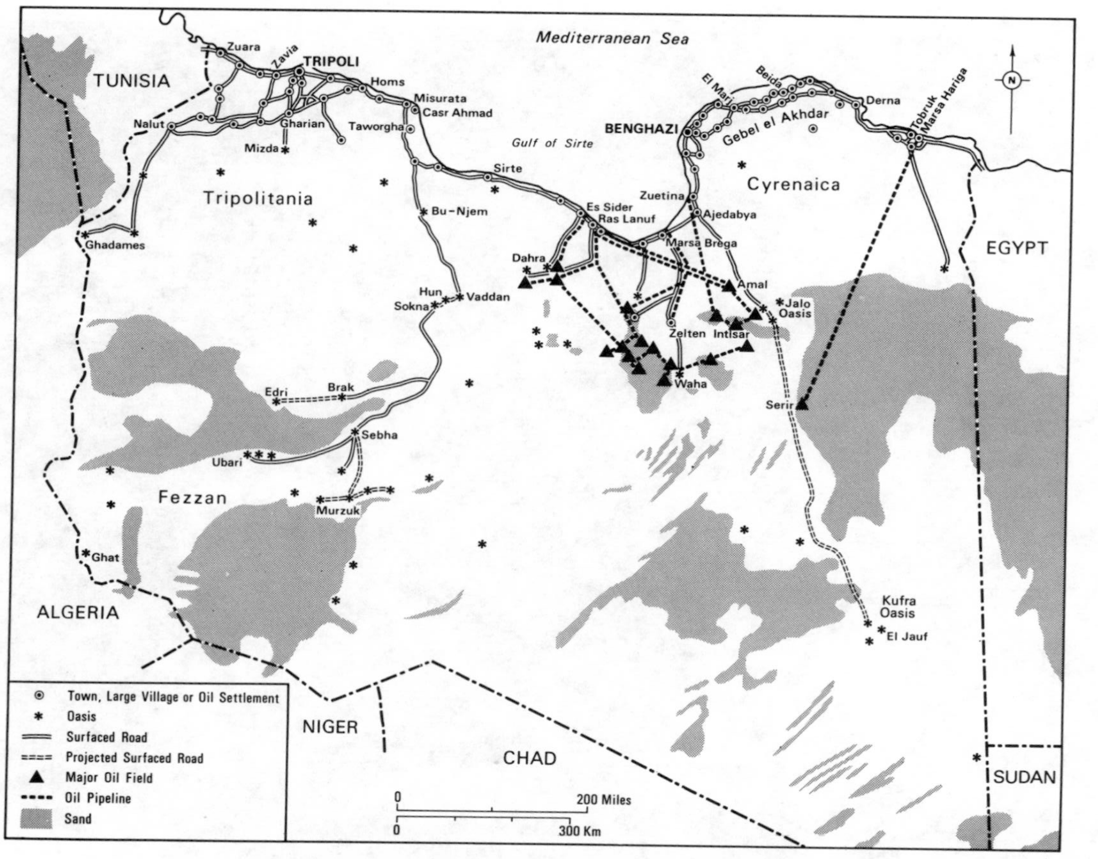

Figure 20.1 Libya: settlement and communications

An exploration well drilled in the Libyan desert by Esso (an Esso photograph)

sulphur content (one per cent compared with 3.5 per cent from the Gulf) and a relatively low residue (29 per cent compared with 36 per cent from the Gulf).

Apart from proximity to Europe, Libya's oilfields enjoy the advantage of being relatively near the Mediterranean coast where the Gulf of Sirte penetrates far inland (Figure 20.1). Thus most pipelines are relatively short, a notable exception being from Serir to Tobruk (560 km). The rivalry of the nine major groups of companies and their desire to begin exports as quickly as possible explains the existence of so many pipelines, not all of which take the shortest route to the sea. The complex of parallel and interlacing pipelines might look very different if based on subsequent knowledge of oilfield locations. Each of the five main oil terminals was originally established by groups of producers for their own use.

20.2 Urban growth and industrial development

Libya's urban population increased steadily from the early years of the twentieth century, but only at about the same rate as the population as a whole. When oil revenue began to stimulate economic growth in the 1960s urban growth rates accelerated to more than double the rate of natural increase, rising to over 10 per cent in some years. There were several reasons for the high rate of urban population increase. First, natural rates of increase persisted at high levels of around 4 per cent per annum. Secondly, large numbers of rural migrants were drawn to the towns to fill thousands of new jobs created by the oil boom. Thus at the end of the first decade of development at least one third of the labour force consisted of individuals who had migrated within Libya.[12] Thirdly, Libya attracted a substantial proportion of its labour force, including skilled workers, from abroad, and these tended to concentrate in the towns. The highest number of expatriates was probably reached in the late 1970s; over 530,000 were reported in 1975, almost one in five of the population of Libya. Expatriate labour made up at least one third of the labour force and in certain sectors such as construction (77.5 per cent) and manufacturing (41.9 per cent) it was even higher.[13] The number had declined to 416,000 by 1985 (see Table 5.5) as lower oil prices took their toll, but even so this represents a considerable increment to the indigenous urban population of Libya.

As a result of these processes, Libya's urban population (defined as those living in towns of over 20,000 inhabitants) grew from 29 per cent in 1965 to 63 per cent in 1984 (Table 6.2) If the smallest urban centres are counted (over 5,000 inhabitants) the proportion grew from 40 per cent in 1964 to 81 per cent in 1980.[14] Apart from rapid growth, the concentration of population in Tripoli and Benghazi should be noted. In 1987 well over two-thirds of Libya's urban dwellers were in these two cities. During the first decade of the oil industry, Tripoli and Benghazi grew largely as a result of the mushrooming of an enormous tertiary sector in response to the needs of the industry. As the chief ports and administrative centres of the country they could hardly escape these benefits. Transport, catering, retailing, construction, light engineering, and government services all created opportunities for employment, and migrants were drawn to the towns by the usual myths of fortunes to be made. In the early years ugly shanty accommodation sprang up on the outskirts, but these had been largely replaced by modern housing estates and high rise blocks by the end of the 1970s. The secondary sector is now well established in these towns, and in spite of government plans to locate industry elsewhere whenever feasible, many of the latest developments are in or near Tripoli and Benghazi. Misurata has emerged as a third important focus of growth, particularly with the construction of a major iron and steelworks and associated port. The population of Misurata has increased almost tenfold in 20 years. According to some estimates Libya's urban population will have tripled or quadrupled by the end of the century, thus creating a continuing need for careful urban planning, the provision of adequate services, and the creation of employment opportunities.

Town planning in Libya in the 1980s compares very favourably with town planning elsewhere in the Middle East. The physical master plans for the urban areas are well-researched and detailed. Moreover the whole of Libya is covered by a series of comprehensive regional plans which provide a national development context for the town plans. One visible result of this integrated approach to physical planning is the emergence of one of Africa's most extensive networks of paved roads (Figure 20.1). This road network is complemented by the provision of electricity, fresh water, sewerage and telecommunications networks which are effectively complete right across the country. Perhaps inevitably many mistakes have been made in the implementation of urban and regional planning strategies in Libya, but it is worth noting that a sound infrastructure is already in place, and the results of planning policies are clearly visible on the ground. Indeed in some of the most densely populated parts, the geographical face of Libya is in a process of total transformation.

During the 1960s relatively short-term plans were prepared by foreign consultants for all the main towns in Libya. Although these plans had some excellent features they were given no regional context, and they tended to underestimate future urban expansion. In 1973 therefore the revolutionary government asked Italconsult to prepare detailed profiles for 36 major towns and associated proposals for settlement planning at regional and national levels to the year 2000.

The consultants made a number of radical recommendations chief of which was the need to link together Libya's eastern and western urban sub-systems by developing the coast of the Gulf of Sirte. A large new national capital would be built near Sirte. In addition, industry should be decentralized, and new growth centres should be identified for manufacturing, commerce and administration. The eventual relocation of the populations of 159 villages lacking resources for future development was also proposed.[15] By no means all the details of Italconsult's plan have been followed, but the development of the Gulf of Sirte coast and decentralization are being energetically implemented. No new capital is being built but new industrial activities and extensive irrigated agriculture are planned along the Sirte coast the latter depending largely on water from the Great Man-made River (GMR), discussed below. Among other measures, decentralization is to be tackled by promoting the rapid growth of Sebha in the southwest and the creation of a new town (60,000 people) near Kufra in the southeast. How far these schemes will succeed in rectifying huge regional disparities remains to be seen, but optimistic expectations should be tempered by bearing in mind the combination of geographical advantages favouring the Tripoli and Benghazi regions.

The creation of a large urban-based industrial sector is one of the development priorities of the revolutionary government of Libya. The aim is to create jobs, substitute local production for imported goods, and above all to lay the foundations for industrial exports as an alternative to dependence on oil exports. By the mid-1980s the results of massive investment in the industrial sector were patchy. Nearly 15 per cent of all capital investment from 1970-1983 was in industry; nearly 150 new industrial units were in operation or under construction,[16] and 7.4 per cent of the labour force was in the industrial sector in 1984, which contributed 4.8 per cent to GDP in the same year.[17] At the same time much production was costly and inefficient, the latter no doubt as a result of poor management, the interference of 'peoples' committees', and the periodic absence of workers on military duties.

Industry was virtually non-existent in Libya before oil revenues began to accrue. The first economic plan (1963–68) sought to create conditions in which the private sector would develop manufacturing industry, including infrastructural projects and financial incentives. A number of small industries were established as a result, chiefly in Tripoli and Behghazi. Following the revolution in 1969 the private sector flourished for a few years, but gradually the new regime introduced its distinctive brand of Arab socialism emphasizing self-reliance and self-sufficiency in food, an expanding public sector and a shrinking private sector. During the 1970s a large number of factories were established and operated by the public sector, often without much thought being given to feasibility.[18] In 1977 factory workers throughout Libya assumed control of their plants, inspired by the ideas of Colonel Qadhafi's 'Green Book'. Although their actions were apparently spontaneous, the takeover of the factories was probably carefully prearranged.[19] By 1979 the private industrial sector had been eliminated in Libya, along with foreign trade and wholesale retail trading.

According to the ambitious 1981–85 Transformation Plan, Libyan industry was to have been massively expanded, with heavy industrial activity strongly emphasized. Iron and steel manufacturing, aluminium smelting, petrochemical manufacture and oil refining all featured prominently, together with a host of smaller industrial projects. Since 1981 the Libyan economy has been badly hit by declining oil revenues. In their best year ever, oil revenues reached US$22,000 million in 1980; by 1986 they had fallen to US$5,000 million.[20] As a result, austerity measures had to be adopted in Libya, and many major projects have been shelved, including industrial projects such as an aluminium complex at Zuara, a fertilizer complex at Sirte, and a petrochemical plant at Ras Lanuf. The huge iron and steel works at Misurata was however completed in 1987, together with its associated port of Casr Ahmad, with a planned capacity of 1,200 million tonnes of steel in the first stage. Plans to bring Libyan iron ore from Wadi Shatti have been postponed because of the cost of the 900 km railway line which would be necessary. Instead, iron ore will be imported via Casr Ahmad. The wisdom of embarking on such large scale industrial projects has been queried by economists, especially at a time of world recession, and in view of Libys's limited internal market. It is also worth noting that the drive towards industrialization in Libya has fuelled the demand for imported goods, largely associated with the construction and equipment of the new plants.

20.3 Agriculture

Previous paragraphs have discussed the remarkable transformation of Libya's urban and industrial scene in the post-revolutionary era since 1969. Perhaps no other Middle Eastern country has experienced such large surpluses of capital combined with such a desire to change the social, political and economic life of the state, as Qadhafi's Libya. The conviction that almost anything can be achieved by capital investment and revolutionary zeal led to some serious mistakes in the agricultural sector, though there have been successes as well.

The agricultural potential of Libya is limited by environmental constraints far more than by mismanagement of the land, or by the legacy of colonialism. Rainfed cultivation is confined to parts of the coastal zone receiving over 200 mm of rainfall per annum, the most reliable areas are confined to the 300 mm isohyet, or 0.7 per cent of the land.[21] Certain parts of the coastal plain are suitable for irrigation and there are groundwater resources available so that some intensive crop cultivation is possible. Soils on the other hand are not good, being sandy, low in organic content, and easily desiccated by high temperatures and winds. Thus yields of cereals and other crops are not high. Outside the 200 mm isohyet small areas of irrigated cultivation are skilfully farmed in a number of scattered oases (Figure 20.1). Besides crop cultivation, Libya was traditionally noted for livestock reared in enormous numbers by nomadic tribes on the fringes of the semi-arid zone. The products of camels, sheep and goats featured prominently in local markets and contributed to Libya's tiny export income. Arable and pastoral farming suffered greatly in the early years of the oil industry in Libya. Agricultural production fell, largely as a result of the movement of farmers to towns in search of higher incomes, and the assumption by government that the country could afford to import its foodstuffs. The food-import bill rose steadily through the 1960s, and even included olive oil which was once produced locally. Although neglect of the agricultural sector was already being tackled before 1969, the revolutionary regime increased allocations to agriculture substantially, particularly for the extensionof irrigation. Their fundamental objective was to make Libya self-sufficient in all basic foodstuffs.

To begin with, efforts were concentrated upon intensification of the old established irrigated farming regions such as the Gefara plain, but by the mid-1970s it was clear that this was having a disastrous effect on the water table.

In places up to 20 km from the coast, the underground water level was falling at 5 m a year, while projected demands for water in future years were very high. The danger was reluctantly recognized, and steps were taken to manage Gefara water more carefully.[22] At the same time the government turned to a large number of new regions and projects in which to develop the ideal of food self-sufficiency. One obvious problem in their implementation was shortage of labour, and foreign workers were heavily engaged in a number of agricultural projects, and the secondary objective of attracting Libyans back to the land from the cities was largely unsuccessful. Since the early 1970s approximately 1.3 million ha of land have been improved for food production in Libya. Probably less than a quarter of this was reclaimed for irrigation, the remaining three quarters being dry farming, pasture and forestry projects. Among the projects there have been both successes and failures, with few schemes coming up to the expectations of government planners.

The Taworgha project near Misurata is an example of a successful irrigation project, where some 300 new farms have been created on 3,000 ha of irrigated land. A central village offers all basic services to the farming community. The Hadaba el Khadra project uses purified sewage water from Tripoli to irrigate 600 ha supporting 100 farms producing fruit and vegetables for the urban market. The Gebel al Akhdar project on the other hand is a successful dry farming and range management project extending over 115,000 ha. Here an integrated regional scheme has improved production on 2,000 farms and helped semi-nomadic tribal groups in their transition to modern cereal farming and orchards. Perhaps the best known of Libya's land reclamation schemes both on account of its scale and location, is the Kufra project. It is also one of the most costly and disappointing adventures so far. The development of the Kufra oasis region was initiated in 1968 by the Occidental Oil Company following their discovery of huge groundwater resources in the area. In 1972 the Libyan government took over the project. The soils of Kufra are sandy but productive when fertilized and irrigated. One hundred wells were dug each of which is the centre of a circle one km in diameter, irrigated by an electronically-controlled rotating sprinkler. Initially 10,000 ha were reclaimed one which fodder crops were grown for sheep, the idea being to supply Libya's coastal cities with mutton. Unfortunately the price of producing and transporting the meat was too high, and in 1977 the Kufra project converted to grain production, but the price of the grain was between 10 and 20 times the world price at the time.[23] The Kufra project was never expanded any further, its associated farm settlement scheme was abandoned, and very few Libyans were ever employed on the project.

The Great Man-Made River Project was inaugurated by Colonel Qadhafi, with his strong personal support, in August 1984. The first phase of this colossal project is to pipe water 1,900 km from wells in the Serir and Tazerbo oases for distribution to the Benghazi and Sirte regions (Figure 20.2). Eventually 180,000 to 200,000 ha will be irrigated, supporting 37,000 new farms for the production of cereals and livestock products. A second phase envisages piping water from the Kufra oasis northwards, and a third phase, a western pipeline drawing on wells in

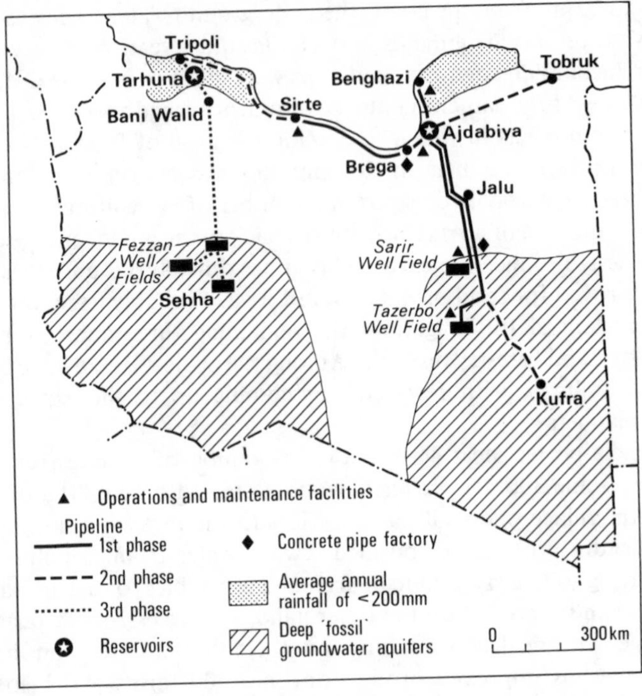

Figure 20.2 Libya's Great Man-Made River Project

the Sebha area. Construction of sections of the four-metre diameter pipeline began in 1986; the first phase of what is one of the world's largest engineering schemes should be complete by 1989. The GMR has been given top priority by the Libyan government, but misgivings have been widely expressed about it. Nobody knows how much 'fossil' groundwater (5000–30,000 years old) lies beneath the southern deserts, though it is certainly enough to sustain the scheme for 50 years. The initial cost of the first phase of the project is US$3500 million, and although the water is forecast to be relatively cheap (18 cents per cu m), the associated costs of land reclamation appear to have been overlooked. These have been put at US$5000–10,000 *per hectare* in the Sirte region.[25] Another fear is that growing urban populations will eventually acquire a disproportionate share of GMR waters, at the expense of agricultural production. According to some estimates Libya's urban population could have quadrupled by the end of the century. As mentioned in Chapter 6, over 5 per cent of the best-watered agricultural land in Libya has already been lost to urban expansion, and the process is clearly not finished yet.[26]

20.4 Conclusion

It is scarcely an exaggeration to say that in two decades the geographical face of Libya has been transformed, while in the next twenty years it could change beyond recognition. Nine-tenths of the land surface, it is true, remains unaffected by the activities of man, but in the populated regions, dramatic change has been the keynote. With the disappearance of nomadism, the rapid growth of the urban population, and schemes for the resettlement of cultivators, the population is undergoing gradual restructuring and redistribution. With the extension of irrigation, tree planting schemes and range management, landscapes are also undergoing visual and functional change. The GMR will add greatly to changes in landscape along the coast of the Gulf of Sirte. Not all visual change has been for the better; some ancient urban centres and historic buildings have been lost in the rush to modernize. The need to create national unity in such a vast country has been met by a network of roads, airports, and telecommunications and broadcasting services. In future there could be a new focus of national unity around the town of Sirte as the GMR project has an effect. Ironically however, the drive to self-sufficiency and true independence has created an ever-increasing dependence upon expatriate know-how and labour which will continue for some time

Libyan oil has given the country considerable political influence, assisted by the favourable geopolitical position of Libya in relation to Europe, the Mediterranean Sea, and the Arab world. Libya has exerted political pressure in countries such as Uganda, Chad, Malta and Sudan, and a number of other groups, notably the Palestinians, are in receipt of Libyan funds. By the mid-1980s however many of Colonel Qadhafi's foreign adventures have cost Libya a great deal of money, and left the country somewhat isolated from the international community.

References

1. International Bank for Reconstruction and Development, *The Economic Development of Libya*, Johns Hopkins Press, Baltimore, 1960, 29.
2. J.A. Allan, *Libya: the Experience of Oil*, Croom-Helm, London and Westview Press,
3. R.I. Lawless and S. Kezeiri, 'Spatial aspects of population change in Libya', *Méditerannée* **4**, 81–86, (1983).
4. R.S. Harrison 'Libya', *Focus* **17**, 1, (November 1966).
5. J.D. Farrell, 'Libya strikes it rich', *Africa Report*, **12**, 8, (April 1967).
6. S. Ghanem, 'The Libyan economy before independence', in *Social and Economic Development of Libya*, (Eds E.G.H. Joffe and K.S. McLachlan), Menas Press, Wisbech, 1982, 148–49.
7. *Ibid*, 144.
8. International Bank for Reconstruction and Development, *The Economic Development of Libya*, Johns Hopkins Press, Baltimore, 1960, 27.
9. W.B. Fisher, 'Problems of modern Libya', *Geogr. J.* **119**, 183–99, (1953).
10. 'Crisis of confidence in Libya', *Petrol. Press Service*, **39**, 282, (August 1972).
11. 'Tough bargaining in Tripoli', *Petrol. Press Service*, **38**, 122 (April 1971).
12. L.L. Bean, 'The labour force and urbanization in the Middle East: analysis and policy', *Ekistics*, **300**, 195–204, (May/June 1983).
13. J.S. Birks and C.A. Sinclair, *Arab Manpower*, Croom Helm, London, 1980, 134.
14. M.O. Attir, 'Libya's pattern of urbanization', *Ekistics*, **300**, 157–62, (May/June 1983).

15. R.I. Lawless and S. Kezeiri 'Spatial aspects of population change in Libya', *Méditerranée*, 81–86 (1983).
16. Administrative Committee for Revolutionary Information, *Achievements of the Great Al Fateh 1969–84* Dar Africa, Rome, 52, (1985).
17. Economist Intelligence Unit, *Country Profile of Libya 1986–87*, Economist Publications, London, (1986).
18. S. Ghanem, 'Changing planning policies in Libya', in *Planning and Development in Modern Libya*, (Eds M.M. Buru, S.M. Ghanem and K.S. McLachlan), Menas Press, Wisbech, 225 (1985).
19. M. Naur, *Political Mobilization and Industry in Libya*, Akademisk Forlag, Denmark, **95**, 1986.
20. Economist Intelligence Unit, *op. cit.*, 10, (1986).
21. J.A. Allan, *Libya: the Experience of Oil*, Croom Helm, London and Westview Press, Boulder, 27, 1981.
22. J.S. Latham, 'A rationale for a "Green River" to supply the Jifarah Plain of Northwest Libya', in *Planning and Development in Modern Libya*, (Eds M.M. Buru, S.M. Ghanem and K.S. McLachlan), Menas Press, Wisbech, 138–150 1985.
23. J.A. Allan, *op. cit.*, 207, 1981.
24. 'Man-made river flows under desert sand', *Middle East Construction*, 27 (January 1986).
25. Economist Intelligence Unit, *op. cit.*, 16, (1986).
26. A.A. Mahmood-Misrati, 'Land conversion to urban use: its impact and character in Libya', *Ekistics*, **300**, 183–94, (May/June 1983).

CHAPTER 21

Conclusion

21.1 Change

The social, political and economic transformation of the Middle East gained momentum during the early years of this century. Change continued unabated after the Second World War and accelerated dramatically in the 1970s and 1980s. Since the first edition of this volume appeared in 1976 events in the Middle East have had major repercussions, of global significance. The 1975–76 civil war in Lebanon, a peace treaty between Egypt and Israel (1978), revolution in Iran (1979), the Iran–Iraq war (since 1980), and the siege of the Grand Mosque in Mecca (1979) are all events which have had, and are having, profound consequences within the region. Directly or indirectly they also brought about the second round of oil price rises in 1979 following those of 1973 (Chapter 9) which Henry Kissinger described as 'one of the pivotal events in the history of this century'.[1] The transformation of the economies of the Middle East and the world which began in 1973 went into top gear in 1979–80.

After 1979 the industrial world redoubled its efforts to sell goods and services to the oil-rich states of the Middle East. The political influence of OPEC soared, and the oil states built up a formidable array of modern weapons. The less populous states with massive revenues to spare were limited in what they could achieve only by the absorptive capacities of their economies and the risk of domestic upheaval. As projects for the construction of ports, roads, airports, hospitals and other largescale amenities multiplied and industrialization got under way, demands for labour increased, and new patterns of international migration for work were created (Chapter 5). Middle East labour migrations have been well studied and documented, though their social implications for both sender countries and host communities have not been fully examined.

As we prepare our second edition, the Middle East is coming to terms with an almost equally sudden slump in the price of oil, notably from 1985. The consequences of this slump for the oil producers are already evident in the form of cancelled, delayed or postponed contracts, and severe cash flow problems for even the richest states.[2] The political power of OPEC is in eclipse for the time being, and little is heard about the once formidable 'oil weapon'. There are signs however, that the price of oil will rise again in the 1990s, and Middle East producers could find themselves centre-stage once more because of their colossal reserves, and their relatively low production costs (Chapter 9). Thus the price of oil rose sharply and then slumped during the life of our first edition; it seems destined to rise again more slowly in the life of this edition. As a result, no doubt, interest in Middle Eastern affairs will revive keenly. Meanwhile, the slowing down of spending on new projects may provide a timely opportunity for stocktaking. There are many who feel that elements of the 'old' Middle East, as typified by the ethos of the historic Islamic cities, was being far too readily swept away in favour of imported ideas less likely to stand the test of time. A breathing space will also allow governments to assess the extent of environmental damage created by rapid development and perhaps provide an opportunity to start putting things right.

Oil revenues have by no means provided the only stimulus to change in the Middle East. Demographic growth has been another major cause, and in many ways it is even more significant because it applies to every state in the region. Astonishingly, since 1976 the population of the region has grown by about 100 million, largely by natural increase, though the Gulf region has received some permanent immigrants. The population of the Middle East is thus youthful, with the potential for high rates of growth in future. Indeed, in terms of manpower, the Arab world could have the status of a mini superpower by the end of the century. Population increase presents no threat to the less populous oil-rich states, such as Saudi Arabia or Libya, but it has extremely serious consequences for the development prospects of the poorer states, with large populations. Egypt provides a classic case study of this problem (Chapter 19), but rapid population growth is the background to economic difficulties elsewhere, as in Iran and Turkey. Israel too, will have to face up to the implications of the rapid increase in the number of Palestinians resident in the occupied areas (Chapter 16).

Demographic increase has helped generate whole new spatial patterns of population, settlement distribution and land use, the true extent of which has been revealed by remote sensing techniques. The outward expansion of towns and cities has been enormous in recent decades. Moreover, populations tend to be increasingly concentrated within major metropolitan regions of the state, creating serious congestion and competition for space (Chapter 6). There is also mounting pressure on land and water resources in the rural Middle East. Land use patterns which have witnessed many transformations in past centuries (Chapter 3) are changing to more intensive uses, as irrigation is more widely used, new types of crop are grown, and greater amounts of chemical fertilizer are applied (Chapter 4). Yet the agricultural sector has not generally been well treated by Middle East governments, in spite of paper commitments in their development plans (Chapter 7). One of the chief concerns of governments is to ensure that the people, especially in towns, are fed, but whilst this is often achieved by imports and subsidized prices rather than through the positive encouragement of domestic agriculture 'food security' has become a worry for major importers. In certain regions the attractions of city life have resulted in an absolute decline in rural population. More commonly, the percentage of rural dwellers is in decline. The effective end of 'horizontal' or desert pastoral nomadism and its replacement by limited-range migrations (or 'truck nomadism') have left vast areas of desert rangeland unproductive of food, especially goat and animal products (Chapters 4 and 11).

Superficially the modern rural landscapes of Middle East lowlands can be deceptive. The spread of irrigation and the greening of large areas creates an impression of agricultural prosperity much favoured by governments, but such prosperity may be very limited in extent (Chapter 13). In certain more traditional agricultural areas, notably in Turkey and Iran productivity remains low in some regions because of the lack of proper price incentives, the persistence of conservative attitudes, unsuitable land tenure systems, and shortage of know-how. Several spectacular irrigation schemes, such as the Great Man-Made River in Libya (Chapter 20) and the Southeast Anatolia Project in Turkey have yet to be completed. As land use patterns change, there is a need to monitor those changes through time, because they are not necessarily permanent. For example one estimate suggested that 40 per cent of desert land reclaimed for agriculture in Egypt had been abandoned by the 1970s.[3]

It seems clear that freshwater supplies are likely to be inadequate in almost every Middle Eastern state by the end of the century. The potential for conflict between states over river basin management should not be exaggerated, neither should it be ignored. There are about 20 shared river basins involving states discussed in this volume, few of which are subject to any agreement.[4] Competition for groundwater could trigger problems between states and neighbouring communities.[5] Desalination on a massive scale could be the solution, but it is costly and the risks of sabotage or pollution are considerable. Much can be done to conserve and recycle more water, but in the end irrigated agriculture, especially close to cities may have to be abandoned, and irrigation schemes modified (Chapter 2). Meanwhile, a decline in irrigated land in Iraq is forecast, as increasing amounts of Euphrates water are used in Syria and Turkey. Chapters 4,13,16,19 and 20 have all touched upon the problem of water supply. Bearing in mind the prevalence of seasonal aridity in the northern part and annual aridity in the southern part of the Middle East, water shortages could place severe constraints upon future economic plans.

Almost half the people in the Middle East are now urban dwellers and the proportion is rising inexorably. While most towns register growth, it is the large cities which attract migrants and grow most rapidly. Immigrants are drawn by the prospects of work (if only in the informal sector of the economy) and most cities now contain a significant amount of modern industrial and commercial activity. All kinds of strategies have been adopted to channel economic growth into peripheral regions and small towns, but without great success. The attempt can lead to conflict between national and regional interests as in Turkey (Chapter 17). Nevertheless the growing gap between living standards in different parts of Middle Eastern states is a worrying problem, as are those between urban and rural dwellers. The Iranian experience however does not hold out much encouragement for improving matters through orthodox agricultural modernization. While modernization was a qualified success, growing numbers of landless peasants moved to the cities where they became disaffected during the downturn in economic fortunes in the 1970s, and played a significant role in the fall of the Shah (Chapter 18).

21.2 Continuity

Against this background of change and development in the Middle East it is perhaps important to recall those geographical elements of the region which remain essentially the same. The disposition of mountains, plains, river valleys and deserts still imposes basic controls on the possibilities for permanent settlement and communications (Chapter 1). The climatic regime and location of groundwater resources clearly limit the potential for agriculture, while water shortages seem likely to become critical in future urban planning and rural development (Chapters 2, 12, and 13). The local geographical environment should be far better understood; failure to take it into proper account has led to many errors of development planning in recent years. Similarly, the basic facts of distances, coastal configuration and global location would reward far more serious examination than they receive. Talk of political and economic integration will remain glib and unrealistic unless placed in the geographical context of regional population distributions, communications and accessibility, political core areas and peripheries. Coastal configuration, especially the location and ownership of islands, will be a key element in future offshore boundary delimitations.

The geopolitical significance of the Middle East is guaranteed to continue by virtue of its global location.[6] Even in an era of declining oil prices and alternative supplies, it remains one of the prime arenas for superpower competition. The concept of the Middle East as one of the world's 'shatterbelt' regions where superpowers compete for influence over client states has recently been reassessed and confirmed.[7] Two geographical factors underly the strategic role of the Middle East. First, it lies adjacent to the Soviet Union, which shares 2200 km of boundaries with Turkey and Iran. What happens in these lands contiguous to the USSR, with its large Muslim population, is of great concern in Moscow. The United States fears Soviet military and economic expansion into the region, these fears being reinforced by the Soviet invasion of Afghanistan in 1979. It is arguable that mutual suspicion between the superpowers would live on even without the presence of large oil reserves in the Gulf region. Secondly, the Middle East remains one of the world's great crossroads for international shipping and aircraft, so that numerous states outside the region are deeply interested in the security of sensitive passageways such as Hormuz, Bab al Mandab, the Suez Canal, and the Strait of Gibraltar. The Turkish Straits are of vital importance to the Soviet Union for international trade and as an outlet for the Black Sea fleet. Thus, superpower interest in the region will remain strong though they may find Middle Eastern client states more difficult to manipulate in future. Among other factors, there are several states whose resources, size, location, and manpower might entitle them to the status of 'emergent regional powers', able to exercise an increasingly independent stance.[8] Certainly, the interplay of geopolitics at global, regional, and local levels is becoming more complex than ever, with the number and variety of geopolitical regions multiplying markedly at different scales.

One fundamental element of the political landscape certain to survive into the next century is the basic outline of the system of states which is of such recent origin (Chapter 10). The anomalies and shortcomings of this pattern of the partitioning of political space have been fully discussed by Drysdale and Blake.[9] In spite of these defects, the states themselves and their boundaries are seemingly being ever more firmly established. Land boundaries are being progressively agreed and delimited, while the internal communications infrastructure and the emergence of core regions, usually around the capital city, are generating centripetal functions and patterns of circulation within established political boundaries. The delimitation of offshore zones of state sovereignty will only serve to increase the permanence of present arrangements (Chapter 10). Two states are experiencing strains and stresses resulting from the poor territorial delimitation drawn up by Britain and France after the First War. Lebanon has effectively disintegrated as a state (Chapter 14), while the economy of Jordan remains very shakey largely because of a poor resource base and poor access (Chapter 15). Yet both states have powerful supporters within and outside the Middle East, and they will not easily disappear from the political map.

Given that the basic political framework seems set to last into the foreseeable future, with some minor adjustments no doubt, what are the prospects for greater economic integration and political cooperation? Not very great, it seems. Trade between states of the region is poorly developed, Japan and the European Community being the dominant partners. Limited raw materials and the low level of manufacturing industry, except in Turkey and Israel inhibit local trade, (Chapter 8). Until recently road and railway systems provided very poor links between states, and there is still much to be done. Political cooperation also has a poor record in the region, though the success of the Gulf Co-operation Council since 1980 (bringing together the Arab states of the Gulf and Saudi Arabia) is a notable exception. Environmental conservation and river basin management may yet prove to be useful media for inter-state collaboration, regardless of political differences, simply because of the environmental issues due to surface in future. Thus many states of the region have already joined with their neighbours to fight marine pollution. The Barcelona Convention of 1976 whose signatories pledged themselves to join forces against pollution in the Mediterranean Sea has been followed by similar agreements between the Red Sea states and Gulf States. Many Middle Eastern states have embarked upon programmes of industrialization which could pose local coastal environmental threats, necessitating regional agreements on acceptable standards and practices. In the arena of river management the example of the Nile may be influential in any future decision to apportion the waters of the Tigris–Euphrates.

It might have been assumed that advances in education, the spread of materialism, and the influences of the West would have undermined the crucial role of Islam in the Middle East. While this may be true among certain groups in the larger urban communities, it has not been the experience of the region as a whole. In fact there is evidence that the reverse may be the case, with people rediscovering their Islamic roots and traditions in a reaction against the westernization and modernization which appear to undermine traditional values. Islamic revolution, as seen in Iran in 1979 is unlikely to provide a model to be repeated elsewhere, although this is not impossible. Rather, Islamic movements will make themselves increasingly felt in the political life of communities – for example in Turkey, Syria, and Egypt. If a revived and militant form of Islam does spread in the Middle East it could follow a diffusion pattern reflecting proximity to Iran and the location of Shī'ite Muslim communities. Meanwhile, Islam remains directly and indirectly one of the oldest and most powerful influences in creating the distinctive cultural geography of the Middle East. Thus, continuity of tradition is to be found alongside changes as far reaching as any occurring in the world today. We have tried to give a balanced view of both in this volume.

Chapter 7 and Chapters 11 to 20, the national case studies, amply demonstrate the variety of development problems to be found in the Middle East. States differ greatly in geographical personality, economic potential, development strategies and political traditions, confirming the diversity of a region which nevertheless displays common characteristics. The earliest comprehensive regional geography of the Middle East emphasized themes of unity and diversity.[10] In the mid-1980s it seems most appropriate to emphasise the diversity of the region in political as well as physical terms. When the present spate of destructive conflicts dies down, elements which unite rather than divide could well come to the fore. Meanwhile, it seems fair to conclude that the Middle East is becoming a region of deepening diversity and regional inequalities as member states embark upon a variety of divergent paths whose final destination remains rather obscure.

References

1. Henry Kissinger, *Years of Upheaval*, Weidenfeld and Nicolson and Michael Joseph, London, 1982.
2. A.M. Farid (Ed) *The Decline of Arab Oil Revenues*, Croom Helm, London, 1986.
3. 'Egypt: desert dreams', *The Economist*, May 12–18 1984, 48–49.
4. United Nations, Department of Economic and Social Affairs, *Register of International Rivers*, Pergamon Press, Oxford, 1978, pp 21–30, 42–45.
5. E.W. Anderson, 'Water resources and boundaries in the Middle East' in *Boundaries and State Territory in the Middle East and North Africa*, (Eds G.H. Blake and R.N. Schofield), Menas Press, Cambridge, 1987, pp 85–97.
6. A. Braun (Ed), *The Middle East in Global Stratgy*, Westview Press, Godstone, 1986.
7. P.L. Kelly, 'Escalation of regional conflict: testing the shatterbelt concept', *Political Geog. Quarterly* **5** (2), 1986. 161–180.
8. S.B. Cohen, 'American foreign policy for the eighties', in *Pluralism and Political Geography*, (Eds N. Kliot and S. Waterman) Croom Helm, London and St Martins Press, New York, 1983, pp 295–310.
9. A.D. Drysdale and G.H. Blake, *The Middle East and North Africa: a Political Geography*, Oxford University Press, New York, 1985.
10. W.B. Fisher, *The Middle East*, Methuen, London, 7th edn, 1978.

Bibliography

Introduction

Ayoob, M. (Ed) (1981). *The Middle East in World Politics*, Croom Helm, London.

Bacharach, J. L. (1985). *A Middle East Studies Handbook*, Cambridge University Press.

Association for Peace (1968). *The Middle East in the Year 2000*, Tel Aviv, pp 1–40.

Bacon, E. E. (1946). 'A preliminary attempt to determine the culture areas of Asia', *Southwestern Journal of Anthropology*, **2**, 117–32.

Bacon, E. E. (1953). 'Problems relating to delimiting the culture areas of Asia', *Memoirs of the Society for American Archaeology*, **9**, 17–23.

Beaumont, P. (1976). 'The Middle East – environmental management problems', *Built Environment Quarterly*, **2**, 104–112.

Beaumont, P. (1980). 'Urban water problems' in *The Changing Middle Eastern City*, (Eds G. H. Blake and R. Lawless), Croom Helm, London, pp 230–50.

Birot, P. and J. Dresch (1955). *La Méditerranée et la Moyen Orient*, 2 vols, Presses Universitaires de France, Paris.

Blake, G. H., Dewdney, J. C. and Mitchell, J. K. (1987). *The Cambridge Atlas of the Middle East and North Africa*, Cambridge University Press.

Blake, G. H. and R. Lawless (1980). *The Changing Middle Eastern City*, Croom Helm, London.

Braun, A. (Ed) (1986). *The Middle East in Global Strategy*, Westview Press, Boulder, Colorado.

Brice, W. C. (1966) *South-West Asia*, University of London Press, London.

Bulliet, R. W. (1979). *Conversion to Islam in the Medieval Period: An Essay in Quantitative History*, Harvard University Press, Cambridge, Mass and London.

Chirol, V. (1903). *The Middle East Question, or Some Political Problems of Indian Defence*, John Murray, London.

Chisholm, M. (1971). *Research in Human Geography*, Social Science Research Council, Heinemann, London, pp 71–72.

Clarke, J. I. and Bowen-Jones, H. (Eds) (1981). *Change and Development in the Middle East. Essays in Honour of W. B. Fisher*, Methuen, London.

Coon, C. S. (1951). *Caravan: The Story of the Middle East*, Henry Holt and Co. New York.

Cressey, C. B. (1960). *Crossroads: Land and Life in South-West Asia*, Lippincott and Co, Chicago.

Daniel, N. (1966). *Islam, Europe and Empire*, Edinburgh University Press.

Davison, R. H. (1960). 'Where is the Middle East?' *Foreign Affairs*, **38**, 665–675.

Drysdale, A. D. and Blake, G. H. (1985). *The Middle East and North Africa: A Political Geography*, Oxford University Press, New York.

Encyclopaedia of Islam. E. J. Brill, Leiden and Luzac and Co, London, 1913–38; new edn 1960-.

English, P. W. (1973) 'Geographical perspectives on the Middle East: the passing of the ecological trilogy'. In *Geographers Abroad*, (Ed M. W. Mikesell), Department of Geography, Research Paper 152, Chicago, pp 134–64.

Farid, A. M. (1984) *The Red Sea: Prospects for Stability*, Croom Helm, London and St Martin's Press, New York.

Fisher, W. B. (1947). 'Unity and diversity in the Middle East', *Geogrl Rev.*, **37**, 414–35.

Fisher, W. B. (1978). *The Middle East*, 7th edn, Methuen, London.

Gillard, D. (1977). *The Struggle for Asia, 1828–1914. A Study of British and Russian Imperialism*, Methuen, London.

Gordon, T. E. (1900). 'The problem of the Middle East', *The Nineteenth Century*, **47**, 413–24.

Hogarth, D. G. (1902). *The Nearer East*, William Heinemann, London.

Hopwood, D. and D. Grimwood-Jones (Eds) (1972). *Middle East and Islam. A Bibliographical Introduction*, Bibliotheca Asiatica, 9.

Hourani, A. (1981). *The Emergence of the Modern Middle East*, Macmillan, London.

Keddie, N. R. (1973). 'Is there a Middle East?' *International Journal of Middle Eastern Studies*, **4**, 255–71.

Koppes, C. R. (1976). 'Captain Mahan, General Gordon and the origin of the term "Middle East" ', *Middle Eastern Studies*, **12**, 95–8.

Landen, R. G. (1970). *The Emergence of the Modern Middle East: Selected Readings*, Van Nostrand-Reinholt, Princeton.

Le Lannou, M. (1966). 'L'isthme du Proche et Moyen-Orient', *Revue Géogr. Lyon*, **41**, 289–302.

Laqueur, W. (1972). *The Struggle for the Middle East. The Soviet Union and the Middle East*, Penguin, Harmondsworth.

Lewis, B. (1964). *The Middle East and the West*, Weidenfeld and Nicolson, London.

Longrigg, S. H. and J. Jankowski (1970). *The Middle East: A Social Geography*, Duckworth, London.

Lorraine, P. (1943). 'Perspectives of the Near East', *Geogrl J.*, **102**, 6–13.

Luciani, G. (Ed) (1984). *The Mediterranean Region*, Croom Helm, London and St Martin's Press, New York.

Lutfiyya, A. M. and C. W. Churchill (Eds) (1970). *Readings in Arab Middle East Society and Culture*, Mouton, Paris and The Hague.

MacMunn, G. (1928). History of the Great War. Military Operations: Egypt and Palestine Vol. 1, HMSO, London, pp 94–97.

Manners, I. R. (1971). 'The desert and the sown: an ecological appraisal of the Middle East', *Focus*, **22(2)**, 1–8.

Mansfield, P. (Ed) (1973). *The Middle East: A Political and Economic Survey*, Oxford University Press.

Martin, L. (1947). 'The miscalled Middle East', *Geogrl Rev.*, **37**, 355–356.

Patai, R. (1952). 'The Middle East as a culture area', *Middle East Journal*, **6**, 1–21. Reprinted in *Readings in Arab Middle East Society and Culture*, (Eds A. M. Lutfiyya and C. W. Churchill), Mouton, Paris and The Hague, 1970, pp 187–204.

Patai, R. (1969). *Golden River to Golden Road. Society, Culture and Technical Change in the Middle East*, 2nd edn, University of Philadelphia Press, Philadelphia.

Peretz, D. (1971). *The Middle East Today*, 2nd edn, Holt, Rinehart and Winston, New York.

Reifenberg, A. (1958). *The Struggle between the Desert and the Sown. Rise and Fall of Agriculture in the Levant*, Government Press, Jerusalem.

Said, E. W. (1978). *Orientalism*, Routledge and Kegan Paul, London.

Said, E. W. (1981). *Covering Islam*, Routledge and Kegan Paul, London.

Smith, C. G. (1968). 'The emergence of the Middle East', *Journal of Contemporary History*, **3**, 3–17.

Steadman, J. M. (1969). *The Myth of Asia*, Simon and Schuster, New York.

Yapp, M. E. (1980). *Strategies of British India: Britain, Iran and Afghanistan 1798–1850*, Clarendon Press, Oxford.

Chapter 1: Relief, Geology, Geomorphology and Soils

Asfia, S., E. B. Bailey and R. C. B. Jones (1948). 'Notes on the geology of the Elburz mountains, north-east of Tehran', *Q. Jl. geol. Soc. Land.*, **104**, 1–42.

Atkinson, K. (1969). 'The dynamics of Terra Rossa soils', *Bulletin of the Faculty of Arts, University of Libya*, **3**, 15–35.

Atkinson, K. (1970). 'Fossil limestone soils in north-west Turkey', *Palaeogeography, Palaeoclimatology, Palaeoecology*, **8**, 29–35.

Barton, D. C. (1938). 'The disintegration and exfoliation of granite in Egypt', *J. Geol.*, **46**, 109–11.

Beaumont, P. (1968). 'Salt weathering on the margin of the Great Kavir, Iran', *Bull. geol. Soc. Am.*, **79**, 1683–4.

Beaumont, P. (1972). 'Alluvial fans along the foothills of the Elburz Mountains, Iran', *Palaeogeography, Palaeoclimatology, Palaeoecology*, **12**, 251–73.

Berry, L. (1961). 'Large scale alluvial islands in the White Nile', *Revue Géomorph. dyn.*, **12**, 105–109.

Beug, H. J. (1967). 'Contributions to the postglacial vegetational history of northern Turkey'. In *Quarternary Palaeoecology*, (Eds E. J. Cushing and H. E. Wright, Jnr) Yale University Press, New Haven and London, pp 349–56.

Birman, J. H. (1968). 'Glacial reconnaissance in Turkey', *Bull. geol. Soc., Am.*, **79**, 109–26.

Blandford, W. T. (1873). 'On the nature and probable origin of the superficial deposits in the valleys and deserts of Central Persia', *Q. Jl geol. Soc. Lond.*, **29**, 495–503.

Bobeck, H. (1937). 'Die Rolle der Eiszeit in Nordwestiran', *Z. Gletscherk. Glacial geol.*, **25 S**, 130–83.

Bobek, H. (1959). *Features and Formation of the Great Kavir and Masileh*, Arid Zone Research Centre, University of Tehrān, Publication No 2, Tehrān, 63 pages.

Bobeck, H. (1963). 'Nature and implications of Quarternary climatic changes in Iran'. In UNESCO & WMO Symposium, *Changes of Climate*, UNESCO, Rome, pp 403–13.

Brice, W. C. (Ed) (1978). *The environmental history of the Near and Middle East since the last Ice Age*, Academic Press, London.

Brown, G. F. (1960). 'Geomorphology of western and central Saudi Arabia', *Report XXI International Geological Congress, Copenhagen*, Pt 21, pp 150–9.

Brown, G. F. (1970). 'Eastern margin of the Red Sea and the coastal structures in Saudi Arabia', *Phil. Trans. R. Soc., A*, **267**, 75–87.

Buringh, P. (1960). *Soils and soil conditions in Iraq*, Ministry of Agriculture, Republic of Iraq, Baghdād, 322 pages.

Butzer, K. W. (1958). *Quarternary stratigraphy and climate in the Near East*, Bonner Geogr. Abhandl., **24**, 157 pages.

Butzer, K. W. (1959). 'Contribution to the Pleistocene geology of the Nile Valley', *Erdkunde*, **2**, 46–67.

Butzer, K. W. (1960). 'Archaeology and geology in ancient Egypt', *Science, N. Y.*, **132**, 1617–24.

Butzer, K. W. (1960). 'On the Pleistocene shorelines of Arabs' Gulf, Egypt', *J. Geol.*, **66**, 626–37.

Butzer, K. W. (1965). 'Desert landforms at the Kurkur Oasis, Egypt', *Ann. Ass. Am. Geogr.*, **55**, 578–691.

Butzer, K. W. (1967). 'Late Glacial and post Glacial climatic variation in the Near East', *Erdkunde*, **11**, 21–35.

Butzer, K. W. and C. L. Hansen (1965). 'On Pleistocene evolution of the Nile Valley in Southern Egypt', *Canadian Geographer*, **9**, 74–83.

Butzer, K. W. and C. L. Hansen (1967). 'Upper Pleistocene stratigraphy in Southern Egypt'. In *Background to African Evolution*, (Eds W. W. Bishop and J. D. Clark) University of Chicago Press, Chicago, pp 329–56.

Butzer, K. W. and C. L. Hansen (1968). *Desert and River in Nubia*, The University of Wisconsin Press, Madison, 562 pages.

Clapp, F. G. (1930). 'Tehran and the Elburz', *Geogrl Rev.*, **29**, 69–85.

Cooke, R. U. (1970). 'Stone pavements in Deserts', *Ann. Ass. Am. Geogr.*, **60,(3)**, 560–77.

Dan, J. and H. Koyumdjisky (1963). 'The soils of Israel and their distribution', *J. Soil Sci.*, **14**, 12–20.

De Ridder, N. A. (1965). 'Sediments of the Konya Basin, central Anatolia, Turkey', *Palaeogeography, Palaeoclimatology, Palaeoecology*, **1**, 225–54.

Dewan, M. L. and J. Famouri (1964). *The Soils of Iran*, Food and Agricultural Organization, United Nations, Rome, 319 pages.

Dietz, R. S. and J. C. Holden (1970). 'The Breakup of Pangaea', *Scient. Am.*, **223**, 30–41.

Dresch, J. (1959). 'Le Piémont de Tehrān'. In Expedition de 1958 en Iran, (Eds J. Dresch *et al.*) *Bull. Ass. Géogr. fr.*, **284–5**, 35–64.

Elgabaly, M. M., I. M. Gewaifel, N. N. Hassan and B. G. Rozanov (1969). *Soil map and land resources of U.A.R.*, Institute of Land Reclamation, Alexandria University, Research Bulletin, No 22, 14 pages.

Emery, K. O. (1960). 'Weathering of the Great Pyramid', *J. sedim. Petrol.*, **30**, 140–3.

Evans, G. (1966). 'The recent sedimentary facies of the Persian Gulf region', *Phil. Trans. R. Soc., A*, **259**, 291–8.

Evans, I. S. (1969–70). 'Salt crystallization and rock weathering: a review', *Revue Géomorph. dyn.*, No 4, 19 Année, 153–77.

FAO (1968). *Definitions of Soil Units for the Soil Map of the World*, Soil Map of the World: FAO/UNESCO Project, World Soil Resources Office, Land and Water Development Division, FAO, Rome, 72 pages.

FAO (1970). *Key to Soil Units for the Soil Map of the World*, FAO/UNESCO Project, Soil Resources, Development and Conservation Service, Land and Water Development Division, FAO, Rome, 16 pages.

Federov, P. V. (1969). 'The marine terraces of the Black Sea Coast of the Caucasus and the problem of the most recent vertical movements', *Dokl. Acad. Nauk. USSR*, **144**, 431–4, (in Russian).

Fleisch, H. and M. Gigout (1966). 'Révue du Quaternaire marin Libanais', *Bull. Geol. Soc. France*, **8**, 10–16.

Fookes, P. G. and J. L. Knill (1969). 'The application of engineering geology in the regional development of northern and central Iran', *Engineering Geology*, **8**, 81–120.

Freznel, B. (1959–60). 'Die Vegetations und Landschaftszonen Nord-Eurasien während der letzten Eiszeit und während der post-glazialen Wärmezeit', *Abh. Akad. Wiss. Liter, (Mainz) Math.-Naturw. Kl. 1959*, No 13,164 pages and 1960 No 6, 167 pages.

Ghaith, A. M. and M. Tanious (1965). *Preliminary Soil Association Map of the United Arab Republic*, Soil Department, Soil Survey Division, Ministry of Agriculture, Cairo, Egypt, 9 pages.

Guilcher, A. (1955). 'Géomorphologie de l'extrémité septentrionale du Banc Farsan (Mer Rouge)', *Ann. Inst. Océanog.*, **30**, 55–100.

Hamdi, A. (1972). *The Soils of Egypt*, Soil Science Society of Egypt, International Symposium on Salt Affected Soils, 4 pages.

Harrison, J. V. (1943). 'The Jaz Murian depression, Persian Baluchistan', *Geogrl J.*, **101**, 206–25.

Harrison, J. V. and N. L. Falcon (1937). 'The Saidmarreh landslip, Southwest Iran', *Geogrl J.*, **89**, 42–57.

Hey, R. W. (1971). 'Quaternary shorelines of the Mediterranean and Black Seas', *Quaternaria*, **15**, (VIII Congrés INQUA – Les Niveaux Marins Quaternaires II – Pleistocene), 273–84.

Hutchinson, G.E. and U.M. Cowgill (1963). 'Chemical examination of a core from Lake Zeribar, Iran', *Science N.Y.*, **140**, 67.

Jelgersma, S. (1966). 'Sea-level changes during the last 10,000 years'. In *World Climate from 8000 to 0 B.C.*, Royal Meteorological Society, London, 54–71.

Jenny, H. (1941). *Factors of Soil Formation*, McGraw-Hill Book Co, New York, 281 pages.

Jewitt, T.N. (1966). 'Soils of arid lands'. In *Arid Lands – A Geographical Appraisal*, (Ed E.S. Hills) Methuen-UNESCO, pp 103–25.

Khalaf, F.I., I.M. Gharib and M.Z. Al-Hashash (1984). 'Types and characteristics of the recent surface deposits of Kuwait, Arabian Gulf', *Journal of Arid Environments*, **7(1)**, 9–33.

Kinsman, D.J.J. (1966). 'Gypsum and anhydrite of recent age, Trucial Coast, Persian Gulf', *Second Symposium on Salt*, 1, Northern Ohio Geological Society, Cleveland, Ohio, pp 302–25.

Krinsley, D.B. (1968). 'Geomorphology of Three Kavirs in Northern Iran'. In *Playa Surface Morphology: Miscellaneous Investigations*, (Ed J.T. Neal) USAF, Office of Aerospace Research, Environmental Research Papers, No 283, pp 105–30.

Krinsley, D.B. (1970). *A Geomorphological and Palaeoclimatological Study of the Playas of Iran*, Parts I and II, Geological Survey, United States Department of the Interior, Washington DC, 486 pages.

Krinsley, D.B. (1972). 'The palaeoclimatic significance of the Iranian playas', *Palaeoecol. Afr.*, **6**, 114–20.

McKee, E.D. (1962). 'Origin of the Nubian and similar sandstones', *Geol. Rdsch.*, **52**, 551–87.

McKenzie, D.P. (1970). 'Plate tectonics of the Mediterranean Region', *Nature, Lond.*, **226**, 239–43.

McKenzie, D.P., D. Davies and P. Molnar (1970). 'Plate tectonics of the Red Sea and East Africa', *Nature, Lond.*, **226**, 243–8.

Messerli, B. (1966). 'Das Problem der eiszeitlichen Vergletscherung am Libanon und Hermon', *Z. Geomorph.*, **10 (1)**, 36–68.

Messerli, B. (1967). 'Die eiszeitliche und die gegenwärtige Vergletscherung im Mittelmerraum', *Geographica helv.*, **3**, 105–228.

Moorman, F. (1959). *The Soil of East Jordan*, FAO, Rome, Report, No 1132.

Oakes, H. (1957). *The Soils of Turkey*, Republic of Turkey, Ministry of Agriculture, Soil Conservation and Farm Irrigation Division, Ankara, Division Publication No 1, 180 pages.

Oberlander, T.M. (1965). *The Zagros Streams: A New Interpretation of Transverse Drainage in an Orogenic Zone*, Syracuse Geographical Series, Syracuse, No 1, 168 pages.

Quennell, A.M. (1958). 'The structural and geomorphic evolution of the Dead Sea Rift', *Q. Jl geol. Soc. Lond.*, **114**, 1–24.

Ravikovitch, S. (1960). *Soils of Israel – Classification of the Soils of Israel*, (In Hebrew – English summary), The Hebrew University of Jerusalem, Faculty of Agriculture, Rehovot, Israel, 89 pages.

Reifenberg, A. (1947). *The Soils of Palestine*, (translated by C.L. Whittles), Thomas Murby, London, 179 pages.

Rieben, E.H. (1955). 'The geology of the Tehrān Plain', *Am. J. Sci.*, **253**, 627–39.

Rieben, E.H. (1960). Les Terrains Alluviaux de la Région de Tehrān, Arid Zone Research Centre, University of Tehran, Publication No 4, 41 pages.

Rieben, E.H. (1966). *Geological Observations on Alluvial Deposits in Northern Iran*, Geol. Surv. Iran, Report 9, 41 pages.

Sanlaville, P. (1967). 'Sur les niveaux marins quaternaires de la région de Tabarja (Liban)', *Comptes Rendues Somm. Soc. Geol. Fr.*, 157–8.

Sanlaville, P. (1971). 'Sur le Tyrrhènien libanais', *Quaternaria*, **15**, (VIII, Congrès INQUA – Les Niveaux Marins Quaternaires, II – Pleistocene), pp 239–48.

Scharlau, K. (1958). 'Zum Problem der Pluvialzeiten in nordost Iran', *Z. Geomorph.*, **2**, 258–77.

Sharon, D. (1962). 'Hammadas in Israel', *Z. Geomorph.*, **6**, 12–147.

Soil Survey Staff, Soil Conservation Service, United States Department of Agriculture (1960). *Soil Classification: A Comprehensive System (7th Approximation)*, US Government Printing Office, Washington DC.

Stevens, J. H. (1969). 'Quarternary events and their effect on soil development in an arid environment – The Trucial States', *Quaternaria*, **10**, 73–81.

Takin, M. (1972). 'Iranian geology and continental drift in the Middle East', *Nature, Lond.*, **235**, 147–50.

Van Liere, W. J. (1961). 'Observations on the Quaternary of Syria', *Ber. Rijkdienst oudheidk. Bodemonderz*, **10–11**, 1–69.

Van Zeist, W. and H. E. Wright Jnr (1963). 'Preliminary pollen studies at Lake Zeribar, Zagros Mountains, southwestern Iran', *Science, N.Y.*, **140**, 65–9.

Vita-Finzi, C. (1968). 'Late Quaternary alluvial chronology of Iran', *Geol. Rdsch.*, **58**, 951–73.

Vita-Finzi, C. (1969). *Mediterranean Valleys*, Cambridge University Press, 140 pages.

Wellman, H. W. (1966). 'Active wrench faults of Iran, Afghanistan and Pakistan', *Geol. Rdsch.*, **55**, 716–35.

Wright, H. E. (1962). 'Pleistocene glaciation in Kurdistan', *Eiszeitalter Gegenw.*, **12**, 131–64.

Wright, H. E. (1962). 'Late Pleistocene geology of coastal Lebanon', *Quaternaria*, **6**, 525–40.

Yaalon, D. H. (1970). 'Parallel stone cracking, a weathering process on desert surfaces', *Geological Institute Technical and Economical Bulletins*, Series C, Pedology, **18**, Bucharest, 107–11.

Chapter 2: Climate and Water

Aberbach, S. H. and A. Sellinger (1967). 'Review of artificial groundwater recharge in the coastal plain of Israel', *Bull. int. Ass. scient. Hydrol.*, **12**, 65–77

Abu-Hilal, A. H. (1985). 'Phosphate pollution in the Jordan Gulf of Aqaba', *Marine Pollution Bulletin*, **16(7)**, 281–5.

Abul-Ata, A. A. (1979). After the Aswan, *Mazingira*, **11**, 21–7.

Ackerman, W. C., G. F. White, E. B. Worthington, J. L. Young (1973). *Man-Made Lakes: their Problems and Environmental Effects*, American Geophysical Union, Geophysical Monograph, **17**.

Addison, H. (1959). *Sun and Shadow at Aswan*, Chapman & Hall, London, 166 pages.

Aelion, E. (1958). 'A report on weather types causing marked storms in Israel during the cold season', *Israel Meteorological Service*, Series C, Miscellaneous Papers 10, 7 pages.

ALESCO/UNEP (1975). *Re-greening of Arab Deserts*, Arab League Educational, Cultural and Scientific Organization and United Nations Environment Programme, Cairo, UNEP Project No 0206–74–002, Final Report.

Al-Khashab, W. H. (1958). *The Water Budget of the Tigris–Euphrates Basin*, University of Chicago, Department of Geography, Research Paper, No 54, 105 pages.

Al-Ruwaih, F. (1984). 'Groundwater chemistry of Dibdiba formation, North Kuwait', *Groundwater*, **22(4)**, 412–17.

Amiran, D. H. K. and M. Gilead (1954). 'Early excessive rainfall and soil erosion in Israel', *Israel Explor. J.*, **4**, 286–95.

Ambroggi, R. P. (1966). 'Water under the Sahara', *Sci. Am.*, **214**, 21–9.

Annesley, T. J., M. J. Hall, and N. A. Hill (1983). 'A master water resources and agricultural development plan for the State of Qatar. Part 2: the systems model and its application', *International Journal of Water Resources Development*, **1(1)**, 31–49.

Anonymous (1984) 'Jeddah water reclamation project', *Middle East Water & Sewage*, **8 (3)**, 22–4.

Anonymous (1984). 'The Riyadh water transmission system', *Middle East Water & Sewage*, **8(3)**, 10–21.

Arab Report and Record, (1970). 'Dam devastates Mediterranean fishing', *Arab Report and Record*, Issue 23, 1–15 December, p 677.

Arab Republic of Egypt, Ministry of Culture and Information, (1972). *The High Dam*, State Information Office, Cairo, Egypt, 54 pages.

Arad, A. (1966). 'Hydrogeochemistry of groundwater in central Israel', *Bull. int. Ass. scient. Hydrol.*, **11**, 122–46.

Arad, A. and U. Kafri (1980). 'Hydrogeological interrelationships between the Judea Group and the Nubian Sandstone aquifers in Sinai and the Negev', *Israeli J. Earth Sci.*, **29**, 67–72.

Atkinson, K., M. Bovis and D. Johnson (1972). 'Man-made oases of Libya', *Geogrl. Mag.*, **45**, 112–15.

Austin, E. E. and D. Dewar (1953). '*Upper Winds Over Mediterranean and Middle East*, Great Britain, Meteorological Research Committee, MRP 811, 10 pages.

Awerbach, L. (1984). 'Desalination – the state of the art', *Asian Water & Sewage*, **1**, 18–22.

Bakiewicz, W., D. M. Milne and M. Noori (1982). 'Hydrogeology of the Umm Er Radhuma aquifer, Saudi Arabia, with reference to fossil gradients', *Q. J. eng. Geol. London*, **15**, 105–26.

Barbour, K. M. (1957). 'A new approach to the Nile Waters problems', *International Affairs*, **33**, 319–30.

Baum, W. A. and L. B. Smith (1953). 'Semi-monthly mean sea-level pressure maps for the Mediterranean area', *Arch. Met. Geophys. Biokilm., Serie A: Meterologie und Geophysik*, **5**, 326–345.

Beaumont, P. (1968). 'Qanats on the Varamin Plain, Iran', *Trans. Inst. Br. Geogr.*, **45**, 169–79.

Beaumont, P. (1968). 'The Road to Jericho: A climatological traverse across the Dead Sea lowlands', *Geography*, **53**, 170–4.

Beaumont, P. (1971). 'Qanat systems in Iran', *Bull. int. Ass. scient. Hydrol.*, **16**, 39–50.

Beaumont, P. (1973). *River Regimes in Iran*, Department of Geography, University of Durham, Occasional Publications, New Series No 1, 29 pages.

Beaumont, P. (1973). 'A Traditional method of groundwater extraction in the Middle East', *Ground Water*, **11**, 23–30.

Beaumont, P. (1974). 'New uses for ancient water'. In Saudi Arabia – Financial Times Survey, *The Financial Times*, London, 10 June.

Beaumont, P. (1974). 'Water resource development in Iran', *Geogrl. J.*, **140**, 418–31.

Beaumont, P. (1976). 'The Middle East – environmental management problems', *Built Environment Quarterly*, **2(2)**, 104–12.

Beaumont, P. (1977). 'Water in Kuwait, *Geography*, **62**, 187–97.

Beaumont, P. (1977). 'Water and development in Saudi Arabia', *Geographical Journal*, **143**, 42–60.

Beaumont, P. (1980). 'Urban water problems'. In *The Changing Middle Eastern City*, (Eds G. H. Blake and R. I. Lawless), Croom Helm, London, pp 230–50.

Beaumont, P. (1981). 'Water resources and their management in the Middle East'. In *Change and development in the Middle East*, (Eds J. I. Clarke and H. Bowen-Jones), Methuen, London, pp 40–72.

Beaumont, P. (1985). 'Irrigated agriculture and ground-water mining on the High Plains of Texas, USA', *Environmental Conservation*, **12(2)**, 119–30.

Beckett, P. H. T. (1951). 'Waters of Persia', *Geogrl Mag.*, **24**, 230–40.

Beckett, P. H. T. (1952). 'Qanats in Persia', *Iran Society Journal*, London, **1**, 125–33.

Beckett, P. H. T. (1953). 'Qanats around Kirman', *Jl R. cent. Asian Soc.*, **40**, 47–58.

Beckett, P. H. T. and E. D. Gordon (1956). 'The climate of Kerman, south Persia', *Q. Jl R. met. Soc.*, **82**, 503–14.

Bell, B. (1970). 'The oldest records of the Nile floods', *Geogrl J.*, **136**, 569–573.

Bémont, F. (1961). 'L'irrigation en Iran', *Annls Gèogr.*, **70**, 597–620.

Birks, J. S. (1984). 'The falaj: modern problems and some possible solutions', *Waterlines*, **2(4)**, 28–31.

Bleeker, W. (Ed), (1960). *The UNESCO/WHO Seminar on Mediterranean Synoptic Meteorology*, Rome 24th November – 13th December 1958, Frien Universität, Berlin, Institut für Meteorologie und Geophysik, Meteorologische, Abhandlungen, Vol 9, 226 pages.

Brawer, M. (1968). 'The geographical background of the Jordan water dispute'. In *Essays in Political Geography*, (Ed C. A. Fisher), Methuen, London, pp 225–42.

Bromehead, C. E. N. (1942). 'The early history of water supply', *Geogrl J.*, **99**, 142–51 and 183–95.

Budyko, M. I., N. A. Yefimova, L. I. Aubenok and L. A. Strokina (1962). 'The heat balance of the surface of the earth', *Soviet Geogr.*, **3**, 3–16.

Burdon, D. J. (1973). 'Groundwater resources of Saudi Arabia', *Groundwater Resources in Arab Countries*, ALESCO Science Monographs, No 2.

Burdon, D. J. (1977). 'Flow of fossil groundwater', *Q. J. eng. Geol. London*, **10**, 97–124.

Burdon, D. J. (1981). 'Infiltration conditions of a major sandstone aquifer around Ghat, Libya', Proc. Second Symp. *Geology of Libya*, Tripoli, Academic Press, Salem, pp 595–609.

Burdon, D. J. (1982). 'Hydrogeological conditions in the Middle East', *Q. J. eng. Geol. London*, **15**, 71–82.

Burdon, D. J. and C. Safadi (1964). 'The karstic groundwaters of Syria', *J. Hydrol.*, **2**, 324–47.

Butzer, K. W. (1960). 'Dynamic climatology of large-scale European circulation patterns in the Mediterranean area', *Met. Rdsch.*, **13**, 97–105.

Cantor, L. M. (1967). *A World Geography of Irrigation*, Praeger, New York and Washington, 252 pages.

Caponera, D. A. (1954). *Water Laws in Moslem Countries*, FAO, Rome, Agricultural Development Paper, No 43, 202 pages.

Carter, D. B., C. W. Thornthwaite and J. R. Mather (1958). *Three Water Balance Maps of Southwest Asia*, Publs. Clim. Drexel. Inst. Technol., **11**, 57 pages.

Caton-Thompson, G. and E. W. Gardner (1939). 'Climate, irrigation and early man in the Hadhramaut', *Geogrl. J.*, **93**, 18–38.

Cederstrom, D. J. (1971). 'Ground water in the Aden Sector of Southern Arabia', *Ground Water*, **9(2)**, 29–34.

Clapp, G. R. (1957). 'Iran, a TVA for the Khuzestan Region', *Middle East Journal*, **11**, 1–11.

Clarke, J. I. and H. Bowen-Jones (Eds) (1981). *Change and Development in the Middle East*, Methuen, London.

Colvin, I. (1964). 'Sharing the water of the Jordan', *Jl R. cent. Asian Soc.*, **51**, 245–50.

Cressey, G. B. (1957). 'Water in the desert', *Ann. Ass. Am. Geogr.*, **47**, 105–24.

Cressey, G. B. (1958). 'Qanats, karez and foggaras', *Geogrl Rev.*, **48**, 27–44.

Dijon, R. (Ed) (1980). *Groundwater in the eastern Mediterranean and western Asia*, UN Water Resources Service Report.

Dincer, T. (1978). 'The use of environmental isotopes in arid zone hydrology', Proc. IAEA Symp. *Application of isotope techniques to arid zone hydrology*, Vienna, pp 23–30.

Dincer, T., A. Al-Mugrin and U. Zimmerman (1974). 'Study of the infiltration and recharge through the sand dunes in arid zones with specific reference to the stable isotopes and thermonuclear tritium', *Journal of Hydrology*, **23**, 79–109.

Duvdevani, S. (1964). 'Dew in Israel and its effects on plants', *Soil Sci.*, **98**, 14–21.

Eagles, B. G. (1985). 'Baghdad sewerage', *Water Pollution Control*, **3**, 299–308.

Ebert, C. H. V. (1965). 'Water resources and land use in the Qatif oasis of Saudi Arabia', *Geogrl Rev.*, **55**, 496–509.

Edmunds, W. M. and E. P. Wright (1979). 'Groundwater recharge and palaeoclimate in the Sirte and Kufra basins', Libya, *J. Hydrol.*, **40**, 215–41.

Eldblom, L. (1967). 'Notes on the problems of irrigation in three Libyan Oases', *Ekistics*, **24 (137)**, 199–202.

Elewa, A. S. (1985). 'Effect of flood water on salt content of Aswan High Dam reservoir', *Hydrobiologia*, **128 (3)**, 249–54.

El-Fandy, M. G. (1952). 'Forecasting thunderstorms in the Red Sea', *Bull. Am. met. Soc.*, **33**, 332–8.

El-Moghraby, A. I. and M. O. El-Sammani (1985). 'On the environmental and socio-economic impact of the Jonglei Canal Project, Southern Sudan', *Environmental Conservation*, **12(1)**, 41–8.

English, P. W. (1968). 'The origin and spread of qanats in the Old World', *Proc. Am. phil. Soc,*, **112**, 170–81.

Field, M. (1973). 'Developing the Nile', *Wld Crops*, **25**, 11–15.

Fisher, W. B. (Ed). (1968). *The Land of Iran – Vol 1 – The Cambridge History of Iran*, Cambridge University Press.

Fitt, R. L. (1953). 'Irrigation development in central Persia', *Jl R. cent. Asian Soc.*, **40**, 124–33.

Flohn, H. (1965). *Contributions to a Synoptic Climatology of the Red Sea Trench and Adjacent Territories*, University of Bonn, Institute of Meteorology, US Department of Army Contract No DA 91-591-EUC-3201, Technical Report, 33 pages.

Flower, D. J. (1968). 'Water use in North-East Iran', In *The Cambridge History of Iran*, (Ed W. B. Fisher), Vol 1, Cambridge University Press, pp 599–610.

French, G.E. and A.G. Hill (1971). *Kuwait: Urban and Medical Ecology*, **4**, Geomedical Monograph Series, Geomedical Research Unit of the Heidelberg Academy of Sciences, Springer—Verlag, Berlin, 124 pages.

Frevert, T. (1985). 'Heavy metals in Lake Kinneret (Israel): total copper and cupric ion concentrations in Lake Kinneret and the river Jordan', *Archiv fur Hydrobiologie*, **104 (4)**, 527–42.

Gabaly, M. El (1977). 'Problems and effects of irrigation in the Near East region'. In *Arid Land Irrigation and Developing Countries*, (Ed E. B. Worthington), Pergamon Press, Oxford, England, 463 pages.

Gabriel, K. R. (1965). *The Israel Artificial Rainfall Stimulation Experiment, An interim Statistical Evaluation of Results*, Hebrew University, Department of Statistics, Jerusalem, Israel, 23 pages.

Galnoor, I. (1980). 'Water policy making in Israel'. In *Water Quality Management under Conditions of Scarcity – Israel as a Case Study*, (Ed H. I. Shuval) Academic Press, New York, pp 287–314.

Ganji, M. H. (1955). 'The climates of Iran', *Société Royale de Géographique d'Egypt Bulletin*, **28**, 195–299.

Ghanem, Y. and J. Eighmy (1984). 'Hema and traditional land use management among arid zone villagers of South Arabia', *Journal of Arid Environments*, **7**, 287–97.

Gilead, M. and N. Rosenan (1954). 'Ten years of dew observation in Israel', *Israel Explor. J.*, **4**, 120–3.

Gischler, C. (1979). *'Water resources in the Arab Middle East and North Africa*, Middle East & North African Studies Press, Wisbech, Cambridge, 132 pages.

Gleeson, T. A. (1953). 'Cyclogenesis in the Mediterranean region', *Arch. Met. Geophys. Bioclim., Serie A. Meteorologie und Geophysik*, **6**, 153–71.

Gleeson, T. A. (1956). 'A comparison of cyclone frequencies in the North Atlantic and Mediterranean regions', *Arch. Met. Geophys. Bioclim., Serie A. Meteorologie und Geophysik*, **9**, 185–90.

Goblot, H. (1962). 'Le problème de l'eau en Iran', *Orient*, **23**, 43–60.

Goldberg, I. A. (1972). *World agroclimatological atlas*, Hydromet Publishers, Moscow/Leningrad.

Goldsmith, E. and Hildyard, N. (Eds) (1986). *The social and environmental effects of large dams*, Wadebridge Ecological Centre.

Gordon, H. (1984). *Iran: Struggle for Power*, Menas Press, Wisbech, 100 pages.

Gordon, A. H. and J. G. Lockwood (1970). 'Maximum one day falls of precipitation in Tehrān', *Weather, Lond.*, **25**, 2–8.

Guerre, A. (1981). 'Study of the Karstic spring of Ayn Zayana, Libya', Proc. Second Symp. *Geology of Libya* Tripoli, Academic Press, Salem, pp 685–701.

Hall, M. J. and N. A. Hill (1983). 'A master water resources and agricultural plan for the State of Qatar. Part 1. Physical setting and resources', *International Journal of Water Resources Development*, **1 (1)**, 15–30.

Hammad, H. Y. (1970). *Ground Water Potentialities in the African Sahara and the Nile Valley*, Beirūt Arab University, Beirūt.

Hare, F. K. (1961). 'The causation of the arid zone'. In *A History of Land-use in Arid Lands*, Arid Zone Research, Vol 17, UNESCO, Paris, pp 35–50.

Hasan, M. R. (1984). 'Hydrology of the Northern and Eastern United Arab Emirates', *Middle East Water & Sewage*, **8 (1)**, 13–16.

Haynes, C. V. and H. Haas (1980). 'Radiocarbon evidence for Holocene recharge of groundwater, western desert, Egypt', *Radiocarbon*, **22**, 705–17.

Haynes, K. E. and D. Whittington (1981). 'International management of the Nile: Stage Three?' *Geographical Review*, **71 (1)**, 17–32.

Hidore, J.J. and Y. Albokhair (1982). 'Sand encroachement in Al-Hasa Oasis, Saudi Arabia, *Geographical Review*, **72**, 350–6.

Hillel, D. I. (1982). *Negev: Land, water and life in a desert environment*, Praeger, London, 288 pages.

Himmida, I. H. (1970). 'The Nubian Artesian Basin, its regional hydro-geological aspects and palaeohydrological reconstruction', *Journal of Hydrology, New Zealand*, **9**, 89–116.

Hollingworth, C. (1971). 'Egypt's Aswan balance-sheet', *The Times*, London, 15 January.

Holz, R. K. (1968). 'The Aswan High Dam', *Prof. Geogr.*, **20**, 230–7.

Houseman, J. (1961). 'Dust haze at Bahrain', *Met. Mag.*, **19**, 41–51.

Hurst, H. E. (1952). *The Nile*, London, Constable, 326 pages.

Ionides, M. G. (1937). *The Regime of the Rivers Euphrates and Tigris*, E. and F. Spon, London, 255 pages.

Ismail, A. M. A. (1984). 'The quality of irrigation ground-water in Qatar and its effect on the development of agriculture', *Journal of Arid Environments* **7(1)**, 101–6.

Israel Embassy (1964). *Israel Water project*, Washington DC.

Issar, A., A. Bein and A. Michaeli (1972). 'On the ancient waters of the Upper Nubian Sandstone aquifer in coastal Sinai and southern Israel', *J. Hydrol.*, **17**, 353–74.

Jabar Al-Mossani, M. A. and M. Salama (1983). 'Monitoring the coastal waters of Kuwait for chemical mutagens', *Environment International*, **9 (5)**, 343–8.

Jiabajee, N. A. (1957). 'Saudi Arabia - water supply of important towns', *Pakist. J. Sci.*, **9**, 189–201.

Johns, R. (1970). 'The Aswan High Dam - power for war torn Egypt', *The Financial Times*, London, 23 July.

Kanter, H. (1967). *Libya 1*, Geomedical Monograph Series, Geomedical Research Unit of the Heidelberg Academy of Sciences, Springer-Verlag, Berlin, 188 pages.

Kashaf, A-A. I. (1981). 'Technical and ecological impacts of the High Aswan dam', *Journal of Hydrology*, **53**, 73–84.

Kashaf, A-A. I. (1983). 'Salt-water intrusion in the Nile delta', *Groundwater*, **21(2)**, 160–7.

Kassas, M. (1971). 'The river Nile ecological system: A study towards an international programme', *Biological Conservation*, **4**, 19–25.

Katnelson, J. (1964). 'The variability of annual precipitation in Palestine', *Archiv für Meteorologie, Geophysik und Bioklimatologie, Serie B*, **13**, 163–72.

Kenyon, K. (1969). 'The origins of the Neolithic', *Advmt. Sci., Lond.*, **26**, 144–60.

Khalaf, J.M. (1956). 'The climate of Iran', *Union Géographique International Congrès International de Géographie, 18e, Rio de Janeiro, 1956, Comptes Rendus*, **2**, 507–25.

Khouri, J. (1972). 'Groundwater in the arid region of the Syrian Arab Republic', *Conf. Reclamation and Development of Desert Land*, Cairo, pp 61–93.

Khouri, J. (1982). 'Hydrogeology of the Syrian steppe and adjoining arid areas, *Q. J. eng. Geol. London*, **15**, 135–54.

Kirby, T.H. and P.R. James (1985). 'Public health engineering in Saudi Arabia – a case study', *Water Pollution Control*, **3**, 347–55.

Kie, G.W. (1984). 'Regulation of the White Nile', *Hydrological Sciences Journal*, **29 (2)**, 191–201.

Klitzsch, E., K. Weistroffer, C. Sonntag and E.M. El Shazly (1977). 'Fossil reserves of groundwater in the central Sahara', *Nat. Resources and Development*, **5**, 19–45.

Knill, J.L. and K.S. Jones (1968). 'Groundwater conditions in greater Tehrān', *Journal of Engineering Geology*, **1**, 181–94

Korzun, U.I. (1974). *Atlas of world water balances*, Hydromet Publishers, Moscow/Leningrad.

Lamb, H.H. (1972). *Climate: Present, Past and Future, Vol. 1, Fundamentals of Climate Now*, Methuen, London, 613 pages.

Lambton, A.K.S. (1969). *The Persian Land Reform*, Clarendon Press, Oxford, 386 pages.

Lavergne, M. (1986). 'The seven deadly sins of Egypt's Aswan High Dam', In *The Social and Environmental Effects of Large Dams*, (Eds E.Y. Goldsmith and N. Hildyard), Wadebridge Ecological Centre, pp 181–3.

Levi, M. (1963). The dry winter of 1962–63: a synoptic analysis', *Israel Explor. J.*, **13**, 229–41.

Little, T. (1965). *High Dam at Aswan*, Methuen, London, 242 pages.

Little, T. (1971). 'Why cry havoc at Aswan? *Middle East International*, June No 3, 5–6.

Lloyd, J.W. (1965). 'The hydrochemistry of the aquifers of North-Eastern Jordan', *Journal of Hydrology*, **3**, 319–30.

Lloyd, J.W. (1980). 'Aspects of environmental isotope chemistry in groundwater in eastern Jordan', Proc. IAEA Symp. *Application of isotope techniques to arid zone hydrology*, Vienna, pp 193–204.

Lloyd, J.W. and M.H. Farag (1978). 'Fossil groundwater gradients in arid regional sedimentary basins', *Groundwater*, **16**, 388–93.

Lomas, J. (1972). 'Forecasting wheat yields from rainfall data in Iran', *World Meteorological Bulletin*, **21**, 9–14.

Lowdermilk, W.C. (1944). *Palestine – Land of Promise*, Victor Gollancz, London, 167 pages.

Mageed, Y.A. (1984). The Jonglai Canal: a conservation project of the Nile', *International Journal of Water Resources Development*, **2 (2/3)**, 85–101.

Main, C.T. (1953). *The Unified Development of the Water Resources of the Jordan Valley Basin*, Boston, Massachusetts.

Marwick, R. and J.P. Germond (1975). 'The River Lar multi-purpose project in Iran', *Water Power and Construction*, **27**, 133–41.

McCaull, J. (1969). 'Conference on the ecological aspects of international development', *Nature and Resources*, (UNESCO), **5**, 5–12.

Meigs, P. (1964). 'Classification and occurrence of Mediterranean-type dry climate'. In *Land Use in Semi-arid Mediterranean Climates*, UNESCO, Paris, Arid Zone Research, **26**, 17–21.

Memon, B. A., A. Kazi and A. S. Bazuhair (1984). 'Hydrogeology of Water Al-Yammaniyah Saudi Arabia', *Ground Water*, **22 (4)**, 406–11.

Meteorological Office (1962). *Weather in the Mediterranean, Vol 1, (Second edition), General Meteorology*, Her Majesty's Stationery Office, London, 362 pages.

Ministry of Information, Tehrān, Iran (1970). *Nationalisation of Water Resources of Iran*, Tehrān, 20 pages.

Murray, G. (1951). 'The water beneath the Egyptian desert', *Geogrl J.*, **117**, 422–34.

Murray, G. W. (1951). 'The Egyptian climate: an historical outline', *Geogrl J.*, **117**, 422–34.

Murray, R. (1960). 'Mean 200-millibar winds at Aden in January 1958', *Met. Mag.*, **89**, 156–7.

Noel, E. (1944). 'Qanats', *Jl R. cent. Asian Soc.*, **31**, 191–202.

Noor-Abboud, A. and J. Crusoe (1984). 'Iraq: saline water goes dowm the drain', *Middle East Economic Digest*, **28 (33)**, 10–11.

Oppenheimer, H. R. (1951). 'Summer drought and water balance of plants growing in the Near East', *J. Ecol.*, **39**, 356–62.

Orsenan, N. (1963). 'Climatic fluctuations in the Middle East during the period of instrumental record', In *Changes of Climate*, UNESCO, Arid Zone Research, Vol 20, 67–74.

Otkun, G. (1969). '. . . and the search for supplies underground is extensive', *The Financial Times*, London, 23 June.

Otkun, G. (1970). 'A thirsty nation looks underground', *The Financial Times*, London, 28 December.

Otkun, G. (1969). *Outlines of Ground Water Resources of Saudi Arabia*, Presented at International Conference on Arid Lands in a Changing World, Tucson, Arizona, USA, 3 June, 1969, 16 pages.

Özis, U. (1983). 'Development plan of the western Tigris Basin in Turkey', *International Journal of Water Resources Development*, **1(4)**, 343–52.

Pachur, H. J. and G. Braun (1980). 'The Palaeoclimate of the Central Sahara, Libya and the Libyan Desert'. In *The Palaeoecology of Africa and the surrounding islands*, A. A. Balkema, Rotterdam, pp 351–63.

Pallas, P. (1981). 'Water Resources of the Socialist People's Libyan Arab Jamahiriya'. In *Geology of Libya*, (Eds M. J. Salam and M. T. Busrewil) Academic Press, London, pp 539–94.

Palmer, C. F. (1986). 'The problems of an urban water supply in Khartoum, Sudan', *Middle East Water & Sewage*, **10(2)**, 47–50.

Parkinson, H. L. and G. F. Worts (1977). *A brief evaluation of groundwater and soils potential for irrigated agriculture, western desert, Egypt*, US Geological Survey, Open File Report.

Parks, Y. P. and P. J. Smith (1983). 'Forces affecting the flow of Aflaj in Oman: a modelling study', *Journal of Hydrology*, **65**, 293–312.

Pedgley, D. E. and P. M. Symmons (1968). 'Weather and the locust upsurge', *Weather, Lond.*, **23**, 484–92.

Penman, H. L. (1948). 'Natural evaporation from open water, bare soil and grass', *Proc. R. Soc. Series A.*, **193**, 120–45.

Perrin de Brichambaut, G. and C. C. Wallén (1963). *A Study of Agroclimatology in Semi-arid and Arid zones of the Near East*, World Meteorological Organization, Technical Note No 56, Geneva, 64 pages.

Pike, J. G. (1970). 'Evaporation of groundwater from coastal playas (Sabkhah) in the Arabian Gulf', *Journal of Hydrology*, **11**, 79–88.

Plan Organisation (1969). *Dam Construction in Iran*, Bureau of Information and Reports, Tehran, 87 pages.

Powers, R. W., L. F. Ramiercz, C. D. Redmond and E. Elberg (1966). *Sedimentary geology of Saudi Arabia*, Prof. Pap. U.S. Geol. Surv. 560 D.

Prunshansky, Y. (1967). *Water Development*, Israel Digest, Israel Today, No 11, Jerusalem, 40 pages.

Raikes, R. L. (1967). *Water, Weather and Prehistory*, John Baker. London, 208 pages.

Ramage, C. S. (1966). 'The summer atmospheric circulation over the Arabian Sea', *Journal of Atmospheric Sciences*, **23**, 144–50.

Ramamurthy, L. A. and J. W. Holmes (1985). 'In search of a characteristic signature for groundwater aquifers – a case study from Israel', *Journal of Hydrology*, **79(3/4)**, 389–400.

Ramaswamy, C. (1965). 'On a remarkable case of dynamical and physical interaction between middle and low latitude weather systems over Iran', *Indian J. Met. Geophys.*, **16**, 177–200.

Raphael, C. N. and H. T. Shairi (1984). 'Water resources for At Tauf, Saudi Arabia: a study of alternative sources for an expanding urban area', *Geographical Journal*, **150(2)**, 183–91.

Ray Choudhuri, A. N., Y. H. Subramanyan and R. Cehllappa (1959). 'A climatological study of storms and depressions in the Arabian Sea', *Indian J. Met. Geophys.*, **10**, 283–90.

Rieben, E. H. (1953). 'Les Resources en Eaux Souterraines de la Plaine Alluviale de Tehran, FAO, Rapport No 168, Rome, 18 pages.

Ritchie, M. (1985). 'Libya: taking the plunge with the GMR', *Middle East Economic Digest*, **29 (29)**, 14–16.

Robinson, J. (1984). 'The treatment of sewage for use in municipal irrigation in Abu Dhabi, UAE, *Water Pollution Control*, **83 (3)**, 387–99.

Rosenthal, E. and S. Mandel (1985). 'Hydrological and hydrogeochemical methods for the delineation of complex groundwater flow systems as evidenced in the Beit-Shean Valley, Israel', *Journal of Hydrology*, **79 (3/4)**, 231–60.

Rosenthal, S. L. and T. A. Gleeson (1959). 'Sea-level anticylogenesis affecting Mediterranean weather', *Arch. Met. Geophys. Bioklim., Serie A, Meteorologie und Geophysik*, **9**, 185–90.

Royal Meteorological Society, (1966). *World Climate from 8000 to 0 B.C.*, Royal Meteorological Society, London, 229 pages.

Salam, M. J. and M. T. Busrewil (Eds) (1981). *Geology of Libya*, Academic Press, London.

Scott, K. F., W. T. N. Reeve and J. P. Germond (1968). 'Farahnaz Pahlavi dam at Latiyan', *Proc. Istn. civ. Engrs*, **39**, 353–95.

Serjeant, R. B. (1964). 'Some irrigation systems in Hadramaut', *Bulletin of the School of Oriental and African Studies*, **27**, 33–76.

Shata, A. A. (1982). 'Hydrology of the Great Nubian Sandstone basin, Egypt', *Q. J. eng. Geol. London*, **15**, 127–33.

Sherbrook, W. C. and P. Paylore (1973). *World Desertification: Cause and Effect*, Arid Lands Resource Information Paper No 3, University of Arizona, Office of Arid Lands Studies, Tucson, Arizona, 168 pages.

Shuval, H.I. (1977). *Water Renovation and Reuse*, Academic Press, New York.

Sivall, T. (1957). 'Sirocco in the Levant', *Geogr. Annlr.*, **39**, 114–42.

Smith, C. G. (1966). 'The disputed waters of the Jordan', *Trans. Inst. Br. Geogr*, **40**, 111–28.

Smith, C. G. (1970). 'Water resources and irrigation development in the Middle East', *Geography*, **55**, 407–25.

Stevens, J. H. (1972). 'Oasis agriculture in the central and eastern Arabian Peninsula', *Geography*, **57**, 321–6.

Stone & Associates (1984). *Projected world market for seawater desalination equipment*, Stone & Associates, Oak Ridge, Tennessee, Oak Ridge National Laboratory, 121 pages.

Storke, C. (1959). 'Evapotranspiration problems in Iraq', *Neth. J. agric. Sci.*, **7**, 269–82.

Striem, H. L. (1967). 'Rainfall groupings in the Middle East', *Bull. inst. Ass. scient. Hydrol.*, **11**, 59–64.

Swailem, F. M., M. S. Hamza and A. I. M. Aly (1983). 'Isotopic composition of groundwater in Kufra, Libya', *International Journal of Water Resources Development*, **1 (4)**, 331–41.

Taylor, G. (1955). *Australia*, Methuen, London, 490 pages.

Thatcher, L., M. Rubin and G. Brown (1961). 'Dating desert ground water', *Science, N.Y.*, **134**, 105.

Thornthwaite, C. W. and J. R. Mather (1957). *Instructions and Tables for Computing the Potential Evapotranspiration and the Water Balance*, Publs. Clim. Drexel. Inst, Technol., Vol 10, 311 pages.

Thornthwaite, C. W., J. R. Mather and D. B. Carter (1958). *Three Water Balance Maps of Southwest Asia*, Publs. Clim. Drexel. Inst. Technol., Vol 11, 57 pages.

Twistleton-Wykenham-Fiennes, D. (1970). 'Sweet water for the hottest land', *Geogrl Mag.*, **42**, 889–93.

Twitchell, K. S. (1944). 'Water resources of Saudi Arabia', *Geogrl Rev.*, **34**, 365–86.

UNESCO/FAO (1963). *Biochemical map of the Mediterranean Zone*, Explanatory notes, UNESCO. Paris, Arid Zone Research, Vol 21, 58 pages.

UNESCO/FAO (1963). *Environmental Physiology and Psychology in Arid Regions – Reviews of Research*, Arid Zone Research, Vol 22, UNESCO, Paris, 345 pages.

UNESCO/FAO (1963). *Changes of Climate*, Arid Zone Research, Vol 20, UNESCO, Paris, 488 pages.

UNESCO (1969). *Discharge of Selected Rivers of the World (Vol 1)*, UNESCO–IASH, Paris, 70 pages.

UNESCO (1971). *Discharge of Selected Rivers of the World (Vol 2), Monthly and Annual Discharges Recorded at Various Selected Stations (from start of observations up to 1964)*, UNESCO, Paris, 194 pages.

UNESCO (1971). *Discharges of Selected Rivers of the World (Vol 3), Mean Monthly and Extreme Discharges (1965–1969)*, (Part I), UNESCO, Paris, 98 pages.

United Arab Republic, Ministry of the High Dam, Aswân High Dam Authority (1972). *Aswân High Dam – Commissioning of the First Units – Transmission of Power to Cairo*, Ministry of the High Dam, Aswân, Egypt, 76 pages.

United Arab republic, Information Department (1963). *The High Dam – Bulwark of our Future*, Information Department, Cairo, 36 pages.

Vahidi, M. (1969). *Water and Irrigation in Iran*, Plan Organisation and Bureau of Information and Reports, Tehrān, 79 pages.

Van Riper, J. E. (1971). *Man's Physical World*. McGraw-Hill, New York, 713 pages.

Wafer, T. A. and A. H. Labib (1973). 'Seepage from Lake Nasser', In *Man-made Lakes: their Problems and Environmental Effects*, (Eds W.C. Ackerman, G.F. Waite, E.B. Worthington and J.L. Young), American Geophysical Union, Geophysical Monograph, **17**, pp 287–291.

Waterbury, J. (1979). *Hydropolitics of the Nile Valley*, Syracuse University Press, New York.

Weickmann, L. (1961). *Some Characteristics of the Sub-tropical Jet Stream in the Middle East and Adjacent Regions*, Meteorological Department, Tehrān, Iran, Meteorological Publications, Series A, No 1, 29 pages.

Weickmann, L. (1963). 'Meteorological and hydrological relationships in the drainage area of the Tigris River north of Baghdad', *Met. Rdsch.*, **16**, 33–8.

Whittington, D. and K. H. Haynes (1985). 'Nile water for whom? Emerging conflicts on water allocation for agricultural expansion in Egypt and Sudan'. In *Agricultural Development in the Middle East*, (Eds P. Beaumont and K. S. McLachlan), Wiley, Chichester, pp 125–149.

Wiener, A. (1972). *The Role of Water in Development*, McGraw-Hill, New York, 483 pages.

Wiener, A. (1977). 'Coping with water deficiency in arid and semi-arid countries through high efficiency water management', *Ambio*, **6**, 77–82.

Wilcox, L. V. (1955). *Classification and Use of Irrigation Waters*, US Dept. Agric. Circular No 969, Washington, 29 pages.

Wittfogel, K. A. (1956). 'The hydraulic civilisation'. In *Man's Role in Changing the Face of the Earth*, (Ed W. L. Thomas). University of Chicago Press, pp 152–64.

World Meteorological Organisation (1964). *High-level forecasting for turbine-engined aircraft operations over Africa and the Middle East*, Proc. of the Joint ICAO/WMO Seminar, Cairo-Nicosia 1961, World Meteorological Organisation, Technical Note 64.

Wright, E. P., A. C. Benfield, W, M. Edmunds and R. Kitching (1982). 'Hydrology of the Kufra and Sirte basins, eastern Libya', *Q. J. eng. Geol. London*, **15**, 83–103.

Wulff, H. E. (1968). 'Qanats of Iran', *Scient. Am.*, **218**, 94–105.

Chapter 3: Landscape Evolution

Abu-Lughod, J. (1972). *Cairo: 1001 Years of the City Victorious*, Princeton University Press, Princeton.

Adams, R. M. (1960). 'Factors influencing the rise of civilisation in the alluvium illustrated by Mesopotamia'. In *City Invincible. A Symposium on Urbanism and Cultural Development in the Ancient Near East*, (Eds C. H. Kraeling and R. M. Adams) University of Chicago Press, Chicago and London, 1960, pp 24–34.

Adams, R. M. (1962). 'Agriculture and urban life in early southwestern Iran', *Science, N. Y.*, **136**, 109–22.

Adams, R. M. (1965). *Land Behind Baghdad. A History of Settlement on the Diyala Plains*, University of Chicago Press, Chicago and London.

Adams, R. M. (1981). *Heartland of Cities*, University of Chicago Press, Chicago and London.

Adams, R. M. and H. J. Nissen, (1972). *The Uruk Countryside. The Natural Setting of Urban Societies*, University of Chicago Press, Chicago and London.

Aharoni, Y. (1968). *The Land of the Bible: An Historical Geography*, Burns and Oates, London.

Amin, S. (1966). *L'économie du Maghreb*, **1**, *Colonisation et la Dècolonisation*, Editions de Minuit, Paris.

Atlas des Centuriations romaines de Tunisie. Institut Geographiques National, Paris, 1954.

Baradez, J. (1949). *Fossatum Africae*, Arts et Métiers Graphiques, Paris.

Barbour, N. (Ed) (1962). *A Survey of North-West Africa*, 2nd edn, Oxford University Press, London.

Bender, B. (1975). *Farming in Prehistory: From Hunter-gatherer to Food Producer*, John Baker, London.

Benevisti, M. (1970). *Crusaders in the Holy Land*, Israel Universities Press, Jerusalem.

Bintliff, J. L. and W. Van Zeist (1982). *Palaeoclimates, Palaeoenvironments and Human Communities in the Eastern Mediterranean Region in Later Prehistory*, British Archaeological Reports, International Series 133, Oxford.

Braidwood, R. J. and C. A. Reed, (1957). 'The achievement and early consequences of food production: a consideration of the archaeological and natural-historical evidence'. *Cold Springs Symposium on Quantitative Biology*, **22**, 19–31.

Brice, W. C. and A. N. Balci (1955). The history of forestry in Turkey, *Orman Fakültesi Dergisi İstanbul Ünversitesi*, **5**, 19–42.

Buchanan, R. H., E. Jones and D. McCourt (Eds) (1971). *Man and his Habitat. Essays Presented to Emyr Estyn Evans*, Routledge and Kegan Paul, London.

Bulliet, R. W. (1975). *The Camel and the Wheel*, Harvard University Press, Cambridge, Mass.

Butzer, K. W. (1970). 'Physical conditions in eastern Europe, western Asia and Egypt before the period of agricultural and urban settlement'. In *The Cambridge Ancient History*, **1**, 3rd edn, Cambridge University Press.

Butzer, K. W. (1972). *Environment and Archaeology*, 2nd edn, Methuen, London

Butzer, K. W. (1976). *Early Hydraulic Civilisation in Egypt: A Study in Cultural Ecology*, University of Chicago Press, Chicago and London.

Caillemer, A. et R. Chevalier, (1954). 'Les centuriations romaines de l'Africa vetus', *Annales, E.S.C.*, **9**, 433–60.

Caillemer, A. et R. Chevalier, (1957). 'Centuriations romaines de Tunisie', *Annales, E.S.C.*, **12**, 275–286.

Carlstein, T. (1982). *Time Resources, Society and Ecology*, **1**, George Allen and Unwin, London.

Clark, J. D. (1962). 'The spread of food production in sub-Saharan Africa', *Journal of African History*, **3**, 211–28.

Clark, J. D. (1964). 'The prehistoric origins of African culture', *Journal of African History*, **5**, 161–83.

Clark, J. D. (1967). *Atlas of African Prehistory*, University of Chicago Press, Chicago and London.

Clarke, J. I. and H. Bowen-Jones. (Eds) (1981). *Change and Development in the Middle East*, Methuen, London.

Cuinet, V. (1890–94). *La Turquie d'Asie, Géographie Administrative, Statistique, Déscriptive et Raisonée de Chaque Provence de l'Asie Mineure*, 4 vols, Ernest Leroux, Paris.

Despois, J. (1961). 'Development of land use in northern Africa'. In *A History of Land Use in Arid Regions, Arid Zone Research*, (Ed L. D. Stamp), Vol 17, UNESCO, pp 219–37.

Edwards, I. E. S., C. J. Gadd, and N. G. L. Hammon, (Eds) (1970). *The Cambridge Ancient History*, 3rd edn, Vol 1, Cambridge University Press, London.

Eisma, D. (1962). 'Beach ridges near Selçuk, Turkey', *Tijdschr.K.ned. aardrijksk. Genoot.*, **79**, 234–246.

English, P. W. (1966). *City and Village in Iran. Settlement and Economy in the Kirman Basin*, University of Wisconsin Press, Madison.

Fisher, S. N. (1968). *The Middle East: A History*, 2nd edn, Knopf, New York.

Fowler, G. L. (1972). Italian colonisation of Tripolitania, *Ann. Ass. Am. Geogr*, **62**, 627–40.

Frank, T. (Ed). (1933–40). *An Economic History of Ancient Rome*, 5 vols, Johns Hopkins, Baltimore.

Garrod, D. A. E. (1970). 'Primitive man in Egypt, western Asia and Europe in Palaeolithic times'. In *The Cambridge Ancient History*, **1**, 3rd edn, Cambridge University Press, pp 70–89.

Goodwood, R. G. (1952a). 'Farming in Roman Libya', *Geogrl Mag.*, **25**, 70–80.

Goodwood, R. G. (1952b). 'The mapping of Roman Libya', *Geogrl J.*, **118**, 142–152.

Grünebaum, G. E. von, (1955). The Muslim town and the Hellenistic town, *Scientia*, **90**, 364–70.

Hachicho, M. A. (1964). 'English travel books about the Arab Near East in the eighteenth century', *Die Welt der Islam*, **9**, 1–206.

Hamdan, G. (1961). 'Evolution of irrigation agriculture in Egypt'. In *A History of Land Use in Arid Regions, Arid Zone Research*, (Ed L. D. Stamp), Vol 17, UNESCO, 119–42.

Harding, A. F. (ed) (1982). *Climatic Change and Later Prehistory*, Edinburgh University Press.

Harris, D. R. (1967). 'New light on plant domestication and the origins of agriculture: a review', *Geogrl Rev.*, **57**, 90–107.

Hasan, M. S. (1958). 'Growth and structure of Iraq's population', 1867–1947, *Bull. Oxf. Univ. Inst. Statist.*, **20**, 339–52.

Hershlag, Z. Y. (1964). *Introduction to the Economic History of the Middle East*, E. J. Brill, Leiden.

Heyd, U. (1960). *Ottoman Documents on Palestine, 1552–1615*, Oxford University Press.

Holt, P. M., A. K. S. Lambton, and B. Lewis (Eds) (1970–71). *The Cambridge History of Islam*, 2 vols, Cambridge University Press.

Hourani, A. (1957). 'The changing face of the Fertile Crescent in the eighteenth century', *Studia Islamica*, **8**, 89–122.

Hutchinson, J., G. Clark, E. M. Jope and R. Riley (Eds) (1977). *The Early History of Agriculture*, Oxford University Press.

Hütteroth, W. (1962). 'Getreidekonjunktur und jüngerer Siedlungsausbau im südlichen Inneranatolien', *Erdkunde*, **16**, 249–71.

Hütteroth, W. (1969). 'Schwankungen von Siedlungsdichte und Siedlungsgrenze in Palästina und Transjordanien seit dem 16 Jahrhundert', *Deutscher Geographentag. Kiel 21–26 Juli*, pp 463–75.

Issawi, C. (Ed) (1966a). *The Economic History of the Middle East, 1800–1914*, University of Chicago Press.

Issawi, C. (1966b). *Egypt in Revolution: An Economic Analysis*, Oxford University Press.

Jones, A. H. M. (1966). *The Decline of the Ancient World*, Longmans, London.

Knight, M. M. (1952–53). Economic space for Europeans in French North Africa, *Economic Development and Cultural Change*, **1**, 360–75.

Kraeling, C. H. and R. M. Adams (Eds) (1960). *City Invincible. A Symposium on Urbanism and Cultural Development in the Ancient Near East*, University of Chicago Press, Chicago and London.

Kurmuş, O. (1974). 'The Role of British Capital in the Development of Western Anatolia', (unpublished PhD thesis, University of London).

Lamb, H. H. (1968). 'Climatic background to the birth of civilisation', *Advt. Sci. Lond.*, **25**, 103–20.

Lampl, P. (1968). *Cities and Planning in the Ancient Near East*, Studio Vista, London.

Lander, D. S. (1958). *Bankers and Pashas. International Finance and Economic Imperialism in Egypt*, Heinemann, London.

Le Strange, G. (1890). *Palestine under the Muslims: A Description of Syria and the Holy Land from A.D. 650 to 1500*, Palestine Exploration Fund, London.

Lewis, N. N. (1949). 'Malaria, irrigation and soil erosion in central Syria', *Geogrl Rev.*, **39**, 278–90.

Lewis, N. N. (1955). 'The frontier of settlement in Syria, 1800–1950', *International Affairs*, **31**, 48–60.

Lombard, M. (1959). Les bois dans la Méditerranée musulmane, *Annales, E.S.C.*, **14**, 234–54.

Marchese, R. T. (1986). *The Lower Meander Flood Plain: A Regional Settlement Study*, British Archaeological Reports, International Series 292. Oxford.

Margalit, H. (1964). 'Some aspects of the cultural landscape of Palestine during the first half of the nineteenth century', *Israel Explor. J.*, **13**, 208–23.

Marthelot, P. (1965). 'Baghdād: notes de geographie humaine', *Annls. Géogr.*, **74**, 24–37.

Matson, F. R. (1966). 'Power and fuel resources in the ancient Near East, *Advmt Sci., Lond.*, **23**, 146–153.

Mikesell, M. W. (1969). The deforestation of Mount Lebanon, *Geogrl Rev.*, **58**, 1–28.

Owen, E. R. J. (1969). *Cotton and the Egyptian Economy, 1820–1914: A Study in Trade and Development*, Oxford University Press.

Owen, R. (1981). *The Middle East in the World Economy, 1800–1914*, Methuen, London.

Quataert, D. (1977). 'Limited revolution: the impact of the Anatolian Railway on Turkish transportation and the provisioning of Istanbul, 1890–1908', *Business History Review*, **51**, 139–60.

Quataert, D. (1980). 'The commercialisation of agriculture in Ottoman Turkey', 1800–1914, *International Journal of Turkish Studies*, **1**, 38–55.

Reed, C. A. (Ed) (1977). *Origins of Agriculture*, Mouton, The Hague and Paris.

Reifenberg, A. (1958). *Struggle between the Desert and the Sown: Rise and Fall of Agriculture in the Levant*, Government Press, Jerusalem.

Renfrew, J. M. (1973). *Palaeoethnobotany: The Prehistoric Food Plants of the Near East and Europe*, Methuen, London.

Rosenam, N. (1963). 'Climatic fluctuations in the Middle East during the period of instrumental record'. In *Changes in Climate. Proceedings of the Rome Symposium Organised by UNESCO and the World Meteorological Organisation. Arid Zone Research*, vol 20, UNESCO, pp 67–73.

Rostovtzeff, M. I. (1941). *The Social and Economic History of the Hellenistic World*, 3 vols. Oxford University Press.

Rowton, M. B. (1967). 'The woodlands of ancient Asia', *Journal of Near Eastern Studies*, **26**, 261–77.

Russell, R. J. (1954). 'Alluvial morphology of Anatolian rivers', *Ann. Ass. Am. Geogr.*, **44**, 363–91.

Smith, C. G. (1970). 'Water resources and irrigation development in the Middle East', *Geography*, **55**, 407–25.

Stamp, L. D. (Ed) (1961). *A History of Land Use in Arid Regions, Arid Zone Research*, Vol 17, UNESCO.

Stigler, R., R. Holloway, R. Solecki, D. Perkins and P. Daly, (1974). *The Old World. Early Man to the Development of Agriculture*, St Martins Press, New York.

Thomas, W. L. (Ed) (1956). *Man's Role in Changing the Face of the Earth*, Chicago University Press, Chicago and London.

Tignor, R. L. (1966). *Modernisation and British Colonial Rule in Egypt, 1882–1914*, Princeton University Press, Princeton.

Ucko, P. J. and G. W. Dimbleby (Eds) (1970). *The Domestication and Exploitation of Plants and Animals*, Duckworth, London.

Ucko, P. J., R. Tringham and G. W. Dimbleby, (Eds) (1972). *Man, Settlement and Urbanism*, Duckworth, London.

Van Zeist, W. and S. Bottema, (1982). 'Vegetational history of the eastern Mediterranean and the Near East during the last 20,000 years'. In *Palaeoclimates, Palaeoenvironments and Human Communities in the Eastern Mediterranean Region in Later Prehistory*, (Eds J. L. Bintliff and W. Van Zeist), British Archaeological Reports, International Series 133, Oxford, pp 277–321.

Vita-Finzi, C. (1969). *The Mediterranean Valleys. Geological Changes in Historical Times*, Cambridge University Press.

Volney, M. C. F. (1786). *Voyage en Syrie et en Egypte pendant les années 1783, 1784 et 1785*, 2 vols. Desenne and Volland, Paris.

Wagstaff, J. M. (1985). *The Evolution of Middle Eastern Landscapes: An Outline to A.D. 1800*, Croom Helm, London.

Watson, A. M. (1983). *Agricultural Innovation in the Early Islamic World*, Cambridge University Press.

Whyte, R. O. (1961). 'Evolution of land use in south-western Asia'. In *A History of Land Use in Arid Regions, Arid Zone Research*, (Ed L. D. Stamp), Vol 17, UNESCO, pp 57–118.

Wright, H. E. (1968). Natural environment of early food production north of Mesopotamia, *Science, N. Y.*, **161**, 334–339.

Wright, H. E. and G. A. Johnson, (1977). 'Population, exchange and early state formation in south-western Iran', *American Anthropological Journal*, **77**, 267–89.

Zohary, D. and P. Spiegel-Roy, (1975). 'The beginnings of fruit-growing in the Old World', *Science N. Y.*, **187**, 319–27.

Chapter 4. Rural Land Use: Patterns and Systems

Abu'Adhirah, S. S. (1978). 'Sedentarisation and the settlement of the bedouin', *Arabian Studies*, **4**, 1–5.

Allan, J. A., K. S. McLachlan and E. T. Penrose (Eds) (1973). *Libya: Agriculture and Economic Development*, Frank Cass, London.

Aresvik, O. (1975). *The Agricultural Development of Turkey*, Praeger, New York, Washington and London.

Arnon, I. and M. Raviv, (1981). *From Fellah to Farmer. A Study on Change in Arab Villages*, Rehovot, Tel Aviv.

Askari, H., J. T.Cummings and B. Harik (1977). 'Land reform in the Middle East', *International Journal of Middle East Studies*, **8**, 437–51.

Baer, G. (1964). *Population and Society in the Arab East*, Routledge and Kegan Paul, London.

Barth, F. (1959–60). 'The land use pattern of migratory tribes of south Persia', *Norsk Geografisk Tidsskrift*, **17**, 1–11.

Bates, D. G. (1973). *Nomads and Farmers: A Study of the Yörük of Southeastern Turkey*, University of Michigan Press, Ann Arbor.

Beaumont, P. (1968). 'Qanats in the Varamin Plain', *Trans. Inst. Br. Geogr.*, **45**, 169–80.

Beaumont, P. and K. McLachlan (Eds) (1985). *Agricultural Development in the Middle East*, John Wiley, Chichester.

Beckett, P. H. T. (1953). 'Qanāts around Kerman', *Jl R. cent. Asian Soc.*, **40**, 47–57.

Benedick, R. E. (1979). 'The High Dam and the transformation of the Nile', *Middle East Journal*, **33**, 119–45.

Benedict, P., E. Tümertekin and F. Mansur (Eds) (1974). *Turkey: Geographic and Social Perspectives*, E. J. Brill, Leiden.

Birks, J. S. (1977). 'The reactions of rural populations to drought: a case study from south-east Arabia', *Erdkunde*, **31**, 299–308.

Birks, J. S. (1981). 'The impact of economic development on pastoral nomadism in the Middle East: an inevitable eclipse?'. In *Change and Development in the Middle East*, (Eds J. I. Clarke and H. Bowen-Jones) Methuen, London, 82–94.

Birks, J. S. (1984). 'The falaj: modern problems and some possible solutions', *Water Lines*, **2**, 28–31.

Caponera, D. A. (1954). *Water Laws in Moslem Countries*, FAO Development Paper No 43, FAO, Rome.

Clarke, J. I. and H. Bowen-Jones (Eds) (1981). *Change and Development in the Middle East*, Methuen, London.

Clawson, M., H. H. Landsberg and L. T. Alexander (1971). *The Agricultural Potential of the Middle East*, American Elsevier Publishing Co, New York, London, Amsterdam.

Cole, D. P. (1975). *Nomads of the Nomads: The al-Murrah Bedouin of the Empty Quarter*, Aldine-Atherton, Inc, Chicago.

Drysdale, A. and G. H. Blake (1985). *The Middle East: A Political Geography*, Oxford University Press.

Eldblom, L. (1961). *Quelques Points de Vue Comparatifs sur les Problèmes d'Irrigation dans les Trois Oases Libyennes de Brâk, Ghadamès et particulièrement Mourzouk*, Lund Studies in Geography No 22, Royal University of Lund, Sweden.

Encyclopaedia of Islam (1934). Volume 4, E. J. Brill, Leyden and Luzac and Co, London.

English, P. E. (1967). 'Urbanites, peasants and nomads: the Middle Eastern ecological trilogy', *J. Geogr.*, **61**, 54–9.

English, P. W. (1968). 'The origin and spread of qanats in the Old World', *Proceedings of the American Philosophical Society*, **112**, 170–81.

Eshag, E. and M. A. Kamal (1968). 'Agrarian reform in the United Arab Republic (Egypt)', *Bulletin of the Oxford Institute of Economics and Statistics*, **30**, 73–104.

Filali, M. Al, (1967). 'Sedentarisation and land problems'. In *Land Policy in the Near East*, (Ed M. R. El Ghonemy), FAO, Rome, pp 38–52.

Fattah, F. (1972). 'Farming cooperatives in Egypt', *World Marxist Review*, **18**, 96–9.

Fisher, W. B. (1968). 'The Land of Iran', Vol 1, *The Cambridge History of Iran*, Cambridge University Press.

George, A. R. (1973). 'Sedentarisation of nomads in Egypt, Israel and Syria: a comparison', *Geography*, **58**, 167–9.

Ghonemy, M. R. El (1967). 'The economic and social development of nomadic populations before and after settlement'. In *Land Policy in the Near East*, (Ed M. R. El Ghonemy), FAO, Rome, pp 317–26.

Ghonemy, M. R. El (1968). 'Land reform and economic development in the Near East', *Land Econ.*, **44**, 36–49.

Goblot, H. (1963). 'Dans l'ancien Iran, les techniques de l'eau la grande histoire', *Annales, E.S.C.*, **18**, 499–520.

Hinderink, J. and M. B. Kiray (1970). *Social Stratification as an Obstacle to Development. A Study of Four Turkish Villages*, Praeger, New York, Washington and London.

Hütteroth, W-D. (1974). 'The influence of social structure on land division and settlement in inner Anatolia', In *Turkey: Geographic and Social Perspectives*, (Eds P. Benedict, E. Tümertekin and F. Mansur), E. J. Brill, Leiden, 19–47.

Ibrahim, A. (1968). 'Classification and characteristic patterns of rural settlements (Egypt)', *Mediterranea*, **23-24**, 332–45.

Issawi, C. (1971). 'Growth and structural change in the Middle East', *Middle East Journal*, **25**, 309–24.

Johnson, D. L. (1969). *The Nature of Nomadism. A Comparative Study of Pastoral Migrations in Southwestern Asia and Northern Africa*, Department of Geography, Research Paper No 118, Chicago.

Katouzian, M. A. (1974). 'Land reform in Iran – case study in the political economy of social engineering', *Journal of Peasant Studies*, **1**, 220–39.

Keilany, Z. (1980). 'Land reform in Syria', *Middle Eastern Studies*, **16**, 209–24.

Kolars, J. (1966). 'Locational aspects of cultural ecology: the case of the goat in non-western agriculture', *Geogrl Rev.*, **56**, 577–84.

Kolars, J. F. (1967). 'Types of rural development'. In *Four Studies on the Economic Development of Turkey*, (Eds F. C. Shorter, J. F. Kolars, D. A. Rustow and O. Yenal), Frank Cass, London, pp 63–87.

Lambton, A. K. S. (1969). *The Persian Land Reform, 1962–66*, Oxford University Press.

Latron, A. (1936). *La Vie Rurale en Syrie et au Liban*, Beirūt.

Marx, E. (1977). 'The tribe as unit of subsistence: nomadic pastoralism in the Middle East', *American Anthropologist*, **79**, 343–63.

Mitchell, W. A. (1971). 'Turkish villages in interior Anatolia and von Thünen's "Isolated State"; a comparative analysis', *Middle East Journal*, **25**, 355–69.

Nattagh, N. (1986). *Agricultural and Regional Development in Iran, 1962–1978*, Menas Press, Wisbech.

Nelson, C. (Ed) (1973). *The Desert and the Sown: Nomads in the Wider Society*, California University Press, Berkeley.

Noel, E. (1944). 'Qanāts', *Royal Central Asian Journal*, **31**, 191–202.

O'Brien, P. (1966). *The Revolution in Egypt's Economic System, from Private Enterprise to Socialism, 1952–1965*, Oxford University Press, London.

Perrin de Brichambaut, G. and C. C. Wallén (1963). *A Study of the Agroclimatology in Semi-Arid and Arid Zones of the Near East*, Technical Note No 56, World Meteorological Organisation, Geneva.

Planhol, X. de (1968). 'Geography of settlement'. In *The Cambridge History of Iran, the Land of Iran*, (Ed W. B. Fisher), vol 1, Cambridge University Press.

Radwan, S. and Lee, E. (1986). *Agrarian Change in Egypt: An Anatomy of Rural Poverty*, Croom Helm, London.

Raphaeli, N. (1966). 'Agrarian reform in Iraq: some political and administrative problems', *Journal of Administration Overseas*, **5**, 102–11.

Richards, A. (1980). 'The agricultural crisis in Egypt', *Journal of Development Studies*, **16**, 303–21.

Ron, Z. (1966). 'Agricultural terraces in the Judean mountains', *Israel Explor. J.*, **16**, 111–22.

Shanin, T. (Ed) (1971). *Peasants and Peasant Societies*, Penguin Books, Harmondsworth.

Shorter, F. C., J. F. Kolars, D. A. Rustow and O. Yenal (Eds) (1967). *Four Studies on the Economic Development of Turkey*, Frank Cass, London.

Stauffer, T. R. (1965). 'The economics of nomadism in Iran', *Middle East Journal*, **22**, 284-302.

Tanoğlu, A. (1954). 'The geography of settlement', *Rev. geogr. Inst. Univ. Istanb.*, **1**, 3-27.

Tuma, E. H. (1970). 'Agrarian reform and urbanisation in the Middle East', *Middle East Journal*, **24**, 163-77.

Turkowski, L. (1969). 'Peasant agriculture in the Judean hills', *Palestine Exploration Quarterly*, **101**, 21-33, 101-12.

Tute, R. C. (1927). *The Ottoman Land Laws*, Jerusalem.

Vieille, P. (1972). 'Les paysans et l'état après le reforme agraire en Iran', *Annales, E.C.S.*, **27**, 347-72.

Van Nieuwenhuijze, C. A. O. (1962). 'The Near Eastern village: a profile'. *Middle East Journal*, **16**, 295-308.

Warriner, D. (1962). *Land Reform and Development in the Middle East*, 2nd Edn Oxford University Press.

Weulersse, J. (1946). *Paysans de Syrie et du Proche Orient*, Gallimard, Paris.

Wilkinson, J. C. (1977). *Water and Settlement in South-East Arabia. A Study of the Aflaj of Oman*, Oxford University Press.

Wilkinson, J. C. (1983). 'Traditional concepts of territory in southeast Arabia', *Geogrl. J.*, **149**, 301-15.

Wulff, H. E. (1966). *The Traditional Crafts of Persia*, MIT Press, Cambridge, Mass and London.

Chapter 5: Population

Abu Jaber, K. (Ed) (1980). *Levels and trends of fertility and mortality in selected Arab countries of West Asia*, University of Jordan, Amman.

Abu-Lughod, J. (1983). 'Social implications of labour migration in the Arab world', in *Arab Resources: the Transformation of a Society* (Ed I. Ibrahim), Croom Helm, London.

Abu-Lughod, J. (1984). 'Theories of development and population: a reassessment and an application to the Arab world'. *Population Bulletin of ECWA*, No 25, 21-48.

Allman, J. (Ed) (1978). Women's Status and Fertility in the Muslim World, Praeger, New York.

Allman, J. (1980). 'The demographic transition in the Middle East and North Africa', *International Journal of Middle East Studies*, **12(3)**, 277-301.

Arnold, F. and N. M. Shah (1984). 'Asian labour migration in the Middle East', *International Migration Review*, **18**, 302.

Barkan, O. L. (1958). 'Essai sur les données statistiques des registres de recensement dans l'Empire Ottoman aux XVe et XVIe siècles', *J. Econ. Soc. Hist. Orient*, **1**, 9-36 (1958).

Beck, L. and N. Keddie (1978). *Women in the Muslim World*, Harvard University Press, Cambridge, Mass.

Birks, J. S. (1986). 'The demographic challenge in the Arab Gulf', *Arab Affairs*, **1**, 72-86 (1986).

Birks, J. S. and C. A. Sinclair (1980). *Arab Manpower*, Croom Helm, London.

Birks, J. S. and C. A. Sinclair (1980). *International Migration in the Arab Region*, International Labour Office, Geneva, 137.

Birks, J. S. and C. A. Sinclair (1981). 'Demographic settling amongst migrant workers', IUSSP, Manila.

Blake, G. H. (1972). 'Israel: immigration and dispersal of population', in *Populations of the Middle East and North Africa*, (Eds J. I. Clarke and W. B. Fisher), University of London Press, London.

Bucht, B. and M. El-Badry (1986). 'Reflections on recent levels and trends of fertility and mortality in Egypt', *Population Studies,* **40(1)**, 101–13.

Caldwell, J. C. and P. Caldwell (1982). 'Fertility transition with special reference to the ECWA region', in *Population and Development in the Middle East*, ECWA, Baghdad, 97–118.

Chanie, J. (1977). 'Religious differentials in fertility: Lebanon', *Population Studies*, **31**, 365–82.

Clarke, J. I. (1985). 'Islamic populations: limited demographic transition', *Geography*, **70**, 118–28.

Clarke, J. I. and W B. Fisher (eds) (1972). *Populations of the Middle East and North Africa*, University of London Press.

Dewdney, J. C. (1972). 'Turkey: recent population trends', in *Populations of the Middle East and North Africa* (Eds J. I. Clarke and W. B. Fisher), University of London Press, London, 42.

Durand, J. D. (1967). 'The modern expansion of world population', *Proc. Amer. Phil. Soc.*, **111**, 151.

ECWA, United Nations (1982). *Population and Development in the Middle East*, UN Economic Commission for Western Asia, Baghdād.

Galaleldin, M. (1984). 'Demographic consequences of alternative types of migration in the Middle East'. In *Proceedings of the Workshop on the Consequences of International Migration*, IUSSP, Canberra.

George, A. R. (1973). 'Processes of sedentarisation of nomads in Egypt, Israel and Syria', *Geography*, **48**, 167–9.

Hassouna, M. T. (1980). 'Assessment of family planning service delivery in Egypt', *Studies in Family Planning*, **11**, 159–66.

HRD base (1987). *Socio-Demographic Profiles of Key Arab Countries* HRD base Ltd, Newcastle upon Tyne.

Ibrahim, I. (Ed) (1983). *Arab resources: the Transformation of a Society*, Croom Helm, London.

Hurewitz, J. C. (1963). 'The politics of rapid population growth in the Middle East', *J. Int. Affairs*, **19**, 27.

International Planned Parenthood Federation (1979). 'The Muslim world', *People*, **6(4)**, 1–30.

Jordan Demographic Survey 1981 (1983). Department of Statistics, Amman.

Lawless, R. I. (1972). 'Iraq: changing population patterns', in *Populations of the Middle East and North Africa* (Eds J. I. Clarke and W. B. Fisher), University of London Press.

Omran, A. R. (1981). *Arab Population*, Croom Helm, London.

Owen, R. (1985). *Migrant workers in the Gulf*, Minority Rights Group, London.

Rizk, H. (1978). 'Fertility trends and differentials in Jordan', in *Women's Status and Fertility in the Muslim World* (Ed J. Allman), Praeger, New York, 113.

Russell, J. C. (1958). 'Late ancient and medieval population', *Trans. Amer. Phil. Soc.*, **48**, 89.

Ryan, M. (1984). *Health Services in the Middle East*, Economist Intelligence Unit, London.

Samman, M. L. (1978). *La population de la Syrie. Etude geo-démographique*, ORSTOM, Paris, 1978, 260.

Seccombe, I. J. (1986). 'Economic recession and international labour migration', *Arab Gulf Journal*, **6**, 43–52.

Seccombe, I. J., C. H. Bleaney and B. Al-Najjar (1984). *International Migration for Employment in the Middle East: An Introductory Biblography*, Centre for Middle East and Islamic Studies, University of Durham.

Shaw, R. P. (1983). *Mobilizing Human Resources in the Arab world*, Kegan Paul, London.

Shorter, F. (1985). *Population Factors in Developing Planning in the Middle East*, Population Council, New York.

State of Israel (1972). Statistical Abstract, **23**, Jerusalem, 127.

Suchindran, C. M. and A. L. Adlakha (1985). 'Levels, trends and differentials of infant and child mortality in Yemen', *Population Bulletin of ESCWA*, **27**, 43–71.

Smart, J. E. (1986). 'Worker circulation between Asia and the Middle East', *Pacific Viewpoint*, **27(1)**, 1–28.

Tamari, S. (1979). 'Minorities in the Middle East', *Middle East Yearbook 1979*, International Communications, London, pp 41–5.

United Nations (1962). *Demographic Yearbook*, New York.

United Nations (1971). Population Distribution and Urbanisation in Selected Countries of the Middle East. In *Studies on Selected Development Problems in Various Countries of the Middle East*, New York, pp 59–78.

Wagstaff, J. M. (1986). 'A note on some nineteenth-century population statistics for Lebanon', *Bull. Bnt. Soc. Middle East Stud.*, **13**, 36–44.

Chapter 6: Towns and Cities

Abd-el-Aziz Nour (1981). 'Factors underlying traditional Islamic urban design', *Planning Outlook*, **24(1)**, 29–32.

Abu-Lughod, J. (1980). 'The growth of Arab cities', *Middle East Yearbook 1980*, International Communications, London, pp 39–44.

Aga Khan Award for Architecture (1978). *Toward an Architecture in the Spirit of Islam*, Seminar at Aiglement, France, April 1978, pp 1–119

Al-Hathloul and A. Ur-Rahmaan (1985). 'The evolution of urban and regional planning in Saudi Arabia', *Ekistics*, **52(312)**, 206–12.

Amin, S. (1966). *L'économie du Maghreb*, **1**, Editions de Minuit, Paris.

Azeez, M. M. (1968). 'Geographical aspects of rural migration from Amara Province, Iraq, 1955-1964'. Unpublished PhD thesis, University of Durham. (Source of Figure 6.8).

Baedeker, K. (1908) *Egypt and the Sudan*, 6th edn, Baedeker, Leipzig.

Berger, M. (Ed) (1963). *The New Metropolis in the Arab World*, Allied Publishers, New York.

Blake, G. H. (1968). *Misurata: a Market Town in Tripolitania*, University of Durham, Department of Geography.

Blake, G. H. and R. I. Lawless, (Eds) (1980). *The Changing Middle Eastern City*, Croom Helm, London.

Brice, W. C. (1966). *Southwest Asia*, University of London Press.

Brown, L. C. (Ed) (1972). *From Medina to Metropolis: Heritage and Change in the Near Eastern City*, Darwin Press, Princeton, New Jersey.

Chandler, T. and G. Fox (1974). *3,000 Years of Urban Growth*, Academic Press, New York and London.

Clark, B. D. and V. F. Costello (1973). 'The urban system and social patterns in Iranian Cities', *Trans. Inst. Br. Geogr.*, 49, 99–128.

Clarke, J. I. (1963). *The Iranian City of Shīrāz*, University of Durham, Department of Geography.

Clarke, J. I. and B. D. Clark (1969). *Kermānshāh: an Iranian Provincial city*, University of Durham, Centre for Middle Eastern and Islamic Studies.

Costa, F. J. and A. G. Noble (1986). 'Planning Arabic towns' *Geog. Review*, **76(2)**, 160–172.

Costello, V. (1980). 'Tehran'. In *Problems and Planning in Third World Cities*, (Ed M. Pacione), Croom Helm, London, pp 156–86.

Costello, V. F. (1977). *Urbanisation in the Middle East*, Cambridge University Press.

Cressey, G. B. (1960). *Crossroads*, J. P. Lippincott Co, International University Edition, New York.

Darwent, D. F. (1965). 'Urban growth in relation to socio-economic development and westernisation; Mashad, Iran'. PhD thesis, University of Durham. (Source of Figure 6.5).

Daryll Forde, C. (1932). 'The ancient cities of the Indus', *Geography*, **17**, 186.

Encyclopaedia Britannica (1902). Tenth edition, Edinburgh and London.

Findlay, A. and R. Paddison (1986). 'Planning the Arab city: the cases of Tunis and Rabat', *Progress in Planning*, **26**, 1–82.

Findlay, A. (1986). 'Amman: urbanization in a "charity state" ', *Bull. de la Soc Languedocienne de Géographie*, **20(2–3)**. 211–20.

Fisher, W. B. (1963). *The Middle East*, 5th edn, Methuen, London.

Fisher, W. B. (1968). 'The Land of Iran' Vol 1, The Cambridge History of Iran, Cambridge University Press.

Fogg, W. (1932). 'The sūq: a study in the human geography of Morocco', *Geography*, **17**, 257–67.

Garrett, J. (1936). 'The site of Damascus', *Geography*, **21**, 283–96.

Gibb, H. A. R. and H. Bowen (1950). *Islamic Society and the West*, Royal Institute of International Affairs, Oxford University Press, London.

Glubb, J. B. (1963). *The Great Arab Conquests*, Hodder and Stoughton, London.

Gordon-Childe, V. (1950). 'The urban revolution', *Town Planning Review*, **21**, 3–17.

Grill, N. C. (1984). *Urbanisation in the Arabian Peninsula*, Centre for Middle East and Islamic Studies, University of Durham.

Hershlag, Z. Y. (1964). *Introduction to the Modern Economic History of the Middle East*, E. J. Brill, Leiden.

Hiorns, F. R. (1956). *Town-building in History*, Harrap, London.

Hourani, A. H. and S. M. Stern (Eds) (1970). *The Islamic City*, Bruno Cassirer, Oxford.

Ismail, A. A. (1972). 'Origin, ideology and physical patterns of Arab urbanisation', *Ekistics*, **33**, 113–23.

Jacobs, J. (1969). *The Economy of Cities*, Jonathan Cape, London.

Karan, P. P. and W. A. Bladen (1983). 'Arab cities', *Focus*, **33(3)**, 1–8.

Kezeiri, S. K. (1985). 'The Middle East small towns', *Planning Outlook*, **28(2)**, 82–5.

Khalaf, S. (1983). 'Some salient features of urbanization in the Arab world', *Ekistics*, **50**, (300) 219–22.

Kliot, N. and A. Soffer (1986). 'The emergence of a metropolitan core area in a new state – the case of Jordan, *Asian and African Studies*, **24(2)**, 217–32.

Lapidus, I. M. (Ed) (1969). *Middle Eastern Cities*, Universities of California, Berkeley and Los Angeles.

Lebon, J. H. G. (1956). 'The site and modern development of Baghdād', *Bull. Soc. Géogr. Egypt*, **24**, 7–33.

Lewis, P. G. (1983). 'Iranian cities', *Focus*, **33(3)**, 12–16.

Lombard, M. (1969). 'Une carte, du bois dans la Mediterranean musalmane (VIIe–IXe siècles)', *Annales – Economies – Societies – Civilisations*, 234–54.

Mahmoud-Mirati, A. A. (1983). 'Land conversion to urban use: its character and impact in Libya', *Ekistics*, **300**, May/June, 183–94.

Mallowan, M. E. L. (1967). 'The development of cities from Al 'Ubaid to the end of Uruk 5', *Cambridge Ancient History*, Vol 1, Cambridge University Press, Chapter VIII.

McAdams, R. (1965). *Land behind Baghdād*, University of Chicago.

Marçais, G. (1957). L'urbanisme musulman', *Mélanges d'histoire et d'archéologie de l'occident musulman*, **1**, Gouvernement Géneral d'Algérie, Algiers, 224–6.

Mellaart, J. (1967). Çatal Huyuk: a neolithic town in Anatolia, Thames and Hudson, London.

Misru, R. P. and O. El Agraa (1983). 'Urbanisation and national development: the guest for appropriate human settlement policies in the Arab world', *Ekistics*, **300**, May/June, 210–18.

Middleton, N. J. (1986). 'Dust storms in the Middle East', *Jnl. of Arid Environments*, **10(2)**, 83–96.

New Geographical Digest (1986). George Philip, London.

Pope, A. U. (1939). *A Survey of Persian Art*, Oxford University Press.

Population Reference Bureau (1984). *World Population Data Sheet 1984*, New York.

Ragette, F. (Ed) (1983). *Beirut of Tomorrow*, American University of Beirut.

Raphael, C. N. and H. T. Shaibu (1984). 'Water resources for At Taif, Saudi Arabia: a study of alternative sources for an expanding urban area' *Geog. Jnl.*, **150(2)**, 183–91.

Roberts, M. H. P. (1979). *An Urban Profile of the Middle East*, St Martin's Press, New York.

Sauvaget, J. (1941). *Alep*, (Album), Librarie Orientaliste Paul Geunther, Paris (Source of Figure 6.3).

Serjeant, R. B. (Ed) (1980). *The Islamic City*, UNESCO, Paris.

Shiber, G. (1964). *The Kuwait Urbanisation*, Kuwait Planning Board, Kuwait, 15–39.

Smit, J. (1977). 'Which future for Alexandria?' *Geoforum*, **8(3)**, 135–40.

Speigal, E. (1967). *New Towns in Israel*, Praeger, Stuttgart and Bern.

Tyrwhitt, J. and P. Psomopoulos (Eds) (1983). 'Urbanisation and social change in the Arab world', *Ekistics*, **50(300)**, 1–239.

Van der Heyden, A. A. M. and H. H. Scullard (Eds) (1959). *Atlas of the Classical World*, Nelson, London.

Waltz, S. E. (1985). 'Women's housing needs in the Arab cultural context of Tunisia', *Ekistics*, **52(310)**, 23–34.

Waterman, S. (1971). 'Pre-Israeli planning in Palestine', *Town Planning Rev.*, **42(1)**, 85–99.

Chapter 7: Problems of Economic Development

Abdel-Malek, A. (1970). 'Sociology and economic history: an essay on mediation'. In *Studies in the Economic History of the Middle East*, (Ed M. A. Cook), Oxford University Press, pp 268–82.

Adams, M. (Ed) (1971). *The Middle East: A Handbook*, Anthony Blond, London.

Albaum, M. and C. S. Davies (1973). 'The spatial structure of socio-economic attributes of Turkish provinces', *International Journal of Middle East Studies*, **4**, 288–310.

Alonso, W. (1963–69). 'Urban and regional imbalances in economic development', *Economic Development and Cultural Change*, **17**, 1–14.

Amiran, D. H. K. (1965). 'Arid zone development: a reappraisal under modern technological conditions', *Econ. Geogr.*, **41**, 189–210.

Baali, F. (1966). Social factors in Iraqi rural-urban migration, *American Journal of Economics and Sociology*, **25**, 359–84.

Baali, F. (1969). 'Agrarian reform in Iraq. some socio-economic aspects', *American Journal of Economics and Sociology*, **28**, 61–76.

Beaumont, P. and K. S. McLachlan (Eds) (1985). *Agricultural Development in the Middle East*, Wiley, Chichester.

Boserup, E. (1965). *The Conditions of Agricultural Growth: The Economics of Agrarian Change under Population Pressure*, Allen and Unwin, London.

Chisholm, M. and B. Rogers (Eds) (1973). *Studies in Human Geography*, Heinemann, London.

Chouri, N. and R. S. Eckaus (1979). 'Interactions of economic and political change; the Egyptian case', *World Development*, **7**, 783–98.

Clark, C. and M. Haswell (1964). *The Economics of Subsistence Agriculture*, Macmillan, London.

Clarke, J. I. (1973). 'Population in movement'. In *Studies in Human Geography*, (Eds M. Chisholm and B. Rogers), Heinemann, London, pp 85–124.

Clarke, J. I. and W. B. Fisher (Eds) (1972). *Populations of the Middle East and North Africa*, University of London Press.

Clarke, J. I. and H. Bowen-Jones (Eds) (1981). *Change and Development in the Middle East. Essays in Honour of W. B. Fisher*, Methuen, London.

Cook, M. A. (Ed) (1970). *Studies in the Economic History of the Middle East*, Oxford University Press, London.

Cooper, C. A. and S. S. Alexander (Eds) (1972). *Economic Development and Population Growth in the Middle East*, American Elsevier Publishing Co, New York, London, Amsterdam.

Dessouki, A. E. H. and A. Al-Labban 'Arms race, defence expenditures and development: the Egyptian case 1952–1973', *Journal of South Asian and Middle Eastern Studies*, **4**, 65–77.

Dresch, J. (1966). 'Utilisation and human geography of the deserts', *Trans. Inst. Br. Geogr.*, **40**, 1–10.

Drysdale, A. (1981). 'The regional equalisation of health care and education in Syria since the Ba'thi Revolution', *International Journal of Middle East Studies*, **13**, 93–111.

Dwyer, D. J. (1968). 'The city in the developing world and the example of South East Asia', *Geography*, **53**, 353–64.

Elkan, W. (1973). *An Introduction to Development Economics*, Penguin Books, Harmondsworth.

Flinn, P. (1968). 'The impact of the technological era', *Journal of Contemporary History*, **3**, 53–68.

Forbes, D. (1984). *The Geography of Underdevelopment*, Croom Helm, London.

Grigg, D. (1985). *The World Food Problems, 1950–1980*, Blackwell, Oxford.

Hershlag, Z. Y. (1964). *Introduction to the Modern Economic History of the Middle East*, E. J. Brill, Leiden.

Hershlag, Z. Y. (1970). *Contemporary Economic Structure of the Middle East*, 2nd edn, E. J. Brill, Leiden.

Holler, J. E. (1964). *Population Growth and Social Change in the Middle East*, George Washington University, Washington, DC.

Hütteroth, W. (1962). 'Getreidkonjunktur und jüngerer Siedlungsausbau im südlichen Inneranatolien', *Erdkunde*, **16**, 249–71.

Issawi, C. (1971). 'Growth and structural change in the Middle East', *Middle East Journal*, **25**, 309–24.

Jansen, A. C. M. (1970). 'The value of the growth pole theory for economic geography', *Tijdsch. econ. soc. Geogr.*, **61**, 67–76.

Joyce, J. (1972). 'Planning prospects in the Middle East', *Contemporary Review*, **220**, 113–17.

Kammash, M. el (1968). *Economic Development and Planning in Egypt*, Praeger, New York, Washington and London.

Kanovsky, E. (1968). 'The economic aftermath of the Six Day War', *Middle East Journal*, **22**, 131–43, 278–96.

Kuznets, S. (1971–72). 'Problems of comparing recent growth rates for developed and less-developed countries', *Economic Development and Cultural Change*, **20**, 185–209.

Lever, H. and C. Huhne (1985). *Debt and Danger. The World Financial Crisis*, Penguin, Harmondsworth.

Magnarella, P. J. (1970). 'From villagers to townsmen in Turkey', *Middle East Journal*, **24**, 229–40.

Maull, H. (1986). 'The arms trade with the Middle East and North Africa'. In *The Middle East and North Africa 1987*, Europa Publications, London, pp 148–53.

Myrdal, G. (1957). *Economic Theory and Underdeveloped Regions*, Duckworth, London.

Ortiz, S. R. de (1972). *Uncertainties in Peasant Farming*, Athlone Press, London.

Sayigh, Y. A. (1982). *The Arab Ecomony: Past Performance and Future Prospects*, Oxford University Press.

Shahshahani, A. and M. Kadhim (1979). 'Development problems of an energy-based economy': Iran, *Journal of South Asian and Middle Eastern Studies*, **2**, 57–83.

Shorter, F. C. (1966). 'The application of development hypotheses in Middle Eastern studies', *Economic Development and Cultural Change*, **14**, 340–54.

Swanson, J. C. (1979). 'Some consequences of emigration for rural economic development in the Yemen Arab Republic', *Middle East Journal*, **33**, 34–43.

Todaro, M. P. (1985). '*Economic Development in the Third World*, 3rd edn, Longmans, London.

Weinbaum, M. G. (1982). *Food Development and Politics in the Middle East*, Westview Press, Boulder; Croom Helm, London.

Wickwar, W. H. (1965). 'Food and social development in the Middle East', *Middle East Journal*, **19**, 177–95.

Wilson, R. (1979). *The Economies of the Middle East*, Brill, Leiden.

Chapter 8: Industry, Trade and Finance

Abdo, A. S. (1970). 'Domestic passenger air transport in Saudi Arabia', *Bulletin of the Faculty of Arts, University of Riyadh*, **1**, 21–39.

Adams, M. (Ed) (1971) *The Middle East: A Handbook*, Anthony Blond, London.

Agwah, A. (1978); 'Import distribution, export expansion and consumption liberalisation: the case of Egypt', *Development and Change*, **9**, 299–329.

Aliboni, R. (1979). *Arab Industrialisation and Economic Integration*, Croom Helm, London.

Alnasrawi, A. (1971). 'The changing pattern of Iraq's foreign trade', *Middle East Journal*, **25**, 481–90.

Askari, H. and J. T. Cummings. (1977). 'The future of economic integration within the Arab world', *International Journal of Middle East Studies*, **8**, 289–315.

Bagley, F. R. C. (1976). 'A bright future after oil: dams and agro-industry in Khuzistan', *Middle East Journal*, **30**, 25–35.

Barbour, K. M. (1972). *The Growth, Location and Structure of Industry in Egypt*, Praeger, New York, Washington and London.

Bardan, A. and B. Khader, (Eds) (1986). *The Economic Development of Jordan*, Croom Helm, London.

Bartsch, W. H. (1971). 'The industrial labor force of Iran: problems of recruitment, training and productivity', *Middle East Journal*, **25**, 15–30.

Boxer, B. (1967). *Israeli Shipping and Foreign Trade*, University of Chicago, Department of Geography, Research Papers, No 48, Chicago.

Carey, J. P. C. and A. G. Carey, (1975). 'Industrial growth and development planning in Iran', *Middle East Journal*, **29**, 1–15.

Chapman, A. S. (1957). 'The economic regions of Turkey as characterised by railway shipments', *Northwestern University Studies in Geography*, (Evanston. Ill), **2**, 71–5.

Cohen, E. J. (1970). *Turkish Economic, Social and Political Change: The Development of a More Prosperous and Open Society*, Praeger, New York, Washington and London.

Collard, E. (1972). 'Trade relations between the E.E.C. and the Arab World', *Middle East International*, **14**, 16–20.

Costello, V. F. (1977). *Urbanization in the Middle East*, Cambridge University Press.

Dewdney, J. C. (1971). *Turkey*, Chatto and Windus, London.

Edwards, C. H. (1986). *Future Energy Propects in the Middle East States*, Special Report on Energy, Economist Intelligence Unit, London.

Edens, D. G. and W. P. Snavely, (1970). 'Planning for economic development in Saudi Arabia', *Middle East Journal*, **24**, 16–30.

Eldem, V. (1953). 'Turkey's transportation', *Middle Eastern Affairs*, **4**, 324–36.

Encle, W. (1966). 'Iraks Industrieentwicklung in Problemen und Zielen', *Orient (Hamburg)*, **7**, 115–20.

Fisher, S. N. (1955). *Social Forces in the Middle East*, Cornell University Press, Ithaca, New York.

Gaube, H., E. Grötzbach, E. Niewöhner-Eberhard, B. Oettinger and E. Wirth (1976). 'Wochenmarkte, Marktorte und Marktzyklen in Vorderasien', *Erdkunde*, **30**, 9–44.

Garnick, D. H. (1961). 'Regional integration and economic development in the Middle East', *Middle Eastern Affairs*, **12**, 294–300.

Gray, A. L. (1976). 'Egypt's Ten Year Economic Plan, 1973–1982', *Middle East Journal*, **30**, 36–48.

Grunwald, K. and J. O. Ronall, (1960). *Industrialisation in the Middle East*, Council for Middle Eastern Affairs Press, New York.

Hajjar, S. G. (Ed) (1985). *The Middle East: From Transition to Development*, E. J. Brill, Leiden.

Hamilton, H. G. (1982). 'The Saudi petrochemical industry: its rationale and effectiveness'. In *State, Society and Economy in Saudi Arabia* (Ed T. Niblock), Croom Helm, London, pp 235–77.

Hershlag, Z. Y. (1964). *Introduction to the Modern Economic History of the Middle East*, E. J. Brill, Leiden.

Hughes, A. M. (1979). 'The future of the Gulf ports', *Geography*, **64**, 54–6.

Hunter, G. (1952). 'The Middle East Supply Centre'. In *The Middle East in the War*, (Ed G. Kirk) Oxford University Press, pp 163–93.

Issawi, C. (1963). *Egypt in Revolution: An Economic Analysis*, Oxford University Press.

Issawi, C. (Ed) (1966). *The Economic History of the Middle East, 1800–1914*, University of Chicago Press, Chicago and London.

Issawi, C. (1967). Iran's economic upsurge, *Middle East Journal*, **21**, 447–61.

Jalal, F. (1971). *The Role of Government in the Industrialisation of Iraq*, Frank Cass, London.

Kammash, M. M. El (1968). *Economic Development and Planning in Egypt*, Praeger, New York, Washington and London.

Kanovsky, E. (1967). 'Arab economic unity', *Middle East Journal*, **21**, 213–235.

Kerwin, R. W. (1950). 'The Turkish roads programme', *Middle East Journal*, **4**, 196–208.

Kirk, G. (1952). *The Middle East in the War*, Oxford University Press.

Kleiman, E. (1967). 'The place of manufacturing in the growth of the Israeli economy', *Journal of Development Studies*, **3**, 226–48.

Kolars, J. F. (1964). 'Types of rural development'. In *Four Studies on the Economic Development of Turkey*, (Eds F. C. Shorter, J. F. Kolars, D. A. Rustow and O. Yenal), Frank Cass, London, pp 63–88.

Kolars, J. F. and H. J. Malin, (1970). 'Population and accessibility: an analysis of Turkish railroads', *Geogrl Rev.*, **60**, 229–46.

Korby, W. (1977). *Probleme der industriellen Entwickslung und Konzentration in Iran*, Dr Ludwig Reichert Verlag, Wiesbaden.

Kreinin, M. (1968). 'Israel and the European Economic Community', *Quart. J. of Econ.*, **82**, 297–312.

Langley, K. M. (1961). *The Industrialisation of Iraq*, Harvard University Press, Cambridge, Mass.

Mabro, R. E. (1971). 'Industrialisation'. In *The Middle East: A Handbook*, (Ed M. Adams), Anthony Blond, London.

McConnell, J. E. (1967). 'The Middle East: competitive or complementary?, *Tijdsch. econ. soc. Geogr.*, **58**, 82–93.

Mallakh, R. El- and M. Kadhim (1976). 'Arab institutionalised development aid: an evaluation', *Middle East Journal*, **30**, 471–84.

Mallakh, R. El-, B. W. Poulson and M. Kadhim (1977). *Capital Investment in the Middle East: Use of Surplus Funds for Regional Development*, Praeger, New York.

Meir, A. (1980). 'The diffusion of industry adoption by kibbutz rural settlements in Israel', *Journal of Developing Areas*, **14**, 539–52.

Meissner, F. and W. Ruby (1978). 'Industrial villages in Israel', *Ekistics*, **45**, 43–8.

Montasser, E. (1974). 'Egypt's pattern of trade and development – a model of import substitution growth', *L'Egypte Contemporaine*, **356**, 141–245.

Musrey, A. G. (1969). *An Arab Common Market: A Study in Inter-Arab Trade Relations, 1920–67*, Praeger, New York, Washington and London.

Niblock, T. (1980). 'The Prospects for Integration in the Arab Gulf'. In *Social and Economic Development in the Arab Gulf* (Ed T. Niblock), Croom Helm, London; St Martin's Press, New York, 187–209.

Niblock, T. (Ed) (1982). *State, Society and Economy in Saudi Arabia*, Croom Helm, London.

Owen, R. (1981). *The Middle East in the World Economy 1800–1914*, Methuen, London.

Péchoux, P-Y. (1984). 'Le démarrage de la grande industrie en Jordanie', *Annales de Géographie*, **93**, 457–61.

Planhol, X. de (1966). 'Small-scale industry and crafts in arid regions'. In *Arid Lands: A Geographic Appraisal*, (Ed E. S. Hills), Methuen, London, pp 273–285.

Pomfret, R. W. T. (1975). 'Manufactured export expansion in a semi-developed economy: the Israeli case', *Economia Internationale*, **28**, 464–77.

Potter, D. (1955). 'The bazaar merchant'. In *Social Forces in the Middle East*, (Ed S. N. Fisher), Cornell University Press, Ithaca, New York, pp 99–115.

Poulson, B. W. and M. Wallace (1979). 'Regional integration in the Middle East: the evidence for trade and capital flows', *Middle East Journal*, **33**, 467–78.

Preston, L. E. (1979). *Trade Patterns in the Middle East*, American Enterprise Institute for Public Policy Research, Washington DC.

Ramazani, R. K. (1964). *The Middle East and the European Common Market*, University of Virginia, Charlottesville.

Rotblat, J. (1975). 'Social organisation and development in an Iranian Persian bazaar', *Economic Development and Cultural Change*, **23**, 292–305.

Shorter, F. C., J. F. Kolars, D. A. Rustow and O. Yenal (Eds) (1964). *Four Studies on the Economic Development of Turkey*, Frank Cass, London.

Underwood, A. M. (1974). *Inter-Arab Financial Flows*, Economic Research Paper 1, Centre for Middle Eastern and Islamic Studies, University of Durham.

Wilson, R. (1977). *Trade and Investment in the Middle East*, Macmillan, London.

Wilson, R. (1978). *The Arab Common Market and Inter-Arab Trade*, Economic Research Paper 4, Centre for Middle Eastern and Islamic Studies, University of Durham.

Wilson, R. (1979a). *The Economies of the Middle East*, E. J. Brill, Leiden.

Wilson, R. (1979b). 'Regular and permanent markets in the San'ā region', *Arabian Studies*, **5**, 189–91.

Wilson, R. (1985). 'Egypt's exports: supply constraints and marketing problems', *British Society for Middle Eastern Studies: Bulletin*, **12**, 135–56.

Wulff, H. E. (1966). *The Traditional Crafts of Persia*, MIT Press, Cambridge, Mass, and London.

Chapter 9: Petroleum

Assah, A. (1969). *Miracle of the Desert Kingdom*, Johnson Publications Ltd, London, 330 pages.

Alnasrawi, A. (1967). *Financing Economic Development in Iraq – The Role of Oil in a Middle Eastern Economy*, Praeger, New York, 188 pages.

Arabian American Oil Company (1965). *Aramco's Role in the Development of the Eastern Province*, Dammam.

Arabian American Oil Company (1970). *Aramco 1970 – A Review of Operations by the Arabian American Oil Company*, Dammam, 22 pages and tables.

Arabian American Oil Company (1971). *Aramco 1971 – A Review of Operations by the Arabian American Oil Company*, Dammam, 7 pages and tables and foldouts.

Berry, J. A. (1972). 'Oil and Soviet policy in the Middle East', *Middle East Journal*, **26**, 149–61.

Bill, J. A. (1971). 'The challenge of change: petroleum and planning in the Middle East', *Focus*, **22(1)**, 1–5.

British Petroleum Company plc (1970). *Our Industry Petroleum*, British Petroleum Company, London, 528 pages.

British Petroleum Company Ltd (1972). *BP Statistical Review of the World Oil Industry 1971*, British Petroleum Company, London, 24 pages.

British Petroleum Company plc (1986). *BP Statistical Review of World Energy*, British Petroleum Company, London.

Caroe, O. (1951). *Wells of Power, the Oil Fields of S.W. Asia*, Macmillan, London, 240 pages.

Cattan, H. (1967). *The Evolution of Oil Concessions in the Middle East and North Africa*, Dobbs-Ferry Oceana Publications, New York, 173 pages.

Cheney, M. S. (1958). *Big Oil Man from Arabia*, Heinemann, London, 320 pages.

Darmstadter, J., P. D. Teitelbaum and J. G. Polach (1971). *Energy in the World Economy: A Statistical Review of Trends in Output, Trade and Consumption Since 1925*, Johns Hopkins Press for Resources of the Future, Baltimore, 876 pages.

Echo of Iran (1971). *Iran Almanac 1971*, Echo of Iran, Tehrān, 808 pages.

Finnie, D. H. (1958). *Desert Enterprise*, Harvard University Press, Cambridge (USA), 224 pages.

Finnie, D. H. (1958). 'Recruitment and training of labour. The Middle East oil industry', *Middle East Journal*, **12**, 127–43.

Fisher, W. B. (Ed) (1968). *The Cambridge History of Iran, Vol 1, – The Land of Iran*, Cambridge University Press.

Frank, H. J. (1966). *Crude Oil Prices in the Middle East*, Praeger, New York, 209 pages.

Frenkel, P. H. (1962). *Oil: The Facts of Life*, Weidenfeld and Nicolson, London.

Ghalayini, A. K. (1970). 'Drilling techniques and cost in Saudi Arabia', *Seventh Arab Petroleum Congress, Beirūt*, **2**, 267–84.

Gordon, R. L. (1970). *The Evolution of Energy Policy in Western Europe: The Reluctant Retreat from Coal*, Praeger, New York, 330 pages.

Hamilton, C. W. (1962). *Americans and oil in the Middle East*, Gulf Publishing Company, Houston, 307 pages.

Hartshorn, J. E. (1967). *Politics and World Oil Economics*, Praeger, New York.

Hirst, D. (1966). *Oil and Public Opinion in the Middle East*, Faber and Faber, London, 127 pages.

Hillmore, P. (1972). 'Oil producers will take 25 pc stake', *The Guardian*, Manchester, 17 December 1972.

Hubbert, M. K. (1969). 'Energy Resources'. In *Resources and Man: A Study and Recommendations of the Division of Earth Sciences of the United States National Academy of Sciences National Research Council*, Freeman, San Francisco, 157–242.

Institute of Petroleum Information Service (1971). *Oil – World Statistics*, Institute of Petroleum, London, 8 pages.

Institute of Petroleum Information Service (1971). *Oil – The Middle East*, Institute of Petroleum, London, 13 pages.

Iranian Oil Operating Companies (1971). *Annual Review 1971*, Iranian Oil Operating Companies, Tehrān, 40 pages.

Iraq, Basrah and Mosul Petroleum Companies (1971). *Review for 1970*, Iraq Petroleum Company, London, 32 pages.

Issawi, C. and M. Yeganeh (1962). *The Economics of Middle Eastern Oil*, Faber and Faber, London, 230 pages.

Jenkins, Iain (1985). 'Iran acts in pipeline war', *Middle East Economic Digest*, 16 November, pp 4–6.

Khatib, A., H, Munif and F. Ruwayha (1963). 'Aramco's participation in Saudi Arabian development', *Fourth Arab Petroleum Congress, Beirūt*, **1**, 53 (A–1).

Kubbah, A. A. Q. (1964). *Libya – its Oil Industry and Economic System*, The Arab Petro-Economic Research Centre, Baghdād, 274 pages.

Kuwait Oil Company (1971). *1971 Review of Operations*, Kuwait Oil Company, Kuwait, 32 pages.

Ledkicher R., G. Rentz, M. Steineke *et al.* (1960). *Aramco Handbook*, Arabian American Oil Company, Dhahran, 343 pages.

Leeman, W. A. (1962). *The Price of Middle East Oil, an Essay in Political Economy*, Cornell University Press, Ithaca, 274 pages.

Lenczowski, G. (1960). *Oil and State in the Middle East*, Cornell University Press, Ithaca, 379 pages.

Lockhart, L. (1953). 'The causes of the Anglo-Persian oil dispute', *Jl R. cent. Asian Soc.*, **40**, 134–50.

Longrigg, S. H. (1949). 'The liquid gold of Arabia', *Jl R. cent. Asian Soc.*, **36**, 20–33.

Longrigg, S. H. (1968). *Oil in the Middle East: its Discovery and Development*, Oxford University Press, London, 519 pages.

Lubbell, H. (1962). *Middle East Oil Crises and Western Europe's Energy Supplies*, A RAND Corporation Study, Johns Hopkins Press, Baltimore, 233 pages.

Lutfi, A. (1968). *OPEC Oil*, The Middle East Research and Publishing Centre, Beirūt, Middle East Oil Monographs No 6, 120 pages.

Lutfi, A. (1965). 'Royalty oil economics – key to Arab participation in the international oil industry', *Fifth Arab Petroleum Congress*, Cairo, **1**, 1(A–4).

McLachlan, K. S. (1972). *Spending Oil Revenues – Development Prospects in the Middle East to 1975*, QER Special No 10, Economist Intelligence Unit Ltd, London, 36 pages.

Manners, G. (1964). *The Geography of Energy*, Hutchinson University Library, London, 205 pages.

Marlowe, J. (1962). *The Persian Gulf in the Twentieth Century*, The Cresset Press, London, 278 pages.

Masters, C. P., H. P. Klemme and A. B. Coury (1982). *Assessment of undiscovered conventionally recoverable petroleum resources of the Arabian–Iranian basin*, United States Geological Survey, Circular No 881, 12 pages.

Melamid, A. (1959). 'Geographical pattern of Iranian oil development', *Econ. Geogr*, **35**, 199–218.

Melamid, A. (1968). 'Industrial activities'. In *The Cambridge History of Iran, Vol 1 – The Land of Iran*, (Ed W. B. Fisher), Cambridge University Press, Cambridge, pp 517–551.

Mikdashi, Z. (1966). *A Financial Analysis of Middle Eastern Oil Concessions 1901–1965*, Praeger, New York, 341 pages.

Moody, J. D. (1970). 'Petroleum demands of future decades', *Bul. Am. Ass. Petrol. Geol.*, **54**, 2239–45.

Netton, I. R. (Ed) (1986). *Arabia and the Gulf: from traditional society to modern states*, Croom Helm, Beckenham, 259 pages.

O'Dell, P. R. (1963). *An Economic Geography of Oil*, G. Bell and Sons, London, 219 pages.

O'Dell, P. R. (1970). *Oil and World Power – A Geographical Interpretation*, Penguin Books, Harmondsworth, 188 pages.

Oxford Regional Economic Atlas (1960). *The Middle East and North Africa*, Oxford University Press, London, 135 pages.

Penrose, E. T. (1968). *The Large International Firm in Developing Countries: The International Petroleum Industry*, George Allen and Unwin, London, 311 pages.

Petroleum Information Bureau (1967). *Oil – Drilling Techniques*, Petroleum Information Bureau, London, 7 pages.

Petroleum Information Bureau (1967). *Oil – The World's Reserves*, Petroleum Information Bureau, London, 5 pages.

Petroleum Information Bureau (1968). *Oil – Refining*, Petroleum Information Bureau, London, 5 pages.

Petroleum Information Bureau (1968). *Oil – Pipelines*, Petroleum Information Bureau, London, 8 pages.

Petroleum Information Bureau (1969). *Oil– Africa*, Petroleum Information Bureau, London, 9 pages.

Petroleum Publishing Co (1970). *International Petroleum Encyclopedia 1971*, Tulsa, Oklahoma, 367 pages.

Petroleum Publishing Co (1972). International Petroleum Encyclopedia 1972, Tulsa, Oklahoma, 448 pages.

Pratt, W, E. and D. Good (Eds) (1950). *World Geography of Petroleum*, American Geographical Society, Special Publication 31, Princeton University Press, 134 pages.

Roosevelt, K. (1949). *Arabs, Oil and History*, Gollancz, London, 271 pages.

Rouhani, F. (1971). *A History of O.P.E.C.*, Praeger, New York, 281 pages.

Sayegh, K. S. (1968). *Oil and Arab Regional Development*, Praeger, New York, 359 pages.

Schurr, S. H. and P. T. Homan (1971). *Middle Eastern Oil and the Western World – Prospects and Problems*, American Elsevier, New York, 206 pages

Shwadran, B. (1956). *The Middle East, Oil and the Great Powers*, Atlantic Press, London, 500 pages.

Snow, C. (1972). 'The emerging giant of world oil', *The Financial Times, London*, 19 December, p 17.

Stevens, G. P. (1949). 'Saudi Arabia's petroleum resources', *Econ. Geogr.*, **25**, 216–25.

Stevens, P. (1986). *The impact of oil on the role of the state in economic development – a case study of the Arab World*, Surrey Energy Economics Centre, University of Surrey, 34 pages.

Stocking, G. W. (1970). *Middle East Oil: a Study in Political and Economic Controversy*, Vanderbilt University Press, Nashville, 485 pages.

Tayim, H. A. (1960). 'Utilisation of natural gas in Saudi Arabia', *Second Arab Petroleum Congress*, Beirūt, **2**, 308–21.

Tetreault, M. A. (1982). *The Organisation of Arab Petroleum Exporting Countries: History, Policies and Prospects*, Greenwood Press, London, 256 pages.

United States, Department of the Interior (1968). *United States Petroleum Through 1980*, Washington, July 1968.

Warman, H. R. (1971). 'Future problems in petroleum exploration', *Petroleum Review*, **25**, 96–101.

Warman, H. R. (1972). 'The future of oil', *Geogrl J.*, **38**, 287–97.

Weeks, L. G. (1968). 'The gas, oil and sulphur potentials of the sea', *Ocean Industry*, **3(6)**, 43–51.

Weeks, L. G. (1971). 'Marine geology and petroleum resources', *World Petroleum Congress, 8, Moscow, Proceedings*, **2**, 99–106.

Yorke, V. and L. Turner (1986). *European interests and Gulf Oil*, Gower Publishing Company, Aldershot, 125 pages.

Chapter 10: The Political Map

Amin, S. H. (1982). 'The Iran-Iraq war, legal implications', *Marine Policy*, **6(3)**, 193–218.

Allan, J. A. (Ed) (1982). *Libya since Independence*, Croom Helm, London.

Bastianelli, F. (1983). 'Boundary delimination in the Mediterranean Sea', *Marine Policy Reports*, **5(4)**, University of Delaware College of Marine Studies.

Beaumont, P. (1978). 'The Euphrates river – an international problem of water resource development', *Environmental Conservation*, **5(1)**, 35–43.

Blake, G. H. (1987). 'Maritime boundaries of the Middle East and North Africa'. In *Boundaries and State Territory in the Middle East and North Africa*, (Eds G. H. Blake and R. N. Schofield) Menas Press, Wisbech.

Blake, G. H. and R. N. Schofield (Eds) (1987). *Boundaries and State Territory in the Middle East and North Africa*, Menas Press, Wisbech, pp 121–34.

Brawer, M. (1968). 'The geographical background of the Jordan water dispute'. In *Essays in Political Geography*, (Ed C. A. Fisher) Methuen, London, pp 225–42.

Chaliand, G. and J.-P. Rageau (1983). *Atlas Stratègique*, Lib. Arthème Fayard, Paris.

Clarke, J. I. and H. Bowen-Jones (Eds) (1981). *Change and Development in the Middle East*, Methuen, London.

Cohen, S. B. (1973). *Geography and Politics in a World Divided*, 2nd edn, Oxford University Press.

Drury, M. P. (1981). 'The political geography of Cyprus'. In *Change and Development in the Middle East*, (Eds J. I. Clarke and H. Bowen-Jones), Methuen, London, pp 289–304.

Drysdale, A. D. and G. H. Blake (1985). *The Middle East and North Africa: a Political Geography*, Oxford University Press, New York.

East, W. G., O. H. K. Spate and C. A. Fisher (Eds) (1971). *The Changing Map of Asia*, Methuen, London.

El-Hakim, A. A. (1979). *The Middle Eastern States and the Law of the Sea*, Manchester University Press, Manchester.

Farid, A. M. (Ed) (1984). *The Red Sea Region*, Croom Helm.

Haupert, G. S. (1969). 'Political geography of the Israel–Syrian boundary dispute 1949–67', *Prof. Geogr.*, **21**, 163–71.

Hawley, D. (1970). *The Trucial States*, George Allen and Unwin, London.

Hourani, A. H. (1970). 'Race, religion and nation-state in the Near East'. In *Arab Middle East Societies and Cultures*, (Eds A. M. Lutifiyya and C. W. Churchill). Mouton, The Hague, pp 1–19.

Ismael, J. Y. (1982). *Iraq and Iran: Roots of Conflict*, Syracuse University Press, Syracuse.

Issawi, C. (1970). 'Political disunity in the Arab World'. In *Readings in Arab Middle East Societies and Cultures*. (Eds A. M. Lutifiyya and C. W. Churchill), Mouton, The Hague, pp 278–84.

Jones, S. B. (1954). 'A unified field theory of political geography', *Ann. Ass. Amer. Geogr.*, **44**, 111–23.

Kelly, J. B. (1964). *Eastern Arabian Frontiers*, Faber and Faber, London.

Kirk, G. E. (1964). *A Short History of the Middle East*, University Paperbacks, London.

Kliot, N. (1986). 'Lebanon: a geography of hostages', *Political Geog. Quarterly*, **5(3)**, 199–220.

Lapidoth-Eschelbacher, R. (1982). *The Red Sea and the Gulf of Aden*, Martinus Nijhoff, The Hague.

Laqueur, W. (Ed) (1969). *The Israel–Arab Reader*, Weidenfeld, London.

Laqueur, W. (1969). *The Struggle for the Middle East*, Routledge and Kegan Paul, London.

Lebon, J. H. G. (1971). 'South-West Asia and Egypt'. In *The Changing Map of Asia*, (Eds W. G. East, O. H. K. Spate and C. A. Fisher), Methuen, London, 5th edn, pp 53–126.

Lutifiyya, A. M. and C. W. Churchill (Eds) (1970). *Readings in Arab Middle East Societies and Cultures*, Mouton, The Hague.

McConnell, J. E. (1962). 'The Middle East: competitive or complementary?' *Tijdschr. econ. Geogr.*, **2**, 82–93.

Melamid, A. (1953). 'Political geography of Trucial Oman and Qatar', *Geogr. Rev.*, **423**, 194–206.

Melamid, A. (1956). 'The Buraimi oasis dispute', *Middle East Affairs*, **7**, 56–63.

Melamid, A. (1968). 'The Shaṭṭ al'Arab boundary dispute', *Middle East Journal*, **22**, 350–57.

Naff, J. and R. C. Matson (Eds) (1984). *Water in the Middle East: Conflict or Cooperation?* Westview, Boulder.

Pounds, N. J. G. (1972). *Political Geography*, 2nd edn, McGraw-Hill, New York.

Prescott, J. R. V. (1985). *The Maritime Political Boundaries of the World*, Methuen, London.

Rozakis, C. (1975). *The Greek–Turkish dispute over the Aegean Continental Shelf*, Law of the Sea Institute, Kingston, Rhode Island.

Schofield, R. N. (1986). *The Evolution of the Shatt al-Arab Boundary Dispute*, Menas Press, Cambridge.

Smith, C. G. (1966). 'The disputed waters of the Jordan', *Trans. Inst. Br. Geogr.*, **40**, 111–28.

Waterbury, J. (1979). *Hydropolitics of the Nile Valley*, Syracuse Press, Syracuse.

Wilkinson, J. C. (1971). 'The Oman question: the background to the political geography of south east Arabia', *Geogrl J.*, **137**, 361–71.

Wilkinson, J. C. (1983). 'Traditional concepts of territory in South East Arabia', *Geogrl. J.*, **149(3)**, 301–15.

Young, H. F. (1985). *Atlas of United States Foreign Relations*, US Department of State Bureau of Public Affairs, Washington DC (2nd edn).

Chapter 11: Tradition and Change in Arabia

Abdul-Ela, M. T. (1965). Some geographical aspects of Al Riyadh, *Bull. Soc. Géogr. Egypte*, **38**, 31–72.

Abdulfattah, K. (1981). *Mountain Farmer and Fellah in 'Asir, South-west Saudi Arabia*, Erlangen Geographische Arbeiten 12, Erlangen.

Abu 'Adirah, S. S. (1978). 'Sedentarisation and settlement of the bedouin', *Arabian Studies*, **4**, 1-5.

Azhary, M. S., El (Ed) (1984). *The Impact of Oil Revenues on Arab Gulf Development*, Croom Helm, London; Westview Press, Boulder.

Beaumont, P. (1977a). 'Water and development in Saudi Arabia', *Geogrl. J.*, **143**, 42–60.

Beaumont, P. (1977b). 'Water in Kuwait', *Geography*, **62**, 187–97.

Beaumont, P. and K. S. McLachlan (Eds) (1985). *Agricultural Development in the Middle East*, Wiley, Chichester.

Belgrave, C. D. (1934). 'Pearl diving in Bahrain', *Jl. R. cent. Asian Soc.*, **21**, 450–2.

Belgrave, C. D. (1968). 'Persian Gulf: past and present', *Jl. R. cent. Asian Soc.*, **55**, 28–34.

Bidwell, R. (1983). *The Two Yemens*, Longmans, London.

Birks, J. S. (1978). 'Development or decline of pastoralists', *Arabian Studies*, **4**, 7–20.

Birks, J. S. and S. E. Letts (1977). 'Diqal and Muqayda: dying oases in Arabia', *Tijdschr. econ. soc. geogr.*, **68**, 145–51.

Bowen, R. L. (1951a). 'Marine industries of eastern Arabia', *Geogrl. Rev.*, **41**, 384–400.

Bowen, R. L. (1951b). 'The pearl fisheries of the Persian Gulf', *Middle East Journal*, **5**, 161–80.

Bowen, R. L. and F. P. Albright (Eds) (1958). *Archaeological Discoveries in South Arabia*, American Foundation for the Study of Man, Johns Hopkins, Baltimore.

Burckhardt, J. L. (1831). *Notes on the Bedouins and Wahabys*, Colburn and Bentley, London.

Carapico, S. (1985). 'Yemeni agriculture in transition', In *Agricultural Development in the Middle East* (Eds P. Beaumont and K. McLachlan), Wiley, Chichester, pp 241–54.

Cole, D. P. (1973). 'The enmeshment of nomads in Sa'ūdi Arabian society: the case of Āl Murrah'. In *The Desert and the Sown: Nomads in the Wider Society* (Ed C. Nelson), Institute of International Studies, Research Series No 21, University of California, Berkeley, pp 113–28.

Cole, D. P. (1975). 'Nomads of the Nomads: The Āl Murrah', *Bedouin of the Empty Quarter*, AHM Publishing Corporation, Arlington Heights, Illinois.

Crary, D. D. (1955). 'Recent agricultural developments in Saudi Arabia', *Geogr. Rev.*, **41**, 366–83.

Dequin, H. (1963). *Die Landwirtschaft Saudisch-Arabiens und ihre Entwicklungs möglichkeiten*, DLG–Verlag–GMBH, Frankfurt am Main.

Dickson, H. H. P. (1951). *The Arab of the Desert: A Glimpse of Badawin Life in Kuwait and Saudi Arabia*, 2nd edn, *Allen and Unwin*, London.

Dowson, V. H. W. (1949). 'The date and the Arab', *Jl. R. cent. Asian Soc.*, **36**, 34–41.

Drysdale, A. and G. H. Blake (1985). *The Middle East and North Africa: A Political Geography*, Oxford University Press.

Dutton, R. (1985). Agricultural policy and development: Oman, Bahrain, Qatar and the United Arab Emirates'. In *Agricultural Development in the Middle East* (Eds P. Beaumont and K. S. McLachlan), Wiley, Chichester, pp 227–40.

Ebert, C. H. V. (1965). 'Water resources and land use in the Qatif oasis of Saudi Arabia', *Geogrl. Rev.*, **55**, 496–509.

Edens, D. G. and W. P. Snavely, (1970). 'Planning for economic development in Saudi Arabia', *Middle East Journal*, **24**, 16–30.

El Attar, M. S. (1964). *Le Sous-développement économique et social du Yémén. Perspective de la révolution yéménite*. Editions Tiers Monde, Algiers.

Fabietti, U. (1982). 'Sedentarisation as a means of detribalisation: some policies of the Saudi Arabian government towards the nomads'. In *State, Society and Economy in Saudi Arabia*, (Ed T. Niblock), Croom Helm, London, pp 186–97.

Feel, M. R., Al- (1985). 'Kuwait: small-scale agriculture in an oil economy'. In *Agricultural Development in the Middle East* (Eds P. Beaumont and K. McLachlan), Wiley, Chichester, pp 279–88.

Fenelon, K. G. (1971). *The United Arab Emirates: An Economic and Social Survey*, 3rd edn, Longmans, London.

Grill, N. C. (1984). *Urbanisation in the Arabian Peninsula*, Centre for Middle Eastern and Islamic Studies, Occasional Papers Series 25, University of Durham.

Habib, J. S. (1978). *Ibn Sa'ud's Warriors of Islam. The Ikhwan of Najd and their Role in the Creation of the Sa'udi Kingdom, 1910–1930*, Social, Economic and Political Studies of the Middle East 27, E. J. Brill, Leiden.

Heard-Bey, F. (1982). *From Trucial States to United Arab Emirates*, Longmans, London and New York.

Helaissi, A. S. (1959). 'The bedouins and tribal life in Saudi Arabia', *International Social Science Journal*, **11**, 532–8.

Hendy, H. F. (1972). 'Ecological consequences of bedouin settlement in Saudi Arabia'. In *The Careless Technology*, (Eds H. T. Farvar and J. P. Milton), Natural History Press, Garden City, New York, pp 683–93.

Hogarth, D. G. (1902). *The Nearer East*, William Heinemann, London.

Hopwood, D. (Ed) (1972). *The Arabian Peninsula*, Allen and Unwin, London.

Hornell, J. (1942). A tentative classification of Arab seacraft, *Mariners' Mirror*, **28**, 11–40.

Hussain, Z. (1978). 'Land and water use in Saudi Arabia', *World Crops*, **30**, 58–61.

Ingrams, H. (1966). *Arabia and the Isles*, John Murray, London.

Joffe, E. G. H. (1985). 'Agricultural development in Saudi Arabia: the problematic path to self-sufficiency'. In *Agricultural Development in the Middle East* (Eds P. Beaumont and K. S. McLachlan), Wiley, Chichester, pp 209–25.

Johnstone, T. M. and J. C. Wilkinson (1960). 'Some geographical aspects of Qatar', *Geogrl. J.*, **126**, 442–450

Katakura, M. (1977). *Bedouin Village: A Study of a Saudi Arabian People in Transition*, University of Tokyo Press.

Kelly, J. B. (1964). *Eastern Arabian Frontiers*, Faber and Faber, London.

King, R. (1972). 'The pilgrimage to Mecca: some geographical and historical aspects, *Erdkunde'*, **26**, 61–72.

Kopp, H. (1985). 'Land usage and its implications for Yemeni agriculture'. In *Economy, Society and Culture in Contemporary Yemen* (Ed B. R. Pridham), Croom Helm, London, pp 41–50.

Lancaster, W. (1981). *The Rwala Bedouin Today*, Cambridge University Press.

Lander, R. G. (1969). 'The modernisation of the Persian Gulf: the period of British dominance'. In *The Princeton University Conference and Twentieth Annual Near Eastern Conference on Middle East Focus: The Persian Gulf*, (Ed T. C. Young), Princeton University Conference, Princeton, pp 1–29.

Leeds, A. and A. P. Vayda (Eds) (1965). *Man, Culture and Animals: The Role of Animals in Human Ecological Adjustments*, American Association for the Advancement of Science, No 78, Washington DC.

Levy, R. (1957). *The Social Structure of Islam*, Cambridge University Press.

Lewis, B., C. Pellart and J. Schaht (Eds) (1965). *The Encyclopaedia of Islam*, **2**, new edn., E. J. Brill, Leiden and Luzac and Co, London.

Machie, J. M. (1924). 'Hasa: an Arabian oasis', *Geogrl. J.*, **63**, 189–207.

Mallakh, R., El, (1966a). 'Kuwait's economic development and her foreign aid programmes', *World Today*, **22**, 13–22.

Mallakh, R., El, (1966b). 'Planning in a capital surplus economy: Kuwait', *Land Econ.*, **42**, 425–440.

Mallakh, R., El, (1979) *Qatar: Development of an Oil Economy*, Croom Helm, London.

Mallakh, R., El, (1981). *The Economic Development of the United Arab Emirates*, Croom Helm, London.

Mallakh, R., El, (1982). *Saudi Arabia: Rush to Development*, Croom Helm, London.

Moosa, A. A., Al-, (1984). 'Kuwait: changing environment in a geographical perspective', *British Society for Middle Eastern Studies: Bulletin*, **11**, 45–57.

Mosely, F. (1966). 'Exploration for water in the Aden Protectorate', *R. Engrs' J.*, **80**, 124–142.

Mundy, M. (1985). 'Agricultural development in the Yemeni Tihama: the past ten years'. In *Economy, Society and Culture in Contemporary Yemen* (Ed B. R. Pridham), Croom Helm, London, pp 22–40.

Musil, A. (1928). *The Manners and Customs of the Rwala Bedouins*, American Geographical Society, New York.

Nakhleh, E. A. (1977). 'Labour markets and citizenship in Bahrayn and Qatar', *Middle East Journal*, **31**, 143–56.

Nelson, C. (Ed) (1973). *The Desert and the Sown: Nomads in the Wider Society*, Institute of International Studies, Research Series No 21, University of California, Berkeley.

Niblock, T. (Ed) (1980). *Social and Economic Development in the Arab Gulf*, Croom Helm, London.

Niblock, T. (Ed) (1982). *State, Society and Economy in Saudi Arabia*, Croom Helm, London.

Ochsenwald, W. (1981). 'Saudi Arabia and the Islamic revival', *International Journal of Middle East Studies*, **13**, 271–86.

Peppelenbosch, P. G. N. (1968). 'Nomadism in the Arabian peninsula: a general appraisal', *Tijdschr. econ. soc. geogr.*, **59**, 335–46.

Pike, J. G. (1979). 'Water resources and agriculture in Qatar', *Arabian Studies*, **5**, 67–86.

Pourcelet, F. (1968). 'Notes de géographie urbaine: l'expansion recente de la ville de Kuwayt', *Cahiers de l'Orient Contemporain*, **71**, 4–8.

Pridham, B. R. (Ed) (1985). *Economy, Society and Culture in Contemporary Yemen*, Croom Helm, London.

Raswan, C. R. (1930). 'Tribal areas and migration lines of the North Arabian bedouins', *Geogrl. Rev.*, **20**, 494–502.

Ronall, J. O. (1970). 'Banking developments in Kuwait', *Middle East Journal*, **24**, 87–90.

Saigh, Y. A. (1971). 'Problems and prospects of development in the Arabian peninsula', *International Journal of Middle Eastern Studies*, **2**, 40–58.

Sanger, R. H. (1954). *The Arabian Peninsula*, Cornell University Press, New York.

Satchell, J. E. (1978). 'Ecology and environment in the United Arab Emirates', *Journal of Arid Environments*, **1**, 201–06.

Seccombe, I. J. (1983). 'Labour migration to the Arabian Gulf: evolution and characteristics, 1920–1950', *British Society for Middle Eastern Studies: Bulletin*, **10**, 3–20.

Serjeant, R. B. (1964). 'Some irrigation systems in Hadramawt', *Bulletin of the School of Oriental and African Studies*, **27**, 32–76.

Serjeant, R. B. (1968). Fisher-folk and fish-traps in al-Bahrain, *Bulletin of the School of Oriental and African Studies*, **31**, 486–514.

Shamekh, A. A. (1975). *Spatial Patterns of Bedouin Settlement in Al-Qasim Region, Saudi Arabia*, Department of Geography, University of Kentucky, Lexington.

Steffen, H. (1979). *Population Geography of the Yemen Arab Republic*, Dr Ludwig Reichert Verlag, Wiesbaden.

Stevens, J. H. (1969). 'Arid zone agricultural development in the Trucial States', *J. Soil Wat. Conserv.*, **24**, 181–3.

Stevens, J. H. (1970). 'The changing agricultural practice of an Arabian oasis', *Geogrl. J.*, **136**, 410–418.

Stevens, J. . H. (1972). 'Oasis agriculture in the central and eastern Arabian peninsula', *Geography* **57**, 321–6

Swanson, J. C. (1979a). 'Some consequences of emigration for rural development in the Yemen Arab Republic', *Middle East Journal*, **33**, 34–44.

Swanson, J. C. (1979b). *Emigration and Economic Development: the Case of the Yemen Arab Republic*, Westview Press, Boulder.

Sweet, L. E. (1964). 'Pirates or politics? Arab societies of the Persian or Arabian Gulf, eighteenth century', *Ethnohistory*, **11**, 262–80.

Sweet, L. E. (1965). 'Camel pastoralism in north Arabia and the minimal camping unit'. In *Man, Culture and Animals: The Role of Animals in Human Ecological Adjustments*, (Eds A. Leeds and A. P. Vayda), American Association for the Advancement of Science. No 78, Washington DC, pp 129–52.

Tan, K. (1970). 'Agricultural problems in southern Arabia', *Wld Crops*, **22**, 397–400.

Townsend, J. (1977). *Oman. The Making of a Modern State*, Croom Helm, London.

Twitchell, K. S. (1958). *Saudi Arabia, with an Account of the Development of its Natural Resources*, 3rd edn, Oxford University Press, New York, 1958.

'Utaibah, M. S., Al- (1977). *Petroleum and the Economy of the United Arab Emirates*, Croom Helm, London.

Vidal, F. S. (1955). *The Oasis of Al-Hasa*, Arabian-American Oil Co, New York.

Wallen, I. F. (1969). 'Non-oil trade and resources'. In *The Princeton University Conference and Twentieth Annual Near Eastern Conference on Middle East Focus: The Persian Gulf*, (Ed T, C. Young), Princeton University Conference, Princeton, pp 107–10.

Weir, S. (1985a). 'Economic aspects of the qat industry in north-west Yemen'. In *Economy, Society and Culture in Contemporary Yemen* (Ed B. R. Pridham), Croom Helm, London, pp 64–82.

Weir, S. (1985b). *Qat in Yemen*, British Museum, London.

Young, A. N. (1983). *Saudi Arabia: The Making of a Financial Giant*, New York University Press.

Young, T. C. (Ed) (1969). *The Princeton University Conference and Twentieth Annual Near Eastern Conference on Middle East Focus: The Persian Gulf*, Princeton University Conference, Princeton, NJ.

Chapter 12: Iraq: Man, Land and Water in an Alluvial Environment

Aart, V. R. (1974). 'Drainage and land reclamation in the Lower Mesopotamian plain', *Nature and Resources*, **10(2)**, 11–17.

Adams, R. M. (1958). 'Survey of ancient water courses and settlements in central Iraq', *Sumer*, **14**, 101–03.

Adams, R. M. (1965). *Land Behind Baghdad*, The University of Chicago Press, 187 pages.

Al-Barazi, N. K. (1961). *The Geography of Agriculture in Irrigated Areas of the Middle Euphrates Valley*, Vol 1 and 2, College of Arts Baghdād University, Baghdād, 183 pages.

Al-Khashab, W. H. (1958). *The Water Budget of the Tigris and Euphrates Basin*, University of Chicago, Department of Geography, Research Paper No 54, 105 pages.

Baali, F. (1969). 'Agrarian reform in Iraq: some socio-economic aspects', *The American Journal of Economics and Sociology*, **28**, 61–76.

Beaumont, P. (1978). 'The Euphrates river – an international problem of water resources development', *Environmental Conservation*, **5(1)**, 35–44.

Beaumont, P. (1985). 'The agricultural environment: an overview'. In *Agricultural Development in the Middle East*, (Eds P. Beaumont and K. S. McLachlan), Wiley, Chichester, pp 3–26.

Beaumont, P. and K. S. McLachlan (Eds) (1985). Agricultural Development in the Middle East, Wiley, Chichester, 349 pages.

Brice, W. C. (1966). *South-West Asia*, University of London Press.

Buringh, P. (1957). 'Living conditions in the lower Mesopotamian plain in ancient times', *Sumer*, **13**, 30–57.

Buringh, P. (1960). *Soils and Soil Conditions in Iraq*, Directorate General of Agricultural Research and Projects, Baghdād, Republic of Iraq, Ministry of Agriculture.

Clarke, J. I. and W. B. Fisher (Ed) (1972). *Populations of the Middle East and North Africa*, University of London Press.

Clawson, M., H. H. Landsberg and L. T. Alexander (1971). *The Agricultural Potential of the Middle East*, Elsevier, New York, 312 pages.

Davies, D. H. (1957). 'Observations on land use in Iraq', *Econ. Geogr.*, **33**, 122–34.

El-Ashry, M. T., J. V. Schilfgaarde and S. Schiffman (1985). 'Salinity pollution from irrigated agriculture', *Journal of Soil & Water Conservation*, **40(1)**, 48–52.

Elgabaly, M. M. (1977). 'Water in arid agriculture: salinity and water logging in the Near East region', *Ambio*, **6**, 36–9.

FAO (1983). *FAO Fertilizer Yearbook 1982*, FAO, Rome.

FAO (1985). *FAO Production Yearbook 1984*, FAO, Rome.

Fernea, R. A. (1959). *Irrigation and Social Organisation Among the El Shabana – a Group of Tribal Cultivators in Southern Iraq*, PhD thesis, University of Chicago.

Fernea, R. A. (1969). 'Land reform and ecology in post-revolutionary Iraq', *Economic Development and Cultural Change*, **17**, 356–81.

Fernea, R. A. and E. W. Fernea (1969). 'Land reform in modern Iraq', *Focus*, **20(2)**, 9–12.

Fernea, R. A. and E. W. Fernea (1969). 'Iraq', *Focus*, **20(2)**, 1–8.

Gelburd, D. (1985). 'Managing salinity: lessons from the past', *Journal of Soil & Water Conservation*, **40(4)**, 329–31.

Gulick, J. (1967). 'Baghdad, portrait of a city in physical and cultural change', *Journal of the American Institute of Planners*, **34**, 339–50.

Harris, S. A. and R. M. Adams (1957). 'A note on canal and marsh stratigraphy near Zubediya', *Sumer*, **13**, 157–63.

Harris, S. A. (1958). 'The Gilgaied and bad-structured soils of central Iraq', *J. Soil Sci.*, **9**, 169–85.

Ionides, M. G. (1937). *The Regime of the Rivers Euphrates and Tigris*, London, 255 pages.

Issawi, C. (1969). 'Economic change and urbanisation in the Middle east'. In *Middle Eastern Cities*, (Ed I. M. Lapidus), University of California Press, Berkeley, 102–21.

Jacobsen, T. and R. McC. Adams (1958). 'Salt and silt in ancient Mesopotamian agriculture', *Science, N. Y.*, **128**, 1251–8.

Jones, L. (1969). 'Rapid population growth in Baghdad and Amman', *Middle East Journal*, **23**, 209–15.

Kaul, R. N. and D. C. P. Thalen (1971). 'Range ecology at the Institute for Applied Research on Natural Resources', Iraq, *Nature and Resources*, **7(2)**, 10–15.

Kharrufa, N. S., G. M. Al-Kaway and H. N. Ismail (1980). *Studies on Crop Consumption Use of Water in Iraq*, Pergamon Press, Oxford.

Kingsman, J. (1970). 'Kurds and Iran: Iraq's changing balance of power', *The New Middle East*, **22**, 25–7.

Laessøe, J. (1953). 'Reflection on modern and ancient oriental waterworks', *Journal of Cuneiform Studies*, **7**, 5–26.

Langley, K. M. (1964). 'Iraq: some aspects of the economic scene', *Middle East Journal* **18**, 180–8.

Langley, K. M. (1967) *The Industrialization of Iraq*, Harvard Middle East Monographs, Cambridge, Mass, 313 pages.

Lawless, R. I. (1972). 'Iraq – changing population patterns'. In *Populations of the Middle East and North Africa*, (Eds J. I. Clarke and W. B. Fisher), University of London Press, 97–129.

Lebon, J. H. G. (1953). 'Population distribution and the agricultural regions of Iraq', *Geogrl. Rev.*, **43**, 223–8.

Lees, G. M. and N. L. Falcon (1952). 'The geographical history of the Mesopotamian plains', *Geogrl. J.*, **118**, 24–39.

Longrigg, S. H. and F. Stokes (1958). *Iraq*, Praeger, New York, 256 pages.

Millon, R. (1962). 'Variations in social responses to the practice of irrigation and agriculture'. In *Civilisations and Desert Lands*, (Ed R. B. Woodbury), University of Utah, Anthropological Papers, Salt Lake City, **62**, pp 56–88.

Mitchell, R. C. (1957). 'Recent tectonic movement in the Mesopotamian plains', *Geogrl. J.*, **123**, 569–71.

Mitchell, C. W. (1959). 'Investigations into the soils and agriculture of the lower Diyala area of eastern Iraq', *Geogrl. J.*, **125**, 390–7.

Mitchell, C. W. and P. E. Naylor (1960). 'Investigations into the soils and agriculture of the middle Diyala region of eastern Iraq', *Geogrl. J.*, **126**, 469–75.

Naval Intelligence Division (Great Britain) (1944). *Iraq and the Persian Gulf*, Geographical Handbook Series BR 524, London, 524 pages

Ockerman, H. W. and S. G. Samano (1985). 'The agricultural development of Iraq'. In *Agricultural Development of the Middle East*, (Eds P. Beaumont and K. S. McLachlan), Wiley, Chichester, pp 189–207.

Phillips, D. G. (1959). 'Rural to urban migration in Iraq', *Economic Development and Cultural Change*, **7**, 405–21.

Population Reference Bureau (1986). *Population Data Sheet 1985*, Washington, DC.

Quint, M. N. (1958). 'The idea of progress in an Iraqi village', *Middle East Journal*, **12**, 369–84.

Simmons, J. L. (1965). 'Agricultural development in Iraq, planning and management failures', *Middle East Journal*, **19**, 129–40.

Smith, C. G. (1970). 'Water resources and irrigation development in the Middle East', *Geography*, **55**, 407–25.

Smith, H. H. *et al.* (1971). *Areas Handbook for Iraq*, The American University, Washington DC, 413 pages.

Smith R. and V. C. Robertson (1962). 'Soil and irrigation classification of shallow soils overlying gypsum beds, northern Iraq', *J. Soil Sci.*, **13**, 106–15.

Storke, C. (1959). 'Evapotranspiration problems in Iraq', *Neth. J. agric. Sci.*, **7**, 269–82.

Tamini, S. A. and M. A. Younis (1972). 'Effect of CEC and irrigation on wheat yields in Iraq', *Wld Crops*, **26(6)**, 310–11.

Thesiger, W. (1964). *The Marsh Arabs*, Longmans, London, 242 pages.

Treakle, H. C. (1966). *The Agricultural Economy of Iraq*, Washington Foreign Regional Analysis Division, Economic Research Service, USDA, Washington, DC, 4 pages.

Ubell, K. (1971). 'Iraq's water resources', *Nature and Resources*, **7(2)**, 3–9

Warriner, D. (1969). 'Revolutions in Iraq'. In *Land Reform in Principle and Practice*, (Ed D. Warriner), Oxford University Press, London, pp 78–108.

Weickmann, L. (1963). 'Meteorological and hydrological relationships in the drainage area of the river Tigris north of Baghdad', *Met. Rdsch.*, **16**, 33–8.

Willcocks, W. (1908). *The Restoration of Ancient Irrigation Work on the Tigris*, Cairo.

Willcocks, W. (1917). *The Irrigation of Mesopotamia*, Spon, London.

Wittfogel, K. A. (1965). 'The hydraulic civilisations'. In *Man's Role in Changing the Face of the Earth*, (Ed W. L. Thomas Jnr), University of Chicago Press, 152–64.

Woodbury, R. B. (Ed) (1962). *Civilisations and Desert Lands*, University of Utah, Anthropological Papers, Salt Lake City.

Wright, H. E. Jnr. (1968). 'Natural environment of early food production north of Mesopotamia', *Science, N.Y.*, **161**, 334–9.

Wright, H. E. Jnr. (1970). 'Environmental change and the origin of agriculture in the Near East', *Bioscience*, **20**, 210–12.

Chapter 13: Agricultural Expansion in Syria

Ashton, B. L. (1928). 'The geography of Syria', *J. Geogr*, **27**, 164–180.

Bourgey, A. (1974). 'Le Barrage de Tabqa et l'améngement du bassin de l'Euphrate en Syrie', *Revue Géogr. de Lyon*, **49**, 434–54.

Beaumont, P. and K. McLachlan (Eds) (1985). *Agricultural Development in the Middle East*, Wiley, Chichester.

Dickie, P. M. and D. B. Noursi (1975). 'Dual markets: the case of the Syrian Arab Republic', *Staff Papers International Monetary Fund*, **22**, 456–68.

Dresch, J. (1963). 'Observations sur la région Palmyre en Syrie', *Bull. Ass. Géogr, fr.*, 2–18

El-Zaim, I. (1968). 'La réforme agraire en Syrie, *Tiers-Monde'*, **9**, 508–17.

Garrett, J. (1936). 'The site of Damascus', *Geography*, **21**, 288–96.

Garzouzi, E. (1963). 'Land reform in Syria', *Middle East Journal*, **17**, 83–90.

Gilbert, A. and M. Fevret (1953). 'La Djezirah Syrienne et son reveil économique', *Revue Géogr. Lyon*, **38**, 1–15, 83–99.

Hamide, A. H. (1959). *La Région d'Alep*, Paris.

Helbaoui, Y. (1963). 'La Population et la population active en Syrie', *Population, Paris*, **18**, 697–714.

Hudson, J. (1968a). 'Syria', *Focus*, **18**, 1–8.

Hudson, J. (1968b). 'The role of irrigation (in Syria)', *Focus*, **18**, 8–11.

International Bank for Reconstruction and Development (1955). *The Economic Development of Syria*, Johns Hopkins, Baltimore.

Issawi, C. (Ed) (1955). *The Economic History of the Middle East 1800–1914*, University of Chicago, Chicago and London.

Kanovsky, E. (1977). *Economic Development of Syria*, University Publishing Projects, Tel Aviv.

Keilany, Z. (1970). 'Economic planning in Syria, 1960–65: an evaluation', *Journal of Developing Areas*, **4**, 361–73.

Keilany, Z. (1973). 'Socialism and economic change in Syria', *Middle Eastern Studies*, **9**, 61–72.

Keilany, Z. (1980). 'Land reform in Syria', *Middle Eastern Studies*, **16**, 209–24.

Latron, A. (1936). *La Vie rurale en Syrie et au Liban*, Beirūt.

Lewis, N. N. (1949). 'Malaria, irrigation and soil erosion in central Syria', *Geogrl. Rev.*, **39**, 278–90.

Lewis, N. N. (1955). 'The frontier of settlement in Syria: 1800–1950', *International Affairs*, **31**, 48–60. Reprinted in Issawi C. (Ed) (1966). *The Economic History of the Middle East, 1800–1914*, University of Chicago, Chicago and London, pp 259–68.

Mahhouk, A. (1956). 'Recent agricultural development and Bedouin settlement in Syria', *Middle East Journal*, **10**, 167–76.

Manners, I. R. and T. Sagafi-Nejad (1985). 'Agricultural Development in Syria'. In *Agricultural Development in the Middle East*, (Eds P. Beaumont and K. McLachlan), Wiley, Chichester.

Ma'oz, M. and Yaniv, A. (Eds) (1986). *Syria under Assad*, Croom Helm, London.

Métral, J. and P. Sanlaville (Eds) (1979). 'Problèmes agraires en Syrie', *Revue Géogr. de Lyon*, **54**, 229–325.

Money-Kyrle A. F. (1956). *Agricultural Development and Research in Syria*. American University of Beirūt, Faculty of Agricultural Science, Publication No 2.

Muir, A. (1951). 'Notes on the soils of Syria', *J. Soil Sci.*, **2**, 163–82.

Orgels, B. (1963). *Contribution a l'etude des problèmes Agricoles de la Syria*, Centre Pour L'Etude des Problèmes du Monde Musulman Contemporain, Brussels.

Petran, T. (1972). *Syria*, Nations of the Modern World, Ernest Benn, London.

Peretz, D. (1964). 'River schemes and their effect on economic development in Jordan, Syria and Lebanon', *Middle East Journal*, **18**, 293–305.

Raymond, A. (Ed) (1980). *La Syrie d'Aujourd'hui*, Centre National de la Recherche Scientifique, Paris.

Reifenberg, A. (1952). 'The soils of Syria and the Lebanon', *J. Soil Soc.*, **3**, 68–88.

Rolley, J. (1948). 'Forest conditions in Syria and Lebanon', *Unasylva*, **2**, 77–80.

Rosciszewski, M. (1965). 'Quelques remarques sur la géographie agraire de la Syrie', *Méditerranée*, **6**, 171–84.

Samman, M. L. (1983). 'Le recensement syrien de 1981', *Population*, **38**, 184–88.

Shakir, K. A. (1965). *Planning for a Middle Eastern Economy: Model for Syria*, Chapman and Hall, London.

Smilianskaya, I. M. (1958). 'Razlozhenie feodalnikh otnoshenii v Sirii i Livane v seredine xix v (The disintegration of feudal relations in Syria and Lebanon in the middle of the nineteenth century)', *Peredneaziatskii Etnosgraficheskii Sbornik*, **1**, 156–179. Translated in Issawi, C. (Ed) (1966). *The Economic History of the Middle East, 1800–1914*, University of Chicago, Chicago and London, pp 227–47.

Thoumin, R. (1936). *Géographie Humaine de la Syrie Centrale*, Ernest Leroux, Paris.

Tresse, R. (1929). 'L'irrigation dans la Ghouta de Damas', *Revue des Etudes Islamiques*, **3**, 459-73.

USAID (United States Agency for International Development) (1980). *Syria: Agricultural Sector Assessment*, Washington, DC.

Vaumas, E. de (1955). 'La population de la Syrie', *Annls. Géogr.*, **64**, 74-80.

Vaumas, E. de (1956). Le Djéziré syrienne, *Annls. Géogr.*, **65**, 64-80.

Warriner, D. (1962). *Land Reform and Development in the Middle East. A Study of Egypt, Syria and Iraq*, 2nd edn, Oxford University Press.

Weulersse, J. (1936). 'Damas, étude de développement urbain', *Bull. Ass. Géogr. fr.*, 5-9.

Weulersse, J. (1938). 'La primauté des cités dans l'économie syrienne, *Congres inter. de Géogr., Amsterdam*, **2**, sect. 3A, pp 233-39.

Weulersse, J. (1946). *Paysans de Syrie et du Proche-Orient*, Gallimard, Paris.

Wirth, E. (1966). 'Damaskus-Aleppo-Beirut; en geographischer Verleich dreier nahöstlicher Städte in Spiegel inrer sozial wirtschaftlich tanangebenden Schichten', *Erde*, **97**, 96-137; 166-202.

Wofast, R. (1967). *Geologie von Syria und der Libanon*, Gebrüder Borntraegerm, Berlin.

Zuckermann, B. (1971) 'Das Euphratprojekt in der Syrischen Arabischen Republik und sein Einflusch auf die Territorialstruktur der syrischen Volkswirtschaft', *Petermanns geogr. Mitt.*, **115**, 98-101.

Chapter 14: Religion, Community and Conflict in Lebanon

Basili, K. (1984). *Memories of Lebanon, 1839-47*, Yad Ben Zvi Jerusalem (Hebrew).

Borgey, A. (1985). 'The war and its geographical consequences in the Lebanon', *Annals de Géographie*, **94**, 1-37.

Chaib, A. E. (1980). 'Analysis of Lebanon's merchandise exports, 1951-1974', *Middle East J.*, **34**, 438-55.

Dawisha, A. I. (1980). 'Syria and the Lebanese Crisis, Macmillan, London.

Deeb, M. K. (1984). 'Lebanon: prospects for national reconciliation in the mid-1980s', *Middle East J.*, **38**, 267-83.

Entelis, J. P. (1973). 'Structural change and organisational development in the Lebanese Kata'ib party', *Middle East Journal*, **27**, 21-35.

Gaube, H. (1977). 'The historical and social background to the Lebanese civil war', *Geographische Rundschau*, **20**, 286-90.

Gilmour, D. (1984). *'Lebanon: The Fractured Country*, revised edn, Sphere, London.

Gordon, D. C. (1983). *The Republic of Lebanon: Nation in Jeopardy*, Westview Press, Boulder; Croom Helm, London.

Gray, A. J. (1983). 'The Crisis in Lebanon, 1975-1983 and Its Geopolitical background'. Unpublished MA dissertation, Durham University.

Gubser, P. (1975). 'The politics of economic interest groups in a Lebanese town', *Middle Eastern Studies*, **11**, 262-83.

Harik, I. (1980). 'Voting participation and political integration in Lebanon, 1943-1974', *Middle Eastern Stud.*, **16**, 27-48.

Heller, P. B. (1980). 'The Syrian factor in the Lebanese civil war', *Journal of South Asia and Middle Eastern Studies*, **4**, 56-76.

Hudson, M. C. (1978). 'The Palestinian factor in the Lebanese civil war', *Middle East J.*, **32**, 261-78.

Jabbra, J. G. and N. W. (1978). 'Local political dynamics in Lebanon: the case of 'Ain al-Qasis', *Anthropological Qtly.*, **51**, 137-51.

Karpat, K.H. (1984). *Ottoman Population, 1830–1914: Demographic and Social Characteristics*, University of Wisconsin Press, Madison and London.

Karpat, K.H. (1985). 'The Ottoman emigration to America', *Int. J. Middle East Stud.*, **17**, 175–209.

Kasyanov, O. (1984). 'Crimes of imperialism and Zionism in Lebanon', *International Affairs* (Moscow), **9**, 53–9.

Khalidi, R. (1984). 'The Palestinians in Lebanon: social repercussions of Israel's invasion', *Middle East J.*, **38**, 255–66.

Khalidi, W. (1979). *Conflict and Violence in Lebanon*, Harvard University Press, Cambridge, Mass.

Kisirwani, M. (1980). 'Foreign interference and religious animosity in Lebanon', *Journal of Contemporary History*, **15**, 685–700.

Kliot, N. (1986). 'Lebanon – a geography of hostages', *Political Geography Qtly.*, **5**, 199–220.

McDowall, D. (1982). *Lebanon: A Conflict of Minorities*, Minority Rights Group Report No 61, London.

Makolisi, SA. (1977). 'An appraisal of Lebanon's postwar economic development and a look at the future', *Middle East J.*, **31**, 267–80.

Muir, J. (1984). 'Lebanon: area of conflict, crucible of peace', *Middle East J.*, **38**, 204–19.

Ryan, S. (1982). 'Israel's invasion of Lebanon: Background to the crisis', *Journal of Palestine Studies*, **11**, **12**, 23–37.

Salibi, K. (1965). *The Modern History of Lebanon*, Caravan Books, New York; Praeger, New York; Weidenfeld and Nicolson.

Salibi, K.S. (1971). 'The Lebanese identity', *Journal of Contemporary History*, **6**, 76–86.

Sayigi, Y. (1983). 'Israel's military performance in Lebanon, 1982', *Journal of Palestine Studies*, **13**, 24–65.

Sirhan, B. (1975). 'Palestinian refugee camp life in Lebanon', *Journal of Palestine Studies*, **4**, 91–107

Toubi, J. (1980). 'Social dynamics in war-torn Lebanon', *Jerusalem Qtly.*, **17**, 83–109.

Vaumas, E. de (1955). 'La répartition confessionelle au Liban et l'équilibre de l'état libanais', *Revue Géogr. alp.*, **43**, 511–604.

Chapter 15: Jordan: The Struggle for Economic Survival

Aharoni, Y. (1966). *The Land of the Bible*, Burns & Oates, London (Translated from the Hebrew by A.F. Rainey), 409 pages.

Atkinson, K. and P. Beaumont (1967). 'Watershed management in northern Jordan', *Wld Crops*, **19**, 63–5.

Atkinson, K. and P. Beaumont (1971). 'The forests of Jordan', *Econ. Bot,*, **25**, 305–11.

Baly, D. (1958). *The Geography of the Bible*, Butterworth Press, London, 303 pages.

Beaumont, P. (1968). 'The Road to Jericho: A climatological traverse across the Dead Sea lowlands', *Geography*, **53**, 170–4.

Beaumont, P. (1985). 'Trends in Middle Eastern agriculture'. In *Agricultural development in the Middle East*, (Eds P. Beaumont and K.S. McLachlan), Wiley, London, pp 305–22.

Beaumont, P. and K. Atkinson (1969). 'Soil erosion and conservation in northern Jordan', *J. Soil Wat. Conserv.*, **24**, 144–7.

Birch, B.P. (1971). 'Jordan's geography after the 1967 war', *Tijdschr. econ. soc. Geogr.*, **72**, 45–52.

Birch, B.P. (1973). 'Recent developments in agriculture, land and water use in Jordan', *Wld Crops*, **25**, 66–76.

Blake, G, S. and M. J. Goldschmidt (1947). *Geology and Water Resources of Palestine*, Government Printer, Jerusalem, 413 pages.

Brawer, M. (1968). 'The geographical background of the Jordan water dispute'. In *Essays in Political Geography*, (Ed C. A. Fisher), Methuen, London, pp 225–42.

Browning, I. (1973). *Petra*, Chatto and Windus, London, 256 pages.

Burdon, D. J. (1959). *Handbook of the Geology of Jordan*, Government of the Hashemite Kingdom of Jordan, Amman, 82 pages.

Casto, E. R. (1937). 'Economic geography of Palestine', *Econ. Geogr.*, **13**, 235–59.

Casto, E. R. and O. W. Dotson (1938). 'Economic geography of Trans-Jordan', *Econ. Geogr.*, **14**, 121–30.

Clarke, J. I. and W. B. Fisher (Eds) (1972). *Populations of the Middle East and North Africa*, University of London Press.

Copeland, R. W. (1965). *The Land and People of Jordan*, J. B. Lippincott Company, Philadelphia, 160 pages.

Davies, H. R. J. (1958). 'Irrigation in Jordan', *Econ. Geogr.*, **34**, 264–71.

Dees, J. L. (1959). 'Jordan's East Ghor canal project', *Middle East J.*, **13**, 357–71.

Fisher, W. B. (1972). 'Jordan: a demographic shatter-belt'. In *Populations of the Middle East and North Africa*, (Eds J. I. Clarke & W. B. Fisher), University of London Press, 202–19.

Garbell, M. A. (1965). 'The Jordan valley plan', *Scient. Am.*, **212(3)**, 23–31.

Gregory, J. W. (1930). 'Palestine and the stability of climate in historic times', *Geogrl J.*, **76**, 487–94.

Hacker, J. M. (1960). *Modern Amman – A Social Study*, Department of Geography, Research Paper Series No 3, University of Durham, 144 pages.

Hindle, P. (1964). 'The population of the Hashemite Kingdom of Jordan, 1961', *Geogrl. J.*, **130**, 261–4.

Haddard, S. M. (1966). 'Principles and procedures used in planning and execution of East Ghor Irrigation project', *Sixth NESA Irrigation Practices Seminar, Amman*, 2–6.

Hare, V. C. (1954). 'The Jordan valley of the future: desert or garden?', *The Near East*, **7**, 8–15.

Hashemite Kingdom of Jordan, Central Water Authority (1962). 'Irrigation in Jordan', *Fourth NESA Irrigation Practices Seminar, Ankara*, 56–153.

Hashemite Kingdom of Jordan (no date). *Five Year Plan for Economic and Social Development 1986–1990*, Ministry of Planning, Amman, Jordan, 574 pages.

Haupert, J. S. (1966). 'Recent progress on Jordan's East Ghor canal', *Prof. Geogr.*, **18**, 9–13.

International Bank for Reconstruction and Development (1957). *Economic Developmentof Jordan*, Johns Hopkins, Baltimore, 488 pages.

Ionides, M. G. and G. S. Blake (1939). *Report on the Water Resources of Trans-Jordan and their Development: Incorporating a Report on Geology, Soils, and Minerals and Hydrogeological Correlations*, Crown Agents for the Colonies, London, 372 pages.

Ionides, M. G. (1946). 'The perspective of water development in Palestine and Trans-jordan', *Jl. R. cent. Asian Soc.*, **33**, 271–80.

Ionides, M. G. (1951). 'The Jordan valley', *Jl. R. cent. Asian Soc.*, **38**, 217–25.

Ionides, M. G. (1953). 'The disputed waters of Jordan', *Middle East Journal*, **7**, 153–64.

Johnston, E. (1954). 'Arab–Israel tension and the Jordan valley', *World Affairs*, **117**, 38–41.

Jones, W. E. (1965). 'The Jordan river valley: a problem in political geography', *Swansea Geographer*, **3**, 77–90.

Jordan Development Board (1965). *Programme for Economic Development 1964–1970*, Amman, 360 pages.

Jordan Valley Commission (1976). *Jordan Valley Development Plan 1975–82 (Summary)*, Jordan Valley Commission, Amman, 38 pages.

Kanovsky, E. (1968). 'The economic aftermath of the Six Day War: Part II, *Middle East Journal*, **22**, 278–96.

Khouri, F. J. (1964). 'The Jordan river, the U.S., and the U.N.', *Middle East Forum*, **40**, 20–24.

Khouri, F. J. (1965). 'The Jordan river controversy', *Révue Politique*, **27**, 32–57.

Kielmans, M. (1987). 'Jordan: oozing confidence about oil', *Middle East Economic Journal*, 14 February 1987.

Kirk, G. (1954). *The Middle East in the War*, Oxford University Press, London, 511 pages.

Kirk, G. (1954). *The Middle East 1945–1950*, Oxford University Press, London, 338 pages.

Long, G. A. (1957). *The Bioclimatology and Vegetation of Eastern Jordan*, FAO, Rome.

Mackenzie, M. (1946). 'Transjordan', *Jl. R. cent. Asian Soc.*, **33**, 260–270.

Manners, I. R. (1970). 'The East Ghor irrigation project', *Focus N. Y.*, **20 (8)**, 8–11.

Mazur, M. P. (1979). *Economic Growth and Development in Jordan*, Croom Helm, London, 314 pages.

Mehdi, M. (1973). 'Israel settlements in the occupied territories', *Middle East International*, **19**, 21–7.

Middle East Economic Digest (1973). 'Economic recovery in Jordan', *MEED*, **17 (24)**, 675–9.

Mitchell, J. E. (1967). 'Planning of water development in the Hashemite Kingdom of Jordan'. In *Water for Peace – Vol 7 – Planning and Developing Water Programs*, Washington, DC, US Gov. Printing Office.

Mountfort, G. (1965). *Portrait of a Desert*, Collins, London, 192 pages.

National Planning Council (1982). *The Five Year Plan for Economic & Social Development 1981–85*, National Planning Council, Amman, 372 pages.

Natur, F. (1962). 'Farm unit layout, distribution and development in East Ghor canal project', *Fourth, NESA Irrigation Practices Seminar, Ankara*, 300–12.

Naval Intelligence Division (Great Britain) (1943). *Palestine and Transjordan*, Geographical handbook Series, BR 514, British Admiralty, 621 pages.

Notestein, F. W. and E. Jurkat (1945). 'Population problems of Palestine', *The Milbank Memorial Fund Quarterly (New York)*, **23**, 307–52.

Nuttonson, M. Y. (1947). 'Agroclimatology and crop ecology of Palestine and Transjordan and climatic analogues in the United States', *Geogrl. Rev.*, **37**, 436–56.

Patai, R. (1958). *The Kingdom of Jordan*, Princeton University Press, Princeton, New Jersey, 315 pages.

Peretz, D. (1964). 'River schemes and their effects on economic development in Jordan, Syria and Lebanon', *Middle East Journal*, **18**, 293–305.

Phillips, P. G. (1954). *The Hashemite Kingdom of Jordan: Prolegomena to a Technical Assistance Program*, Department of Geography, Research Paper No 34, University of Chicago, 191 pages.

Poore, M. E. D. and V. C. Robertson (1964). *An Approach to the Rapid Description and Mapping of Biological Habitats*, Sub-Commission on Conservation of Territorial Biological Communities of the International Biological Programme, 68 pages.

Randall, R. (1968). *Jordan and the Holy Land*, Frederick Muller, London, 243 pages.

Reece, H. C., T. D. Roberts, J. P. Coury, S. Cooper, A. E. Farrier, T. Tompkins and N. B. Turk (1969). *Area Handbook for the Hashemite Kingdom of Jordan*, The American University, Washington, DC, 372 pages.

Schattner, I. (1962). *The Lower Jordan Valley*, Scripta Hierosolymitana, Publications of the Hebrew University, Jerusalem, vol 11, 123 pages.

Smith, C. G. (1966). 'The disputed waters of the Jordan', *Trans. Inst. Br. Geogr.*, **40**, 111–28.

Smith, R. A. and B. P. Birch (1963). 'The East Ghor irrigation project in the Jordan valley', *Geography*, **48**, 407–09.

Sparrow, J. G. (1961). *Modern Jordan*, Allen and Unwin, London, 180 pages.

Sutcliffe, C. R. (1973). 'The East Ghor Canal Project: a case study of refugee resettlement, 1961–1966', *The Middle East Journal*, **27**, 471–82.

Talal, H. (1967). 'Growth and stability in the Jordanian economy', *Middle East Journal*, **17**, 92–100.

UNDP/FAO (1970). *Investigation of the Sandstone Aquifers of East Jordan – Jordan – The Hydrology of the Mesozorie-Cainozorie Aquifers of the Western Highlands and Plateau of East Jordan*, FAO, Rome.

Van Valkenburg, S. (1954). 'The Hashemite Kingdom of the Jordan: a study in economic geography', *Econ. Geogr.*, **30**, 101–16.

Vouras, P. P. (1967). 'Jordan', *Focus*, **17(6)**, 1–6.

Whyte, R. O. (1950). 'The phytogeographical zones of Palestine', *Geogrl. Rev.*, **40**, 600–14.

Chapter 16: Israel and the Occupied Areas

Adam-Smith, G. (1900). *Historical Geography of the Holy Land*, 7th edn, Hodder and Stoughton, London.

Bachi, R. (1967). 'Effects of migration on the geographic distribution of population in Israel', *International Union for Scientific Study of Population, Conference Papers*, Sydney, pp 737–51.

Ben-Arieh, Y. (1985). *Jerusalem in the Nineteenth Century*, Croom Helm, London.

Baly, D. (1957). *The Geography of the Bible*, Lutterworth, London.

Baratz, J. (1960). *A Village by the Jordan*, Ichud Habonim, Tel Aviv.

Ben-Arieh, Y. (1968). 'The changing landscape of the central Jordan valley', *Scripta Hierosolymitana*, Jerusalem, **15**, Pamphlet 3, 1–131.

Benvenisti, M. (1986). *1986 Report: Demographic, Economic, Legal, Social and Political Developments in the West Bank*. West Bank Data Base Project, Jerusalem.

Benvenisti, M., Z. Abu-Zayed and D. Rubinstein, (1986). *The West Bank Handbook: a Political Lexicon* The Jerusalem Post, Jerusalem.

Benvenisti, M. (1986). *West Bank Data Project: a Survey of Israel's Policies*, The Jerusalem Post, Jerusalem.

Ben-Yosef, Z. (1985). *Map of Settlement in Eretz Israel*, Settlement Division of the Zionist Organisation, Jerusalem.

Central Bureau of Statistics (1967–68). *Population Census of 1967: West Bank, Golan Heights, Gaza Strip, Northern Sinai*, 2 vols, Jerusalem.

Cohen, A. (1964). *Arab Border Villages in Israel*, Manchester University Press.

Cohen, S. B. (1987). *The Geopolitics of Israel's Border Question*, Westview Press, Godstone.

Elazar, D. J. (Ed) (1982). *Judea, Samaria and Gaza: Views on the Present and Future*, American Enterprise Institute, Washington DC.

Eisenstadt, S. N. (1958). 'Israel'. In *The Institutions of Advanced Societies*, (Ed M. E. Rose), University of Minnesota Press, Minneapolis.

Farid, A. M. and H. Sirriyeh (Eds) (1985). *Israel and Arab Water*, Arab Research Centre, London and Ithaca Press, London.

Gilbert, M. (1979). *The Arab-Israeli Conflict: Its History in Maps*. 3rd edn, Weidenfeld and Nicolson, London.

Granott, A. (1951). *The Land System in Palestine*, Eyre and Spottiswoode, London.

HMSO (1937). Cmd 5479: *Palestine Royal Commission Report*, HMSO, London (The 'Peel Commission').

HMSO (1938). Cmd 5854: *Palestine Partition Commission Report*, HMSO, London. (The 'Woodhead Commission').

HMSO (1939). Cmd 6019: *Palestine – Statement of Policy*, HMSO, London (The 'Macdonald White Paper').

Hale, G. A. (1982). 'Diaspora versus ghourba: the territorial restructuring of Palestine'. In *Tension Areas of the World*, (Ed D. Gordon Bennett), Park Press, Delray Beach, pp 130–54.

Halperin, H. (1957). *Changing Patterns in Israel Agriculture*, Routledge and Kegan Paul, London.

Hansell, H. J. (1978). Letter from State Department Legal Adviser concerning legality of Israeli Settlements in the Occupied Territories, April 21, 1978, *Americans for Middle East Understanding*, New York.

Harris, W. W. (1980). *Taking Root: Israeli Settlement in the West Bank, the Golan, and Gaza-Sinai 1967–1980*, Research Studies Press, John Wiley, Chichester.

Jewish Encyclopaedia (1901). Vol 1, Funk and Wagnalls, New York and London.

Karmon, Y. (1971). *Israel – a Regional Geography*, John Wiley, Chichester.

Kendall, H. and K. H. Baruth (1949). *Village Development in Palestine During the British Mandate*, Crown Agent, London.

Klatzmann, (1963). *Les enseignements des l'expérience Israélienne*, Presses Universitaires de France, Paris.

Kliot, N. (1987). 'Here and there – the phenomenology of settlement removal from northern Sinai', *J. Applied Behavioural Science*, No 1, pp 35–51.

Kliot, N. and S. Waterman (Eds) (1983). *Pluralism and Political Geography: People Territory and State,* Croom Helm, London and St Martins Press, New York.

Locke, R. and A. Stewart, (1985). *Bantustan Gaza*, Zed Books and Council for the Advancement of Arab-British Understanding, London.

Lowdermilk, W. C. (1946). *Palestine Land of Promise*, Victor Golancz, London.

Murray, G. W. (1953). 'The land of Sinai', *Geogrl. J.*, **119**, 140–54.

Nakhleh, E. A. (Ed) (1980). *A Palestinian Agenda for the West Bank and Gaza*, American Enterprise Institute, Washington DC.

Naval Intelligence Division (Great Britain) (1943). *Palestine and Transjordan*, Geographical Handbook Series, BR 514.

Newman, D. (Ed) (1985). *The Impact of Gush Emunim: Politics and Settlement in the West Bank*, Croom Helm, London.

Newman, D. (1984). 'Ideological and political influences on Israeli urban colonization: the West Bank and Galilee mountains', *Canadian Geographer*, **28 (2)**, 142–55.

Newman, D. (1984). 'The development of the Yishuv Kehillati in Judea and Samaria: political process and settlement form', *Tijdschrift voor Econ. en. Soc. Geografie*, **75 (2)**, 140–50.

Nir, D. (1968). *La vallée de Beth-Chéane*, Libraire Armand Colin, Paris.

Orni, E. and E. Efrat (1964). *Geography of Israel*, Israel Program for Scientific Translations, Jerusalem.

Orni, E. and E. Efrat (1971). *Geography of Israel*, Israel Universities Press, Jerusalem.

Paran, U. (1970). 'Kibbutzim in Israel: their development and distribution', *Jerusalem Studies in Geography*, vol 1, Hebrew University, Jerusalem.

Portugali, J. (1986). 'Arab labour in Tel Aviv: a preliminary study', *Int. Jnl. of Urban and Regional Research*, **10 (3)**, 352–76.

Peretz, D. (1985). *The West Bank*, Westview Special Studies on the Middle East, Westview Press, Godstone.

Ron, Z. Y. D. (1985). 'Development and management of irrigation systems in mountain regions of the Holy Land', *Trans. Inst. Br. Geogr.*, N.S.10, 149–69.

Roth, C and G. Wigoder (Eds) (1971). *Encyclopaedia Judaica*, 16 vols. Keter, Jerusalem.

Rowley, G. (1984). *Israel Into Palestine*, Mansell Publishing, London.

Rubner, A. (1960). *The Economy of Israel*, Frank Cass, London.

Sayigh, Y. A. (1968). *Palestine in Focus*, Palestine Research Center, Beirūt.

Soffer, A., and J. V. Minghi (1986). 'Israel's security landscapes: the impact of military considerations on land uses', *Professional Geographer*, **38 (1)**, 28–41.

State of Israel (1972). *Statistical Abstract of Israel*, **23**, Jerusalem.

Survey of Israel (1985). *Atlas of Israel*, 3rd edn, Collier Macmillan, Basingstoke.

UNESCO (1964). 'Agricultural planning and village community in Israel', *Arid Zone Research*, **23**, United Nations, New York.

Waterman, S. (1985). 'Not just milk and honey: Israeli human geography since the Six Day War', *Progress in Human Geography*, **9 (2)**, 194–234.

Chapter 17: The Industrialization of Turkey

Aktan, R. (1966). 'Turkish agricultural problems', *Mediterranea*, **12**, 266–75.

Albaum, M. and C. S. Davies (1973). 'The spatial structure of socio-economic attributes of Turkish provinces'. *International Journal of Middle Eastern Studies*, **4**, 288–310.

Alexander, A. P. (1960). 'Industrial entrepreneurship in Turkey: origins and growth', *Economic Development and Cultural Change*, **8**, 349–65.

Aresvik, O. (1975). *The Agricultural Development of Turkey*, Praeger, New York, Washington, London.

Barchard, D. (1985). *Turkey and the West*, Chatham House Papers 27, Royal Institute of International Affairs; Routledge and Kegan Paul, London.

Beeley, B. W. (1983). *Migration – The Turkish Case*, Open University, Milton Keynes.

Benedict, P., E. Tümertekin and F. Mansur (Eds) (1974). *Turkey: Geographic and Social Prespectives*, E. J. Brill, Leiden.

Burgel, G. (1967). 'Note sur le development récent de l'agglomeration d'Istanbul', *Bull. Ass. Géogr. fr.*, Nos 355 to 361, 51–63.

Chapman, A. S. (1957). 'The economic regions of Turkey as characterized by railway shipments', *Northwestern University Studies in Geography* (Evanston, Ill.), **2**, 71–5.

Clark, E. C. (1974). 'The Ottoman industrial revolution', *International Journal of Middle East Studies*, **5**, 65–76.

Clark, J. (1970–71). 'The growth of Ankara', 1961–69, *Rev. geogr. Inst. Univ. Istanb.*, **13**, 119–40.

Clarke, J. I. and H. Bowen-Jones (Eds) (1981). *Change and Development in the Middle East. Essays in Honour of W. B. Fisher*, Methuen, London.

Cohen, E. J. (1970). *Turkish Economic, Social and Political Change: Development of a More Prosperous and Open Society*, Praeger, New York, Washington and London.

Crabbe, G. (1944). 'Turkey: a record of industrial and commercial progress in the last quarter of a century', *Jl. R. cent. Asian Soc.*, **31**, 48–63.

Cuinet, V. (1890–94). *La Turquie d'Asie. Géographie Administrative, Déscriptive et Raisonée de Chaque Provence de l'Asie Mineure*, 4 vols, Ernest Leroux, Paris.

Danielson, M. N. and R. Keleş (1985). *The Politics of Rapid Urbanisation: Government and Growth in Modern Turkey*, Holmes and Meier, New York.

Darkot, B. (1958). *Türkiye Iktisadî Coğrafasi*, Bermet, Istanbul.

Dewdney, J. C. (1971). *Turkey*, Chatto and Windus, London

Dewdney, J. C. (1981). 'Agricultural development in Turkey'. In *Change and Development in the Middle East. Essays in Honour of W. B. Fisher* (Eds J. I. Clarke and H. Bowen-Jones), Methuen, London, pp 213–23.

Diamond, W. (1950). 'The Industrial Development Bank of Turkey', *Middle East Journal*, **4**, 349–51.

Dodd, C. H. (1969). *Politics and Government in Turkey*, Manchester University Press.

Dodd, C. H. (1979). *Democracy and Development in Turkey*, Eothen Press, Beverley.

Dodd, C. H. (1983). *The Crisis of Turkish Democracy*, Eothen Press, Beverley.

Dooren, P. J. van (1969). Structural and institutional obstacles facing Turkey's peasant farmers, *Tropical Man*, **2**, 107–61.

Dubetsky, (1976). 'Kinship, primordial ties and factory organisation in Turkey: an anthropological view', *International Journal of Middle East Studies*, **7**, 433–51.

Dulgarian, M. and E. Tümertekin (1962). The population of Istanbul: patterns and changes, 1955–60, *Rev. geogr. Inst. Univ. Istanb.*, **8**, 251–8.

Eldem, V. (1953). Turkey's transportation, *Middle Eastern Affairs*, **4**, 324–36.

Eraydin, A. (1981). 'Foreign investment, international labour migration and the Turkish economy'. In *Spatial Analysis and the Industrial Environment*, vol 2, *International Industrial Systems* (Eds F. E. I. Hamilton and G. J. R. Linge), Wiley, Chichester, pp 225–64.

Eren, N. (1966). 'Financial aspects of Turkish planning', *Middle East J.*, **20**, 187–95.

Erinç, S. (1950). 'Climatic types and the variation of moisture regions in Turkey', *Geogrl Rev.*, **40**, 224–35.

Erinç, S. and N. Tunçdilek (1952). 'The agricultural regions of Turkey', *Geogrl Rev.*, **42**, 189–203.

Fry, M. J. (1971). 'Turkey's first Five-Year Development Plan: an assessment', *Economic Journal*, **81**, 306–26.

Hale, W. (1981). *The Political and Economic Development of Modern Turkey*, Croom Helm, London.

Hamilton, F. E. I. and G. J. R. Linge (Eds) (1981). *Spatial Analysis and the Industrial Environment, vol 2, International Industrial Systems*, Wiley, Chichester.

Helburn, N. (1955). 'A stereotype of agriculture in semi-arid Turkey', *Geogrl Rev.*, **45**, 375–84.

Hershlag, Z. Y. (1964). *Introduction to the Modern Economic History of the Middle East*, E. J. Brill, Leiden.

Hershlag, Z. Y. (1968a). *Turkey: The Challenge of Growth*, E. J. Brill, Leiden.

Hershlag, Z. Y. (1968b). *Economic Planning in Turkey*, Economic Research Foundation, İstanbul.

Hiltner, J. (1962). 'The distribution of Turkish manufacturing', *J. Geogr., N.Y.*, **61**, 251–58.

Hinderink, J. and M. B. Kiray (1970). *Social Stratification as an Obstacle to Development. A Study of Four Turkish Villages*, Praeger, New York, Washington and London.

Hirsch, E. (1971). *Poverty and Plenty on the Turkish Farm: A Economic Study of Turkish Agriculture in the 1950s*, Middle East Institute, Columbia University, New York.

Hirsch, E. and A. Hirsch (1963). 'Changes in agricultural output per capita of rural population in Turkey', 1927–60, *Economic Development and Cultural Change*, **11**, 372–94.

Hirsch, E. and A. Hirsch (1966). Changes in terms of trade to farmers and their effect on real farm income per capita of rural population in Turkey', 1927–60, *Economic Development and Cultural Change*, **14**, 440–57.

Hudson, R. and J. Lewis (Eds) (1985). *Uneven Development in Southern Europe*, Methuen, London and New York.

Hütteroth, W-D. (1982). *Türkei*, Wissenschaftliche Buchgesellschaft, Darmstadt.

Issawi, C. (Ed) (1966). *The Economic History of the Middle East, 1800–1914*, Chicago University Press, Chicago and London.

Karataş, C. (1977). 'Contemporary economic problems in Turkey', *British Society for Middle East Studies: Bulletin*, **4**, 3–20.

Karpat, K. H. (1977). *The Gecekondu*, Cambridge University Press.

Karpat, K. H. (1985). *Ottoman Population, 1830–1914*, University of Wisconsin Press, Madison.

Keleş, R. Y. (1961). *Türkiyede Şehirleşme Haraketleri (1927–1960)*, Faculty of Political Science, Ankara, mimeographed.

Keleş, R. Y. (1963). 'Regional disparities in Turkey', *Ekistics*, **15**, 331–35.

Keleş, R. Y. (1966). Urbanisation and balanced regional development in Turkey, *Ekistics*, **22**, 163–8.

Keleş, R. (1985). 'The effects of external migration on regional development in Turkey'. In *Uneven Development in Southern Europe* (Eds R. Hudson and J. Lewis), Methuen, London and New York, pp 54–75.

Kepenek, Y. (1984). *Türkiye Ekonomisi*, 2nd edn, Savaş Yayınları, Ankara.

Kerwin, R. W. (1950). 'The Turkish roads programme', *Middle East J.*, **4**, 196–208.

Kerwin, R. W. (1951). 'Private enterprise in Turkish industrial development', *Middle East J.*, **5**, 21–38.

Keyder, C. (1981). *The Definition of a Peripheral Economy: Turkey, 1923–1929*, Maisons des Sciences de l'Homme, Paris; Cambridge University Press.

Kolars, J. F. and H. J. Malin (1970). 'Population and accessibility: an analysis of Turkish railroads', *Geogrl Rev.*, **60**, 229–46.

Kolodny, Y. (1968). 'Données récentes sur la population urbaine de la Turquie', *Méditerranée*, **9**, 165–80.

Kroner, G. (1969). 'Der Bau des Euphrat-Dammes bei Keban (Ostananatolien): Möglichkeiten und Grenzen einer raumplanerischen Lösung', *Raumforschung und Raumordung*, **27**, 156–62.

Küçük, Y. and Aksoy, A. (1981). The development of the planning concept in Turkey', *Etudes Balkaniques*, **4**, 51–7.

Kündig-Steiner, W. (1968). 'Neueste kulturlandschaftliche Veränderungen in Ostanatolien, speziell in der Region Kars', *Geographica helv.*, **23**, 129–131.

Kurmuş, O. (1974). 'The role of British Capital in the Development of Western Anatolia, 1850–1913', unpublished PhD thesis, University of London.

Landau, J. M. (Ed) (1984). *Atatürk and the Modernisation of Turkey*, E. J. Brill, Leiden.

Levine, N. (1980). 'Anti-urbanisation: an implicit development policy in Turkey', *Journal of Developing Areas*, **14**, 513–37.

Lewis, B. (1968). *The Emergence of Modern Turkey*, 2nd edn, Oxford University Press, London.

Louis, H. (1972). 'Die Bevölkerungsverteilung in der Türkei 1965 und ihre Entwicklung seit 1935', *Erdkunde*, **26**, 161–77.

Merriam, G. P. (1926). 'The regional geography of Anatolia', *Econ. Geogr.*, **2**, 86–107.

Mitchell, J. K. M. (1986). *The South East Anatolia Project – Commercial Opportunities: Preliminary Report*, (mimeographed) Ankara.

Morris, J. A. (1960). 'Recent problems of economic development in Turkey', *Middle East J.*, **14**, 1–14.

Ökçün, G. (1984). *Osmanlı Sanayü: 1913–1915 İstatistikleri*, partial revision, Hil Yayin, İstanbul.

Okyas, O. (1965). 'The concept of Etatism', *Economic Journal*, **75**, 98–111.

Owen, R. (1981). *The Middle East and the World Economy 1800–1914*, Methuen, London.

Poroy, I. I. (1972). 'Planning with a large public sector: Turkey (1963–1967)', *International Journal of Middle Eastern Studies*, **3**, 348–60.

Quartet, D. (1977). 'Limited revolution: the impact of the Anatolian Railway on Turkish transportation and the provisioning of Istanbul, 1890–1908', *Business History Review*, **51**, 139–60.

Quartet, D. (1980). 'The commercialisation of agriculture in Ottoman Turkey, 1800–1914', *International Journal of Turkish Studies*, **1**, 38–55.

Ritter, G. (1972). 'Landflucht und Städtewachstum in der Türkei', *Erdkunde*, **26**, 177–96.

Rivkin, M. D. (1965). *Area Development for National Growth. The Turkish Precedent*, Praeger, New York, Washington and London.

Robinson, R. D. (1967). *High-level Manpower in Economic Development: The Turkish Case*, Middle East Monographs No 17, Harvard University Press, Cambridge, Mass.

Sarc, O. C. (1941). 'Tanzimat ve sanayimiz'. In *Tanzimat*, İstanbul, pp 423–40. Translated as The Tanzimat and our industry. In *The Economic History of the Middle East, 1800–1914*, (Ed C. Issawi), Chicago University Press, Chicago and London, 1966, pp 48–59.

Simpson, D. J. (1965). 'Development as a process: the Menderes phase in Turkey', *Middle East J.*, **19**, 141–52.

Snyder, W. W. (1969). 'Turkish economic development: the first Five Year Plan', *Journal of Development Studies*, **6**, 58–71.

Soysal, M. (1976). *Die Siedlungs-und Landschaftsentwicklung der Çukorova*, Erlangen Geographische Arbeiten 4, Erlangen.

Stewig, R. (1972). 'Die industrialisierung in der Türkei', *Erde*, **103**, 21–47.

Tanoğlu, A., S. Erinç and E. Tümertekin (1961). *Turkiye Atlasi*, Istanbul Üniversitesi Edebiyat Fakültesi Yayinlau, No 903, Istanbul.

Toepfer, H. (1980). 'Mobilität und Investitionsverhalten Turkischer Gastarbeiter nach der Remigration', *Erdkunde*, **34**, 206–14.

Treadway, R. C. (1972). 'Gradients of metropolitan dominance in Turkey; alternative models', *Demography*, **9**, 13–34.

Tümertekin, E. (1955). 'The iron and steel industry in Turkey', *Econ. Geogr.*, **31**, 174–84.

Tümertekin, E. (1961). 'L'activité industrielle à Istanbul', *Rev. geogr. Inst. Univ. Istanb.*, **7**, 35–52.

Tümertekin, E. (1970–71). 'Manufacturing and suburbanisation in İstanbul', *Rev. geogr. Inst. Univ. Istanb.*, **13**, 1–40.

Tümertekin, E. (1970–71). 'Gradual internal migration in Turkey: a test of Ravenstein's hypothesis', *Rev. geogr. Inst. Univ. Istanb.*, **13**, 157–69.

Walker, D. F. (Ed) (1980). *Planning Industrial Development*, Wiley, Chichester.

Wålstedt, B. (1980). *State Manufacturing Enterprise in a Mixed Economy: The Turkish Case*, Johns Hopkins University Press, Baltimore and London.

Wary, R. E. and D. A. Rustow (Eds) (1964). *Political Modernization in Japan and Turkey*, Princeton University Press, Princeton.

Yarruz, J. (1952). 'The development of Ankara', *J. T. Plann. Inst. Lond.*, **38**, 251–252.

Chapter 18: Iran: Agriculture and its Modernization

Adams, R. M. (1962). 'Agriculture and urban life in early southwestern Iran', *Science, N.Y.*, **136**, 109–22.

Ajami, I. (1973). 'Land reform and modernisation of the farming structure in Iran', *Oxford Agrarian Studies*, **2**, 120–31.

Amini, S. (1983). 'The origin, function and disappearance of collective production units (Harasch) in rural areas of Iran', *Der Tropenlandwirt*, **84**, 47–61.

Aresvik, O. (1976). *The agricultural development of Iran*, Praeger, New York.

Arfa, H. (1963). 'Land reform in Iran', *Jl. R. cent. Asian Soc.*, **50**, 132–7.

Avery, P. (1965). *Modern Iran*, Ernest Benn, London, 527 pages.

Ayazi, M. (1961). 'Drainage and reclamation problems in the Garmsar area'. In UNESCO, *Salinity Problems in the Arid Zone*, Arid Zone Research, vol 14, 285–90.

Banami, A. (1961). *The Modernization of Iran 1921–41*, Stanford University Press, Stanford, 191 pages.

Barth, F. (1964). *Nomads of South Persia, The Basseri Tribe of the Khamseh Confederacy*, Universitetsforlaget, Oslo, 159 pages.

Barth, F. (1962). 'Nomadism in the mountain and plateau areas of South West Asia'. In UNESCO, *The Problems of the Arid Zone*, Arid Zone Research, Paris, vol 18, 341–55.

Beaumont, P. (1968). 'Qanats on the Varamin Plain, Iran', *Trans. Inst. Br. Geogr.*, **45**, 169–79.

Beaumont, P. (1971). 'Qanat systems in Iran', *Bull. int. Ass. scient. Hydrol.*, **16**, 39–50.

Beaumont, P. (1973). 'A traditional method of ground water extraction in the Middle East', *Ground Water*, **11**, 23–30.

Beaumont, P. (1973). *River Regimes in Iran*, Occasional Publications (New Series), No 1, Department of Geography, University of Durham, 29 pages.

Beaumont, P. (1974). 'Water resource development in Iran', *Geogrl. J.*, **140**, 100–10.

Beaumont, P. (1976). 'Environmental management problems – the Middle East', *Built Environment Quarterly*, **2**, 104–12.

Beaumont, P. (1985). 'The agricultural environment: an overview'. In *Agricultural development in the Middle East*, (Eds P. Beaumont and K. S. McLachlan), Wiley, Chichester, pp 3–26.

Beaumont, P. and K. S. McLachlan (Eds) (1985). *Agricultural Development in the Middle East*, Wiley, Chichester.

Beaumont, P. and J. H. Neville (1968). 'Rice cultivation in Iran's Caspian lowlands', *Wld Crops*, **20**, 70–3.

Beckett, P. H. T. (1953). 'Qanats around Kerman', *Jl. R. cent. Asian Soc.*, **40**, 47–58.

Beckett, P. H. T. (1957). 'Tools and crafts in south central Persia', *Man*, *57*, 145–8.

Beckett, P. H. T. (1958). 'The soils of Kerman, south Persia', *J. Soil Sci.*, **9**, 20–32.

Beckett, P. H. T. and E. D. Gordon (1956). 'The climate of Kerman, South Persia', *Q. Jl R. met. Soc.*, **82**, 503–14.

Beckett, P. H. T. and E. D. Gordon (1966). 'Land use and settlement round Kerman in southern Iran', *Geogrl J.*, **132**, 476–91.

Bémont, F. (1961). 'L'irrigation en Iran', *Annls Gèogr.*, **70**, 597–620.

Bharier, J. (1968). 'A note on the population of Iran, 1900–1966', *Population Studies*, **22**, 273–9.

Bharier, J. (1971). *Economic Development in Iran 1900–1970*, Oxford University Press, London, 314 pages.

Bharier, J. (1972). 'The growth of towns and villages in Iran, 1900–1970', *Middle Eastern Studies*, **8**, 51–61.

Bill, J. A. (1963). 'The social and economic foundations of power in contemporary Iran', *Middle East Journal*, **17**, 400–18.

Black, A. G. (1948). 'Iranian agriculture – present and prospective', *Journal of Farm Economics*, **30**, 422–42.

Bobek, H. (1968). 'Vegetation'. In *The Cambridge History of Iran, Vol 1 – The Land of Iran*, (Ed W. B. Fisher), Cambridge University Press.

Bonine, M. E. and N. Keddie (1981). *Modern Iran – the dialectics of continuity and change*, State of New York Press, Albany, 464 pages.

Bowen-Jones, H. (1968). 'Agriculture'. In *The Cambridge History of Iran, Vol 1 – The Land of Iran*, (Ed W. B. Fisher), Cambridge University Press, Cambridge, pp 565–98.

Caponera, D. (1954). *Water Laws in Moslem Countries*, FAO Agricultural Development Papers, No 43, 202 pages.

Clark, B. D. (1972). 'Iran: changing population patterns'. In *Populations of the Middle East and North Africa*, (Eds J. I. Clarke and W. B. Fisher), University of London Press, 68–96.

Clark, B. D. and V. Costello (1973). 'The urban system and social patterns in Iranian cities', *Trans. Inst. Br. Geogr.*, **59**, 99–128.

Clarke, J. I. (1963). *The Iranian City of Shiraz*, Department of Geography, Research Paper Series No 7, University of Durham, 55 pages.

Clarke, J. I. and B. D. Clarke (1969). *Kermanshah, an Iranian Provincial City*, Department of Geography, Research Paper Series No 10, University of Durham, 137 pages.

Clarke, J. I. and W. B. Fisher (Eds) (1972). *Populations of the Middle East and North Africa*, University of London Press.

Development and Resources Corporation (1959). *The Unified Development of the Natural Resources of the Khurzestan Region*, New York, 162 pages.

Dewan, M. L. and J. Famouri (1968). 'Soils'. In *The Cambridge History of Iran, Vol 1 – The Land of Iran*, (Ed W. B. Fisher), Cambridge University Press, Cambridge, pp 250–63.

Djavid-Pour, E. (1968). 'Condition of agriculture and farmers before execution of the Iranian Law of Land Reforms', *Mediterranea*, **25**, 572–5.

Echo of Iran (1971). *Iran Almanac 1971*, Echo of Iran, Tehrān, Iran, 808 pages.

Ehlers, E. (1985). 'The Iranian village: a socio-economic microcosm'. In *Agricultural development in the Middle East*, (Eds P. Beaumont and K. S. McLachlan), Wiley, Chichester, pp 151–70.

Ehlers, E. (1977). 'Social and economic consequences of large-scale irrigation development – the Dez Irrigation project, Khuzestan, Iran'. In *Arid Land Irrigation in Developing Countries*, (Ed E. B. Worthington), Pergamon Press, Oxford, pp 85–97.

Ehlers, E. (1983). 'Rent capitalism and unequal development in the Middle East: the case of Iran'. In *Work, Income and Inequality: Payment Systems in the Third World*, (Ed F. Stewart), Macmillan, London, pp 32–61.

Elwell-Sutton, L. P. (1958). 'Nationalism and neutralism in Iran', *Middle East Journal*, **12**, 20–33.

English, P. W. (1966). *City and Village in Iran*, University of Wisconsin Press, Madison, 204 pages.

Field, M. (1972). 'Agro-business and agricultural planning in Iran', *Wld Crops*, **24**, 68–72.

Fisher, W. B. (Ed) (1968). *Cambridge History of Iran, Vol 1 – The Land of Iran*, Cambridge University Press, 784 pages.

Fitt, R. L. (1953). 'Irrigation development in central Persia', *Jl. R. cent. Asian Soc.*, **40**, 124–33.

Flower, D. J. (1968). 'Water use in north-east Iran'. In *The Cambridge History of Iran, Vol 1 – The Land of Iran*, (Ed W. B. Fisher), Cambridge University Press, pp 599–610.

Freivalds, J. (1972). 'Farm corporations in Iran: an alternative to traditional agriculture', *Middle East Journal*, **26**, 185–93.

Ganji, M. H. (1960). *Iranian Rainfall Data*, University of Tehrān, Arid Zone Research Centre, Publication No 3, 191 pages.

Ganji, M. H. (1968). 'Climate'. In *The Cambridge History of Iran, Vol 1 – The Land of Iran*, (Ed W. B. Fisher), Cambridge University Press, pp 212–49.

Gastil, R. D. (1958). 'Middle class impediments to Iranian modernisation, *Public Opinion Quarterly*, **22**, 325–9.

Gittinger, J. P. (1965). *Planning for Agricultural Development: the Iranian Experience*, National Planning Association, Washington DC, 121 pages.

Gittinger, J. P. (1967). 'Planning and agricultural policy in Iran – program effects and indirect effects', *Economic Development and Cultural Change*, **16**, 107–17.

Goblot, H. (1962). 'Le Problème de l'eau en Iran', *Orient*, **23**, 43–59.

Goodell, G. (1975). 'Agricultural production in a traditional village of northern Khuzestan'. In *Tradionelle und moderne Formen der Landwirtschaft in Iran*, (E. Ehlers and G. Goodell), Marburger Geogr. Schnriftan 64, Marburg, pp 243–89.

Hadary, G. (1951). 'The agrarian reform problem in Iran', *Middle East Journal*, **5**, 181–96.

Halliday, F. (1979). *Iran – dictatorship and development*, Penguin, Harmondsworth, 348 pages.

Harrison, J. V. (1932). 'The Bakhtiari Country, S.W. Persia', *Geogrl J.*, **80**, 193–210.

Hayden, L. (1949). 'Living standards in rural Iran, a case study', *Middle East Journal*, **3**, 140–50.

Hooglund, E. (1973). 'The *Khwushnishin* population of Iran', *Iranian Studies*, **6**, 229–45.

Hooglund, E. (1982). *Land and Revolution in Iran 1960–1980*, University of Texas Press, Austin.

Hooglund, E. J. (1981). 'Rural socio-economic organisation in transition: the case of Iran's bonehs'. In *Modern Iran*, (Eds M. E. Bonine and N. Keddie), State University of New York Press, Albany, pp 191–207.

Issawi, C. (1967). 'Iran's economic upsurge', *Middle East Journal*, **17**, 447–61.

Johnson, V. W. (1960). 'Agriculture in the economic development of Iran', *Land Economics*, **36**, 314–21.

Keddie, N. R. (1968). 'The Iranian village before and after land reform', *Journal of Contemporary History*, **3**, 69–91.

Keddie, N. R. (1980). *Iran, religion, politics and society*, Cass, London.

Kernan, H. S. (1957). 'Forest management in Iran', *Middle East Journal*, **11**, 199–202.

Khamsi, F. (1969). 'Land reform in Iran', *Monthly Review*, **21(2)**, 20–8.

Lambton, A. K. S. (1953). *Landlord and Peasant in Persia*, Oxford University Press, London, 459 pages.

Lambton, A. K. S. (1957). 'Impact of the West on Iran', *International Affairs*, London, **33**, 12–25.

Lambton, A. K. S. (1969). *The Persian Land Reform*, Clarendon Press, Oxford, 386 pages.

Lambton, A. K. S. (1969). 'Land reform and co-operative societies in Persia', *Jl. R. cent. Asian Soc.*, **56**, 142–55 and 245–58.

McLachlan, K. S. (1968). 'Land reform in Iran'. In *The Cambridge History of Iran, Vol 1 – The Land of Iran*, (Ed W. B. Fisher), Cambridge University Press, Cambridge, 684–716.

Ministry of Information (1970). *Farm Corporations in Iran*, Ministry of Information, Tehrān, 10 pages.

Ministry of Land Reform and Rural Cooperation (1970). *Land Reform Programme in Iran*, Rural Research Centre, Ministry of Land Reform and Rural Cooperation, Tehrān, 20 pages.

Naval Intelligence Division (Great Britain) (1945). *Persia*, Geographical Handbook Series, BR 525, 638 pages.

Noel, E. (1944). 'Qanats', *Jl. R. cent. Asian Soc.*, **31**, 191–202.

Oberlander, T. M. (1968). 'Hydrography'. In *The Cambridge History of Iran, Vol 1 – The Land of Iran*, (Ed W. B. Fisher), Cambridge University Press, pp 264–79.

Okazaki, S. (1985). 'Agricultural mechanisation in Iran'. In *Agricultural Development in the Middle East*, (Eds P. Beaumont and K. S. McLachlan), Wiley, Chichester, pp 171–87.

Plan and Budget Organisation (1982). *The First Five Year Economic Social and Cultural Macro-Development Plan of the Islamic Republic of Iran 1362–1366 (1983/84–1987/88)*, Tehrān, Iran.

Plan Organisation, Iran (1968). *Fourth National Development Plan, 1968–1972*, Plan Organisation, The Imperial Government of Iran, Tehrān.

Plan Organisation, Iran (1969). *Dam Construction in Iran*, Bureau of Information and Reports, Tehrān, 87 pages.

Planhol, X. de (1966). 'Aspects of mountain life in Anotolia and Iran'. In *Geography as Human Ecology*, (Eds S. R. Eyre and G. R. J. Jones), Arnold, London, pp 291–308.

Population Reference Bureau (1986). *World Population Data Sheet 1985*, Washington, DC.

Sahebdiam-Bunodière, C. (1962). 'L'agriculture en Iran', *Orient*, **21**, 33–47.

Salmanzadeh, C. (1980). *Agricultural change and rural society in southern Iran*, Menas Press, Wisbech.

Smith, A. (1953). *Blind White Fish in Persia*, E. P. Dutton and Co, New York, 256 pages.

Spooner, B. (1963). 'The function of religion in Persian Society', *Iran*, **1**, 83–95.

Spooner, B. (1966). 'Iranian kinship and marriage', *Iran*, **4**, 51–9.

Stewart, F. (Ed) (1983). *Work, Income and Inequality: Payment Systems in the Third World*, Macmillan, London.

Sunderland, E. (1968). 'Pastoralism, nomadism and the social anthropology of Iran'. In *The Cambridge History of Iran, Vol 1 – The Land of Iran*, (Ed W. B. Fisher), Cambridge University Press, 611–83.

Sykes, C. (1957). 'Persian gardens', *Geogrl Mag.*, **30**, 326–9.

Vahidi, M. (1968). *Water and Irrigation in Iran*, Plan Organisation, published by Bureau of Information and Reports, Tehrān, 79 pages.

Worthington, E. B. (Ed) (1977). *Arid Land Irrigation in Developing Countries*, Pergamon Press, Oxford.

Wulff, H. E. (1968). 'Qanats of Iran', *Scient. Am.*, **218 (4)**, 94–105.

Young, T. C. (1948). 'The problem of westernization in Modern Iran', *Middle East Journal*, **2**, 47–59.

Chapter 19: Egypt: Population Growth and Agricultural Development

Abdel-Fadil, M.(1975). *Development, Income Distribution and Social Change in Rural Egypt (1952–1970)*, Cambridge University Press.

Abdel Hakim, M. S. and W. Abdel Hamid (1982). *Some Aspects of Urbansiation in Egypt*, Centre for Middle Eastern and Islamic Studies, Occasional Papers Series No 1, University of Durham.

Abou Zeid, A. M. (1979). 'New Towns and Rural Development in Egypt', *Africa*, **49**, 283–90.

Abu-Lughod, J. L. (1963–64). 'Rural–Urban differences as a function of the demographic transition: Egyptian data and analytical model', *American Journal of Sociology*, **69**, 476–90.

Adamowicz, M. (1970). 'Transformation of agricultural structure in the United Arab Republic', *Africana Bulletin*, **43**, 76, University of Warsaw.

Allan, J. A. (1981). 'High Aswan Dam is a Success Story', *Geographical Magazine*, **53(6)**, 393–6.

Ammar, H. M. (1954). *Growing Up in an Egyptian Village*, Routledge and Kegan Paul, London.

Anon, (1981). 'Housing in the Suez Canal towns: an introduction', *Third World Planning Review*, **3**, 141.

Ayrout, H. H. (1963). *The Egyptian Peasant*, Beacon Press, Boston.

Bailey, R. (Ed), (1983). *Egypt: Development Targets and Realities*, MEED Special Report (July), London.

Ball, J. (1939). *Contributions to the Geography of Egypt*, Government Press, Cairo.

Beaumont, P. and K. McLachlan (Eds) (1985). *Agricultural Development in the Middle East*, Wiley, Chichester.

Benedick, R. E. (1979). 'The High Dam and the transformation of the Nile', *Middle East Jnl.*, **33(1)**, 119–44.

Beshara, A. (1981). 'Planning New Development Regions in Egypt', *Third World Planning Review*, **3**, 234–49.

Blackman, W. S. (1971). *The Fellāhīn of Upper Egypt, their Religious, Social and Industrial Life*, Frank Cass, London.

Butter, D. (Ed) (1985). *Egypt*, MEED Special Report (November), London.

Cooper, M. N. (1982). *The Transformation of Egypt*, Croom Helm, London.

Daniels, C. (1983). *Egypt in the 1980s: The Challenge*, Economist Intelligence Unit Special Report No 158, London.

Economist Intelligence Unit, (1986). *Country Profile - Egypt: 1986*, London.

Economist Intelligence Unit (1970). *Annual Supplement on Egypt*, London.

Eshag, E. and M. A. Kamal (1968). 'Agrarian reform in the United Arab Republic', *Bull. Oxf. Univ. Inst. Statist.*, **30**, 73–104.

Granott, A. (1956). *Agrarian Reform and the Record of Israel*, Eyre and Spottiswoode, London.

Grove, D. (1982). Egypt has too much water, *Geographical Magazine*, **54(8)**, 437–41.

Harik, I. F. (1974). *The Political Mobilization of Peasants. A Study of an Egyptian Community*, Indiana University Press, Bloomington.

Holt, P.M. (Ed) (1968). *Political and Social Change in Modern Egypt*, Oxford University Press, London.

Hopkins, H. (1969). *Egypt, the Crucible*, Secker and Warburg, London.

Hyland, A.D.C., A.G. Tipple, and N. Wilkinson. (1984). *Housing in Egypt*, University of Newcastle upon Tyne, School of Architecture Housing Course Working Paper Series No 1, Newcastle.

Jenner, M. (1985). 'Cairo in Peril', *Geographical Magazine*, **57(9)**, 474–80.

Kumawat, R.R. (1979). 'Settlements in Newly Irrigated Areas in Egypt', *Ekistics*, **46**, 231–4.

May, J.M. and I.S. Jarcho (1961). *The Ecology of Malnutrition in the Far and Near East*, Hafner, New York.

Meyer, G. (1980). 'Effects of the 'New Valley' Project upon the development of the Egyptian Oases', *Applied Geography and Development*, (Tubingen), **15**, 96–116.

Middle East and North Africa Yearbook: 1986 Europa, London 1986.

Ministry of Information (1983). *Agriculture in the Arab Republic of Egypt*, Cairo.

Ministry of Land Reclamation (1972). *Land Reclamation Development in the Arab Republic of Egypt*, Cairo.

Mountjoy, A.B. (1972). 'Egypt cultivates her deserts', *Geogr. Mag.*, **44**, 241–50.

Platt, R.R. and M.B. Hefny (1958). *Egypt: a Compendium*, American Geographical Society, New York.

Radwan, S. (1974). *Capital Formation in Egyptian Industry and Agriculture (1982–1967)*, Ithaca Press, London.

Radwan, S. and E. Lee (1986). *Agrarian Change in Egypt: An Anatomy of Rural Poverty*, Croom Helm, London.

Richards, A. (1980). 'The agricultural crisis in Egypt', *Jnl. Development Studies*, **16(3)**, 303–21.

Saab, G.S. (1967). *The Egyptian Agrarian Reform 1952–1962*, Oxford University Press, London.

Stamp, L.D. (1964). *The Geography of Life and Death*, Fontana, London.

Vatikiotis, P.J. (1980). *The History of Egypt*, Wiedenfeld and Nicolson.

Voll, S.P. (1980). 'Egyptian land reclamation since the revolution', *Middle East Journal*, **34(1)**, 127–48.

Wallach, B. (1986). 'The Nile Valley', *Focus*, **36(1)**, 16–19.

Waterbury, J. (1979). *The Hydropolitics of the Nile Valley*, Syracuse University Press, New York.

Whittington, D. and K.E. Haynes (1980). 'Valuing water in the agricultural environment of Egypt: some estimations and policy considerations', *Regional Science Perspectives*, **10**, 109–26.

Whittington, D. and K.E. Haynes (1985). 'Nile water for whom? emerging conflicts in water allocation for agricultural expansion in Egypt and Sudan'. In *Agricultural Development in the Middle East*, (Eds P. Beaumont and K.S. McLachlan), Wiley, Chichester.

Wilber, D.N. (Ed) (1969). *The United Arab Republic, its People, its Society, its Culture*, Human Relations Area Files, New Haven.

World Development Report (1985). Oxford University Press, London.

Chapter 20: Libya

Allan, J.A. (1981). *Libya: The Experience of Oil*, Croom Helm, London, and Westview Press, Boulder.

Allan, J.A. (Ed) (1982). *Libya Since Independence: Economic and Political Development*, Croom Helm, London, and St Martin's Press, New York.

Attir, M.O. (1983). 'Libya's pattern of urbanization', *Ekistics 300*, 157–62.

Bean, L.L. (1983). 'The labour force and urbanization in the Middle East', *Ekistics*, **300**, 195–204.

Birks, J.S. and C.A. Sinclair (1980). *Arab Manpower*, Croom Helm, London.

Blake, G.H. (1968). *Misurata: a Market Town in Tripolitania*, Department of Geography, University of Durham, pp 1–34.

Blake, G.H. (1979). 'Urbanisation and development planning in Libya'. In *Development of Urban Systems in Africa*, (Eds R.A. Obudho and S. El-Shakhs), Praeger, New York, 99–115.

Buru, M.M., S.M. Ghanem and K.S. McLachlan (Eds) (1985). *Planning and Development in Modern Libya*, Society for Libyan Studies, London, and Menas Press, Wisbech.

Economist Intelligence Unit (1985). *Country Profile: Libya 1986–87*, Economist Publications, London.

Farrell, J.D. (1967). 'Libya strikes it rich', *Africa Report*, **12**, 8.

Fisher, W.B. (1953). 'Problems of modern Libya', *Geogr. J.*, **119**, 183–99.

Fowler, G.L. (1972). 'Italian colonisation in Tripolitania', *Ann. Ass. Am. Geogr.*, **62**, 627–40.

Fowler, G.L. (1973). 'Decolonisation of rural Libya', *Ann. Ass. Am. Geogr.*, **63**, 490–506.

Habib, H. (1986). 'Changing patterns in Libyan Foreign policy', *J. South Asian and Middle Eastern Studies*, **10(2)**, 3–15.

Harrison, R.S. (1966). 'Libya' *Focus*, **17**, 1.

International Bank for Reconstruction and Development (1960). *The Economic Development of Libya*, Johns Hopkins Press, Baltimore.

Joffé, E.G.H., and K.S. McLachlan (Eds) (1982). *Social and Economic Development of Libya*, Menas Press, Wisbech.

Kezeiri, S.K. (1983). 'Urban planning in Libya', *Libyan Studies*, **14**, 9–15.

Kezeiri, S.K. (1986). 'Population growth of the Libyan small towns', *Libyan Studies*, **17**, 115–63.

Khader, B. and El-Wifati (1986). *The Economic Development of Libya*, Croom Helm, London.

Lawless, R.I. and S. Kezeiri (1983). 'Spatial aspects of population change in Libya', *Méditerranée*, **4**, 81–6.

Mahmood-Misrati, A.A. (1983). 'Land conversion to urban use: its impact and character in Libya', *Ekistics*, **300**, 183–94.

Mason, J.P. (1982). 'Qa dhafi's "revolution" and change in a Libyan oasis community', *Middle East Journal*, **36(3)**, 319–35.

Naur, M. (1986). *Political Mobilization and Industry in Libya*, Akademisk Forlag, Denmark.

Obudho, R.A. and S. El-Shakhs (Eds) (1979). *Development of Urban Systems in Africa*, Praeger, New York.

Penrose, E., J.A. Allan and K.S. McLachlan (Eds) (1970). *Agriculture and the Economic Development of Libya*, Universities of Libya and London, and British Petroleum, London.

Petrol. Press Service (1971). 'Tough bargaining in Tripoli' *Petrol. Press Service*, **38**, 122.

Petrol. Press Service (1972). 'Crisis of confidence in Libya', *Petrol Press Service*, **39**, 282.

Chapter 21: Conclusion

Anderson, E.W. (1987). 'Water resources and boundaries in the Middle East'. In *Boundaries and State Territory in the Middle East and North Africa*, (Eds G.H. Blake and R.N. Schofield), Menas Press, Cambridge, pp 85–97.

Beaumont, P. and K.S. McLachlan (Eds) (1985). *Agricultural Development in the Middle East*, John Wiley, Chichester.

Blake, G.H., J.C. Dewdney and J.K. Mitchell (1987). *Cambridge Atlas of the Middle East and North Africa*, Cambridge University Press.

Blake, G.H. and Schofield, R.N. (Eds) (1987). *Boundaries and State Territory in the Middle East and North Africa*, Menas Press, Cambridge.

Braun, A. (Ed) (1986). *The Middle East in Global Strategy*, Westview Press, Godstone.

Drysdale, A.D. and G.H. Blake (1985). *The Middle East and North Africa: a Political Geography*, Oxford University Press, New York.

Farid, A.M. (Ed) (1986). *The Decline of Arab Oil Revenues*, Croom Helm, London.

Fisher, W.B. (1978). The Middle East, 7th edn, Methuen, London.

Kelly, P.L. (1986). 'Escalation of regional conflict: testing the shatterbelt concept', *Political Geog. Quarterly*, **5(2)**, 161–80.

Kissinger, H. (1982). *Years of Upheaval*, Weidenfeld and Nicolson; Michael Joseph, London.

Kliot, N. and S. Waterman (Eds) (1983). *Pluralism and Political Geography: People, State and Territory*, Croom Helm, London and St Martin's Press, New York.

Tuma, E.H. (1980). 'The rich and the poor in the Midle East', *Middle East Journal*, **34(4)**, 413–37.

United Nations, Department of Economic and Social Affairs, (1978). *Register of International Rivers*, Pergamon Press, Oxford.

Wise, G.S. and C. Issawi (Eds) (1982). *Middle East Perspectives: the Next Twenty Years*, Darwin Press, Princeton, NJ.

Index